Statistics and Probability

Raghu Raj Bahadur : b.1924

Statistics and Probability

A
Raghu Raj Bahadur
Festschrift

Edited by

J.K. Ghosh • S.K. Mitra
K.R. Parthasarathy • B.L.S. Prakasa Rao

PUBLISHING FOR ONE WORLD
WILEY EASTERN LIMITED
NEW DELHI • BANGALORE • BOMBAY • CALCUTTA • GUWAHATI
HYDERABAD • LUCKNOW • MADRAS • PUNE

WILEY EASTERN LIMITED

New Delhi : 4835/24 Ansari Road, Daryaganj, New Delhi 110 002
Bangalore : 27, Bull Temple Road, Basavangudi, Bangalore 560 004
Bombay : Post Box No. 4124, Saraswati Mandir School, Kennedy Bridge, Nana Chowk, Bombay 400 007
Calcutta : 40/8, Ballygunge Circular Road, Calcutta 700 019
Guwahati : Pan Bazar, Rani Bari, Guwahati 781 001
Hyderabad : 1-2-412/9, Gaganmahal, Near A V College, Domalguda, Hyderabad 500 029
Lucknow : 18, Pandit Madan Mohan Malviya Marg, Lucknow 226 001
Madras : No. 6, First Main Road, Gandhi Nagar, Madras 600 020
Pune : Flat No. 2, Building No. 7, Indira Co-op Housing Society Ltd., Indira Heights, Paud Fatta, Erandawane, Karve Road, Pune 411 038

ISBN: 81-224-0541-X

Published by H.S. Poplai for Wiley Eastern Limited, 4835/24, Ansari Road, Daryaganj, New Delhi 110 002. Typeset by Jay Compusoft, H-296, Uday Niketan, Narain Vihar, New Delhi 110 028, and printed at Baba Barkha Nath Printers, N-48, Kirti Nagar, New Delhi 110 015.

Printed in India.

Production : M.I. Thomas

Preface

We have great pleasure in bringing out this volume of essays in honour of Raghu Raj Bahadur, an outstanding statistician and a noble man, on the occasion of his seventieth birthday.

Bahadur was born on April 30, 1924 at the city of Delhi where he completed his school and college education. He received his doctor's degree in 1950 from the University of North Carolina at Chapel Hill in the U.S.A. During the years 1956–61 he served as a Professor in the Research and Training School of the Indian Statistical Institute at Calcutta where, along with P.C. Mahalanobis and C.R. Rao, he played a major role in leading a whole generation of young probability theorists and mathematical statisticians to the frontiers of research by his inspiring lectures on several topics including, among others, inference, sequential analysis, measure-theoretic probability and limit theorems. In 1961 he settled down permanently in the U.S.A. as a Professor of Statistics at the University of Chicago and left an unbridgeable gap in statistical education at Calcutta.

Bahadur's research has left an indelible imprint in the areas of sufficiency and asymptotics by an exemplary combination of statistical insight and mathematical skill. Bearing his name there now exists a whole theory of comparison of statistical tests on the one hand and a representation of quantiles on the other. Statistical literature is strewn with several major contributions from the pen of Bahadur among which we mention a few: results on large deviations; the Markov-like property of transitivity obeyed by minimal sufficient statistics under independent experimentation: a tight set of sufficient conditions for the consistency of the maximum likelihood estimate in a very general and abstract setting where the violation of any one of the conditions leads to a counter example; a deep and delightful converse to the famous Rao-Blackwell theorem on the existence of a minimum variance unbiased estimate under the assumption of completeness i.e. existence, under certain regularity conditions, of an estimable parametric function with no minimum variance unbiased estimate whenever the minimal sufficient statistic is not complete. Taken as a whole Bahadur's work is deep, austere, elegant and of lasting value. In a world obsessed with the pressure to publish or perish he has stuck to the discipline of publishing only what is worthy of appearing in print. Indeed, he has offered his advice and admonition to younger colleagues in his inimitable style: "publish and perish".

Recognition of Bahadur's contributions to mathematical statistics came first in the form of John Simon Guggenheim Fellowship during 1968–69. He was President of the Institute of Mathematical Statistics and Wald Lecturer in 1974. He is a Fellow of the Indian National Science Academy and the Indian Academy of Sciences and a Member of the American Academy of Arts and Sciences. The late Professor Jerzy Neyman, in an address delivered during the inauguration of the Delhi Campus of the Indian Statistical Institute in 1974, described Bahadur as the brightest star of Indian origin shining over the U.S. in the statistical horizon.

The papers included in this volume are works in several areas, including sufficiency, asymptotics, reliability, linear models, stochastic processes and group representations. The diversity of topics as well as affiliations of the authors reflect the respect and affection that are widely accorded to Bahadur.

In conclusion we wish to thank all the authors who have contributed to this volume and the scores of referees who rendered valuable assistance in processing the papers. We acknowledge with gratitude the generous help that we have received from R.L. Karandikar, Isha Dewan, S. Hande, V.P. Sharma, B. Ganesan and Meena Ramachandran in this endeavour.

26th April, 1993

J.K. GHOSH
S.K. MITRA
K.R. PARTHASARATHY
B.L.S. PRAKASA RAO

Contents

Publications of Raghu Raj Bahadur

Book

Some Limit Theorems in Statistics. NBS-CBMS Monograph, No. 4 SIAM, 1971.

Research Papers

On a problem in the theory of k populations, *Ann. Math. Statist.* **21** (1950), 362-375.

The problem of the greater mean (with H. Robbins), *Ann. Math. Statist.* **21** (1950), 469-487.

A property of the t statistic, *Sankhyā* **12** (1952), 78-88.

Impartial decision rules and sufficient statistics (with Leo A. Goodman), *Ann. Math. Statist.* **23** (1952), 553-562.

Sufficiency and statistical decision functions, *Ann. Math. Statist.* **25** (1954), 423-462.

Two comments on sufficiency and statistical decision functions (with E. L. Lehmann), *Ann. Math. Statist.* **26** (1955), 139-142.

A characterization of sufficiency, *Ann. Math. Statist.* **26** (1955), 286-293.

Statistics and subfields, *Ann. Math. Statist.* **26** (1955), 490-497.

Measurable subspaces and subalgebras, *Proc. Amer. Math. Soc.* **6** (1955), 565-570.

The nonexistence of certain statistical procedures in non-parametric problems (with L. J. Savage), *Ann. Math. Statist.* **27** (1956), 1115-1122.

On unbiased estimates of uniformly minimum variance, *Sankhyā* **18** (1957), 211-224.

Examples of inconsistency of maximum likelihood estimates, *Sankhyā* **20** (1958), 207-210.

A note on the fundamental identity of sequential analysis, *Ann. Math. Statist.* **29** (1958), 534-543.

Some approximations to the binomial distribution function, *Ann. Math. Statist.* **31** (1960), 43-54.

Simultaneous comparison of the optimum and sign tests of a normal mean, *Contributions to Probability and Statistics: Essays in Honor of Harold Hotelling* (Eds. I. Olkin *et al.*), Stanford University Press (1960), 79-88.

Stochastic comparison of tests, *Ann. Math. Statist.* **31** (1960), 276-295.

On the asymptotic efficiency of tests and estimates, *Sankhyā* **22** (1960), 229-252.

On deviations of the sample mean (with R. R. Rao), *Ann. Math. Statist.* **31** (1960), 1015-1027.

On the number of distinct values in a large sample from an infinite discrete distribution, *Proc. Nat. Inst. Sciences, India*, **A 26**, (Supp. II), (1960), 67-75.

A representation of the joint distribution of n dichotomous items, *Studies in Item Analysis and Prediction*, (Ed. H. Solomon), Stanford University Press (1961), 158-168.

On classification based on responses to n dichotomous items, *Studies in Item Analysis and Prediction*, (Ed. H. Solomon), Stanford University Press (1961), 169-176.

Classification into two multivariate normal distributions with unequal covariances (with T. W. Anderson). *Ann. Math. Statist.* **33** (1962), 420-431.

On Fisher's bound for asymptotic variances, *Ann. Math. Statist.* **35** (1964), 1545-1552.

An optimal property of the likelihood ratio statistic, *Proc. Fifth Berkeley Symp. Math. Statist. Prob.*, **1**, (1965), 13-26.

A note on quantiles in large samples, *Ann. Math. Statist.* **37** (1966), 577-580.

Rates of convergence of estimates and test statistics, *Ann. Math. Statist.* **38** (1967), 303-324.

Substitution in conditional expectation (with P. J. Bickel), *Ann. Math. Statist.* **39** (1968), 377-378.

On conditional test levels in large samples (with P. J. Bickel), *Essays in Probability and Statistics*, S.N. Roy memorial volume, (Eds. R.C. Bose *et al.*) University of North Carolina Press (1969), 25-34.

Some asymptotic properties of likelihood ratios on general sample spaces (with M. Ragavachari), *Proc. Sixth Berkeley Symp. Math. Statist. Prob.*, **1** (1970), 129-152.

Examples of inconsistency of the likelihood ratio statistic, *Sankhyā* **34** (1972), 81-84.

A note on UMV estimates and ancillary statistics. *Contributions to Statistics* (Hájek Memorial Volume) (Ed. J. Jurečková), D. Riedel Pub. Co. (1979), 19-24.

On large deviations of the sample mean in general vector spaces, (with S. L. Zabell), *Ann. Probab.*, **7** (1979), 587-621.

Large deviations, tests and estimates, (with J. C. Gupta and S. L. Zabell). *Asymptotic Theory of Statistical Tests and Estimation*, Hoeffding Volume, (Ed. I.M. Chakravarti), Academic Press (1980), 33-64.

On large deviations of maximum likelihood and related estimates. Tech. Report No. 121, Department of Statistics, University of Chicago, 1980.

A note on the effective variance of a randomly stopped mean. *Statistics and Probability : Essays in Honor of C. R. Rao*, (Eds. G. Kallianpur *et al.*), North Holland (1982), 39-43.

Hodges Superefficiency, *Encyclopedia of Statistical Sciences*, Vol. 3. John Wiley (1983), 645-646.

Large deviations of the maximum likelihood estimate in the Markov chain case. *Recent Advances in Statistics: Papers in Honor of H. Chernoff* (Eds. M.H. Rizvi *et al.*), Academic Press (1983), 273-280.

Some further properties of the LR statistic in general sample spaces (with T. K. Chandra and D. Lambert), Proc. Ind. Stat. Inst. Golden Jub. Int. Conference, (Eds. J.K. Ghosh and J. Roy), Indian Statistical Institute (1984), 1-19.

Distributional optimality and second-order efficiency of test statistics (with J. C. Gupta), 1985. In *Adaptive Statistical Procedures and Related Topics*, IMS Lecture Notes — Monograph Series No. 8 (Ed. John Van Ryzin), 315-331.

Remarks on transitive sufficiency. In *Glimpses of India's Statistical Heritage* (Eds. J.K. Ghosh *et al.*) Wiley Eastern Ltd. (1992), 3-10.

J.K. Ghosh, S.K. Mitra, K.R. Parthasarathy and B.L.S. Prakasa Rao (*eds.*),
Statistics and Probability : A Raghu Raj Bahadur Festschrift
Published by Wiley Eastern Limited, Publishers, (1993), 1 – 12

Bahadur Efficiency and Local Optimality of a New Nonparametric Test of Symmetry

V.L. ABBAKUMOV
Petrozavodsk State University

and

YA. YU. NIKITIN
St. Petersburg University

Abstract

We introduce and study some new nonparametric statistics for testing symmetry that are the centered versions of the well-known Smirnov statistic and its one-sided counterparts. We find their limit distributions, large deviations, Bahadur efficiency and conditions of local optimality in the Bahadur sense. It turns out that the centering has an effect on the properties of Smirnov's statistic. In the case of the location alternative the latter is Bahadur locally optimal only for observations with the Laplace distribution and the new statistic is locally optimal only in the case of Cauchy distribution.

1 Introduction

Let $X_1, X_2, ..., X_n$ be independent one-dimensional observations with common distribution function (d.f.) F_0. The hypothesis of symmetry

$$H : \Delta F_0(x) \equiv F_0(x) + F_0(-x) - 1 = 0 \quad \forall\, x \in \mathbb{R}^1$$

AMS Subject Classifications : 60F10, 62G10, 62G20.
Key words : testing of symmetry, large deviations, U-statistics, Bahadur efficiency, local optimality.

is one of the most interesting and important examples in classical nonparametric statistics. In this paper we consider for simplicity only the most frequent alternative to symmetry - the hypothesis of shift in location. We suppose that under the alternative hypothesis A the d.f. of X_i is $G(x+\theta)$ where $G(x)$ is some completely known continuous and symmetric d.f. and $\theta > 0$.

A number of tests have been proposed for testing H against A. Most known of them are the sign test and various signed rank tests such as the Wilcoxon test and the Fraser test (see, for example, Hájek and Šidak (1967), Hollander and Wolfe (1973) or Huškova (1984)).

Another class of test statistics is based directly on the empirical d.f. F_n, corresponding to the sample $X_1, ..., X_n$. Smirnov (1947) proposed the statistic

$$K_n = \sup_x |\Delta F_n(x)| \equiv \sup_x |F_n(x) + F_n(-x) - 1|$$

Butler (1969) and Chatterjee and Sen (1973) have studied some of its properties. In his unpublished dissertation Chentsov (1958) introduced the ω^2-type sttistic

$$R_n^2 = \int_{-\infty}^{\infty} (\Delta F_n(x))^2 dF_n(x).$$

Various properties and variants of it were investigated by Orlov (1972), Rothman and Woodroofe (1972), Srinivasan and Godio (1974), Gregory (1977), Koziol (1979) and Nikitin (1982). A centered version of this statistic was proposed and studied by Hill and Rao (1981) and later by Abbakumov (1986).

The aim of the present paper is to introduce the centered variant of the Kolmogorov-Smirnov statistic K_n, namely

$$H_n = \sup_x \left| \Delta F_n(x) - \int_{-\infty}^{\infty} (\Delta F_n(y)) dF_n(y) \right|$$

and to study its limit distribution, Bahadur efficiency and conditions of local optimality in Bahadur sense. We consider also one-sided statistics of this type:

$$H_n^+ = \sup_x \left[\Delta F_n(x) - \int_{-\infty}^{\infty} (\Delta F_n(y)) dF_n(y) \right]$$

$$H_n^- = \sup_x \left[\int_{-\infty}^{\infty} (\Delta F_n(y)) dF_n(y) - \Delta F_n(x) \right]$$

It should be mentioned that the statistics similar to H_n and $H_n^\pm$ for goodness of fit were proposed by Watson (1976) in the context of testing uniformity on the circle. Later limit distributions of these statistics were studied by Darling (1983a), (1983b), their large deviations and Bahadur efficiency by Nikitin (1987). Further on we obtain similar results for statistics H_n, H_n^+ and H_n^- though it turns out to be somewhat more laborious due to their more complicated structure. Specifically we need the large deviation result for non-degenerate U-statistics of Dasgupta (1984).

2 Limit Distributions

Let $D(0,1)$ be the space of continuous from the left (in the point 0 from the right) functions having only finite number of jumps endowed with the uniform metric ρ. A functional L, defined on $D(0,1)$, is said to be continuously differentiable of order $k \geq 1$ at the point F_0 if there exists a functional $l(F_0, v)$ for which the following holds:

There exists a $v \in C[0,1]$ such that for any sequence $\{v_h\}, h \in I\!R^1, v_h \in D[0,1]$ satisfying $\lim_{h\to 0} \rho(v_h, v) = 0$ one has

$$(1) \qquad \lim_{h\to 0} \frac{L(F_0 + hv_h) - L(F_0)}{h^k} = l(F_0, v);$$

$$(2) \qquad \lim_{h\to 0} l(F_0, v_h) = l(F_0, v).$$

PROPOSITION 1 : (Borovkov (1984, §8)). Let F_n be the empirical d.f. corresponding to a sample $X_1, ..., X_n$ with continuous d.f. F_0 and L a continuously differentiable functional of order k on $D(0,1)$. Then

$$n^{k/2}[L(F_n) - L(F_0)] \to l(F_0, w^0)$$

in distribution, where w^0 is the Brownian bridge process.

As all three statistics H_n, H_n^+ and H_n^- are obviously distribution-free under H we may suppose that $F_0(x) = x, 0 \leq x \leq 1$. It is easy to verify that these statistics considered as functionals of F_n are continuously differentiable of order 1. Applying Proposition 1 we get convergence in distribution

$$\sqrt{n} H_n \to \sup_{0\leq t\leq 1} |Z(t)|$$

$$\sqrt{n} H_n^{\pm} \to \sup_{0\leq t\leq 1} Z(t)$$

where $Z(t)$ is a Gaussian process with zero mean and covariance

$$EZ(t)Z(s) = 2[\min(t, s, 1-t, 1-s) - t(1-t) - s(1-s)] + \frac{1}{3}.$$

The question of obtaining explicit formulas for corresponding limit distributions will not be discussed here.

3 Large deviations under H

We introduce the auxiliary function

$$f(a,t) = -at + \ln\left(\frac{\sinh t}{t}\right), a \geq 0, t \geq 0$$

and put

$$k(a) = \inf\{f(a,t) : t \geq 0\}$$

It follows immediately from the implicit function theorem that the function $k(a)$ is continuous and that

$$k(a) = -\frac{3}{2}a^2(1 + o(1)), \quad a \to 0. \tag{1}$$

We shall show below that the logarithmic asymptotics of large deviations under H is the same for all three statistics under consideration. So we shall study only the statistic H_n^+. The main result of this section is:

THEOREM 1 : Under hypothesis H for sufficiently small positive a

$$\lim_{n\to\infty} n^{-1}\ln P(H_n^+ > a) = -k(a).$$

We need the following result concerning large deviations of U-statistics. Let $\Phi : I\!R^2 \to I\!R^1$ be a symmetric function and consider the U-statistic

$$U_n = \binom{n}{2}^{-1} \sum_{1\leq i<j\leq n} \Phi(X_i, X_j),$$

where $X_1, ..., X_n$ are i.i.d. random variables. Let

$$\Psi(t) = E[\Phi(X_1, X_2)|X_1 = t].$$

PROPOSITION 2 : (Dasgupta (1984)). Suppose that $E\Phi(X_1, X_2) = 0$, $E\Psi^2(X_1) > 0, E\ \exp\{t\Phi^2(X_1, X_2)\} < \infty$ for all $t \in I\!R^1$, and that for some $\varepsilon_0 > 0$

$$\lim_{n\to\infty} n^{-1}\ln P\left(\sum_{i=1}^{n}\Psi(X_i) > n\varepsilon_0\right) < \infty.$$

Then for any $\varepsilon > 0$ and $n \to \infty$

$$P(U_n > 2\varepsilon) = P\left(\sum_{i=1}^{n}\Psi(X_i) > n\varepsilon\right)(1 + o(1))$$

Now we proceed to the proof of Theorem 1 that resembles the proofs in Abrahamson (1967) and Bahadur (1971) where classical Kolmogorov-Smirnov statistics were considered. Without loss of generality we may assume that the disribution of X_i is uniform on $[-1, 1]$. Denote for any $u \in [-1, 1]$

$$H_n^+(u) = F_n(u) + F_n(-u) - \int_{-1}^{1}(F_n(u) + F_n(-u))dF_n(u)$$

and observe that

$$H_n^+(u) = n^{-2} \sum_{i,j=1}^{n} \varphi_u(X_i, X_j) + n^{-1},$$

where the symmetric kernel φ_u is

$$\begin{aligned}\varphi_u(X_i, X_j) &= \frac{1}{2}\{I(X_i < u) + I(X_i < -u) + I(X_j < u) + I(X_j < -u) \\ &\quad -2I(X_i + X_j < 0) - 1\},\end{aligned}$$

and $I(A)$ denotes the indicator of the event A.

In order to apply the Proposition 2 we introduce

$$\overline{H}_n^+ = \binom{n}{2}^{-1} \sum_{1 \leq i < j \leq n} \varphi_u(X_i, X_j)$$

and notice that for any u

$$|H_n^+(u) - \bar{H}_n^+(u)| \leq 7n^{-1} \tag{2}$$

It is easy to verify that

$$\Psi_u(t) \equiv E[\varphi_u(X_1, X_2)|X_1 = t] = \frac{1}{2}\{I(t < u) + I(t < -u) + t - 1\}.$$

Calculating the moment generating function of the random variable $Y = \Psi_u(X_1)$ we obtain for $t \geq 0$

$$\lambda_u(t) \equiv Ee^{tY} = \frac{4}{t} \sinh \frac{t}{4} \cosh \left(\frac{t}{4} - \frac{t|u|}{2}\right).$$

As the function $\lambda_u(t)$ is monotonic in u for $u > 0$ and $u < 0$ it follows that

$$\sup_{-1 \leq u \leq 1} \inf_{t \geq 0} [-at + ln\lambda_u(t)] = \inf_{t \geq 0} [-at + ln\lambda_1(t)] = k(2a) \tag{3}$$

Now for every $u \in [-1, 1]$ we have because of (2)

$$P(H_n^+ > a) \geq P(H_n^+(u) > a) \geq P(\bar{H}_n^+(u) > a + \delta) \tag{4}$$

for arbitrary $\delta > 0$ and sufficiently large n. Applying Proposition 2 for $\Phi = \varphi_n$ to inequality (4) we obtain

$$\lim_{n \to \infty} \inf n^{-1} ln\, P(H_n^+ > a) \geq \lim_{n \to \infty} n^{-1}\, ln\, P\left(n^{-1} \sum_{i=1}^{n} \Psi_u(X_i) > \frac{a + \delta}{2}\right).$$

The limit in the right-hand side may be evaluated by the Chernoff theorem (Chernoff (1952), Bahadur (1971)):

$$\lim_{n \to \infty} n^{-1} ln\, P\left(n^{-1} \sum_{i=1}^{n} \Psi_u(X_i) > \frac{a + \delta}{2}\right) = \inf_{t \geq 0}\left[-\frac{a + \delta}{2} t + ln\lambda_u(t)\right].$$

As $u \in [-1, 1]$ is arbitrary, we have, taking $u = 1$:

$$\lim_{n\to\infty} \inf n^{-1} \ln P(H_n^+ > a) \geq \inf_{t\geq 0}\left[-\frac{a+\delta}{2}t + \ln \lambda_1(t)\right] \equiv k(a+\delta).$$

By continuity of k, taking $\delta \to 0$, we obtain

$$\lim_{n\to\infty} \inf n^{-1} \ln P(H_n^+ > a) \geq k(a) \tag{5}$$

To prove the opposite inequality we fix some large natural number N and consider integers $m, -N \leq m \leq N$. We may write

$$\begin{aligned}
P(H_n^+ > a) &= P\left(\sup_{-1\leq u\leq 1} n^{-2}\sum_{i,j=1}^{n} \varphi_u(X_i, X_j) > a - n^{-1}\right) \\
&\leq P\left(\max_{-N\leq m\leq N}\left[n^{-2}\sum_{i,j=1}^{n}\varphi_{\frac{m}{N}}(X_i, X_j)\right.\right. \\
&+ \left.\left.\frac{1}{2n}\sum_{i=1}^{n} I\left(-\frac{m}{N} < X_i < -\frac{m-1}{N}\right)\right] > a - \frac{1}{n}\right) \\
&\leq \sum_{m=-N}^{N} P\left(H_n^+\left(\frac{m}{N}\right)\right. \\
&+ \left.\frac{1}{2n}\sum_{i=1}^{n} I\left(-\frac{m}{N} < X_i < -\frac{m-1}{N}\right) > a\right) \\
&\leq \sum_{m=-N}^{N} P\left(H_n^+\left(\frac{m}{N}\right) > a - \tau_N\right) \\
&+ \sum_{m=-N}^{N} P\left(\frac{1}{2n}\sum_{i=1}^{n} I\left(-\frac{m}{N} < X_i < -\frac{m-1}{N}\right) > \tau_N\right) \\
&= \Sigma_1 + \Sigma_2,
\end{aligned}$$

say, where $\tau_N > 0$ will be specified below.

To estimate Σ_1 we shall use the so-called Bernstein theorem (Bahadur (1971)):

$$P(Y \geq 0) \leq \inf_{t\geq 0} Ee^{tY}. \tag{6}$$

Using (3) and (6) we have for any $\delta_1 > \tau_N + 7n^{-1}$

$$\begin{aligned}
\Sigma_1 &\leq 2N \sup_{-1\leq u\leq 1} P(H_n^+(u) > a - \tau_N) \\
&\leq 2N \sup_{-1\leq u\leq 1} P(\bar{H}_n^+(u) > a - \delta_1)
\end{aligned}$$

$$\begin{aligned}
&= 2N \sup_{-1\le u\le 1} P\left(n^{-1}\sum_{i=1}^{n}\Psi_u(X_i) > \frac{a-\delta_1}{2}\right)(1+o(1)) \\
&\le 2N \exp\left(n\left[\sup_{-1\le u\le 1}\inf_{t\ge 0}\left(-\frac{a-\delta_1}{2}t + ln\,\lambda_u(t)\right)\right]\right)(1+o(1)) \\
&= 2N\exp(nk(a-\delta_1))(1+o(1)).
\end{aligned}$$

As to the sum $\sum_2$ we have with the aid of (6)

$$\begin{aligned}
\Sigma_2 &\le 2N \sup_{-1\le u\le 1} P\left(n^{-1}\sum_{i=1}^{n} I\left(-u < X_i < -u + \frac{1}{N}\right) > 2\tau_N\right) \\
&\le 2N\exp(-n[(1-2\tau_N)ln(1-2\tau_N) + 2\tau_N ln(2\tau_N) + 2\tau_N\, ln\,(N-1) \\
&\le 2N\exp(-n(-\ln 2 + 2\tau_N\, ln(N-1))).
\end{aligned}$$

If we take now $2\tau_N = (ln(N-1))^{-1/2}$, then we see that $\sum_2$ is exponentially decreasing much faster than the first sum $\sum_1$. It follows that

$$\lim_{n\to\infty}\ \sup n^{-1}\, ln\, P(H_n^+ > a) \le k(a-\delta_1)$$

and consequently because of continuity of k

$$\lim_{n\to\infty}\ \sup n^{-1}\, ln\, P(H_n^+ > a) \le k(a). \tag{7}$$

Inequalities (5) and (7) lead to the conclusion of Theorem 1.

It can be immediately seen that the proof of Theorem 1 is valid with minor changes also for the statistic H_n^- and gives in this case the same result. As $H_n = \max(H_n^+, H_n^-)$, the same is true for H_n, too.

4 Bahadur efficiency and conditions of local Bahadur optimality

We know already from the section 2 that all three statistics under consideration are continuous functionals of empirical d.f. in the uniform topology. It follows from the Glivenko-Cantelli theorem that under the alternative A

$$H_n^+ \to b_{H^+}(\theta) = \sup_x[\Delta G(x+\theta) - \int_{-\infty}^{\infty}\Delta G(y+\theta)dG(y+\theta)] \text{ a.s.}$$

Similar formulas are valid for $b_{H-}(\theta)$ and $b_H(\theta)$. Supposing that the d.f. G has the bounded density g, we can find the main part of $b_{H+}(\theta)$ as $\theta \to 0$, namely

$$b_{H^+}(\theta) \sim 2\sup_x[g(x) - \int_{-\infty}^{\infty} g^2(y)dy]\cdot\theta$$

Analogously

$$b_{H^-}(\theta) \sim 2\int_{-\infty}^{\infty} g^2(y)dy\cdot\theta$$

$$b_H(\theta) \sim 2\sup_x |g(x) - \int_{-\infty}^{\infty} g^2(y)dy| \cdot \theta$$

Large deviation results from the previous section enable now to find local exact Bahadur slopes using Bahadur (1967), p. 309, or Theorem 7.2 of Bahadur (1971). We obtain for $\theta \to 0$

$$\left.\begin{array}{lcl} \bar{C}_{H^+}(\theta) & \sim & 12\{\sup_x[g(x) - \int_{-\infty}^{\infty} g^2(y)dy]\}^2 \cdot \theta^2 \\ \bar{C}_{H^-}(\theta) & \sim & 12(\int_{-\infty}^{\infty} g^2(y)dy)^2\theta^2 \\ \bar{C}_H(\theta) & \sim & 12\sup_x|g(x) - \int_{-\infty}^{\infty} g^2(y)dy|^2\theta^2 \end{array}\right\} \tag{8}$$

It follows from Ho (1973) that the exact slope $\bar{C}_T(\theta)$ of any sequence of statistics $\{T_n\}$ used for testing H against A satisfies the inequality

$$\bar{C}_T(\theta) \le 2K_g(\theta) \equiv 2\int_{-\infty}^{\infty} ln\frac{2g(x+\theta)}{g(x+\theta)+g(-x+\theta)}g(x+\theta)dx \tag{9}$$

that goes back to general results of Bahadur (1967), (1971) and Raghavachari (1970). In typical cases with more or less regular densities g the following asymptotics holds as $\theta \to 0$:

$$K_g(\theta) \sim \frac{1}{2}I(g)\theta^2, \tag{10}$$

where $I(g) = \int_{-\infty}^{\infty} \frac{g'^2(x)}{g(x)}dx$ is the usual Fisher information. So we can compare for such g the local Bahadur indices (that are coefficients at θ^2 in the right-hand side of (8)) with their upper bound $I(g)$. Their ratio may be called the local Bahadur efficiency. Simple calculations show for example that in the case of Laplace distribution this efficiency equals 0.75 for all three sequences of statistics, whereas in the normal case the efficiency of $\{H_n^+\}$ is 0.164, but the efficiency of $\{H_n^-\}$ and $\{H_n\}$ is 0.955.

An interesting question consists in finding for what density g from some *a priori* given class $\mathcal{G}$ these sequences are locally optimal in the Bahadur sense so that (see (9))

$$\bar{C}_H(\theta) \sim 2K_g(\theta), \quad \theta \to 0$$

with similar formulas for one-sided statistics. The statement of the problem seems to go back to Bahadur (1967), p. 313, but the initial results were obtained by Nikitin (1981), (1982), (1984). The solution for statistics $\{H_n^\pm\}$ and $\{H_n\}$ formulated below is very close to results in Nikitin (1987), where he studied the counterparts of these for testing goodness of fit. Consider the class $\mathcal{G}$ of positive, bounded, differentiable and symmetric densities g on $\mathbb{R}^1$ that satisfy (10) with $0 < I(g) < \infty$. We emphasize that the condition (10) is quite natural and is often used in different parts of asymptotic statistics (see, e.g., Borovkov (1984), Kullback (1959) or Ibragimov and Hasminskii (1981)).

Suppose that the alternative distribution of observations has a density g from this class. The condition of local Bahadur optimality of the sequence $\{H_n^+\}$ in $\mathcal{G}$ is fulfilled if

$$12\left[\sup_x\left(g(x)-\int_{-\infty}^{\infty}g^2(y)dy\right)\right]^2=I(g) \tag{11}$$

Denoting $\phi(x)=g(G^{-1}(x))$ we may replace (11) by

$$12\left[\sup_x\left(\phi(x)-\int_0^1\phi(y)dy\right)\right]^2=\int_0^1\phi'^2(x)dx \tag{12}$$

with boundary conditions $\phi(0)=\phi(1)=0$. It is clear that by Cauchy-Schwarz inequality for any x

$$\left[\phi(x)-\int_0^1\phi(y)dy\right]^2=\left(\int_0^1\phi'(y)\left(I(y<x)+y-x-\frac{1}{2}\right)dy\right)^2\leq$$

$$\leq\frac{1}{12}\int_0^1\phi'^2(y)dy)$$

Therefore the equality (12) is valid if

$$\phi'(y)=C\left(I(y<x)+y-x-\frac{1}{2}\right),\quad 0\leq x\leq 1.$$

As it should be $\phi'(0)\geq 0$ and $\phi'(1)\leq 0$, we obtain easily that $x=1/2$. Then we have

$$\phi(y)=C\min{}^2(y,1-y),\quad C>0 \tag{13}$$

and consequently that (11) takes place iff

$$g(x)=\frac{1}{\sigma}p_1\left(\frac{x}{\sigma}\right),\quad \sigma>0,$$

where

$$p_1(x)=\frac{1}{2(1+|x|)^2} \tag{14}$$

The density p_1 is quoted in Hájek, Šidak (1967), Chapter 1, as one of typical examples of strongly unimodal densities.

The local optimality of $\{H_n^-\}$ in $\mathcal{G}$ takes place iff

$$12\left(\int_{-\infty}^{\infty}g^2(y)dy\right)^2=I(g)$$

or

$$12\left(\int_0^1 \phi(y)dy\right)^2 = \int_0^1 \phi'^2(y)dy \tag{15}$$

under zero boundary conditions once again. As

$$\left(\int_0^1 \phi(y)dy\right)^2 = \left(\int_0^1 \phi'(y)\left(y-\frac{1}{2}\right)dy\right)^2 \leq \frac{1}{12}\int_0^1 \phi'^2(y)dy,$$

we conclude that the equality (15) holds iff

$$\phi'(y) = C\left(y-\frac{1}{2}\right)$$

or

$$\phi(y) = Cy(1-y), \quad C>0.$$

It follows that (15) is valid only if g is the logistic density:

$$g(x) = \frac{1}{\sigma}p_2\left(\frac{x}{\sigma}\right), \quad \sigma > 0,$$

where

$$p_2(x) = \frac{e^x}{(1+e^x)^2} \tag{16}$$

It is obvious that the local optimality of $\{H_n\}$ in $\mathcal{G}$ takes place only if g is of the form (15) or (17) (up to the scale factor). The results of this section may by summarized as follows.

THEOREM 2. The sequence $\{H_n^+\}$ is Bahadur locally optimal in the class $\mathcal{G}$ only in the case when the underlying density g of observations under the alternative has the form (14) (up to the scale factor), the sequence $\{H_n^-\}$– only in the case when g is logistic (16) and the sequence $\{H_n\}$ only in these two cases.

Theorem 2 may be considered as a new example of characterization of distributions by the property of local Bahadur optimality of nonparametric statistics.

REFERENCES

[1] Abbakumov, V.L. (1986), *Large deviations and asymptotic efficiency of Watson-type test for symmetry.*, Vestnik Leningrad Univ. N 4, 98-100 (in Russian).

[2] Abrahamson, I.G. (167), *Exact Bahadur efficiencies for the Kolmogorov-Smirnov and Kuiper one - and two - sample statistics.*, Ann. Math. Stat. **38, N 5**, 1475-1490.

[3] Bahadur, R.R. (1967), *Rates of convergence of estimates and test statistics.*, Ann. Math. Stat. **38, N 2**, 303-324.

[4] Bahadur, R.R. (1971), *Some limit theorems in statistics.*, SIAM, Philadelphia.

[5] Borovkov, A.A. (1984), *Mathematical Statistics.*, Moscow, Nauka (in Russian).

[6] Butler C. (1969), *A test for symmetry using the sample distribution function.*, Ann. Math. Stat. **40, N 6**, 2209-2210.

[7] Chatterjee S.K., Sen P.K. (1973), *On Kolmogorov-Smirnov type tests for symmetry.*, Ann. Inst. Stat. Math. **25, N 2**, 287-300.

[8] Chentsov, N.N. (1958), *Substantiation of statistical tests by the methods of random processes.*, Candidate thesis (in Russian).

[9] Chernoff, H. (1952), *A measure of asymptotic efficiency for tests of a hypothesis based on sums of observations.*, Ann. Math. Stat. **23, N 4**, 493-507.

[10] Darling, D.A. (1983a), *On the asymptotic distribution of Watson's statistic.*, Ann. Stat. **11, N 6**, 1263-1266.

[11] Darling, D.A. (1983b), *On the supremum of a certain Gaussian process.*, Ann. Probab. **11, N 4**, 803-806.

[12] Dasgupta R. (1984), *On large deviation probabilities of U-statistics in non-i.i.d. case.* Sankhya. Indian J. Statist. **A46, N 1**, 110-116.

[13] Gregory, G.G. (1977), *Cramér-von Mises type tests for symmetry.*, South Afr. Stat. J. **11, N 1**, 49-61.

[14] Hájek J., Šidak Z.(1967), *Theory of rank tests*, Prague, Academia.

[15] Hill, D.L., Rao, P.V. (1981), *Tests of symmetry based on Watson statistics.*, Commun. Statist. Theor. Meth. **A10, N11**, 1111-1125.

[16] Ho, Nguen Van (1973), *Asymptotic efficiency in the Bahadur sense for the signed rank tests.*, Proc. Prague Symp. on Asympt. Stat. **2**, 127-156.

[17] Hollander, M., Wolfe, D.A. (1973), *Nonparametric statistical methods*, Wiley, N.Y..

[18] Huškova, M. (1984), *Hypothesis of symmetry.*, In: Handbook of Statistics. v.4, Nonparametric Methods, 63-78, North-Holland, Amsterdam.

[19] Ibragimov, I.A., Has'minskii, R.Z. (1981), *Statistical Estimation. Asymptotic Theory.*, Springer-Verlag.

[20] Koziol, J.A. (1980), *On a Cramér-von Mises type statistic for testing symmetry.*, Journ. Amer. Stat. Assoc. **75, N 369**, 161-167.

[21] Kullback, S. (1959), *Information Theory and Statistics.*, Wiley, N.Y..

[22] Nikitin, Ya.Yu.(1981), *Characterization of distributions by the property of local asymptotic optimality of test statistics.*, Zapiski Nauchn. Semin. LOMI **108**, 120-134 (in Russian).

[23] Nikitin, Ya.Yu. (1982), *Bahadur efficiency of integral tests for symmetry.*, Zapiski Nauchn. Semin. LOMI **119**, 181-194(in Russian).

[24] Nikitin, Ya.Yu. (1984), *Bahadur local asymptotic optimality and characterization problems*, Theory Probab. Appl. **29, N 1**, 79-92.

[25] Nikitin, Ya. Yu. (1987), *Bahadur efficiency of Darling-Watson goodness of fit tests.*, Zapiski Nauchn. Semin. LOMI **158**, 138-145 (in Russian).

[26] Orlov A.I. (1972), *On testing symmetry of a distribution.*, Theory Probab. Appl. **17, N 2**, 372-377.

[27] Raghavachari M. (1970), *On a theorem of Bahadur on the rate of convergence of test statistics.*, Ann. Math. Stat. **41, N 6**, 1695-1699.

[28] Rothman, E.D., Woodroofe M. (1972), *A Cramér-von Mises type statistic for testing symmetry.*, Ann. Math. Stat. **43, N 6**, 2035-2038.

[29] Smirnov N.V. (1947), *On a test of symmetry of a distribution of random variable.*, Doklady Acad. Sci. USSR **56, N 1**, 13-16.

[30] Srinivasan, R., Godio, P.B. (1974), *A Cramér-von Mises type statistic for testing symmetry.*, Biometrika **61, N 1**, 196-198.

[31] Watson G.S. (1976), *Optimal invariant tests for uniformity.*, Studies in Probability and Statistics., 121-127, North Holland.

Department of Mathematics of
Petrozavodsk State University
185000 Petrozavodsk
RUSSIA

Department of Mathematics
and Mechanics
St. Petersburg University
198904 RUSSIA

of $X(1|m)$ and $X(2|m)$, given X_m, implies that the conditional distibution of $(X'(1|m),\ X'(2|m))'$, given X_m is $(n-1)$-variate normal distribution with mean vector $\mu^*(m)$ and dispersion matrix $\Sigma^*(m)$ for almost all X_m. It follows by Theorem 4.13 of FKN that X has $N_n(\mu, \Sigma)$ distribution. The conditional independence of $X(1|m)$ and $X(2|m)$ also implies by Lemma 2.1, that

$$\Sigma^*_{12}(m) = 0 \text{ for all } m,\ 2 \leq m \leq n-1,$$

which in turn implies that Σ is a Markov matrix. Thus $(i) \Rightarrow (ii)$. The reverse implication being trivial, the theorem follows.

The following counter-example shows that the condition in Theorem 2.1 that Σ be a positive definite matrix cannot be relaxed. Suppose $n > 2$ and let $X_1 = X_2 = \cdots = X_{n-2} = 0$ almost surely. Suppose $(X_{n-1}, X_n)'$ has the non-normal $EC_2(0,\ \Sigma_2,\ \phi)$ distribution, where Σ_2 is a 2×2 positive definite matrix. Then $(X_1, \cdots, X_n)'$ has the $EC_n\ (0, \Sigma, \phi)$ distribution where

$$\Sigma = \begin{bmatrix} O_{n-2} & O \\ O & \Sigma_2 \end{bmatrix},$$

O_{n-2} is a square null matrix of order $n-2$ and the orders of other null matrices, generically denoted by O are easily determined. It is now readily verified that $(X_1, \cdots, X_n)'$ has the Markov property, although it has non-normal elliptically contoured distribution.

The lemma 2.2 below is a simple generalization of theorem 4.11 of FKN. Although it is not connected with the main result of this section, we state it because it provides a characterization of normality of X in the class Φ_n without requiring that Σ be a diagonal matrix or that it be positive definite.

LEMMA 2.2. Suppose X has $EC_n\ (\mu,\ \Sigma,\ \phi)$ distribution with $\Sigma_{12} = 0$. Suppose there exist at least one j, $1 \leq j \leq m$ and at least one k, $(m+1) \leq k \leq n$, such that σ_{jj} and σ_{kk} are both positive. Then the following statements are equivalent:

(i) The random vector X has $N_n(\mu, \Sigma)$ distribution.

(ii) $X^{(1)}$ and $X^{(2)}$ are independent random vectors.

(iii) If $1 \leq u \leq m$ and $m+1 \leq v \leq n$, then X_u and X_v are independent r.v.s.

We omit the proof as it is similar to that of theorem 4.11 of FKN. The counter example in the preceding paragraph also shows that one cannot relax the condition that σ_{jj} and σ_{kk} be positive in Lemma 2.2.

3 Characterization of the $E_n(\lambda)$ Distribution

Let the $n \times 1$ random vector $Z = (Z_1, \cdots, Z_n)'$ have the l_1-norm symmetric distribution $L_n(F)$. Define

$$Z^{(1)} = (Z_1, \cdots, Z_m)',$$

$$Z^{(2)} = (Z_{m+1}, \cdots, Z_n)' = Z(2|m)$$

and

$$Z(1|m) = (Z_1, \cdots, Z_{m-1})',$$
$$2 \leq m < n.$$

The following theorem provides the characterization stated in the introduction.

THEOREM 3.1: Let $Z = (Z_1, \cdots, Z_n)'$ have the distribution $L_n(F)$ and let F be a non-degenerate d.f. Then the following statements are equivalent.

(i) Z has the Markov property.

(ii) The components $Z_1, \cdots, Z_n$ of Z are i.i.d. exponential r.v.s.

PROOF. Suppose (i) holds. Then the conditional distribution of $(Z'(1|m), Z'(2|m))'$ given Z_m is $L_{n-1}(F_{z_m})$, with the d.f. F_{z_m} depending on Z_m; cf. Theorem 5.7 of FKN. By the Markov property, $Z(1|m)$ and $Z(2|m)$ are conditionally independent, given Z_m. Hence by Theorem 5.18 of FKN, the components of $Z(1|m)$ and $Z(2|m)$ are conditionally i.i.d. exponential $E(\lambda(Z_m))$ r.v.s, whose parameter $\lambda(Z_m)$ may depend on Z_m. However, by the Markov property, the conditional distribution of $Z^{(2)}$ given $Z^{(1)}$ is same as its conditional distribution given Z_m. Moreover, since the distribution F generating $L_n(F)$ is non-degenerate, this conditional distribution of $Z^{(2)}$ given $Z^{(1)}$ is non-degenerate for at least one m, $2 \leq m < n$. Thus, $Z^{(2)}$, conditionally given $Z^{(1)}$ or equivalently given Z_m, has an exponential distribution. Hence by Theorem 5.19 of FKN, Z has i.i.d. $E(\lambda)$ components. This proves that if (i) holds (ii) holds. The reverse implication $(ii) \Rightarrow (i)$, being trivial, the proof of the theorem is complete.

Acknowledgement

The author thanks the referee for his critical comments on the first version of this paper.

REFERENCES

Fang, K.T.; Kotz, S. and Ng, K.W. (1990). *Symmetric multivariate and related distributions*, Chapman and Hall, New York.

Feller, W. (1969). *An introduction to probability theory and its applications*, Vol II, Wiley Eastern, New Delhi.

Lawrence, A.J. (1991). Directionality and reversibility in time series, *International Statist. Review*, **59**, 67-79.

Department of Statistics
University of Poona
Pune - 411 007, India

J.K. Ghosh, S.K. Mitra, K.R. Parthasarathy and B.L.S. Prakasa Rao (*eds.*),
Statistics and Probability : A Raghu Raj Bahadur Festschrift
Published by Wiley Eastern Limited, Publishers, (1993), 19 – 35

Iterative Procedures for Exact Maximum Likelihood Estimation in the First-Order Guassian Moving Average Model

T. W. ANDERSON[1]
Stanford University

and

R. P. MENTZ[2]
University of Tucumán

Abstract

This paper deals with the estimation of the two parameters of a first-order zero-mean Gaussian moving average model by maximizing the exact likelihood function. "Exact" here means that the multivariate normal density function is considered without approximations in its functional form, in the determinant, or in the inverse of the covariance matrix. A general approach proposed for statistical problems in which the covariance matrix

[1] Research supported by U. S. Army Research Office Contract No. DAAL03-89-K-0033.

[2] Research partially supported by the Argentine Research Council (CONICET).

AMS Subject Classification: Primary 62M10.

Key words and Phrases: Moving average, maximum likelihood estimation, iterative procedures.

(or its inverse) has a linear structure with a finite number of parameters was already specialized to ARMA (p, q) models; here it is specialized further to the MA(1) model. The analysis leads to iterative procedures. Consideration of these involves the choice among four dichotomous alternatives: model parameters vs. autocorrelations, likelihood vs. concentrated likelihood, Newton-Raphson vs. scoring, and time vs. frequency domains. Different choices yield different iterative procedures. The procedures involve certain traces and quadratic forms, whose evaluation is also considered. Connection with other approaches in the literature and references are given.

1 Introduction

Maximum likelihood estimates of the parameters in Gaussian time series maximize the likelihood function

$$L(\boldsymbol{\theta}) = (2\pi)^{-T/2}|\boldsymbol{\Sigma}|^{-1/2}\exp\left\{-\frac{1}{2}\boldsymbol{y}'\boldsymbol{\Sigma}^{-1}\boldsymbol{y}\right\}, \tag{1.1}$$

where $\boldsymbol{y}$ is the vector of observations, $\mathcal{E}\boldsymbol{y} = \mathbf{0}$, $\mathcal{E}\boldsymbol{y}\boldsymbol{y}' = \boldsymbol{\Sigma} = \boldsymbol{\Sigma}(\boldsymbol{\theta})$. Exact maximum likelihood estimation is complicated because even for standard models the needed determinants and inverses are either not known in closed forms or, if they are known, they involve complicated parametric functions. Thus only for the first-order autoregressive model are the explicit forms of the maximum likelihood estimators of its two parameters known (Hasza, 1980).

One procedure that is often used is to operate mathematically with the likelihood function, so that it can be evaluated numerically at points of the parameter space, and then to optimize the function by varying values of the parameters, using an efficient computer program. For example, the IMSL Library (1979) package of Fortran subroutines uses a "modified steepest descent algorithm" to compute estimates of the parameters of autoregressive moving average (ARMA) models; the statistical package BMDP (1985) uses "Gauss-Marquardt methods" to perform linear and nonlinear estimations. One idea in this area by Box and Jenkins (1976) was the "backcasting" procedure used to evaluate the approximate likelihood function in some time series models. Useful suggestions have been the Cholesky decomposition of the covariance matrix and "Woodbury's formula," as in Phadke and Kedem (1978), for example. Ansley (1979) studied several approaches and proposed a new algorithm; he reviewed earlier work by Newbold (1974), Dent (1977), Ali (1977), Osborn (1977), and Hillmer and Tiao (1979). He showed the equivalence of the proposals by Newbold (1974) and Dent (1977), and related this approach to Ali (1977). See also Nicholls and Hall (1979).

Another approach is to derive an iterative procedure. Some iterative procedures use the likelihood equations deduced by setting the derivatives of the likelihood function equal to 0 to obtain a procedure of the form $\widehat{\boldsymbol{\theta}}_i = g(\widehat{\boldsymbol{\theta}}_{i-1})$ for some function g depending on $\boldsymbol{y}$, where $\widehat{\boldsymbol{\theta}}_i$, $i = 1, 2, \ldots$ are successive numerical values of the estimates, with $\widehat{\boldsymbol{\theta}}_0$ as a starting value. Anderson (1975, 1977) considered several procedures for exact maximum likelihood estimation and some approximations. Godolphin and de Gooijer (1982) presented a procedure for the first-order moving average model. In general, deriving iterative procedures tends to involve more mathematical elaborations of, and more knowledge about, the likelihood function than the approaches mentioned in the previous paragraph. The Box-Jenkins (1976) "backcasting" procedure is iterative; in this procedure, to deal with the initial values, a "backward" form of the given process is used, in which "time" is reversed. A general iterative procedure for estimation purposes is the so-called "EM-algorithm" (Dempster, et al., 1977), proposed initially for some missing value statistical problems; in our context the initial values can be thought of as missing values, and when applicable the EM algorithm will suggest a sequence of conditional expectation-optimization steps for the iterative computation of the maximum likelihood estimates. Another approach involves Kalman filtering techniques; see, for example, Harvey (1981). Kalman filtering ideas in the present context are often related to incorporating in the likelihood function the error of the optimal (mimimum mean square) prediction of an observation and its variance. Some considerations about these three ideas and procedures ("backcasting", EM-algorithm, and optimal prediction), and in particular their relations to the present analysis, are given in Anderson and Mentz (1990).

The purpose of this paper is to consider in detail iterative procedures for exact maximum likelihood estimation in the first-order Gaussian moving average model. Section 2 introduces the model; Section 3 deals in general with iterative procedures, and the specific procedures are derived in Section 4. Sections 5 contains some remarks about computation and Section 6 a general discussion about the paper.

2 The First-Order Moving Average Model

The Gaussian autoregressive moving average process of order (p, q), denoted here by ARMA (p, q), is defined by

$$y_t + \beta_1 y_{t-1} + \ldots + \beta_p y_{t-p} = u_t + \alpha_1 u_{t-1} + \ldots + \alpha_q u_{t-q}, \quad t = \ldots, -1, 0, 1, \ldots, \tag{2.1}$$

where the y_t are observable, the u_t are unobservable independent normal random variables with $\mathcal{E}u_t = 0$, $\mathcal{E}u_t^2 = \sigma^2$, $0 < \sigma^2 < \infty$. The parameters of the process are $\beta_1, \ldots, \beta_p, \alpha_1, \ldots, \alpha_q$ and σ^2. For stationarity we require that the polynomial equation $z^p + \beta_1 z^{p-1} + \ldots + \beta_p = 0$ has all roots less than 1 in absolute value, in which case (2.1) can be expressed as ("inverted" into) an infinite moving average. The process is stationary for any choice of the moving average parameters; however, if the polynomial equation $w^q + \alpha_1 w^{q-1} + \ldots + \alpha_q = 0$ has all roots less than 1 in absolute value, (2.1) can be inverted into an infinite autoregression. We further assume that the two polynomial equations have no root in common (since otherwise the polynomials would include common factors that could be cancelled out).

In this paper we deal with the special case of a Gaussian first-order moving average process, MA(1), defined by

$$y_t = u_t + \alpha u_{t-1}, \quad t = \ldots, -1, 0, 1, \ldots, \tag{2.2}$$

where the u_t are as those in (2.1), and α and σ^2 are the parameters of the process. The Gaussian MA(1) is stationary for any value of α and is invertible into an infinite autoregression if $|\alpha| < 1$.

The autocovariance (or covariance) sequence of the process is

$$\begin{aligned} \sigma_s &= \sigma^2(1+\alpha^2), & s &= 0, \\ &= \sigma^2\alpha, & |s| &= 1, \\ &= 0, & |s| &> 1. \end{aligned} \tag{2.3}$$

The autocorrelation (or correlation) sequence of the process is

$$\begin{aligned} \rho_s &= 1, & s &= 0, \\ &= \tfrac{\alpha}{1+\alpha^2}, & |s| &= 1, \\ &= 0, & |s| &> 1. \end{aligned} \tag{2.4}$$

For convenience, we write ρ for ρ_1. The covariance sequence satisfies the Fourier inversion formula

$$\sigma_s = \int_{-\pi}^{\pi} e^{i\lambda s} f(\lambda)\, d\lambda, \tag{2.5}$$

where $f(\lambda)$ is the spectral density of the process, given by

$$f(\lambda) = \frac{\sigma^2}{2\pi}|e^{i\lambda} + \alpha|^2 = \frac{\sigma^2}{2\pi}(1 + \alpha^2 + 2\alpha\cos\lambda), \quad -\pi \le \lambda \le \pi. \tag{2.6}$$

If $y_1, \ldots, y_T$ is a sample from (2.2), $\boldsymbol{y} = (y_1, \ldots, y_T)'$ has a multivariate normal distribution with expectation $\mathcal{E}\boldsymbol{y} = \mathbf{0}$ and covariance matrix $\mathcal{E}\boldsymbol{y}\boldsymbol{y}' = \boldsymbol{\Sigma} = (\sigma_{|i-j|})$.

Let us introduce the $T \times T$ matrices $\boldsymbol{P}$ and $\boldsymbol{R}$ (the correlation matrix) defined by

$$\boldsymbol{\Sigma} = \sigma^2\boldsymbol{P} = \sigma^2(1+\alpha^2)\boldsymbol{R} = \sigma_0\boldsymbol{R}. \tag{2.7}$$

We note that $\boldsymbol{\Sigma}$, $\boldsymbol{P}$, and $\boldsymbol{R}$ can be written as linear combinations of the identity matrix $\boldsymbol{I}$ and the matrix $\boldsymbol{G}$ that is symmetric, has 1's along the diagonals immediately above and below the main diagonal, and 0's elsewhere.

In this notation

$$
\begin{aligned}
\boldsymbol{\Sigma} &= \sigma_0 \boldsymbol{I} + \sigma_1 \boldsymbol{G} = \sigma^2(1+\alpha^2)\boldsymbol{I} + \sigma^2 \alpha \boldsymbol{G}, & (2.8)\\
\boldsymbol{P} &= (1+\alpha^2)\boldsymbol{I} + \alpha \boldsymbol{G}, & (2.9)\\
\boldsymbol{R} &= \boldsymbol{I} + \frac{\alpha}{1+\alpha^2}\boldsymbol{G} = \boldsymbol{I} + \rho \boldsymbol{G}. & (2.10)
\end{aligned}
$$

For any α and T, $\boldsymbol{\Sigma}$, $\boldsymbol{P}$, and $\boldsymbol{R}$ are positive definite, the first because we also assume that $\sigma^2 > 0$. As a function of ρ, $\boldsymbol{R}$ is positive definite for $-a < \rho < a$, where $a = 1/\{2\cos[\pi/(T+1)]\}$; see Anderson and Takemura (1986).

The likelihood function can be written as a function of α and σ^2 as

$$
\begin{aligned}
L^*(\alpha, \sigma^2) &= (2\pi)^{-T/2} |\boldsymbol{\Sigma}|^{-1/2} \exp\left\{ -\frac{1}{2} \boldsymbol{y}' \boldsymbol{\Sigma}^{-1} \boldsymbol{y} \right\} & (2.11)\\
&= (2\pi)^{-T/2} (\sigma^2)^{-T/2} |\boldsymbol{P}|^{-1/2} \exp\left\{ -\frac{1}{2\sigma^2} \boldsymbol{y}' \boldsymbol{P}^{-1} \boldsymbol{y} \right\}. & (2.12)
\end{aligned}
$$

Instead of operating with (2.12) for purposes of maximum likelihood estimation, we can separate the analysis into two parts: we maximize (2.12) with respect to σ^2 at

$$\widehat{\sigma}^2 = \frac{1}{T} \boldsymbol{y}' \boldsymbol{P}^{-1} \boldsymbol{y}, \tag{2.13}$$

and then substitute $\sigma^2 = \widehat{\sigma}^2$ in (2.12) to derive the "concentrated likelihood function," which is a constant times the square root of

$$n^*(\alpha) = \frac{1}{|\boldsymbol{P}|(\boldsymbol{y}'\boldsymbol{P}^{-1}\boldsymbol{y})^T}. \tag{2.14}$$

Since $\rho = \alpha/(1+\alpha^2) = (1/\alpha)/[1+(1/\alpha)^2]$, the likelihood function attains all possible values on the set for which $|\alpha| \le 1$, $0 < \sigma^2 < \infty$. Hence, without loss of generality we restrict attention to this set. Note that $|\alpha| < 1$ is the condition for invertibility.

In terms of σ_0 and ρ the likelihood function can be written as

$$L(\sigma_0, \rho) = (2\pi)^{-T/2} (\sigma_0)^{-T/2} |\boldsymbol{R}|^{-1/2} \exp\left\{ -\frac{1}{2\sigma_0} \boldsymbol{y}' \boldsymbol{R}^{-1} \boldsymbol{y} \right\}. \tag{2.15}$$

If the analysis is separated into two parts, we maximize (2.15) with respect to σ_0 at

$$\widehat{\sigma}_0 = \frac{1}{T} \boldsymbol{y}' \boldsymbol{R}^{-1} \boldsymbol{y}, \tag{2.16}$$

and then maximize with respect to ρ the function

$$n(\rho) = \frac{1}{|\boldsymbol{R}|(\boldsymbol{y}'\boldsymbol{R}^{-1}\boldsymbol{y})^T}. \tag{2.17}$$

Note that $n^*(\alpha) = n[\rho(\alpha)]$, where $\rho(\alpha) = \alpha/(1+\alpha^2)$.

For $|\alpha| \leq 1$ and $|\rho| \leq \frac{1}{2}$ the following three pairs provide alternative equivalent parametrizations for (2.2): α and σ^2, σ_0 and σ_1, ρ and σ_0. For example,

$$\alpha = \frac{1 - \sqrt{1 - 4(\sigma_1/\sigma_0)^2}}{2(\sigma_1/\sigma_0)} = \frac{1 - \sqrt{1 - 4\rho^2}}{2\rho}. \tag{2.18}$$

Hence, for purposes of maximum likelihood estimation we can operate with (2.11) as a function of σ_0 and σ_1, with (2.12) as a function of α and σ^2, or with (2.15) as a function of ρ and σ_0. Similarly, in (2.14) we operate with a function of α, and in (2.17) with a function of ρ.

For further details about the moving average model see, for example, Anderson (1971) or Anderson and Mentz (1980).

3 Some Approaches to the Iterative Estimation by Maximum Likelihood

The method of maximum likelihood proposes to estimate the parameters by maximizing (2.11), (2.12) or (2.15); alternatively, we can use the pair (2.13) and (2.14) or the pair (2.16) and (2.17). A basic difficulty comes from the complicated nature of the parametric functions that are involved. Let $\Delta_T = |\boldsymbol{P}|$; then as functions of α and ρ we have for $|\alpha| < 1$,

$$\Delta_T = \frac{1 - \alpha^{2(T+1)}}{1 - \alpha^2} \tag{3.1}$$

$$= \frac{1}{2^T} \sum_{j=0}^{[T/2]} \binom{T+1}{2j+1} (1 - 4\rho^2)^j, \quad T = 0, 1, \dots, \tag{3.2}$$

respectively. These results can be verified by showing that Δ_T satisfies the homogeneous difference equation

$$\Delta_T - (1 + \alpha^2)\Delta_{T-1} + \alpha^2 \Delta_{T-2} = 0, \quad T = 2, 3, \dots,$$

with $\Delta_0 = 1$, $\Delta_1 = (1 - \alpha^4)/(1 - \alpha^2)$. See Anderson (1971), Section 6.7, for example. If p^{ij} denotes the i, j-th component of $\boldsymbol{P}^{-1}$, then

$$p^{ij} = (-\alpha)^{j-i} \frac{\Delta_{i-1}\Delta_{T-j}}{\Delta_T}, \quad j \geq i, \tag{3.3}$$

so that as a function of α,

$$p^{ij} = (-\alpha)^{j-i} \frac{(1 - \alpha^{2i})(1 - \alpha^{2(T-j+1)})}{(1 - \alpha^2)(1 - \alpha^{2(T+1)})}, \quad j \geq i. \tag{3.4}$$

See Shaman (1969).

We consider four aspects of maximum likelihood estimation.

I. Likelihood vs. concentrated likelihood functions

As indicated in Section 2, operating with (2.11) or (2.12) in terms of α and σ^2 is mathematically equivalent to operating with (2.13) and with the concentrated likelihood function or with (2.14); similarly, operating with (2.15) in terms of σ_0 and ρ is mathematically equivalent to operating with (2.16) and (2.17), in the sense that the solutions are the same. However, for the same set of parameters, the estimating equations are not the same for the likelihood and concentrated likelihood functions.

II. α and σ^2 vs. ρ and σ_0

We have a choice of parameters to consider. As indicated in Section 2, the three pairs that we discussed provide equivalent alternative parametrizations for the model (2.2). Since the maximum likelihood estimation procedure is invariant under such type of transformations of the parameters, it follows that from a mathematical point of view it is immaterial with which set we choose to operate. However, different sets of parameters may lead to different estimating equations. For a study of the relations between the α and ρ parametrizations under the concentrated likelihood approach see Anderson and Mentz (1992).

III. Time vs. frequency domains

We presented our results so far (except for (2.5) and (2.6)) in the "time domain". We can consider the effect of a transformation. Let $\boldsymbol{K}$ be the orthogonal $T \times T$ matrix with components

$$\sqrt{\frac{2}{T+1}} \sin \frac{\pi j k}{T+1}, \qquad j, k = 1, \ldots, T, \tag{3.5}$$

and let $\boldsymbol{D}$ be the diagonal matrix with diagonal elements

$$d_j = 2 \cos \frac{\pi j}{T+1}, \qquad j = 1, \ldots, T. \tag{3.6}$$

Then

$$\boldsymbol{K}'\boldsymbol{K} = \boldsymbol{K}\boldsymbol{K}' = \boldsymbol{I}, \qquad \boldsymbol{K}'\boldsymbol{G}\boldsymbol{K} = \boldsymbol{D}. \tag{3.7}$$

We have a way to diagonalize the matrix $\boldsymbol{G}$ appearing in (2.8), (2.9), and (2.10). This transformation (except for a factor of 1/2) is given in detail in Anderson (1971), Section 6.5.4. If $\boldsymbol{y} = \boldsymbol{K}\boldsymbol{z}$, then $\boldsymbol{z} = \boldsymbol{K}'\boldsymbol{y}$, that is,

$$z_j = \sqrt{\frac{2}{T+1}} \sum_{k=1}^{T} y_k \sin \frac{\pi j k}{T+1}, \qquad j = 1, \ldots, T; \tag{3.8}$$

then $\boldsymbol{z}$ is multivariate normal with $\mathcal{E}\boldsymbol{z} = \boldsymbol{0}$ and

$$\mathcal{E}\boldsymbol{z}\boldsymbol{z}' = \sigma_0 \boldsymbol{I} + \sigma_1 \boldsymbol{D} = \sigma_0(\boldsymbol{I} + \rho \boldsymbol{D}). \tag{3.9}$$

The transformation given by (3.5) and (3.6) can be compared with the standard Fourier representation or transform, as given, for example, in Anderson (1971) Section 4.2.

This then provides an alternative approach that we may call a "frequency domain approach": any expression in terms of $\boldsymbol{G}$ can be translated into an expression in terms of $\boldsymbol{D}$, and any method formally presented in terms of $\boldsymbol{y}$ can be translated into a method presented in terms of $\boldsymbol{z}$.

These results will be used in Section 4.3 below to derive expressions for certain traces and quadratic forms. See also the remarks on computations in Section 5.

IV. Scoring vs. Newton-Raphson

To maximize the L or n functions (or the L^* or n^* functions) we have available two procedures based on a Taylor's expansion. We illustrate this with

$$\log n(\rho) = -\log|\boldsymbol{R}| - T\log(\boldsymbol{y}'\boldsymbol{R}^{-1}\boldsymbol{y}) \tag{3.10}$$

coming from (2.17). The expansion of its derivative with respect to ρ around a value ρ_0 is

$$\frac{d}{d\rho}\log n(\rho) = \frac{d}{d\rho}\log n(\rho)\bigg|_{\rho=\rho_0} + (\rho-\rho_0)\frac{d^2}{d\rho^2}\log n(\rho)\bigg|_{\rho=\rho_0} + R(\rho,\rho_0), \tag{3.11}$$

where $R(\rho, \rho_0)$ is a remainder. The estimating equation is obtained by setting this derivative equal to 0.

The Newton-Raphson procedure consists in replacing the remainder by 0 and setting $\rho = \widehat{\rho}^{(i)}$ and $\rho_0 = \widehat{\rho}^{(i-1)}$. The iterative procedure is then

$$\left\{-\frac{d^2}{d\rho^2}\log n(\rho)\right\}\widehat{\rho}^{(i)} = \frac{d}{d\rho}\log n(\rho) + \left\{-\frac{d^2}{d\rho^2}\log n(\rho)\right\}\widehat{\rho}^{(i-1)}, \tag{3.12}$$

where all derivatives are evaluated at $\rho = \widehat{\rho}^{(i-1)}$.

In the method of scoring the second derivative is replaced by its expectation, where the random vector $\boldsymbol{y}$ is taken with distribution having parameter $\rho = \rho_0$.

About these alternative approaches we note that the dichotomies likelihood vs. concentrated likelihood functions and scoring vs. Newton-Raphson procedures, arise from theoretical considerations. The other dichotomies, α and σ^2 vs. ρ and σ_0, and time vs. frequency domains, are motivated on computational grounds.

From this analysis it follows that to estimate the parameters of the model (2.2) by maximum likelihood under normality, we have sixteen alternative approaches, which may lead to different iterative procedures. Some of these have already been presented in the literature, as will be noted in Section 4.

Anderson (1977) emphasized these dichotomies, while operating with the likelihood function.

4 Iterative Procedures Derived from Newton-Raphson and Scoring Methods

In this section we derive several iterative procedures to estimate the parameters of model (2.2), with emphasis on the estimation of ρ. After some preliminaries and the introduction of some notation, we operate successively in the time and frequency domains.

4.1. Notation and general rules

To estimate by maximum likelihood the covariances of the moving average part and the coefficients of the autoregressive part of an ARMA(p,q) model, Anderson (1977), Section 4.1, derived the equations that in general correspond to the iterative procedures in the Gaussian case. In the case of a MA(q) model these equations are

$$\widehat{\boldsymbol{\Lambda}}_{i-1}(\widehat{\boldsymbol{\sigma}}_i - \widehat{\boldsymbol{\sigma}}_{i-1}) = \widehat{\mathbf{s}}_{i-1}, \tag{4.1}$$

where $\widehat{\boldsymbol{\sigma}}_i = (\widehat{\sigma}_0^{(i)}, \widehat{\sigma}_1^{(i)}, \ldots, \widehat{\sigma}_q^{(i)})'$ is the vector of estimated covariances at step i, $\widehat{\boldsymbol{\Lambda}}_{i-1}$ is an estimate of the information matrix for the covariances, and $\widehat{\mathbf{s}}_{i-1}$ is composed of estimates of $\partial \log L/\partial\sigma_j$, $j = 0, 1, \ldots, q$. From (4.1) we deduce the iterative procedure

$$\widehat{\boldsymbol{\Lambda}}_{i-1}\widehat{\boldsymbol{\sigma}}_i = \widehat{\mathbf{s}}_{i-1} + \widehat{\boldsymbol{\Lambda}}_{i-1}\widehat{\boldsymbol{\sigma}}_{i-1} \equiv \widehat{\mathbf{r}}_{i-1}, \tag{4.2}$$

that for the MA(1) model is written in terms of components as

$$\begin{pmatrix} \widehat{\lambda}_{00}^{(i-1)} & \widehat{\lambda}_{01}^{(i-1)} \\ \widehat{\lambda}_{10}^{(i-1)} & \widehat{\lambda}_{11}^{(i-1)} \end{pmatrix} \begin{pmatrix} \widehat{\sigma}_0^{(i)} \\ \widehat{\sigma}_1^{(i)} \end{pmatrix} = \begin{pmatrix} \widehat{r}_0^{(i-1)} \\ \widehat{r}_1^{(i-1)} \end{pmatrix}. \tag{4.3}$$

Solving for $\widehat{\sigma}_0^{(i)}$ and $\widehat{\sigma}_1^{(i)}$ and using $\widehat{\rho}^{(i)} = \widehat{\sigma}_1^{(i)}/\widehat{\sigma}_0^{(i)}$, we obtain the iterative procedure

$$\left\{\widehat{\lambda}_{01}^{(i-1)}\widehat{r}_1^{(i-1)} - \widehat{\lambda}_{11}^{(i-1)}\widehat{r}_0^{(i-1)}\right\}\widehat{\rho}^{(i)} = \widehat{\lambda}_{10}^{(i-1)}\widehat{r}_0^{(i-1)} - \widehat{\lambda}_{00}^{(i-1)}\widehat{r}_1^{(i-1)}. \tag{4.4}$$

The needed λ_{ij} and r_i can be evaluated by using the scoring or Newton-Raphson procedures, as will be shown next.

The derivation of these procedures will lead to certain quadratic forms and traces that we now consider. We use the fact that if $\boldsymbol{A}$ and $\boldsymbol{B}$ are square matrices with $\boldsymbol{B}$ nonsingular, then $\boldsymbol{AB} = \boldsymbol{BA}$ implies that $\boldsymbol{AB}^{-1} = \boldsymbol{B}^{-1}\boldsymbol{A}$. We use this result with $\boldsymbol{A} = \boldsymbol{G}$ and $\boldsymbol{B} = \boldsymbol{R}$, for example. From (2.10) we use $\boldsymbol{R} = \boldsymbol{I} + \rho\boldsymbol{G}$, and hence,

$$\boldsymbol{RG} = (\boldsymbol{I} + \rho\boldsymbol{G})\boldsymbol{G} = \boldsymbol{G} + \rho\boldsymbol{G}^2 = \boldsymbol{G}(\boldsymbol{I} + \rho\boldsymbol{G}) = \boldsymbol{GR}. \tag{4.5}$$

We now define a set of quadratic forms by

$$q_{jk} = \boldsymbol{y}'\boldsymbol{R}^{-(j+1)}\boldsymbol{G}^k\boldsymbol{y}, \quad j = -1, 0, 1, \ldots; \quad k = 0, 1, 2, \ldots, \tag{4.6}$$

and a set of traces by

$$t_{jk} = \operatorname{tr} \boldsymbol{R}^{-\mathrm{j}}\boldsymbol{G}^{\mathrm{k}}, \quad \mathrm{j}, \mathrm{k} = 0, 1, 2, \ldots \tag{4.7}$$

We note that q_{jk} is a random variable, a function of $\boldsymbol{y}$, and that

$$\begin{aligned} \mathcal{E}q_{jk} &= \mathcal{E}\boldsymbol{y}'\boldsymbol{R}^{-(j+1)}\boldsymbol{G}^k\boldsymbol{y} = \mathcal{E}\operatorname{tr}\boldsymbol{y}'\boldsymbol{R}^{-(\mathrm{j}+1)}\boldsymbol{G}^{\mathrm{k}}\boldsymbol{y} \\ &= \operatorname{tr}\boldsymbol{R}^{-(\mathrm{j}+1)}\boldsymbol{G}^{\mathrm{k}}\mathcal{E}\boldsymbol{y}\boldsymbol{y}' = \operatorname{tr}\boldsymbol{R}^{-(\mathrm{j}+1)}\boldsymbol{G}^{\mathrm{k}}\boldsymbol{\Sigma} \\ &= \operatorname{tr}\boldsymbol{R}^{-(\mathrm{j}+1)}\boldsymbol{G}^{\mathrm{k}}\sigma_0\boldsymbol{R} \\ &= \sigma_0\operatorname{tr}\boldsymbol{R}^{-\mathrm{j}}\boldsymbol{G}^{\mathrm{k}} = \sigma_0 \mathrm{t}_{\mathrm{jk}}, \end{aligned} \tag{4.8}$$

where we used (2.7) and the fact that $\boldsymbol{G}$ and $\boldsymbol{R}$ commute.

In our operations we shall find quadratic forms like $\boldsymbol{y}'\boldsymbol{R}^{-1}\boldsymbol{G}\boldsymbol{R}^{-1}\boldsymbol{y}$, $\boldsymbol{y}'\boldsymbol{R}^{-1}\boldsymbol{G}\boldsymbol{R}^{-1}\boldsymbol{G}\boldsymbol{y}$, etc., and traces like $\operatorname{tr}\boldsymbol{R}^{-2}\boldsymbol{G}$, $\operatorname{tr}\boldsymbol{R}^{-1}\boldsymbol{G}\boldsymbol{R}^{-1}$, etc., so that frequent use of this commutative property will be made.

4.2. Time domain considerations

4.2.1. Procedures based on the likelihood function

ITERATIVE PROCEDURE 1 (*Time, Likelihood Function, Scoring*)

In Anderson (1977) Section 4.2, it is shown that (4.1) above leads to a system that in the case of a MA(1) model is

$$\begin{aligned} \operatorname{tr}\widehat{\boldsymbol{\Sigma}}_{\mathrm{i}-1}^{-2}\widehat{\sigma}_0^{(\mathrm{i})} + \operatorname{tr}\widehat{\boldsymbol{\Sigma}}_{\mathrm{i}-1}^{-2}\boldsymbol{G}\widehat{\sigma}_1^{(\mathrm{i})} &= \boldsymbol{y}'\widehat{\boldsymbol{\Sigma}}_{i-1}^{-2}\boldsymbol{y}, \\ \operatorname{tr}\widehat{\boldsymbol{\Sigma}}_{\mathrm{i}-1}^{-1}\boldsymbol{G}\widehat{\boldsymbol{\Sigma}}_{\mathrm{i}-1}^{-1}\widehat{\sigma}_0^{(\mathrm{i})} + \operatorname{tr}\widehat{\boldsymbol{\Sigma}}_{\mathrm{i}-1}^{-1}\boldsymbol{G}\widehat{\boldsymbol{\Sigma}}_{\mathrm{i}-1}^{-1}\boldsymbol{G}\widehat{\sigma}_1^{(\mathrm{i})} &= \boldsymbol{y}'\widehat{\boldsymbol{\Sigma}}_{i-1}^{-1}\boldsymbol{G}\widehat{\boldsymbol{\Sigma}}_{i-1}^{-1}\boldsymbol{y}. \end{aligned} \tag{4.9}$$

These equations can also be written with $\boldsymbol{R}$ instead of $\boldsymbol{\Sigma}$, using the fact that $\boldsymbol{\Sigma} = \sigma_0\boldsymbol{R}$, and σ_0^{-2} cancelled throughout. We can then write the resulting linear system as

$$\begin{pmatrix} \widehat{t}_{20}^{(i-1)} & \widehat{t}_{21}^{(i-1)} \\ \widehat{t}_{21}^{(i-1)} & \widehat{t}_{22}^{(i-1)} \end{pmatrix} \begin{pmatrix} \widehat{\sigma}_0^{(i)} \\ \sigma_1^{(i)} \end{pmatrix} = \begin{pmatrix} \widehat{q}_{10}^{(i-1)} \\ \widehat{q}_{11}^{(i-1)} \end{pmatrix}. \tag{4.10}$$

Solving this system for $\widehat{\sigma}_0^{(i)}$ and $\widehat{\sigma}_1^{(i)}$, and using the definition of $\widehat{\rho}^{(i)}$, we obtain the following iterative procedure:

$$\left\{\widehat{t}_{22}^{(i-1)}\widehat{q}_{10}^{(i-1)} - \widehat{t}_{21}^{(i-1)}\widehat{q}_{11}^{(i-1)}\right\}\widehat{\rho}^{(i)} = \widehat{t}_{20}^{(i-1)}\widehat{q}_{11}^{(i-1)} - \widehat{t}_{21}^{(i-1)}\widehat{q}_{10}^{(i-1)}. \tag{4.11}$$

ITERATIVE PROCEDURE 2 (*Time, Likelihood Function, Newton-Raphson*)
Equations (4.2) in Anderson (1975) become, in the case of a MA(1) model,

$$\left(\boldsymbol{y}'\widehat{\boldsymbol{\Sigma}}_{i-1}^{-3}\boldsymbol{y} - \frac{1}{2}\operatorname{tr}\widehat{\boldsymbol{\Sigma}}_{i-1}^{-2}\right)\widehat{\sigma}_0^{(i)} + \left(\boldsymbol{y}'\widehat{\boldsymbol{\Sigma}}_{i-1}^{-2}\boldsymbol{G}\widehat{\boldsymbol{\Sigma}}_{i-1}^{-1}\boldsymbol{y} - \frac{1}{2}\operatorname{tr}\widehat{\boldsymbol{\Sigma}}_{i-1}^{-2}\boldsymbol{G}\right)\widehat{\sigma}_1^{(i)} = \frac{3}{2}\boldsymbol{y}'\widehat{\boldsymbol{\Sigma}}_{i-1}^{-2}\boldsymbol{y} - \operatorname{tr}\widehat{\boldsymbol{\Sigma}}_{i-1}^{-1},$$

$$\left(\boldsymbol{y}'\widehat{\boldsymbol{\Sigma}}_{i-1}^{-1}\boldsymbol{G}\widehat{\boldsymbol{\Sigma}}_{i-1}^{-2}\boldsymbol{y} - \frac{1}{2}\operatorname{tr}\widehat{\boldsymbol{\Sigma}}_{i-1}^{-1}\boldsymbol{G}\widehat{\boldsymbol{\Sigma}}_{i-1}^{-1}\right)\widehat{\sigma}_0^{(i)} - \left((\boldsymbol{y}'\widehat{\boldsymbol{\Sigma}}_{i-1}^{-1}\boldsymbol{G}\widehat{\boldsymbol{\Sigma}}_{i-1}^{-1}\boldsymbol{G}\widehat{\boldsymbol{\Sigma}}_{i-1}^{-1}\boldsymbol{y} - \frac{1}{2}\operatorname{tr}\widehat{\boldsymbol{\Sigma}}_{i-1}^{-1}\boldsymbol{G}\widehat{\boldsymbol{\Sigma}}_{i-1}^{-1}\boldsymbol{G}\right)\widehat{\sigma}_1^{(i)} = \frac{3}{2}\boldsymbol{y}'\widehat{\boldsymbol{\Sigma}}_{i-1}^{-1}\boldsymbol{G}\widehat{\boldsymbol{\Sigma}}_{i-1}^{-1}\boldsymbol{y} - \operatorname{tr}\widehat{\boldsymbol{\Sigma}}_{i-1}^{-1}\boldsymbol{G}. \tag{4.12}$$

Since $\boldsymbol{G}$ commutes with $\boldsymbol{\Sigma}$ ($\boldsymbol{\Sigma} = \sigma_0\boldsymbol{R}$), it also commutes with $\boldsymbol{\Sigma}^{-1}$. Introducing the notation

$$q_{jk}^* = \boldsymbol{y}'\boldsymbol{\Sigma}^{-(j+1)}\boldsymbol{G}^k\boldsymbol{y}, \qquad t_{jk}^* = \operatorname{tr}\boldsymbol{\Sigma}^{-j}\boldsymbol{G}^k, \tag{4.13}$$

so that $\mathcal{E}q_{jk}^* = t_{jk}^*$, we write (4.12) in matrix form as

$$\begin{pmatrix} \widehat{q}_{20}^{*(i-1)} - \frac{1}{2}\widehat{t}_{20}^{*(i-1)} & \widehat{q}_{21}^{*(i-1)} - \frac{1}{2}\widehat{t}_{21}^{*(i-1)} \\ \widehat{q}_{21}^{*(i-1)} - \frac{1}{2}\widehat{t}_{21}^{*(i-1)} & \widehat{q}_{22}^{*(i-1)} - \frac{1}{2}\widehat{t}_{22}^{*(i-1)} \end{pmatrix}\begin{pmatrix} \widehat{\sigma}_0^{(i)} \\ \widehat{\sigma}_1^{(i)} \end{pmatrix} = \begin{pmatrix} \frac{3}{2}\widehat{q}_{10}^{*(i-1)} & \widehat{t}_{10}^{*(i-1)} \\ \frac{3}{2}\widehat{q}_{11}^{*(i-1)} & -\widehat{t}_{11}^{*(i-1)} \end{pmatrix} \tag{4.14}$$

Since $\widehat{\boldsymbol{\Sigma}}^{-1}$ appears in the coefficients of the system (4.12) raised to different powers, substitution of $\boldsymbol{\Sigma} = \sigma_0\boldsymbol{R}$ will not produce the cancellation of all powers of σ_0^{-1}. Hence, we do not write an expression for ρ as we did in Iterative Procedure 1, but leave the iterations to be carried out for σ_0 and σ_1 as indicated by (4.14).

4.2.2. Procedures based on the concentrated likelihood function

ITERATIVE PROCEDURE 3 (*Time, Concentrated Likelihood Function, Scoring*)

We operate with (3.10), that is,

$$\log n(\rho) = -\log|\boldsymbol{I} + \rho\boldsymbol{G}| - T\log\{\boldsymbol{y}'(\boldsymbol{I} + \rho\boldsymbol{G})^{-1}\boldsymbol{y}\}. \tag{4.15}$$

We have

$$\begin{aligned} \frac{d}{d\rho}\log n(\rho) &= -\operatorname{tr}\boldsymbol{R}^{-1}\boldsymbol{G} + T\frac{\boldsymbol{y}'\boldsymbol{R}^{-1}\boldsymbol{G}\boldsymbol{R}^{-1}\boldsymbol{y}}{\boldsymbol{y}'\boldsymbol{R}^{-1}\boldsymbol{y}} \\ &= -\frac{(\operatorname{tr}\boldsymbol{R}^{-1}\boldsymbol{G})(\boldsymbol{y}'\boldsymbol{R}^{-1}\boldsymbol{y}) - T\boldsymbol{y}'\boldsymbol{R}^{-1}\boldsymbol{G}\boldsymbol{R}^{-1}\boldsymbol{y}}{\boldsymbol{y}'\boldsymbol{R}^{-1}\boldsymbol{y}} \\ &= 0. \end{aligned} \tag{4.16}$$

Since

$$I = R^{-1}R = R^{-1}(I + \rho G) = R^{-1} + \rho R^{-1}G, \tag{4.17}$$

we have

$$\begin{aligned} \operatorname{tr} R^{-1}G &= \operatorname{tr} R^{-1}G(R^{-1} + \rho R^{-1}G) \\ &= \operatorname{tr} R^{-1}GR^{-1} + \rho \operatorname{tr} R^{-1}GR^{-1}G. \end{aligned} \tag{4.18}$$

Substitution in the numerator of (4.16) gives the estimating equation

$$\begin{aligned} \operatorname{tr} \widehat{R}_{i-1}^{-1}G\widehat{R}_{i-1}^{-1}G(y'\widehat{R}_{i-1}^{-1}y)\widehat{\rho}^{(i)} &= Ty'\widehat{R}_{i-1}^{-1}G\widehat{R}_{i-1}^{-1}y - \\ &\qquad \operatorname{tr} \widehat{R}_{i-1}^{-1}G\widehat{R}^{-1}(y'\widehat{R}_{i-1}^{-1}y), \end{aligned} \tag{4.19}$$

or, in the notation introduced in (4.6) and (4.7),

$$\widehat{t}_{22}^{(i-1)}\widehat{q}_{00}^{(i-1)}\widehat{\rho}^{(i)} = T\widehat{q}_{11}^{(i-1)} - \widehat{t}_{21}^{(i-1)}\widehat{q}_{00}^{(i-1)}. \tag{4.20}$$

The equations (4.9) can be obtained by the substitution (4.18) or by replacing second derivatives by their expectations; (4.19) is obtained here in analogy and that is why we call the procedure scoring.

ITERATIVE PROCEDURE 4 (*Time, Concentrated Likelihood Function, Scoring*)

To derive Iterative Procedure 3 we used (4.18). Another approach is to go back to the general structure given by (3.12). The first derivative of $\log n(\rho)$ with respect to ρ is given by (4.16), and the second derivative is

$$\begin{aligned} &\frac{d^2}{d\rho^2}\log n(\rho) \\ &= \operatorname{tr} R^{-1}GR^{-1}G - 2T\frac{y'R^{-1}GR^{-1}G^{-1}Ry}{y'R^{-1}y} + T\frac{(y'R^{-1}GR^{-1}y)^2}{(y'R^{-1}y)^2}. \end{aligned} \tag{4.21}$$

The method of scoring consists of replacing (4.21) by its expected value; instead, we can replace each quadratic form in (4.21) by its expected value, using (4.8). In the notation of (4.6) and (4.17) this is

$$t_{22} + T\frac{-2\sigma_0 t_{22}\sigma_0 T + (\sigma_0 t_{11})^2}{(\sigma_0 T)^2} = -t_{22} + \frac{t_{11}^2}{T}. \tag{4.22}$$

Finally, Iterative Procedure 4 to estimate ρ is given by

$$\begin{aligned} \widehat{q}_{00}^{(i-1)}\left\{\widehat{t}_{22}^{(i-1)} - \frac{1}{T}\left[\widehat{t}_{11}^{(i-1)}\right]^2\right\}\widehat{\rho}^{(i)} &= T\widehat{q}_{11}^{(i-1)} - \widehat{t}_{11}^{(i-1)}\widehat{q}_{00}^{(i-1)} + \widehat{q}_{00}^{(i-1)} \\ &\qquad \left\{\widehat{t}_{22}^{(i-1)} - \frac{1}{T}\left[\widehat{t}_{11}^{(i-1)}\right]^2\right\}\widehat{\rho}^{(i-1)}. \end{aligned} \tag{4.23}$$

ITERATIVE PROCEDURE 5 (*Time, Concentrated Likelihood Function, Newton-Raphson*)

The iterative procedure is (3.12). From (4.16)

$$\begin{aligned}(\boldsymbol{y}'\boldsymbol{R}^{-1}\boldsymbol{y})^2\frac{d}{d\rho}\log n(\rho) &= T(\boldsymbol{y}'\boldsymbol{R}^{-1}\boldsymbol{G}\boldsymbol{R}^{-1}\boldsymbol{y})(\boldsymbol{y}'\boldsymbol{R}^{-1}\boldsymbol{y}) \\ &\quad -\operatorname{tr}\boldsymbol{R}^{-1}\boldsymbol{G}(\boldsymbol{y}'\boldsymbol{R}^{-1}\boldsymbol{y})^2 \\ &= Tq_{11}q_{00} - t_{11}q_{00}^2. \end{aligned} \tag{4.24}$$

From (4.21)

$$-(\boldsymbol{y}'\boldsymbol{R}^{-1}\boldsymbol{y})^2\frac{d^2}{d\rho^2}\log n(\rho) = 2Tq_{22}q_{00} - t_{22}q_{00}^2 - Tq_{11}^2. \tag{4.25}$$

It follows that the iterative procedure for ρ can be written as

$$\begin{aligned}&\left\{2T\widehat{q}_{22}^{(i-1)}\widehat{q}_{00}^{(i-1)} - \widehat{t}_{22}^{(i-1)}\left[\widehat{q}_{00}^{(i-1)}\right]^2 - T\left[\widehat{q}_{11}^{(i-1)}\right]^2\right\}\widehat{\rho}^{(i)} \\ &\quad = T\widehat{q}_{11}^{(i-1)}\widehat{q}_{00}^{(i-1)} - \widehat{t}_{11}^{(i-1)}\left[\widehat{q}_{00}^{(i-1)}\right]^2 \\ &\quad + \left\{2T\widehat{q}_{22}^{(i-1)}\widehat{q}_{00}^{(i-1)} - \widehat{t}_{22}^{(i-1)}\left[\widehat{q}_{00}^{(i-1)}\right]^2 - T\left[\widehat{q}_{11}^{(i-1)}\right]^2\right\}\widehat{\rho}^{(i-1)}.\end{aligned}$$

4.3. Frequency domain considerations

In this section we write all quadratic forms and traces appearing in the iterative procedures presented in Section 4.2., in terms of the elements introduced in paragraph III of Section 3.

From (3.7) we have $\boldsymbol{I} = \boldsymbol{K}\boldsymbol{K}'$ and $\boldsymbol{G} = \boldsymbol{K}\boldsymbol{D}\boldsymbol{K}'$, where $\boldsymbol{D}$ is diagonal with diagonal elements $d_j = 2\cos[\pi j/(T+1)]$, $j = 1, 2, \ldots, T$. Then

$$\boldsymbol{R} = \boldsymbol{I} + \rho\boldsymbol{G} = \boldsymbol{K}\boldsymbol{K}' + \rho\boldsymbol{K}\boldsymbol{D}\boldsymbol{K}' = \boldsymbol{K}(\boldsymbol{I} + \rho\boldsymbol{D})\boldsymbol{K}', \tag{4.26}$$

so that $\boldsymbol{R}$ is also diagonalized by the orthogonal matrix $\boldsymbol{K}$, and $\boldsymbol{I} + \rho\boldsymbol{D}$ has diagonal elements $1 + 2\rho\cos[\pi j/(T+1)]$. It then follows that

$$\boldsymbol{R}^{-s} = \boldsymbol{K}(\boldsymbol{I} + \rho\boldsymbol{D})^{-s}\boldsymbol{K}', \qquad \boldsymbol{G}^s = \boldsymbol{K}\boldsymbol{D}^s\boldsymbol{K}', \qquad s = 0, 1, \ldots. \tag{4.27}$$

Putting these results together we find that the quadratic forms introduced in (4.6) can be written in terms of the z_j defined in (3.8) as

$$\begin{aligned} q_{jk} &= \boldsymbol{y}'\boldsymbol{R}^{-(j+1)}\boldsymbol{G}^k\boldsymbol{y} = \boldsymbol{y}'\boldsymbol{K}(\boldsymbol{I} + \rho\boldsymbol{D})^{-(j+1)}\boldsymbol{K}'\boldsymbol{K}\boldsymbol{D}^k\boldsymbol{K}'\boldsymbol{y} \\ &= (\boldsymbol{K}'\boldsymbol{y})'(\boldsymbol{I} + \rho\boldsymbol{D})^{-(j+1)}\boldsymbol{D}^k(\boldsymbol{K}'\boldsymbol{y}) = \boldsymbol{z}'(\boldsymbol{I} + \rho\boldsymbol{D})^{-(j+1)}\boldsymbol{D}^k\boldsymbol{z} \\ &= \sum_{s=1}^{T}\frac{d_s^k}{(1 + \rho d_s)^{j+1}}z_s^2, \quad j = -1, 0, 1, \ldots; \; k = 0, 1, \ldots, \end{aligned} \tag{4.28}$$

while the traces introduced in (4.7) become

$$\begin{aligned} t_{jk} &= \operatorname{tr} \boldsymbol{R}^{-j}\boldsymbol{G}^{k} = \operatorname{tr} \boldsymbol{K}(\boldsymbol{I}+\rho\boldsymbol{D})^{-j}\boldsymbol{D}^{k}\boldsymbol{K}' \\ &= \operatorname{tr}(\boldsymbol{I}+\rho\boldsymbol{D})^{-j}\boldsymbol{D}^{k}\boldsymbol{K}'\boldsymbol{K} = \operatorname{tr}(\boldsymbol{I}+\rho\boldsymbol{D})^{-j}\boldsymbol{D}^{k} \\ &= \sum_{s=1}^{T} \frac{d_s^k}{(1+\rho d_s)^j}, \qquad j,k = 0,1,\ldots . \end{aligned} \tag{4.29}$$

For an interesting connection of these results with the analysis of variance see the paper by Speed (1987) and the comments by Anderson (1987).

5 Remarks on Computations

The iterative procedures presented in Section 4 involve certain traces $t_{jk} = \operatorname{tr} \boldsymbol{R}^{-j}\boldsymbol{G}^{k}$ and quadratic forms $q_{jk} = \boldsymbol{y}'\boldsymbol{R}^{-(j+1)}\boldsymbol{G}^{k}\boldsymbol{y}$ for small values of j and k. From the definition $\boldsymbol{R} = \boldsymbol{I} + \rho\boldsymbol{G}$ we deduce that $\boldsymbol{G} = \rho^{-1}(\boldsymbol{R} - \boldsymbol{I})$ for $\rho \neq 0$. Then for use in all procedures presented in Section 4 it suffices to compute t_{10}, t_{20}, q_{00}, q_{10}, and q_{20}; for example,

$$t_{11} = \rho^{-1}(T - t_{10}),\ t_{21} = \rho^{-1}(t_{10} - t_{20}),\ t_{22} = \rho^{-2}(T - 2t_{10} + t_{20}), \tag{5.1}$$

and

$$q_{11} = \rho^{-1}(q_{00} - q_{10}),\ q_{21} = \rho^{-1}(q_{10} - q_{20}),\ q_{22} = \rho^{-2}(q_{00} - 2q_{10} + q_{20}). \tag{5.2}$$

Anderson and Mentz (1990) considered several alternative procedures to compute these traces and quadratic forms. From their analyses it follows that evaluation of frequency domain expressions (4.29) for the traces, namely,

$$t_{10} = \sum_{s=1}^{T} \frac{1}{1+\rho d_s}, \qquad t_{20} = \sum_{s=1}^{T} \frac{1}{(1+\rho d_s)^2} \tag{5.3}$$

involve the least number of operations, namely in the order of $2T$ additions and subtractions and $2T$ multiplications and divisions per iteration, for a given value of ρ.

Similar expressions coming from (4.28) could be used for the quadratic forms; in such a case one must first transform the available observations $y_1, \ldots, y_T$ by means of (3.8), which can be done by the Fast Fourier Transform. However, for quadratic forms the necessary computations in the time domain correspond to solving the linear systems

$$\boldsymbol{y} = \boldsymbol{R}\boldsymbol{x}, \qquad \boldsymbol{x} = \boldsymbol{R}\boldsymbol{v} \tag{5.4}$$

for $\boldsymbol{x}$ and $\boldsymbol{v}$, respectively, and this can be done very efficiently by the method of successive elimination. These calculations involve in the order of $7T$ additions and subtractions and $10T$ multiplications and divisions per iteration for a given value of ρ and a fixed set $y_1, \ldots, y_T$ of observations.

This will then provide a computationally effective procedure, whose details are given in Anderson and Mentz (1990), together with a comparison with alternative computational procedures. Questions of speed of convergence and behavior of the procedures for particular values of the parameters and of the sample size remain to be studied.

6 Discussion

In this paper the method of maximum likelihood is applied to the estimation of the two parameters in a Gaussian MA(1) model when observations $y_1, \ldots, y_T$ are available. In spite of the long-standing interest among writers on time series, the problem has proved difficult to analyze.

Many of the known approaches do not deal with the *exact* likelihood function (1.1); sometimes the determinant of the covariance matrix (the Jacobian of the transformation from independent errors) is omitted; sometimes it is approximated; often the inverse covariance matrix in the quadratic form is approximated.

To deal with the exact likelihood function we explore in this paper the possibility of deriving iterative procedures of the general form $\hat{\boldsymbol{\theta}}_i = g(\hat{\boldsymbol{\theta}}_{i-1})$; an alternative procedure is to evaluate the likelihood function at points of the parameter space, and then use an optimization algorithm; however, this approach is not pursued here.

A general approach for the statistical inference in problems where the corresponding covariance matrix or its inverse has a linear structure (is a finite linear combination of known matrices, the coefficients being the unknown parameters) was given in particular for ARMA (p, q) models by Anderson (1977). In the present paper we consider in detail the MA(1) model, and derive several iterative procedures to estimate ρ, the first-order autocorrelation of the process, except for one procedure in which we estimate σ_0 and σ_1, the 0th- and 1st-order autocovariances of the process.

One such procedure, namely (4.20), can be written in full detail as

$$\hat{\rho}^{(i)} = \frac{T\boldsymbol{y}'\hat{\boldsymbol{R}}_{(i-1)}^{-2}G\boldsymbol{y} - \boldsymbol{y}'\hat{\boldsymbol{R}}_{(i-1)}^{-1}\boldsymbol{y}\left(\operatorname{tr}\hat{\boldsymbol{R}}_{(i-1)}^{-2}\boldsymbol{G}\right)}{\boldsymbol{y}'\hat{\boldsymbol{R}}_{(i-1)}^{-1}\boldsymbol{y}\left(\operatorname{tr}\hat{\boldsymbol{R}}_{(i-1)}^{-2}\boldsymbol{G}^2\right)}, \qquad i = 1, 2, \ldots, \quad (6.1)$$

and interpreted as follows: $\boldsymbol{y} = (y_1, \ldots, y_T)'$ is the vector of observations and is given; with an arbitrary starting value $\hat{\rho}^{(0)}$ (which may be a simple inefficient estimate coming from some procedure), we form $\hat{\boldsymbol{R}}_0 = \boldsymbol{I} + \hat{\rho}^{(0)}\boldsymbol{G}$ and then compute all traces and quadratic forms appearing in (6.1); they then determine $\hat{\rho}^{(1)}$, which is used as the first value generated by the iterative procedure, to form $\hat{\boldsymbol{R}}_1 = \boldsymbol{I} + \hat{\rho}^{(1)}\boldsymbol{G}$, and then continue with the iterations. Some stopping rule, like $\left|\hat{\rho}^{(i)} - \hat{\rho}^{(i-1)}\right| < \delta$ can be used to terminate the iterations and to declare $\hat{\rho}$ as the last value generated iteratively.

If ρ is estimated as $\hat{\rho}$ by maximum likelihood, then $\hat{\boldsymbol{R}} = \boldsymbol{I} + \hat{\rho}\boldsymbol{G}$ is the maximum likelihood estimate of the correlation matrix, and

$$\hat{\sigma}_0 = \frac{1}{T} \boldsymbol{y}' \hat{\boldsymbol{R}}^{-1} \boldsymbol{y} \tag{6.2}$$

is the maximum likelihood estimate of σ_0. Then $(\hat{\sigma}_0, \hat{\rho})$ is the maximum likelihood estimate of a pair of parameters, and from them estimates of the other pairs can be deduced; see, however, Anderson and Mentz (1992).

Even when many of the results of Anderson (1977) and related work are valid for general ARMA (p, q) models, the results presented here for the MA(1) model depend in some key ways on the fact that the covariance matrix, or the correlation matrix, of T consecutive observations, can be expressed as a linear combination of the identity matrix $\boldsymbol{I}$ and a matrix $\boldsymbol{G}$, and they are simultaneously diagonalizable by an orthogonal matrix $\boldsymbol{K}$ defined in (3.5). This property will not extend to more complicated models, even to MA(q) models for $q > 1$. It remains to be seen whether the present analysis can be extended to other models, and if so, how the resulting procedures are.

REFERENCES

Ali, M. M. (1977). Analysis of autoregressive-moving average models: estimation and prediction, *Biometrika* **64**, 535-545.

Anderson, T. W. (1971), *The Statistical Analysis of Time Series*, New York: John Wiley and Sons.

Anderson, T. W. (1975). Maximum likelihood estimation of parameters of an autoregressive process with moving average residuals and other covariance matrices with linear structure, *Ann. Statist.* **3**, 1283-1304.

Anderson, T. W. (1977). Estimation for autoregressive moving average models in the time and frequency domains, *Ann. Statist.* **5**, 842–865.

Anderson, T. W. (1987). Discussion of T. P. Speed, What is an analysis of variance? *Ann. Statist.* **15**, 911-913.

Anderson, T. W., and R. P. Mentz (1980). On the structure of the likelihood function of autoregressive and moving average models, *J. Time Series Analysis* **1**, 83-94.

Anderson, T. W., and R. P. Mentz (1990). Iterative procedures for exact maximum likelihood estimation in the first-order Gaussian moving average model. Technical Report No. 25, U.S. Army Research Office Contract DAAL03-89-K-0033, Stanford University.

Anderson, T. W., and R. P. Mentz (1992). A note on maximum likelihood estimation in the first-order Gaussian moving average model, *Statistics and Probability Letters*, **16**, 205-211.

Anderson, T. W., and Akimichi Takemura (1986). Why do noninvertible estimated moving averages occur? *J. Time Series Analysis* **7**, 235–254.

Ansley, C. F. (1979). An algorithm for the exact likelihood of a mixed autoregressive moving average process, *Biometrika* **66**, 59-65.

BMDP (1985). *BMDP Statistical Software*, Los Angeles, California.

Box, G. E. P., and G. M. Jenkins (1976). *Time Series Analysis: Forecasting and Control*, Second edition. San Francisco: Holden-Day.

Dempster, A. P., N. M. Laird, and D. B. Rubin (1977). Maximum likelihood from incomplete data via the EM algorithm, *J. Roy. Statist. Soc. Series B* **39**, 1–38.

Dent, W. (1977). Computation of the exact likelihood function of an ARIMA process, *J. Statistical Computation and Simulation* **5**, 193-206.

Godolphin, E. J., and J. G. de Gooijer (1982). On the maximum likelihood estimation of the parameters of Gaussian moving average process, *Biometrika* **69**, 443–451.

Hasza, David P. (1980). A note on the maximum likelihood estimation for the first-order autoregressive process, *Commun. Statist. A: Theory and Methods* **A9**, 1411-1415.

Harvey, A. C. (1981). *Time Series Models*, Oxford: Philip Allan.

Hillmer, S. G., and G. C. Tiao (1979). Likelihood function of stationary multiple autoregressive moving average models, *J. Amer. Statist. Assoc.* **74**, 652-661.

IMSL (1979). *The IMSL Library*, Houston, Texas.

Newbold, P. (1974). The exact likelihood function for a mixed auto-regressive moving average process, *Biometrika* **61**, 423-426.

Nicholls, D. F., and A. D. Hall (1979). The exact likelihood function of multivariate autoregressive moving average models, *Biometrika* **66**, 259-264.

Osborn, D. R. (1977). Exact and approximate maximum likelihood estimators for vector moving average processes, *J. Roy. Statist. Soc. Series B* **39**, 114-118.

Phadke, M. S., and G. Kedem (1978). Computation of the exact likelihood function of multivariate moving average models, *Biometrika* **65**, 511–520.

Shaman, P. (1969). On the inverse of the covariance matrix of a first-order moving average, *Biometrika* **56**, 595–600.

Speed, T. P. (1987). What is an analysis of variance? *Ann. Statist.* **15**, 885-910.

Department of Statistics
Stanford University
Stanford, CA 94305
U.S.A.

Institute of Statistics
University of Tucumán
4000 Tucumán, ARGENTINA

J.K. Ghosh, S.K. Mitra, K.R. Parthasarathy and B.L.S. Prakasa Rao (*eds.*),
Statistics and Probability : A Raghu Raj Bahadur Festschrift
Published by Wiley Eastern Limited, Publishers, (1993), 37 – 62

Parameter Consistency of Invariant Tests for MANOVA and Related Multivariate Hypotheses

T. W. ANDERSON[1]
Stanford University

and

MICHAEL D. PERLMAN[2]
University of Washington

Abstract

A statistical test is said to be parameter consistent (PC) if, for fixed sample size, its power approaches one for any sequence of alternatives whose distance (in the Kullback-Leibler metric) from the null hypothesis approaches infinity. Necessary and sufficient conditions are given for the parameter consistency of invariant tests for the multivariate analysis of variance (MANOVA). The Roy maximum-root test, the Lawley-Hotelling trace test, and the likelihood ratio test (Wilks' criterion) are PC, whereas the Bartlett-

[1]Research supported by U.S. National Science Foundation Grant No. DMS-89-04851 and by U.S. Army Research Office Contract No. DAAG29-89-k-OO33.
[2]Research supported in part by U.S. National Science Foundation Grant No. DMS-89-02211 and by U.S. National Security Agency Grant No. M,DA 904-92-H-3083.
AMS Subject Classifications : Primary 62H15; Secondary 62F05.
Key Words and Phrases: multivariate analysis of variance, MANOVA, invariant tests, consistency, noncentral Wishart matrix, characteristic roots, stochastic bounds, tests for independence, multivariate components of variance, equality of covariance matrices.

Nanda-Pillai trace test, although consistent in the usual sense as the sample size approaches infinity, fails to be PC unless the significance level or the error degrees of freedom n is sufficiently large. These results extend to the problems of testing independence of variates, testing equality of two covariance matrices against one-sided alternatives, and to other multivariate hypotheses.

1 Introduction

The general multivariate analysis of variance (MANOVA) testing problem can be reduced to the following canonical form by means of sufficiency and invariance (cf. Anderson (1984), §8.3.3; Lehmann (1986), §8.1-8.2): observe the independent random matrices[3]

$$X(p\times r) \sim N(\xi, \Sigma\otimes I_r), \qquad Y(p\times n) \sim N(0,\Sigma\otimes I_n) \tag{1.1}$$

and test

$$H_0 : \xi = 0 \quad \text{vs.} \quad H_1 : \xi \neq 0 \tag{1.2}$$

with Σ unknown. Thus the $r+n$ columns of X and Y are mutually independent, normally distributed column vectors $(p\times 1)$ with common covariance matrix Σ. It is assumed that Σ is nonsingular and $n \geq p$, so that the unbiased estimate $n^{-1}YY'$ of Σ is nonsingular with probability one. The matrix $\xi(p\times r)$ is the matrix of unknown mean vectors of X.

For any matrix with all characteristic roots real S let $\mathrm{ch}_1(S) \geq \mathrm{ch}_2(S) \geq \cdots$ denote its ordered characteristic roots. Define

$$\begin{aligned} t = p\wedge r, &\qquad s = p\vee r, \\ c_i \equiv c_i(X,Y) &= \mathrm{ch}_i[XX'(YY')^{-1}], \quad 1\leq i\leq t, \\ c \equiv c(X,Y) &= (c_1,\ldots,c_t), \end{aligned} \tag{1.3}$$

so that $c_1 \geq \cdots \geq c_t \geq 0$. The random vector c is a representation of the maximal invariant statistic under the group of transformations

$$(X,Y) \to (BX\Psi_1, BY\Psi_2), \tag{1.4}$$

where $B(p\times p)$ is nonsingular and $\Psi_1(r\times r)$ and $\Psi_2(n\times n)$ are orthogonal. The testing problem (1.2) remains invariant under this group. Now define

$$\begin{aligned} \lambda_i &\equiv \lambda_i(\xi,\Sigma) = \mathrm{ch}_i(\xi\xi'\Sigma^{-1}), \quad 1\leq i\leq t, \\ \lambda &\equiv \lambda(\xi,\Sigma) = (\lambda_1,\ldots,\lambda_t), \end{aligned} \tag{1.5}$$

so that $\lambda_1 \geq \cdots \geq \lambda_t \geq 0$ and λ is a representation of the maximal invariant parameter. Then the MANOVA problem (1.2) is equivalent to that of testing

$$H_0 : \lambda = 0 \quad \text{vs.} \quad H_1 : \lambda \neq 0. \tag{1.6}$$

[3] I_r denotes the $r\times r$ identity matrix.

An invariant test for (1.2) ≡ (1.6) depends on (X, Y) only through $c_1, \ldots, c_t$, while its power function depends on (ξ, Σ) only through the noncentrality parameters $\lambda_1, \ldots, \lambda_t$. Since c_i estimates λ_i, the appropriate invariant tests are those that reject H_0 for large values of $c_1, \ldots, c_t$. More precisely, Schwartz (1967b) showed that every admissible invariant test must have a *monotone acceptance region* in terms of $c_1, \ldots, c_t$. In the present paper, therefore, we restrict attention to this class of monotone invariant tests and study their power functions as one or more $\lambda_i \to \infty$.

Perlman and Olkin (1980) showed that every monotone invariant test is *unbiased* for the testing problem (1.2)≡(1.6), so this criterion does not differentiate among admissible invariant tests. Likewise, neither does the usual notion of *consistency*, which we shall call *sample size consistency* (SSC), which requires the power to approach one for every fixed alternative as the sample size approaches infinity. Here we introduce the notion of *parameter consistency* (PC) and show that this property does in fact differentiate among admissible tests.

In a general hypothesis-testing problem, a level α test is defined to be PC if, for fixed sample size, its power approaches one for all sequences of alternatives whose Kullback-Leibler distance from the null hypothesis approaches infinity. For the canonical MANOVA problem this definition is equivalent to the following:

DEFINITION 1.1. For fixed p, r, n, and α, an invariant level α test with acceptance region A for (1.2)≡(1.6) is *parameter consistent* (PC) if its power $P_\lambda\{A^c\}$ approaches 1 as $\|\lambda\| \to \infty$, where $\|\lambda\| = \sum \lambda_i \equiv \mathrm{tr}(\xi\xi'\Sigma^{-1})$.[4] For $i = 1, \ldots, t$, such a test is *parameter consistent of degree i* (PC(i)) if its power at λ approaches 1 as $\lambda_i \to \infty$. Since $\lambda_1 \geq \cdots \geq 0$, PC ⇔ PC(1) ⇒ $\cdots$ ⇒ PC(t). □

Consider a sequence of alternatives $\{(\xi^{(m)}, \Sigma^{(m)})|m = 1, 2, \ldots\}$ such that $\|\lambda^{(m)}\| \to \infty$ as $m \to \infty$, where $\lambda^{(m)} = \lambda(\xi^{(m)}, \Sigma^{(m)})$. For $i = 0, 1, \ldots, t$, such a sequence is said to have *asymptotic rank i* if $(\lambda^{(m)})_j \to \infty$ as $m \to \infty$ for $j = 1, \ldots, i$ but $(\lambda^{(m)})_j$ remains bounded as $m \to \infty$ for $j = i+1, \ldots, t$. Thus, $(\xi^{(m)}, \Sigma^{(m)})$ has asymptotic rank $\geq i$ iff $(\lambda^{(m)})_i \to \infty$ as $m \to \infty$. Clearly, an invariant test is PC(i) iff its power approaches 1 for all sequences of asymptotic rank $\geq i$. (Also see Remark 3.7.)

EXAMPLE 1.2. *(One-way MANOVA, equal sample sizes — cf. Muirhead (1982), §10.7).* Suppose that $\{X_{jk}|1 \leq j \leq J,\ 1 \leq k \leq K\}$ is an array of independent random p-vectors with $X_{jk} \sim N(\xi_j, \Sigma)$. Consider the problem of testing

$$H_0 : \xi_1 = \cdots = \xi_J \quad \text{vs.} \quad H_1 : \text{not } H_0$$

[4] It can be shown that $\|\lambda\| \equiv \mathrm{tr}(\xi\xi'\Sigma^{-1})$ is twice the minimum Kullback-Leibler distance from the alternative distribution $N(\xi, \Sigma \otimes I_r)$ to the null hypothesis $H_0 \equiv \{N(0, \Lambda \otimes I_r)|\Lambda \text{ positive definite}\}$.

with Σ unknown. This can be reduced to the canonical MANOVA model (1.1) with $r = J - 1$, $n = J(K-1)$,

$$X = K^{1/2}(X_1, \cdots, X_J)\Gamma,$$

$$\xi = K^{1/2}(\xi_1, \cdots, \xi_J)\Gamma,$$

where $\Gamma(J \times (J-1))$ satisfies $\Gamma'\Gamma = I_{J-1}$, $(1, \cdots, 1)\Gamma = (0, \ldots, 0)$. For $i = 1, \ldots, t \equiv p \wedge (J-1)$, the matrix ξ has rank i iff $\xi_1, \ldots, \xi_J$ lie in some i-dimensional hyperplane $\mathcal{H}$ in $\mathbb{R}^p$.

A sequence of alternatives $(\xi^{(m)}, \Sigma^{(m)})$ has asymptotic rank i iff for each m there exists an i-dimensional hyperplane $\mathcal{H}^{(m)}$ in $\mathbb{R}^p$ such that the sum of the Euclidean distances from $(\mu_1)^{(m)}, \cdots (\mu_J)^{(m)}$ to $\mathcal{H}^{(m)}$ remains bounded as $m \to \infty$ while the dispersion among $(\mu_1)^{(m)}, \cdots, (\mu_J)^{(m)}$ along any direction within $\mathcal{H}^{(m)}$ approaches ∞, where

$$(\mu_j)^{(m)} = (\Sigma^{(m)})^{-1/2}(\xi_j)^{(m)}.$$

If a monotone invariant test fails to be PC(i), then for fixed p, J, K, and α, its power remains strictly below 1 for every such sequence of alternatives, even though it may be admissible in the decision-theoretic sense. Injudicious or routine use of such a test (for example, in a statistical computer package) could result in failure to detect a sizeable departure from the null hypothesis H_0. □

In Section 3 we present necessary and sufficient conditions for the parameter consistency of monotone invariant tests for MANOVA. These results are based on a series of stochastic bounds for the characteristic roots of a noncentral Wishart matrix, presented first in Section 2. Specific tests are discussed in Section 4. Whereas the Roy maximum-root test, the Lawley-Hotelling trace test, and the likelihood ratio test (Wilks' criterion) are PC, the Bartlett-Nanda-Pillai trace test, although admissible, proper Bayes, locally most powerful invariant, and consistent in the usual sense as the sample size approaches infinity, fails to be PC unless the significance level α or the error degrees of freedom n is sufficiently large. Thus, caution is advised before this latter test criterion is chosen.

These results and conclusions also apply to monotone invariant tests for the independence of two sets of variates (Section 5) and for the equality of two covariance matrices vs. a one-sided alternative in a multivariate components-of-variance model (Section 6). Extensions to the GMANOVA (growth curves) testing problem will be made in a subsequent paper. Some of the results in the present paper were announced in Anderson and Perlman (1987).

2 Stochastic bounds for the characteristic roots of a noncentral Wishart matrix

In order to study the power of an invariant test for the MANOVA problem (1.2)≡(1.6) as a function of the noncentrality parameters $\lambda \equiv (\lambda_1, \ldots, \lambda_t)$, we may assume that $(\xi, \Sigma) = (\mu, I_p)$, where $\mu : p \times r$ satisfies

$$\lambda_i = \text{ch}_i(\mu\mu'), \quad 1 \le i \le t.$$

Under this assumption XX' and YY' have standard Wishart distributions, noncentral and central, respectively, and are independent. Let

$$l_i \equiv l_i(X) = \text{ch}_i(XX'), \quad 1 \le i \le t,$$

so that $l_1 \ge \ldots \ge l_t \ge 0$. Our results on parameter consistency are based on a series of stochastic bounds for $l_1, \ldots, l_t$. In Lemmas 2.1–2.3, p, r, and n remain fixed, while $\mathcal{X}^2_m(u)$ denotes a noncentral chi-square random variable with m degrees of freedom and noncentrality parameter u.

LEMMA 2.1. *For $1 \le k \le t$ and $a > 0$,*
(i) $P\{l_1 + \cdots l_k \le a\} \le P\{\mathcal{X}^2_{ks}(\lambda_1 + \cdots + \lambda_k) \le a\}$,
(ii) $P\{l_{t-k+1} + \cdots + l_t \ge a\} \le P\{\mathcal{X}^2_{ks}(\lambda_{t-k+1} + \cdots + \lambda_t) \ge a\}$.
(Note that equality holds for $k = t$, since $l_1 + \cdots + l_t = \text{tr}(XX') \sim \mathcal{X}^2_{pr}(\lambda_1 + \cdots + \lambda_t)$.)

PROOF. Because $\text{ch}_i(XX') = \text{ch}_i(X'X)$ for $1 \le i \le t$, we may suppose that $p \le r$, so $t = p$, $s = r$. Since

$$l_1 + \cdots + l_k = \sup_\Gamma \text{tr}(\Gamma XX'\Gamma'), \tag{2.1}$$

where $\Gamma : k \times p$ ranges over all row-orthogonal matrices (i.e., $\Gamma\Gamma' = I_k$), and since $\mathcal{X}^2_n(u)$ is stochastically increasing in u, we have

$$\begin{aligned} P[l_1 + \cdots + l_k \le a] &\le \inf_\Gamma P\{\text{tr}(\Gamma XX'\Gamma') \le a\} \\ &= \inf_\Gamma P\{\mathcal{X}^2_{kr}(\text{tr}(\Gamma\mu\mu'\Gamma')) \le a\} \\ &= P\{\mathcal{X}^2_{kr}(\sup_\Gamma \text{tr}(\Gamma\mu\mu'\Gamma')) \le a\} \\ &= P\{\mathcal{X}^2_{kr}(\lambda_1 + \cdots + \lambda_k) \le a\}. \end{aligned}$$

The desired stochastic bound for $l_{t-k+1} + \cdots + l_t$ follows similarly, starting from the relation

$$l_{t-k+1} + \cdots + l_t = \inf_\Gamma \text{tr}(\Gamma XX'\Gamma') \tag{2.2}$$

Now set

$$V = X - \mu \sim N(0, I_p \otimes I_r),$$
$$v_k = \text{ch}_1(VV') + \cdots + \text{ch}_k(VV').$$

Note that the distribution of v_k depends on p and r but not on λ, since VV' has the central Wishart distribution $W_p(I_p, r)$. For $u, v \geq 0$ define

$$G(u,v) = (u^{1/2} + v^{1/2})^2,$$
$$H(u,v) = [(u^{1/2} - v^{1/2}) \vee 0]^2.$$

LEMMA 2.2. *For $1 \leq k \leq t$ and $a > 0$,*

(i) $P\{l_1 + \cdots + l_k \geq a\} \leq P\{G(\lambda_1 + \cdots + \lambda_k,\ v_k) \geq a\}$,

(ii) $P\{l_{t-k+1} + \cdots + l_t \leq a\} \leq P\{H(\lambda_{t-k+1} + \cdots + \lambda_t,\ v_k) \leq a\}$.

PROOF. (i) Write $X = V + \mu$ and apply (2.1). From the Cauchy-Schwarz inequality,

$$\begin{aligned} \text{tr}(\Gamma XX'\Gamma') &= \text{tr}[\Gamma(V+\mu)(V+\mu)'\Gamma'] \\ &= \text{tr}(\Gamma VV'\Gamma') + \text{tr}(\Gamma\mu\mu'\Gamma') + 2\text{tr}(\Gamma V\mu'\Gamma') \\ &\leq \text{tr}(\Gamma VV'\Gamma') + \text{tr}(\Gamma\mu\mu'\Gamma') + 2[\text{tr}(\Gamma VV'\Gamma')\text{tr}(\Gamma\mu\mu'\Gamma')]^{1/2} \\ &\equiv G(\text{tr}(\Gamma\mu\mu'\Gamma'),\ \text{tr}(\Gamma VV'\Gamma')). \end{aligned}$$

Since $G(u,v)$ is increasing in u and v, it follows from (2.1) that

$$\begin{aligned} l_1 + \cdots + l_k &\leq G(\sup_\Gamma \text{tr}(\Gamma\mu\mu'\Gamma'),\ \sup_\Gamma \text{tr}(\Gamma VV'\Gamma')) \\ &= G(\lambda_1 + \cdots + \lambda_k,\ v_k), \end{aligned}$$

from which (i) follows.

(ii) Again by the Cauchy-Schwarz inequality,

$$\begin{aligned} \text{tr}(\Gamma XX'\Gamma') &\geq \text{tr}(\Gamma VV'\Gamma') + \text{tr}(\Gamma\mu\mu'\Gamma') - 2[\text{tr}(\Gamma VV'\Gamma')\text{tr}(\Gamma\mu\mu'\Gamma')]^{1/2} \\ &= [(\text{tr}(\Gamma\mu\mu'\Gamma'))^{1/2} - (\text{tr}(\Gamma VV'\Gamma'))^{1/2}]^2 \\ &\geq H(\text{tr}(\Gamma\mu\mu'\Gamma'),\ \text{tr}(\Gamma VV'\Gamma')). \end{aligned}$$

Since $H(u,v)$ is increasing in u and decreasing in v, it follows from (2.2) that

$$\begin{aligned} l_{t-k+1} + \cdots + l_t &\geq H(\inf_\Gamma \text{tr}(\Gamma\mu\mu'\Gamma'),\ \sup_\Gamma \text{tr}(\Gamma VV'\Gamma')) \\ &= H(\lambda_{t-k+1} + \cdots + \lambda_t,\ v_k), \end{aligned}$$

which implies (ii). □

The stochastic bounds in Lemmas 2.1 and 2.2 are fairly sharp. They lead to the following stochastic bounds for individual characteristic roots that are not as sharp, but that suffice to determine parameter consistency.

LEMMA 2.3. *For $1 \le k \le t$ and $a > 0$,*

$$P\{H(\lambda_{t-k+1}, v_k) \ge ka\} \le P\{l_{t-k+1} \ge a\} \le P\{\mathcal{X}^2_{ks}(k\lambda_{t-k+1}) \ge a\}.$$

PROOF. Begin with the inequalities

$$\frac{1}{k}(l_{t-k+1} + \cdots + l_t) \le l_{t-k+1} \le l_{t-k+1} + \cdots + l_t. \tag{2.3}$$

Lemma 2.1(ii) then implies that

$$\begin{aligned} P\{l_{t-k+1} \ge a\} &\le P\{l_{t-k+1} + \cdots + l_t \ge a\} \\ &\le P\{\mathcal{X}^2_{ks}(\lambda_{t-k+1} + \cdots + \lambda_t) \ge a\} \\ &\le P\{\mathcal{X}^2_{ks}(k\lambda_{t-k+1}) \ge a\}, \end{aligned}$$

while Lemma 2.2(ii) yields

$$\begin{aligned} P\{l_{t-k+1} \ge a\} &\ge P\{l_{t-k+1} + \cdots + l_t \ge ka\} \\ &\ge P\{H(\lambda_{t-k+1} + \cdots + \lambda_t,\ v_k) \ge ka\} \\ &\ge P\{H(\lambda_{t-k+1},\ v_k) \ge ka\}. \end{aligned}$$

□

REMARK 2.4. The inequality $l_k \le (l_1 + \cdots l_k)/k$ and Lemma 2.2(i) together imply that

$$P\{l_k \ge a\} \le P\{G(\lambda_1 + \cdots + \lambda_k,\ v_k) \ge ka\},$$

but this does not lead to an upper bound in terms of λ_k alone. Similarly, Lemma 2.1(i) does not yield a lower bound in terms of λ_k alone. □

Because $H(u,v) \to \infty$ as $u \to \infty$ and $\mathcal{X}^2_m(u)$ is stochastically increasing in u, and since the distribution of v_k does not involve λ, Lemma 2.3 implies that $l_i \xrightarrow{P} \infty$ if and only if $\lambda_i \to \infty$, uniformly in $\{\lambda_j | j \ne i\}$. In order to determine the stochastic behavior of $c_i \equiv \text{ch}_i[XX'(YY')^{-1}]$, apply the inequalities

$$\frac{l_i}{\text{ch}_1(YY')} \le c_i \le \frac{l_i}{\text{ch}_p(YY')}.$$

Since $l_i \equiv \text{ch}_i(XX')$ is independent of Y, Lemma 2.3 yields

$$\begin{aligned} &P\left\{\frac{H(\lambda_i, v_{t-i+1})}{\text{ch}_1(YY')} \ge (t-i+1)a\right\} \\ &\le P\{c_i \ge a\} \le P\left\{\frac{\mathcal{X}^2_{(t-i+1)s}((t-i+1)\lambda_i)}{\text{ch}_p(YY')} \ge a\right\}, \end{aligned} \tag{2.4}$$

where each numerator is independent of the corresponding denominator. Furthermore, $YY' \sim W_p(I_p, n)$, so its distribution does not involve λ. Finally, YY' is nonsingular with probability 1 since $n \geq p$, and $\text{ch}_1(YY') \leq \text{tr}(YY') \sim \mathcal{X}^2_{pn}$, so

$$P\{0 < \text{ch}_p(YY') < \text{ch}_1(YY') < \infty\} = 1. \tag{2.5}$$

It follows that $c_i \xrightarrow{\text{P}} \infty$ if and only if $\lambda_i \to \infty$, uniformly in $\{\lambda_j | j \neq i\}$.

3 Parameter consistency of monotone invariant tests for MANOVA

In order to avoid consideration of points at infinity in the sample space, it is convenient to characterize parameter consistent invariant tests for the MANOVA problem (1.2) ≡ (1.6) in terms of

$$\begin{aligned} d_i &\equiv d_i(X,Y) = \text{ch}_i[XX'(XX'+YY')^{-1}], \quad 1 \leq i \leq t, \\ d &\equiv d(X,Y) = (d_1, \ldots, d_t), \end{aligned} \tag{3.1}$$

where the range of d is the bounded set

$$\mathcal{D}_t \equiv \{x \equiv (x_1, \ldots, x_t) | 1 > x_1 \geq \cdots \geq x_t \geq 0\}. \tag{3.2}$$

Since

$$d_i = c_i(c_i+1)^{-1}, \tag{3.3}$$

d_i is a strictly increasing function of c_i, hence d is equivalent to c as a representation of the maximal invariant statistic and $d_i \xrightarrow{\text{P}} 1$ if and only if $c_i \xrightarrow{\text{P}} \infty$. Furthermore the central ($\lambda = 0$) and noncentral ($\lambda \neq 0$) distributions of d are absolutely continuous with respect to Lebesgue measure on the interior of $\mathcal{D}_t$.

For any subset $A \subseteq \mathcal{D}_t$, let $\bar{A}$ denote its closure in $\bar{\mathcal{D}}_t$, where

$$\bar{\mathcal{D}}_t = \{x | 1 \geq x_1 \geq \cdots \geq x_t \geq 0\}. \tag{3.4}$$

A subset $A \subseteq \mathcal{D}_t$ ($A \subseteq \bar{\mathcal{D}}_t$) is said to be *monotone* in $\mathcal{D}_t$ (in $\bar{\mathcal{D}}_t$) if

$$x \in A,\ y \in \mathcal{D}_t\ (y \in \bar{\mathcal{D}}_t),\ x \geq y \Rightarrow y \in A. \tag{3.5}$$

Here, $x \geq y$ means that $x_i \geq y_i$ for $i = 1, \ldots, t$. If A is monotone in $\mathcal{D}_t$, then $\bar{A}$ is monotone in $\bar{\mathcal{D}}_t$. Schwartz (1967b, Theorem 2) noted that the acceptance region (in terms of d) of any admissible invariant test is Lebesgue-equivalent to a monotone subset A of $\mathcal{D}_t$; hence, since $\text{Leb}\{\bar{A} \backslash A\} = 0$ by Lemma 3.2(ii) (to follow), such an acceptance region is Lebesgue-equivalent to a closed monotone subset $\bar{A} \subseteq \bar{\mathcal{D}}_t$. Therefore Theorem 3.1, our main result, although stated for *closed* monotone invariant acceptance regions, in fact determines necessary and sufficient conditions for the parameter consistency of *any* monotone acceptance region in $\mathcal{D}_t$, hence for the parameter consistency of *any* admissible invariant test.[5]

Denote the vertices of $\bar{\mathcal{D}}_t$ by $e_0(\equiv 0)$, $e_1, \ldots, e_t$, where

$$e_i = (\underbrace{1, \ldots, 1}_{i}, \underbrace{0, \ldots, 0}_{t-i}).$$

For $0 \leq \eta \leq 1$ and $0 \leq i \leq t$ define $e_i(\eta) \in \bar{\mathcal{D}}_t$ by

$$e_i(\eta) = e_i + \eta(e_t - e_i) \equiv (\underbrace{1, \ldots, 1}_{i}, \underbrace{\eta, \ldots, \eta}_{t-i}).$$

Note that $e_0(\eta) \leq e_1(\eta) \leq \cdots \leq e_t(\eta) \equiv e_t$, so that if $A \subseteq \bar{\mathcal{D}}_t$ is monotone, then $e_i(\eta) \in A \Rightarrow e_{i-1}(\eta) \in A$. For convenience we define $\lambda_0 = \infty$, $\lambda_{t+1} = 0$, and say that a test is PC(0) if its power function is 1 for every λ, that is, if its rejection region coincides with $\bar{\mathcal{D}}_t$ a.e.

THEOREM 3.1. *Let $A \subseteq \bar{\mathcal{D}}_t$ be a closed monotone invariant level α acceptance region for the testing problem (1.2) $\equiv$ (1.6). Fix α, p, r, n, and i, $0 \leq i \leq t$.*

(i) If $e_i(\eta) \notin A$ for all $\eta > 0$ (in particular, if $e_i \notin A$), then

$$P_\lambda\{d \notin A\} \to 1$$

as $\lambda_i \to \infty$, i.e., the test with acceptance region A is PC(i).

(ii) If $e_i(\eta) \in A$ for some $\eta > 0$, then the test is not PC(i).

(iii) Hence, for $1 \leq i \leq t$ a necessary and sufficient condition that the invariant test with acceptance region A be PC(i) *is that $e_i(\eta) \notin A$ for all $\eta > 0$. A sufficient condition is that $e_i \notin A$.*

(iv) For $1 \leq i \leq t$, if $e_i(\eta) \notin A$ for all $\eta > 0$ (in particular, if $e_i \notin A$) but $e_{i-1}(\eta) \in A$ for some $\eta > 0$, then $P_\lambda\{d \notin A\} \to 1$ if and only if $\lambda_i \to \infty$, i.e., the test is PC(i) *but not* PC($i-1$).

PROOF. (i) The result is trivial for $i = 0$, so assume that $1 \leq i \leq t$. We must show that for $\varepsilon > 0$, there exists $N_\varepsilon \equiv N(\varepsilon, A)$ such that

$$P_\lambda\{d \notin A\} \geq 1 - \varepsilon \tag{3.6}$$

[5] In fact, Schwartz also proved that the acceptance region of an admissible invariant test is not only a.e. monotone but is also convex in $d_1^{\frac{1}{2}}, \ldots, d_t^{\frac{1}{2}}$. Therefore, Theorem 3.1 also applies to some inadmissible invariant tests.

whenever $\lambda_i \geq N_\varepsilon$. For $0 < \beta < \eta < 1$ define $e_i(\eta, \beta) = e_i(\eta) - \beta e_t \in \mathcal{D}_t$. Since A is closed and $e_i(\eta) \notin A$, there exists $0 < \beta(\eta) < \eta$ such that $e_i(\eta, \beta(\eta)) \in \mathcal{D}_t \backslash A$. Therefore, by the monotonicity of A,

$$\begin{aligned} P_\lambda\{d \notin A\} &\geq P_\lambda\{d \geq e_i(\eta, \beta(\eta))\} \quad (3.7)\\ &= P_\lambda\{d_i \geq 1 - \beta(\eta),\ d_t \geq \eta - \beta(\eta)\} \\ &\geq P_\lambda\{d_i \geq 1 - \beta(\eta)\} + P_\lambda\{d_t \geq \eta - \beta(\eta)\} - 1 \\ &\geq P_\lambda\{d_i \geq 1 - \beta(\eta)\} + P_0\{d_t \geq \eta - \beta(\eta)\} - 1. \end{aligned}$$

The last inequality is a consequence of Theorem 4.2 of Perlman and Olkin, (1980). Since $\eta - \beta(\eta) \to 0$ as $\eta \to 0$ and $P_0\{d_t > 0\} = 1$, given $\varepsilon > 0$ there exists $\eta_\varepsilon > 0$ such that

$$P_0\{d_t \geq \eta_\varepsilon - \beta(\eta_\varepsilon)\} \geq 1 - \frac{\varepsilon}{2}.$$

Also, from (2.4) and (3.3),

$$P_\lambda\{d_i \geq 1 - \beta(\eta_\varepsilon)\} \geq P\left\{\frac{H(\lambda_i, v_{t-i+1})}{\mathrm{ch}_1(YY')} \geq (t - i + 1)\frac{1 - \beta(\eta_\varepsilon)}{\beta(\eta_\varepsilon)}\right\}. \quad (3.8)$$

Because $H(u, v) \to \infty$ as $u \to \infty$, there exists N_ε such that

$$P_\lambda\{d_i \geq 1 - \beta(\eta_\varepsilon)\} \geq 1 - \frac{\varepsilon}{2}$$

whenever $\lambda_i \geq N_\varepsilon$. Thus, $\lambda_i \geq N_\varepsilon$ implies that (3.6) holds.

(ii) The result is trivial for $i = t$, so assume that $0 \leq i \leq t - 1$. We must show that for any $0 < M < \infty$ there exists $\delta_M \equiv \delta(M; A)$ such that $P_\lambda\{d \notin A\} \leq 1 - \delta_M < 1$ whenever $\lambda_{i+1} \leq M$. Since A is monotone and $e_i(\eta) \in A$, $A \supseteq \{d \in \bar{\mathcal{D}}_t | d_{i+1} \leq \eta\}$. Then as above, it again follows from (2.4) and (3.3) that

$$\begin{aligned} P_\lambda\{d \notin A\} &\leq P_\lambda\{d_{i+1} > \eta\} \quad (3.9)\\ &= P_\lambda\left\{c_{i+1} > \frac{\eta}{1 - \eta}\right\} \\ &\leq P\left\{\frac{\chi^2_{(t-i)s}((t - i)\lambda_{i+1})}{\mathrm{ch}_p(YY')} > \frac{\eta}{1 - \eta}\right\}, \end{aligned}$$

from which the desired result follows immediately.

(iii), (iv). These facts follow directly from (i) and (ii). □

If the upper boundary of A is not too irregular, Theorem 3.1 states that the test with acceptance region A is PC(i) if either $e_i \notin A$ (cf. Fig. 3.1a) or e_i lies in the upper boundary of A (Fig. 3.1b), whereas it fails to be PC(i) if $e_i \in A \backslash$ (upper boundary of A) (Fig. 3.1c). Thus, the test is PC(i) if e_i lies either in the *rejection* region or in its lower boundary. These three cases are illustrated in Figures 3.1a,b,c, where $i = 1$ and $t = 2$. Note that if $\alpha > 0$, every monotone invariant level α test must be PC(t).

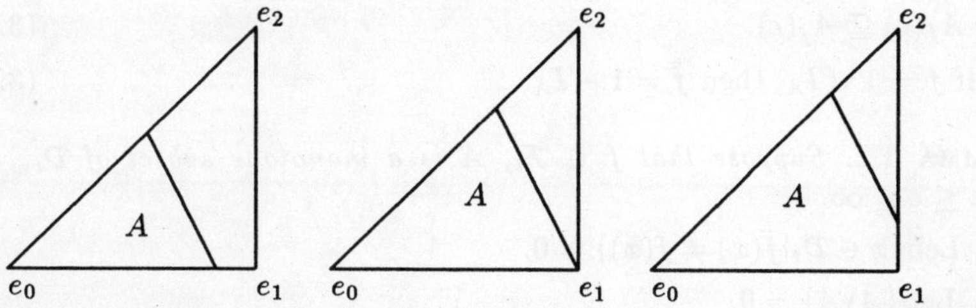

Fig. 3.1a: test is PC(1). **Fig. 3.1b**: test is PC(1). **Fig. 3.1c**: test is *not* PC(1).

An invariant level α acceptance region $A \subseteq \mathcal{D}_t$ usually is defined in terms of an invariant test statistic $f \equiv f(d)$ as follows:

$$A = A_f(c_\alpha) \equiv \{d \in \mathcal{D}_t | f(d) \leq c_\alpha\}, \tag{3.10}$$

where $c_\alpha \equiv c_\alpha(p, r, n; f)$ satisfies

$$P_{\lambda=0}\{f(d) \leq c_\alpha\} = 1 - \alpha. \tag{3.11}$$

If f is *monotone* on $\mathcal{D}_t$ (i.e., nondecreasing in each d_i) then $A_f(c_\alpha)$ is a monotone invariant acceptance region with power function given by $P_\lambda\{f(d) > c_\alpha\}$. In Corollary 3.3 we shall restate Theorem 3.1 in order to characterize PC in terms of the test statistic f. First we show that without loss of generality, we may restrict attention to statistics f that are lower semicontinuous as well as monotone.

Let $\mathcal{F}_t$ denote the collection of all extended real-valued, monotone test statistics f defined on $\mathcal{D}_t$. If $f \in \mathcal{F}_t$, then for every $c \in [-\infty, \infty]$, $A_f(c)$ given by (3.10) is a monotone acceptance region; conversely, every monotone region $A \subseteq \mathcal{D}_t$ is of the form (3.10) for some $f \in \mathcal{F}_t$ (e.g., take $f = 1 - I_A$, $c = 0$).

Let $\bar{\mathcal{F}}_t$ denote the collection of all extended real-valued, monotone, lower semicontinuous functions defined on $\bar{\mathcal{D}}_t$. For $f \in \mathcal{F}_t$, define $\bar{f} \in \bar{\mathcal{F}}_t$ as follows:

$$\bar{f}(x) = f(x-) \equiv \lim_{\varepsilon \downarrow 0} f((1-\varepsilon)x), \quad x \in \bar{\mathcal{D}}_t. \tag{3.12}$$

The following properties are readily verified:

$$\bar{f} \text{ is monotone on } \bar{\mathcal{D}}_t. \tag{3.13}$$

$$\bar{f}(x-) = \bar{f}(x) \text{ for all } x \in \bar{\mathcal{D}}_t. \tag{3.14}$$

$$\bar{f} \text{ is lower semicontinuous on } \bar{\mathcal{D}}_t \text{ (hence } \bar{f} \in \bar{\mathcal{F}}_t). \tag{3.15}$$

$$A_{\bar{f}}(c) \equiv \{x \in \bar{\mathcal{D}}_t | \bar{f}(x) \leq c\} \text{ is closed and monotone.} \tag{3.16}$$

$$A_{\bar{f}}(c) = \overline{A_f(c)}. \tag{3.17}$$

$\bar{f}(x) \leq f(x)$ for all $x \in \mathcal{D}_t$. (3.18)

If $f \in \mathcal{F}_t$, $g \in \bar{\mathcal{F}}_t$, and $g \leq f$ on $\mathcal{D}_t$, then $g \leq \bar{f}$ on $\mathcal{D}_t$. (3.19)

$A_f(c) \subseteq A_{\bar{f}}(c)$. (3.20)

If $f = 1 - I_A$, then $\bar{f} = 1 - I_{\bar{A}}$. (3.21)

LEMMA 3.2. *Suppose that $f \in \mathcal{F}_t$, A is a monotone subset of $\mathcal{D}_t$, and $-\infty \leq c \leq \infty$.*

(i) $\text{Leb}\{x \in \mathcal{D}_t | \bar{f}(x) \neq f(x)\} = 0$.

(ii) $\text{Leb}\{\bar{A} \backslash A\} = 0$.

(iii) $\text{Leb}\{A_{\bar{f}}(c) \backslash A_f(c)\} = 0$.

PROOF. (i) Transform to polar coordinates and apply Fubini's Theorem as follows. For $x \in \mathcal{D}_t$ $(x \neq 0)$ define

$$\begin{aligned} y &\equiv y(x) = (x_1^{-1}x_2, \ldots, x_1^{-1}x_t) \in \bar{\mathcal{D}}_{t-1}, \\ \varphi_y(x_1) &= f(x_1, x_1 y) \equiv f(x), \\ B_y &= \{x_1 \in [0,1] | \varphi_y(x_1-) \neq \varphi_y(x_1)\} \equiv \{x_1 | \bar{f}(x_1, x_1 y) \neq f(x_1, x_1 y))\}, \end{aligned}$$

so that

$$\text{Leb}\{x | \bar{f}(x) \neq f(x)\} = \int_{\bar{\mathcal{D}}_{t-1}} \left\{ \int_{B_y} x_1^{t-1} dx_1 \right\} dy.$$

Since $\varphi_y(\cdot)$ is nondecreasing, B_y is at most countable, so the iterated integral is 0.

(ii) Apply (i) and (3.21).

(iii) Apply (ii) and (3.17). □

Lemma 3.2 shows that for every $f \in \mathcal{F}_t$ there exists $\bar{f} \in \bar{\mathcal{F}}_t$ such that the level α tests determined by f and $\bar{f}$ are equivalent, i.e., have identical power functions. Therefore the following corollary to Theorem 3.1, although stated for a monotone *lower semicontinuous* statistic $f \in \bar{\mathcal{F}}_t$, also determines necessary and sufficient conditions for parameter consistency of a test determined by *any* monotone test statistic $g \in \mathcal{F}_t$ (just apply Corollary 3.3 with f replaced by $\bar{g}$).

COROLLARY 3.3. *Suppose that $f \in \bar{\mathcal{F}}_t$ and that c_α satisfies (3.11). Fix α, p, r, n, and i, $0 \leq i \leq t$.*

(i) If $f(e_i(\eta)) > c_\alpha$ for all $\eta > 0$ (in particular, if $f(e_i) > c_\alpha$), then

$$P_\lambda\{f(d) > c_\alpha\} \to 1$$

as $\lambda_i \to \infty$, i.e., the level α test based on f is PC(i). *If $\eta = 0$ is a point of increase of $f(e_i(\eta))$ (i.e., $\eta > 0 \Rightarrow f(e_i(\eta)) > f(e_i)$), then $f(e_i) \geq c_\alpha$ suffices.*

(ii) If $f(e_i(\eta)) \le c_\alpha$ for some $\eta > 0$, then the test is not PC(i). *If $f(e_i(\eta))$ is continuous at $\eta = 0$, then $f(e_i) < c_\alpha$ suffices.*

(iii) Hence, for $1 \le i \le t$, a necessary and sufficient condition that the level α test based on f be PC(i) *is that $f(e_i(\eta)) > c_\alpha$ for all $\eta > 0$. If $\eta = 0$ is both a point of increase and a continuity point of $f(e_i(\eta))$, then $f(e_i) \ge c_\alpha$ is necessary and sufficient.*

(iv) For $1 \le i \le t$, if $f(e_i(\eta)) > c_\alpha$ for all $\eta > 0$ (in particular, if $f(e_i) > c_\alpha$) but $f(e_{i-1}(\eta)) \le c_\alpha$ for some $\eta > 0$, then

$$P_\lambda\{f(d) > c_\alpha\} \to 1$$

if and only if $\lambda_i \to \infty$, i.e., the level α test based on f is PC(i) *but not* PC($i-1$). *If $f(e_{i-1}(\eta))$ is continuous at $\eta = 0$, then $f(e_i) > c_\alpha > f(e_{i-1})$ suffices. If in addition $\eta = 0$ is a point of increase of $f(e_i(\eta))$, then $f(e_i) \ge c_\alpha > f(e_{i-1})$ suffices.*

PROOF. Since $f(e_i(\eta)) > c_\alpha$ if and only if $e_i(\eta) \in A_f(c_\alpha)$, which by (3.16) is a closed monotone subset of $\bar{\mathcal{D}}_t$, Theorem 3.1 applies directly with $A = A_f(c_\alpha)$. □

For the next remark, if $f \in \bar{\mathcal{F}}_t$ we define

$$f_0(\eta) = f(e_0(\eta)) \equiv f(\eta, \ldots, \eta), \quad 0 \le \eta \le 1. \tag{3.22}$$

REMARK 3.4. Let f and c_α be as in Corollary 3.3.

(i) If $\alpha > 0$ then $c_\alpha < f(e_t) \equiv f_0(1-)$, so by Corollary 3.3(i), a sufficient condition that the level α test determined by f be PC(i) is that $f(e_i(\eta)) = f(e_t)$ for all $\eta > 0$. In particular, $f(e_i) = f(e_t)$ suffices. Consequently, if $f(d) = \varphi(d_1, \ldots, d_i)$ for some $\varphi \in \bar{\mathcal{F}}_i$, then the level α test determined by f is PC(i). (This does not rule out the possibility that it may be PC(i') for some $i' < i$.)

(ii) If $\alpha < 1$ then $c_\alpha \ge f_0(0+)$, so by Corollary 3.3(ii), a necessary condition that the level α test determined by f be PC(i) is that $f(e_i(\eta)) > f_0(0+)$ for all $\eta > 0$. In particular, if $f(e_i(\eta))$ is continuous at $\eta = 0$, then $f(e_i) \ge f_0(0+)$ is necessary for PC(i). If in addition $\eta = 0$ is a point of increase of $f_0(\eta)$, then $c_\alpha > f_0(0+)$, so $f(e_i) > f_0(0+)$ is necessary for PC(i). Also by Corollary 3.3(ii), if $f(d) = \varphi(d_{i+1}, \ldots, d_t)$ for some $\varphi \in \bar{\mathcal{F}}_{t-i}$, then the level α test determined by f is not PC(i). (This does not rule out the possibility that it may fail to be PC(i') for some $i' > i$.)

(iii) If $\alpha > 0$ and $f(e_i(\eta)) = \infty$ for all $\eta > 0$ (in particular, if $f(e_i) = \infty$), then the level α test determined by f is PC(i).

(iv) If $\alpha < 1$ and $f(e_i(\eta)) = -\infty$ for some $\eta > 0$ (or, if $f \ge 0$ on $\bar{\mathcal{D}}_t$ and $f(e_i(\eta)) = 0$ for some $\eta > 0$), then the level α test determined by f is not PC(i). If $f(e_i(\eta))$ is continuous at $\eta = 0$ and $\eta = 0$ is a point of increase of $f_0(\eta)$, then $f(e_i) = -\infty$ implies that this test is not PC(i). This last statement remains true if "$f(e_i) = -\infty$" is replaced by "$f \geq 0$ on $\bar{\mathcal{D}}_t$ and $f(e_i) = 0$." □

It follows from Remark 3.4 (i) and (ii) that if $f(d) = \varphi(d_{k+1}, \ldots, d_l)$ for some $\varphi \in \bar{\mathcal{F}}_{l-k}$, where $0 \leq k < l \leq t$, then for $0 < a < 1$, the level α test based on f is PC(l) but not PC(k). It may or may not be PC(i) for $k < i < l$: for example, the level α test based on $f(d) \equiv \prod_{k+1}^{l} d_j$ is PC(l) but not PC($l-1$), while those based on $\prod_{k+1}^{l} d_j(1-d_j)^{-1}$ and $\sum_{k+1}^{l} d_j(1-d_j)^{-1}$ are PC($k+1$) but not PC(k). Furthermore, parameter consistency may depend on the value of α: when $k < i \leq l$, the level α test based on $f(d) \equiv \sum_{k+1}^{l} d_j$ is PC(i) but not PC($i-1$) for α satisfying $i-k-1 < c_\alpha \leq i-k$. These and other examples are presented in Table 3.1, where Corollary 3.3 is applied to determine the parameter consistency of the level α tests ($0 < \alpha < 1$) determined by several monotone test statistics $f \in \mathcal{F}_t$. (Recall that if $f \in \mathcal{F}_t \backslash \bar{\mathcal{F}}_t$, it is necessary to apply Corollary 3.3 to $\bar{f} \in \bar{\mathcal{F}}_t$.) Furthermore, it is straightforward to apply Corollary 3.3 and Remark 3.4 to determine conditions for test statistics of the more general forms $\sum \varphi_i(d_i)$ and $\prod \varphi_i(d_i)$.

REMARK 3.5. When $\lambda = 0$, each $d_i \equiv \mathrm{ch}_i[XX'(XX'+YY')^{-1}]$ is stochastically increasing in r and stochastically decreasing in n. Arguing as in Perlman (1974, p. 24), it can be shown that $c_\alpha \equiv c_\alpha(p, r, n; f)$ is increasing in p and r and decreasing in α and n. It follows that for a test statistic $f \in \bar{\mathcal{F}}_t$, if the test with acceptance region $A_f(c_\alpha(p, r, n; f))$ is PC(i), then the test with acceptance region $A_f(c_{\alpha'}(p', r', n'; f))$ is also PC(i) whenever $\alpha' \geq \alpha$, $n' \geq n$, $p' \leq p$, $r' \leq r$. □

REMARK 3.6. If $n < p < n + r$ and $\alpha < 1$, no invariant level α test can be parameter consistent ($\equiv$ PC(1)), although, for example, the test based on $f(d) \equiv d_{1+p-n}$ is PC($1+p-n$). If $n + r \leq p$, however, no nontrivial invariant test exists (cf. Lehmann (1986), Chapter 8, problems 1 and 2). □

REMARK 3.7. It follows from the proof of Theorem 3.1 (and the proofs of Lemmas 2.1-2.3) that a monotone invariant test is PC(i) iff it is *uniformly* PC(i), i.e., iff its power at $\lambda = (\lambda_1, \ldots, \lambda_t)$ approaches 1 uniformly in $\{\lambda_j | j \neq i\}$ as $\lambda_i \to \infty$. This equivalence would also follow from the conjecture of Perlman and Olkin (1980) that the power of any monotone test is increasing in each λ_j, for then the power at $(\lambda_1, \ldots, \lambda_i,\ \lambda_{i+1}, \ldots, \lambda_t)$ would be no less than power at $(\lambda_i, \ldots, \lambda_i,\ 0, \ldots, 0)$. Furthermore, this conjecture would imply that a monotone invariant test is PC(i) iff its power approaches 1 for all sequences of alternatives $\{(\xi^{(m)}, \Sigma^{(m)})\}$ of asymptotic rank $= i$. □

Table 3.1: Parameter consistency for monotone test statistics $f \in \mathcal{F}_t$. Here k, l are integers satisfying $0 \le k < l \le t$.

$f(d)$	range of c_α, $0 < \alpha < 1$	$f(e_i)$ (or $f(e_i(\eta))$ if η appears)	conclusion from Corollary 3.3
$\prod_{k+1}^{l} d_j$	$0 < c_\alpha < 1$	$\begin{cases} 0, & i < l \\ 1, & i \ge l \end{cases}$	PC(l), not PC($l-1$)
$\prod_{k+1}^{l} d_j(1-d_j)^{-1}$	$0 < c_\alpha < \infty$	$\begin{cases} \left(\frac{\eta}{1-\eta}\right)^{l-k}, & i \le k \\ \infty, & i > k \end{cases}$	PC($k+1$), not PC(k)
$\sum_{k+1}^{l} d_j$	$0 < c_\alpha < l-k$	$\begin{cases} 0, & i \le k \\ i-k, & k+1 \le i \le l \\ l-k, & i > l \end{cases}$	$\{(i,\alpha) \mid k+1 \le i \le l,\ i-k-1 < c_\alpha \le$ PC(i), not PC($i-1$); PC($k+1$) iff $c_\alpha \le 1$.
$\sum_{k+1}^{l} d_j(1-d_j)^{-1}$	$0 < c_\alpha < \infty$	$\begin{cases} 0, & i \le k \\ \infty, & i > k \end{cases}$	PC($k+1$), not PC(k)
$1-\prod_{k+1}^{l}(1-d_j)$	$0 < c_\alpha < 1$	$\begin{cases} 0, & i \le k \\ 1, & i > k \end{cases}$	PC($k+1$), not PC(k)
$\left(\sum_{k+1}^{l} d_j^{-1}\right)^{-1}$	$0 < c_\alpha < (l-k)^{-1}$	$\begin{cases} 0, & i < l \\ (l-k)^{-1}, & i \ge l \end{cases}$	PC(l), not PC($l-1$)
$\sum^{*}\prod_{1}^{r} d_{j_m}$	$0 < c_\alpha < \binom{l-k}{r}$	$\begin{cases} 0, & i < k+r \\ \binom{i-k}{r}, & k+r \le i \le l \\ \binom{l-k}{r}, & i > l \end{cases}$	$\{(i,\alpha) \mid k+r \le i \le l,\ \binom{i-k-1}{r} <$ PC(i), not PC($i-1$); PC($k+r$) iff $c_\alpha \le 1$.
$\sum^{*}\prod_{1}^{r} d_{j_m}(1-d_{j_m})^{-1}$	$0 < c_\alpha < \infty$	$\begin{cases} \binom{l-k}{r}\left(\frac{\eta}{1-\eta}\right)^{r}, & i \le k \\ \infty, & i > k \end{cases}$	PC($k+1$), not PC(k)
$\binom{l-k}{r} - \sum^{*}\prod_{1}^{r}(1-d_{j_m})$	$0 < c_\alpha < \binom{l-k}{r}$	$\begin{cases} 0, & i \le k \\ \binom{l-k}{r} - \binom{l-i}{r}, & k+1 \le i \le l-r+1 \\ \binom{l-k}{r}, & i > l-r+1 \end{cases}$	$\{(i,\alpha) \mid k+1 \le i \le l-r+1,\ \binom{l-k}{r}$ $< c_\alpha \le \binom{l-k}{r} - \binom{l-i}{r}\}$: PC; PC($k+1$) iff $c_\alpha \le \binom{l-k}{r} - \binom{l-k}{r}$

The summation $\sum^{*}$ extends over all r-tuples $j_1, \ldots, j_r$ such that $k+1 \le j_1 < \cdots < j_r \le l$, where $1 \le r \le l-k$. We define $\binom{s}{r} = 0$ if $s < r$.

4 Parameter consistency of specific invariant tests for MANOVA

We now examine the parameter consistency of the tests based on six familiar statistics:

$$f_1(d) = d_1 \qquad \equiv \mathrm{ch}_1[XX'(XX'+YY')^{-1}] \qquad \text{(S.N. Roy)}$$

$$f_2(d) = \textstyle\sum_1^t \frac{d_j}{1-d_j} \qquad \equiv \mathrm{tr}[XX'(YY')^{-1}] \qquad \text{(Lawley-Hotelling)}$$

$$f_3(d) = \textstyle\prod_1^t \frac{1}{1-d_j} \qquad \equiv \frac{\det(XX'+YY')}{\det(YY')} \qquad \text{(Likelihood ratio (Wilks))}$$

$$f_4(d) = \textstyle\sum_1^t d_j \qquad \equiv \mathrm{tr}[XX'(XX'+YY')^{-1}] \qquad \text{(Bartlett-Nanda-Pillai)}$$

$$f_5(d) = \textstyle\prod_1^t d_j \qquad \equiv \left\{ \begin{array}{ll} \frac{\det(XX')}{\det(XX'+YY')} & \text{if } p \le r \\ \frac{\det(X'X)}{\det(XX'+YY')} & \text{if } p > r \end{array} \right\}$$

$$f_6(d) = \textstyle\prod_1^t \frac{d_j}{1-d_j} \qquad \equiv \left\{ \begin{array}{ll} \frac{\det(XX')}{\det(YY')} & \text{if } p \le r \\ \frac{\det(X'X)}{\det(YY')} & \text{if } p > r \end{array} \right\}.$$

The statistics f_1, f_2, f_3, f_4 are those most commonly recommended for the MANOVA testing problem (1.2) ≡ (1.6) (cf. Anderson (1984), Ch. 8), while the statistics f_5, f_6 are also familiar (e.g. Olson (1974)). Each of these six statistics is monotone, in fact belongs to the class $\bar{\mathcal{F}}_t$, so each of the six corresponding tests is unbiased (Perlman and Olkin (1980)). From the results of Schwartz (1967b), the tests based on f_1, f_2, f_3, f_4 are admissible while those based on f_5, f_6 are inadmissible. Each of the six tests is consistent in the usual sense, i.e., SSC (cf. Anderson and Perlman (1987)).

The Bartlett Nanda-Pillai (BNP) test, based on f_4, frequently has been advocated on the basis of power and robustness (eg. Olson (1974, 1976)). Kiefer and Schwartz (1965) showed that the BNP test is proper Bayes, Schwartz (1967a) proved that the level α BNP test is *locally* most powerful among all invariant level α tests, while numerical studies have shown that the BNP test has high power against certain (fixed) alternatives. In fact, Olson (1976) wrote: "A review of statistical literature concerning the power and robustness of the four most promising tests leads to the recommendation of the Pillai-Bartlett trace statistic for general use. ... It is sufficiently powerful to detect population differences in any noncentrality structure."

Unfortunately, this last statement is not completely accurate. As will be seen below, the results in Section 3 show that the BNP test is not PC unless α or n is sufficiently large. This failure implies that there exist alternatives arbitrarily far from the null hypothesis H_0 which, with positive probability, the BNP test will fail to detect.

The PC properties of the tests based on f_1–f_6 may be determined from Table 3.1. The (admissible) tests based on f_1, f_2, f_3 and the (inadmissible) test based on f_6 are PC(1). The (inadmissible) test based on f_5 is PC(t) but not PC($t-1$). Finally, the PC property of the BNP test based on f_4 depends nontrivially on p, r, n, and α: for $1 \le i \le t$, the level α BNP test is PC(i) but not PC($i-1$) iff

$$i - 1 < c_\alpha(p, r, n; f_4) \le i. \tag{4.1}$$

In particular, for $0 < \alpha < 1$ the level α BNP test is PC $\equiv$ PC(1) iff

$$c_\alpha(p, r, n; f_4) \le 1. \tag{4.2}$$

It is of some interest to determine those values of (p, r, n, α) for which (4.2) fails and, in this case, to determine the maximum shortcoming in power of the BNP test for distant alternatives of asymptotic rank 1. In the remainder of this section some approximate answers to these questions are obtained.

Because $c_\alpha(p, r, n; f_4)$ is decreasing in α and n,[6] for fixed (p, r, n) [respectively, fixed (p, r, α)] there exists $\alpha^*(p, r, n) > 0$ [resp., $n^*(\alpha, p, r) \ge p$] such that (4.2) holds iff

$$\alpha \ge \alpha^*(p, r, n) \quad [\text{resp., } n \ge n * (\alpha, p, r)]. \tag{4.3}$$

If (4.3) does not hold, the BNP test fails to be PC.

Values of $\alpha^*(p, r, n)$ and $n^*(\alpha, p, r)$ are determined from the defining equation

$$c_{\alpha^*}(p, r, n; f_4) = 1, \tag{4.4}$$

or equivalently (see (3.1))

$$\alpha^* = P_{\lambda=0}\{f_4(d(p, r, n)) > 1\}, \tag{4.5}$$

where $d(p, r, n)$ indicates the dependence of the eigenvalues d on the dimensions p, r, n of X and Y. It is well known that when $\lambda = 0$ (i.e., when $\xi \equiv EX = 0$),

$$d(p, r, n) \sim d(r, p, n + r - p) \tag{4.6}$$

(cf. Anderson (1984, Ch. 9) or Perlman (1974, §2)), so that

$$c_\alpha(p, r, n; f_4) = c_\alpha(r, p, n + r - p;\ f_4), \tag{4.7}$$

$$\alpha^*(p, r, n) = \alpha^*(r, p, n + r - p), \tag{4.8}$$

[6] See Perlman (1974, p. 24), where $c_\alpha(p, r, n; f_4)$ is denoted as $k_4(\alpha, p, r, n)$. (In the second displayed expression on p. 23 of Perlman (1974), $\notin$ should be $\in$.)

$$n^*(\alpha, p, r) = n^*(\alpha, r, p) - r + p. \tag{4.9}$$

It can be deduced from (4.5)–(4.9) that $\alpha^*(p, r, n)$ is increasing in p and r but decreasing in n, while $n^*(\alpha, p, r)$ is increasing in p and r but decreasing in α. For $\alpha = .05$ and .01, respectively, Tables I and II in Perlman (1974) list values of $n^*(\alpha, p, r)$ for $2 \le p, r \le 16$; typical values for $\alpha = .05$ are $n^*(.05, 2, 2) = 7$, $n^*(.05, 4, 4) = 20$, $n^*(.05, 8, 4) = 40$, $n^*(.05, 8, 8) = 73$, and $n^*(.05, 16, 8) = 145$, while values for $\alpha^* = .01$ are slightly larger. In fact, $n^*(\alpha, p, r) \to \infty$ as $\alpha \to 0$.

When $\lambda = 0$ and p, r are fixed,

$$n^{-1}\, XX' \xrightarrow{P} 0 \text{ and } n^{-1}\, YY' \xrightarrow{P} I_p \tag{4.10}$$

as $n \to \infty$, so $d(p, r, n) \xrightarrow{P} 0$. By (4.5), this implies that $\alpha^*(p, r, n) \to 0$ as $n \to \infty$ with p and r fixed. If, however, p, r, and n are all of the same order of magnitude then the approximations (Pillai (1955))

$$\begin{aligned}
\frac{1}{t} f_4(d(p,r,n)) \quad &\approx \quad \text{Beta}\left(\frac{pr}{2}, \frac{t(n-p+t)}{2}\right), \\
E[f_4(d(p,r,n))] \quad &= \quad \frac{pr}{(n+r)} \text{ (exact)}, \\
Var[f_4(d(p,r,n))] \quad &\approx \quad \frac{2prt(n-p+t)}{(n+r)^2(t(n+r)+2)},
\end{aligned}$$

for the central ($\lambda = 0$) distribution of $f_4(d)$ suggest that $\alpha^*(p, r, n) \to 1$ as p, r, and n increase at the same rate. In fact, these approximations suggest that $\alpha^*(p, r, n)$ remains substantially above 0 as p, r, and n increase with $n \le pr(1 - \varepsilon)$ for $\varepsilon > 0$.

Some approximate values of $\alpha^*(p, r, n)$, obtained from the tables of Schuurmann et al. (1973) for the central distribution of $f_4(d)$, are presented in Table 4.1. This table can be extended by means of the relation (4.8), e.g., $\alpha^*(3, 2, 8) = \alpha^*(2, 3, 7) \gtrsim .05$. Furthermore, when $r = n \ge p$, $\text{tr}[XX'(XX' + YY')^{-1}]$ and $\text{tr}[YY'(XX' + YY')^{-1}]$ have identical central distributions, while

$$\text{tr}[XX'(XX' + YY')^{-1}] = p - \text{tr}[YY'(XX' + YY')^{-1}], \tag{4.11}$$

so their common central distribution must be symmetric about $p/2$. Thus, by (4.5),

$$\begin{aligned}
\alpha^*(2, n, n) \quad &= \quad \alpha^*(n, 2, 2n-2) = .5 \quad (n \ge 2) & (4.12)\\
\alpha^*(p, n, n) \quad &= \quad \alpha^*(n, p, 2n-p) > .5 \quad (n \ge p \ge 3). & (4.13)
\end{aligned}$$

Together with Table 4.1, this paragraph indicates that in many cases the BNP test fails to be PC for the standard values of α used in practice.

When the BNP test fails to be PC, i.e., when (4.2) fails, it is of interest to examine the maximum power shortcoming for distant alternatives of

Table 4.1: Approximate values for $\alpha^*(p,r,n)$. The level α Bartlett-Nanda-Pillai trace test fails to be parameter consistent when $\alpha < \alpha^*(p,r,n)$.

α^* (p, r, n) $\gtrsim$ .05	α^* (p, r, n) $\gtrsim$ .10	α^* (p, r, n) $\gtrsim$ .25
p, r, n	p, r, n	p, r, n
2, 3, 7	2, 3, 6	2, 3, 4
2, 5, 10	2, 5, 9	2, 5, 7
2, 7, 13	2, 7, 12	2, 7, 10
3, 4, 14	3, 4, 12	3, 4, 10
3, 6, 20	3, 6, 18	3, 6, 15
3, 8, 26	3, 8, 23	3, 8, 19
4, 5, 24	4, 5, 21	4, 5, 18
4, 7, 32	4, 7, 29	4, 7, 25
4, 9, 39	4, 9, 36	4, 9, 32

asymptotic rank 1. If the power of the BNP test is monotonically increasing in each λ_i, which would follow from the conjecture of Perlman and Olkin (1980) that every monotone invariant test has this power property[7], then this maximum shortcoming would be given by

$$1 - \lim_{\lambda_1 \to \infty} P_{(\lambda_1,0,\cdots,0)}\{f_4(d(p,r,n)) > c_\alpha(p,r,n,f_4)\} \equiv s_\alpha(p,r,n). \tag{4.14}$$

Note that (4.6) and (4.7) imply that

$$s_\alpha(p,r,n) = s_\alpha(r,p,n+r-p). \tag{4.15}$$

While the proof of Theorem 3.1(ii) provides a general lower bound for $s_\alpha(p,r,n)$, a tighter lower bound may be obtained in terms of the central ($\lambda = 0$) distribution of $f_4(d(p,r-1,n)$ as follows. Partition X as (X_1, X_2) where X_1 is $p \times 1$ and X_2 is $p \times (r-1)$. Then

$$\begin{aligned} & \operatorname{tr}[XX'(XX'+YY')^{-1}] \\ = \; & \operatorname{tr}[X_1X_1'(XX'+YY')^{-1}] + \operatorname{tr}[X_2X_2'(XX'+YY')^{-1}] \\ \le \; & 1 + \operatorname{tr}[X_2X_2'(X_2X_2'+YY')^{-1}]. \end{aligned} \tag{4.16}$$

When $\lambda = (\lambda_1, 0, \ldots, 0)$ we may set $(\xi, \Sigma) = (\mu, I_p)$ where $\mu_{11} = \lambda_1^{\frac{1}{2}}$ and $\mu_{ij} = 0$ otherwise. Since $E(X_2) = 0$, it follows that

$$\begin{aligned} s_\alpha(p,r,n) & \ge P_{\lambda=0}\{f_4(d(p,r-1,n)) \le c_\alpha(p,r,n;f_4) - 1\} \\ & \equiv \tilde{s}_\alpha(p,r,n). \end{aligned} \tag{4.17}$$

By (4.11) we have the alternative expression

$$\tilde{s}_\alpha(p,r,n) = P_{\lambda=0}\{f_4(d(p,n,r-1)) \ge p+1-c_\alpha(p,r,n;f_4)\}. \tag{4.18}$$

[7] Perlman and Olkin (1980) do show that the probability in (4.14) is monotonically increasing in λ_1.

Table 4.2: Approximate values of $\tilde{s}_{.05}(p,r,n)$, a lower bound for the maximum power shortcoming $s_{.05}(p,r,n)$ of the level .05 Bartlett-Nanda-Pillai test.

$.10 \leq \tilde{s}_{.05}(p,r,n) \leq .25$	$.05 \leq \tilde{s}_{.05}(p,r,n) \leq .10$	$.01 \leq \tilde{s}_{.05}(p,r,n) \leq .05$
p, r, n	p, r, n	p, r, n
2, 3, 3	2, 3, 6	2, 3, 7
2, 3, 5	2, 5, 5	2, 5, 7
		2, 7, 7
4, 5, 5	4, 5, 11	4, 5, 13
4, 5, 9	4, 7, 11	4, 5, 15
4, 7, 7	4, 9, 9	4, 7, 13
4, 7, 9	4, 9, 11	4, 7, 17
		4, 9, 13
		4, 9, 17
6, 7, 7	6, 7, 17	6, 7, 21
6, 7, 15	6, 7, 19	6, 7, 23
6, 9, 9	6, 9, 17	6, 9, 21
6, 9, 15	6, 9, 19	6, 9, 25

Some approximate values for the lower bound $\tilde{s}_{.05}(p,r,n)$ are given in Table 4.2, again obtained from Schuurmann, et al. (1973). These values indicate that the maximum power shortcoming for the level .05 BNP test might not be too large if n is moderately large, but this cannot be stated conclusively since $\tilde{s}_{.05}$ is only a lower bound for $s_{.05}$. Bondar (1991) recently has obtained more accurate values for $s_\alpha(p,r,n)$ which indicate that the power shortcoming of the BNP test can be substantial. It should be stressed that the competing Roy, Lawley-Hotelling, and likelihood ratio (Wilks) tests always have zero power shortcoming for distant alternatives since they are PC.

5 Invariant tests for independence of variates

Because the problem of testing independence of two sets of variates in a multivariate normal population can be reduced to the canonical MANOVA problem by conditioning on one set of variates, all of the preceding results regarding parameter consistency (and sample size consistency) of invariant tests for MANOVA remain valid for the corresponding tests for independence. This correspondence and the conditioning argument are now outlined. The reader is referred to Anderson (1984, §9.8 and §9.10), Anderson and Das Gupta (1964a), and Perlman (1974, §2) for further details.

Based on a sample of size m from the (p_1+p_2)-variate normal population

$$N\left[\begin{pmatrix}0\\0\end{pmatrix}, \begin{pmatrix}\Sigma_{11} & \Sigma_{12}\\ \Sigma_{21} & \Sigma_{22}\end{pmatrix} \equiv \Sigma\right], \quad \Sigma_{ij} : p_i \times p_j,$$

we may construct a (central) Wishart random matrix

$$S \equiv \begin{pmatrix} S_{11} & S_{12} \\ S_{21} & S_{22} \end{pmatrix} \sim W_{p_1+p_2}\left[\begin{pmatrix} \Sigma_{11} & \Sigma_{12} \\ \Sigma_{21} & \Sigma_{22} \end{pmatrix}, \; m\right]$$

with m degrees of freedom. Assume that Σ is nonsingular and that $m \geq p_1 + p_2$, so that S is nonsingular with probability 1. Without loss of generality, also assume that $p_1 \leq p_2$. The problem of independence is equivalent to testing

$$H_0 : \Sigma_{12} \quad \text{vs.} \quad H_1 : \Sigma_{12} \neq 0 \tag{5.1}$$

with Σ_{11} and Σ_{22} unknown.

For $1 \leq i \leq t \equiv p_1 \wedge p_2$ $(= p_1$ here), define

$$r_i^2 = \operatorname{ch}_i(S_{12}S_{22}^{-1}S_{21}S_{11}^{-1}), \qquad r^2 = (r_1^2, \ldots, r_t^2) \tag{5.2}$$

$$\rho_i^2 = \operatorname{ch}_i(\Sigma_{12}\Sigma_{22}^{-1}\Sigma_{21}\Sigma_{11}^{-1}), \qquad \rho^2 = (\rho_1^2, \ldots, \rho_t^2) \tag{5.3}$$

$$\gamma_i = \operatorname{ch}_i(\Sigma_{12}\Sigma_{22}^{-1}\Sigma_{21}\Sigma_{11\cdot 2}^{-1}) \tag{5.4}$$

$$\equiv \rho_i^2/(1-\rho_i^2), \qquad \gamma = (\gamma_1, \ldots, \gamma_t).$$

Thus, $1 > r_1 > \cdots > r_t > 0$ (resp., $1 > \rho_1 \geq \cdots \geq \rho_t \geq 0$) are the sample (population) canonical correlation coefficients and represent the maximal invariant statistic (parameter) under the group of transformations

$$S \to \begin{pmatrix} B_1 & 0 \\ 0 & B_2 \end{pmatrix} S \begin{pmatrix} B_1 & 0 \\ 0 & B_2 \end{pmatrix}', \tag{5.5}$$

where $B_i(p_i \times p_i)$ is nonsingular, which group leaves the testing problem (5.1) invariant. Any invariant test depends on S only through r^2, while the distribution of r^2 depends on Σ only through ρ^2, or equivalently, only through γ. Then (5.1) is equivalent to the problem of testing

$$H_0 : \gamma = 0 \quad \text{vs.} \quad H_1 : \gamma \neq 0. \tag{5.6}$$

As in the MANOVA problem, every admissible invariant test for (5.1)≡(5.6) must have a *monotone* acceptance region A in terms of $r^2 \equiv (r_1^2, \ldots, r_t^2)$ and every such monotone invariant test is unbiased. (The range of r^2, like that of d in Section 3, is the set $\mathcal{D}_t$.) For fixed p_1, p_2, m, and α, an invariant level α test with acceptance region A is PC if its power $P_\gamma\{r^2 \notin A\} \to 1$ as $\sum \gamma_i \equiv \operatorname{tr}(\Sigma_{12}\Sigma_{22}^{-1}\Sigma_{21}\Sigma_{11\cdot 2}^{-1}) \to \infty$.[8] For $i = 1, \ldots, t$ $(\equiv p_1)$, such a test is said to be PC(i) if its power at the alternative γ approaches 1 as $\gamma_i \to \infty$. Again, PC ⇔ PC(1) ⇒ ⋯ ⇒ PC(t).

All results in Sections 3 and 4 regarding PC properties of invariant tests with monotone acceptance regions of the forms $\{d \in A\}$ and $\{f(d) \leq c_\alpha\}$

[8]It can be shown that the minimum Kullback-Leibler distance from the alternative $N(0, \Sigma \otimes I_m)$ to the null hypothesis $H_0 \equiv \{N(0, \operatorname{Diag}(\Lambda_{11}, \Lambda_{22}) \otimes I_m | \Lambda_{11}, \Lambda_{22}$ positive definite$\}$ is $(m/2)\Sigma \log(1+\gamma_i)$, which $\to \infty$ iff $\Sigma\gamma_i \to \infty$.

for the MANOVA problem carry over to invariant tests with monotone acceptance regions of the forms $\{r^2 \in A\}$ and $\{f(r^2) < c_\alpha\}$ for the independence problem. For example, in the statements of Theorem 3.1 and Corollary 3.3, simply replace p, r, n by p_1, p_2, m, $P_\lambda\{d \notin A\}$ by $P_\gamma\{r^2 \notin A\}$, $P_\lambda\{f(d) > c_\alpha\}$ by $P_\gamma\{f(r^2) > c_\alpha\}$, and λ_i by γ_i.

We sketch the proof of this modified version of Theorem 3.1. By invariance, the distribution of r^2 under Σ is the same as when

$$\Sigma = \begin{pmatrix} I_{p_1} & \Delta(\rho) \\ \Delta(\rho)' & I_{p_2} \end{pmatrix}, \tag{5.7}$$

where $\Delta(\rho) : p_1 \times p_2$ is defined by $\Delta_{ij}(\rho) = \rho_i \delta_{ij}$ with δ_{ij} the Kronecker delta function. Under (5.7), it can be shown that the *conditional* distribution of r^2 given S_{22} is the same as the distribution of d in (3.1) under the following correspondences:

$$(p, r, n) \leftrightarrow (p_1, p_2, m - p_2) \tag{5.8}$$

$$d_i \leftrightarrow r_i^2 \tag{5.9}$$

$$\lambda_i \leftrightarrow \text{ch}_i[D(\gamma)\tilde{S}_{22}], \tag{5.10}$$

where $D(\gamma) = \text{Diag}(\gamma_1, \cdots, \gamma_t)$ is $p_1 \times p_1$ and where $\tilde{S}_{22}$ is the upper left-hand $p_1 \times p_1$ block of S_{22}. The unconditional distribution of r^2 is determined by this conditional distribution and by the marginal Wishart distribution of $\tilde{S}_{22}$:

$$\tilde{S}_{22} \sim W_{p_1}(I_{p_1}, m). \tag{5.11}$$

In particular, the central ($\gamma = 0$) distribution of $r^2 \equiv r^2(p_1, p_2, m)$ does not depend on S_{22} and is identical to the central ($\lambda = 0$) distribution of $d \equiv d(p_1, p_2, m - p_2)$.

Now condition on S_{22} to obtain

$$P_\gamma\{r^2 \notin A\} = E[P_\gamma\{r^2 \notin A | S_{22}\}]. \tag{5.12}$$

By (5.10), the original version of Theorem 3.1 gives necessary and/or sufficient conditions for the implication

$$\text{ch}_i[D(\gamma)\tilde{S}_{22}] \to \infty \Rightarrow P_\gamma\{r^2 \notin A | S_{22}\} \to 1 \tag{5.13}$$

to hold for every fixed S_{22}. But for $i = 1, \ldots, t\ (\equiv p_1)$,

$$\gamma_i \text{ch}_{p_1}(\tilde{S}_{22}) \le \text{ch}_i[D(\gamma)\tilde{S}_{22}] \le \gamma_i \text{ch}_1(\tilde{S}_{22}), \tag{5.14}$$

while, by (5.11) and the assumption that $m \ge p_1 + p_2\ (\ge p_1)$,

$$P\{0 < \text{ch}_{p_1}(\tilde{S}_{22}) < \text{ch}_1(\tilde{S}_{22}) < \infty\} = 1, \tag{5.15}$$

so for every fixed S_{22},

$$\mathrm{ch}_i[D(\gamma)\tilde{S}_{22}] \to \infty \;\Leftrightarrow\; \gamma_i \to \infty. \tag{5.16}$$

Thus the same necessary and/or sufficient conditions on A are valid for the implication

$$\gamma_i \to \infty \;\Rightarrow\; P_\gamma\{r^2 \notin A|S_{22}\} \to 1 \tag{5.17}$$

to hold for every fixed S_{22}. These facts suffice to establish the modified version of Theorem 3.1, and similarly Corollary 3.3, for the independence problem.

Thus the PC properties of an invariant test based on a monotone statistic $f(r^2)$ for the independence problem (5.1)≡(5.6) are the same as those of an invariant test based on $f(d)$ for the MANOVA problem (1.2)≡(1.6). In particular, by (4.2) and (5.8), the BNP test based on

$$f_4(r^2) \equiv \mathrm{tr}(S_{12}S_{22}^{-1}S_{21}S_{11}^{-1}) \tag{5.18}$$

is PC(1) ≡ PC only if

$$c_\alpha(p_1, p_2, m - p_2; f_4) \le 1, \tag{5.19}$$

and its maximum power shortcoming is given by $s_\alpha(p_1, p_2, m-p_2)$ in (4.14).[9] By contrast, the Roy, Lawley-Hotelling, and likelihood ratio (≡ Wilks) tests, based on $f_1(r^2)$, $f_2(r^2)$, and $f_3(r^2)$, respectively, are always PC(1).

6 Multivariate components of variance

We return to the MANOVA testing problem (1.1)-(1.2) but now assume that ξ is itself random. That is, replace (1.1) with the following distributional assumptions: $X(p \times r)$, $\xi(p \times r)$, and $Y(p \times n)$ are mutually independent, and

$$X|\xi \sim N(\xi, \Sigma \otimes I_r), \quad \xi \sim N(0,\ \Phi \otimes I_r), \quad Y \sim N(0,\ \Sigma \otimes I_n). \tag{6.1}$$

Here X, Y are observed but ξ is unobservable, Σ is positive definite and Φ is positive semidefinite. Thus, unconditionally, X and Y are independent with

$$X \sim N(0, (\Sigma + \Phi) \otimes I_r), \qquad Y \sim N(0, \Sigma \otimes I_n), \tag{6.2}$$

so we have a multivariate components of variance (≡ random effects) model (cf. Anderson (1984), §10.6). We assume that $p \le \min(r, n)$ to insure that Σ and Φ are estimable, so now $t \equiv p \wedge r = p$.

[9] Again assuming the validity of the conjecture of Perlman and Olkin (1980).

The testing problem (1.2) is replaced by that of testing

$$H_0 : \Phi = 0 \quad \text{vs.} \quad H_1 : \Phi \neq 0 \tag{6.3}$$

with Σ unknown. As in Section 1, this problem is invariant under the group of transformations given by (1.4). The statistics $c \equiv (c_1, \ldots, c_t)$ in (1.3) and $d \equiv (d_1, \ldots, d_t)$ in (3.1) again are equivalent representations of the maximal invariant statistic, but now the maximal invariant parameter is given by ω, defined as follows:

$$\begin{aligned} \omega_i &\equiv \omega_i(\Phi, \Sigma) = \text{ch}_i(\Phi\Sigma^{-1}), \quad 1 \leq i \leq t, \\ \omega &\equiv \omega(\Phi, \Sigma) = (\omega_1, \ldots, \omega_t). \end{aligned} \tag{6.4}$$

Any invariant test depends on (X, Y) only through d, whose distribution depends on (Φ, Σ) only through ω. The testing problem (6.2) is equivalent to that of testing

$$H_0 : \omega = 0 \quad \text{vs.} \quad H_1 : \omega \neq 0. \tag{6.5}$$

As before, every admissible invariant test for (6.3)≡(6.5) must have a monotone acceptance region A in terms of $d \equiv (d_1, \ldots, d_t) \in \mathcal{D}_t$, and every such monotone invariant test is unbiased (Anderson and Das Gupta (1964b)). For fixed p, r, n, and α, an invariant level α test is PC if its power $P_\omega\{d \notin A\} \to 1$ as $\sum \omega_i \equiv \text{tr}(\Phi\Sigma^{-1}) \to \infty$.[10] For $i = 1, \ldots, t \ (\equiv p)$, such a test is said to be PC(i) if its power at the alternative ω approaches 1 as $\omega_i \to \infty$.

Once again, all results in Sections 3 and 4 regarding parameter consistency of tests with monotone invariant acceptance regions of the forms $\{d \in A\}$ and $\{f(d) \leq c_\alpha\}$ for the MANOVA problem (1.2)≡(1.6) carry over to the present problem (6.3)≡(6.5); now the only notational change required is to replace λ by ω. We briefly sketch the proof of Theorem 3.1 in the present context.

By invariance, the distribution of d under (Φ, Σ) is the same as under $(D(\omega), I_p)$, where $D(\omega) = \text{Diag}(\omega_1, \ldots, \omega_t) : p \times p$. In this case, by (6.1) the *conditional* distribution of d given ξ is the same as the distribution of d in the MANOVA problem with noncentrality parameters $\lambda \equiv (\lambda_1, \ldots, \lambda_t)$, where $\lambda_i = \text{ch}_i(\xi\xi'\Sigma^{-1})$ (cf. (1.5)). In particular, since $P_\omega\{\xi = 0\} = 1$ when $\omega = 0$, the central distribution of d is the same as its central distribution in MANOVA.

Theorem 3.1 gives necessary and/or sufficient conditions on A for the implication

$$\text{ch}_i(\xi\xi'\Sigma^{-1}) \to \infty \Rightarrow P_\omega\{d \notin A|\xi\} \to 1 \tag{6.6}$$

[10] It can be shown that the minimum Kullback-Leibler distance from the alternative $N(0, (\Sigma+\Phi)\otimes I_r) \times N(0, \Sigma\otimes I_n)$ to the null hypothesis $H_0 \equiv \{N(0, \Lambda\otimes I_r) \times N(0, \Lambda\otimes I_n) | \Lambda$ positive definite $\}$ is $(1/2)[r\Sigma \log \alpha_i + n\Sigma \log \beta_i]$, where $\alpha_i = (1 + h\omega_i)$, $\beta_i = \alpha_i/(1 + \omega_i)$, and $h = n/(r + n)$. Since $h \leq \beta_i \leq 1$, it follows that this distance $\to \infty$ iff $\Sigma\omega_i \to \infty$.

to hold for every fixed ξ. Furthermore, by (6.1) and the assumption $(\Phi, \Sigma) = (D(\omega), I_p)$, $\text{ch}_i(\xi\xi'\Sigma^{-1})$ has the same distribution as $\text{ch}_i[D(\omega)UU']$, where

$$U(p \times r) \sim N(0, I_p \otimes I_r). \tag{6.7}$$

But for $i = 1, \ldots, t\ (\equiv p)$

$$\omega_i \text{ch}_p(UU') \leq \text{ch}_i[D(\omega)UU'] \leq \omega_i \text{ch}_1(UU'), \tag{6.8}$$

while, since $p \leq r$,

$$P\{0 < \text{ch}_p(UU') < \text{ch}_1(UU') < \infty\} = 1, \tag{6.9}$$

so for every fixed U (i.e., for every fixed ξ),

$$\text{ch}_i[D(\omega)UU'] \to \infty \;\Leftrightarrow\; \omega_i \to \infty. \tag{6.10}$$

Thus the same necessary and/or sufficient conditions on A are valid for the implication

$$\omega_i \to \infty \;\Rightarrow\; P_\omega\{d \notin A|\xi\} \to 1 \tag{6.11}$$

to hold for every fixed ξ. These facts are sufficient to establish Theorem 3.1, and similarly Corollary 3.3, for the testing problem (6.3)≡(6.5), provided that λ is replaced by ω.

Therefore, the PC properties of an invariant test based on a monotone statistic $f(d)$ for the components-of-variance problem (6.3)≡(6.5) are the same as for the MANOVA problem (1.2)≡(1.6). In particular, the BNP test based on

$$f_4(d) \equiv \text{tr}[XX'(XX' + YY')^{-1}]$$

is PC(1) only if (4.2) holds, or equivalently (4.3). The maximum power shortcoming of the BNP test is again given by $s_\alpha(p, r, n)$ in (4.14).[11] The Roy, Lawley-Hotelling, and Wilks tests, based on $f_1(d)$, $f_2(d)$, and $f_3(d)$, respectively, are always PC(1).

It should be noted, however, that the test based on the Wilks statistic $f_3(d)$ is *not* the likelihood ratio test for the present problem (6.3)≡(6.5). The likelihood ratio test is given, for example, in eqn. (14) of Anderson (1984, §10.6) and is easily shown to be PC(1) by applying Corollary 3.3.

REFERENCES

Anderson, T. W. (1984). *An Introduction to Multivariate Statistical Analysis* (second edition). John Wiley & Sons, New York.

[11] In fact, whereas the validity of (4.14) for the MANOVA problem depends on the monotone power conjecture of Perlman and Olkin (1980), this conjecture has been established for the components-of-variance problem by Anderson and Das Gupta (1964b), so in the present context the validity of (4.14) is certain.

Anderson, T. W. and Das Gupta, S. (1964a). Monotonicity of the power functions of some tests of independence between two sets of variates. *Ann. Math. Statist.* **35**, 206-208.

Anderson, T. W. and Das Gupta, S. (1964b). A monotonicity property of the power functions of some tests of the equality of two covariance matrices. *Ann. Math. Statist.* **35**, 1059-1063.

Anderson, T. W. and Perlman, M. D. (1987). Consistency of invariant tests for the multivariate analysis of variance. *Proc. Second Int. Conf. in Statist.* (T. Pukkila, S. Puntanen, eds.), 225-243.

Bondar, J.V. (1991). On the power of Pillai's test for means in MANOVA. In preparation.

Kiefer, J. and Schwartz, R. E. (1965). Admissible Bayes character of T^2, R^2, and other fully invariant tests for classical multivariate normal problems. *Ann. Math. Statist.* **36**, 747-770.

Lehmann, E. L. (1986). *Testing Statistical Hypotheses* (second edition). John Wiley & Sons, New York.

Muirhead, R.J. (1982). *Aspects of Multivariate Statistical Theory.* John Wiley & Sons, New York.

Olson, C. L. (1974). Comparative robustness of six tests in multivariate analysis of variance. *J. Amer. Statist. Assoc.* **69**, 894-908.

Olson, C.L. (1976). On choosing a test statistic in multivariate analysis of variance. *Psych. Bulletin* **83**, 579-586.

Perlman, M. D. (1974). On the monotonicity of the power functions of tests based on traces of multivariate beta matrices. *J. Multivariate Analysis* **4**, 22-30.

Perlman, M. D. and Olkin, I. (1980). Unbiasedness of invariant tests for MANOVA and other multivariate problems. *Ann. Statist.* **8**, 1326-1341.

Pillai, K.C.S. (1955). Some new test criteria in multivariate analysis. *Ann. Math. Statist.* **26**, 117-121.

Schuurmann, F.J., Krishnaiah, P.R., and Chattopadhyay, A.K. (1973). Exact percentage points of the distribution of the trace of $S_1(S_1 + S_2)^{-1}$. Report No. ARL 73-008, Aerospace Research Laboratories, Wright-Patterson AFB, Ohio.

Schwartz, R. E. (1967a). Locally minimax tests. *Ann. Math. Statist.* **38**, 340-360.

Schwartz, R. E. (1967b). Admissible tests in multivariate analysis of variance. *Ann. Math. Statist.* **38**, 698-710.

Department of Statistics
Stanford University
Stanford, California 94305
U.S.A.

Department of Statistics
University of Washington
Seattle, Washington 98195
U.S.A.

J.K. Ghosh, S.K. Mitra, K.R. Parthasarathy and B.L.S. Prakasa Rao (*eds.*),
Statistics and Probability : A Raghu Raj Bahadur Festschrift
Published by Wiley Eastern Limited, Publishers, (1993), 63 – 71

Edgeworth Expansions in Non-regular Cases and Their Applications to Bootstrap[1]

GUTTI JOGESH BABU
Pennsylvania State University, University Park, U. S. A.

Abstract

Edgeworth expansions for sums of non-identically distributed random vectors and of random vectors with lattice and non-lattice coordinates are reviewed. Their applications to errors-in-variables models, least absolute deviation estimators *etc.*, are presented. Applications to bootstrap are also discussed.

1 Introduction

Let f be the Fourier transform or characteristic function of a distribution F and ψ be the characteristic function of another distribution Ψ. Consider the formal expansion

$$\begin{aligned} f(t) &= (\exp[\log f(t) - \log\psi(t)])\psi(t) \\ &= \left(\exp\left(\sum_{r=1}^{\infty}(\kappa_r - \gamma_r)\frac{(it)^r}{r!}\right)\right)\psi(t). \end{aligned}$$

[1]Research supported in part by NSA Grant MDA904–90–H–1001 and NSF Grants DMS-9007717 and DMS-9208066.

AMS Subject Classifications : Primary 62E20; Secondary 60F05

Key words and Phrases: mean absolute deviation; non-lattice distribution; errors-in-variables; t-type statistics

Let κ_r and γ_r denote cummulants of F and Ψ respectively. The Fourier transform of $(-D)^r\Psi$ is given by the function g_r, where D denotes the differential and $g_r(t) = (it)^r\psi(t)$. By formally inverting the right hand side of the above equation term by term one obtains

$$F(x) = \left(\exp\left(\sum_{r=1}^{\infty}(\kappa_r - \gamma_r)\frac{(-D)^r}{r!}\right)\right)\Psi(x).$$

This is called Charlier differential series (Charlier, 1906). The formal expansion, with normal distribution as the developing function, was considered by Chebyshev (1890), Edgeworth (1905) and Charlier (1905). The choice of Ψ has no effect on convergence, but it clearly has great influence on the quality of approximation by the first few terms. For details on the history of such expansions see Wallace (1958). The approximation by finite number of terms have come to be known as Edgeworth expansions.

In their fundamental paper, Bhattacharya and Ghosh (1978) show the validity of formal Edgeworth expansions for a wide class of statistics. Their paper is to a great extent, responsible for the renewed interest in Edgeworth expansions. Applications to the asymptotic theory of bootstrap can be cited as another reason for the recent spurt in work on Edgeworth expansions. In a series of papers, Babu and Singh (1983, 1984) and Singh and Babu (1990) have shown superiority of the bootstrap method for a wide class of statistics in approximating the sampling distribution, using the techniques and results of Bhattacharya and Ghosh (1978) on Edgeworth expansions. This class includes statistics which can be written as a smooth function of sample mean of *i.i.d.* random vectors.

To state the main result of Bhattacharya and Ghosh (1978), let $\bar{Z}_n$ denote the sample mean of n independent identically distributed random vectors $Z_1, \ldots, Z_n$ in R^k with mean μ. Let P_n denote the distribution of $\sqrt{n}(\bar{Z}_n - \mu)$ and let g be a measurable function. Under some conditions on Z_1, they show that

$$\int gd(\mathrm{P}_n - \Psi) = o(n^{-s/2}) + \delta_n(g),$$

where $\delta_n(g)$ goes to zero at certain rate,

$$\Psi_{s,n}(x) = \Phi_\Sigma(x) + \phi_\Sigma(x)\left(\sum_{i=1}^{s} n^{-i/2}p_i(x)\right),$$

and p_i are linear combinations of *Chebyshev-Hermite* polynomials with coefficients depending on the moments of Z_1. Using this result they establish that

$$\begin{aligned}\mathrm{P}(\sqrt{n}(H(\bar{Z}_n) - H(\mu)) \le x) &= \Phi_\sigma(x) + \phi_\sigma(x)\left(\sum_{i=1}^{s} n^{-i/2}q_i(x)\right)\\ &\quad + o(n^{-s/2}),\end{aligned}$$

holds under some conditions on Z_1 and on the smooth function H, where q_i are polynomials.

In this paper we review some of the recent work done at Pennsylvania State University on Edgeworth Expansions.

In addition to conditions on the moments of Z_1, these results are proved assuming:

a) Z_1 satisfies Cramér's condition

$$\limsup_{\|t\|\to\infty} |\mathrm{E}(e^{it'Z})| < 1,$$

and

b) the random variables $Z_1, \ldots, Z_n$ are independent and identically distributed.

For one term expansions condition **a)** can be relaxed to,

a') the distribution of Z_1 satisfies strongly non-lattice condition,

$$|\mathrm{E}(e^{it'Z})| < 1 \quad \text{for all} \quad t \neq 0,$$

i.e., the distribution of Z_1 is not concentrated on countably many parallel hyperplanes.

For several statistics of interest, some of these conditions are violated. For example, condition **a)** or **a')** is violated, if a statistic is normalized using L_1*-estimate* of scale rather than the traditional L_2*-estimate*. Condition **b)** does not hold in the context of errors-in-variables regression.

For such statistics the standard results on Edgeworth expansions are not applicable. These non-regular cases are discussed in the next few sections.

2 Presence of Lattice Structure

Ratio Estimator

The ratio estimator $\hat{\theta} = \bar{Y}_n/\bar{X}_n$ of $\theta = \mathrm{E}(Y)/\mathrm{E}(X)$, is often encountered in sample surveys, where $\bar{V}_n$ denotes the sample mean of $V_1, \ldots, V_n$. Quite often the auxilliary variable X has lattice distribution and the variable of interest Y has a continuous distribution. The difference $\hat{\theta} - \theta$ can be rewritten as

$$\hat{\theta} - \theta = H(\bar{Z}_n) = \frac{\bar{W}_n}{\mathrm{E}(X)}\left(1 + \frac{\mathrm{E}(X) - \bar{X}_n}{\bar{X}_n}\right)$$

where $Z_i = (W_i, X_i)$, H is a smooth function and the random variable $W_i = Y_i - \theta X_i$ has continuous distribution. Notice that the leading term has continuous distribution and the lattice variable X occurs only in quadratic or higher order terms in the Taylor series expansion

$$H(\bar{Z}_n) - H(\mu) = (\bar{Z}_n - \mu)'\ell(\mu) + (\bar{Z}_n - \mu)'L(\bar{Z}_n - \mu) + \cdots,$$

where $\mu = (\mathrm{E}(W), \mathrm{E}(X))$. In this case Edgeworth expansions are valid by the Corollary of Babu and Singh (1989). Babu (1991) generalized the results of Babu and Singh (1989) to s-term Edgeworth expansions, when the lattice structure does not appear in the first few terms of the Taylor series expansion of the function H at $\bar{Z}_n$ around μ. This is achieved by convoluting the lattice variable with a 'small smooth' random variable and removing its effect later.

Mean absolute deviations

Easy access to inexpensive computing facilities of late, has renewed interest in the statistical methodology based on L_1-norm, instead of the traditional methods based on L_2-norm or least squares. Recent developments in robust procedures in statistical inference, is also to some extent responsible for this trend. The use of L_1-norm in estimation and tests of significance is not new, the idea goes back to Galileo, but the complexity of the distributions involved stood in the way of its use in practical applications. The least-squares (or L_2) estimation, which also has its roots in celestial mechanics gained popularity due to its mathematical tractability. In this section we consider t-type statistics, where the usual least-squares scaling factor is replaced by the one based on mean absolute deviations.

To be precise, let $X_1, \ldots, X_n$ be independent random variables with a common continuous distribution F having mean μ and variance σ^2. For scale parameter we use the mean absolute deviation

$$M_n = \frac{1}{n}\sum_{i=1}^{n} |X_i - \bar{X}_n|,$$

instead of the classical root mean square deviation

$$s_n = \left(\frac{1}{n}\sum_{i=1}^{n}(X_i - \bar{X}_n)^2\right)^{1/2}.$$

Consider the robust student's t-type statistic, $H_n = \sqrt{n}(\bar{X}_n - \mu)/M_n$. Suppose F has a derivative f in a neighborhood of μ and suppose $\mathrm{f}(\mu)$ is positive. If for some $a > 0$ and $b > 0$,

$$|\mathrm{f}(x) - \mathrm{f}(\mu)| \le a|x - \mu|^b,$$

for x near μ, then it can be shown (see Babu and Rao, 1992) that

$$H_n = \sqrt{n}(\mathrm{H}(\bar{Z}_n) - \mathrm{H}(\mathrm{E}(Z_1))) + R_n,$$

where

$$Y_i \;=\; \sigma^{-1}(X_i - \mu), \quad \gamma^{-1} = E|Y_1|,$$

$$\begin{aligned} Z_i &= (Y_i, |Y_i| - \gamma^{-1}, \quad I(Y_i \le 0) - \mathrm{F}(\mu)), \\ \mathrm{H}(x,y,z) &= \gamma x(1 - y\gamma - x\gamma(2\mathrm{F}(\mu) - 1) \\ &\quad - x(2z + x\sigma \mathrm{f}(\mu)) + \gamma^2(y + x(2\mathrm{F}(\mu) - 1))^2), \end{aligned}$$

and for some $K > 0$ and $\epsilon > 0$,

$$\mathrm{P}(|R_n| > Kn^{-1-\epsilon}) = o(n^{-1}).$$

Note that the last component of Z_1 is a lattice variable. The standard results on Edgeworth expansions are not applicable. The parameter γ^{-1} is called the shape parameter and it is generally known. For example, its value for the normal family is $\sqrt{2/\pi}$ and is $1/\sqrt{2}$ for the double exponential family. The results of Babu and Singh (1989) lead to valid one term Edgeworth expansion for H_n. A similar result holds for bootstrapped version of H_n. These results establish that the bootstrap automatically corrects for skewness of the sampling distribution of H_n provided the shape parameter is known. The case when the shape parameter is unknown is presently under investigation.

In the same spirit, Bai and Rao (1991) obtained s-term Edgeworth expansions for $\sqrt{n}(\ell\sigma\ell')^{-1/2}(H(\bar{Z}_n) - H(\mu))$ through a conditional argument, where $\ell = \mathrm{grad}(H(\mu)) \neq 0$. In addition to the conditions on the moments $\mathrm{E}|Z_1|^s < \infty, \quad s \ge 3$, and $\ell_1 > 0$, they assume the partial (conditional) Cramér's condition

$$\limsup_{\|t\| \to \infty} \mathrm{E}\left|\mathrm{E}\left(e^{itZ_{11}} \mid Z_{21}, \cdots, Z_{k1}\right)\right| < 1,$$

where $Z_1 = (Z_{11}, \cdots, Z_{k1})$ and ℓ_1 is the first coordinate of ℓ.

However, this method does not lead to expansions for the multivariate means. Another disadvantage of using the partial Cramér's condition is that it does not hold in general, even in the limit, for the bootstrapped sample. This prevents applications to bootstrap.

Conditions on the Moments

In all the results mentioned above the assumption on the moments is too restrictive, needing existence of moments higher than those required to define the expansions. Consider the simple example of Student's t-statistic,

$$\hat{t} = \sqrt{n}(\bar{X}_n - \mu)/s_n,$$

which can be written as

$$\hat{t} = H\left(\left(\bar{X}_n - \mu, \ \overline{(X_n - \mu)^2}\right)\right),$$

where H is a smooth function on R^2. Standard proof of one term expansion requires at least *three* moments of (X_1, X_1^2), *i.e.*, *six* moments of X_1. But only three moments of X_1 appear in the actual expansion. Hall (1987) showed that Edgeworth expansions for Student's t-statistic are valid under minimal moment conditions. Babu and Bai (1993) generalized this and obtained s- term expansions under minimal moment conditions and *conditional* Cramér's condition for statistics which are smooth functions of means of random vectors. See also Bhattacharya and Ghosh (1988). Chibisov (1980 and 1981) obtained similar results under Cramér's condition, for statistics which can be written as a polynomial in sample means. However, there is a major flaw in the proof given by Chibisov, which we could not fix.

3 Errors in Variables Regression

In this section we consider Edgeworth expansions for statistics, based on independent but not identically distributed random vectors. Second order correctness for bootstrap methods for sums of independent but not identically distributed random variables are considered by Liu (1988) in the univariate case, using Edgeworth expansions. This situation occurs in errors-in-variables (EIV) regression context, even when the residuals are known to be *i.i.d.* The EIV models have been studied extensively in the literature. See Deeming (1968), Fuller (1987), Gleser (1985) and Jones (1979) among others. Consider the simple linear EIV model (X_i, Y_i):

$$X_i = u_i + \delta_i, \quad Y_i = v_i + \epsilon_i,$$

where (δ_i, ϵ_i) are independent mean zero random vectors and u_i and v_i are unknown nuisance parameters. If u_i are known, then the analysis becomes simpler, though it leads to consideration of sums of independent but not identically distributed random vectors. In many applications to fields like ecology and astronomy, very little is known about the values of u_i except that on the average they behave well. In this paper u_i are assumed to be unknown. We are interested in the linear relation $v_i = \alpha + \beta u_i$. Let (δ_i, ϵ_i) be independent copies of (δ, ϵ), and σ_δ and σ_ϵ respectively denote the standard deviations of δ_1 and ϵ_1. Further let $\lambda = \sigma_\epsilon^2/\sigma_\delta^2$. It is well known (see for example Jones, 1979) that the least squares estimators of β and α are given by

$$\hat{\beta}_1 = h + \text{sign}\,(S_{XY})(\lambda + h^2)^{1/2} \quad \text{and} \quad \hat{\alpha}_1 = \bar{Y}_n - \hat{\beta}_1 \bar{X}_n,$$

when λ is known, where $h = (S_{YY} - \lambda S_{XX})/2S_{XY}$,

$$S_{XY} = \sum_{i=1}^n (X_i - \bar{X}_n)(Y_i - \bar{Y}_n),$$

$$S_{XX} = \sum_{i=1}^n (X_i - \bar{X}_n)^2, \quad \text{and} \quad S_{YY} = \sum_{i=1}^n (Y_i - \bar{Y}_n)^2.$$

Instead, if σ_δ alone is known, $S_{XX} > n\sigma_\delta^2$ and $S_{YY}(S_{XX} - n\sigma_\delta^2) > S_{XY}^2$, then the least squares estimators of β and α are given by

$$\hat{\beta}_2 = S_{XY}/(S_{XX} - n\sigma_\delta^2), \quad \text{and} \quad \hat{\alpha}_2 = \bar{Y}_n - \hat{\beta}_2 \bar{X}_n.$$

On the other hand, if σ_ϵ alone is known, $S_{XX}\,(S_{YY} - n\sigma_\epsilon^2) > S_{XY}^2$ and $S_{YY} > n\sigma_\epsilon^2$, then the least squares estimators of β and α are given by

$$\hat{\beta}_3 = (S_{YY} - n\sigma_\epsilon^2)/S_{XY}, \quad \text{and} \quad \hat{\alpha}_3 = \bar{Y}_n - \hat{\beta}_3\bar{X}_n.$$

Simple algebra leads to the fact that $\hat{\beta}_r$ and $\hat{\alpha}_r$, $r = 1, 2, 3$ can be written as smooth functions of the mean of

$$\xi_{jn} = (\epsilon_j^2, \delta_j^2, \epsilon_j\delta_j, \epsilon_j, \delta_j, u_j\epsilon_j, u_j\delta_j).$$

Edgeworth expansions for ξ_{jn} lead to those of $\hat{\beta}_r$. But standard results on Edgeworth expansions are not applicable for two reasons. The first one being that ξ_{jn} are not identically distributed and the second one being that components of ξ_{jn} are linearly dependent. But on the average, the vectors ξ_{jn} behave very well, provided for some $\nu \geq 3$,

$$\sum u_{jn} = 0, \quad \tfrac{1}{n}\sum u_{jn}^2 \to \eta > 0 \quad \text{and} \quad \sup_n \tfrac{1}{n}\sum |u_{jn}|^\nu < \infty.$$

Babu and Bai (1992) have shown that if ϵ and δ are independent and continuous random variables with finite sixth moment, and if the conditions on u_{jn} hold with $\nu = 3$, then $\sqrt{n}(\hat{\beta}_r - \beta)$ has valid one-term Edgeworth expansion, for $r = 1, 2, 3$. In fact, ϵ and δ need not be independent, but satisfy weak continuity assumptions on the conditional distributions of ϵ and δ. If the conditions on u_{jn} hold with $\nu > 3$, then higher order Edgeworth expansions can be obtained. EIV model illustrates the case where both conditions **a′)** and **b)** are violated.

As is demonstrated in Babu and Singh (1983, and 1984), bootstrap approximations work best for studentized statistics in general. To establish such a result, Babu and Bai (1992) have obtained Edgeworth expansions for studentized statistics $\sqrt{n}(\hat{\beta}_r - \beta)/\hat{\sigma}_r$, $r = 1, 2, 3$, and their bootstrapped versions. The scaling factors $\hat{\sigma}_r$ are obtained using jackknife type arguments and are given by,

$$\hat{\sigma}_1^2 = n\hat{\beta}_1^2\psi\sum_{i=1}^{n}\left((Y_i - \bar{Y}_n)^2 - \lambda(X_i - \bar{X}_n)^2 - 2h(X_i - \bar{X}_n)(Y_i - \bar{Y}_n)\right)^2,$$

$$\hat{\sigma}_2^2 = n(\hat{\beta}_2/S_{XY})^2\sum_{i=1}^{n}\left((X_i - \bar{X}_n)(Y_i - \bar{Y}_n - \hat{\beta}_2(X_i - \bar{X}_n)) + \hat{\beta}_2\sigma_\delta^2\right)^2,$$

$$\hat{\sigma}_3^2 = nS_{XY}^{-2}\sum_{i=1}^{n}\left((Y_i - \bar{Y}_n)(Y_i - \bar{Y}_n - \hat{\beta}_3(X_i - \bar{X}_n)) - \sigma_\epsilon^2\right)^2,$$

where $\psi^{-1} = 4S_{XY}^2(h^2 + \lambda)$.

REFERENCES

Babu, G. J. (1991). Edgeworth expansions for statistics, which are functions of lattice and non-lattice variables, *Statistics & Probability Letters*, **12**, 1-7.

Babu, G. J. and Bai, Z. D. (1992). Edgeworth expansions for errors - in - variables models, *J. Multivariate Analysis*, **42**, 226-244..

Babu, G. J. and Bai, Z. D. (1993). Edgeworth expansions of a function of sample means under minimal moment conditions and partial Cramér's condition, To appear in *Sankhyā* A.

Babu, G. J. and Rao, C. R. (1992). Expansions for statistics involving the mean absolute deviations, *Annals of the Institute of Statistical Mathematics*, **44**, 387-403.

Babu, G. J. and Singh, K. (1983). Inference on means using the bootstrap, *Annals of Statistics*, **11**, 999-1003.

Babu, G. J. and Singh, K. (1984). On one term Edgeworth correction by Efron's Bootstrap, *Sankhyā*, Series A, **46**, 219-232.

Babu, G. J. and Singh, K. (1989). On Edgeworth expansions in the mixture cases, *Annals of Statistics*, **17**, 443-447.

Bai, Z. D. and Rao, C. R. (1991). Edgeworth expansion of a function of sample means, *Annals of Statistics*, **19**, 1295-1315.

Bhattacharya, R.N. and Ghosh, J.K. (1978). On the validity of the formal Edgeworth expansion, *Annals of Statistics*, **6**, 434–451.

Bhattacharya, R.N. and Ghosh, J.K. (1988). On moment conditions for valid formal Edgeworth expansion, *J. Multivariate Analysis*, **27**, 68-79.

Charlier, C. V. L. (1905). Über das Fehlergestz, *Ark. Mat. Astr. och Fys.*, **2**, No. 8, 9.

Charlier, C. V. L. (1906). Über die Darstellung willkürlicher Funktionen, *Ark. Mat. Astr. och Fys.*, **2**, No. 20, 35.

Chebyshev, P. L. (1890). Sur deux théorèmes relatifs aux probabilités, *Acta Math.*, **14**, 305-315.

Chibisov, D. M. (1980). An asymptotic expansion for the distribution of a statistic admitting a stochastic expansion - I, *Theory of Probability and its Applications*, **25**, 732-744.

Chibisov, D. M. (1981). An asymptotic expansion for the distribution of a statistic admitting a stochastic expansion - II, *Theory of Probability and its Applications*, **26**, 1-12.

Deeming, T. J. (1966). The Analysis of Linear Correlation in Astronomy, *Vistas in Astronomy*, **10**, 125-142.

Edgeworth, F. Y. (1905). The law of error, *Cambridge Philos. Trans.*, **20**, 36-66 and 113-141.

Fuller, W. A. (1987). *Measurement error models*, John Wiley, New York.

Gleser, L. J. (1985). A note on G. R. Dolby's unreplicated ultrastructural model, *Biometrika*, **72**, 117-124.

Hall, P. (1987). Edgeworth expansion for Student's *t* statistic under minimal moment conditions, *Annals of Probability*, **15**, 920-931.

Jones, T. A. (1979). Fitting straight lines when both variables are subject to error, *Math. Geology*, **11**, 1-25.

Liu, R. Y. (1988). Bootstrap procedures under some non-i.i.d. models, *Ann. Statist.*, **16**, 1696-1708.

Singh, K and Babu, G. J. (1990). On Asymptotic optimality of the bootstrap, *Scandinavian Journal of Statistics*, **17**, 1-9.

Wallace, D. L. (1958). Asymptotic approximations to distributions, *Ann. Math. Stat.*, **29**, 635-654.

Department of Statistics
319 Classroom Building
Pennsylvania State University
University Park, PA 16802
U. S. A.

J.K. Ghosh, S.K. Mitra, K.R. Parthasarathy and B.L.S. Prakasa Rao (*eds.*).
Statistics and Probability: A Raghu Raj Bahadur Festschrift
Published by Wiley Eastern Limited, Publishers, (1993), 73 - 81

The Range of the Infinitesimal Generator of an Ergodic Diffusion

GOPAL K. BASAK
University of California at Berkeley

and

RABI N. BHATTACHARYA [1]
Indiana University

Abstract

Sufficient conditions are given for a function f to belong to the range of the infinitesimal generator of an ergodic nonsingular diffusion X_t. Functional central limit theorems and laws of iterated logarithm hold for time integrals $\int_0^t f(X_s)ds$ of such functions f.

1 Introduction

In this article a broad class of functions on the state space $\Re^k$ of an ergodic *nonsingular diffusion* is determined for which the functional versions of the central limit theorem and the law of the iterated logarithm hold. It is known from Bhattacharya [4] (Theorems 2.1, 2.6, 2.7, Remark 2.6.2), with a supplement from Touati [9] (Theorem 6) that these limit theorems are valid for all functions in the range of the infinitesimal generator on $L^2(\Re^k, \pi)$, where π is the unique invariant probability of the diffusion. This approach, based on martingale central limit theorems and semigroup theory, has an

[1] Research of this author is partially supported by NSF Grant DMS 9206937.
AMS Subject Classifications : Primary 60F17; secondary 60J60.
Key words and Phrases: nonsingular diffusion, invariant probability, semigroup of transition operators.

advantage over the renewal method used in Bhattacharya and Ramasubramanian [5] in that the variance parameter of the limiting Brownian motion is explicitly given in terms of the solution of an elliptic equation and the inner product on $L^2(\Re^k, \pi)$. An additional advantage is that the results in Bhattacharya [4] are applicable to general ergodic Markov processes including ergodic *singular diffusions*. The latter class is analyzed in Basak and Bhattacharya [2] and Basak [1].

An open problem is to explicitly characterize, in terms of the coefficients of the diffusion, those ergodic diffusions for which the range of the infinitesimal generator is all of $1^{\perp}$, which is the set of all mean zero, square integrable functions on $L^2(\Re^k, \pi)$. Indeed, a broad sufficient condition itself will be quite useful. Note that the range is $1^{\perp}$ if and only if zero is an isolated point of the spectrum of the generator (Bhattacharya [4], p. 201). The multivariate Ornstein–Uhlenbeck process (see Bhattacharya and Waymire [6], pp. 597, 598) is an example in which zero is an isolated point of the spectrum.

2 Main Results

Consider a diffusion $\{X_t\}$ on $\Re^k (k \geq 1)$ generated by an elliptic operator (see Ikeda and Watanabe [8])

$$L = \frac{1}{2}\sum_{i,j=1}^{k} a_{ij}(x)\partial_i\partial_j + \sum_{i=1}^{k} b_i(x)\partial_i \tag{2.1}$$

Assume that the coefficients $a_{ij}(x)$ are continuous, $b_i(x)$ are measurable functions which are bounded on compacts, and that for each x, the $k \times k$ matrix $((a_{ij}(x)))$ is symmetric and positive definite. Under the condition (2.4) imposed below, the diffusion is conservative (indeed recurrent) and may be constructed on the canonical space $\Omega = C([0,\infty) : \Re^k)$, the set of all continuous functions on $[0,\infty)$ into $\Re^k$. We let X_t be the coordinate projection: $X_t(\omega) = \omega_t (\omega \in \Omega)$. For each $x \in \Re^k$, let P_x denote the distribution of the diffusion starting at x. By randomizing over x, one may construct P_μ as the distribution of the diffusion having initial distribution μ. Let E_x, E_μ denote expectations under P_x, P_μ respectively.

For a given $z \in \Re^k$ and $r_0 > 0$ define, for $x \neq z$,

$$\begin{gathered} A(x) = \frac{\sum a_{ij}(x)(x_i - z_i)(x_j - z_j)}{|x-z|^2}, \\ B(x) = \sum a_{ii}(x), \quad C(x) = 2\sum (x_i - z_i) b_i(x), \\ \bar{\alpha}(r) = \sup\nolimits_{|x-z|=r} A(x), \quad \underline{\alpha}(r) = \inf\nolimits_{|x-z|=r} A(x), \\ \bar{\beta}(r) = \sup\nolimits_{|x-z|=r} \frac{B(x) - A(x) + C(x),}{A(x)}, \quad \underline{\beta}(r) = \inf\nolimits_{|x-z|=r} \frac{B(x) - A(x) + C(x)}{A(x)}, \\ \bar{I}(r) = \int_{r_0}^{r} \frac{\bar{\beta}(u)}{u} du, \quad \underline{I}(r) \int_{r_0}^{r} \frac{\underline{\beta}(u)}{u} du \end{gathered} \tag{2.2}$$

It is known that the *transition probability density* $p(t; x, y)$ of the diffusion admits an invariant probability (necessarily unique) if and only if the diffusion is positive recurrent (Has'minskii [7], Bhattacharya [3]), i.e., if and only if $E_x \tau_r < \infty$ for all $x, z \in \Re^k$ and $r > 0$, where

$$\tau_r = \inf\{t \geq 0 : |X_t - z| = r\} . \tag{2.3}$$

A proof of the following criterion of Has'minskii may be found in Bhattacharya [3].

LEMMA 1. *The diffusion* $\{X_t\}$ *is positive recurrent if for some* $z \in \Re^k$ *and some* $r_0 > 0$ *one has*

$$\int_{r_0}^{\infty} \exp\{-\bar{I}(r)\} dr = \infty \, , \quad \int_{r_0}^{\infty} \frac{\exp\{\bar{I}(r)\}}{\underline{\alpha}(r)} \, dr < \infty \tag{2.4}$$

For a positive recurrent diffusion let π denote the unique *invariant probability.* On the Hilbert space $L^2(\Re^k, \pi)$ define the *transition operators* $T_t (t > 0)$,

$$(T_t f)(x) = E_x f(X_t) = \int f(y) p(t; x, y) dy. \tag{2.5}$$

The *infinitesimal generator* of this semigroup is $\hat{L}$ defined by

$$\hat{L} f = \lim_{t \downarrow 0} \frac{T_t f - f}{t} \tag{2.6}$$

defined on the doamin $D_{\hat{L}}$ comprising all f in $L^2(\Re^k, \pi)$ such that the limit in (2.6) exists in the L^2–norm. For smooth f vanishing outside a compact set one may show that $\hat{L} f = Lf$. Let $\Re_{\hat{L}}$ denote the *range of* $\hat{L}$,

$$\Re_{\hat{L}} = \{\hat{L} f : f \in D_{\hat{L}}\}. \tag{2.7}$$

The main technical problem now is to identify a broad subset of $\Re_{\hat{L}}$. For any nonnegative measurable function h on $[r_0, \infty)$ define

$$\phi(r; h) = \int_{r_0}^{r} \exp\{-\bar{I}(u)\} \left(\int_{u}^{\infty} \frac{h(v) \exp\{\bar{I}(v)\}}{\underline{\alpha}(v)} \, dv \right) du \; (r \geq r_0). \tag{2.8}$$

For any f define the function h_f on $[0, \infty)$ by

$$h_f(r) = \sup_{|x-z|=r} |f(x)|. \tag{2.9}$$

LEMMA 2. *Let f be a real valued measurable function on $\Re^k$ which is bounded on compacts.*

(a) *For every* $z \in \Re^k, r > 0$ *and positive integer* n *one has*

$$\sup_{x:|x-z| \leq r} E_x \left(\int_0^{\tau_r} |f(X_t)| dt \right)^n < \infty. \tag{2.10}$$

(b) *If* (2.4) *holds, then*

$$\sup_{x:|x-z|=r} E_x \int_0^{\tau_{r_0}} |f(X_t)|dt \leq \phi(r; h_f) \; (r > r_0), \tag{2.11}$$

and

$$\sup_{x:|x-z|=r} E_x \left(\int_0^{\tau_{r_0}} |f(X_t)|dt\right)^2 \leq 2\phi(r; h_f\phi(\cdot; h_f)) \; (r > r_0), \tag{2.12}$$

The proof of Lemma 2 may be found in Bhattacharya and Ramasubramanian [5] (Corollary 1.7, relation (2.9), and the last paragraph of the proof of Theorem 2.9).

LEMMA 3. *Assume* (2.4). *If f is measurable and bounded on compacts, and satisfies (for some $r_0 > 0$)*

$$\int_{r_0}^{\infty} \frac{h_f^2(r)\exp\{\bar{I}(r)\}}{\underline{\alpha}(r)}\, dr < \infty, \tag{2.13}$$

then $f \in L^2(\Re^k, \pi)$.

PROOF. Choose $z \in \Re^k, r_0 > 0$. Fix r_1 such that $0 < r_0 < r_1$. Define the following sequence of stopping times relative to the family of sigmafields $\mathcal{F}_t := \sigma\{X_s : 0 \leq s \leq t\}$ $(t \geq 0)$,

$$\begin{aligned}
\tau^{(1)} &= \inf\{t \geq 0 : |X_t - z| = r_1\}, \\
\tau^{(2)} &= \inf\{t > \tau^{(1)} : |X_t - z| = r_0\}, \\
\tau^{(2n+1)} &= \inf\{t \geq \tau^{(2n)} : |X_t - z| = r_1\}, \\
\tau^{(2n+2)} &= \inf\{t > \tau^{(2n+1)} : |X_t - z| = r_0\}, \\
& \qquad (n = 1, 2, \ldots,).
\end{aligned} \tag{2.14}$$

Then $\{X_{\tau^{(2n)}} : n = 1, 2, \ldots\}$ is a discrete parameter Markov process on the state space $\{x : |x - z| = r_0\}$. Its transition probability admits a unique invariant probability m and one has (see Has'minskii [7] or Bhattacharya [3])

$$\int_{\Re^k} g(x)\pi(dx) = \frac{E_m \int_0^{\tau^{(2)}} g(X_t)dt}{E_m \tau^{(2)}}, \tag{2.15}$$

for every integrable g. Now, by (2.10) and (2.11) applied to f^2,

$$\begin{aligned}
E_m \int_0^{\tau^{(2)}} f^2(X_t)dt &= E_m \int_0^{\tau^{(1)}} f^2(X_t)dt + E_m \int_{\tau^{(1)}}^{\tau^{(2)}} f^2(X_t)dt \\
&\leq \sup_{\{x:|x-z|=r_0\}} E_x \int_0^{\tau_{r_1}} f^2(X_t)dt \\
&\quad + \sup_{\{x:|x-z|=r_1\}} E_x \int_0^{\tau_{r_0}} f^2(X_t)dt \\
&\leq c + \phi(r_1; h_{f^2}) = c + \phi(r_1; h_f^2) < \infty,
\end{aligned} \tag{2.16}$$

where c is a positive constant. Note that the finiteness of $\phi(r_1; h_f^2)$ follows from (2.13). The integrability of f^2 with respect to m now follows from (2.15). □

THEOREM 4. *Assume* (2.4) *and*

$$\int_{r_0}^{\infty} \frac{\phi(r;1)}{\underline{\alpha}(r)} \exp\{\bar{I}(r)\} dr < \infty, \tag{2.17}$$

where $\phi(r;1)$ *is given by* (2.8) *with* $h \equiv 1$. *If* f *is measurable, and bounded on compacts,* (2.13) *holds and*

$$\int_{r_0}^{\infty} \frac{\phi^2(r;h_f)}{\underline{\alpha}(r)} \exp\{\bar{I}(r)\} dr < \infty, \tag{2.18}$$

then $\tilde{f} := f - \int f(y)\pi(dy) \in \Re_{\hat{L}}$. *In particular, under* (2.4) *and* (2.17) *all bounded measurable centered functions belong to* $\Re_{\hat{L}}$.

PROOF. Define the *resolvent operator* $R_\lambda (\lambda > 0)$,

$$(R_\lambda g)(x) = \int_0^{\infty} e^{-\lambda t}(T_t g)(x) dt \qquad (g \in L^2(\Re^k, \pi)) . \tag{2.19}$$

By Lemma 3, $\tilde{f} := f - \int f(y)\pi(dy) \in L^2(\Re^k, \pi)$. Using the notation (2.14), one has

$$\begin{aligned} R_\lambda \tilde{f} &= \int_0^{\infty} e^{-\lambda t}(T_t \tilde{f})(x) dt = E_x \int_0^{\infty} e^{-\lambda t} \tilde{f}(X_t) dt \\ &= E_x \int_0^{\tau^{(2)}} e^{-\lambda t} \tilde{f}(X_t) dt + \sum_{n=1}^{\infty} E_x \int_{\tau^{(2n)}}^{\tau^{(2n+2)}} e^{-\lambda t} \tilde{f}(X_t) dt. \end{aligned} \tag{2.20}$$

We will show that $\{R_\lambda \tilde{f} : \lambda > 0\}$ is a bounded subset of $L^2(\Re^k, \pi)$. First,

$$\begin{aligned} \left|E_x \int_0^{\tau^{(2)}} e^{-\lambda t} \tilde{f}(X_t) dt\right| &\leq E_x \int_0^{\tau^{(2)}} |\tilde{f}(X_t)| dt \\ &\leq E_x \int_0^{\tau^{(1)}} |\tilde{f}(X_t)| dt \\ &+ \sup_{\{y:|y-z|=r_1\}} E_y \int_0^{\tau_{r_0}} |\tilde{f}(X_t)| dt \\ &\leq E_x \int_0^{\tau_{r_1}} |\tilde{f}(X_t)| dt + \phi(r_1; h_{\tilde{f}}) \end{aligned} \tag{2.21}$$

using the strong Markov property for the second inequality, and (2.11) for the third. Since $\phi(r_1; h_{\tilde{f}}) \leq \phi(r_1; h_f) + c\phi(r_1; 1)$, where $c = |\int f(y)\pi(dy)|$,

$\phi(r_1; h_{\tilde{f}}) < \infty$ under (2.4) and (2.13). Also, by (2.10),

$$C_1 := \sup_{\{x:|x-z|\leq r_1\}} E_x \int_0^{\tau_{r_1}} |\tilde{f}(X_t)|dt < \infty \tag{2.22}$$

and, by (2.11),

$$\begin{aligned} \sup_{\{x:|x-z|=r\}} E_x \int_0^{\tau_{r_0}} |\tilde{f}(X_t)|dt &\leq \phi(r; h_{\tilde{f}}) \\ &\leq c\phi(r;1) + \phi(r; h_f) \; (r > r_0). \end{aligned} \tag{2.23}$$

By (2.15), (2.21) - (2.23),

$$\int_{\Re^k} \left(E_x \int_0^{\tau^{(2)}} |\tilde{f}(X_t)|dt \right) \pi(dx) \leq C_2 + 2 \int_{\{x:|x-z|>r_1\}} \phi^2(|x-z|; h_f)\pi(dx). \tag{2.24}$$

By Lemma 3, and (2.18), the last integral is finite. Thus the first summand on the extreme right side of (2.20) is uniformly bounded in $L^2(\Re^k, \pi)$. It remains to show that the infinite sum in (2.20) is also uniformly (for $\lambda > 0$) bounded. Let us compare this sum with

$$g(x) = \sum_{n=1}^{\infty} g_n(x), \tag{2.25}$$

where

$$g_n(x) = E_x \int_{\tau^{(2n)}}^{\tau^{(2n+2)}} \tilde{f}(X_t)dt. \tag{2.26}$$

By the strong Markov property,

$$g_n(x) = \int_{\{y:|y-z|=r_0\}} w(y)\mu_{x,n}(dy), \tag{2.27}$$

where $\mu_{x,n}$ is the P_x – distribution of $X_{\tau^{(2n)}}$, and

$$w(y) = E_y \int_0^{\tau^{(2n)}} \tilde{f}(X_t)dt \qquad (|y-z| = r_0). \tag{2.28}$$

But the transition probability of the Markov process $\{X_{\tau^{(2n)}}\}$ satisfies Doeblin's condition (Bhattacharya [3], proof of Theorem 3.5), and $w(y) \leq C_1 + \phi(r_1; h_{\tilde{f}})$. Therefore, $g_n(x)$ converges exponentially fast, uniformly for all x, to

$$\begin{aligned} \int_{\{y:|y-z|=r_0\}} w(y)m(dy) &= E_m \int_0^{\tau^{(2)}} \tilde{f}(X_t)dt \\ &= (E_m\tau^{(2)}) \int_{\Re^k} \tilde{f}(x)\pi(dx) = 0. \end{aligned} \tag{2.29}$$

Hence $g(x)$ is bounded on $\Re^k$. Now

$$\left.\begin{array}{l} |E_x \int_{\tau^{(2n)}}^{\tau^{(2n+2)}} e^{-\lambda t}\tilde{f}(X_t)dt - g_n(x)| \\ \quad = |E_x\{e^{-\lambda\tau^{(2n)}}(E_{X_{\tau^{(2n)}}} \int_0^{\tau^{(2)}} (e^{-\lambda t}-1)\tilde{f}(X_t)dt)\}| \\ \quad \le |E_x\{e^{-\lambda\tau^{(2n)}}(E_{X_{\tau^{(2n)}}}[(1-e^{-\lambda\tau^{(2)}})]\cdot\int_0^{\tau^{(2)}} |\tilde{f}(X_t)|dt)\}| \end{array}\right\} \tag{2.30}$$

But, by Lemma 2,

$$\left.\begin{array}{l} E_y[1-e^{-\lambda\tau^{(2)}}\int_0^{\tau^{(2)}} |\tilde{f}(X_t)|dt] \\ \quad \le [E_y(1-e^{-\lambda\tau^{(2)}})^2]^{1/2}\cdot[E_y(\int_0^{\tau^{(2)}} |\tilde{f}(X_t)|dt)^2]^{\frac{1}{2}} \\ \quad \le [E_y(1-e^{-\lambda\tau^{(2)}})^2]^{\frac{1}{2}}\cdot[C_3+4\phi(r_1;h_f\phi(\cdot;h_f))]^{\frac{1}{2}} \\ \quad = C_4[E_y(1-e^{-\lambda\tau^{(2)}})^2]^{\frac{1}{2}} \ (|y-z|=r_0). \end{array}\right\} \tag{2.31}$$

For $\phi(r_1;h_f\phi(\cdot;h_f))$ is finite by (2.13) and (2.18) (using the Schwarz inequality). Hence

$$\left.\begin{array}{l} |E_x \int_{\tau^{(2n)}}^{\tau^{(2n+2)}} e^{-\lambda t}\tilde{f}(X_t)dt - g_n(x)| \\ \le C_4(\sup_{\{y:|y-z|=r_0\}}[E_y(1-e^{-\lambda\tau^{(2)}})^2]^{\frac{1}{2}}) \qquad (E_x e^{-\lambda\tau^{(2n)}}). \end{array}\right\} \tag{2.32}$$

Next, by Lemma 2 (with $f\equiv 1$) and (2.17),

$$E_y(1-e^{-\lambda\tau^{(2)}})^2 \le \lambda^2 E_y(\tau^{(2)})^2 \le C_5\lambda^2 \ (|y-z|=r_0). \tag{2.33}$$

Also, by repeated applications of the strong Markov property,

$$E_x e^{-\lambda\tau^{(2n)}} = E_x\{e^{-\lambda\tau^{(2)}} E_{X_{\tau^{(2)}}} e^{-\lambda\tau^{(2n-2)}}\} \le E_x e^{-\lambda\tau^{(2)}}(\delta(\lambda))^{n-1} \tag{2.34}$$

where

$$\left.\begin{array}{l} \delta(\lambda) = \sup_{\{y:|y-z|=r_0\}} E_y e^{-\lambda\tau^{(2)}} \le 1-\lambda\inf_{|y-z|=r_0} E_y\tau^{(2)} \\ \quad +\frac{\lambda^2}{2}\sup_{|y-z|=r_0}(E_y\tau^{(2)})^2 \\ \quad \le 1-C_6\lambda \quad (\lambda\le\varepsilon). \end{array}\right\} \tag{2.35}$$

for some $C_6>0$ and some $\varepsilon>0$ (use, e.g., relation (2.8) in Bhattacharya and Ramasubramanian [5]). From (2.32) – (2.35) one obtains, for $0\le\lambda\le\varepsilon$,

$$|E_x\int_{\tau^{(2n)}}^{\tau^{(2n+2)}} e^{-\lambda t}\tilde{f}(X_t)dt - g_n(x)| \le C_7\lambda(1-C_6\lambda)^{n-1}E_x e^{-\lambda\tau^{(2)}} \tag{2.36}$$

Summing over n, one has

$$\begin{aligned} |\sum_{n=1}^{\infty} E_x\int_{\tau^{(2n)}}^{\tau^{(2n+2)}} e^{-\lambda t}\tilde{f}(X_t)dt| &\le |g(x)| + \frac{C_7\lambda}{C_6\lambda}E_x e^{-\lambda\tau^{(2)}} \\ &\le C_8+\frac{C_7}{C_6}=C_9. \end{aligned} \tag{2.37}$$

Hence $\{R_\lambda \tilde{f} : \lambda > 0\}$ is a bounded subset of $L^2(\Re^k, \pi)$. Therefore, there exists a sequence $\lambda_n \downarrow 0$ such that $R_{\lambda_n} \tilde{f}$ converges *weakly* to some $u \in L^2(\Re^k, \pi)$. But $R_{\lambda_n} = (\lambda_n - \hat{L})^{-1}$. Hence $u_n := (\lambda_n - \hat{L})^{-1}\tilde{f}$ converges weakly to u. Now $(\lambda_n - \hat{L})u_n = \tilde{f}$ for all n, $\lambda_n u_n$ converges (strongly) to zero as $n \to \infty$. Therefore, $\hat{L}(-u_n) \to \tilde{f}$ and $n \to \infty$. The graph of $\hat{L}$ is closed, strongly as well as weakly (Yosida [10], pp. 79, 125), and $(-u_n, \hat{L}(-u_n))$ belongs to this graph (for all n) and converges weakly in the space $L^2 \times L^2$ to $(-u, \tilde{f})$. Therefore, $-u \in D_{\hat{L}}$ and $\hat{L}(-u) = \tilde{f}$. □

The following corollary is an immediate consequence of Theorem 4 above, Theorem 2.6, Theorem 2.8 and Remark 2.6.2 in Bhattacharya [4], and Theorem 6 in Touati [9].

COROLLARY 5. *Consider a diffusion, satisfying* (2.4) *and* (2.17). *Then for every f, satisfying the hypothesis of Theorem 4, and $x \in \Re^k$,* (a) *the processes $\{n^{-1/2} \int_0^{nt} \tilde{f}(X^x(s))ds : t \geq 0\}$ converge in distribution, as $n \longrightarrow \infty$, to a Brownian Motion with mean zero and variance parameter $\sigma^2 = -2 \int f(x)h(x)\pi(dx)$, where h is a function in the domain of $\hat{L}$ such that $\hat{L}h = f - \bar{f} = \tilde{f}$, and* (b) *the sequence of random functions $\{(2n \log\log n)^{-1/2} \int_0^{nt} \tilde{f}(X^x(s))ds : 0 \leq t \leq 1\}$ is relatively compact in $C[0,1]$ with probability one and the set of limit points is the set of all absolutely continuous functions θ on $[0,1]$ satisfying $\int_0^1 (\theta'(t))^2 dt \leq \sigma^2, \theta(0) = 0$.*

REFERENCES

[1] Basak, G.K. (1991) A class of limit theorems for singular diffusions. *J. Multivariate Anal.*, **39** 44–59.

[2] Basak, G.K. and Bhattacharya, R.N. (1992) Stability in distribution for a class of singular diffusions. *Ann. Probab.* **20** 312–321.

[3] Bhattacharya, R.N. (1978) Criteria for recurrence and existence of invariant probability measures for multidimensional diffusions. *Ann. Probab.* **6** 541–553. (1980). Correction Note, Ibid 8 1194–95.

[4] Bhattacharya, R.N. (1982). On the functional central limit theorem and the law of iterated logarithm for Markov processes. *Z. Wahrsch. verw. Gebiete* **60** 185–201.

[5] Bhattacharya, R.N. and Ramasubramanian, S. (1982) Recurrence and ergodicity of diffusions. *J. Multivariate Anal.* **12** 95–122.

[6] Bhattacharya, R.N. and Waymire, E.C. (1990). *Stochastic Processes with Applications*. Wiley, New York.

[7] Has'minskii, R.Z. (1960). Ergodic properties of recurrent diffusion processes and stabilization of the solution of the Cauchy problem for parabolic equations. *Theor. Probability Appl.* **5** 179–196. (English Translation).

[8] Ikeda, N. and Watanabe, S. (1981) *Stochastic Differential Equations and Diffusion Processes.* North Holland, Amsterdam.

[9] Touati, P.A. (1990) Loi fonctionnelle du logarithme itere pour les processus de Markov recurrents. *Ann. Probab.* **18** 140–159.

[10] Yosida, K. (1965) *Functional Analysis.* Springer-Verlag, New York.

Department of Statistics
University of California
Berkeley, California 94720
U.S.A.

Department of Mathematics
Indiana University
Bloomington, Indiana 47405
U.S.A.

J.K. Ghosh, S.K. Mitra, K.R. Parthasarathy and B.L.S. Prakasa Rao (*eds.*),
Statistics and Probability : A Raghu Raj Bahadur Festschrift
Published by Wiley Eastern Limited, Publishers, (1993), 83 – 96

Asymptotic Distribution of the Likelihood Ratio Statistic in a Prototypical Non Regular Problem

P. BICKEL[1]
University of California at Berkeley, Berkeley

and

H. CHERNOFF[2]
Harvard University, Cambridge

Abstract

This paper addresses the asymptotic behavior of $M_n^* = \sup_t S_n^*(t)$ where

$$S_n^*(t) = n^{-1/2} \sum_{i=1}^{n} y^*(X_i, t)$$

$$y^*(x,t) = (e^{tx - t^2/2} - 1 - tx)/(e^{t^2} - 1 - t^2)^{1/2}$$

and $X_1, X_2, \ldots, X_n$ are i.i.d. $N(0,1)$ random variables.

[1]Supported in part by NSF Grant DMS-8505550 at the Mathematical Sciences Research Institute

[2]Supported in part by NSF Grant DMS-8505550 and ONR contract N00014-91-J1005.

AMS Subject Classifications: Primary 60G70; secondary 60G05, 62M99.

Key words and Phrases: Asymptotic distribution, supremum of stochastic process, Slepian theorem, Iterated logarithm, Kolmogorov bound, Hungarian construction, likelihood-ratio tests, mixtures of distributions.

It will be shown that

$$M_n^* = (\log_2 n)^{1/2} + (V_n - \log(\sqrt{2\pi}))(\log_2 n)^{-1/2}$$

where $\log_2 n = \log(\log n)$ and, as $n \to \infty$,

$$P\{V_n \le v\} \to \exp(-e^{-v}).$$

This result gives the asymptotic behavior of the likelihood-ratio test statistic for testing whether a mixture of two normal distributions with specified variance is simply a pure normal with that variance. We expect that the analysis for this problem will be relevant to a class of testing problems, including mixture problems and change point problems, characterized by a singularity in the natural parametrization when the null hypothesis is true, that leads to a loss of indentification under the null hypothesis.

1 Introduction

This paper addresses the asymptotic behavior of $M_n^* = \sup_t S_n^*(t)$ where

$$S_n^*(t) = n^{-1/2} \sum_{i=1}^{n} y^*(X_i, t) \tag{1}$$

$$y^*(x,t) = (e^{tx-t^2/2} - 1 - tx)/(e^{t^2} - 1 - t^2)^{1/2} \tag{2}$$

and $X_1, X_2, \ldots, X_n$ are i. i. d. $N(0,1)$ random variables. It will be shown that

$$M_n^* = (\log_2 n)^{1/2} + (V_n - \log(\sqrt{2\pi}))(\log_2 n)^{-1/2} \tag{3}$$

where $\log_2 n = \log(\log n)$ and, as $n \to \infty$,

$$P\{V_n \le v\} \to \exp(-e^{-v}). \tag{4}$$

Hartigan (1984) has shown that the logarithm of the likelihood-ratio test statistic for the test that a mixture of two normal distributions with known variance is a pure normal with that variance, can be approximated by an expression stochastically equal to M_n^{*2} and that $M_n^* \to \infty$ in probability. His proof uses the fact that the $y^*(X_i, t)$ have mean 0 and variance 1, and hence $S_n^*(t)$ is asymptotically normal with mean 0 and variance 1 for each value of t. The covariances of S_n^* at s and t become small for s and t far enough apart, and hence M_n^* relates to the maximum of a large number of almost independent normals, and approaches infinity. He conjectures correctly that $M_n^* = O((\log_2 n)^{1/2})$.

There is a large class of problems, including mixture problems and change point problems, that are characterized by a natural parameterization which degenerates under the null hypothesis (See Ghosh and Sen (1985)). For example, in Hartigan's problem, the underlying parameters are α, μ_1 and μ_2, the mixture proportion and the two means. A single distribution under the null hypothesis, e.g. $N(0,1)$, can be represented by infinitely many combinations of these parameters. In other words, there is a loss of identification under the null hypothesis.

Heuristically, what happens in such problems is the following (compare Chernoff (1954).) We suppose we have $X_1, \dots, X_n$ i.i.d. $p(\cdot, \theta)$, $\theta = (\phi, \psi)$, Euclidean. Suppose also that ϕ is unidentifiable if $\psi = 0$, $p(\cdot, \phi, 0) = p(\cdot, 0, 0)$. We expect that, for $\psi = O(n^{-1/2})$,

$$\sum_{j=1}^{n} \log \frac{p(X_i, \theta)}{p(X_i, \phi, 0)} \simeq \Psi^T \sum_{i=1}^{n} T(X_i, \phi)$$

$$-\frac{n}{2} \Psi^T \sum(\phi) \Psi + o_p(1)$$

where $E_0 T = 0$, $\mathrm{Var}_0 T = \sum(\phi)$, and hence that,

$$2 \sup_{\phi, \psi \neq 0} \sum_{i=1}^{n} \log \frac{p(X_i, \theta)}{p(X_i, \phi, 0)} = \sup_{\phi} \|Z_n(\phi)\|^2 + o_p(1)$$

where

$$Z_n(\phi) = n^{-1/2} \sum_{i=1}^{n} \sum{}^{-1/2}(\phi) T(X_i, \phi).$$

It is reasonable to expect that $\sup_\phi \|Z_n(\phi)\|^2$ behaves to first order like $\sup_\phi \|Z(\phi)\|^2$ where Z is Gaussian, $EZ(\phi) = 0$ and $\mathrm{Var} Z(\phi) = I$ (the identity). One possibility now is that $P[\sup_\phi Z(\phi) < \infty] = 1$. We expect this to happen if the domain of ϕ is compact, for instance, if X is distributed as $(1-\psi)\,\mathrm{Bin}(m, \frac{1}{2}) + \psi\,\mathrm{Bin}(m, \phi)$, a case considered in Chernoff and Lander (1989). It can even happen if ϕ is unbounded. Take X distributed as bivariate $\mathcal{N}(\psi, \phi\psi, I)$, in which case, $\sup_\phi \|Z_n(\phi)\|^2$ is χ^2_2, rather than the naively expected χ^2_1. However, in the normal mixture problem, $\sup_\phi \|Z_n(\phi)\|^2 = \infty$. We may still hope, however, that $P[\sup_\phi \|Z_n(\phi)\|^2 < \infty] = 1$. Suppose further, $\sup_\phi \|Z_n(\phi)\|^2 = \sup\{\|Z_n(\phi)\|^2 : \phi \in K_n\}$ plus lower order terms where K_n is compact, K_n tends, as $n \to \infty$, to the range of ϕ (which can be a cone rather than the whole Euclidean space). Then we may expect that the theory of extrema of Gaussian processes — see Leadbetter, Lindgren and Rootzen (1983) will be a guide to higher order behavior of $\sup_\phi \|Z_n(\phi)\|^2$. That is the prescription that we carry out in this paper for the special case of normal mixtures. We expect the approach to be applicable to most other mixture and change point problems.

2 Outline

Instead of attacking $S_n^*(t)$ directly, we will first deal with the simpler expression

$$S_n(t) = n^{-1/2} \sum_{i=1}^{n} (e^{tX_i - t^2/2} - 1) e^{-t^2/2} \tag{5}$$

with supremum

$$M_n = sup_t S_n(t). \tag{6}$$

Then

$$\begin{aligned} S_n(t) &= n^{1/2} \int_0^1 y(x,t)[dF_n(u) - dF(u)] \\ &= \int_0^1 y(x,t)\, dB_n(u) \end{aligned} \tag{7}$$

where

$$y(x,t) = e^{tx - t^2} \tag{8}$$

and

$$\begin{gathered} u = \Phi(x) = \textstyle\int_{-\infty}^{x} \phi(v)\, dv = 1 - \Phi^c(x), \\ \phi(v) = (2\pi)^{-1/2} \exp(-v^2/2), \\ F_n(u) = \text{sample c. d. f. of } U_i = \Phi(X_i) = 1 - U_i^c, \\ F(u) = u = 1 - u^c, \end{gathered}$$

and

$$B_n(u) = n^{1/2}[F_n(u) - F(u)]. \tag{9}$$

We will relate $S_n(t)$ to the Gaussian Process

$$S_0(t) = \int_0^1 y(x,t) dB_0(u) \tag{10}$$

where B_0 is the Brownian Bridge. The Hungarian Construction (Komlos, Major, Tusnady, (1975)) gives us

$$sup_{0 \le u \le 1} |B_n(u) - B_0(u)| = O_p(n^{-1/2} \log n) \tag{11}$$

on a suitable probability space.

Both S_0 and S_n are zero mean stochastic processes with common covariance function

$$\rho(s,t) = \exp(-(s-t)^2/2) - \exp(-s^2/2 - t^2/2).$$

By adjoining to S_0 and S_n, the relatively trivial additional term $\tilde{X} e^{-t^2/2}$ where $\tilde{X}$ is independent of S_0 and S_n and $\mathcal{L}(\tilde{X}) = N(0,1)$, we have the processes

$$\tilde{S}_0(t) = S_0(t) + \tilde{X} e^{-t^2/2} \tag{12}$$

and

$$\tilde{S}_n(t) = S_n(t) + \tilde{X} e^{-t^2/2} \tag{13}$$

with common covariance function

$$\tilde{\rho}(s,t) = e^{-(s-t)^2/2} \tag{14}$$

Since $\tilde{S}_0$ is a Gaussian and stationary process its maximal behavior is well known (see Leadbetter, 1983) i.e.,

$$\begin{aligned}\tilde{M}(T) &= sup_{-T\leq t\leq T}\tilde{S}_0(t)\\ &\doteq (2\log 2T)^{1/2}+(V_T-\log(2\pi))(2\log 2T)^{-1/2} \qquad (15)\end{aligned}$$

where, as $T\to\infty$,

$$P\{V_T\leq v\}\to \exp(-e^{-v}). \qquad (16)$$

Furthermore, the location of the value of t for which $\tilde{S}_0$ attains its maximum is uniformly distributed over the range $(-T,T)$.

The main idea of the derivation is to show that $S_n(t)$ behaves like $S_0(t)$ and $\tilde{S}_0(t)$ for $|t|\leq\sqrt{\log n/2}$, and becomes relatively small for $|t|\geq\sqrt{\log n/2}$.

For $t>0$, we shall decompose each of the integrals S_0 and S_n into two parts in one of two ways. Then, for $i=1,2$

$$\begin{aligned}S^u_{ni}(t) &= \int_{x>x_{ni}} y(x,t)dB_n(u),\\ S^u_{0i}(t) &= \int_{x>x_{ni}} y(x,t)dB_0(u),\\ S^l_{ni}(t) &= \int_{x\leq x_{ni}} y(x,t)dB_n(u),\\ S^l_{0i}(t) &= \int_{x\leq x_{ni}} y(x,t)dB_0(u)\end{aligned}$$

We will also find it convenient to define, for $i=1,2$,

$$\begin{aligned}\tilde{S}^u_{0i}(t) &= \tilde{X}e^{-t^2/2}\Phi^c(x_{ni}-t)+S^u_{0i}(t)\\ \tilde{S}^u_{ni}(t) &= \tilde{X}e^{-t^2/2}\Phi^c(x_{ni}-t)+S^u_{ni}(t)\\ \tilde{S}^l_{0i}(t) &= \tilde{X}e^{-t^2/2}\Phi(x_{ni}-t)+S^l_{0i}(t)\\ \tilde{S}^l_{ni}(t) &= \tilde{X}e^{-t^2/2}\Phi(x_{ni}-t)+S^l_{ni}(t)\end{aligned}$$

since the corresponding covariance functions are

$$\tilde{\rho}^u_i(s,t)=\Phi^c(x_{ni}-(s+t))e^{-(s-t)^2/2} \qquad (17)$$

for $\tilde{S}^u_{0i}$ and $\tilde{S}^u_{ni}$, and

$$\tilde{\rho}^l_i(s,t)=\Phi(x_{ni}-(s+t))e^{-(s-t)^2/2} \qquad (18)$$

for $\tilde{S}^l_{0i}$ and $\tilde{S}^l_{ni}$.

Our main immediate goal is to show that M_n is stochastically asymptotically equivalent to $\tilde{M}(\sqrt{\log n/2})$. The argument involves several ranges of t and two different levels of x_{ni}, and five basic tools or established theorems. The levels of x_{ni} and the crucial t values are given by

$$\begin{aligned}
x_{n1}^2 &= 2\log n - 4\log_2 n \\
x_{n2}^2 &= 2\log n - 2\log_2 n \\
t_{n0} &= (2\log_3 n)^{1/2} \\
t_{n1} &= x_{n1}/2 - 2(\log_2 n)^{1/2} \\
t_{n2} &= x_{n1}/2 - 2(\log_3 n)^{1/2} \\
t_{n3} &= x_{n2}/2 + 2(\log_3 n)^{1/2} \\
t_{n4} &= x_{n2}/2 + 2(\log_2 n)^{1/2} \\
t_{n5} &= (\log n)^{1/2},
\end{aligned}$$

where $\log_3 n \equiv \log(\log_2 n)$. Note that $0 < x_{n2} - x_{n1} = o(1)$.

We shall abbreviate the basic tools with letters. Thus the Hungarian Construction, referred to previously, will be labeled H. The law of the iterated logarithm will be labeled I. Slepian's theorem states that if $\rho_1(s,t)$ and $\rho_2(s,t)$ are the autocovariances for two Gaussian processes with mean 0 and variance 1, and $\rho_1(s,t) \geq \rho_2(s,t)$ for all s, t, then the supremum of the first process is stochastically smaller than the supremum of the second, (see Leadbetter, 1983), and will be labeled S.

Another basic tool, to be labeled T for Tail, is the fact that

$$\sup_{0\leq u\leq 1} [F_n(u)/F(u)] = O_p(1) \tag{19}$$

and

$$\sup_{0\leq u\leq 1} [(1 - F_n(u))/(1-u)] = O_p(1) \tag{20}$$

Finally, a fifth tool to be labeled K is the Kolmogorov Bound, (see Billingsley (1968)), which states that if $Z(t)$ is a stochastic process which satisfies

$$E[Z(s) - Z(t)]^2 \leq c(s-t)^2, \tag{21}$$

for $0 \leq s \leq t \leq 1$, then

$$P[\sup_{0<t<1} |Z(t) - Z(0)| \geq z] \leq Kc/z^2 \tag{22}$$

where K is an absolute constant.

The main argument will show that $S_n(t) - S_0(t) = o_p(\log_2 n)^{-1/2}$ for $|t| \leq t_{n1}$. Since the difference between $\sup_{|t|\leq t_{n1}} S_0(t)$ and $\sup_{|t|\leq t_{n1}} \tilde{S}_0(t)$ is negligible, that will establish that $\sup_{|t|\leq t_{n1}} S_n(t)$ behaves asymptotically like $\tilde{M}(\sqrt{\log n/2})$. But we want this result for $\sup_t S_n(t)$. To achieve this it suffices then to show that $\sup_{|t|>t_{n1}} S_n(t) = o_p(\log_2 n)^{1/2}$, and hence the supremum over all t is achieved for $|t| \leq t_{n1}$ with probability approaching one. Finally the difference between M_n and M_n^* will be shown to be negligible.

We now outline in more detail the arguments used to show that $S_n(t) - S_0(t) = o_p(\log_2 n)^{-1/2}$ for $0 \leq t \leq t_{n1}$ and $S_n(t) = o_p(\log_2 n)^{1/2}$ for $t \geq t_{n1}$. Recall that $S_n(t) = S_{n1}^l(t) + S_{n1}^u(t) = S_{n2}^l(t) + S_{n2}^u(t)$ and that $S_0(t)$ can be decomposed similarly.

(1) $0 \leq t \leq t_{n1}$

$$H1: \quad S_{n1}^l(t) - S_{01}^l(t) = o_p(\log_2 n)^{-1/2}$$
$$I1: \quad S_{01}^u(t) = o_p(\log_2 n)^{-1/2}$$
$$T1: \quad S_{n1}^u(t) = o_p(\log_2 n)^{-1/2}$$

(2) $t_{n1} \leq t \leq t_{n2}$

$$H2: \quad S_{n1}^l(t) - S_{01}^l(t) = o_p(\log_2 n)^{1/2}$$
$$S2: \quad S_{01}^l(t) = o_p(\log_2 n)^{1/2}$$
$$K2: \quad S_{n1}^u(t) = o_p(\log_2 n)^{1/2}$$

(3) $t_{n2} \leq t \leq t_{n3}$

$$K3: \quad S_n(t) = o_p(\log_2 n)^{1/2}$$

(4) $t_{n3} \leq t \leq t_{n4}$

$$K4: \quad S_{n2}^l(t) = o_p(\log_2 n)^{1/2}$$
$$T4: \quad S_{n2}^u(t) = o_p(\log_2 n)^{1/2}$$

(5) $t_{n4} \leq t \leq t_{n5}$

$$K5: \quad S_{n2}^l(t) = o_p(\log_2 n)^{1/2}$$
$$T5: \quad S_{n2}^u(t) = o_p(\log_2 n)^{1/2}$$

(6) $t_{n5} \leq t$

$$T6: \quad S_n(t) = o_p(\log_2 n)^{1/2}$$

3 The Kolmogorov Bound argument

First we note a simple corollary of the Kolmogorov Bound result. If the interval over which Z is defined, is replaced by one of length L, then the bound cK/z^2, for the probability of the maximum deviation, is replaced by $cK(L/z)^2$. Consequently, we have $\sup(|Z(s) - Z(t)|) = (Lc^{1/2})O_p(1)$.

3.1 $K3$ Since $\tilde{\rho}(s,t) = \exp[-(s-t)^2/2]$,

$$\begin{aligned} E\{[\tilde{S}_n(s) - \tilde{S}_n(t)]^2\} &= 2(1 - e^{-(s-t)^2/2}) \\ &\leq (s-t)^2 \end{aligned}$$

and it follows that

$$\begin{aligned} \sup_{t_{n2} \leq t \leq t_{n3}} |\tilde{S}_n(t) - \tilde{S}_n(t_{n2})| &= O_p(t_{n3} - t_{n2}) \\ &= o_p(\log_2 n)^{1/2}. \end{aligned}$$

Then

$$\begin{aligned}\sup_{t_{n2}\le t\le t_{n3}} S_n(t) &\le S_n(t_{n2}) + \sup_{t_{n2}\le t\le t_{n3}} |\tilde{S}_n(t) - \tilde{S}_n(t_{n2})| \\ &\quad + \sup_{t_{n2}\le t\le t_{n3}} |e^{-t^2/2} - e^{-t_{n2}^2/2}| O_p(1) \\ &= O_p(1) + o_p(\log_2 n)^{1/2} + o_p(1) \\ &= o_p(\log_2 n)^{1/2} \qquad (23)\end{aligned}$$

3.2 $K2$ From Equation 17 it follows that

$$\begin{aligned}E\{[\tilde{S}^u_{n1}(s) - \tilde{S}^u_{n1}(t)]^2\} &= \Phi^c(x_{n1} - 2s) + \Phi^c(x_{n1} - 2t) \\ &\quad -2\Phi^c(x_{n1} - (t+s)) + 2\Phi^c(x_{n1} - (t+s))[1 - e^{-(s-t)^2/2}]\end{aligned}$$

For $0 \le t \le s \le x_{n1}/2 - 1$,

$$\begin{aligned}\Phi^c(x_{n1} - 2s) + \Phi^c(x_{n1} - 2t) - 2\Phi^c(x_{n1} - (t+s)) &\le \\ (x_{n1} - 2s)\phi(x_{n1} - 2s)(s-t)^2\end{aligned}$$

and thus

$$E\{[\tilde{S}^u_{n1}(s) - \tilde{S}^u_{n1}(t)]^2\} \le \{\Phi^c(x_{n1} - 2s) + (x_{n1} - 2s)\phi(x_{n1} - 2s)\}(s-t)^2,$$

and for $t_{n1} \le t \le s \le t_{n2}$, the coefficient of $(s-t)^2$ is bounded by $O[(x_{n1} - 2t_{n2})\phi(x_{n1} - 2t_{n2})]$. It follows that

$$\begin{aligned}\sup_{t_{n1}\le t\le t_{n2}} S^u_{n1}(t) &\le O_p(1) + O_p(t_{n2} - t_{n1}) O[(x_{n1} - 2t_{n2})\phi(x_{n1} - 2t_{n2})]^{1/2} \\ &= O_p(1) + O_p((\log_2 n)^{1/2}) o(1) \\ &= o_p((\log_2 n)^{1/2}). \qquad (24)\end{aligned}$$

3.3 $K4$ Essentially the same argument as that used for K2 applies to $S^l_{n2}(t)$. Here $\Phi^c(x_{n1} - 2t)$ in the proof for K2 is replaced by $\Phi(x_{n2} - 2t)$ and the argument $x_{n2} - 2t$ is negative.

3.4 $K5$ Here again, the same argument applies, except that in place of Equation 24, we have

$$\begin{aligned}\sup_{t_{n4}\le t\le t_{n5}} S^\ell_{n2}(t) &= O_p(1) + O_p(\log n)^{1/2} O[(x_{n2} - 2t_{n4})\phi(x_{n2} - 2t_{n4})]^{1/2} \\ &= O_p(1) + O_p(\log n)^{1/2} O[(\log_2 n)^{1/2}\, e^{-16\log_2 n/2}]^{1/2} \\ &= O_p(1) \qquad (25)\end{aligned}$$

4 The Slepian argument, $S2$

The process $\tilde{S}^l_{01}(t)[\Phi((x_{n1} - 2t))]^{-1/2}$ has covariance function

$$\frac{\tilde{\rho}^l_{01}(s,t)}{[\tilde{\rho}^l_{01}(s,s)\tilde{\rho}^l_{01}(t,t)]^{1/2}} = \frac{\Phi(x_{n1} - (t+s))}{[\Phi(x_{n1} - 2s)\Phi(x_{n1} - 2t)]^{1/2}} e^{-(s-t)^2/2}.$$

It is easy to see that $\log \dot{\Phi}(x)$ is concave, and hence this covariance function is no less that $e^{-(s-t)^2/2}$, and, by Slepian's theorem

$$\begin{aligned}\sup_{t_{n1}\le t\le t_{n2}} \tilde{S}^l_{01}(t) &= O_p(\log(t_{n2}-t_{n1}))^{1/2}[\Phi(x_{n1}-2t_{n1})]^{1/2}\\ &= O_p[(\log_2 n)^{1/4}]O(1),\end{aligned}$$

and

$$\sup_{t_{n1}\le t\le t_{n2}} S^l_{01}(t) = o_p(\log_2 n)^{1/2}. \tag{26}$$

5 The Hungarian Construction argument, $H1, H2$

The following argument applies for $0 \le t \le t_{n2}$, i. e., for both $H1$ and $H2$. We have

$$S^l_{n1}(t) - S^l_{01}(t) = \int_{x\le x_{n1}} e^{tx-t^2}[dB_n(u) - dB_0(u)].$$

and after integration by parts,

$$\begin{aligned}|S^l_{n1}(t) - S^l_{01}(t)| &\le \sup_u |B_n(u) - B_0(u)|e^{tx_{n1}-t^2}\\ &= O_p(n^{-1/2}\log n)\exp[-(t-x_{n1}/2)^2 + x^2_{n1}/4]\\ &= O_p(n^{-1/2}\log n)(\log_2 n)^{-4}(n^{1/2}(\log n)^{-1})\\ &= o_p(\log_2 n)^{-1/2}.\end{aligned} \tag{27}$$

6 Iterated Logarithm argument, $I1$

The Law of the Iterated Logarithm implies that

$$B_0(u) = O_p(1)[(1-u)\log_2(1-u)^{-1}]^{1/2}. \tag{28}$$

Integrating by parts, and concentrating on $0 \le t \le t_{n1}$, we have

$$\begin{aligned}S^u_{01}(t) &= \int_{x>x_{n1}} e^{tx-t^2}\, dB_0(u)\\ &= -\int_{x>x_{n1}} B_0(u)\, d(e^{tx-t^2}) - B_0[\Phi(x_{n1})]e^{tx_{n1}-t^2}.\end{aligned} \tag{29}$$

$$\begin{aligned}B_0[\Phi(x_{n1})]e^{tx_{n1}-t^2} &= O_p(1)[\Phi^c(x_{n1})\log_2[\Phi^c(x_{n1})]^{-1}]^{1/2}e^{tx_{n1}-t^2}\\ &= O_p(1)[x_{n1}^{-1/2}\exp(-x^2_{n1}/4)(\log x_{n1})^{1/2}e^{tx_{n1}-t^2}\\ &= O_p(1)\exp(-(t-x_{n1}/2)^2)\left[\frac{\log x_{n1}}{x_{n1}}\right]^{1/2}\end{aligned}$$

$$\begin{aligned}\sup_{0\le t\le t_{n1}} |B_0[\Phi(x_{n1})]e^{tx_{n1}-t^2}| &= O_p(1)(\log n)^{-4}\left[\frac{\log_2 n}{\sqrt{\log n}}\right]^{1/2} \\ &= o_p(\log_2 n)^{-1/2}. \qquad (30)\end{aligned}$$

Also

$$\begin{aligned}\int_{x>x_{n1}} B_0(u)\,d(e^{tx-t^2}) &= t\int_{x_{n1}}^{\infty} B_0(u)e^{tx-t^2}\,dx \\ &= tO_p(1)\int_{x_{n1}}^{\infty}\frac{\exp(-x^2/4)}{x^{1/2}}(\log x)^{1/2}e^{tx-t^2}\,dx \\ &\quad \text{by (28) and the usual estimate for } \Phi^c \\ &= tO_p(1)\int_{x_{n1}}^{\infty}\left(\frac{\log x}{x}\right)^{1/2}\exp[-(t-x/2)^2]\,dx \\ &= O_p(1)t\left[\frac{\log x_{n1}}{x_{n1}}\right]^{1/2}\Phi^c\left(\frac{x_{n1}-2t}{\sqrt{2}}\right).\end{aligned}$$

$$\begin{aligned}\sup_{0\le t\le t_{n1}}\left|\int_{x>x_{n1}} B_0(u)\,d(e^{tx-t^2})\right| &= O_p(1)t_{n1}\left(\frac{\log x_{n1}}{x_{n1}}\right)^{1/2}\Phi^c\left(\frac{x_{n1}-2t_{n1}}{\sqrt{2}}\right) \\ &= O_p(1)(\log n)^{1/2}\frac{(\log_2 n)^{1/2}}{(\log n)^{1/4}}\frac{(\log n)^{-4}}{(\log_2 n)^{1/2}} \\ &= o_p(\log_2 n)^{-1/2}. \qquad (31)\end{aligned}$$

It follows that

$$\sup_{0\le t\le t_{n1}} |S^u_{01}(t)| = o_p(\log_2 n)^{-1/2} \qquad (32)$$

7 Tail argument

7.1 $T1$ We have

$$S^u_{n1}(t) = \int_{x>x_{n1}} e^{tx-t^2}n^{1/2}[dF_n(u) - dF(u)].$$

Since $F^c_n(u) = O_p(1)(1-u)$ for $0\le u\le 1$

$$\begin{aligned}S^u_{n1}(t) &= O_p(1)n^{1/2}\int_{x>x_{n1}} e^{tx-t^2}\,du \\ &= O_p(1)n^{1/2}e^{-t^2/2}\int_{x_{n1}}^{\infty} e^{-(x-t)^2/2}\,dx \\ &= O_p(1)n^{1/2}e^{-t^2/2}\Phi^c(x_{n1}-t). \qquad (33)\end{aligned}$$

Then

$$\sup_{0\le t\le t_{n1}} |S^u_{n1}(t)| = O_p(1)n^{1/2}\frac{e^{-(x_{n1}-t_{n1})^2/2}}{x_{n1}-t_{n1}}e^{-t^2_{n1}/2}$$

$$\begin{aligned} &= O_p(1)n^{1/2}\exp[-x_{n1}^2/4 - (t_{n1} - x_{n1}/2)^2]/(x_{n1} - t_{n1}) \\ &= O_p(1)n^{1/2}\frac{n^{-1/2}(\log n)}{(\log n)^{1/2}(\log n)^4} \\ &= o_p(\log_2 n)^{-1/2}. \end{aligned} \tag{34}$$

7.2 $T4, T5$ The same argument yields

$$\sup_{t_{n3}\le t\le t_{n5}} |S_{n2}^u(t)| = O_p(1)n^{1/2} \sup_{t\ge t_{n3}} \Phi^c(x_{n2}-t)e^{-t^2/2}.$$

But it is not hard to show that $\Phi^c(x_{n2}-t)e^{-t^2/2}$ attains its maximum value for $t \approx x_{n2}/2 + 1/x_{n2}$, and that maximum value satisfies

$$\sup_t \Phi^c(x_{n2}-t)e^{-t^2/2} = \sqrt{2/\pi}e^{-x_{n2}^2/4}(x_{n2})^{-1}(1+o(1)).$$

Thus

$$\begin{aligned} \sup_{t\ge t_{n3}} S_{n2}^u(t) &= O_p(1)n^{1/2}\frac{n^{-1/2}(\log n)^{1/2}}{(\log n)^{1/2}} \\ &= O_p(1). \end{aligned} \tag{35}$$

7.3 $T6$ Replacing x_{n1} in Equation 33 by $-\infty$, we have

$$\begin{aligned} \sup_{t_{n5}\le t} S_n(t) &= O_p(1)n^{1/2}\, e^{-t_{n5}^2/2} \\ &= O_p(1). \end{aligned} \tag{36}$$

8 Assembly

We shall derive our result for $M(t) = \sup_t S_n(t)$ by assembling the results of Sections 3 to 7 and showing that $\sup_t S_n(t)$, $\sup_{|t|\le t_{n1}} S_n(t)$, and $\sup_{|t|\le t_{n1}} S_0(t)$ are stochastically equivalent to $\tilde{M}(t_{n1}) + o_p(\log_2 n)^{-1/2}$, where we recall that $\tilde{M}(T) = \sup_{|t|\le\, T} \tilde{S}_0(t)$. But first we shall show that

$$\sup_{|t|\le t_{n1}} S_0(t) = \tilde{M}(t_{n1}) + o_p(\log_2 n)^{-1/2}. \tag{37}$$

The difficulty in this first step is that the difference $\tilde{X}e^{-t^2/2}$ between $S_0(t)$ and $\tilde{S}_0(t)$ is of the order $O_p(1)$ for t small. Thus it is useful to show that the region $|t| \ge t_{n0} = (2\log_3 n)^{1/2}$, where $\tilde{X}e^{-t^2/2} = o_p(\log_2 n)^{-1/2}$, is very likely to contain the maximizing values of t in the above suprema.

Because the location of the maximizing value of the stationary process $\tilde{S}_0(t)$ is uniformly distributed in the range of t considered, the probability that the maximizing value of t for the range $|t| \le t_{n1}$ is located within

$[-t_{n0}, t_{n0}]$ approaches zero. Moreover the fact that $\tilde{M}(t_{n0}) = O_p(\log_4 n)^{1/2}$ while $\tilde{M}(t_{n1}) = (\log_2 n)^{1/2} + o_p(1)$ implies that $\sup_{|t|\le t_{n0}} S_0(t) = o_p(\log_2 n)^{1/2}$, and the supremum of $S_0(t)$ over $[-t_{n1}, t_{n1}]$ will take place for $|t| \ge t_{n0}$ where $S_0(t)$ and $\tilde{S}_0(t)$ differ by $o_p(\log_2 n)^{-1/2}$, thus establishing Equation 37.

Now we proceed to combine the results of Sections 3 to 7. It is clear that

$$\sup_{t\ge t_{n1}} S_n(t) = o_p(\log_2 n)^{1/2}. \tag{38}$$

For $0 \le t \le t_{n1}$,

$$\begin{aligned} S_n(t) &= S_{n1}^u(t) + [S_{n1}^l(t) - S_{01}^l(t)] - S_{01}^u(t) + S_0(t) \\ &= o_p(\log_2 n)^{-1/2} + S_0(t) \end{aligned} \tag{39}$$

By symmetry, Equation 39 holds also for $-t_{n1} \le t \le 0$, and Equation 38 holds for the supremum over $t \le -t_{n1}$. Thus

$$\sup_t S_n(t) = \sup_{|t|\le t_{n1}} S_0(t) + o_p(\log_2 n)^{-1/2}$$

is stochastically equivalent to $\tilde{M}(t_{n1}) + o_p(\log_2 n)^{-1/2}$, and therefore also to $\tilde{M}(\sqrt{\log n}/2) + o_p(\log_2 n)^{-1/2}$. It follows that

$$\sup_t S_n(t) = (\log_2 n)^{1/2} + (V - \log(\sqrt{2}\pi) + o_p(1))(\log_2 n)^{-1/2}. \tag{40}$$

This is the desired result for $S_n(t)$.

In the next section, where we extend our result to hold for S_n^*, we will use the fact that the the supremum of $S_n(t)$ is attained for $|t| > t_{n0}$.

9 Extension to $S_n^*(t)$

We may write

$$y^*(x,t) = (e^{tx-t^2/2} - 1 - tx)e^{-t^2/2}\ h(t)$$

where

$$h(t) = t^{-2} h_1(t)$$

and, as $t \to \infty$,

$$h(t) = 1 + O(t^2 e^{-t^2})$$

and the derivatives of $h(t)$ approach zero, $h_1(t) > 0$, and $h_1(t)$ is analytic. For $t \ne 0$,

$$y^*(x,t) = [y(x,t) - e^{-t^2/2}] + [y(x,t) - e^{-t^2/2}][h(t) - 1] - xte^{-t^2/2}\ h(t).$$

Thus,

$$S_n^*(t) = S_n(t) + S_n(t)[h(t) - 1] - n^{1/2}\bar{X}te^{-t^2/2}\ h(t) \tag{41}$$

and

$$\begin{aligned}\sup_{t^2>2\log_3 n} S_n^*(t) &= \sup_{t^2>2\log_3 n} S_n(t) + O_p(\log_2 n)^{1/2}(\log_3 n)(\log_2 n)^{-2} \\ &\quad + O_p(1)(\log_3 n)^{1/2}(\log_2 n)^{-1} \\ &= \sup_t S_n(t) + o_p(\log_2 n)^{-1/2}. \end{aligned} \tag{42}$$

The covariance function of $S_n^*(t)$ is

$$\begin{aligned}\rho^*(s,t) &= [e^{st} - 1 - st]e^{-s^2/2}h(s)e^{-t^2/2}\,h(t) \\ &= e^{-(s-t)^2/2}h(s)h(t)[1-(1+st)e^{-st}].\end{aligned}$$

If s and t have the same sign,

$$\begin{aligned}E\{[S_n^*(s) - S_n^*(t)]^2\} &= 2[1-\rho^*(s,t)] \\ &\leq c(s-t)^2\end{aligned} \tag{43}$$

for some constant c. Hence the Kolmogorov bound gives

$$\sup_{t^2\leq 2\log_3 n} |S_n^*(t) - S_n^*(0)| = O_p(\log_3 n)^{1/2},$$

and with probability approaching one,

$$\begin{aligned}\sup_t S_n^*(t) &= \sup_{t^2>2\log_3 n} S_n^*(t) \\ &= \sup_t S_n(t) + o_p(\log_2 n)^{-1/2}\end{aligned} \tag{44}$$

which yields the desired result, Equation 3, for S_n^*.

REFERENCES

Billingsley, P. (1968). Convergence of probability measures. J. Wiley. New York.

Chernoff, H. (1954). On the distribution of the likelihood ratio. *Ann. Math. Statist.* **25**, 573-578.

Chernoff, H. and Lander, E. (1989). Asymptotic distribution of the likelihood ratio test that a mixture of two binomials is a single binomial. Technical Report. *Harvard University.*

Ghosh, J.K. and Sen, P.K. (1985). On the asymptotic performance of the log likelihood ratio statistic for the mixture model and related results. *Proc. Berkeley Conference in honor of J. Neyman and J. Kiefer*, Vol. II, pp 789-906.

Hartigan, J.A. (1985). A failure of likelihood ratio asymptotics for normal mixtures. *Proc. Berkeley Conference in Honor of Jerzy Neyman and Jack Kiefer.* Wadsworth Advanced Books, Monterey, CA and Institute of Mathematical Statistics. Hayward, CA.

Komlos, J., Major, P., Tusnady, G. (1975). An approximation of partial sums of independent r.v.s and the sample distribution function. *Z. Wahrsch. verw. Geb.* **32**, 111-131.

Leadbetter, M.R., Lindgren, G., Rootzen, H. (1983). Extremes and related properties of random sequences. *Springer Verlag.* New York.

Department of Statistics
Statistical Laboratory
University of California, Bekeley
Berkeley, California 94720
USA

Science Center
Department of Statistics
Harvard University
1 Oxford Street
Cambridge, MA 02138, USA

J.K. Ghosh, S.K. Mitra, K.R. Parthasarathy and B.L.S. Prakasa Rao (*eds.*),
Statistics and Probability : A Raghu Raj Bahadur Festschrift
Published by Wiley Eastern Limited, Publishers, (1993), 97 – 134

Hausdorff Dimension: Self-Similarity and Independent Processes; Cross-Similarity and Markov Processes

PATRICK BILLINGSLEY
The University of Chicago

Abstract

It is shown how the Hausdorff dimensions of certain self-similar sets in k-space can be determined by means of entropies of independent processes. A more general notion of cross-similarity is introduced; dimensions of cross-similar sets can be determined by studying the entropies of Markov processes. Although standard probability theory is used in the paper, the treatment of dimension is self-contained.

1 Hausdorff dimension: The idea

Imagine a sheet of paper in 3-space. It may be flat, perhaps canted in some way, or it may be curved. In any case, we regard the sheet as representing a surface, a two-dimensional geometric object. One way to make mathematics of this concept of dimension is through the idea of *measure*.

We think of the sheet as having *finite, positive area.* Since we think of the sheet as having no thickness, we also think of it as having volume *zero.* What about its length? Although we do not usually attach a length to a surface, if it is to have one at all, this length should be at least as great as that of any geometric figure the surface contains. And of course we can

AMS Subject Classifications : Primary 11K55, 60J10; secondary, 28A80.
Key words and Phrases : Hausdorff measure, Hausdorff dimension, Markov chains.

Figure 1

Figure 2

inscribe on our sheet a curve (a rectifiable curve) of arbitrarily great length (even a curve of infinite length). Thus our surface has *infinite length.* In sum,

$$\left.\begin{array}{l}\text{length } = \infty,\\ 0 < \text{ area } < \infty,\\ \text{volume } = 0.\end{array}\right\} \tag{1.1}$$

If we think of length, area, and volume as one-dimensional, two- dimensional, and three-dimensional measures, (1.1) becomes,

$$\left.\begin{array}{l}\text{1-dimensional measure} = \infty,\\ 0 < \text{2-dimensional measure} < \infty,\\ \text{3-dimensional measure} = 0.\end{array}\right\} \tag{1.2}$$

As the dimension attached to the measure passes the value 2, the measure jumps from ∞ down to 0, and the idea now is to make this the basis of a definition of geometric dimension. If this program is to succeed, if it is not to be an exercise in circularity, we need to frame an independent definition of s-dimensional measure. This is the next step.

2 Hausdorff measure

A ball (sphere) of diameter 1 in s-dimensional Euclidean space R^s has ordinary volume $c_s = \pi^{s/2}/2^s\Gamma(1+s/2)$. (It will turn out in the end that the actual value of c_s is of no importance at all for our theory.) Magnifying a set in R^s by the factor θ multiplies its volume by the factor θ^s: a ball of diameter d in R^s has ordinary volume $c_s d^s$. This is our starting point. Although we certainly need no definition of dimension in order to know what R^s *is* (namely, the set of s-tuples of reals), this is the source of our intuition: We take R^s as the prototype of a geometric object of dimension s.

Now suppose that B is a ball of diameter d in R^k. Even though s and k may differ (focus on the case $s \leq k$), we take $c_s d^s$ as the s-dimensional measure of B. Here k is an integer, but s can be any positive real number. A disc (ball) of diameter d in the plane ($k = 2$) has 2-dimensional measure $\pi d^2/4$ and 1-dimensional measure d ($c_2 = \pi/4$ and $c_1 = 1$). Suppose now that A is a set in R^k. As a first attempt at a definition of the s-dimensional measure of A we imagine it covered by balls in an efficient way (the balls do not extend much beyond A and do not overlap much) and add up the s-dimensional measures of these balls. This leads to the standard measure-theoretic device of taking the infimium of covering-sums. Provisionally, take the s-dimensional measure of A to be

$$H^s(A) = \inf \sum_n c_s(\text{diam } B_n)^s, \tag{2.1}$$

where the infimum extends over coverings of A by sequences $B_1, B_2, \ldots$ of balls.

If A is a line segment of ordinary length l in the plane ($k = 2$), then $H^1(A) \leq l$ (cover A by a single disc of diameter l). Since $A \cap B_n$ is a segment of length at most diam B_n, $l \leq \sum_n \text{diam} B_n$ if $A \subset \cup_n B_n$. Thus A has 1-dimensional measure l, which is what we want. But now there arises a problem. Let A be the union of two parallel segments of length l, at a distance of δ from one another, as in Figure 3. (assume $0 < \delta < l$). Since

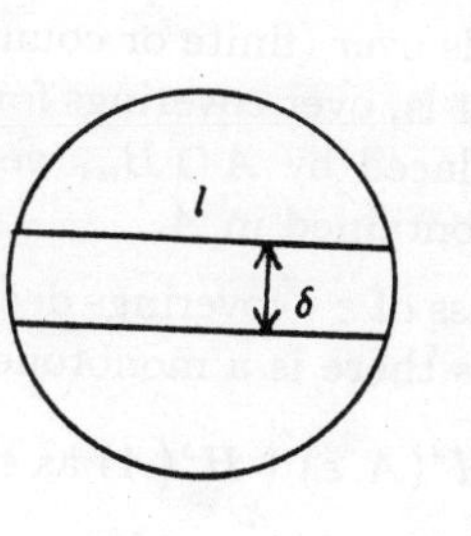

Figure 3

the lengths of the segments composing A add to $2l$, we want $H^1(A) = 2l$. But A can be covered by a single disc B of diameter $\sqrt{l^2 + \delta^2}$, so that by (2.1), $H^1(A) = \sqrt{l^2 + \delta^2}$ — essentially l for small δ. The trouble is that the disc B does double duty, covering both segments at once.

The remedy for this is to consider only coverings by balls of radius at

Figure 4

most ε and then let $\varepsilon \to 0$. Discard (2.1) and replace it by

$$H^s(A,\varepsilon) = \inf \sum_n c_s(\text{diam } B_n)^s, \tag{2.2}$$

where now the infimum extends over coverings of A by balls B_n satisfying diam $B_n \leq \varepsilon$ ($\varepsilon > 0$ in all that follows). For the union of the two segments considered above, if $\varepsilon < \delta/2$, then an efficient covering will look like the one in Figure 4, for which the sum in (2.2) comes to $2l$, and this is what we want. For $H^s(A)$ we will take the limit ($\varepsilon \to 0$) of (2.2), and for the dimension of A we will take the number s_0 (we will show that one exists) for which $H^s(A) = \infty$ if $s < s_0$ and $H^s(A) = 0$ if $s > s_0$.

In this definition of dimension we will need to distinguish only the three cases (i) $H^s(A) = \infty$, (ii) $0 < H^s(A) < \infty$, and (iii) $H^s(A) = 0$. Therefore, we can simplify (2.2) by suppressing the factor c_s. For the same reason, we can take the B_n to be arbitrary sets instead of balls (a set of finite diameter d can be enclosed in a ball of diameter $2d$, which increases the covering sum by an irrelevant factor of at most 2^s). Writing $|B_n|$ for diam B_n, we therefore replace (2.2) by

$$H^s(A,\varepsilon) = \inf \sum_n |B_n|^s, \tag{2.3}$$

where the infimum extends over (finite or countable) ε *– coverings* of A by (arbitrary) sets B_n — that is, over coverings for which $|B_n| \leq \varepsilon$. Since (2.3) is unaffected if B_n is replaced by $A \cap B_n$, we can, when it is convenient, assume that each B_n is contained in A.

As ε decreases, the class of ε - coverings decreases, and so (2.3) increases (or stays the same). Thus there is a monotone limit $H^s(A)$:

$$H^s(A,\varepsilon) \uparrow H^s(A) \text{ as } \varepsilon \downarrow 0. \tag{2.4}$$

This limit is the *s-dimensional Hausdorff measure* of A. We can assume $s > 0$ here, although it is natural to take $H^0(A)$ as the number of points in A ($H^0(A) = \infty$ if A is infinite). It is important to understand that s need not be an integer. The definition applies to sets in arbitrary metric spaces, but here we will be concerned only with sets in R^k.

Obviously $H^s(A,\varepsilon)$ increases with A and hence so does $H^s(A)$. Thus H^s is *monotone*:

$$H^s(A) \leq H^s(B) \text{ if } A \subset B \tag{2.5}$$

Further, H^s is *subadditive:*

$$H^s(\bigcup_m A_m) \le \Sigma_m H^s(A_m). \tag{2.6}$$

Since this is obvious if some $H^s(A_m)$ is infinite, assume they are all finite and for each m let $B_{m1}, B_{m2}, \ldots$ be an ε-covering of A_m satisfying $\sum_n |B_{mn}|^s < H^s(A_m, \varepsilon) + \delta/2^m (\delta > 0)$. Since the B_{mn} taken all together form an ε-covering of A.

$$H^s(\bigcup_m A_m, \varepsilon) \le \sum_m \sum_n |B_{mn}|^s < \sum_m (H^s(A_m, \varepsilon) + \delta/2^m) \le \sum_m H^s(A_m) + \delta.$$

Let $\varepsilon \to 0$ and then $\delta \to 0$: (2.6) follows.

Suppose that $A \subset R^k$ and $B \subset R^j$, and that B is A magnified (or contracted) by the factor θ. Suppose, that is, that there is a one-to-one map f of A onto B with the property that the distance in R^j from $f(x)$ to $f(y)$ is θ times the distance in R^k from x to y. Thus B is similar to A with similarity ratio θ. To an ε-covering of A there clearly corresponds a $\theta\varepsilon$-covering of B, and the new covering sum is θ^s times the old one. This can be reversed, and so $H^s(A, \varepsilon) = \theta^s H^s(B, \theta\varepsilon)$. Therefore,

$$H^s(B) = \theta^s H^s(A). \tag{2.7}$$

In particular, $H^s(A) = H^s(B)$ if A and B are isometric ($\theta = 1$) — for example if B is a translation of A.

Hausdorff measure has other properties we will not need here. It can be shown that H^s is countably additive on the class of Borel sets in R^k: If A_m are disjoint Borel sets, then

$$H^s(\bigcup_m A_m) = \sum_m H^s(A_m). \tag{2.8}$$

See Section 8. If A is a Borel set in R^k, it can also be shown that $H^k(A)$ coincides with the ordinary Lebesque measure of A (although this requires restoring the factor c_s). See Section 19 of [4].

3 Hausdorff dimension: Definition

We have still to define the dimension of A. Suppose that $H^s(A)$ is finite and choose K so that $H^s(A) < K < \infty$. Suppose that δ is positive. By (2.4), $H^s(A, \varepsilon) < K$, and whatever ε may be, there is an ε-covering $B_1, B_2, \ldots$ of A for which $\Sigma_n |B_n|^s < K$. But then $H^{s+\delta}(A, \varepsilon) \le \Sigma_n |B_n|^{s+\delta} \le \Sigma_n |B_n|^s \varepsilon^\delta \le K\varepsilon^\delta \le K\varepsilon^\delta$, and letting $\varepsilon \to 0$ shows that $H^{s+\delta}(A) = 0$. Thus $H^s(A) < \infty$ implies $H^{s+\delta}(A) = 0$.

Therefore, there exists a unique s_0 such that

$$H^s(A) = \begin{cases} \infty & \text{if } s < s_0, \\ 0 & \text{if } s > s_0. \end{cases} \tag{3.1}$$

Here $H^{s_0}(A)$ itself may be 0, finite and positive, or infinite. This s_0 we take as the *Hausdorff dimension of* A, denoted dim A. Compare (3.1) with (1.2), which was our starting point. The interesting thing is that the s_0 in (3.1) need not be an integer (recall that $H^s(A)$ is defined for all positive s). For this reason, Hausdorff dimension is sometimes called *fractional* dimension.

If it happens that $0 < H^s(A) < \infty$, then A is called an *s-set*; of course dim $A = s$ in this case.

There are two extreme cases of (3.1). In the first, $H^s(A) = 0$ for all $s > 0$, so that dim $A = 0$. This happens, for example, if A is finite or countable. In the second case, $H^s(A) = \infty$ for all $s > 0$, so that dim $A = \infty$. Although this can happen for sets in Hilbert space, for example, it cannot happen for sets in Euclidean space.

The first properties of Hausdorff dimension are easy to prove. If $A \subset B$, then because of (2.5), the place where $H^s(B)$ jumps from ∞ down to 0 must lie to the right of (or else coincide with) the place where $H^s(A)$ does:

$$\dim A \leq \dim B \text{ if } A \subset B. \tag{3.2}$$

Further,

$$\dim \bigcup_m A_m = \sup_m \dim A_m. \tag{3.3}$$

To see this, notice first that if $\sup_m A_m < s$, then $H^s(A_m) = 0$ for all m, and hence $H^s(\bigcup_m A_m) = 0$ by (2.6), so that dim $\bigcup_m A_m \leq s$. Therefore the left side of (3.3) is at most the right. But the reverse inequality follows immediately from (3.2).

It is not hard to see that if $\lambda_k(A)$ is the ordinary k-dimensional Lebesque measure of a Borel set A in R^k, then $\lambda_k(A)$ and $H^k(A)$ are both 0, both finite and positive, or both infinite. A bounded set therefore has dimension at most k, and by (3.3), every set has dimension at most k. And if $\lambda_k(A) > 0$, in particular if A has nonempty interior, then dim $A = k$, as intuition requires.

4 Self-similar sets

There are interesting sets whose Hausdorff dimensions can be guessed at from their properties of self-similarity. The prime example is the Cantor set. Split the Cantor set C into $C_1 = C \cap [0, 1/3]$ and $C_2 = C \cap [2/3, 1]$

Figure 5

Because of the relation (2.8), additivity, $H^s(C) = H^s(C_1) + H^s(C_2)$. (We have not proved (2.8), but we will see that the arguments in this section are incomplete on other grounds anyway.) Because of the way C is defined, C_1 is C contracted by the factor 1/3, and by (2.7), $H^s(C_1) = (\frac{1}{3})^s H^s(C)$. Since the same is true of C_2, we arrive at

$$H^s(C) = 2\left(\frac{1}{3}\right)^s H^s(C). \tag{4.1}$$

If $s = \dim C$, and if we are lucky and C is an s-set, then (4.1) leads to $2(\frac{1}{3})^s = 1$, the solution of which is $s = \log 2/\log 3$. This leads to the conjecture that $\dim C = \log 2/\log 3$. The conjecture is correct, but this argument alone does not prove it: we need somehow to exclude the possibilities $H^s(C) = 0$ and $H^s(C) = \infty$.

The *Sierpinski* set is a planar analogue of the Cantor set. Split a closed equilateral triangle into four equilateral triangles (Figure 6) and remove

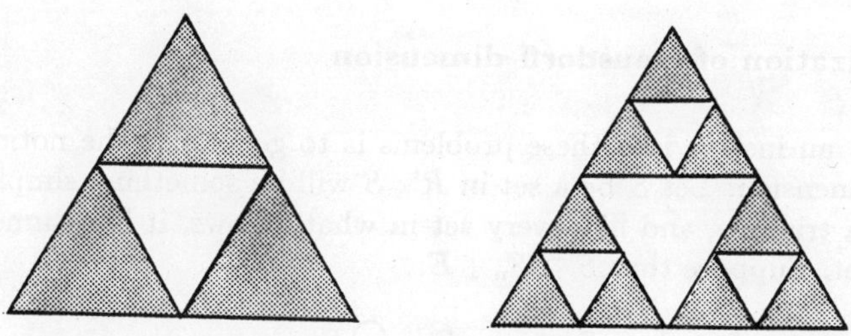

etc.

4

Figure 6

the open middle one. Split each of the three remaining triangles in the same way and remove the middle piece. Continue as in the case of the Cantor set, arriving at the Sierpinski set W. If W_1, W_2, W_3 are the intersections of W with the three triangles retained at the first stage, then $W = W_1 \cup W_2 \cup W_3$, and each W_i is W scaled down by the factor $\frac{1}{2}$, so that $H^s(W_i) = (\frac{1}{2})^s H^s(W)$. If we ignore the fact that the W_i are not quite disjoint (they share vertices) and use additivity, we obtain $H^s(W) = 3(\frac{1}{2})^s H^s(W)$. If $s = \dim W$, and if again we are lucky and W is an s-set, we can cancel $H^s(W)$ and solve for s: $\dim W = \log 3/\log 2$, which is the correct dimension.

These two arguments can be completed, and the essential reason is that the Cantor and Sierpinski sets are compact. In other cases, the argument simply leads to the wrong answer.

For x in the interval [0,1], let $\Sigma_{n=1}^{\infty} d_n(x)/r^n$ be its r-adic expansion — the nonterminating one, say, except for $x = 0$. Let $N_n(x, i)$ be the number of occurrences of the digit i ($i = 0, \ldots, r-1$) among $d_1(x), \ldots, d_n(x)$. If

$p_0, \ldots, p_{r-1}$ are nonnegative and add to 1, consider the *Eggleston* set V, the set of x for which

$$\lim_n \tfrac{1}{n} N_n(x,i) = p_i, \quad 0 \le i \le r. \tag{4.2}$$

Write $V = V_0 \cup \ldots \cup V_{r-1}$, where V_i is V intersected with $[i/r, (i+1)/r]$. Since the limits here do not depend on the value of $d_1(x)$, it is clear that each V_i is V contracted by the factor $1/r$, and so $H^s(V) = rr^{-s}H^s(V)$. If again V is an s-set, again we can cancel and solve for s: $\dim E = 1$. But this answer is wrong. The correct answer is more interesting.

$$\dim V = -\sum_{i=0}^{r-1} p_i \log p_i / \log r. \tag{4.3}$$

This implies that V is *not* an s-set (unless $p_i \equiv r^{-1}$).

There are similar interesting examples in R^k, and the theory of the following sections can be used to derive their Hausdorff dimensions.

5 A generalization of Hausdorff dimension

One way to get an insight into these problems is to generalize the notion of Hausdorff dimension. Let S be a set in R^k; S will be something simple, like a cube or a triangle, and like every set in what follows, it is assumed to be a Borel set. Suppose that $S \supset S_n \downarrow E$:

$$S \supset S_1 \supset S_2 \supset \cdots, E = \bigcap_n S_n.$$

Suppose that $\mathcal{A}_n$ is a finite or countable partition of S_n, $n = 1, 2, \ldots$, and for $n \ge 2$ suppose that $\mathcal{A}_n$ refines $\mathcal{A}_{n-1}$ in the sense that each element of $\mathcal{A}_n$ is contained in some element of $\mathcal{A}_{n-1}$. We will find the dimensions of certain subsets of the intersection E. The results here stem from those in [2].

EXAMPLE 1. Suppose ($k = 1$) that $S = S_n = E = [0,1]$ and that $\mathcal{A}_n$ consists of the r-adic intervals

$$[0, r^{-n}], (r^{-n}, 2r^{-n}], \ldots, (ir^{-n}, (i+1)r^{-n}], \cdots, (1 - r^{-n}, 1]. \tag{5.1}$$

This example is connected in a natural way with the base-r digits $d_n(x)$ and the Eggleston set. □

EXAMPLE 2. Again take $S = [0,1]$. Let $\mathcal{A}_1$ consist of the intervals $[0, \frac{1}{3}], [\frac{2}{3}, 1]$; let $\mathcal{A}_2$ consist of the intervals $[0, \frac{1}{9}], [\frac{2}{9}, \frac{1}{3}], [\frac{2}{3}, \frac{7}{9}], [\frac{8}{9}, 1]$, and so on, so that $\mathcal{A}_n$ consists of the intervals $[i3^{-n}, (i+1)3^{-n}]$ which are contained in some interval of $\mathcal{A}_{n-1}$ and for which i is even. Let S_n be the union of the 2^n intervals in $\mathcal{A}_n$. In contrast to Example 1, here S_n is a proper subset of S_{n-1}. The intersection $E = \cap_n S_n$ is the Cantor set. □

Let $\mathcal{A} = \{\emptyset\} \cup \bigcup_{n=1}^{\infty} \mathcal{A}_n$, let $\mathcal{S}$ consist of the Borel sets in S, and let μ be a measure on $(S, \mathcal{S})$. For $A \subset E$, define, a *μ-ε-covering* of A as a covering (necessarily finite or countable) by $\mathcal{A}$-sets with $\mu(v_i) \leq \varepsilon$. For $s > 0$, define

$$\mu^s(A, \varepsilon) = \inf \sum_i \mu(v_i)^{s/k}, \tag{5.2}$$

where the infimum extends over all *μ-ε-coverings* of $\mathcal{A}$. Note that if $\mu = \lambda_k$ and v_i is a ball, then $\lambda_k(v_i)^{s/k}$ is of the order $|v_i|^s$. Define $\mu^s(A)$ (it must be distinguished from $\mu(A)^s$) as the limit:

$$\mu^s(A, \varepsilon) \uparrow \mu^s(A) \text{ as } \varepsilon \downarrow 0. \tag{5.3}$$

The set function $\mu^s(\cdot)$ is monotone and subadditive, and the proofs go through exactly as in the case of $H^s(\cdot)$ (see (2.5) and (2.6)):

$$\mu^s(A) \leq \mu^s(B) \text{ if } A \subset B \tag{5.4}$$

and

$$\mu^s(\bigcup_m A_m) \leq \sum_m \mu^s(A_m). \tag{5.5}$$

Since $\mu^s(\emptyset) = 0, \mu^s(\cdot)$ is an outer measure on the class of subsets of E.

Just as in the case of $H^s(\cdot), \mu^s(A) < \infty$ implies $\mu^{s+\delta}(A) = 0$ for $\delta > 0$. Therefore, there exists a number $\dim_\mu A$ $(0 \leq \dim_\mu A \leq \infty)$ such that

$$\mu^s(A) = \begin{cases} \infty & \text{if } 0 < s < \dim_\mu A, \\ 0 & \text{if } \dim_\mu A < s. \end{cases} \tag{5.6}$$

This is the dimension of A with respect to μ. Note that the definitions depend on the S_n and $\mathcal{A}_n$ as well as on μ. It can happen that $\dim_\mu A = \infty$, for example if A contains a point of positive μ-mass.

It is easy to prove the analogues of (3.2) and (3.3):

$$\dim_\mu A \leq \dim_\mu B \quad \text{if} \quad A \subset B \tag{5.7}$$

and

$$\dim_\mu \bigcup A_m = \sup_m \dim_\mu A_m. \tag{5.8}$$

Note. The definitions (5.2), (5.3), and (5.6) make sense if $\mathcal{A}$ is a perfectly arbitrary collection of subsets of R^k and μ is an arbitrary nonnegative, real function on $\mathcal{A}$. Then (5.4), (5.5), (5.7), and (5.8) are all valid. If $\mathcal{A}$ contains all the subsets of R^k and $\mu(v) = |v|^k$, we are back in the standard case. But we will use in an essential way the further conditions imposed above on $\mathcal{A}$ and μ.

The program now consists of two steps: Suppose that $A \subset E = \bigcap_n S_n$.

(i) Find conditions on the $\mathcal{A}_n$ ensuring that if μ is λ_k, then $\dim_\mu$ A is the ordinary Hausdorff dimension of A: $\dim_{\lambda_k} A = \dim A$.

(ii) Study how, for fixed A, $\dim_\mu A$ varies with μ.

For $x \in E$ and $n \geq 1$, let $u_n(x)$ be the element of A_n that contains x. In connection with step (ii), we will be able in certain cases to prove that

$$\lim_n \frac{\log \mu(u_n(x))}{\log \nu(u_n(x))} = \rho \tag{5.9}$$

($\rho \geq 0$) for every x in A, where $A \subset E$ and μ and ν are two measures on $(S, \mathcal{S})$. To understand the implications of this condition, suppose that the ratio in (5.9) acutally equals ρ for every x in A and every n. Then $\mu(v) = \nu(v)^\rho$ for every A-set v that meets A, and these are the only ones that affect $\mu^s(A), \nu^s(A), \dim_\mu A, \dim_\nu A$. But it then follows by (5.2) and (5.3) that $\mu^s(A) = \nu^{\rho s}(A)$. Therefore, the t at which $\nu^t(A)$ jumps from ∞ to 0 is ρ times the s at which $\mu^s(A)$ does so: $\dim_\nu A = \rho \dim_\mu A$.

To see the point of all this, return to Example 1, and let ν be λ_1, 1-dimensional Lebesgue measure. Since $\lambda_1(v) = r^{-n}$ for each v in A_n, $\log \lambda_1 u_n(x)) = -n \log r$. Let μ be the probability measure on $(S, \mathcal{S})$ under which $d_1(x), d_2(x), \cdots$ are independent random variables that assume the value i with probability p_i (see Lemma 2 in Section 6). If $N_n(x, i)$ is the number of occurrences of i among $d_1(x), \cdots, d_n(x)$, then by independence, $\mu(u_n(x)) = \Sigma_i N_n(i, x) \log p_i$ (assume the p_i are strictly positive). If x is in the Eggleston set V, so that (4.2) holds, then $\log \mu(u_n(x))/\log \lambda_1(u_n(x)) \to h = -\Sigma_i p_i \log p_i / \log r$. It will follow by the argument after (5.9) (in its precise form) that $\dim_{\lambda_1} V = h \dim_\mu V$. It will turn out, as a consequence of step (i) above, that $\dim_{\lambda_1} V = \dim V$, so that $\dim V = h \dim V$. Finally, $\mu(V) = 1$ by the strong law of large numbers, and from this it follows easily that $\dim_\mu V = 1 : \dim A = h$, as (4.3) asserts.

This is the approach taken in [2] and [3]. Here it will be extended from the line to R^k.

We turn to step (i). We need two further conditions on the $\mathcal{A}_n$.

CONDITION 1. *There is a constant C_1 such that*

$$|v|^k \leq C_1 \lambda_k(v) \tag{5.10}$$

for all v in $\mathcal{A}$.

If the elements of $\mathcal{A}$ are balls, as in Examples 1 and 2, then Condition 1 holds ($C_1 = 1/c_k$, where as before c_k is the volume of a ball of diameter 1 in R^k). An inequality opposite to (5.10) always holds. If B is the closed ball of radius $|v|$ about some point of v, then $\lambda_k(v) \leq \lambda_k(B) = 2^k c_k |v|^k$.

$$\lambda_k(v) \leq 2^k c_k |v|^k . \tag{5.11}$$

But suppose v is a rectangle ($k = 2$) of width δ and length d; then $|v|$ exceeds d, while $\lambda_k(v)$ can be made arbitrarily small by taking δ small. The inequality (5.10) is therefore a real restriction on the sets in $\mathcal{A}$.

Condition 1 holds if and only if there is a constant C such that each v in $\mathcal{A}$ can be enclosed in a ball B for which $\lambda_k(B) \leq C\lambda_k(v)$, a condition useful in the theory of differentiation in R^k (see [13], p. 163). A stronger condition is that for each v in $\mathcal{A}$ there are balls B_0 and B_1 such that $B_0 \subset v \subset B_1$ and $|B_1|/|B_0| \leq C$.

As before, for $x \in E$ let $u_n(x)$ be the $\mathcal{A}_n$-set containing x.

CONDITION 2. *The diameter $|v|$ is bounded away from 0 and ∞ for v in $\mathcal{A}_1$, and there is a constant C_2 such that*

$$|u_{n-1}(x)|/|u_n(x)| \leq C_2 \tag{5.12}$$

for $x \in E$ and $n > 1$; further,

$$|u_n(x)| \to 0 \tag{5.13}$$

for $x \in E$.

Condition 2 obviously holds in Examples 1 and 2. In the next section this condition will be replaced by a stronger one.

THEOREM 1. *If Conditions 1 and 2 hold, then*

$$\dim A = \dim_{\lambda_k} A \tag{5.14}$$

for all $A \subset E$.

For the proof we need a preliminary result.

LEMMA 1. *If Conditions 1 and 2 hold, then there exist an integer N and positive constants C_3 and ε_0 such that, if B is any open ball with $|B| < \varepsilon_0$, then $E \cap B$ can be covered by N or fewer $\mathcal{A}$-sets each having λ_k-measure at most $C_3\lambda_k(B)$.*

PROOF. Choose ε_0 so that $\varepsilon_0 < |v|$ for $v \in \mathcal{A}_1$, and suppose that B is an open ball with $|B| < \varepsilon_0$. For $x \in E$, choose $n = n(x)$ so that $|u_n(x)| \leq |B| < |u_{n-1}(x)|$; this is possible by (5.13), and it follows by (5.12) that $C_2^{-1}|B| < |u_n(x)| \leq |B|$. This gives a covering of E by $\mathcal{A}$-sets v_i satisfying

$$C_2^{-1}|B| < |v_i| \leq |B|. \tag{5.15}$$

Since each $\mathcal{A}_n$ refines the preceding one, for any two sets in $\mathcal{A}$, either one of them contains the other or else they are disjoint. Therefore, if we remove each v_i that is contained in some other v_j, the remaining v_i still cover E and they are now *disjoint.*

Let B' be the closed ball with the same center x_0 as B and with diameter $3|B|$. If $x \in v_i \cap B$ and $y \in v_i$, then $|y-x_0| \leq |y-x|+|x-x_0| \leq |v_i|+\frac{1}{2}|B| \leq \frac{3}{2}|B|$. In other words, if v_i meets B, then $v_i \subset B'$. Let Σ' denote summation over those i for which v_i does meet B. Since the v_i are disjoint, it follows by (5.10) and the left-hand inequality in (5.15) that

$$\begin{aligned} 3^k c_k |B|^k &= \lambda_k(B') \geq \Sigma' \lambda_k(v_i) \geq C_1^{-1} \Sigma' |v_i|^k \\ &\geq C_1^{-1} \Sigma' C_2^{-k} |B|^k. \end{aligned}$$

The number of v_i that meet B is therefore at most the integer part N of $3^k c_k C_1 C_2^k$. These v_i cover $E \cap B$, and by (5.11) and the right-hand inequality in (5.15), each satisfies $\lambda_k(v_i) \leq 2^k c_k |v_i|^k \leq 2^k c_k |B|^k = 2^k \lambda_k(B)$. Take $C_3 = 2^k$. □

PROOF OF THEOREM 1. By (3.1) and (5.6), $\dim A \leq \dim_{\lambda_k} A$ will follow if (for $s > 0$)

$$H^s(A) \leq K_{k,s}\, \lambda_k^s(A) \tag{5.16}$$

for some constant $K_{k,s}$; and $\dim_{\lambda_k} A \leq \dim A$ will follow if

$$\lambda_k^s(A) \leq K'_{k,s} H^s(A) \tag{5.17}$$

for some constant $K'_{k,s}$. In proving these inequalities, we can assume the measures on the right are finite.

First, (5.16). Suppose that $\lambda_k^s(A,\varepsilon) \leq \lambda_k^s(A) < L < \infty$ and choose a λ_k-ε-covering $\{v_i\}$ of A with $\Sigma_i \lambda_k(v_i)^{s/k} \leq L$. By (5.10), $|v_i| \leq C_1^{1/k} \lambda_k(v_k)^{1/k} \leq C_1^{1/k} \varepsilon^{1/k}$, so that $\{v_i\}$ is an ordinary (that is to say, in the sense of Section 2) $C_1^{1/k}\varepsilon^{1/k}$-covering of A, and hence

$$H^s(A, C_1^{1/k}\varepsilon^{1/k}) \leq \sum_i |v_i|^s \leq \sum_i C_1^{s/k} \lambda_k(v_i)^{s/k} \leq C_1^{s/k} L.$$

Therefore (let $\varepsilon \to 0$), $H^s(A) \leq C_1^{s/k} L$. Let $L \downarrow \lambda_k^s(A)$: $K_{k,s} = C_1^{s/k}$ will work.

Next, (5.17). Suppose that $H^s(A,\varepsilon) \leq H^s(A) < H < \infty$. Suppose that $\varepsilon < \varepsilon_0/4$ and choose an ordinary ε-covering $\{A_i\}$ of A with $\sum_i |A_i|^s \leq H$. Enclose A_i in an open ball B_i of radius $2|A_i|$ (diameter 4 $|A_i|$) — the extra factor of 2 makes an open B_i work) and cover (Lemma 1) $A \cap B_i$ by N or fewer sets $v_{i1} v_{i2}, \cdots$ of $\mathcal{A}$ chosen so that $\lambda_k(v_{ij}) \leq C_3 \lambda_k(B_i)$. The v_{ij} taken all together cover A, and

$$\lambda_k(v_{ij}) \leq C_3 \lambda_k(B_i) = C_3 c_k (4|A_i|)^4 \leq 4^k C_3 c_k \varepsilon^k.$$

Therefore,

$$\lambda_k^s(A, 4^k C_3 c_k \varepsilon^k) \;\leq\; \sum_{ij} \lambda_k(v_{ij})^{s/k}$$

$$\leq \sum_{ij} 4^s C_3^{s/k} c_k^{s/k} |A_i|^s$$

$$\leq \sum_i N 4^s C_3^{s/k} c_k^{s/k} |A_i|^s \leq N 4^s C_3^{s/k} c_k^{s/k} H.$$

Let $\varepsilon \downarrow 0$ and $H \downarrow H^s(A)$: $K'_{s,k} = N4^s C_3^{s/k} c_k^{s/k}$ will work. □

From (5.16) and (5.17) it follows that A is an s-set if and only if $0 < \lambda_k^s(A) < \infty$.

We turn now to step (ii). How does $\dim_\mu A$ vary with μ? Consider a one-sided version of the condition (5.9):

$$\left.\begin{array}{c} \liminf_n \dfrac{\log \mu(u_n(x))}{\log \nu(u_n(x))} \geq \rho \quad \text{or} \\ \forall \delta \exists N \forall n \geq N : \mu(u_n(x)) \leq \nu(u_n(x)))^{\rho-\delta}. \end{array}\right\} \tag{5.18}$$

Here δ is positive, and N depends on x and δ. If

$$0 < \nu(u_n(x)) < 1, \quad 0 < \mu(u_n(x)) \tag{5.19}$$

holds for all sufficiently large n, then the first condition in (5.18) implies the second one. The second condition is the "official" one — the one needed in the following proof. It has the advantage that it avoids the assumption (5.19); but the first condition is intuitively clearer and is the one to be verified in practice.

THEOREM 2. *If* (5.18) *holds for all x in A, then*

$$\dim_\nu A \geq \rho \dim_\mu A. \tag{5.20}$$

Here, as in all that follows, ρ is nonnegative. Although $\dim_\mu A = \infty$ is possible, in all the interesting cases $\dim_\mu A < \infty$ and even $\dim_\mu A \leq k$. *In Theorems 2 through 5, all dimensions are therefore assumed finite.* And we assume $A \subset E$, since $v_n(x)$ is defined for all n only if $x \in E$.

PROOF. We may assume in the proof that $\rho > 0$. Also assume at first that $\nu(v) > 0$ for every $\mathcal{A}$-set v that meets A.

Fix a δ satisfying $0 < \delta < \rho$. For positive ε, let A_ε be the set of x in A such that for every n, either $\nu(u_n(x)) \geq \varepsilon$ or $\mu(u_n(x)) \leq \nu(u_n(x))^{\rho-\delta}$. Clearly A_ε increases as ε decreases. Suppose that $x \in A$ and let $\varepsilon = \nu(u_N(x))$, which is positive by our temporary assumption (recall that N depends on x as well as on δ). Whatever n may be, if $\nu(u_n(x)) < \varepsilon$, then $n > N$ and therefore $\mu(u_n(x)) \leq \nu(u_n(x))^{\rho-\delta}$ by (5.18); thus $x \in A_\varepsilon$. Therefore, $A_\varepsilon \uparrow A$ as $\varepsilon \downarrow 0$, and by (5.8) (let ε go to 0 through the rationals) it suffices to prove that

$$\dim_\nu A \geq \rho \dim_\mu A_\varepsilon. \tag{5.21}$$

Let $\dim_\nu A = d$ (assumed finite), and suppose that $0 < \eta < \varepsilon$. Since $\dim_\nu A_\varepsilon < d + \delta, \nu^{d+\delta}(A_\varepsilon, \eta) = \nu^{d+\delta}(A_\varepsilon) = 0$. Therefore, there exists a ν-η-covering $\{v_i\}$ of A_ε for which $\Sigma_i \nu(v_i)^{(d+\delta)/k} \leq 1$. We can assume that each v_i meets A_ε. But then, since $\nu(v_i) \leq \eta < \varepsilon$ and $v_i = u_n(x)$ for some x in A_ε and some n, we have $\mu(v_i)^{1/(\rho-\delta)} \leq \nu(v_i)$, and hence

$$\begin{aligned}\mu^{(d+\delta)/(\rho-\delta)}(A_\varepsilon, \eta^{\rho-\delta}) &\leq \sum_i \mu(v_i)^{(d+\delta)/(\rho-\delta)k} \\ &\leq \sum_i \nu(v_i)^{(d+\delta)/k} \leq 1.\end{aligned}$$

Let $\eta \to 0$: $\mu^{(d+\delta)/(\rho-\delta)}(A_\varepsilon) \leq 1$ and $\dim_\mu A_\varepsilon \leq (d+\delta)/(\rho-\delta)$. Let $\delta \to 0$: $\dim_\mu A_\varepsilon \leq d/\rho$, which is (5.21).

It remains to remove the assumption that $\nu(v) > 0$ for every $\mathcal{A}$-set v that meets A. Let F_ν be the union of all $\mathcal{A}$-sets of ν-measure 0. Then $\dim_\nu F_\nu = 0$, and hence by (5.7) and (5.8), $\dim_\nu A = \dim_\nu B$ if the symmetric difference of A and B is contained in F_ν. Suppose of x that $\nu(u_n(x)) = 0$ for some n; then $\nu(u_m(x)) = 0$ for $m \geq n$ and hence, if x satisfies (5.18) (and $\rho - \delta > 0$), $\mu(u_m(x)) = 0$ for some m. Thus $A \cap F_\nu \subset A \cap F_\mu$, and so

$$\begin{aligned}\dim_\nu A &= \dim_\nu(A \cap F_\nu^c) \geq \rho \dim_\mu(A \cap F_\nu^c) \geq \rho \dim_\mu(A \cap F_\mu^c) \\ &= \rho \dim_\mu A.\end{aligned}$$

where the first inequality follows from the case of (5.20) already covered.□

Consider next the condition

$$\left.\begin{aligned}&\limsup_n \frac{\log\ \mu(u_n(x))}{\log \nu(u_n(x))} \leq \rho \ \text{ or} \\ &\forall\delta\exists N\forall n \geq N : \nu(u_n(x))^{\rho+\delta} \leq \mu(u_n(x)).\end{aligned}\right\} \tag{5.22}$$

Again, if (5.19) holds for large n, the first condition implies the second. If $\rho > 0$, then (5.22) is effectively (5.18) with μ and ν interchanged and ρ replaced by its reciprocal. If (5.22) holds on A, it therefore follows by Theorem 2 that $\dim_\mu A \geq \rho^{-1} \dim_\nu A$, or $\dim_\nu A \leq \rho \dim_\mu A$. But ρ here need not be positive: If (5.22) holds for $\rho = 0$, it also holds for positive ρ, and if $\dim_\nu A \leq \rho \dim_\mu A$ for all positive ρ then $\dim_\nu A = 0$.

THEOREM 3. *If* (5.22) *holds for all x in A, then*

$$\dim_\nu A \leq \rho \dim_\mu A. \tag{5.23}$$

If $\log \mu(u_n(x))/ \log \nu(u_n(x)) \to \rho$ on A, so that (5.18) and (5.22) both hold there, it follows that $\dim_\nu A = \rho \dim_\mu A$.

A condition weaker than (5.22) is

$$\left.\begin{aligned}&\liminf_n \frac{\log_\mu(u_n(x))}{\log \nu(u_n(x))} \leq \rho \ \text{ or} \\ &\forall\delta\forall N\exists n \geq N : \nu(u_n(x))^{\rho+\delta} \leq \mu(u_n(x)).\end{aligned}\right\} \tag{5.24}$$

THEOREM 4. *If, for all x in A, $\nu(u_n(x)) \to 0$ and (5.24) holds, and if $\mu(S) < \infty$, then*

$$\dim_\nu A \leq \rho k. \tag{5.25}$$

PROOF. Given ε and δ, consider the sets $u_n(x)$ for which $x \in A, \nu(u_n(x)) \leq \varepsilon$, and $\nu(u_n(x))^{\rho+\delta} \leq \mu(u_n(x))$. By the hypothesis, these sets cover A: we have for A a ν-ε-covering $\{v_i\}$ for which $\nu(v_i)^{\rho+\delta} \leq \mu(v_i)$. Since each $\mathcal{A}_n$ refines the preceding one, for any two sets in $\mathcal{A}$, one of them contains the other or else they are disjoint. Therefore, if we remove each v_i that is containted in some other v_j, the union is unchanged and the sets are now disjoint:

$$\mu(S) \geq \sum_i \mu(v_i)^{k/k} \geq \sum_i \nu(v_i)^{(\rho+\delta)k/k} \geq \nu^{(\rho+\delta)k}(A, \varepsilon).$$

Hence (let $\varepsilon \to 0$) $\nu^{(\rho+\delta)k}(A) \leq \mu(S) < \infty$, hence $\dim_\nu A \leq (\rho + \delta)k$, and hence (let $\delta \to 0$) $\dim_\nu A \leq \rho k$. □

It is possible to construct an example showing (i) that Theorem 3 becomes false if the hypothesis is weakened by the substitution of a limit inferior for the limit superior, and (ii) that Theorem 4 becomes false if the conclusion is strengthened by the substitution of (5.23) for (5.25). See Example 10 at the end of the paper.

Now take $\nu = \lambda_k$ in Theorems 2 and 4:

THEOREM 5. *Suppose that Conditions 1 and 2 hold and that $A \subset E$ and $\rho \geq 0$.*

(i) *If*

$$\left.\begin{array}{c} \liminf_n \dfrac{\log \mu(u_n(x))}{\log \lambda_k(u_n(x))} \geq \rho \text{ or} \\ \forall\delta\exists N \forall n \geq N : \mu(u_n(x)) \leq \lambda_k(u_n(x))^{\rho-\delta} \end{array}\right\} \tag{5.26}$$

holds for all x in A, and if $\mu(A) > 0$, then

$$\dim A \geq \rho k \tag{5.27}$$

(ii) *If*

$$\left.\begin{array}{c} \liminf_n \frac{\log \mu(u_n(x))}{\log \lambda_k(u_n(x))} \leq \rho \text{ or} \\ \forall\delta\forall N \exists n \geq N : \lambda_k(u_n(x))^{\rho+\delta} \leq \mu(u_n(x)) \end{array}\right\} \tag{5.28}$$

holds for all x in A, and if $\mu(S) < \infty$, then

$$\dim A \leq \rho k. \tag{5.29}$$

In (5.26) and (5.28), the condition needed for the proofs is the second one. If

$$0 < \lambda_k(u_n(x)) < 1, \quad 0 < \mu(u_n(x)) \tag{5.30}$$

holds for sufficiently large n, the other condition suffices.

PROOF. By Theorem 1, $\dim A = \dim_{\lambda_k} A$. By Theorem 2, if (5.26) holds on A, then $\dim A \geq \rho \dim_\mu A$. But if $\{v_i\}$ is a μ-ε-covering of A, then from $\sum_i \mu(u_i)^{k/k} \geq \mu(A)$ follows $\mu^k(A) \geq \mu^k(A,\varepsilon) \geq \mu(A) > 0$. Therefore $\dim_\mu A \geq k$, which proves part (i).

By (5.11) and (5.13), $\lambda_k(u_n(x)) \leq 2^k c_k |u_n(x)|^k \to 0$. Part (ii) is therefore an immediate consequence of Theorem 4. □

In (5.29) we can take A to consist of *all* the x in E that satisfy (5.28). If (5.26) and (5.28) both hold on A, so that the limit inferior is exactly ρ, and if $0 < \mu(A) \leq \mu(S) < \infty$, then the two parts of the theorem together give $\dim A = \rho k$.

Results somewhat similar to Theorem 5 can be descried among the Gothic wonders of [9]; see paragraphs 2.10.19 (1) and 2.10.19 (3) there.

6 Self-similarity and independent processes

We now put further restrictions on $\mathcal{A}$. Assume that $\mathcal{A}_1$ contains r sets and each element of $\mathcal{A}_n$ contains exactly r elements of $\mathcal{A}_{n+1}$, so that $\mathcal{A}_n$ contains r^n sets in all. Let $D = \{0, 1, \cdots, r-1\}$ be the set of base-r digits. Label the elements of $\mathcal{A}_1$ by these digits in some order. If an element of $\mathcal{A}_n$ has the label $i_1, \cdots, i_n$ (an element of D^n), to the r elements of $\mathcal{A}_{n+1}$ contained in it assign the labels $i_1, \cdots, i_n$, $i (i \in D)$ in some order. Let $v(i_1, \cdots, i_n)$ be the element of $\mathcal{A}_n$ with label $i_1, \cdots, i_n$, and as before, for $x \in E$ let $u_n(x)$ be the element of $\mathcal{A}_n$ that contains x. This is the set-up in symbolic dynamics.

For $x \in E$, let $d_1(x), \cdots, d_n(x)$ be the label assigned to $u_n(x)$, so that $u_n(x) = v(d_1(x), \ldots, d_n(x))$. If in Example 1 of the preceding section the intervals are labeled in the usual order, then the $d_n(x)$ are the digits in the base-r expansion of x (the nonterminating one unless $x = 0$). In the general case, the labels may be assigned arbitrarily, and in fact D may be arbitrary. Let

$$E_{i_1 \cdots i_n} = E \cap v(i_1, \cdots, i_n).$$

CONDITION 3. *The set S is compact, and if*

$$u \in \mathcal{A}, v \in \mathcal{A}_n, u \cap v = \emptyset, \tag{6.1}$$

then there is a w such that

$$w \subset v, w \in \mathcal{A}_{n+1}, u \cap w^- = \emptyset. \tag{6.2}$$

The bar in (6.2) denotes closure. The significance of (6.1) and (6.2) will become clear in the proof of Lemma 2 below.

CONDITION 4. *There is a vector* $q = (q_0, \cdots, q_{r-1})$ *such that* $q_i > 0, \sum_i q_i \leq 1$, *and*

$$\lambda_k(v(i_1, \cdots, i_n)) = q_{i_1} \cdots q_{i_n}, \; i_1, \cdots, i_n \in D^n; \tag{6.3}$$

and $\lambda_k(S) = 1$.

The requirement $\lambda_k(S) = 1$ is merely a convenient normalization. If $0 < \lambda_k(S) < \infty$, applying an appropriate similarity transformation will lead to $\lambda_k(S) = 1$, and Hausdorff dimension is invariant under similarity transformations. What (6.3) says is that each element v of $\mathcal{A}_n$ splits into $r + 1$ pieces, namely the r elements of $\mathcal{A}_{n+1}$ contained in v, which have measures $\lambda_k(v)q_i$, and the part left over, a piece (perhaps empty) of measure $\lambda_k(v)(1 - \sum_i q_i)$. The labeling must be consistent with this condition.

From (5.10) and (5.11) it follows that $|u_{n-1}(x)|^k/|u_n(x)|^k \leq C_1 c_k 2^k$ $\lambda_k(u_{n-1}(x))/\lambda_k(u_n(x))$; if q is the minimum q_i, then $\lambda_k(u_n(x)) \geq q\lambda_k$ $(u_{n-1}(x))$. Therefore, Conditions 1 and 4 imply (5.12) with $C_2^k = q^{-1}C_1 c_k 2^k$ And (6.3) and (5.10) together imply (5.13). *Condition* 2 *therefore follows from Conditions* 1 *and* 4, so that the results of Section 5 apply.

EXAMPLE 1 *(continued).* In addition to Condition 1 (and Condition 2), Example 1 of the preceding section satisfies Condition 3 and 4: If (6.1) holds and v lies to the left [right] of u, take w to be the leftmost [rightmost] element of $\mathcal{A}_{n+1}$ contained in v. In this example, $q_i = r^{-1}$ and $\sum_i q_i = 1$. □

EXAMPLE 2 *(continued).* In Example 2, Condition 3 holds because if two elements of $\mathcal{A}$ are disjoint, so are their closures. In condition 4, $q_i = 1/3$ and $\sum_i q_i = 2/3$. □

EXAMPLE 3. As a slight generalization of Example 2, take $S = [0, 1]$ and $q_0 + q_1 < 1$ $(k = 1, r = 2)$. Let $\mathcal{A}_1$ consist of the intervals $[0, q_0]$ and $[1 - q_1, 1]$, and continue to subdivide as shown in Figure 7 (where $q_0 = 1/2$ and $q_1 = 1/4$). Example 2 is the case $q_0 = q_1 = 1/3$. The intersection E is a Cantor-like set; it is similar to each E_i, the similarity ratio being q_i. By

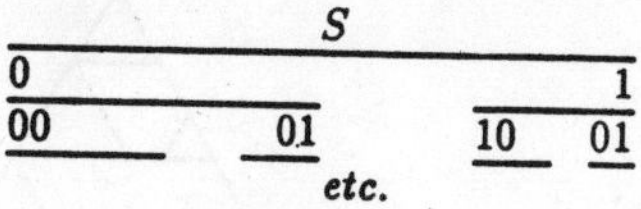

Figure 7

self-similarity, $H^s(E) = q_0^s H^s(E) + q_1^s H(E)$, and if E is an s-set, it follows that $\dim S$ is the solution of $q_0^s + q_1^s = 1$.

But suppose E is constructed as in Figure 8, where the order of the successive subintervals is determined in some haphazard way. Although the destroys the self-similarity, Conditions 1 through 4 still hold, and as we will see, dim E is unchanged. □

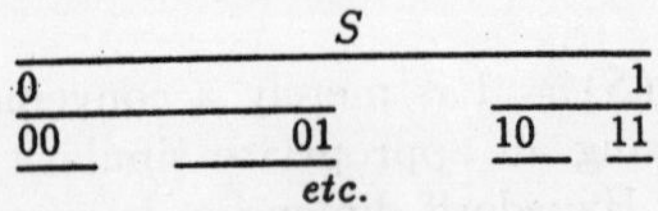

Figure 8

EXAMPLE 4. Let S and the elements of $\mathcal{A}$ be nonoverlapping equilateral triangles, as shown in Figure 9 ($r = 3$). If two triangles share a vertex, as in the figure, it can be assigned to either of the two, arbitrarily. If $q_i \equiv 1/3$, E is the Sierpinski set. Condition 1 is easy to check. Condition 3 holds as well: For example, $v(2)$ is disjoint from $v(0)$, and $v(2,2)^-$ is disjoint from $v(0)$. Condition 4 holds by construction.

The self-similarity in this example can be destroyed by using triangles that are not similar (but have the same areas as before). A little care is required to ensure that Conditions 1 through 4 still hold. □

Let J_1 be the set of probability vectors $p = (p_0, \cdots, p_{r-1}) : p_i \geq 0$ and $\sum_i p_i = 1$. Let J_1° be the subset where $p_i > 0$ for all i; J_1° is the interior of J_1 relative to the hyperplane in R^r determined by the constraint $\sum_i p_i = 1$.

LEMMA 2. *If Conditions* 1, 3, *and* 4 *hold and* $p \in J_1^0$, *then there exists a probability measure* μ_p *on* $(S,\mathcal{S})$ *such that*

$$\mu_p(v(i_1, \cdots, i_n)) = p_{i_1} \cdots p_{i_n}, \quad i_1, \cdots, i_n \in D^n. \tag{6.4}$$

Under $\mu_p, d_1(x), d_2(x), \cdots$ are independent random variables assuming the value i with probability p_i. The discussion following (5.9) shows the

Figure 9

role μ_p will play. It follows from (6.4) that $\mu_p(S_n) = \mu_p(E) = 1$ and hence $\mu_p(E_{i_1\cdots i_n}) = p_{i_1}\cdots p_{i_n}$. In the interesting cases, $\lambda_k(E) = 0$, so that μ_p is singular.

PROOF OF THE LEMMA. Consider, on some probability space $(\Omega, \mathcal{F}, \mathcal{P})$, independent random variables ξ_n that assume only the values $0, 1, \cdots, r-1$ and satisfy $P[\omega : \xi_n(\omega) = i] = p_i$; if $V(i_1, \cdots, i_n) = [\omega : \xi_1(\omega) = i_1, \cdots, \xi_n(\omega) = i_n]$, this is the same as requiring

$$\mathcal{P}(V(i_1, \cdots, i_n)) = p_{i_1}\cdots p_{i_n}. \tag{6.5}$$

By (6.3), $v(i_1, \cdots, i_n)$ is nonempty, and by (5.10) and (6.3), $\max[|v| : v \in \mathcal{A}_n] \to 0$ (Conditions 1 and 4 enter the proof only here). Therefore, $\bigcap_n v(\xi_1(\omega), \cdots, \xi_n(\omega))^-$ consists of a single point, call it $T\omega$. If $x(i_1, \cdots, i_n)$ is some arbitrarily chosen point of $v(i_1, \cdots, i_n)$ and $T_n(\omega) = x(\xi_1(\omega), \cdots, \xi_k(\omega))$, then $T_n\omega \to T\omega$ for all ω, and since T_n is measurable $\mathcal{F}/\mathcal{S}$, so is T. Define $\mu_p = \mathcal{P}T^{-1}$.

Let Ω_0 be the set of ω such that

$$T\omega \in v(\xi_1(\omega), \cdots, \xi_n(\omega)) \tag{6.6}$$

for all n. Assume $\Omega_0 \in \mathcal{F}$ and $\mathcal{P}(\Omega_0) = 1$, which we will prove in a moment. If (6.6) holds, then $T\omega \in v(i_1, \cdots, i_n)$ if and only if $\xi_j(\omega) = i_j$ for $j \le n$, that is, if and only if $\omega \in V(i_1, \cdots, i_n)$. In other words, the symmetric difference of $V(i_1, \cdots, i_n)$ and $T^{-1}v(i_1, \cdots i_n)$ is contained in Ω_0^c. If $\mathcal{P}(\Omega_0^c) = 0$, then $\mu_p(v(i_1, \cdots, i_n)) = \mathcal{P}(T^{-1}v(i_1, \cdots, i_n))$ coincides with $\mathcal{P}(V(i_1, \cdots, i_n))$, so that (6.4) follows from (6.5).

Since (6.6) holds for all n if and only if it holds for infinitely many $n, \omega \in \Omega_0^c$ implies that there exists an n_0 such that

$$T\omega \in v(\xi_1(\omega), \cdots, \xi_n(\omega))^- - v(\xi_1(\omega)), \cdots, \xi_n(\omega))$$

for all $n \ge n_0$. Assume this is true and take $u = u_{n_0}(T\omega)$. Since $T\omega$ lies in u but not in $v(\xi_1(\omega), \cdots, \xi_{n_0}(\omega))$, these two sets are different elements of $\mathcal{A}_{n_0}$. It follows that $v(\xi_1(\omega), \cdots, \xi_n(\omega))$ is disjoint from u for $n \ge n_0$; but since $v(\xi_1(\omega), \cdots, \xi_{n+1}(\omega))^- \cap u$ contains $T\omega$, it is nonempty. Therefore, $\omega \in \Omega_0^c$ implies that there exist an n_0 and a set $u \in \mathcal{A}$ such that

$$\left.\begin{array}{ll} v(\xi_1(\omega), \cdots, \xi_n(\omega)) \cap u = \emptyset, & n \ge n_0, \\ v(\xi_1(\omega), \cdots, \xi_{n+1}(\omega))^- \cap u \ne \emptyset, & n \ge n_0. \end{array}\right\} \tag{6.7}$$

If Ω_{u,n_0} is the set where (6.7) holds, then $\Omega_0^c \subset \bigcup \Omega_{u,n_0}$. If we assume that $(\Omega, \mathcal{F}, \mathcal{P})$ is complete, as we may, then it is enough to prove that $\mathcal{P}(\Omega_{u,n_0}) = 0$ for each u and n_0. This is where (6.1) and (6.2) (Condition 3) come in.

If $v(i_1, \cdots, i_n)$ is disjoint from u, choose i in D so that $v(i_1, \cdots, i_n, i)^- \cap u = \emptyset$ (possible if (6.1) implies (6.2)) and set $f_{u,n}(i_1, \cdots, i_n) = i$; otherwise (to be definite) set $f_{u,n}(i_1, \cdots, i_n) = 0$ ($f_{u,n}$ maps D^n into D). Now (6.7) implies that

$$\xi_{n+1}(\omega) \neq f_{u,n}(\xi_1(\omega), \cdots, \xi_n(\omega)), \quad n \geq n_0. \tag{6.8}$$

Since the p_i are all positive.

$$\mathcal{P}[\xi_{n+1} \neq f_{u,n}(\xi_1, \cdots, \xi_n) | \xi_i = i_j, j \leq n] \leq 1 - \min p_i < 1, \tag{6.9}$$

and it follows that the event (6.8) has probability 0. □

For $x \in E$ and $i \in D$, let $N_n(x, i)$ be the number of occurrences of i among $d_1(x), \cdots, d_n(x)$, let $f_n(x, i) = N_n(x, i)/n$ be the relative frequency of i, and let $f_n(x)$ be the probability vector $(f_n(x, 0), \cdots, f_n(x, r-1))$. Now

$$\log \lambda_k(u_n(x)) = \sum_i N_n(x, i) \log q_i$$

(recall that $q_i > 0$), and if $p \in J_1^\circ$, then

$$\log \mu_p(u_n(x)) = \sum_i N_n(x, i) \log p_i. \tag{6.10}$$

For $p \in J_1$, define

$$h(p : q) = \sum_i p_i \log p_i / \sum_i p_i \log q_i,$$

where in the numerator the convention is that $0 \log 0 = 0$.

For $p \in J_1$, let $L(P)$ be the set of x in E for which

$$f_n(x) \to p. \tag{6.11}$$

If $p \in J_1^\circ$ and $x \in L(p)$, then

$$\frac{\log \mu_p(u_n(x))}{\log \lambda_k(u_n(x))} \to h(p : q). \tag{6.12}$$

Since $\mu_p(L(p)) = 1$ by the strong law of large numbers, it follows by Theorem 5 that

$$\dim L(p) = kh(p : q). \tag{6.13}$$

Now suppose that some of the p_i may be 0 — that p lies in J_1 but perhaps not in J_1°. Define $L_0(p)$ as the set of x in E for which (6.11) holds and for which, if $p_i = 0$, then $d_n(x)$ is never i ($N_n(x, i) \equiv 0$). If $p \in J_1 - J_1^\circ$, then $L_0(p)$ is a proper subset of $L(p)$; if $p \in J_1^\circ$, then $L_0(p) = L(p)$. If $N_n(x, i) = p_i = 0$, the convention 0 log 0 applies to (6.10) (which is undefined if $N_n(x, i) > 0 = p_i$ for some i). Then (6.12) holds on $L_0(p)$, and since clearly $\mu_p(L_0(p)) = 1$, we have

$$\dim L_0(p) = kh(p : q). \tag{6.14}$$

But what about (6.13) in the case $p \in J_1 - J_1^\circ$? Suppose that π lies in J_1° and put $\xi_\pi = \sum_i p_i \log \pi_i / \sum_i p_i \log q_i$. Then $\log \mu_\pi(u_n(x))/\log \lambda_k(u_n(x)) \to \xi_\pi$ for $x \in L(p)$, and therefore $\dim L(p) \le k\xi_\pi$ by Theorem 5 (ii). Let $\pi \to p$: $\dim L(p) \le kh(p:q)$ by continuity, and from $L_0(p) \subset L(p)$ it follows that

$$\dim L_0(p) = \dim L(p) = kh(p:q), \quad p \in J_1. \tag{6.15}$$

EXAMPLE 1 *(continued)*. Put

$$h(p) = -\sum_i p_i \log p_i.$$

Since $k = 1$ and $q_i \equiv r^{-1}$, (6.15) reduces to

$$\dim L_0(p) = \dim L(p) = h(p)/\log r,$$

which is Eggleston's theorem. □

For $G \subset J_1$, let $L(G) = \bigcup_{p \in G} L(p)$ and $L_0(G) = \cup_{p \in G} L_o(p)$. Let $U(G)$ be the set of x in E such that for each ε and n_0 there exist a $p \in G$ and an $n \ge n_0$ such that $|f_n(x) - p| < \varepsilon$. Thus L(G) (L for "lower") is the x-set where the sequence $\{f_n(x)\}$ converges to some point of G and $U(G)$ (U for "upper") is the x-set where some subsequence converges to some point of G^{-1}. Thus

$$L_0(G) \subset L(G) \subset U(G) = U(G^-).$$

We want next to prove

$$\dim L_0(G) = \dim U(G) = kh(G:q), \tag{6.16}$$

where

$$h(G:q) = \sup[h(p:q) : p \in G]. \tag{6.17}$$

If $p \in G$, then $\dim L_0(G) \ge \dim L_0(p) = kh(p:q)$ by (6.14), and so it suffices to prove

$$\dim U(G) \le kh(G:q). \tag{6.18}$$

The argument is similar to one in [5]. If $q = \max_i q_i$, then $q < 1 (q_i > 0$ and $\sum_i q_i \le 1)$ and $-\sum_i \pi_i \log q_i \ge \log q^{-1}$ if $\pi \in J_1$. For $\pi \in J_1^\circ$ and $0 < \varepsilon < \log q^{-1}$, let V_π^ε be the set of p in J_1 for which

$$-\sum_i |p_i - \pi_i| \log \pi_i < \varepsilon \tag{6.19}$$

and

$$-\sum_i |p_i - \pi_i| \log q_i < \varepsilon \tag{6.20}$$

If $x \in J_1^\circ$ and $f_n(x) \in V_\pi^\varepsilon$, then

$$\frac{\log \mu_\pi(u_n(x))}{\log \lambda_k(u_n(x))} \leq \frac{-\sum_i \pi_i \log \pi_i + \varepsilon}{-\sum_i \pi_i \log q_i - \varepsilon};$$

the denominator here is positive, because of the constraint on ε.

By Theorem 5(ii), $\pi \in J_1^\circ$ implies

$$\dim[x \in E : f_n(x) \in V_\pi^\varepsilon \ i.o.] \leq k \frac{-\sum_i \pi_i \log \pi_i + \varepsilon}{-\sum_i \pi_i \log q_i - \varepsilon}, \tag{6.21}$$

where "i.o." means "for infinitely many n." If π lies in J_1° and converges to a point p of J_1, then the sums in (6.19) and (6.20) converge to 0. Since J_1° is dense in J_1, if $p \in J_1$, then there exists for positive ε and δ a π in J_1° such that $|\pi - p| < \delta$ and $p \in V_\pi^\varepsilon$. Let $G^\delta = [p \in J_1\colon \operatorname{dist}(p, G) < \delta]$. Then the V_π^ε for π in $G^\delta \cap J_1^\circ$ cover G^-. Since V_π^ε is open relative to J_1, some finite subcollection $V_{\pi(1)}^\varepsilon, \cdots, V_{\pi(a)}^\varepsilon$ covers G^-, and so $G^\eta \subset \bigcup_{\alpha=1}^a V_{\pi(\alpha)}^\varepsilon$ for some positive η. If $x \in U(G)$, then $f_n(x) \in G^\eta$ for infinitely many n. Therefore,

$$U(G) \subset \bigcup_{\alpha=1}^{a} [x \in E : f_n(x) \in V_{\pi(\alpha)}^\varepsilon \text{ i.o.}].$$

By (6.21)

$$\dim U(G) \leq k \sup \left[\frac{-\sum_i \pi_i \log \pi_i + \varepsilon}{-\sum_i \pi_i \log q_i - \varepsilon} : \pi \in G^\delta \cap J_1^\circ \right].$$

Let δ and then ε tend to 0: (6.18) follows.

This proves (6.16), which is much stronger than (6.15), even for the case $G = \{p\}$:

THEOREM 6. *Conditions* 1, 3, *and* 4 *together imply* (6.16).

Since it implies $\dim \bigcup_{p \in G} L(p) = \sup_{p \in G} \dim L(p)$, this theorem in part has to do with extending (5.8) to uncountable unions. For a general investigation of this question, see Cajer [6].

Since $U(J_1) = E$, the theorem implies

$$\dim E = kh(J_1 : q). \tag{6.22}$$

This result can be reformulated:

THEOREM 7. *The supremum $h(J_1 : q)$ is the solution of the equation*

$$\sum_i q_i^t = 1. \tag{6.23}$$

This solution t is of course unique, and $0 < t \leq 1$ ($t = 1$ if and only if $\sum_i q_i = 1$).

PROOF. If t satisfies (6.23) and $\pi_i = q_i^t$, then $\pi = (\pi_0, \cdots, \pi_{r-1}) \in J_1^\circ$ and $h(p : q) = th(p : \pi)$. If $p = \pi$, then $h(p : \pi) = 1$, and for arbitrary p it follows by convexity that (let $\sum'$ denote summation over those i for which $p_i > 0$)

$$\sum_i p_i \log \pi_i - \sum_i p_i \log p_i = \sum{}' p_i \log(\pi_i/p_i) \leq \log \sum{}' \pi_i \leq 0, \tag{6.24}$$

so that $h(p : \pi) \leq 1$. The suprema over p of $h(p, \pi)$ and $h(p : q)$ are therefore 1 and t, respectively. □

EXAMPLE 2 *(continued)*. Here $r = 2, E = U(J_1)$ is the Cantor set, $q_1 = q_2 = 1/3$, and $\dim E = h(J_1 : q) = \log 2/\log 3$. □

EXAMPLE 1 *(continued)*. Suppose $r = 3$ and let G be the set of (p_0, p_1, p_2) for which $p_1 = 0$. Then $(q_i \equiv 1/3)\, h(G : q) = \log 2/\log 3$, and $U(G)$ is the set of x in [0,1] for which $\liminf_n f_n(x, 1) = 0$. This set is larger than the Cantor set but has the same dimension. And $L_0(G)$ is smaller than the Cantor set but has the same dimension. □

EXAMPLE 3 *(continued)*. In this case, since $k = 1$, $\dim E$ is the supremum of

$$(p_0 \log p_0 + p_1 \log p_1)/(p_0 \log q_0 + p_1 \log q_1)$$

over all (p_0, p_1). Alternatively, it is the solution s of $q_0^1 + q_1^t = 1$. And (6.15) gives an Eggleston- like result. □

EXAMPLE 4. *(continued)*. Here $\dim E = 2t$, where t solves $q_0^t + q_1^t + q_2^t = 1$. In the Sierpinsky case, $q_1 \equiv 1/4$, $t = \log 3/\log 4$, and $\dim E = \log 3/\log 2$. □

EXAMPLE 5. Let S be an isosceles triangle with base angles of $\pi/6$. Divide the base into three equal parts and remove the equilateral triangle over middle one, as in the figure, which shows $v(0)$ and $v(1)$. Continue as shown. The intersection E is the Koch curve, Figure 11. Since $r = 2$ and $q_0 = q_1 = 1/3, \dim S$ is $2t$, where t is the solution of $2(1/3)^t = 1 : \dim S = \log 4/\log 3$. The bars and labels show the partition of E into the sets $E_{i_1 i_2 i_3}$. The set of x on the Koch curve for which i occurs in the sequence $\{d_n(x)\}$ of labels with asymptotic relative frequency p_i has Hausdorff dimension $-2(p_0 \log p_0 + p_1 \log p_1)/\log 3$. □

The changes on this theme can be rung and infinitum. Here is a slightly different example.

EXAMPLE 6. Suppose $k = 2$ and S is the closed unit square. Let $\mathcal{A}_n$ consist

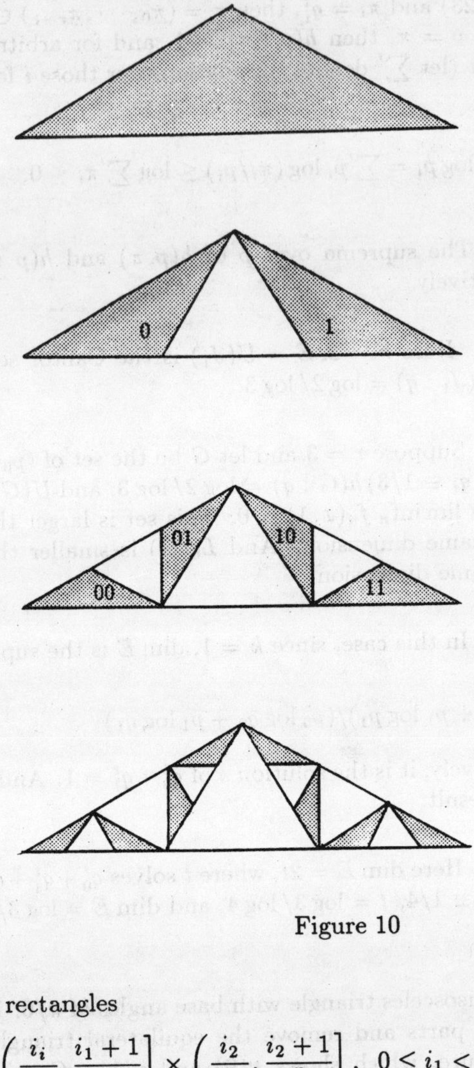

Figure 10

of the rectangles

$$\left(\frac{i_i}{r_1^{n_1}}, \frac{i_1+1}{r_1^{n_1}}\right] \times \left(\frac{i_2}{r_2^{n_2}}, \frac{i_2+1}{r_2^{n_2}}\right], \quad 0 \le i_1 < r_1, 0 \le i_2 < r_2, \tag{6.25}$$

(close the interval on the left for $i_1 = 0$ and for $i_2 = 0$), where

$$n_1 = n, \quad n_2 = \left[n_1 \frac{\log r_1}{\log r_2} \right], \tag{6.26}$$

the brackets denoting integer part. Since $1 \le r_2^{-n_2}/r_1^{-n_1} < r_2$, the rectangles (6.25) are almost square, and Condition 1 holds ($C_1 = r_2 + 1$); this is false if we take $n_2 = n_1$ (unless $r_2 = r_1$). And Condition 2 holds.

For $x = (x_1, x_2)$ in S, let x_l have r_l-adic expansion $\sum_n d_n^{(l)}(x) r_l^{-n}$ (the nonterminating one unless $x_l = 0$). Let μ_l be the measure on [0,1] under

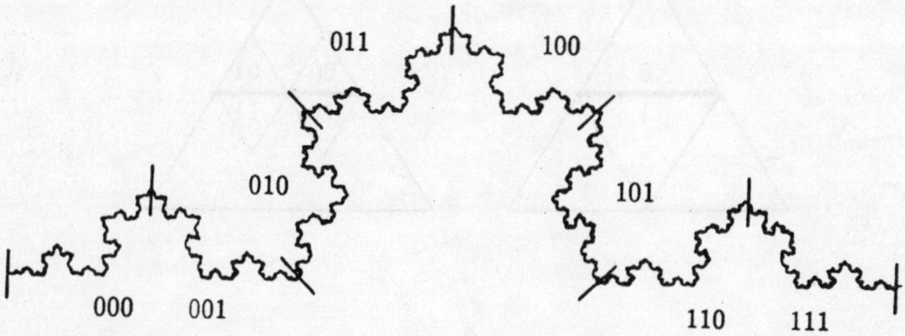

Figure 11

which the digits of the r_l-adic expansion are independent with probabilities $p_i^{(l)}, 0 \leq i < r_l$.

Define μ on S as the product $\mu = \mu_1 \times \mu_2$. Let $N_n^{(l)}(x, i)$ be the number of occurrences of i among $d_1^{(l)}(x), \cdots, d_n^{(l)}(x)$. If n, n_1, and n_2 are related by (6.26), then

$$\log \mu(u_n(x)) = \sum_{l=1}^{2} \sum_{t=0}^{r_l - 1} N_{n_l}^{(l)}(x, t) \log p_i^{(l)}.$$

If $V^{(l)}$ is the Eggleston set on [0,1] corresponding to the probabilities $p_0^{(l)}, \cdots, p_{r_l-1}^{(l)}$, then $x \in V^{(1)} \times V^{(2)}$ implies

$$\frac{\log \mu(u_n(x))}{\log \lambda_2(u_n(x))} \to \frac{1}{2} \sum_{l=1}^{2} \left\{ - \sum_{i=0}^{r_l - 1} p_i^{(l)} \log p_i^{(l)} / \log r_l \right\}.$$

It follows by the usual argument that $\dim(V^{(1)} \times V(2))$ is twice this limit:

$$\dim(V^{(1)} \times V^{(2)}) = \dim V^{(1)} + \dim V^{(2)}.$$

This would be obvious if Hausdorff dimension were always additive for Cartesian products, but it is not (see [8, p. 73]). □

This result for $r_1 = r_2$ is due to Cutler [7]; see also Kenyon and Peres [10]. That it does not follow immediately from Theorem 6 suggests that Condition 4 should be relaxed in some way.

Although the results in this section do not require self- similarity, they do tie in with self-similarity in a natural way. Suppose that $E = \bigcup_{i=0}^{r-1} E_i$, where $E_i = \psi_i E, \psi_i$ being a similarity transformation with similarity ratio $\theta_i < 1$. Then $H^s(E_i) = \theta_i^s H^s(E)$. If the E_i are disjoint, or have intersections of H^s-measure 0, then $H^s(E) = \sum_i \theta_i^s H^s(E)$. If E is an s-set, then its dimension must be the solution of

$$\sum_i \theta_i^s = 1 \tag{6.27}$$

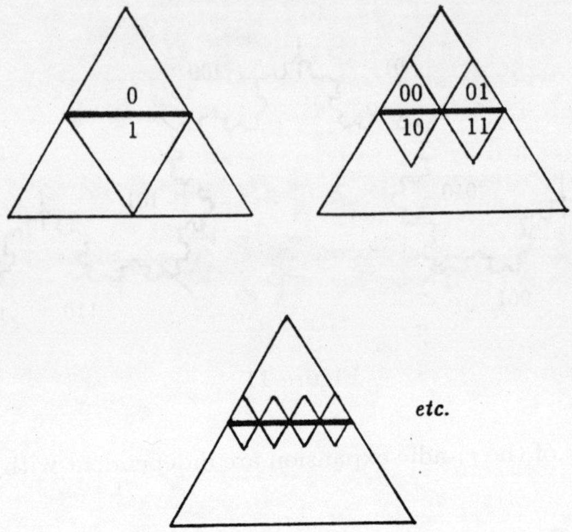

Figure 12

According to a theorem of Moran [12], all this is true if there exists a bounded, open set V and disjoint sets V_i such that $V = \bigcup_i V_i$ and $V_i = \psi_i V$.

If the closures V_i^- happen to be disjoint, this can easily be deduced from Theorem 6: Take $v(i_1, \cdots, i_n) = \psi_{i_n} \cdots \psi_{i_1} V^-$ and $q_i = \theta_i^k$. Note that the solution of (6.27) is k times the solution of (6.23). From Theorem 6, I can "almost" deduce the general case of Moran's theorem — almost but not quite.

EXAMPLE 7. Suppose that $\mathcal{A}$ consists of equilateral triangles, as in Figure 12. If two triangles in $\mathcal{A}_n$ share an edge, the edge belongs to the top one (the edge common to $v(0,0)$ and $v(1,0)$ lies in $v(0,0)$). If two upper triangles share a vertex (as $v(0,0)$ and $v(0,1)$ do), assign it to the left one. Conditions 1,2, and 4 hold, but Condition 3 does not. Since E is the heavy horizontal segment in the figure, for an arbitrary p we have, for example, $\mu_p(v(1,0)) = \mu_p(\emptyset) = 0$ instead of $\mu_p(v(1,0)) = p_1 p_0$. Thus Lemma 2 does not hold.

Since E is in effect the unit interval, we can find the dimensions of all sorts of subsets of E. But we cannot directly apply the theory of this section: the formula (6.17) does not hold as it stands. □

Something like Condition 3 seems to be essential.

7 Cross-similarity and Markov processes

Suppose as in the preceding section that Conditions 1 and 3 hold, but generalize Condition 4 in the following way. Let q be a positive ($q_i > 0$) subprobability ($\sum_i q_i \leq 1$) vector and let $Q = [q_{ij}]\,(i, j \in D)$ be an r by r positive ($q_{ij} > 0$) substochastic ($\sum_j q_{ij} \leq 1$) matrix.

CONDITION 5. *There are a positive subprobability vector q and a positive substochastic matrix Q for which*

$$\lambda_k(v(i_1, \cdots, i_n)) = q_{i_1} q_{i_1 i_2} \cdots q_{i_{n-1} i_n}, \quad i_1, \cdots, i_n \in D^n; \qquad (7.1)$$

and $\lambda_k(S) = 1$.

Thus $\lambda_k(v(i)) = q_i$, as before, but now an element $v(i_1, \cdots, i_n)$ of $\mathcal{A}_n$ splits into the r elements of $\mathcal{A}_{n+1}$ containted in $v(i_1, \cdots, i_n)$ — and these have measures $\lambda_k(v(i_1, \cdots, i_n))q_{i_n i}$ — together with a part left over (perhaps empty) of measure $\lambda_k(v(i_1, \cdots, i_n))(1 - \sum_i q_{i_n i})$. It turns out that the specific values of the q_i have no effect on the dimensions we will calculate.

As in the preceding section, it is easy to see that Conditions 1 and 5 together imply Condition 2.

EXAMPLE 8. Suppose that $S = [0, 1], r = 2$, and the elements of $\mathcal{A}_n$ are closed disjoint intervals. Figure 13 shows the case where

$$Q = \begin{bmatrix} \frac{1}{2}, & \frac{1}{4} \\ \frac{1}{3}, & \frac{1}{3} \end{bmatrix}$$

and $v(i_1, \cdots, i_n, 0)$ $[v(i_1, \cdots, i_n, 1)]$ always has the same left [right] endpoint as $v(i_1, \cdots, i_n)$. In this case there is a kind of *cross- similarity*:

E_0 is similar to E_{00} with ratio $q_0 q_{00}/q_0$,
E_0 is similar to E_{10} with ratio $q_1 q_{10}/q_0$,
E_1 is similar to E_{01} with ratio $q_0 q_{01}/q_1$,
E_1 is similar to E_{11} with ratio $q_1 q_{11}/q_1$.

If we permute the order of the $\mathcal{A}_{n+1}$-sets within an element of $\mathcal{A}_n$, we destroy this cross-similarity. But as we will see, this has no effect on the dimension common to E_0, E_1, and E. □

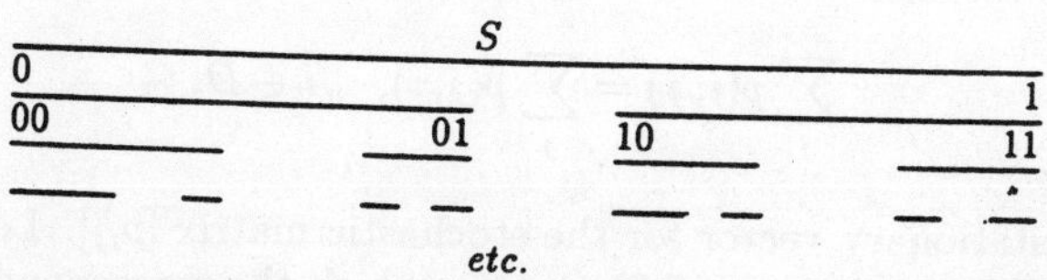

Figure 13

Even though cross-similarity will not figure in our proofs, let us examine the consequences of it. Suppose that for all i and j in D, E_j is similar to E_{ij} with ratio $q_i q_j^{-1} q_{ij}$. Let d_i and d_{ij} be the dimensions of E_i and E_{ij}. Since E_j and E_{ij} are similar, $d_j = d_{ij}$. But then $d_i = \max_j d_{ij} = \max_j d_j$ is independent of i, so that E, E_i, E_{ij} all have the same dimension. And *if* one of them is an s-set, so are they all. *If* the E_{ij} are disjoint, or if their intersections have H^s-measure 0, then $H^s(E_i) = \sum_j H^s(E_{ij}) = \sum_j (q_i q_j^{-1} q_{ij})^s H^s(E_j)$, or

$$q_i^{-s} H^s(E_i) = \sum_j q_{ij}^s q_j^{-s} H^s(E_j). \tag{7.2}$$

If E is an s-set, it follows that 1 is an eigenvalue of the matrix $[q_{ij}^s]$ and that one of the corresponding right eigenvectors has positive components. It will turn out that this condition completely characterizes s. As in the case of self-similarity, none of this applies to cross-similar sets that are not compact.

To find the dimensions of E and subsets of it under the assumption (7.1) requires studying the frequencies not of individual digits but of pairs of adjacent digits, and this leads in a natural way to Markov chains. Let J_2^+ be the set of r by r matrices $P = [p(i,j)](i,j \in D)$ that have nonnegative elements and satisfy

$$\sum_{ij} p(i,j) = 1.$$

Given a P in J_2^+, define a probability vector $p = (p_0, \cdots, p_{r-1})$ by

$$p_i = \sum_j p(i,j)$$

and a stochastic matrix $[p_{ij}]$ by

$$p_{ij} = \begin{cases} p(i,j)/p_i & \text{If } p_i > 0, \\ r^{-1} & \text{if } p_i = 0. \end{cases}$$

(If $p_i = 0$, an arbitrary positive probability vector can be used for the ith row.) Then p and $[p_{ij}]$ are functions of P.

If P also satisfies

$$\sum_j p(i,j) = \sum_j p(j,i), \quad i \in D, \tag{7.3}$$

then p is a stationary vector for the stochastic matrix $[p_{ij}]$. Let J_2 consist of the P in J_2^+ that satisfy (7.3) and for which the corresponding Markov chain does not have two disjoint, nonempty, closed sets of states. For P in J_2, p is the *only* stationary vector for $[p_{ij}]$, and the transient states i are exactly the ones for which $p_i = 0$. (If those states are removed, the chain becomes irreducible.)

We need a generalization of Lemma 2. Let J_2° consist of the elements of J_2 that have positive entries.

LEMMA 3. *If Conditions* 1, 3, *and* 5 *hold and* $P \in J_2^\circ$, *then there exists a probability measure* μ_P *on* (S, S) *such that*

$$\mu_P(v(i_1, \cdots, i_n)) = p_{i_1} p_{i_1 i_2} \cdots p_{i_{n-1} i_n}, \quad i_1, \ldots, i_n \in D^n. \tag{7.4}$$

PROOF. Replace (6.5) by $P(V(i_i, \ldots, i_n)) = p_{i_1} p_{i_1 i_2} \ldots p_{i_{n-1} i_n}$ and in (6.9) replace $\min_i p_i$ by $\min_{ij} p_{ij}$. The old proof then goes through without further change. □

As before, the intersection E supports μ_P.

Assume Conditions 1, 3, and 5 hold. For $x \in E$, let $N_n(x, i, j)$ be the number of l, $1 \leq l < n$, such that $d_l(x) = i$ and $d_{l+1}(x) = j$; let $F_n(x, i, j) = N_n(x, i, j)/(n-1)$, and let $F_n(x)$ be the r by r matrix $[F_n(x, i, j)]$. Although $F_n(x)$ lies in J_2^+, it may not satisfy (7.3), since $\sum_j N_n(x, i, j)$ and $\sum_j N_n(x, j, i)$ may differ by 1.

For $P \in J_2$, define

$$h(P : Q) = \sum_{ij} p_i p_{ij} \log p_{ij} \Big/ \sum_{ij} p_i p_{ij} \log q_{ij}, \tag{7.5}$$

and for $G \subset J_2$, define

$$h(G : Q) = \sup[h(P : Q) : P \in G].$$

For P in J_2, let $L(P)$ be the set of x in E such that

$$F_n(x) \to P,$$

and let $L_0(P)$ be the set of x in $L(P)$ such that $N_n(x, i, j) \equiv 0$ for any pairs i, j with $p_i p_{ij} = 0$. For $G \subset J_2$, let $L_0(G) = \bigcup_{p \in G} L_0(P)$, and $L(G) = \cup_{P \in G} L(P)$, and let $U(G)$ be the set of x in E such that some subsequence of $\{F_n(x)\}$ converges to some point of G^-.

We will prove

$$\dim L_0(P) = \dim L(P) = \dim U(P) = kh(P : Q) \tag{7.6}$$

and

$$\dim L_0(G) = \dim L(G) = \dim U(G) = kh(G : P) \tag{7.7}$$

Of course (7.6) is a special case of (7.7). Since $P \in G$ implies $L_0(P) \subset L_0(G) \subset L(G) \subset U(G)$, it is enough to prove

$$\dim L_0(P) = kh(P : Q) \tag{7.8}$$

for $P \in J_2$ and

$$\dim U(G) \leq kh(G:Q) \tag{7.9}$$

for $G \subset J_2$.

If $P \in J_2$ and $x \in L_0(P)$, then

$$\frac{\log \mu_P(u_n(x))}{\log \lambda_k(u_n(x))} = \frac{\log p_{d_1(x)} + \sum_{ij} N_n(x,i,j)\log p_{ij}}{\log q_{d1(x)} + \sum_{ij} N_n(x,i,j)\log q_{ij}} \to h(P:Q).$$

Since $\mu_P(L_0(P)) = 1$ by the strong law of large numbers of Markov chains (or the ergodic theorem), (7.8) follows from Theorem 5.

The proof for (7.9) is almost the same as that for (6.18). For $\Pi \in J_2^\circ$, let V_Π^ε be the set of P in J_2 for which

$$-\sum_{ij} |p(i,j) - \pi(i,j)| \log \pi_{ij} < \varepsilon$$

and

$$-\sum_{ij} |p(i,j) - \pi(i,j)| \log q_{ij} < \varepsilon$$

If $\Pi \in J_2^\circ$ and $F_n(x) \in V_\Pi^\varepsilon$, then

$$\frac{\log \mu_\Pi(u_n(x))}{\log \lambda_k(u_n(x))} \leq \frac{-n^{-1}\log \pi_{d_1}(x) - \sum_{ij}\pi(i,j)\log \pi_{i,j} + \varepsilon}{-n^{-1}\log q_{d_1}(x) - \sum_{ij}\pi(i,j)\log q_{i,j} - \varepsilon}.$$

By Theorem 5(ii)

$$\dim[x \in E : F_n(x) \in V_\Pi^\varepsilon \text{ i.o.}] \leq k\frac{-\sum_{ij}\pi(i,j)\log \pi_{ij} + \varepsilon}{-\sum_{ij}\pi(i,j)\log q_{ij} - \varepsilon}$$

for $\Pi \in J_2^\circ$. (The denominator on the right is positive if $0 < \varepsilon < \log q^{-1}$, where $q = \max_{ij} q_{ij}$.) As in the previous proof, for positive ε and δ, there exist $V_{\Pi(1)}^\varepsilon, \cdots, V_{\Pi(a)}^\varepsilon$, with $\Pi(\alpha) \in G^\delta \cap J_1^\circ$, and a positive η such that $G^\eta \subset \bigcup_{\alpha=1}^q V_{\Pi(\alpha)}^\varepsilon$. The rest of the proof is as in Section 6.

THEOREM 8. *Conditions* 1, 3 *and* 5 *together imply* (7.7)

Since $E = U(J_2)$, we have in analogy with (6.22),

$$\dim E = kh(J_2 : Q). \tag{7.10}$$

THEOREM 9. *The supremum $h(J_2 : Q)$ is the (unique) t for which the matrix $Q_t = [q_{ij}^t]$ has eigenvalue* 1 *and one of the corresponding right eigenvectors is positive:*

$$y_i = \sum_j q_{ij}^t y_i, \quad y_i > 0, \tag{7.11}$$

for some vector y.

PROOF. Suppose that (7.11) holds and put $\pi_{ij} = q_{ij}^t y_j / y_i$. Then $[\pi_{ij}]$ is a positive stochastic matrix and has a unique stationary vector $\pi = (\pi_1, \cdots, \pi_{r-1})$. Define $\pi(i,j) = \pi_i \pi_{ij}$ and $\Pi = [\pi(i,j)]$. Then $\Pi \in J_2^\circ$ and a calculation shows that $\sum_{ij} \pi_i \pi_{ij} \log \pi_{ij} = \sum_{ij} \pi_i \pi_{ij} \log q_{ij}^t$, so that $h(\Pi : Q) = t$. On the other hand, consider any P in J_2. In analogy with (6.24) we have ($\sum'$ denotes summation over those j for which $p_{ij} > 0$)

$$\sum_j p_{ij} \log \pi_{ij} - \sum_j p_{ij} \log p_{ij} \leq \log \Sigma' \pi_{ij} \leq 0.$$

Therefore, $-\sum_{ij} p_i p_{ij} \log p_{ij} \leq -\sum_{ij} p_i p_{ij} \log \pi_{ij}$. But from another calculation it follows that $-\sum_{ij} p_i p_{ij} \log \pi_{ij} = -\sum_{ij} p_i p_{ij} \log q_{ij}^t$, and hence $h(P : Q) \leq t = h(\Pi : Q)$. Thus if (7.11) holds, then $t = h(J_2 : Q)$. It follows of course that there is at most one such t.

We must still show that $t = h(J_2 : Q)$ does satisfy (7.11) for some y. For fixed $Q, h = h(P : Q)$ is continuous on J_2. We first show that the maximum is attained at a P in J_2°. Write $\phi(x) = -x \log x$ ($\phi(0) = 0$). Fix P. If

$$T = \sum_{ij} \phi(p(i,j)) - \sum_i \phi(p_i) \tag{7.12}$$

and

$$B = \sum_{ij} p(i,j) \log q_{ij}^{-1}, \tag{7.13}$$

then

$$h = T/B. \tag{7.14}$$

Let P' have components $p'(i,j) = (p(i,j)+s)/(1+r^2 s)$, where $0 < s < r^{-2}$ Then we have $P' \in J_2^\circ$, $p'_i = (p_i + rs)/(1 + r^2 s)$, and

$$\begin{aligned} h(P' : Q) = \Big\{ T + \textstyle\sum_{ij} [\phi(p(i,j) + s) - \phi(p(i,j))] \\ - \textstyle\sum_i [\phi(p_i + rs) - \phi(p_i)] \Big\} \Big/ \Big\{ B + s \textstyle\sum_{ij} \log q_{ij}^{-1} \Big\}. \end{aligned} \tag{7.15}$$

If $p_i = 0$, then $p(i,j) = 0$ for all j, and

$$\begin{aligned} \sum_j [\phi(p(i,j) + s) \quad -\phi(p(i,j))] - [\phi(p_i + rs) - \phi(p_i)] \\ = \quad rs \log r = O(s). \end{aligned}$$

If $p_i > 0$, then $\phi(p_i + rs) - \phi(p_i) = O(s)$ by the mean-value theorem. Similarly, $\phi(p(i,j) + s) - \phi(p(i,j))$ is $\phi(s)$ or $O(s)$ according as $p(i,j) = 0$ or $p(i,j) > 0$. Applying these facts to (7.15), we arrive at

$$h(P' : Q) = \frac{T + c\phi(s) + O(s)}{B + O(s)} \tag{7.16}$$

where c is the number of pairs i, j for which $p(i,j) = 0$ and $p_i > 0$.

Suppose now that $p(i,j) = 0$ for some i, j. If $p_i = 0$, then $p(l,i) = 0$ for all l, and for some l we must have $p_l > 0$. Thus there must exist a pair i, j for which $p(i,j) = 0$ and $p_i > 0$. Thus $c > 0$ in (7.16), and it follows that $h(P' : Q) > h(P : Q)$ for small enough s: There is an extreme point in the interior J_2°.

To simplify the notation, let the indices run from 1 to r instead of from 0 to $r-1$. Let $p(i,j)$ for $1 \le i < r$, $1 \le j \le r$ be positive variables with sum less than 1. Successively define the p_i and the other $p(i,j)$ by

$$\left.\begin{aligned} p_i &= \sum_{j=i}^{r} p(i,j), & 1 \le i < r; \\ p_r &= 1 - \textstyle\sum_{i=1}^{r-1} p_i; & \\ p(r,j) &= p_j - \textstyle\sum_{i=1}^{r-1} p(i,j), & 1 \le j < r; \\ p(r,r) &= p_r - \textstyle\sum_{i=1}^{r-1} p(i,r). & \end{aligned}\right\} \tag{7.17}$$

This parameterizes J_2°. Differentiate with respect to $p(\alpha,\beta)$ for $1 \le \alpha < r$, $1 \le \beta \le r$; the derivatives of the four functions in (7.17) are

$$\delta_{i\alpha}, \quad -1, \quad \delta_{j\alpha} - \delta_{j\beta}, -1 - \delta_{r\beta},$$

respectively. The derivatives of (7.12) and (7.13) are

$$\frac{\partial t}{\partial p(\alpha,\beta)} = \log\{p(r,\beta)p(r,r)p_\alpha / p(\alpha,\beta)p(r,\alpha)p_r\}$$

and

$$\frac{\partial B}{\partial p(\alpha,\beta)} = \log\{q_{r\beta}q_{rr}/q_{\alpha\beta}q_{r\alpha}\}$$

(for $\beta = r$ as well as for $\beta < r$). Now (7.14) has derivative

$$\frac{\partial h}{\partial p(\alpha,\beta)} = B^{-1}\left\{\frac{\partial T}{\partial p(\alpha,\beta)} - h\frac{\partial B}{\partial p(\alpha,\beta)}\right\}. \tag{7.18}$$

Suppose the derivatives (7.18) vanish at a point where $h = t$. It follows from the last three equations that

$$p(\alpha,\beta) = \frac{p(r,r)}{q_{rr}^t p_r}\,\frac{q_{r\alpha}^t p_\alpha}{p(r,\alpha)}\,q_{\alpha\beta}^t\,\frac{p(r,\beta)}{q_{r\beta}^t} \tag{7.19}$$

holds for $1 \le \alpha < r$, $1 \le \beta \le r$. But for $\alpha = r$, (7.19) is an algebraic identity. Thus the $p(i,j)$ have the form

$$p(i,j) = x_i q_{ij}^t y_i, \quad 1 \le i, j \le r, \tag{7.20}$$

for positive vectors x and y, where $p_i/x_i y_i = K = q_{rr}^t p_r / p(r,r)$. The constraints $\sum_i p_i = 1, p_i = \sum_j p(i,j)$, and $\sum_i p(i,j) = p_j$ give

$$K\sum_i x_i y_i = 1, \quad K y_i = \sum_j q_{ij}^t y_i, \quad \sum_i x_i q_{ij}^t = K x_j. \tag{7.21}$$

Thus there exists in J_2° a $P = [p(i,j)]$ of the form (7.20), where the constraints (7.21) hold and $t = h = T/B$. (This can of course also be derived by Lagrange multipliers, but the argument is not substantially simpler.)

From $T = tB$, together with (7.20) and (7.21), it follows that $K = 1$. But now the middle constraint in (7.21) is just that in (7.11). Thus there does exist a t such that (7.11) holds for some vector y. □

Perhaps this argument could be simplified by use of the Frobinius theory.

Although the t specified by Theorem 9 is unique, one can ask whether there can be several values of t for which $Q_t = [q_{ij}^t]$ has eigenvalue 1. This is the same as asking whether $F(t) = \det[Q_t - I]$ has more than one zero. Since $0 < q_{ij} < 1$, it is easy to see that F is strictly increasing if $r = 2$; since $F(t)$ increases from -1 to +1 as t runs from 0 to ∞, F has a unique zero in this case.

But suppose that $r = 4$ and

$$Q = \begin{bmatrix} Q' & E \\ E & Q'' \end{bmatrix},$$

where Q' and Q'' are 2 by 2 matrices and all four entries of E are ε. If F' and F'' are the functions corresponding to Q' and Q'', then $F(t) = F'(t)F''(t) + G(t)$ where $G(t)$ is the sum of the 20 terms of the expansion of $\det[Q_t - I]$ containing at least one factor from E. But for $t \geq 0$ and $0 < \varepsilon < 1$, $|G(t)| \leq 20\varepsilon$. Choose Q' and Q'' so that F' and F'' have distinct zeros; then there are points t_1, t_2, and t_3 for which the values of $F'(t)F''(t)$ are positive, negative, and positive, respectively, and for which $0 < t_1 < t_2 < t_3$. For small enough ε, the values of $F(t)$ at t_1, t_2, and t_3 also alternate in sign, so that $F(t)$ has two distinct zeros. (This argument was shown me by William Beckner.)

EXAMPLE 9. Let $\mathcal{O}_1$ be the operation of removing from an equilateral triangle ($k = 2$) the center equilateral triangle, as in Figure 14. Let $\mathcal{O}_2$ be the operation of removing the center regular hexagon, as in Figure 15. Each operation leaves three congruent equilateral triangles. Apply $\mathcal{O}_1$ to an equilateral triangle. To the bottom two remaining triangles apply $\mathcal{O}_1$, and to the other one apply $\mathcal{O}_2$. Continue, as in Figure 16.

The limit E is cross-similar, and the matrix Q is

$$Q = \begin{bmatrix} \frac{1}{9} & \frac{1}{9} & \frac{1}{9} \\ \frac{1}{4} & \frac{1}{4} & \frac{1}{4} \\ \frac{1}{4} & \frac{1}{4} & \frac{1}{4} \end{bmatrix}$$

Since $\det\,[Q_t - I] = 9^{-t} + 2.4^{-t} - 1$, the zero is in this case unique. A calculation shows that dim E as given by Theorems 8 and 9 is between 1.3 and 1.4. One could derive a result of the Eggleston type.

Moran's theorem on self-similar sets (end of Section 6) doubtless has an extension to cross-similar sets.

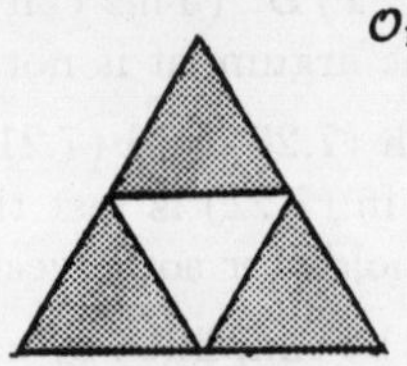

Figure 14

8 Appendix

Suppose that $\alpha_n \leq \mu(v)$ and $|v| \leq \beta_n$ for $v \in \mathcal{A}_n$, where

$$\alpha_n > 0, \quad \beta_n \to 0. \tag{8.1}$$

This follows from Condition 1 together with either of Conditions 4 and 5. We will show that (8.1) *implies that the outer measure $\mu^s(\cdot)$ satisfies Carathéodory's condition*: If A and B are at positive distance, then

$$\mu^s(A \cup B) = \mu^s(A) + \mu^s(B). \tag{8.2}$$

From this it follows (see, for example, Theorems 11.1 and 11.5 of [4]) that *μ^s $(\cdot)$ is countably additive on the class $\mathcal{S}$ of Borel sets in S.*

Because of subadditivity, in proving (8.2) we need only show that the right side is at most the left, and we can assume that $\mu^s(A \cup B, \varepsilon) \leq \mu^s(A \cup B) < K < \infty$. Choose n so that $\beta_m <$ dist (A, B) for $m \geq n$, and then choose ε so that $\varepsilon < \alpha_m$ for $m \leq n$. Let $\{v_i\}$ be a μ-ε-covering of $A \cup B$ for which $\sum_i \mu(v_i)^{s/k} < K$. Since $\mu(v_i) \leq \varepsilon, v_i$ cannot lie in $\mathcal{A}_m$ for any $m \leq n$, hence must lie in $\mathcal{A}_m$ for some $m > n$, and hence must satisfy $|v_i| < \text{dist}(A, B)$. Thus v_i cannot meet both A and B. If $U\,[V]$ consists of the indices i for which v_i meets $A\,[B]$, then

$$\mu^s(A, \varepsilon) + \mu^s(B, \varepsilon) \leq \sum_{i \in U} \mu(v_i)^{s/k} + \sum_{i \in V} \mu(v_i)^{s/k} \leq \sum_i \mu(v_i)^{s/k} < K.$$

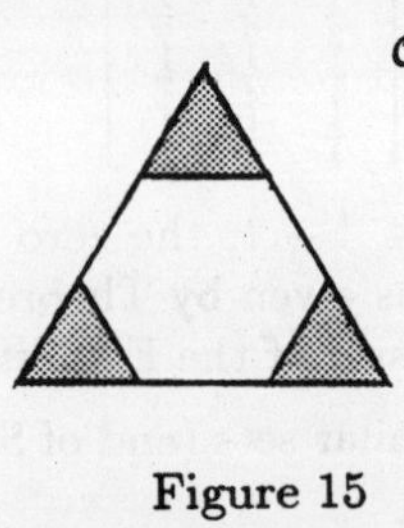

Figure 15

etc.

Figure 16

Let $\varepsilon \to 0$ and then $K \to \mu^s(A \cup B)$: (8.2) follows.

This argument is essentially the same as the proof (p. 248 of [4]) that $H^s(\cdot)$ also satisfies Carathéodory's condition.

Suppose now that $\mu^s(\cdot)$ and $\nu^s(\cdot)$ *are* countably additive on $\mathcal{S}$ for each s. Strengthen (5.18) to

$$\exists N \forall n \geq N : \mu(u_n(x)) \leq \nu(u_n(x))^\rho, \tag{8.3}$$

where $\rho > 0$. *If* (8.3) *holds for all* $x \in A (A \subset E)$, *then*

$$\mu^{t/\rho}(A) \leq \nu^t(A) \tag{8.4}$$

for all t. Of course (8.4) implies (5.20): $\dim_\nu A \geq \rho \dim_\mu A$.

To prove this, assume first that $\nu(v) > 0$ for every $\mathcal{A}$-set v that meets A. Let A_ε be the set of x in A such that for every n, either $\nu(u_n(x)) \geq \varepsilon$ or $\mu(u_n(x)) \leq \nu(u_n(x))^\rho$. It follows as in the proof of Theorem 2 that $A_\varepsilon \uparrow A$ as $\varepsilon \downarrow 0$, and so by countable additivity, (8.4) will follow if

$$\mu^{t/\rho}(A_\varepsilon) \leq \nu^t(A_\varepsilon). \tag{8.5}$$

Suppose that $\nu^t(A_\varepsilon, \eta) \leq \nu^t(A_\varepsilon) < K < \infty$, where $0 < \eta < \varepsilon$. Choose a ν-η-covering $\{v_i\}$ of A_ε such that $\sum_i \nu(v_i)^{t/k} < K$ and each v_i meets A_ε. Then $\mu(v_i)^{1/\rho} \leq \nu(v_i)$ for all i, and hence $\mu^{t/p}(A_\varepsilon, \eta^\rho) \leq \sum_i \mu(v_i)^{t/\rho k} \leq \sum_i \nu(v_i)^{t/k} < K$. Let $\eta \to 0$ and then $K \to \nu^t(A_\varepsilon)$: (8.5) follows.

Define F_ν as the union of all $\mathcal{A}$-sets of ν-measure 0. Then $\nu^s(F_\nu) = 0$ for $s > 0$, and from (8.3) it follows as in the proof of Theorem 2 that $A \cap F_\nu \subset A \cap F_\mu$. By monotonicity and subadditivity,

$$\mu^{t/\rho}(A) = \mu^{t/\rho}(A \cap F_\mu^c) \leq \mu^{t/p}(A \cap F_\nu^c) \leq \nu^t(A \cap F_\nu^c) \leq \nu^t(A),$$

where the next-to-last inequality follows by the case of (8.4) already covered. This proves (8.4) in general.

Next, strengthen (5.24) to

$$\forall N \exists n \geq N : \nu(u_n(x))^\rho \leq \mu(u_n(x)), \tag{8.6}$$

where $\rho > 0$. *If* (8.6)*holds and* $\nu(u_n(x)) \to 0$ *for all* $x \in A (A \subset E)$, *then*

$$\nu^{\rho k}(A) \leq \mu(S). \tag{8.7}$$

If $\mu(S) < \infty$, then of course (8.7) implies (5.25): $\dim_\nu A \leq \rho k$.

To prove this, consider for given ε the sets $u_n(x)$ for which $x \in A$, $\nu(u_n(x)) < \varepsilon$, and $\nu(u_n(x))^\rho \leq \mu(u_n(x))$. It follows as in the proof of Theorem 4 that these sets cover A: we have for A a ν-ε-cover for which $\nu(v_i)^\rho \leq \mu(v_i)$. As before, discard superfluous v_i, so that the remaining ones are disjoint. Then

$$\mu(S) \geq \sum_i \mu(v_i)^{k/k} \geq \sum_i \nu(v_i)^{\rho k/k} \geq \nu^{\rho k}(A, \varepsilon).$$

Let $\varepsilon \to 0$: (8.7) follows.

EXAMPLE 2 *(continued)*. Here $k = 1$ and E is the Cantor set. Let $\mu = \mu_p$ be Cantor measure, the measure (Lemma 2) such that $\mu(u_n(x)) = 1/2^n$. Since $\lambda_1(u_n(x)) = 1/3^n$, if $s = \log 2/\log 3$, then

$$\lambda_1(u_n(x))^s = \mu(u_n(x)) \tag{8.8}$$

for all $x \in E$. Take $A = E$, $\nu = \lambda_1$, and $t = \rho = s$ in (8.4): $\mu(E) \le \lambda_1^s(E)$. Take $A = E, \nu = \lambda_1, k = 1$, and $\rho = s$ in (8.6): $\lambda_1^s(E) \le \mu(S)$. Since $\mu(S) = \mu(E) = 1$, we arrive at $\lambda_1^s(E) = 1$. Thus E is an s-set (see the remark following the proof of Theorem 1). In particular $\dim E = s = \log 2/\log 3$.

This is one way of seeing that the Cantor set must be an s-set (the same argument works for the Sierpinsky set). Now (8.7) can be written $\log \mu(u_n(x))/\log \lambda_1(u_n(x)) = s$, so that in this case (6.12) can be strengthened to

$$\frac{\log \mu_\rho(u_n(x))}{\log \lambda_1(u_n(x))} = h(p:q). \tag{8.9}$$

In the case of the Eggleston set (Example 1), however, (6.12) cannot be strengthened to (8.8); this is one way of seeing why the Eggleston set is not an s-set.

□

Here is the example referred to in the comment following the proof of Theorem 4.

EXAMPLE 10. Put the positive integers into successive blocks $B_1, B_2, \cdots$ of lengths $b_1, b_2, \cdots$, where

$$(b_1 + \cdots + b_{j-1})/b_j \to 0. \tag{8.10}$$

Let ν be the measure on [0,1] under which the dyadic digits $d_1(x), d_2(x), \cdots$ are independent and

$$\nu[x : d_n(x) = 0] = p_n = \begin{cases} 1/2 & \text{for } n \in B_j, \quad j \text{ odd,} \\ 1/n^2 & \text{for } n \in B_j, \quad j \text{ even} \end{cases} \tag{8.11}$$

Since $p_n = 1/2$ for infinitely many n, ν has no points of positive mass. If j is odd and $n = b_1 + \cdots + b_j$, then $\nu(u_n(x)) \le 1/2^{b_j}$ for all x. If μ is Lebesgue measure, then by (8.9),

$$\liminf_n \frac{\log \mu(u_n(x))}{\log \nu(u_n(x))} \le 1 \tag{8.12}$$

for all x.

If A is the x-set where $d_n(x) = 1$ for all but finitely many of the n in $\bigcup_i B_{2i}$, then (Borel-Cantelli) $\nu(A) = 1$ and $\dim_\nu A = 1$. If $x \in A, n =$

$b_1 + \cdots + b_j$, and j is even and large enough, then $N_n(x,1) \geq b_j$. By (8.10), $\limsup_n f_n(x,1) = 1$ on A, so that by (6.16) applied to Example 2, $\dim_\mu A = \dim A = 0$. Thus: (8.12) holds for all x but $\dim_\nu A = 1$ and $\dim_m A = 0$. □

REFERENCES

[1] Patrick Billingsley, "Hausdorff dimension in probability theory," *Ill. J. Math.* **4** (1960), 187-209.

[2] ———, "Hausdorff dimension in probability theory II," *Ill. J. Math* **5** (1961), 594-601.

[3] ———, *Ergodic Theory and Information,* Wiley, 1965.

[4] ———, *Probability and Measure,* second edition, Wiley, 1986.

[5] Patrick Billingsley and Inge Henningsen, "Hausdorff dimension of some continued-fraction sets," *Z. Wahrscheinlichkeitstheorie verw. Gebiete* **31** (1975), 163-173.

[6] Helmut Cajer, *Billingsley Dimension in Probability Spaces,* Volume 892 of *Lecture Notes in Mathematics,* Springer, 1981.

[7] Coleen D. Cutler, "The Hausdorff dimension distribution of finite measures in Euclidean space," *Can. J. Math.* **38**(1986), 1459-1484.

[8] K.J. Falconer, *The Geometry of Fractal Sets,* Cambridge University Press, 1985.

[9] Herbert Federer, *Geometric Measure Theory,* Springer, 1969.

[10] Richard Kenyon and Yuval Peres, "Intersecting random translates of invariant Cantor sets," preprint.

[11] Benoit B. Mandelbrot, *The Fractal Geometry of Nature,* Freeman, 1983.

[12] P.A.P. Moran, "Additive functions of intervals and Hausdorff measure," *Proc. Camb. Phil. Soc.* **42** (1946), 15-23.

[13] Walter Rudin, *Real and Complex Analysis,* second edition, McGraw-Hill, 1974.

Department of Statistics
The University of Chicago
5734 University Avenue
Chicago. Illinois 60637
U.S.A.

J.K. Ghosh, S.K. Mitra, K.R. Parthasarathy and B.L.S. Prakasa Rao (*eds.*),
Statistics and Probability : A Raghu Raj Bahadur Festschrift
Published by Wiley Eastern Limited, Publishers, (1993), 135 – 145

Sharp Probability Bounds for Itô Processes

DONALD L. BURKHOLDER
University of Illinois, Urbana, Illinois

Abstract

Let X and Y be Itô processes such that the initial value and predictable integrands of Y are majorized in absolute value by those of X. A sharp bound on the distribution of the maximal function of Y is obtained under the additional condition that X is a nonnegative local submartingale. If X is a non negative supermartingale, a similar inequality holds but with a different best constant.

Let X and Y be Itô processes defined by

$$X_t = X_0 + \int_0^t \varphi_s \, dB_s + \int_0^t \psi_s \, ds, \tag{1}$$

$$Y_t = Y_0 + \int_0^t \tilde{\varphi}_s \, dB_s + \int_0^t \tilde{\psi}_s \, ds. \tag{2}$$

AMS Subject Classifications. Primary 60H05, 60G48; secondary 60J60.
Key words and Phrases. Stochastic integral, martingale, submartingale, supermartingale, maximal inequality.

The underlying probability space $(\Omega, \mathcal{F}, P)$ is complete and is filtered by an increasing right-continuous family $(\mathcal{F}_t)_{t\geq 0}$ of sub-σ-fields of $\mathcal{F}$ where $\mathcal{F}_0$ contains all $A \in \mathcal{F}$ with $P(A) = 0$. The real Brownian motion B starts at 0, is adapted to $(\mathcal{F}_t)_{t\geq 0}$, and the process $(B_{t+s} - B_s)_{t\geq 0}$ is independent of $\mathcal{F}_s$ for all $s \geq 0$. The real processes φ and ψ are predictable,

$$P\left(\int_0^t \varphi_s^2\, ds < \infty \text{ and } \int_0^t |\psi_s|\, ds < \infty \text{ for all } t > 0\right) = 1, \tag{3}$$

and $\tilde{\varphi}$ and $\tilde{\psi}$ satisfy the same conditions. Both X and Y are adapted and can be taken to be everywhere continuous. For further background on Itô processes, see, for example, Ikeda and Watanabe [7].

Assume throughout the paper that

$$|Y_0| \leq |X_0|, \tag{4}$$
$$|\tilde{\varphi}| \leq |\varphi|, \tag{5}$$
$$|\tilde{\varphi}| \leq |\psi|, \tag{6}$$

and, for simplicity, that X_0 is constant.

NOTATION. Let $\|X\|_1 = \sup_{t\geq 0} \|X_t\|_1$ and $|||X|||_1 = \sup \|X_\tau\|_1$ where the latter supremum is taken over all bounded stopping times τ of $(\mathcal{F}_t)_{t\geq 0}$. Of course, $\|X\|_1 \leq |||X|||_1$ and Doob's optional sampling theorem implies that if X is a martingale or a nonnegative submartingale, then equality holds. The Itô process X is a local martingale if $\psi = 0$ and a nonnegative local submartingale if $X \geq 0$ and $\psi \geq 0$. Strict inequality can occur for such X as in Examples 3 and 4 below. However, if $X \geq 0 \geq \psi$, then X is a nonnegative supermartingale and, in this case,

$$|||X|||_1 = \|X\|_1 = E\,X_0.$$

Throughout the paper Y^* denotes the maximal function of Y:

$$Y^*(\omega) = \sup_{t\geq 0} |Y_t(\omega)|.$$

THEOREM 1 *Let X and Y be Itô processes as above. If (i) $\psi = 0$ or if (ii) $X \geq 0 \geq \psi$, then, for all $\lambda > 0$,*

$$\lambda\, P(Y^* \geq \lambda) \leq 2|||X|||_1 \tag{7}$$

and, for both (i) and (ii), the constant 2 is best possible. If (iii) $X \geq 0$ and $\psi \geq 0$, then

$$\lambda\, P(Y^* \geq \lambda) \leq 3|||X|||_1 \tag{8}$$

and the constant 3 is best possible.

As can be seen from the proof, the left-hand side of (7) and of (8) can be replaced by

$$\lambda P\left(\sup_{t\geq 0}(|X_t|+|Y_t|)\geq\lambda\right). \tag{9}$$

If $|||X|||_1$ is replaced by $\|X\|_1$ in (7) then, under condition (i), the inequality need not hold. The same is true for (8) under condition (iii). See Examples 3 and 4.

That 2 is the best constant in (7) under condition (i) is proved in [1] but our method there does not give (7) with the left-hand side replaced by (9) as does our method here. That 2 and 3 are possible choices and, in fact, are the best possible choices for the constants under conditions (ii) and (iii), respectively, is the main contribution here.

For sharp probability bounds and norm inequalities for semimartingales X and Y that are not necessarily continuous, see [5]. We study here the important special case of Itô processes as a first step. Their continuity leads to simpler proofs that are of interest in their own right.

LEMMA 1 *Let M and N be nonnegative continuous processes adapted to $(\mathcal{F}_t)_{t\geq 0}$ such that $N_0 \leq M_0$ where M_0 is constant, and*

$$E\,N_\tau^2 \leq E\,M_\tau^2 \tag{10}$$

for all bounded stopping times τ. Then, for all $\lambda > 0$,

$$\lambda P\left(\sup_{t\geq 0}(M_t+N_t)\geq\lambda\right)\leq 2|||M|||_1. \tag{11}$$

PROOF: It is enough to prove that $\lambda P(\sup_{t\geq 0}(M_t+N_t)>\lambda)\leq 2|||M|||_1$. We can assume that $\lambda=1$ and $2|||M|||_1<1$ so $M_0+N_0\leq 2M_0<1$.

Define the stopping time μ by

$$\mu=\inf\{t\geq 0 : M_t+N_t>1\}$$

and let $\tau=\mu\wedge t$. The desired inequality follows from

$$P\left(\sup_{0\leq s\leq t}(M_s+N_s)>1\right)\leq P(M_\tau+N_\tau=1)\leq 2|||M|||_1 \tag{12}$$

where

$$\begin{aligned} P(M_\tau+N_\tau=1) &\leq P\left(2M_\tau-M_\tau^2+N_\tau^2=1\right)\\ &\leq E\left[2M_\tau-M_\tau^2+N_\tau^2\right] \end{aligned}$$

in which we have used the nonnegativity of the integrand. Inequality (12) now follows from (10).

PROOF OF THEOREM 1: (i) Under this condition ($\psi = 0$), the proof of

$$\lambda P\left(\sup_{t\geq 0}(|X_t| + |Y_t|) \geq \lambda\right) \leq 2|||X|||_1 \tag{13}$$

follows from Lemma 1. The first step is to reduce the proof of the equivalent inequality $\lambda P(\sup_{t\geq 0}(|X_t| + |Y_t|) > \lambda) \leq 2|||X|||_1$ to the case in which X and Y are L^2-bounded martingales. Let k be a positive integer and

$$\nu_k = \inf\left\{t > 0 : \int_0^t \varphi_s^2 \, ds > k\right\}. \tag{14}$$

By (3), $\lim_{k\to\infty} \nu_k = \infty$ a.s. Using this and the definition of $|||X|||_1$, we can assume that X and Y are the same as the original X and Y stopped at ν_k. Then $M_t = |X_t|$ and $N_t = |Y_t|$ define submartingales M and N satisfying the conditions of Lemma 1. By (4) and (5), the L^2-bounded martingales X and Y satisfy

$$\begin{aligned} EY_\tau^2 &= EY_0^2 + E\int_0^\tau \tilde{\varphi}_s^2 \, ds \\ &\leq EX_0^2 + E\int_0^\tau \varphi_s^2 \, ds \\ &= EX_\tau^2 \end{aligned}$$

for all bounded stopping times τ and this gives (10).

EXAMPLE 1. To see that 2 is best in (7), let $\tau = \inf\{t > 0 : B_t = -1\}$ and $\varphi_s = -\tilde{\varphi}_s = 1(\tau \geq s)$, the indicator function of $\{\tau \geq s\}$. Also, let $X_0 = Y_0 = 1$ and $\psi = \tilde{\psi} = 0$. Then the processes X and Y given by (1) and (2) satisfy

$$\begin{aligned} X_t &= 1 + B_{\tau\wedge t}, \\ Y_t &= 1 - B_{\tau\wedge t}. \end{aligned}$$

Note that X is a nonnegative martingale starting at 1 so $|||X|||_1 = \|X\|_1 = 1$. But almost surely $Y_t = 2$ for all large t, so $P(Y^* \geq 2) = 1$ and, for this example, equality holds in (7) for $\lambda = 2$. Therefore, 2 is the best constant in (7).

(ii) To prove this part of Theorem 1, we can assume, as in the proof of Lemma 1, that $\lambda = 1$ and $2|||X|||_1 < 1$ so $X_0 + |Y_0| \leq 2X_0 < 1$. Let

$$S = \{(x, y) : x \geq 0 \text{ and } x + |y| \leq 1\}.$$

The continuous process (X, Y) starts in S and can be stopped at the boundary. In fact, let

$$\mu = \inf\{t > 0 : X_t + |Y_t| > 1\} \tag{15}$$

and $\tau = \mu \wedge \nu_k \wedge t$ where ν_k is the stopping time defined in (14). Note that $(X_s, Y_s) \in S$ for all $s \leq \tau$ so

$$P\left(\sup_{0\leq s\leq t}(|X_s|+|Y_s|>1\right)\leq \lim_{k\to\infty} P(X_\tau+|Y_\tau|=1).$$

Therefore the proof of

$$\lambda P\left(\sup_{t\geq 0}(|X_t|+|Y_t|)\geq\lambda\right)\leq 2EX_0 \tag{16}$$

under condition (ii) will be complete once we show that

$$P(X_\tau+|Y_\tau|=1)\leq 2EX_0 \tag{17}$$

for all $k \geq 1$ and $t > 0$. This inequality is equivalent to

$$Ev(X_\tau, Y_\tau)\leq 2EX_0 \tag{18}$$

where $v(x,y) = 1$ if $x + |y| = 1$ and $v(x,y) = 0$ elsewhere in S. To prove (18), we shall show that there is a majorant u of v on S such that

$$Eu(X_\tau, Y_\tau)\leq Eu(X_0,Y_0)\leq 2EX_0 \tag{19}$$

It is easy to check that the function u defined by

$$u(x,y)=2x-x^2+y^2 \tag{20}$$

majorizes v on S. The right-hand side of (19) follows from (4). To prove the left-hand, note first that

$$\begin{aligned} u_x(x,y) &= 2-2x,\ u_y(x,y)=2y, \\ u_{xx}(x,y) &= -2,\ u_{xy}(x,y)=0,\ u_{yy}(x,y)=2. \end{aligned}$$

Then, using Itô's formula and writing u_x for $u_x(X_s, Y_s)$, $\cdots$, we obtain

$$\begin{aligned} u(X_\tau,Y_\tau) &= u(X_0,Y_0)+\int_0^\tau [u_x\varphi_s+u_y\tilde{\varphi}_s]\,dB_s \\ &\quad +\int_0^\tau \left[u_x\psi_s+u_y\tilde{\psi}_s\right] ds \\ &\quad +\frac{1}{2}\int_0^\tau [u_{xx}\varphi_s^2+2u_{xy}\varphi_s\tilde{\varphi}_s+u_{yy}\tilde{\varphi}_s^2]\,ds \\ &= u(X_0,Y_0)+\int_0^\tau [(2-2X_s)\varphi_s+2Y_s\tilde{\varphi}_s]\,dB_s \\ &\quad +\int_0^\tau \left[(2-2X_s)\psi_s+2Y_s\tilde{\psi}_s\right] ds+\int_0^\tau [\tilde{\varphi}_s^2-\varphi_s^2]\,ds. \end{aligned}$$

By (5), (6), and the definition of S, the integrand in each of the last two integrals is nonpositive. The integrand in the third from the last integral is bounded by $4|\varphi_s|$. Therefore, by (14), the integral is square-integrable and has expectation 0. This gives the left side of (19) and completes the proof of (16), hence (7), under the condition (ii).

The process X in Example 1 satisfies not only (i) but also (ii), so 2 is also the best constant in (7) under the latter condition.

(iii) To prove (8), we can assume that $X \geq \alpha$ where $\alpha \in (0,1)$: If we can prove the inequality for the pair $(X+\alpha, Y)$, the inequality will follow for (X,Y) by letting α decrease to 0. We can also assume that $\lambda = 1$ and $3|||X|||_1 < 1$ so $X_0 + |Y_0| \leq 2X_0 < 1$ Let

$$S_\alpha = \{(x,y) : x \geq \alpha \text{ and } x + |y| \leq 1\}.$$

With ν_k as in (14) and μ as in (15), define $\tau = \mu \wedge \nu_k \wedge t$ as in (ii). Here $(X_s, Y_s) \in S_\alpha$ for $s \leq \tau$ and the inequality

$$\lambda P\left(\sup_{t\geq 0}(|X_t| + |Y_t|) \geq \lambda\right) \leq 3|||X|||_1 \tag{21}$$

will follow once we show that

$$P(X_\tau + |Y_\tau| = 1) \leq 3EX_\tau \tag{22}$$

for all $k \geq 1$ and $t > 0$. Since $\alpha \leq X_\tau \leq 1$, the expectation EX_τ is finite, so it is enough to show that

$$Ev(X_\tau, Y_\tau) \leq 0 \tag{23}$$

where $v(x,y) = 1 - 3x$ if $x + |y| = 1$ and $v(x,y) = -3x$ elsewhere in S_α. We shall show that there is a majorant u of v on S_α such that

$$Eu(X_\tau, Y_\tau) \leq Eu(X_0, Y_0) \leq 0 \tag{24}$$

It is easy to check that the function u defined by

$$u(x,y) = (|y| - 2x)\sqrt{x + |y|} \tag{25}$$

is twice continuously differentiable on the open right-half plane and majorizes v on S_α. The right-hand side of (24) follows from (4) and the positivity of X. We shall now prove the left-hand side. Note that

$$\begin{aligned}
u_x(x,y) &= -\frac{6x + 3|y|}{2\sqrt{x+|y|}}, \quad u_y(x,y) = \frac{3y}{2\sqrt{x+|y|}}, \\
u_{xx}(x,y) &= -\frac{6x + 9|y|}{4(x+|y|)^{3/2}}, \\
u_{xy}(x,y) &= -\frac{3y}{4(x+|y|)^{3/2}}, \quad u_{yy}(x,y) = \frac{6x + 3|y|}{4(x+|y|)^{3/2}}.
\end{aligned}$$

On S_α, these five derivatives are bounded and

$$|u_y| \leq -u_x \text{ and } 2|u_{xy}| + |u_{yy}| = -u_{xx}. \tag{26}$$

Again writing u_x for $u_x(X_s, Y_s), \cdots$ and using Itô's formula, we obtain

$$\begin{aligned} u(X_\tau, Y_\tau) &= u(X_0, Y_0) + \int_0^\tau [u_x\varphi_s + u_y\tilde{\varphi}_s]\, dB_s \\ &\quad + \int_0^\tau \left[u_x\psi_s + u_y\tilde{\psi}_s\right] ds \\ &\quad + \frac{1}{2}\int_0^\tau \left[u_{xx}\varphi_s^2 + 2u_{xy}\varphi_s\tilde{\varphi}_s + u_{yy}\tilde{\varphi}_s^2\right] ds. \end{aligned}$$

By (5), (6) and (26), the integrands in the second and third integrals are nonpositive. The integrand of the first integral is bounded by a constant multiple of $|\varphi_s|$. Therefore, by (14), the integral is square-integrable and has expectation 0. This gives the left-hand side of (24) and finishes the proof of (21), hence (8), under condition (iii).

We shall complete the proof of Theorem 1 by constructing an example showing that the constant 3 in (8) is best possible.

EXAMPLE 2. Let N be a positive integer. We shall define a pair of Itô processes X^N and Y^N satisfying condition (iii) such that

$$P\left((Y^N)^* \geq 1\right) = 1 \text{ and } |||X^N|||_1 = 1/3 + 1/6N. \tag{27}$$

If $\beta < 3$, then $P((Y^N)^* \geq 1) > \beta|||X^N|||_1$ for all large N showing that 3 is the best possible constant in (8).

Our first step is to construct a pair of Itô processes X and Y that we shall use in the definition of X^N and Y^N. We shall need an increasing sequence of stopping times: $\tau_0 = 0$, $\tau_1 = 1$, $\tau_2 = 2$,

$$\tau_3 = \inf\{t > 2 : B_t - B_2 \notin (-1, 1)\},$$

and, for $n \geq 1$,

$$\begin{aligned} \tau_{3n+1} &= \inf\{t > \tau_{3n} : B_t - B_{\tau_{3n}} \notin (-2n, 1)\}, \\ \tau_{3n+2} &= \tau_{3n+1} + 1, \\ \tau_{3n+3} &= \inf\left\{t > \tau_{3n+2} : B_t - B_{\tau_{3n+2}} \notin (-1, 2n+1)\right\}. \end{aligned}$$

If $n \geq 0$, then $E\tau_n$ is finite and

$$\begin{aligned} P(B_{\tau_{3n+3}} - B_{\tau_{3n+2}} = 2n+1) &= 1/(2n+2). \\ P(B_{\tau_{3n+3}} - B_{\tau_{3n+2}} = -1) &= (2n+1)/(2n+2). \end{aligned}$$

If $n \geq 1$, then

$$\begin{aligned} P(B_{\tau_{3n+1}} - B_{\tau_{3n}} = 1) &= 2n/(2n+1), \\ P(B_{\tau_{3n+1}} - B_{\tau_{3n}} = -2n) &= 1/(2n+1). \end{aligned}$$

Set $X_0 = Y_0 = 0$. If $0 < s \leq 1$, let $\varphi_s = \tilde{\varphi}_s = \psi_s = \tilde{\psi} = 0$. For $1 < s \leq 2$, let $\varphi_s = \tilde{\varphi}_s = 0$ and

$$\begin{aligned} \psi_s &= 1, \\ \tilde{\psi}_s &= \operatorname{sgn}(B_{1/2} - B_{1/3}). \end{aligned}$$

Here we set $\operatorname{sgn}(x) = -1$ for negative x and $+1$ for nonnegative x. If $2 < s \leq \tau_3$, let $\psi_s = \tilde{\psi}_s = 0$ and

$$\begin{aligned} \varphi_s &= 1(Y_2 = 1) + 1(Y_2 = -1), \\ \tilde{\varphi}_s &= -1(Y_2 = 1) + 1(Y_2 = -1). \end{aligned}$$

On the random interval $(0, \tau_3]$, both ψ and X are nonnegative. Furthermore,

$$\begin{aligned} P(X_{\tau_3} = 0,\ Y_{\tau_3} = 2) &= 1/4, \\ P(X_{\tau_3} = 2,\ Y_{\tau_3} = 0) &= 1/2, \\ P(X_{\tau_3} = 0,\ Y_{\tau_3} = -2) &= 1/4. \end{aligned}$$

Now let $n \geq 1$. If $\tau_{3n} < s \leq \tau_{3n+1}$, let $\psi_s = \tilde{\psi}_s = 0$ and

$$\begin{aligned} \varphi_s &= 1(X_{\tau_{3n}} = 2n,\ Y_{\tau_{3n}} = 0), \\ \tilde{\varphi}_s &= \operatorname{sgn}(B_{1/3n} - B_{1/(3n+1)})1(X_{\tau_{3n}} = 2n,\ Y_{\tau_{3n}} = 0). \end{aligned}$$

If $\tau_{3n+1} < s \leq \tau_{3n+2}$, let $\varphi_s = \tilde{\varphi}_s = 0$ and

$$\begin{aligned} \psi_2 &= 1(X_{\tau_{3n+1}} = 0, Y_{\tau_{3n+1}} = 2n) + 1(X_{\tau_{3n+1}} = 0, Y_{\tau_{3n+1}} = -2n), \\ \tilde{\psi}_2 &= 1(X_{\tau_{3n+1}} = 0, Y_{\tau_{3n+1}} = 2n) - 1(X_{\tau_{3n+1}} = 0, Y_{\tau_{3n+1}} = -2n). \end{aligned}$$

Finally, if $\tau_{3n+2} < s \leq \tau_{3n+3}$, let $\psi_s = \tilde{\psi}_s = 0$ and

$$\begin{aligned} \varphi_s &= 1(X_{\tau_{3n+2}} = 1, Y_{\tau_{3n+2}} = 2n+1) \\ &\quad +1(X_{\tau_{3n+2}} = 1, Y_{\tau_{3n+2}} = -2n-1) \\ &\quad -1(X_{\tau_{3n+2}} = 2n+1, Y_{\tau_{3n+2}} = 1) \\ &\quad -1(X_{\tau_{3n+2}} = 2n+1, Y_{\tau_{3n+2}} = -1), \\ \tilde{\varphi}_s &= -1(X_{\tau_{3n+2}} = 1, Y_{\tau_{3n+2}} = 2n+1) \\ &\quad +1(X_{\tau_{3n+2}} = 1, Y_{\tau_{3n+2}} = -2n-1) \\ &\quad +1(X_{\tau_{3n+2}} = 2n+1, Y_{\tau_{3n+2}} = 1) \\ &\quad -1(X_{\tau_{3n+2}} = 2n+1, Y_{\tau_{3n+2}} = -1). \end{aligned}$$

This completes the definition of X and Y. By induction, for $n \geq 1$,

$$\begin{aligned} P(X_{\tau_{3n}} = 0, Y_{\tau_{3n}} = 2n) &= 1/3 - 1/12n, \\ P(X_{\tau_{3n}} = 2n, Y_{\tau_{3n}} = 0) &= 1/3 + 1/6n, \\ P(X_{\tau_{3n}} = 0, Y_{\tau_{3n}} = -2n) &= 1/3 - 1/12n. \end{aligned} \tag{28}$$

Let N be a positive integer and $\tau = \inf\{t > \tau_{3N} : B_t - B_{\tau_{3N}} \notin (-1,1)\}$. On $[0, \tau_{3N}]$, let $X^N = X/2N$ and $Y^N = Y/2N$. If $\tau_{3N} < s \le \tau$, let $\psi_s^N = 0$, $\tilde{\psi}_s^N = 0$ and

$$\varphi_s^N = \tilde{\varphi}_s^N = 1(X_{\tau_{3N}}^N = 1,\ Y_{\tau_{3N}}^N = 0).$$

If $s > \tau$, let $\varphi_s^N = \tilde{\varphi}_s^N = \psi_s^N = \tilde{\psi}_s^N = 0$. This completes the definition of X^N and Y^N. It is easy to check that these are uniformly bounded Itô processes satisfying condition (iii) as well as (4), (5) and (6). In fact, the finiteness of $E\tau$ implies that X^N is a nonnegative submartingale. Also,

$$EX_\tau^N - EX_{\tau_{3N}}^N = E\int_{\tau_{3N}}^{\tau} \varphi_s^N \, dB_s = 0$$

so X^N satisfies

$$|||X^N|||_1 = \|X^N\|_1 = EX_\tau^N = EX_{\tau_{3N}}^N.$$

By (28), the latter expectation is $1/3 + 1/6N$ and this gives the right-hand side of (27). The left-hand side follows from the fact that $P(|Y_\tau^N| = 1) = 1$.

EXAMPLE 3. There is an Itô process X with $\psi = 0$ such that $\|X\|_1 < \infty$ but

$$\sup_{\lambda>0} \lambda P(X^* \ge \lambda) = \infty. \tag{29}$$

This shows that, even with $Y = X$, (7) need not hold under condition (i) if $|||X|||_1$ is replaced by $\|X\|_1$. This implies that here $|||X|||_1 > \|X\|_1$ as can be seen directly. Let

$$X_t = \int_0^t \varphi_s \, dB_s \tag{30}$$

where φ is defined as follows. If $n \ge 1$ is odd, let

$$\tau_n = \inf\left\{ t > n-1 : \int_{n-1}^{t} \frac{1}{n-s} dB_s \notin (-1,1) \right\}$$

and then, for even positive integers n, let

$$\tau_n = \inf\left\{ t > n-1 : \int_{n-2}^{\tau_{n-1}} \frac{1}{n-1-s} dB_s + \int_{n-1}^{t} \frac{1}{n-s} dB_s = 0 \right\}.$$

Since $\int_{n-1}^{n} (n-s)^{-2} ds = \infty$, the process $(\int_{n-1}^t (n-s)^{-1} dB_s)_{n-1<t<n}$ is unbounded from above and from below a.s. so, for $n \ge 1$,

$$n - 1 < \tau_n < n \quad \text{a.s.}$$

Let

$$\varphi_s = \sum_{n=1}^{\infty} 1(n-1 < s \le \tau_n)\frac{1}{n-s}.$$

Then φ is predictable and

$$\int_0^n \varphi_s^2 ds = \sum_{k=1}^{n} \int_{k-1}^{\tau_k} \left(\frac{1}{k-s}\right)^2 ds < \infty \text{ a.s.},$$

so X defined by (30) is a local martingale.

Note that $X_2 = X_4 = \cdots = 0$ a.s. and $(X_t)_{0\le t\le 2}$, $(X_{2+t})_{0\le t\le 2}$, $\cdots$ is an independent sequence of identically distributed processes. If $\lambda > 1$, then $P(\sup_{0\le t\le 2} |X_t| > \lambda) = 1/\lambda > 0$ so $P(X^* \ge \lambda) = 1$ and (29) holds. On the other hand, since $\sup_{0\le t\le 2} E|X_t| = 1$, we have that $\|X\|_1 = 1$.

EXAMPLE 4. There is a nonnegative Itô process X with $\psi \ge 0$ such that both (29) and $\|X\|_1 < \infty$ hold. To see this let $X_0 = 0$ and ψ be the nonrandom function defined by

$$\begin{aligned} \psi_s &= 1 \text{ if } s \in (0,1] \cup (2,3] \cup (4,5] \cup \cdots, \\ &= 0 \text{ elsewhere on } (0,\infty). \end{aligned}$$

Let φ be a predictable process satisfying $\varphi_s = 0$ where $\psi_s = 1$ and

$$\int_1^2 \varphi_s dB_s = \int_3^4 \varphi_s dB_s = \cdots = -1 \text{ a.s.}$$

Such a φ exists, as can be seen from Example 3, and $X_2 = X_4 = \cdots = 0$. The process X is nonnegative, $\|X\|_1 = 1$, and, as in Example 3, it satisfies (29).

REFERENCES

[1] D.L. Burkholder, *A sharp inequality for martingale transforms* Ann. Probab. **7** (1979), 858-863.

[2] ————, *Sharp inequalities for martingales and stochastic integrals* Colloque Paul Lévy, Palaiseau, 1987, Astérisque **157-158** (1988), 75-94.

[3] ————, *Differential subordination of harmonic functions and martingales* Proceedings of the Seminar on Harmonic Analysis and Partial Differential Equations, El Escorial 1987, Lecture Notes in Mathematics **1384** (1989), 1-23.

[4] ————, *Explorations in martingale theory and its applications* Ecole d'Eté de Probabilités de Saint-Flour XIX - 1989, Lecture Notes in Mathematics **1464** (1991), 1-66.

[5] ———, *Strong differential subordination and stochastic integration* manuscript in preparation.

[6] C. Dellacherie and P.A. Meyer, "Probabilités et potentiel: théorie des martingales," Hermann, Paris, 1980.

[7] N. Ikeda and S. Watanabe "Stochastic Differential Equations and Diffusion Processes," North-Holland, Amsterdam, 1981.

Department of Mathematics
University of Illinois
1409 West Green Street
Urbana, Illinois 61801

J.K. Ghosh, S.K. Mitra, K.R. Parthasarathy and B.L.S. Prakasa Rao (*eds.*),
Statistics and Probability : A Raghu Raj Bahadur Festschrift
Published by Wiley Eastern Limited, Publishers, (1993), 147 - 163

Estimating a Sequence from Noisy Copies[1]

HERMAN CHERNOFF
Harvard University, Cambridge, MA
Mathematical Sciences Research Institute, Berkeley, CA
The Hebrew University, Jerusalem

Abstract

There is a proposal to build a machine capable of sequencing relatively large portions of DNA in a few minutes. It is anticipated that, when built, this device will be subject to a substantial error rate. Thus it will yield noisy copies of sequences as large as 40,000 base pairs with error rates possibly as high as 20 percent per base pair. How many noisy copies should be made to yield reliable estimates of the original sequence? An algorithm is outlined, and analyzed in part, for estimating the true sequence, assuming a somewhat unrealistic model for the true sequence and the copies. The main parts of the algorithm consist of using a partial resolution of the registration problem via matches of long *words*, to reduce the problem, in several stages, to one dealing with many short copies of about 20 base pairs each. Multiple matches are then used to refine the alignment, and to set up a template for the short copies on which estimated *letters* are entered using a weighting scheme depending on the partial alignment.

[1]Research supported by NSF grant DMS-850550 at the Mathematical Sciences Research Institute and by the Hebrew University, and in part by ONR under contract N00014-91-J1005.

AMS Subject Classifications : Primary 62M99; secondary 68R10, 92O8.

Keywords and Phrases: DNA, clones, sequencing, alignment, estimation.

1 Introduction

There is a proposal to build a device capable of sequencing relatively large portions of DNA in a relatively short time, e.g. 40,000 nucleotides or base pairs in a few minutes. However it is anticipated that the device will be subject to a relatively high error rate. As many as ten to twenty percent of the readings may be in error. It will be possible to produce several readings, which we shall call *copies*. How many copies will be required to obtain a high degree of accuracy, say one error in 10,000 base pairs, and what algorithm will deliver such performance with a reasonable amount of effort?

In this paper I outline an algorithm and establish some of its properties, assuming somewhat unrealistic preliminary models for the true sequence to be copied and for the errors of copying.

If and when this, or some other, device is constructed for rapidly sequencing clones of DNA, it will be possible to model the error rate more realistically, by applying the device to clones which have been sequenced in the past.

We assume that the sources of error will consist of *substitution* (one letter for another), *transposition* (inverting the order of two successive letters), the occasional *insertion* of a few letters that do not belong in the sequence, or the *deletion* of a short subsequence. Finally parts of the beginning and end of the sequence may be dropped, leading to a *registration* or alignment problem, making it difficult to align the copies properly. The initial registration problem could be serious depending on how large a portion of the sequences may be dropped. However, even if the copies were properly aligned at one point, there would still remain a less important, but nontrivial registration problem because of the random deletions and insertions.

The algorithm outlined here relies heavily on the partial resolution of the registration problem, and uses that resolution to reduce, in several stages, our problem to one of dealing with many short sequences of length about 20. It involves the use of matching *words* or subsequences, long enough to support strongly the inference that they are correct copies of words in the *true* sequences.

At this point the method of dealing with the short sequences is rather *ad hoc* and could bear improvement, although it is not computationally intensive. A dissertation by T. Blackwell is in process where the maximum-likelihood method is used to attack short sequences. It is apparently too computationally intensive to apply to long sequences.

What makes sense for some orders of magnitude of the relevant parameters of the process would not be very useful for others. To help keep these orders of magnitude in mind, we shall often deal with very specific numbers to supplement general algebraic symbols.

In Section 2 we will describe the probability models to be used to describe the DNA sequence and its copies. In Section 3, we will outline an algorithm to be used together with some informal justifications. In the remaining sections we will derive some results useful in evaluating the algorithm.

There exists a considerable literature on alignment and consensus of multiple DNA sequences, (see Waterman, 1989 and Doolittle, 1990). Typically the objective of the alignment is to see how closely related are various DNA sequences, and possibly to sketch the evolutionary tree. Criteria of optimality are used to define algorithms. These criteria are partly *ad hoc* and partly take into account biological issues. For example if a *gap* appears in one strand, one would be highly inclined to expect the same gap to appear in biological descendants of that DNA sequence. Thus, while hash coding is widely used to recognize long subsequences which are equal or almost equal, there is relatively little else of direct applicability to our stated problem in the literature.

2 Models

We assume a *true* sequence of length n_s of i.i.d. *letters* selected with equal probability from the characters A, C, T, G. There are n_c independent copies. Each copy is generated in the following manner. A subsequence of length N_r is dropped at the beginning where N_r has standard deviation σ_r. A corresponding subsequence of length N_r' with standard deviation σ_r is dropped at the end. The remaining symbols are copied in sequence, from beginning to end, with the following errors. On reaching the j-th symbol x_j, there is a probability p_s of a substitution, the substituted letter being selected with equal probability from the three other possibilities. There is a probability p_t of a transposition with the previous letter. There are probabilities p_{ij} of the insertion of j randomly selected letters after the given letter, and there are probabilities p_{dj} of the deletion of the next j letters in the sequence.

We assume that all these events take place independently. There is some ambiguity in the description above unless one takes each operation to occur in the sequence described. For an observer who knows the true sequence, it may be difficult to distinguish between an insertion of two letters followed by the deletion of two letters, from two successive substitutions. If the various values of p are small, the overall error rate is approximately p_e, the sum of the various p values.

For concreteness, assume $n_s = 40,000$, $n_c = 13$, $\sigma_r = 3,000$, $p_s = 0.08$, $p_t = 0.04$, $p_{i1} = p_{d1} = 0.03$, $p_{i2} = p_{d2} = 0.008$, $p_{i3} = p_{d3} = 0.002$, and $p_{ij} = p_{dj} = 0$ for $j \geq 3$. With these numbers the overall error rate $p_e = 1 - q_e \approx 0.2$ per letter. We will associate with the standard deviation σ_r a corresponding *uncertainty in alignment* of $u = 6\sigma_r$.

While the error model described above is slightly ambiguous, the results to be derived will depend on more precisely stated models. The values of the error probabilities presented here were made up solely for the sake of discussion, and have no sound empirical or theoretical support. Once a technology is developed for fast sequencing, experimentation will allow one to estimate the relevant error probabilities. Finally the model for the true sequence is certainly naive, and robustness considerations will be required to ascertain the effects of various known phenomena, such as many repeats of short subsequences in true DNA sequences. However, at this stage of technology it seems premature to attempt to use a more realistic model and to derive precise results.

3 Algorithm

3.1 Registration of long sequences

The probability that a specified long word of length n_w in the true sequence is copied correctly may be approximated by $q_e^{n_w}$. The probability that two specified copies will both agree on this word is $q_e^{2n_w}$. This may be contrasted with 4^{-n_w}, which is the probability that two randomly selected words of length n_w will agree. If we align two sequences by a long matching word of length n_w, we have a likelihood-ratio for *correct registration* (CR) versus random pairing of $q_e^{2n_w}4^{n_w}$, which does not reflect the fact that there are roughly $n_s u$ pairs of words that are candidates in the matching process. Neither does it reflect that there are almost n_s words which are candidates to be copied correctly. Thus a more appropriate likelihood-ratio would be $q_e^{2n_w}4^{n_w}\,u^{-1}$. Taking logarithms to the base 10, we have

$$L = n_w \log_{10}(4q_e^2) - \log_{10}(u). \tag{1}$$

For the tentative numbers selected in Section 2 for concreteness, we have, for $n_w = 20$,

$$L = (0.408)n_w - \log_{10}(u) = 3.91$$

When a word of length n_w, or greater, matches on two copies within the limits of the alignment uncertainty, we identify that word as a *point of contact* of the two copies. The notion of regarding a word of length 20 as a point is sensible when considering copies of length 40,000. That notion will no longer make sense when considering words of length 3 or 4 for short sections of copies of length 20. We now have a graph consisting of n_c sequences, along which are located *vertices* or points of contact where pairs and occasionally triples of copies have a common word of length at least 20. Figure 1 is a simplified diagram, where the vertices are represented by vertical lines connecting dots in order to facilitate comprehension.

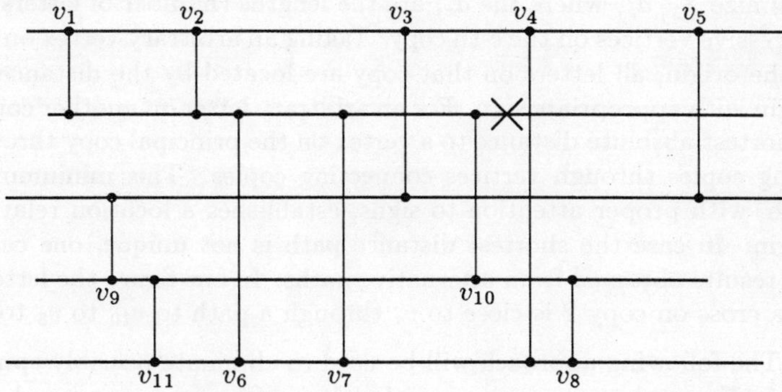

Figure 1: Five Copies With Vertices

Distances along these copies may be measured by the number of letters from one point to another. Because of the error structure, it is possible that two copies contact at v_1 and v_2 and that the distances between v_1 and v_2 differ somewhat from one copy to the other. With our model, and the proposed error probabilities, the measured distance between two points on a copy would have a variance $d[0.03+4(0.008)+9(0.002)+0.03+4(0.008)+9(0.002)] = 0.160d$ where d is the actual distance between the points on the true sequence. More generally, the coefficient of d would be

$$\sum_j (p_{ij} + p_{dj})j^2 - [\sum_j (p_{ij} - p_{dj})j]^2.$$

The method of hash coding (Minsky and Papert, 1969) makes the computational effort, in locating the vertices, to be of the order of magnitude of $n_s n_c n_w$. For the convenience of readers not acquainted with hash coding, a brief description appears in the appendix. In practice, it is probably a good idea to establish the vertices sequentially starting with those corresponding to the longest common words, and going down in word length until we reach a designated value of n_w. When the word length is high, the probability of correct registrations is very high. Nevertheless there is a small probability that a vertex will be spurious. Thus each such contact should be checked out for consistency with the majority of earlier and later vertices. A sign of inconsistency will be an inversion in position, e. g. if v_1 comes before v_2 on one copy and in reverse order on another copy, there is an inconsistency. Another sign would be that the distances between v_1 and v_2 calculated along two copies differ by many standard deviations. Typically, the rare spurious vertex will be discarded as soon as it is observed, but ultimately, the method described below will be used.

We propose now to establish a tentative position for each letter on each copied sequence. This is done to define the algorithm, and a certain amount of arbitrariness is excusable. First a *principal copy* is defined so as to minimize $\sum_j d_{cj}^2$ where the d_{cj} are the lengths (number of letters) between successive vertices on the c-th copy. Taking an arbitrary vertex on that copy as the origin, all letters on that copy are located by the distance from the origin with appropriate sign. For an arbitrary letter on another copy there is a shortest absolute distance to a vertex on the principal copy through paths along copies through vertices connecting copies. This minimum distance path, with proper attention to signs, establishes a location relative to the origin. In case the shortest distance path is not unique, one can average the results obtained from alternative paths. In our figure the letter marked by a cross on copy 2 is close to v_4 through a path to v_{10} to v_8 to v_4.

The following approach will be used to eliminate possibly spurious vertices. For each vertex, pretend that it is not a common word. On each copy mark the location of contact with a cross, and determine its location. If the difference, according to our previous calculation, between the two measurements of location exceeds a certain number of standard deviations, eliminate the vertex. The variance of the difference is 2(0.160) times the shortest absolute distance to the principal copy.

Having estimated positions along each copy, break these copies into about 25 pieces or *sections* aligned among the copies and which overlap somewhat. Here we assume that there will be enough vertices, so that almost all of the sections will have at least two vertices, one near each end. Note that a vertex connecting any two copies, is useful for sectioning all of the copies. Later we shall discuss how many vertices to expect and how they will be distributed. In the meantime, the breaking or sectioning rules have been left rather vague.

Now each section will have a length of about 2,000, and we have a problem similar to our original one if we attempt to *sequence* each of the approximately 25 sections. The main differences are that the length is considerably less than 40,000 and the uncertainty about registration, which may originally have been considerable, and which we assumed to be quite large in our analysis, could now be replaced by a relatively minor amount. As a result the $u = 18,000$ in our calculation of L could be reduced to a small fraction of 2,000. For example with $u = 6\sqrt{(0.160)(2,000)} = 107$, we can still obtain the same value of L after reducing n_w by $(0.408)^{-1}\log_{10}(18,000/107) > 5$.

In Section 5.3 we indicate the feasibility of reducing the problem to sections of length 2,000 and then, in two more stages, to moderately well registered sections of length 160 and then 20.

3.2 Short copies - registration and template

Consider n_c copies of a sequence of approximately 20 letters, where there is an uncertainty in registration of u which averages about 3. Now we intend to establish a finer registration and a *template*, consisting of a number of spaces representing an estimate of the spaces in the true sequence, on which we will put estimates of the letters of the true sequence. For the registration we shall use words of length at least $n_w = 3$ or 4, which are repeated in a substantial number, m of the n_c copies, and in positions consistent with the order and uncertainty. These will be called ***well established***, and fall into two categories. These are the ***spurious*** and the ***well copied words***. These categories correspond to the relationship with the unknown true sequence.

Note that there are 18 successive overlapping words of length 3 in a section of 20 letters and that with errors in copying, we don't need many copies for almost all of the 64 possible words to be very likely to occur in at least one of the copies. On the other hand it is much less likely for 7 out of 13 copies to have a random word in common in roughly the right location. In fact we can roughly measure the inferential impact of m copies of a word of n_w letters by the odds ratio

$$R = \frac{[q_e^{n_w}]^m[1 - q_e^{n_w}]^{n_c - m}}{4^{n_w}[u4^{-n_w}]^m[1 - u4^{-n_w}]^{n_c - m}}. \tag{2}$$

For $u = 3$, $n_w = 3$ we have

$$R \approx (21.3)^m(0.512)^{n_c}/64$$

and for $u = 3$, $n_w = 4$,

$$R \approx (58.6)^m(0.597)^{n_c}/256.$$

Thus one needs about 3/4 as many four letter words in common as three letter words to have the same impact. For example for $m = 7$, $n_c = 13$, $n_w = 3$, $\log_{10}R \approx 3.8$ while for $m = 5$, $n_c = 13$, $n_w = 4$, $\log_{10}R \approx 3.5$.

We establish several words in decreasing order of odds ratio from among these n_c copies, making sure to establish consistency in order and location. Here we have the additional consistency requirement that overlapping words have to be consistent. Thus the word $ACGT$ would be inconsistent with the word $AGTC$ if they were supposed to overlap. To apply this test of consistency, one must first establish the locations of the words, which we assume to start at the leftmost letters.

Suppose w_1 and w_2 are two words with high odds ratio which appear simultaneously on m_s copies. On the c-th copy where both words appear, there is a distance $X_c = d(c; w_1, w_2)$ which is the (signed) number of letters from the leftmost letter of w_1 to that of w_2. Then the estimated distance, $d(w_1, w_2) = \hat{X}$ is obtained as that integer which minimizes

$$S = \sum_c [|X_c - \hat{X}| - 1.5\{X_c = \hat{X}\}]. \tag{3}$$

This is a modification of the median which gives a modal value extra weight, and can be regarded as an approximation to the MLE based on the model of i.i.d. observations on a random variable X for which the distribution is $P_\theta\{X = x\} = f(x - \theta)$ where, for $j > 0$, $f(j) = p_{ij}$ and $f(-j) = p_{dj}$, and $f(0) = 1 - \sum p_{ij} - \sum p_{dj}$.

This estimated distance can be applied to establish a *template* on which well established words are listed and with an estimated number of spaces.

In dealing with short words, one problem that is likely to arise is exemplified by the sequence $ACACAC$ which matches $ACAC$ in two possible ways. In this case both choices should be given a weight of half an observation in the sum defining S. This problem is likely to arise more frequently in real DNA strands than in the model of Section 2 because, in the real case, it often happens that short subsequences are subject to multiple repetitions.

3.3 Final estimation

Once the template is established, each position on that template must be given an estimated letter. The method will be to assign each blank space a vector of weights, one component for each of the four letters of the alphabet. For each copy containing two well established words of the template, we assign weight according to a suitably defined weight function $w(x, x_0; b, b_0)$ as follows. Let b_0 and b be the number of blank spaces, on the template and on the copy, between the two words. Let the blank on the template be a distance x_0 to the right of the end of the first word. Then for $x = 1, 2, \ldots, b$, the letter in position x is assigned a weight $w(x, x_0; b, b_0)$ for the blank in position x_0.

Similarly, a function $w(x, x_0)$ can be used for positions to the right of the last well established word or to the left of the first well established word on each copy containing some well established word. Finally each blank space on the template is assigned, as its estimated letter, that letter which has accumulated the most weight. The author has experimented with a few functions w, but is not especially pleased with them, and leaves it to the reader to propose one more aesthetic with natural monotonicity properties.

4 Evaluation of the algorithm-General remarks

There are two parts of the evaluation. For the long copies, analysis gives some useful results about the frequency of spurious and of true vertices. These support the claim that the problem can be reduced to the analysis of many short copies. The registration of short copies and the final estimation are difficult to handle analytically. On the other hand, Monte Carlo simulations should be feasible for that part. Since the aim is to establish very high reliability, an efficient Monte Carlo would demand some form of variance reduction. It seems appropriate to apply *importance sampling* using an increase in the levels of the various error probabilities. A moderate increase in these probabilities should lead to a substantial rate of errors of estimation which could be well estimated and yield effective estimates for the original error rates, and for that matter, for other alternative sets of error rates, with the same random run. Incidentally, it is plausible that a statistical study of these errors would be useful in improving the short copy part of the algorithm.

In Section 5, we will present some results relevant to the approximate registration of long sequences. These involve estimates of how many spurious and reliable vertices to expect. In Section 6, we shall present results that indicate how many well established words are likely to be spurious, and how close the typical letter is to a well copied word.

One shortcoming, of both the theoretical and Monte Carlo approaches to the evaluation of the algorithm for short copies, is that no sophisticated attempt has been made to consider the conditioning implicit in the derivation of the short copies.

A final caveat is that the implicit assumption that the target sequence being estimated consists of randomly selected letter may lead to inaccurate results. Suppose that the target has 50 successive repetitions of a short word, e.g., ACT. It is possible that our algorithm will tend to yield estimates with fewer repetitions of that word. Should such a phenomenon exist, it will probably be easily recognized, and some method may have to be developed to cope with it. No such attempt is considered here.

5 Distribution of Vertices

5.1 Spurious Vertices

The probability that two random sequences of length n_s will have a specified word of length n_w in common within a distance $u/2$ of each other is approximately $n_s u 4^{-2n_w}$. Since there are n_c copies, there are approximately $n_c^2/2$ pairs of sequences, and considering that there are 4^{n_w} words of length n_w, the expected number of vertices corresponding to words of length n_w among n_c random sequences within the distance $u/2$ is approximately

$$n_f = n_s u 4^{-2n_w}(n_c^2/2)4^{n_w} = 0.5 n_s u n_c^2 4^{-n_w}. \tag{4}$$

For $n_w = 20$, $n_s = 40,000$, $n_c = 13$, and $u = 18,000$, we have $n_f = 0.055$. Thus the probability of a spurious vertex is small under our random model assumption. It is most likely that the few spurious vertices we expect will be in inappropriate locations and easily detected.

Also, n_f represents an overestimate of the expected number of false vertices, because it counts overlapping common words several times. That effect is more important in the next subsection.

Incidentally, if we had good reason to reduce u by a factor of 4, we could achieve the same value of n_f with $n_w = 19$.

5.2 True vertices

Given our model for errors in copying, the probability that a designated word will be copied correctly by a specified copy is at least $q_e^{n_w}$. The probability that one of the other $n_c - 1$ copies will also copy that word correctly is approximately $(n_c - 1)q_e^{n_w}$, and so the expected number of correctly copied words, of length n_w on the specified copy, which form a vertex, is approximately $n_s(n_c - 1)q_e^{2n_w} = 63.8$ for $n_s = 40,000$, $n_w = 20$, and $n_c = 13$. However, when a word of length 20 is copied correctly, it is quite likely that the twenty-first letter will also be copied correctly, giving rise to two words of length 20. Thus the 63.8 above represents somewhat fewer separated vertices or islands of correctly matched words of length 20 or greater.

To deal with this problem, we introduce a process of i.i.d. Bernoulli random variables which assign one with probability q. We form a renewal process which starts at the first one and goes through N sequences of L_1 ones, R_1 zeros, L_2 ones,..., L_N ones and R_N zeros, where N is selected by the condition that L_N is the first $L_j \geq n_w$. Then the process renews. Using $\mathcal{G}(q)$ to represent the geometric distribution with probabilities $q(1-q)^{x-1}$, we shall prove

LEMMA 1
$$\begin{aligned} \mathcal{L}(R_j) &= \mathcal{G}(q), & ER_j &= q^{-1}, \\ \mathcal{L}(L_j) &= \mathcal{G}(1-q), & EL_j &= (1-q)^{-1}, \\ \mathcal{L}(N) &= \mathcal{G}(q^{n_w-1}), \\ EL_N &= n_w - 1 + (1-q)^{-1}, \\ ET &= (1-q)^{-1}q^{-n_w} \end{aligned}$$

where

$$T = \sum_{j=1}^{N}(L_j + R_j). \tag{5}$$

Before proving this lemma, we note that it implies, for our application, that there should be approximately

$$m_c = n_s/ET = p_e n_s q_e^{n_w} \tag{6}$$

renewals representing correctly copied nonoverlapping words of length at least n_w and the average length of these words is $n_w - 1 + p_e^{-1}$. When $n_s = 40,000$, $n_w = 20$, and $p_e = 0.2$, we have $m_c \approx 92.2$.

Proof: The first two lines are obvious, as is the third when we observe that $P\{L_j \geq n_w\} = q^{n_w-1}$. Then $EL_N = E\{L_j | L_j \geq n_w\} = n_w - 1 + EL_j$ which yields the fourth line. Finally, by Wald's Lemma,

$$ET = E(N)E(L_j + R_j) = q^{-(n_w-1)}[q^{-1} + (1-q)^{-1}]$$

and the last line follows.

Given a correctly copied word of length L_N in a specified copy, we ask what is the probability that there is a correct match of length at least n_w

in another specified copy. To address this question we introduce a sequence of n_s Bernoulli random variables with parameter q where $n_w \leq n_s < 2n_w$. We shall derive

LEMMA 2 *Given a sequence of n_s i.i.d. Bernoulli random variables with parameter q, where $n_w \leq n_s < 2n_w$, the probability that this sequence contains a word of at least n_w consecutive ones is*

$$q^{n_w}[1 + (1-q)(n_s - n_w)].$$

For our application of this lemma, the proof of which follows, the role of n_s is played by L_N, with expectation given in the preceding lemma. The probability that L_N is as large as $2n_w$ is slight. Hence this lemma implies that the probability that a correctly copied word, of length at least n_w, will lead to a vertex with a specified one of the other copies, is approximately

$$p_c = (1 + q_e)q_e^{n_w} \tag{7}$$

which is 0.021 for $n_w = 20$ and $p_e = 1 - q_e = 0.2$.
Proof: Let $A_j =$ {the n_w elements of the sequence starting with the j-th are all ones}. We seek the probability of the union of the $n_s - n_w + 1$ events A_j. Every realization where exactly one of the A_j occurs is counted once in $\sum P(A_j) = (n_s - n_w + 1)P(A_1)$. Every realization where exactly two of the A_j occur is counted twice in $\sum P(A_j)$, but only once in $\sum P(A_j A_{j+1}) = (n_s - n_w)P(A_1 A_2)$ because $n_s - n_w < n_w$. Every realization where exactly three of the A_j occur is counted three times in the first sum and twice in the second, etc. Thus

$$\begin{aligned} P(\cup A_j) &= (n_s - n_w + 1)P(A_1) - (n_s - n_w)P(A_1 A_2) \\ &= (n_s - n_w + 1)q^{n_w} - (n_s - n_w)q^{n_w+1}. \end{aligned}$$

and the lemma follows.

The probability that a correctly copied word of length at least n_w, on a specified copy, will lead to a vertex with at least one of the other copies, is then $[1 - (1 - p_c)^{n_c - 1}]$, and thus the expected number of vertices on a specified copy is

$$n_{v1} = m_c[1 - (1 - p_c)^{n_c - 1}] \tag{8}$$

and is 20.7 for $n_s = 40,000$, $p_e = 0.2$, $n_c = 13$, and $n_w = 20$.

If there were no triple vertices the total number of expected vertices would be $n_{v1} n_c / 2 = 134.5$ because each vertex involves two copies. For later stages with somewhat shorter sequences, the effect of triple vertices is no longer negligible. We need to consider Y = the number of copies which form a vertex with a correctly copied word on a specified copy. Let V_j = the number of vertices which touch $j \geq 2$ copies, and the total number of vertices is

$$V = \sum_{j=2}^{\infty} V_j.$$

Then

$$V_j \approx n_c m_c\ P(Y = j - 1)\frac{1}{j}$$

where Y has a binomial distribution with parameters $n_c - 1$ and p_c. We write

$$P\{Y \leq y\} = \mathcal{B}(y; n_c - 1, p_c) = 1 - \mathcal{B}^c(y; n_c - 1, p_c)$$

Thus

$$\begin{aligned} EV &\approx n_c m_c \sum_{j=2}^{n_c} \binom{n_c - 1}{j - 1} p_c^{j-1}(1 - p_c)^{n_c - j}\frac{1}{j} \\ &= m_c\ p_c^{-1} \sum_{j=2}^{n_c} \binom{n_c}{j} p_c^j (1 - p_c)^{n_c - j} \\ &= m_c\ p_c^{-1}[1 - (1 - p_c)^{n_c} - n_c p_c (1 - p_c)^{n_c - 1}]. \end{aligned}$$

Combining these results, we have

THEOREM 1.

$$\begin{aligned} n_f &= .5 n_s u n_c^2 4^{-n_w} \\ m_c &= p_e n_s q_e^{n_w} \\ p_c &= (1 + q_e) q_e^{n_w} \\ n_{v1} &= m_c[1 - (1 - p_c)^{n_c - 1}] \\ EV &= \frac{1 - q_e}{1 + q_e} n_s [1 - (1 - p_c)^{n_c} - n_c p_c (1 - p_c)^{n_c - 1}] \end{aligned}$$

For our choice of $n_s = 40{,}000$, $n_c = 13$, $n_w = 20$, $u = 18{,}000$ and $p_e = 1 - q_e = 0.2$, we have $n_f = 0.06$, $m_c = 92.2$, $p_c = 0.021$, $n_{v1} = 20.7$, and $EV = 131$.

As long as the length of the word is a small fraction of the length of the copy, and the formation of a vertex at any specified place is rather improbable, we expect the locations of the vertices to behave like a Poisson process. That means that the average distance between the vertices involving the principal copy would be less than one twentieth of 40,000 or 2,000, and the average distance between approximately 130 vertices would be less than 400. Since the largest spacing between n uniformly distributed random variables can be expressed asymptotically as $(\ln(n) + W)/n$ where W has the extreme value distribution with *cdf* $\exp(-e^{-w})$, the largest of these approximately 130 distances would be about $(40,000)\ln(130)/130 \approx 1,500$.

One would anticipate that the error in measuring the distances between vertices would have a variance of about $(0.16)(1,500)/2 \approx 120$, or that the registration would be accurate within an uncertainty of $6\sqrt{120} = 66$. This reduction implies that the rare false vertex will be easily detected by virtue of its incorrect location. There should be no difficulty in breaking up the copies into partially registered sections of size about 2,000.

5.3 Second stage

In the second stage, we have a considerable advantage over the first stage in that a reasonable value for u would be about 80. Substituting in Theorem 1 with $n_s = 2,000$, $n_c = 13$, $n_w = 12$, $u = 80$, we have $n_f = 0.81$, $m_c = 27.5$, $p_c = 0.124$, $n_{v1} = 21.8$, and $EV = 109$. In this stage, the effect of triple vertices has become nontrivial. The expected number of false vertices can be reduced by a factor of 4 by taking $n_w = 13$, thereby cutting down somewhat the expected number of true vertices. One could go immediately to a third stage with small copies, but caution suggests first taking another stage with copies of size about 160.

A third stage, with $n_s = 160$, $n_c = 13$, $u = 30$ and $n_w = 9$, yields $n_f = 1.5$, $m_c = 4.3$, $p_c = 0.24$, $n_{v1} = 4.1$ and $EV = 15.2$, and it should be possible to divide these sections into moderately well registered short subsections of size about 20. These section will be called ***short copies.***

6 Short copies

A key to the implementation of our proposed algorithm for short copies is the requirement that there be very few spurious cases of well established words and an abundance of well copied words.

6.1 Spurious cases of established words

To get some handle on the number of spurious well established words, let us assume that all the copies are randomly generated. Given a specified word of n_w letters, the probability that it appears within a specified space of $n_w - 1 + u$ letters is bounded by $u4^{-n_w}$. For most words this is a good bound. For $n_w = u = 3$, the worst case is the word AAA. Here the probability is 0.039 instead of $3/64 = 0.047$.

The probability that the word appears in this region on m or more of the copies is bounded by $B^c(m-1; n_c, u4^{-n_w})$. For $n_w = u = 3$, $n_c = 13$, $m = 5$, this bound is 0.00021, which suggests that such spurious copies will not be very rare (with 64 three letter words in 2,000 short copies), and will bring confusion on occasion. This effect can be reduced by taking suitably larger values of m and n_w, possibly in conjunction with larger values of n_c needed to increase the frequency of ***honest*** well copied words.

6.2 Well copied words

We now present an analysis relevant to the frequency of well copied words.

In a true sequence of length n_s, there are $n_s - n_w + 1$ words of length n_w. Each has probability $q_e^{n_w}$ of being copied correctly, and so the expected number of words, *counting overlaps*, of length n_w, which are copied correctly at least m times is

$$(n_s - n_w + 1)\mathcal{B}^c(m-1; n_c, q_e^{n_w}).$$

There doesn't seem to be a simple way of adjusting for the overlap. While every word of length n_w+1, which is copied correctly m times, leads to two corresponding overlapping words of length n_w, the reverse is not necessarily the case. A similar statement applies for words of length $n_w + 2$ which are counted twice as words of length $n_w + 1$ and three times for length n_w. Thus

$$(n_s - n_w + 1)\mathcal{B}^c(m-1; n_c, q_e^{n_w}) - (n_s - n_w)\mathcal{B}^c(m; n_c, q_e^{n_w+1})$$

should represent an underestimate of the number of words of length at least n_w that are well copied.

Another approach to comprehending the abundance of well copied words is to ask how close will a given letter be to a well copied word. First we address the following question. What is the probability that a specified letter of the true sequence appears in a word of length n_w which is well copied? To attack this question we must consider the $2n_w - 1$ letters with the specified letter in the center. In other words, we ask, what is

$$P\{\cup_{j=1}^{n_w} A_j\}$$

where A_j is the event that the word of length n_w starting with the j-th of the $2n_w - 1$ letters, is correctly copied at least m times. Let B_j be the event that the word is copied correctly on a specified copy.

We shall attack this problem indirectly, by finding $P\{M \leq m\}$ where M is the maximum number of times that any of the n_w words containing the specified letter is copied correctly. We need some notation. With a copy we identify which of the n_w words is copied correctly, defining a vector $\mathbf{t} = (t_1, t_2, \ldots, t_{n_w})$ which consists of ones corresponding to correctly copied words, and zeros elsewhere. No two components of $\mathbf{t}$ which are ones can be separated by zeros. The set H of such vectors may be written as $H = H^* \cup \{\mathbf{0}\}$ where each element of H^* contains at least one one. Since there are $j+1$ elements of H^* with j zeros, H^* contains $n^* = n_w(n_w+1)/2$ elements.

Associated with each $\mathbf{t} \in H$, we have the event

$$E_{\mathbf{t}} = (\cap_{\{t_j=1\}} B_j)(\cap_{\{t_j=0\}} B_j^c),$$

and for $\mathbf{t} \in H^*$, we easily calculate

$$p_{\mathbf{t}} = P(E_{\mathbf{t}}) = q_e^{r_1} p_e^{r_2} \tag{9}$$

where $r_1 = \sum t_j + (n_w - 1)$ and $r_2 = 2 - (t_1 + t_{n_w})$. Here r_1 corresponds to the number of ones in $\mathbf{t}$ and r_2 is the number of sides, of the connected group of ones in $\mathbf{t}$, on which there are zeros. Since the $E_\mathbf{t}$ are disjoint, we can calculate $p_\mathbf{0}$ by complementation, i. e.

$$p_\mathbf{0} = 1 - \sum_{\mathbf{t} \in H^*} p_\mathbf{t}.$$

The conditional probabilities, given that some word is copied correctly, are given for $\mathbf{t} \in H^*$ by

$$p_\mathbf{t}^* = P(E_\mathbf{t} | E_\mathbf{0}^c) = \frac{p_\mathbf{t}}{1 - p_\mathbf{0}}.$$

Let $\mathbf{s} = \{s_\mathbf{t} : \mathbf{t} \in H\}$ be a vector with $n^* + 1$ nonnegative integer elements representing how often $E_\mathbf{t}$ occurs in the n_c copies. Thus, for each $\mathbf{s}$, we have

$$\sum_{\mathbf{t} \in H} s_\mathbf{t} = n_c. \tag{10}$$

Let $\mathbf{s}^* = \{s_\mathbf{t} : \mathbf{t} \in H^*\}$ and $s_\mathbf{0} = s_\mathbf{0}(\mathbf{s}^*) = n_c - \sum_{\mathbf{t} \in H^*} s_\mathbf{t}$. Then, we have

THEOREM 2.

$$P\{M \le m\} = \sum_{\mathbf{s}^*} \frac{n_c!}{s_\mathbf{0}!} p_\mathbf{0}^{s_\mathbf{0}} (1 - p_\mathbf{0})^{n_c - s_\mathbf{0}} \prod_{\mathbf{t} \in H^*} \frac{(p_\mathbf{t}^*)^{s_\mathbf{t}}}{s_\mathbf{t}!} \tag{11}$$

where the sum over the set of $\mathbf{s}^*$ *is subject to the restrictions*

$$\sum_\mathbf{t} t_j s_\mathbf{t} \le m \text{ for } j = 1, 2, \ldots, n_w \tag{12}$$

which do not involve n_c.

Proof: This theorem is an immediate consequence of the multinomial distribution applied to the disjoint events $E_\mathbf{t}$ where $E_\mathbf{0}$ is separated out.

To illustrate the theorem in the case $n_w = 3$, the 7 elements of H are binary representations of the integers, 0,1,2,3,4,6, and 7, and the restrictions are

$$\begin{aligned} s_7 + s_6 + s_4 &\le m, \\ s_7 + s_6 + s_3 + s_2 &\le m, \\ s_7 + s_3 + s_1 &\le m. \end{aligned}$$

Theorem 2 is not very insightful by itself, since the right hand side looks complicated. But it may be implemented in a computationally convenient fashion. The major part of the computational effort lies in the algorithm identifying the vectors $s_\mathbf{t}^*$ which satisfy the inequalities of Equation 12. In this calculation there is little extra effort in dealing with lower values of m and other values of n_c simultaneously.

Table 1 presents the results of these calculations for $n_c = 13(1)20$, $n_w = 3, 4$, and $m = 0(1)7$. The implications of this table can be seen in the following statements directed at $n_c = 13, n_w = 3$, and $m = 5$. Then, the probability that a specified letter does not appear in a well copied word is 0.0185. Consider 3 letters which are spaced 5 letters apart in the true sequence. The probability that none of the three letters appears in a well copied word is $(6.3)10^{-6}$. The probability that at most one of these letters will be in a well copied word is 0.0010. We expect fewer than 30 cases out of 40,000 where a letter fails to be in a well copied word, or between two well copied words within 5 letters.

Table 1. $100P\{M \leq m\}$

	$n_w = 3$							
$n_c\backslash m$	0	1	2	3	4	5	6	7
13	0.00	0.00	0.02	0.28	1.85	7.73	21.83	44.39
14	0.00	0.00	0.01	0.12	0.90	4.27	13.73	31.74
15	0.00	0.00	0.00	0.05	0.42	2.26	8.25	21.56
16	0.00	0.00	0.00	0.02	0.19	1.16	4.76	13.98
17	0.00	0.00	0.00	0.01	0.09	0.58	2.65	8.69
18	0.00	0.00	0.00	0.00	0.04	0.28	1.42	5.19
19	0.00	0.00	0.00	0.00	0.02	0.13	0.74	3.00
20	0.00	0.00	0.00	0.00	0.01	0.06	0.38	1.68
	$n_w = 4$							
$n_c\backslash m$	0	1	2	3	4	5	6	7
13	0.00	0.01	0.21	1.84	8.53	24.53	48.70	72.86
14	0.00	0.00	0.10	0.98	5.16	16.82	37.48	61.99
15	0.00	0.00	0.05	0.51	3.04	11.15	27.79	50.85
16	0.00	0.00	0.02	0.26	1.74	7.16	19.91	40.28
17	0.00	0.00	0.01	0.13	0.98	4.48	13.82	30.85
18	0.00	0.00	0.00	0.06	0.54	2.73	9.32	22.90
19	0.00	0.00	0.00	0.03	0.29	1.62	6.11	16.50
20	0.00	0.00	0.00	0.01	0.15	0.94	3.91	11.57

These calculations suggest that with the error probabilities we have been using, with $p_e = 0.2$, 13 copies will not be adequate to achieve an error rate of one out of 10,000. More likely 20 copies will be needed. In any case a Monte Carlo study will be required to derive an accurate judgement.

Appendix

Consider the information retrieval problem of storing a large library of $n = n_s n_c = 40,000 \times 13 = 520,000$ potential words of length $n_w = 20$ into a memory bank, and deciding when a specified word is in that bank. One way would be to have an available labeled address for each of the $4^{20} \approx 10^{12}$ possible words. Then each word of the library is put into its labeled address.

When the specified word is presented we look in the corresponding address to see if it is occupied. This plan is very demanding of memory space. Hash coding provides an alternative which trades off the cost of making comparisons against the cost of memory.

Suppose that we have a memory bank which holds $2n$ or approximately one million labeled addresses. We select some random uniform numbers $y_1, y_2, \ldots$ and a function $z = f(x, y)$ which maps a 20 letter word x and a random number y into an address z.

Proceeding sequentially, after $x_1, x_2, \ldots, x_{i-1}$ have been stored in positions $z_1, z_2, \ldots, z_{i-1}$, x_i is stored in position $f(x_i, y_1)$ if that position is vacant. If it is occupied, we try $f(x_i, y_2)$, etc. until a vacancy is found, and that position is labeled z_i. If f is not too simple, the z_i will tend to be more or less randomly distributed among the available positions in the memory bank, and two trials will be required on the average to find a vacancy when most of the n items have been stored. Finally, when it is desired to know whether a specified word x is stored in the library, one searches $f(x, y_1), f(x, y_2), \ldots$, until one finds x or an empty position. In the latter case x is not in the library. One requires on the average two trials and comparisons before reaching an empty position.

Increasing the size of the memory leads to a reduction in the expected number of trials and comparisons. The relative costs of memory and comparisons leads to a potential trade off of costs.

In our application we have a variation of the above information retrieval problem where we may have to store or compare $n_s n_c$ words of length n_w. Assuming that each calculation and comparison involves an effort proportional to n_w, the computational effort is proportional to $n_s n_c n_w$ with rather modest memory requirements.

REFERENCES

Blackwell, T. (1992). Unfinished dissertation.

Doolittle, R.F. (1990). *Computer Analysis of Protein and Nucleic Acid Sequences.* Methods in Enzymology, Vol. 183, Academic Press, New York

Minsky, M. and Papert, S. (1969). *Perceptrons.* M.I.T. Press, Cambridge, Mass.

Waterman, M.S. (1989), *Mathematical Methods for DNA Sequences.* CRC Press, Boca Raton, Fla.

Department of Statistics
Harvard University
Cambridge, MA 02138
U.S.A.

J.K. Ghosh, S.K. Mitra, K.R. Parthasarathy and B.L.S. Prakasa Rao (*eds.*),
Statistics and Probability : A Raghu Raj Bahadur Festschrift
Published by Wiley Eastern Limited, Publishers, (1993), 165 – 170

On Stochastic Ordering of Partial Sums with Application to Reliability Theory[1]

ARTHUR COHEN and H. B. SACKROWITZ
Rutgers University, New Brunswick

Abstract

Let $\mathbf{X} = (X_1, \ldots, X_n)$ be a random vector whose density depends on a parameter vector $\boldsymbol{\lambda} = (\lambda_1, \ldots, \lambda_n)$. Assume the density is decreasing in transposition. Let $T_j = \sum_{i=1}^{j} X_i$, $j = 1, 2, \ldots, n$, and $\mathbf{T} = (T_1, \ldots, T_n)$ be the vector of partial sums. Let π be a permutation of $(1, 2, \ldots, n)$ and consider the partial ordering $\pi \geq^{b_1} \pi'$ if either $\pi = \pi'$ or π can be obtained from π' by a sequence of permutations, starting with π', whereby each recursive permutation is obtained by a pairwise correction of out-of-order elements of the preceding permutation. We prove that conditioned on $T_n = t$, the distribution of $\mathbf{T}$ under π' (i.e., the density of $\mathbf{X}$ when the parameter vector is $\boldsymbol{\lambda}_{\pi'}$, a permutation of $(\lambda_1, \ldots, \lambda_n)$, $\lambda_1 \leq \ldots \leq \lambda_n$), is stochastic larger than the distribution of $\mathbf{T}$ under π if $\pi \geq^{b_1} \pi'$. The result has important applications in reliability theory. It generalizes a result of Boland, El-Neweihi, and Proschan (1991) in several directions.

---

[1] Research supported by NSF Grant DMS-9112784.

AMS Subject Classifications : 62N05, 60K10.
Key Words and Phrases: Decreasing in transposition, monotone likelihood ratio, multivariate stochastic ordering, inspection and repair policies, waiting time distributions, permutation, partial orderings.

1 Introduction and Summary

Let $\mathbf{X} = (X_1, \ldots, X_n)$ be a random vector and $\boldsymbol{\lambda} = (\lambda_1, \ldots, \lambda_n)$, $\lambda_1 \le \lambda_2 \le \ldots \le \lambda_n$, be a vector of parameters such that the joint density (or probability mass function) of $\mathbf{X}$ is denoted by $f_{\mathbf{X}}(\mathbf{x}; \boldsymbol{\lambda})$. Assume $f_{\mathbf{X}}(\mathbf{x}; \lambda)$ is decreasing in transposition (DT). That is, (i) $f_{\mathbf{X}}$ is unchanged when the same permutation is applied to $\boldsymbol{\lambda}$ and to $\mathbf{x}$ and (ii) $f_{\mathbf{X}}(\mathbf{x}; \boldsymbol{\lambda}) \ge f_{\mathbf{X}}(\mathbf{x}'; \boldsymbol{\lambda})$ whenever $\mathbf{x}'$ and $\mathbf{x}$ differ in two coordinates only, say i and j, and $(x_j - x_i)(j-i) \ge 0$ and $x_i' = x_j$, $x_j' = x_i$. The notion of DT is discussed in Hollander, Proschan, and Sethuraman (1977). Note that if X_i are independent with densities of the same form (i.e., $f_{X_i}(x_i; \lambda_i) = f_X(x_i; \lambda_i)$, $i = 1, 2, \ldots, n$) that depend on parameters λ_i and the densities are monotone likelihood ratio (MLR), then $f_{\mathbf{X}}(\mathbf{x}; \boldsymbol{\lambda})$ is DT. The important property (ii) above of a DT density is that if $x_i < x_j$ (and $j > i$)

$$f_{\mathbf{X}}(x_1, \ldots, x_i, \ldots, x_j, \ldots, x_n; \boldsymbol{\lambda}) \ge f_{\mathbf{X}}(x_1, \ldots, x_j, \ldots, x_i, \ldots, x_n; \boldsymbol{\lambda}). \tag{1.1}$$

Before stating the main result we need two additional definitions.

DEFINITION 1.1: For two permutations of the integers $(1, 2, \ldots, n)$, π and π', we say that $\pi \ge^{b_1} \pi'$ if either $\pi = \pi'$ or π can be obtained from π' by a sequence of permutations, whereby each successive permutation is obtained by a pairwise correction of out-of-order elements of the preceding permuation. For example, $(2, 1, 4, 3) \ge^{b_1} (2, 3, 4, 1)$.

DEFINITION 1.2: Let $\mathbf{U}$, $\mathbf{V}$ be $1 \times n$ random vectors. We say $\mathbf{V}$ is stochastically greater than or equal to $\mathbf{U}$, $\mathbf{U} \le^{st} \mathbf{V}$, if for every nondecreasing function $h(\mathbf{u})$ for which expectations exist

$$Eh(\mathbf{U}) \le Eh(\mathbf{V}). \tag{1.2}$$

(h is a nondecreasing function of $\mathbf{u}$ if for all arguments fixed except u_i, h is nondecreasing in u_i, $i = 1, 2, \ldots, n$.)

Let $T_j = X_1 + X_2 + \cdots + X_j$, $j = 1, 2, \ldots, n$ be partial sums of the X's and let $\mathbf{T} = (T_1, T_2, \ldots, T_n)$. Let $\boldsymbol{\lambda}_\pi$ be the $1 \times n$ vector of $\boldsymbol{\lambda}$ with coordinates permuted according to π (i.e., if $\pi_1 = (1, 2, \ldots, n)$, then $\boldsymbol{\lambda}_{\pi_1} = (\lambda_1, \lambda_2, \ldots, \lambda_n)$ if $\pi_2 = (2, 1, 3, \ldots, n)$, $\boldsymbol{\lambda}_{\pi_2} = (\lambda_2, \lambda_1, \lambda_3, \ldots, \lambda_n)$, etc.) Let $\mathbf{T}_\pi$ represent the vector of partial sums when the distribution of $\mathbf{X}$ is $f_{\mathbf{X}}(\mathbf{x}; \boldsymbol{\lambda}_\pi)$. We show that conditioned on $T_n = t$, if $\pi \ge^{b_1} \pi'$, then $\mathbf{T}_\pi \le^{st} \mathbf{T}_{\pi'}$.

This result was inspired by the work of Boland, El-Neweihi, and Proschan (BEP) (1991). They studied a special case of this model where X_i are independent Bernoulli variables with parameters p_i, $p_1 \le p_2 \le \cdots \le p_n$. They studied waiting time distributions which are nonincreasing functions of partial sums. They derive a stochastic ordering result. We will demonstrate that such a result is a special case of our result. BEP (1991) indicate the applicability of these types of results to inspection and repair policies. We refer to their paper for motivation and applications in reliability theory. The applications cited there would be appropriate for many of the models that are special cases of our general setup.

There are three ways in which the current result here improves on some of the results of BEP (1991). Namely,

(i) In BEP (1991), the components of $\mathbf{X}$ are independent Bernoulli variables with probabilities p_i, $p_1 \leq p_2 \leq \cdots \leq p_n$. Here we have a much larger class of distributions including independent exponential family distributions of the same form with parameters λ_i as well as indpendent MLR families with the same form with parameters λ_i. Even some models with nonindependent X_i are treated here under the DT assumption.

(ii) In BEP (1991), stochastic ordering is demonstrated for each one-dimensional waiting time distribution. Here we show the stronger result that an entire vector of waiting time distributions has a stochastic ordering.

(iii) The proof here is totally different and considerably shorter than in BEP (1991).

We prove our main result in Section 2. The connection to waiting time distributions is made in Section 3.

2 Stochastic Ordering of Partial Sums

We will appeal to the following slight modification of Theorem B.1 of Marshall and Olkin (1979).

THEOREM 2.1: *Consider random vectors* $(\mathbf{U}, \mathbf{V})$. *The pair satisfy* $\mathbf{U} \leq^{st} \mathbf{V}$ *if and only if there exist random vectors* $(\tilde{\mathbf{U}}, \tilde{\mathbf{V}}, \mathbf{Z})$ *with* $\mathbf{Z} = (Z_1, \ldots, Z_n)$, $Z_i \geq 0$ *with probability 1, such that*

(2.1) $\tilde{\mathbf{V}} = \tilde{\mathbf{U}} + \mathbf{Z}$,

(2.2) $\mathbf{U}$ *and* $\tilde{\mathbf{U}}$ *are identically distributed,*

$\mathbf{V}$ *and* $\tilde{\mathbf{V}}$ *are identically distributed.*

Next we have

LEMMA 2.2: *Suppose the density* $f_{\mathbf{X}}(\mathbf{x}; \lambda)$ *is DT. Then the conditional density of* $\mathbf{X}$ *given* $T_n = \sum X_i = t$ *is also DT.*

PROOF: See Theorem 3.9 of Hollander, Proschan, and Sethuraman (1977).

The main result is

THEOREM 2.3: *Assume the distribution of* $\mathbf{X}$ *depends on* λ *and is DT. Conditioned on* $T_n = t$, *if* $\pi \geq^{b_1} \pi'$, *then* $\mathbf{T}_\pi \leq^{st} \mathbf{T}_{\pi'}$. *Furthermore* $T_\pi \leq^{st} \mathbf{T}_{\pi'}$ *unconditionally as well.*

PROOF: Lemma 2.2 and the assumption of the theorem imply that the conditional distribution of $\mathbf{X}$ given T_n is DT. Thus we have, using the property in (1.1),

(2.3)
$$f_{\mathbf{X}|T_n}(x_1,\ldots,x_i,\ldots,x_j,\ldots,x_n;\lambda) \geq f_{\mathbf{X}|T_n}(x_1,\ldots,x_j,\ldots,x_i,\ldots,x_n;\lambda)$$

for $j > i$, $x_j > x_i$, $\lambda_1 \leq \lambda_2 \leq \cdots \leq \lambda_n$.

Now let

$$r(\mathbf{x},\lambda) = \frac{f_{\mathbf{X}|T_n}(x_1,\ldots,x_j,\ldots,x_i,\ldots,x_n;\lambda)}{f_{\mathbf{X}|T_n}(x_1,\ldots,x_i,\ldots,x_j,\ldots,x_n;\lambda)}, \tag{2.4}$$

and note $0 \leq r(\mathbf{x},\lambda) \leq 1$. At this point, let $\mathbf{Y}$ have the same distribution as $\mathbf{X}$ given $T_n = t$ when $\boldsymbol{\lambda}$ is the parameter (i.e., $\lambda_i < \lambda_j$). Define the random vector $\mathbf{Q}$ conditionally on $\mathbf{Y} = \mathbf{x}$ as follows: For sample points $\mathbf{x} = (x_1,\ldots,x_i,\ldots,x_j,\ldots,x_n)$ where $x_i < x_j$, $\mathbf{Q}$ takes on two possible sets of values, namely $\mathbf{q}_1 = (0,\ldots,0)$, $\mathbf{q}_2 = (0,\ldots,0, x_i - x_j, 0,\ldots,0, x_j - x_i, 0,\ldots,0)$. The $x_i - x_j$ occurs as the ith coordinate while the $x_j - x_i$ occurs as the jth coordinate. The conditioned probability that $\mathbf{Q} = \mathbf{q}_1$ or $\mathbf{q}_2$ will be given as follows: When $x_i < x_j$ and $\mathbf{Y} = \mathbf{y}_1 = (x_1,\ldots,x_i,\ldots,x_j,\ldots,x_n)$, then $P(\mathbf{Q} = \mathbf{q}_1|\mathbf{Y} = \mathbf{y}_1) = r(\mathbf{x};\lambda)$. If $\mathbf{Y} = \mathbf{y}_2 = (x_1,\ldots,x_j,\ldots,x_i,\ldots,x_n)$, $P(\mathbf{Q} = \mathbf{q}_2|\mathbf{Y} = \mathbf{y}_2) = 0$.

When $x_i \geq x_j$, $P(\mathbf{Q} = \mathbf{q}_1|\mathbf{Y} = \mathbf{y}_1) = 1$ and $P(\mathbf{Q} = \mathbf{q}_2|\mathbf{Y}_2 = \mathbf{y}_2) = 1 - r(\mathbf{x};\lambda)$. Defining $\mathbf{Q}$ this way, it follows that $\mathbf{Y} + \mathbf{Q}$ is distributed as $\mathbf{X}$ given $T_n = t$ when λ' is the parameter where recall $\lambda' = (\lambda_1,\ldots,\lambda_j,\ldots,\lambda_i,\ldots,\lambda_n)$. To see this, let $h_{\mathbf{Y}+\mathbf{Q}}(\mathbf{x}) = h_{\mathbf{Y}+\mathbf{Q}}(x_1,\ldots,x_i,\ldots,x_j,\ldots,x_n)$ denote the density of $\mathbf{Y} + \mathbf{Q}$ (T_n is fixed throughout). Then

$$\begin{aligned} h_{\mathbf{Y}+\mathbf{Q}}(\mathbf{x}) &= h_{\mathbf{Y}}(\mathbf{x})P(\mathbf{Q} = \mathbf{q}_1|\mathbf{Y} = \mathbf{y}_1) \\ &\quad + h_{\mathbf{Y}}(x_1,\ldots,x_j,\ldots,x_i\ldots,x_n)P(\mathbf{Q} = \mathbf{q}_2|\mathbf{Y}_2 = \mathbf{y}_2). \end{aligned} \tag{2.5}$$

Now for $x_i < x_j$, (2.5) becomes

$$f_{\mathbf{X}|T_n}(\mathbf{x};\lambda)\cdot\frac{f_{\mathbf{X}|T_n}(\mathbf{x};\lambda')}{f_{\mathbf{X}|T_n}(\mathbf{x};\lambda)} + 0 = f_{\mathbf{X}|T_n}(\mathbf{x};\lambda'). \tag{2.6}$$

For $x_i \geq x_j$, (2.5) becomes

$$f_{\mathbf{X}|T_n}(\mathbf{x};\lambda)\cdot 1 + f_{\mathbf{X}|T_n}(\mathbf{x}';\lambda)\left(1 - \frac{f_{\mathbf{X}|T_n}(\mathbf{x}';\lambda')}{f_{\mathbf{X}|T_n}(\mathbf{x}';\lambda)}\right), \tag{2.7}$$

where $\mathbf{x}' = \mathbf{y}_2$. Note that (2.7) becomes $f_{\mathbf{X}|T_n}(\mathbf{x};\lambda')$ since $f_{\mathbf{X}|T_n}(\mathbf{x}';\lambda) = f_{\mathbf{X}|T_n}(\mathbf{x};\lambda')$ and $f_{\mathbf{X}|T_n}(\mathbf{x}';\lambda') = f_{\mathbf{X}|T_n}(\mathbf{x};\lambda)$.

Finally, define $\mathbf{Z} = (Z_1,\ldots,Z_n)$ by $Z_j = Q_1 + \cdots + Q_j$, $j = 1,2,\ldots,n$. Then $P(\mathbf{Z} \geq 0) = 1$. Furthermore, if $S_j = Y_1 + \cdots + Y_j$, all $j = 1,2,\ldots,n$, then $\mathbf{S} = (S_1, \ldots, S_n)$ has the same distribution as $\mathbf{T}$ given $T_n = t$ and λ is the true parameter. Hence $\mathbf{S} + \mathbf{Z}$ under λ has the same distribution as $\mathbf{T}$ given $T_n = t$ and λ' is the true parameter. Now appeal to Theorem 2.1 to complete the proof when π can be obtained from π' with one transposition. However, if more than one transposition is necessary the argument is easily repeated. ||

3 Waiting Times

If the X_i are Bernoulli with parameters p_i, then BEP (1991) define $W_{k,\pi}$ as the waiting time, under inspection procedure π, until the discovery of the kth failed component. They study the conditional distribution of $W_{k,\pi}$ given $N = n$, where N is the total number of failed components. Note in our terminology N is the same as T_n. Notice that $W_{k,\pi}$, $k = 1, \ldots, n$ is a function of the partial sums in $\mathbf{T}$. The random variable $W_{k,\pi} = \nu$, if and only if $T_\nu = \sum_{i=1}^{\nu} X_i = k$ and $T_{\nu-1} = k - 1$. Thus $W_{k,\pi}$ is a decreasing function of $\mathbf{T}$ which, by virtue of Theorem 2.3 implies the following:

COROLLARY 3.1: *If $\pi \geq^{b_1} \pi'$, then for any one k among $k = 1, \ldots, n$, $W_{k,\pi'}$ $\leq^{st} W_{k,\pi}$, conditioned on T_n. Furthermore $W_{k,\pi'} \leq^{st} W_{k,\pi}$ unconditionally.*

Next note that the vector $\mathbf{W}_\pi = (W_{1,\pi}, W_{2,\pi}, \ldots, W_{n,\pi})$ is a 1-1 function of $\mathbf{T}$, say $\mathbf{g}(\mathbf{T})$, a vector-valued function of a vector. Each component of $\mathbf{W}_\pi$ is a nonincreasing function of $\mathbf{T}$.

COROLLARY 3.2: *If $\pi \geq^{b_1} \pi'$, $\mathbf{W}_{\pi'} \leq^{st} \mathbf{W}_\pi$, conditioned on T_n. Furthermore $W_{\pi'} \leq^{st} W_\pi$ unconditionally.*

PROOF: Let H be a nondecreasing function of $\mathbf{W}$. Then conditioned on T_n,

$$E_{\pi'} H(\mathbf{W}) = E_{\pi'} H(\mathbf{g}(\mathbf{T})) = E_{\pi'} H^*(\mathbf{T}) \tag{3.1}$$

where $H^*(\mathbf{T})$ is a nonincreasing function of $\mathbf{T}$. This latter fact and Theorem 2.3 imply that

$$E_{\pi'} H^*(\mathbf{T}) \leq E_\pi H^*(\mathbf{T}) = E_\pi H(\mathbf{W}). \; || \tag{3.2}$$

REMARK 3.3: Waiting times are appropriately defined for situations other than when X_i are independent Bernoulli variables with varying probabilities p_i. Hence the development of the previous section is felt to be applicable under many additional models. The partial sums themselves are quantities that have practical interpretations.

REFERENCES

Boland, P. J., El-Neweihi, E., and Proschan, F. (1991) Stochastic order for inspection and repair policies. *The Annals of Applied Probability* **1**, 207-218.

Hollander, M., Proschan, F., and Sethuraman, J. (1977) Functions decreasing in transposition and their applications in ranking problems. *Ann. Statist.* **5**, 722-733.

Marshall, A. W. and Olkin, I. (1979) *Inequalities: Theory of Majorization and its Applications.* Academic Press.

Rutgers University
New Brunswick
New Jersey 08903
USA

J.K. Ghosh, S.K. Mitra, K.R. Parthasarathy and B.L.S. Prakasa Rao (*eds.*),
Statistics and Probability : A Raghu Raj Bahadur Festschrift
Published by Wiley Eastern Limited, Publishers, (1993), 171 - 190

Large Deviations for Sequences of Mixtures

I. H. DINWOODIE
Tulane University

and

S. L. ZABELL
Northwestern University

Abstract

Large deviations for sequences of mixtures more general than exchangeable sequences are studied.

1 Introduction

Let Θ be a topological space, and for each $\theta \in \Theta$, let $\{P_\theta^n : n \geq 1\}$ be a sequence of probability measures on the Borel sets of a topological space X such that the function $\theta \to P_\theta^n(A)$ is Borel measurable for every integer $n \geq 1$ and Borel set $A \subset X$. Given a probability measure μ on the Borel sets of Θ, let

$$P^n(A) =: \int_\Theta P_\theta^n(A)\, d\mu(\theta).$$

AMS Subject Classification : 60F10
Key Words : large deviations, mixtures, Markov chains, exponential continuity.

In this paper we continue the study, begun in Dinwoodie and Zabell (1992), of the large deviation behavior of the sequence of mixtures $\{P^n\}$, in terms of the large deviation behavior of the component sequences $\{P_\theta^n\}$. The basic result of our earlier paper was the following:

THEOREM 1.1 Let $A \subset X$ be measurable, let Θ be first countable, let $S(\mu)$ be the support of μ, and for each $\theta \in \Theta$, let $\Lambda_\theta(A)$ be a number, $0 \leq \Lambda_\theta(A) \leq \infty$. If

$$\liminf_{n\to\infty} \frac{1}{n} \log P_{\theta_n}^n(A) \geq -\Lambda_\theta(A) \tag{1.1}$$

whenever $\theta_n \to \theta \in S(\mu)$, then

$$\liminf_{n\to\infty} \frac{1}{n} \log P^n(A) \geq -\inf\{\, \Lambda_\theta(A) : \theta \in S(\mu)\,\}. \tag{1.2}$$

2. If $S(\mu)$ is compact, and

$$\limsup_{n\to\infty} \frac{1}{n} \log P_{\theta_n}^n(A) \leq -\Lambda_\theta(A), \tag{1.3}$$

whenever $\theta_n \to \theta \in S(\mu)$, then

$$\limsup_{n\to\infty} \frac{1}{n} \log P^n(A) \leq -\inf\{\, \Lambda_\theta(A) : \theta \in S(\mu)\,\}. \tag{1.4}$$

In the theory of large deviations the set function $\Lambda_\theta(A)$ ordinarily arises as the infimum of a point function. Thus, let $\{\lambda_\theta : \theta \in \Theta\}$ be a family of functions such that $\lambda_\theta : X \to [0,\infty]$ for every $\theta \in \Theta$, and let $\Lambda_\theta(A) = \inf\{\lambda_\theta(x) : x \in A\}$. Given a family of sequences $\{P_\theta^n : \theta \in \Theta\}$, say that the family $\{P_\theta^n\}$ is *exponentially continuous* with respect to $\{\lambda_\theta\}$ and $S(\mu)$ if for this choice of Λ_θ the lower bound (1.1) is satisfied for every open set A, and the upper bound (1.3) is satisfied for every closed set A. In this case the sequence $\{P_\theta^n : n \geq 1\}$ necessarily satisfies a large deviation principle (LDP) with rate function λ_θ for each $\theta \in S(\mu)$, and the sequence $\{P^n : n \geq 1\}$ an LDP with rate function $\lambda(x) =: \inf\{\lambda_\theta(x) : \theta \in S(\mu)\}$.

For example, if E is a locally convex space, $\{X_j : j \geq 1\}$ a sequence of E-valued random vectors which are independent and identically distributed with respect to a probability measure P_θ, and

$$P_\theta^n(A) =: P_\theta\{\frac{1}{n}(X_1 + X_2 + \cdots + X_n) \in A\},$$

then under conditions of considerable generality it can be shown that the family $\{P_\theta^n\}$ is exponentially continuous with respect to the classical Cramér-Chernoff rate function $\lambda_\theta(v) = \sup_{\xi\in E^*} \langle v, \xi\rangle - \log E_\theta[\exp\langle X_1, \xi\rangle]$, where E^* is the topological dual of E (Dinwoodie and Zabell, 1990, Theorem 3.1);

a corresponding large deviation result for the sequence of sample means arising from an exchangeable random sequence then immediately follows (Dinwoodie and Zabell, 1990, Theorem 4.1).

In the present paper these results are extended in several ways. Sections 2 and 3 discuss to what extent the exponential continuity conditions (1.1) and (1.3) may be weakened without loss of the corresponding bounds (1.2) and (1.4). In general, one would hope that if the rate functions $\lambda_\theta(x)$ vary continuously in an appropriate function space, then at least some large deviation properties would also vary continuously. In Section 4 an appropriate form of such functional continuity is discussed. In Section 5 the requirement in Theorem 1.1 above that the parameter space Θ be compact is eliminated given suitable restrictions on the family $\{P_\theta^n\}$. In Section 6 these results are then applied to the case of parametrized families of Markov chains: an exponential continuity property is established for functionals of a Markov chain, giving a large deviation theorem for Markov exchangeable sequences (see, e.g., Diaconis and Freedman, 1984 and 1987, for the definition and basic properties of such sequences).

2 The Lower Bound for Mixtures

Let $A \subset X$ be a fixed measurable set. Let us say that a nonnegative function $\Lambda : \Theta \to [0, \infty]$ is a (lower) *parametric rate function* (for A) if for every $\theta \in \Theta$,

$$\liminf_{n\to\infty} \frac{1}{n} \log P_\theta^n(A) \geq -\Lambda(\theta); \tag{2.1}$$

and that $\Lambda(\theta)$ is a (lower) *mixture rate function* (for A) if for every probability measure μ on Θ,

$$\liminf_{n\to\infty} \frac{1}{n} \log P^n(A) \geq -\inf\{\Lambda(\theta) : \theta \in S(\mu)\}, \tag{2.2}$$

where P^n is the μ-mixture of $\{P_\theta^n : \theta \in \Theta\}$.

It is immediate that any mixture rate function Λ is a parametric rate function: for each θ, consider the mixing measure $\mu = \delta_\theta$ which concentrates all mass on the point θ. On the other hand, not every function Λ which satisfies (2.1) will also satisfy (2.2). In this section we describe the optimal parametric and mixture rate functions, determine the relationship between the two, and use this relationship to give a simple sufficient condition to insure that a function satisfying (2.1) also satisfies (2.2).

The optimal lower parametric rate function is of course given by the exact lower bound

$$\alpha(\theta) =: -\liminf_{n\to\infty} \frac{1}{n} \log P_\theta^n(A). \tag{2.3}$$

Unfortunately, the function $\alpha(\theta)$ can only be explicitly calculated in very special circumstances; and the bounds provided by the present theory of large deviations specify the actual limits attained only in very special cases. For example, even in the simple i.i.d. setting, the classical lower bound for an open set does not always coincide with the limit inferior in (2.1) if the set is not convex; see the important example constructed by Slaby (1988).

Examples where α is not upper semicontinuous suggest considering the function

$$\alpha^*(\theta) =: \inf_{U \ni \theta} \{\sup\{\alpha(\gamma) : \gamma \in U\}\}. \tag{2.4}$$

The function α^* is the upper envelope (or upper semicontinuous regularization) of α. It is immediate that α^* is upper semicontinuous, that $\alpha^* \geq \alpha$, and that α^* is the least function having this property: i.e., if $\Lambda : \Theta \to [0, \infty]$ is any other upper semicontinuous function such that $\Lambda \geq \alpha$, then $\Lambda \geq \alpha^*$.

THEOREM 2.1 Suppose μ is a probability measure on Θ and P^n is the μ-mixture of $\{P^n_\theta : \theta \in \Theta\}$. Then

1. $$\liminf_{n\to\infty} \frac{1}{n} \log P^n(A) \geq -\inf\{\alpha^*(\theta) : \theta \in S(\mu)\}.$$

2. If $\Lambda : \Theta \to [0, \infty]$ is an upper semicontinuous function such that

$$\liminf_{n\to\infty} \frac{1}{n} \log P^n_\theta(A) \geq -\Lambda(\theta) \quad \text{for every } \theta \in \Theta \tag{2.5}$$

then

$$\liminf_{n\to\infty} \frac{1}{n} \log P^n(A) \geq -\inf\{\Lambda(\theta) : \theta \in S(\mu)\}. \tag{2.6}$$

PROOF Let θ be an arbitrary element of $S(\mu)$; it suffices to show that

$$\liminf_{n\to\infty} \frac{1}{n} \log P^n(A) \geq -\alpha^*(\theta). \tag{2.7}$$

We may assume that $\alpha^*(\theta) < \infty$. Let $\epsilon > 0$ and N_θ a neighborhood of θ such that $\alpha^*(\gamma) < \alpha^*(\theta) + \epsilon$, $\gamma \in N_\theta \subset \Theta$; such a neighborhood N_θ exists by the upper semicontinuity of α^*. Now let

$$B_i = \{\gamma \in N_\theta : P^n_\gamma(A) \geq \exp[-n(\alpha(\gamma) + \epsilon)], \quad n \geq i\}.$$

Clearly $N_\theta = \bigcup_{i\geq 1} B_i$ and since $\theta \in S(\mu)$, $\mu(N_\theta) > 0$; hence one of the B_i, say B_t, must also have positive measure. Thus for $n \geq t$,

$$\begin{aligned} P^n(A) &\geq \int_{B_t} P^n_\gamma(A)\, d\mu(\gamma) \\ &\geq \int_{B_t} e^{-n[\alpha(\gamma)+\epsilon]}\, d\mu(\gamma) \\ &\geq \int_{B_t} e^{-n[\alpha^*(\theta)+2\epsilon]}\, d\mu(\gamma) \\ \text{hence} \qquad \liminf_{n\to\infty} \tfrac{1}{n} \log P^n(U) &\geq -\alpha^*(\theta) - 2\epsilon. \end{aligned}$$

Since $\epsilon > 0$ is arbitrary, the first part of the theorem follows. The second half of the theorem is an immediate consequence of the first: it follows from (2.5) that $\Lambda \geq \alpha$, hence $\Lambda \geq \alpha^*$. □

REMARK 2.1 A simple sufficient condition for the upper semicontinuity of $\Lambda_\theta(A)$ is that the rate function $\lambda_\theta(x)$ be upper semicontinuous in θ for every $x \in X$ (the infimum of a family of upper semicontinuous functions is again upper semicontinuous). This sufficient condition, however, is far from necessary: even in the exchangeable case, where mixtures of i.i.d. sequences with common law $\pi(\theta)$ are involved, where λ_θ is the classical Cramér-Chernoff rate function and the parameterization by Θ is continuous in the weak topology, $\lambda_\theta(x)$ need not be upper semicontinuous in θ, although it can be shown that in this case $\Lambda_\theta(U)$ is *always* upper semicontinuous when U is open. (If, for example, $\Theta = [0,1]$, and $\pi(\theta) = \delta_\theta$, then $\lambda_\theta(0) = 0$ for $\theta = 0$, but $\lambda_\theta(0) = \infty$, $0 < \theta \leq 1$.) The upper semicontinuity of $\lambda_\theta(x)$ in x is known to hold in a number of interesting cases; see, e.g., Section 4 of Bahadur, Gupta, and Zabell (1980).

It is immediate that any (lower) mixture rate function Λ must satisfy the inequality $\Lambda(\theta) \geq \alpha(\theta)$ for all $\theta \in \Theta$, while any function $\Lambda(\theta)$ satisfying the inequality $\Lambda(\theta) \geq \alpha^*(\theta)$ for every $\theta \in \Theta$ is necessarily, by Theorem 2.1, a mixture rate function. The question naturally arises as to the optimality of α^*. Given a family $\{P_\theta^n\}$, let F denote the set of all possible mixture rate functions corresponding to that family, and let $\Lambda_L(\theta) =: \inf\{\Lambda(\theta) : \Lambda \in F\}$.

THEOREM 2.2 1. $\Lambda_L \in F$. 2. If $\Lambda \in F$, then $\Lambda_L \leq \Lambda$. 3. The function Λ_L satisfies the inequalities $\alpha \leq \Lambda_L \leq \alpha^*$, and $\Lambda_L^* = \alpha^*$.

PROOF If $\Lambda_L \notin F$, then there exists a μ-mixture P and a point $\theta \in S(\mu)$ for which (2.2) fails; i.e., such that $\liminf_{n\to\infty} n^{-1} \log P^n(A) < -\Lambda_L(\theta)$. By the definition of Λ_L, there then exists a $\Lambda \in F$ such that (2.2) also fails for that Λ and $\mu = \delta_\theta$, which is impossible. Thus $\Lambda_L \in F$.

That $\Lambda_L \leq \Lambda$ for any $\Lambda \in F$ is immediate. Because $\alpha \leq \Lambda$ for any $\Lambda \in F$, clearly $\alpha \leq \Lambda_L$; it then follows from Theorem 2.1 that α^* is a mixture rate function, and thus $\alpha \leq \Lambda_L \leq \alpha^*$. Since α^* is the least upper semicontinuous function lying above α, it follows that $\Lambda_L^* \geq \alpha^*$, and therefore $\Lambda_L^* = \alpha^*$. □

The last theorem in this section clarifies the role that the exponential continuity condition (1.1) plays vis-à-vis the lower bound. Let $\Lambda : \Theta \to [0,\infty]$ and consider the following four conditions on Λ:

(L1) If $\theta_n \to \theta \in \Theta$, then

$$\liminf_{n\to\infty} \frac{1}{n} \log P_{\theta_n}^n(A) \geq -\Lambda(\theta).$$

(L2) For every $\theta \in \Theta$ and any $\epsilon > 0$, there exists a neighborhood U_θ of θ and an integer $N_\theta \geq 1$ such that for all $\gamma \in U_\theta$ and all $n \geq N_\theta$,

$$P_\gamma^n(A) \geq e^{-n(\Lambda(\theta)+\epsilon)}.$$

(L3) For every $\theta \in \Theta$ and any $\epsilon > 0$, there exists a neighborhood U_θ of θ

such that for all $\gamma \in U_\theta$ there exists an integer N_γ with

$$P_\gamma^n(A) \geq e^{-n(\Lambda(\theta)+\epsilon)}, \quad n \geq N_\gamma.$$

(L4) For every $\theta \in \Theta$, $\Lambda(\theta) \geq \alpha^*(\theta)$.

Example 4.1 below illustrates that (L3) may hold without (L2). The precise relationship between the four conditions is given by the next theorem.

THEOREM 2.3 (L1) $\Leftrightarrow$ (L2) $\Rightarrow$ (L3) $\Leftrightarrow$ (L4).

PROOF To see that (L1) implies (L2), argue by contradiction: if (L2) did not hold for some θ and $\epsilon > 0$, then one can construct a sequence $\theta_n \to \theta$ for which (L1) failed. It is immediate that (L2) implies (L1). That (L2) implies (L3) is trivial. Finally, if (L3) holds, then it follows that for any $\gamma \in U_\theta$

$$-\alpha(\gamma) = \liminf_{n\to\infty} \frac{1}{n} \log P_\gamma^n(A) \geq -\Lambda(\theta) - \epsilon \tag{2.8}$$

and thus that $\alpha^*(\theta) \leq \Lambda(\theta) + \epsilon$. Since $\epsilon > 0$ was arbitrary, (L4) follows. Conversely, if (L4), then, by definition, $\inf_{U\ni\theta}\{\sup_{\gamma\in U}\{\alpha(\gamma)\}\} \leq \Lambda(\theta)$, and thus for any $\epsilon > 0$, there exists a neighborhood U of θ such that for every $\gamma \in U$, (2.8) holds, and thus (L3) follows. □

3 The Upper Bound for Mixtures

Let $A \subset X$ again denote a fixed measurable set. Corresponding to the above analysis for the lower bound, define in parallel fashion upper parametric and mixture rate functions, and the optimal upper parametric rate function $\beta(\theta)$:

$$\beta(\theta) =: -\limsup_{n\to\infty} \frac{1}{n} \log P_\theta^n(A).$$

Let $\beta_*(\theta)$ denote the lower envelope of $\beta(\theta)$, that is, the greatest lower semicontinuous function lying under $\beta(\theta)$, and let $\Lambda^U(\theta) =: \sup\{\Lambda(\theta) : \Lambda \in G\}$ denote the supremum over the family G of all upper mixture rate functions, i.e., the optimal mixture rate function.

In the case of the lower bound the inequality $\alpha \leq \Lambda_L \leq \alpha^*$ holds (and hence that $\Lambda_L^* = \alpha^*$). In the case of the upper bound, however, the analogous inequality $\beta_* \leq \Lambda^U \leq \beta$ need not hold: the possibility $\Lambda^U(\theta) < \beta_*(\theta)$ can occur for some $\theta \in \Theta$. Example 3.1 illustrates this phenomenon.

EXAMPLE 3.1 Let $\Theta = [0,1]$, let $P_0^n(A) = 0$, $n \geq 1$; and for $0 < \theta \leq 1$, let $P_\theta^n(A) = 1$, $1 \leq n \leq \theta^{-1}$, and $P_\theta^n(A) = 0$, $n > \theta^{-1}$. Then $\beta(\theta) \equiv \infty$ and $\beta_* = \beta$, since β is lower semicontinuous. But $\Lambda^U < \beta_*$: if $\Lambda^U(\theta)$ were infinite for all $\theta \in \Theta$, then for any probability measure μ on Θ for which $\lim_{n\to\infty} n^{-1}\mu[0, n^{-1}] = 0$, the μ-mixture P^n would satisfy the inequality

$$\limsup_{n\to\infty} \frac{1}{n} \log P^n(A) = 0 > -\infty = -\inf\{\Lambda^U(\theta) : \theta \in S(\mu)\},$$

which by the definition of Λ^U is impossible. In fact, it follows from Theorem 3.1 below that $\Lambda^U(0) = 0$ and $\Lambda^U(\theta) = \infty$ for $0 < \theta < 1$.

Because of such possibilities, it is necessary to find a replacement for β_* to serve as a lower bound for Λ^U. Let $\Psi(\theta, N) =: -\sup_{n \geq N}[n^{-1} \log P_\theta^n(A)]$ and, with $\Psi_*(., N)$ the lower envelope of $\Psi(., N)$, let $\Xi(\theta) =: \sup\{\Psi_*(\theta, N) : N \geq 1\}$. The lower semicontinuous function $\Xi(\theta)$ plays a role for the upper bound analogous to that of α^* for the lower bound.

THEOREM 3.1 Suppose μ is a probability measure on Θ and P^n is the μ-mixture of $\{P_\theta^n : \theta \in \Theta\}$.

1. If Θ is compact, then $\limsup_{n\to\infty} \dfrac{1}{n} \log P^n(A) \leq -\inf\{\Xi(\theta) : \theta \in S(\mu)\}$.

2. If $\Lambda_*^U(\theta)$ is the lower envelope of the optimal upper mixture rate function $\Lambda^U(\theta)$, then $\Lambda_*^U(\theta) \leq \Xi(\theta)$.
3. If Θ is compact, then $\Lambda_*^U(\theta) = \Xi(\theta)$.

PROOF The first statement is easily proved using the lower semicontinuity of Ξ and the compactness of Θ. To prove the second we must show that any lower semicontinuous mixture rate function Λ is below Ξ. If this were not true there would be a point $\theta \in \Theta$ and number $c > 0$ such that $\Xi(\theta) < c < \Lambda(\theta) \leq \infty$. By the lower semicontinuity of Λ, there is a neighborhood N_θ of θ such that $c < \inf\{\Lambda(\gamma) : \gamma \in N_\theta\}$. By the definition of Ξ, there is a neighborhood $U_\theta \subset N_\theta$ of θ such that

$$\limsup_{n\to\infty} \sup_{\gamma \in U_\theta} \frac{1}{n} \log P_\gamma^n(A) > -c.$$

Let $\{n_k : k \geq 1\}$ be a subsequence such that for each $k \geq 1$,

$$\sup_{\gamma \in U_\theta} \frac{1}{n_k} \log P_\gamma^{n_k}(A) > -c.$$

and choose $\theta_k \in U$ such that $\dfrac{1}{n_k} \log P_{\theta_k}^{n_k} > -c$. Now let $\mu(\theta_k) = \dfrac{1}{\zeta(2)k^2}$. Then

$$\begin{aligned}\limsup_{n\to\infty} \tfrac{1}{n} \log P^n(A) &\geq \limsup_{k\to\infty} \frac{1}{n_k} \log \mu(\theta_k) P_{\theta_k}^{n_k}(A) \geq -c \\ &> -\inf\{\Lambda(\gamma) : \gamma \in N_\theta\} \\ &\geq -\inf\{\Lambda(\gamma) : \gamma \in S(\mu)\}.\end{aligned}$$

Thus Λ is not a mixture rate function under the assumption that Λ is not below Ξ, which proves the assertion. The final assertion follows from the first two: the first assertion shows that $\Xi(\theta)$ is an upper mixture rate function, and hence by definition $\Lambda_*^U(\theta) \geq \Xi(\theta)$; the second shows that $\Lambda_*^U(\theta) \leq \Xi(\theta)$. □

One can deduce from Theorem 3.1 an optimality property for Ξ analogous to that of α^* provided Θ is compact. The next theorem states, in part, that if Θ is compact, then Ξ is an upper mixture rate function and that every lower semicontinuous upper mixture rate function lies below Ξ. Consider the four conditions (U1)–(U4) defined as follows:

(U1) If $\theta_n \to \theta \in \Theta$, then

$$\limsup_{n\to\infty} \frac{1}{n} \log P^n_{\theta_n}(A) \leq -\Lambda(\theta).$$

(U2) For every $\theta \in \Theta$ and any $c < \Lambda(\theta)$, there exists a neighborhood U_θ of θ and an integer $N_\theta \geq 1$ such that for all $\gamma \in U_\theta$ and all $n \geq N_\theta$,

$$P^n_\gamma(A) \leq e^{-nc}.$$

(U3) For every $\theta \in \Theta$ and any $c < \Lambda(\theta)$, there exists a neighborhood U_θ of θ such that for all $\gamma \in U_\theta$ there exists an integer N_γ with

$$P^n_\gamma(A) \leq e^{-nc}, \qquad n \geq N_\gamma.$$

(U4) For every $\theta \in \Theta$, $\Lambda(\theta) \leq \beta_*(\theta)$.

THEOREM 3.2 (U1) $\Leftrightarrow$ (U2) $\Rightarrow$ (U3) $\Leftrightarrow$ (U4).

PROOF *Mutatis mutandis*, the same as that for Theorem 2.3. □

THEOREM 3.3 A function $\Lambda(\theta)$ satisfies condition (U2) if and only if $\Lambda(\theta) \leq \Xi(\theta)$.

PROOF If Λ satisfies condition (U2), then it follows from the definition of Ξ that $\Lambda \leq \Xi$. Conversely, Ξ satisfies condition (U2) and so if $\Lambda \leq \Xi$ then Λ will certainly satisfy condition (U2) as well. □

REMARK 3.1 Theorems 3.1 and 3.3 together imply that when Λ is a lower semicontinuous function and Θ is compact, then Λ is an upper mixture rate function if and only if Λ satisfies the uniform asymptotic condition (U2).

4 Epi-Convergence and Large Deviations

In this section we assume that for each $\theta \in \Theta$ the sequence $\{P^n_\theta\}$ satisfies a large deviation principle with a rate function $\lambda_\theta : X \to [0, \infty]$. Thus, by definition, λ_θ is lower semicontinuous, and for $\theta_n \equiv \theta$, the set function $\Lambda_\theta(A) =: \inf\{\lambda_\theta(x) : x \in A\}$ satisfies (1.1) for every open set A, and (1.3) for every closed set A. It will be seen that certain continuity properties for P^n_θ in θ can be characterized precisely in terms of a form of functional convergence for the family $\{\lambda_\theta : \theta \in \Theta\}$ known as "epi-convergence."

A sequence of extended real valued functions $f_n : X \to [0, \infty]$ is said to be *epiconvergent* to the function f at the point x if

(4.1) there exists a sequence $x_n \to x$ such that $\limsup f_n(x_n) \leq f(x)$;

(4.2) for every sequence $x_n \to x$, $\liminf_{n\to\infty} f_n(x_n) \geq f(x)$.

If X is first countable, the joint lower semicontinuity of $\lambda_\theta(x)$ is obviously equivalent to property (4.2) holding for every family $\{\lambda_{\theta_n}\}$ for which $\theta_n \to \theta \in \Theta$.

The following lemma provides the link between epi-convergence and the theory of large deviations; it was suggested by Remark 2.82 of Attouch (1984), and is a refinement of a result in Vervaat (1981).

LEMMA 4.1 Let X be a first countable Hausdorff topological space, let $\{f_n : n \geq 0\}$, be a sequence of nonnegative extended real-valued functions on X, and let $f_n(A) =: \inf\{f_n(x) : x \in A\}$.

1. For every $x \in X$, there exists a sequence $x_n \to x$ such that $\limsup_{n\to\infty} f_n(x_n) \leq f_0(x) \Leftrightarrow$ for every open set $U \subset X$, $\limsup_{n\to\infty} f_n(U) \leq f_0(U)$.

2. If f_0 is lower semicontinuous, then for every $x \in X$ and sequence $x_n \to x$, $\liminf_{n\to\infty} f_n(x_n) \geq f_0(x) \Leftrightarrow$ for every compact set $K \subset X$, $\liminf_{n\to\infty} f_n(K) \geq f_0(K)$.

PROOF (1) $\Rightarrow$: Let $U \subset X$ be open. Given $\epsilon > 0$, there exists $x \in U$ such that $f_0(x) \leq f_0(U) + \epsilon$. Choose $x_n \to x$ such that $\limsup_{n\to\infty} f_n(x_n) \leq f_0(x)$; clearly $\limsup_{n\to\infty} f_n(U) \leq \limsup_{n\to\infty} f_n(x_n) \leq f_0(x) \leq f_0(U)+\epsilon$, and since $\epsilon > 0$ was arbitrary, it follows that $\limsup_{n\to\infty} f_n(U) \leq f_0(U)$.

(1) $\Leftarrow$: Given $x \in X$, let $\{U_k : k \geq 1\}$ be a decreasing sequence of open neighborhoods of x such that $U_k \downarrow x$. Since $\limsup_{n\to\infty} f_n(U_k) \leq f_0(U_k)$, there exists an increasing sequence of integers $\{n_k : k \geq 1\}$ such that $f_n(U_k) \leq f_0(U_k) + 1/k$ for every $k \geq 1$ and $n \geq n_k$. For n in the range $n_k \leq n < n_{k+1}$, choose $x_n \in U_k$ so that $f_n(x_n) \leq f_n(U_k) + 1/k$, and therefore $f_n(x_n) \leq f_0(x) + 2/k$. Then $\limsup_{n\to\infty} f_n(x_n) \leq f_0(x)$, and since $x_n \to x$, the result follows.

(2) $\Rightarrow$: Let $K \subset X$ be compact. If $a =: \liminf_{n\to\infty} f_n(K) < \beta =: f_0(K)$, then there exists a sequence $\{x_n\} \subset K$ such that $\liminf_{n\to\infty} f_n(x_n) = \alpha$. Because K is compact, and X is first countable, K is sequentially compact, hence there exists a convergent subsequence $\{x_{n_j}\}$ of $\{x_n\}$ such that $\lim_{j\to\infty} f(x_{n_j}) = \alpha$; let $x =: \lim_{j\to\infty} x_{n_j} \in K$. Then $\lim_{j\to\infty} f_{n_j}(x_{n_j}) = \alpha < \beta = f_0(K) \leq f_0(x)$, which by assumption is impossible; thus $\liminf_{n\to\infty} f_n(K) \geq f_0(K)$.

(2) $\Leftarrow$: Suppose there exist $x \in X$ and $x_n \to x$ such that $\alpha =: \liminf_{n\to\infty} f_n(x_n) < \beta =: f_0(x)$; we show that this leads to a contradiction. Choose γ such that $\alpha < \gamma < \beta$. Because f_0 is lower semicontinuous and $x_n \to x$, there exists an open neighborhood U of x such that $f_0(U) \geq \gamma$ and an integer N such that $x_n \in U$ for $n \geq N$. Let $K =: \bigcup\{x_n : n \geq N\} \cup \{x\}$. Then $\liminf_{n\to\infty} f_n(K) \leq \alpha < \gamma \leq f_0(K)$, which by hypothesis is impossible. □

As usual, if a sequence $\{P_\theta^n\}$ satisfies a large deviation principle with rate function $\lambda_\theta : X \to [0, \infty]$, then $\Lambda_\theta(A) =: \inf\{\lambda_\theta(x) : x \in A\}$.

THEOREM 4.1 If X is first countable and $\{P_\theta^n\}$ satisfies a large deviation principle with rate function $\lambda_\theta : X \to [0, \infty]$ for every $\theta \in \Theta$, then the following three conditions are equivalent:

(i) $\{\lambda_\theta : \theta \in \Theta\}$ satisfies the left epi-convergence condition (4.1) for every sequence $\theta_n \to \theta$;

(ii) $\Lambda_\theta(U)$ is upper semicontinuous in θ for every open set $U \subset E$;

(iii) $\Lambda_\theta(U)$ satisfies (L3) for every open set $U \subset E$.

PROOF That (i) $\Rightarrow$ (ii) is Lemma 4.1; and that (ii) $\Rightarrow$ (iii) follows from Theorems 2.1 and 2.2: if $\Lambda_\theta(U)$ is upper semicontinuous in θ, then it satisfies (L4), hence (L3). To see that (iii) $\Rightarrow$ (i), let $(\theta, x) \in \Theta \times X$, let $\theta_n \to \theta$, and let $\{V_k : k \geq 1\}$ be a countable open neighborhood base for x, such that each V_k is open. By (L3), for each V_k there exists an open neighborhood U_k of θ such that for every $\gamma \in U_k$, $P_\gamma^m(A) \geq \exp\{-m(\Lambda_\theta(V_k) + 1/k)\}$ for all m sufficiently large. Since $\theta_n \to \theta$ and the U_k are open, there exists an increasing sequence of positive integers $1 \leq n_1 < n_2 < n_3 < \ldots$ such that $\theta_n \in U_k$ whenever $n \geq n_k$. Again by (L3), for any $k \geq 1$, and every n such that $n_k \leq n < n_{k+1}$, there exists an integer M_k such that

$$-\frac{1}{m} \log P_{\theta_n}^m(V_k) \leq \Lambda_\theta(V_k) + \frac{1}{k}$$

whenever $m \geq M_k$. For every such k and n, there exists an element $x_{n,k} \in \bar{V}_k$ such that $\lambda_{\theta_n}(x_{n,k}) - \frac{1}{k} \leq \Lambda_{\theta_n}(\bar{V}_k)$, and since λ_θ is a rate function for every θ,

$$\Lambda_{\theta_n}(\bar{V}_k) \leq -\frac{1}{m} \log P_{\theta_n}^m(\bar{V}_k)$$

for all m sufficiently large. It follows that for every $k \geq 1$, n such that $n_k \leq n < n_{k+1}$, and x_n chosen as above, $\lambda_{\theta_n}(x_{n,k}) \leq \lambda_\theta(x) + \frac{2}{k}$, and hence $\limsup_n \lambda_{\theta_n}(x_{n,n}) \leq \lambda_\theta(x)$. Since $x_{n,n} \to x$, the result follows. □

Theorem 4.2 establishes a similar equivalence between the right epi-convergence condition (4.2) and the pointwise asymptotic property (U3) for compact sets.

THEOREM 4.2 Suppose X is first countable and that probability measures on X are regular. If $\{P_\theta^n\}$ satisfies a large deviation principle with rate function $\lambda_\theta : X \to [0, \infty]$ for every $\theta \in \Theta$, then the following three properties are equivalent:

(i) $\{\lambda_\theta : \theta \in \Theta\}$ satisfies the right epi-convergence condition (4.2) for every sequence $\theta_n \to \theta$;

(ii) $\Lambda_\theta(K)$ is lower semicontinuous in θ for every compact set $K \subset E$;

(iii) $\Lambda_\theta(K)$ satisfies (U3) for every compact set $K \subset E$.

PROOF That (i) $\Rightarrow$ (ii) is Lemma 4.1; that (ii) $\Rightarrow$ (iii) is nearly immediate. Namely, suppose that $\Lambda_\theta(K)$ is lower semicontinuous in θ. Let $c < \Lambda_\theta(K)$; since $\Lambda_\theta(K)$ is lower semicontinuous, there exists a neighborhood U_θ of θ such that $\Lambda_\gamma(K) > c$ for every $\gamma \in U_\theta$. But P_γ^n satisfies a large deviation principle, hence there exists an integer N_γ such that $P_\gamma^n(K) \le e^{-nc}$ for every $n \ge N_\gamma$, and (U3) follows.

It remains to show that (iii) $\Rightarrow$ (i). Suppose that (4.2) does not hold. Then there exists a pair $(\theta, v) \in \Theta \times X$ and a sequence $(\theta_k, v_k) \to (\theta, v)$ such that $\lambda_{\theta_k}(v_k) + 3\epsilon < \lambda_\theta(v)$ for some $\epsilon > 0$. Since λ_θ is lower semicontinuous on X, there is a neighborhood U of v such that $\Lambda_\theta(U) > \lambda_\theta(v) - \epsilon$. Let $\{U_k^0 : k \ge 1\}$ be a countable neighborhood base of $v \in X$ such that $U_{k+1}^0 \subset U_k^0 \subset U$. By taking a subsequence of $\{v_k : k \ge 1\}$ if necessary we can assume without loss of generality that $v_k \in U_k^0$.

For each $k \ge 1$, let $\{U_k^i : i \ge 1\}$ be a countable open neighborhood base at v_k with $U_k^i \subset U$. For each $k \ge 1$ and $i \ge 1$, there exists n_i^k such that for $n \ge n_i^k$,

$$P_{\theta_k}^n(U_k^i \cap U_k^0) > \exp\{-n(\Lambda_{\theta_k}(U_i^k \cap U_k^0) + \epsilon)\}.$$

We can assume that $n_1^k < n_2^k < \ldots$. Now let K_i^k be a compact set contained in $U_i^k \cap U_k^0$ such that

$$P_{\theta_k}^{n_i^k}(K_i^k) > e^{-n_i^k(\Lambda_{\theta_k}(U_i^k \cap U_k^0) + \epsilon)}.$$

Let $K^k = \bigcup_{i \ge 1} K_i^k \cup \{v_k\}$. Then K^k is compact, and if we set $K = \bigcup_{k \ge 1} K^k \cup \{v\}$ then $K \subset U$, since $U_i^k \subset U$, and K is compact. (If $\{U_\alpha\}$ is an open covering of K, then $v \in U_{\alpha_0}$, and since $v_k \in U_k^0 \subset U_{\alpha_0}$ for k large, $K^k \subset U_k^0 \subset U_{\alpha_0}$.)

It suffices to show that the compact set K does not have property (U3). Since $\lambda_{\theta_k}(v_k) + \epsilon \le \lambda_\theta(v) - 2\epsilon \le \Lambda_\theta(U) - \epsilon \le \Lambda_\theta(K) - \epsilon$, it follows that for $i \ge 1$,

$$\begin{aligned} P_{\theta_k}^{n_i^k}(K) \ge P_{\theta_k}^{n_i^k}(K_i^k) \quad &> e^{-n_i^k}(\Lambda_{\theta_k}(U_i^k \cap U_k^0) + \epsilon) \\ &\ge e^{-n_i^k}(\lambda_{\theta_k}(v_k) + \epsilon) \\ &\ge e^{-n_i^k}(\Lambda_\theta(K) - \epsilon) \end{aligned}$$

which contradicts (U3). This proves the theorem. □

The following example illustrates the difference between properties (L1) and (L3), as well as that between (U1) and (U3).

EXAMPLE 4.1 Let $\Theta = \{1, 2, 3, \ldots \infty\}$ be the topological space of extended positive integers. (a) Let $X = \{x_1, x_2\}$; let $P_\theta^n(x_1) = 1$ along the diagonal $1 \le n = \theta < \infty$; and let $P_\theta^n(x_1) = 0$ otherwise (so that $P_\infty^n(x_1) \equiv 0$). Then $\lambda_\theta(x_1) \equiv \infty$ and $\lambda_\theta(x_2) \equiv 0$, hence $\lambda_\theta(x)$ is jointly continuous in θ and x, and $\Lambda_\theta(A)$ is continuous in θ for any subset $A \subset X$. Thus (4.1), (4.2), (L3), and (U3) are all satisfied, but if $K = \{x_1\}$, $U = \{x_2\}$, and $\theta_n = n$, $n \ge 1$, then $\theta_n \to \infty$, yet $-\infty = n_{-1} \log P_{\theta_n}^n(U) < -\Lambda_\infty(U) = 0$, where U is open, and $0 = \lim_n n^{-1} \log P_{\theta_n}^n(K) > -\Lambda_\infty(K) = -\infty$, where K is compact, so that neither (L1) nor (U1) are satisfied.

If P_θ^n is the distribution of the sample mean $\bar{X}_n$ arising from an independent and identically distributed sequence of random vectors $X_1, X_2, X_3,$

(with respect to P_θ), and if $\lambda_\theta(v)$ is the corresponding Cramér-Chernoff rate function corresponding to P_θ , the above distinctions largely collapse. For in this case, provided the map $\theta \to P_\theta$ is continuous in the weak topology, the exponential continuity condition (1.1) is always satisfied whenever U is open (Dinwoodie and Zabell, 1992, Theorem 3.1), hence (by Theorem 2.3) $\Lambda_\theta(U)$ satisfies L3, hence (by Theorem 4.1) $\Lambda_\theta(U)$ is upper semicontinuous and $\lambda_\theta(v)$ satisfies the epi-convergence condition (4.1). If in addition $\lambda_\theta(v)$ satisfies the epi-convergence condition (4.2), then the exponential continuity condition (1.3) is satisfied for every compact set (Dinwoodie and Zabell, 1992, Theorem 3.1); thus in the i.i.d. setting (4.2) is equivalent to (1.3) for compact sets K; see also Example 5.2.

5 Upper Bounds for Noncompact Parameter Spaces

In this section we investigate to what extent it is possible to dispense with the assumption in Theorem 1.1 that $S(\mu)$ be compact. Our main result views the time at which the large deviation rate takes effect to be a random variable on the space (Θ, μ). Given a measurable set $C \subset X$, the existence of the moment generating function of this random variable on an interval $[0, \beta)$ is shown to be sufficient for the upper bound to hold for the sequence $\{P^n(C) : n \geq 1\}$.

Given a measurable set $C \subset X$, let $N(C, r) : \Theta \to [0, \infty]$ denote the function whose value at θ is

$$N_\theta(C, r) =: \inf\{n \geq 1 : P_\theta^m(C) \leq e^{-mr}, \quad m \geq n\}. \tag{5.1}$$

Let μ be a fixed probability measure on Θ; the function $N_\theta(C, r)$ will be viewed as a random variable on Θ relative to μ.

THEOREM 5.1 Let $C \subset X$ be measurable, and let $0 \leq \Lambda(C) \leq \infty$ be a fixed number. If

$$\int_\Theta e^{tN_\theta(C,r)}\, d\mu < \infty \tag{5.2}$$

for all $0 \leq t < \Lambda(C)$ and all $0 \leq r < \Lambda(C)$, then

$$\limsup_{n\to\infty} \frac{1}{n} \log P^n(C) \leq -\Lambda(C).$$

PROOF If $\Lambda(C) = 0$, the assertion is trivial, so we may assume that $\Lambda(C) > 0$. Fix $0 \leq r < \Lambda(C)$. By (5.2), if $s > 0$ is chosen such that $r < s < \Lambda(C)$, then

$$\mu\{\theta : N_\theta(C, r) > n\} \leq e^{-ns} \int_\Theta e^{sN_\theta(C,r)}\, d\mu.$$

Hence there exists N_s such that for $n \geq N_s$, $\mu\{\theta : N_\theta(C, r) > n\} \leq e^{-nr}$. Let $A =: \{\theta : N_\theta(C, r) > n\}$, and $A^c =: \{\theta : N_\theta(C, r) \leq n\}$. Then for

$n \geq N_s$,

$$P^n(C) = \int_A P^n_\theta(C)\, d\mu + \int_{A^c} P^n_\theta(C)\, d\mu \leq e^{-nr} + e^{-nr},$$

hence $\limsup_{n\to\infty} n^{-1} \log P^n(C) \leq -r$. Since $0 \leq r < \Lambda(C)$ is arbitrary, the theorem follows. □

If Θ is compact, as in Theorem 1.1, the locally uniform onset of asymptotics in (1.3) is sufficient to insure that the onset of asymptotics is in fact globally uniform. If Θ is not compact, as in Theorem 5.1, a global condition relating the onset of asymptotics to the measure μ is required; the resulting bound is thus specific to a given mixture, rather than uniform over all mixtures, as is the case when Θ is compact and (1.3) holds. The next lemma shows that the hypotheses of Theorem 5.1 are weaker than those of Theorem 1.1.

LEMMA 5.1 If Θ is compact and condition (1.3) holds for $\{\Lambda_\theta(C) \in [0,\infty] : \theta \in \Theta\}$, then condition (5.2) holds for $\Lambda(C) =: \inf\{\Lambda_\theta(C) : \theta \in \Theta\}$.

PROOF Fix $0 \leq r < \Lambda(C)$. For each $\theta \in \Theta$, there exists by (1.3) an open neighborhood U_θ and an integer $N_\theta \geq 1$ such that for all $n \geq N_\theta$

$$\sup_{\gamma \in U_\theta} P^n_\gamma(C) \leq e^{-nr}.$$

Let $\{U_{\theta_j} : 1 \leq j \leq k\}$ be a finite subcover of Θ from the covering $\{U_\theta : \theta \in \Theta\}$. Let $M(C,r) = \max_j N_{\theta_j}$. Then the random variable $N_\theta(C,r)$ is bounded above by $M(C,r)$, so condition (5.2) holds. □

The following lemma is needed for Proposition 5.1, which will be applied in Example 5.1.

LEMMA 5.2 Let $a_n \geq 0$ and $\sum a_n < \infty$. If $r_n =: \sum_{k\geq n} a_k$, then

$$\limsup_{n\to\infty} \frac{1}{n} \log r_n = \limsup_{n\to\infty} \frac{1}{n} \log a_n.$$

PROOF Since $a_n \leq r_n$, it suffices to show that

$$\limsup_{n\to\infty} \frac{1}{n} \log r_n \leq \limsup_{n\to\infty} \frac{1}{n} \log a_n.$$

Let $S(t) =: \sum_{n\geq 1} a_n t^n$; and let $R =: \sup\{t \geq 0 : S(t) < \infty\}$ denote the radius of convergence of $S(t)$. Because $\sum a_n < \infty$, clearly $R \geq 1$. Recall the Hadamard formula $R = \left\{\limsup_{n\to\infty} \sqrt[n]{a_n}\right\}^{-1}$. If $R = 1$, then $\limsup_{n\to\infty} \frac{1}{n} \log a_n = 0$, and the desired inequality is immediate. If $R > 1$, choose and fix t such that $1 \leq t < R$. Since $r_n \leq t^{-n} \sum_{k\geq n} a_k t^k \leq t^{-n} S(t)$, it follows that $\limsup_{n\to\infty} n^{-1} \log r_n \leq -\log t$; since $t < R$ was arbitrary, the desired inequality again follows. □

PROPOSITION 5.1 If Y is a nonnegative integer-valued random variable, and $M(t) = E[\exp(tY)]$, then

$$\limsup_{n\to\infty} \frac{1}{n} \log P[Y \geq n] = -\sup\{t \geq 0 : M(t) < \infty\}.$$

PROOF An immediate consequence of Lemma 5.2. □

The following example illustrates the relationship between delayed asymptotics in each component P_θ^n and the relative weight assigned by μ to each component in the mixture P^n.

EXAMPLE 5.1 Let $\Theta = \{1, 2, 3, \ldots\}$, and let $X = P(R)$ be the set of probability measures on the real line endowed with the weak topology. Let (Ω, F) be a measurable space, let $\{P_j : j \in \Theta\}$ be a family of probability measures on (Ω, F), and let $\{X_n : n \geq 1\}$ be a sequence of random variables defined on (Ω, F) which under P_j are independent and uniform on $[(2j+1)^{-1}, (2j)^{-1}]$. Let p_n denote the empirical probability measure based on the sample $\{X_i : 1 \leq i \leq n\}$, and let $P_j^n = P_j \circ p_n^{-1}$; the family $\{P_j^n : j \in \Theta\}$ is a sequence of probability measures on $P(R)$. Let $\mu_j, j \geq 1$, be a probability measure on the positive integers; and let P^n be the probability on X given by

$$P^n(A) = \sum_{j\geq n} P_j^n(A)\mu_j.$$

Let $C_j \subset X$ be the closed set of discrete measures having no more than j mass points all of which lie in the interval $[(2j+1)^{-1}, (2j)^{-1}]$, and let $C = \bigcup_{j\geq 1} C_j \cup \delta_0$; clearly the set C is closed, $P_j^n(C) = 0$ for $j < n$, and $P_j^n(C) = 1$ for $n \leq j$. It follows that $P^n(C) = \sum_{j\geq n} \mu_j$ and $N_j(C, r) = j+1$ for any $r > 0$. Thus for $t \geq 0$,

$$\int e^{tN_j(C,r)}\, d\mu(j) = \sum_{j=1}^{\infty} e^{t(j+1)}\mu_j. \tag{5.3}$$

It then follows from Proposition 5.1 and (5.3) that

$$\begin{aligned}\limsup_{n\to\infty} \tfrac{1}{n} \log P^n(C) &= \limsup_{n\to\infty} \tfrac{1}{n} \log \textstyle\sum_{j\geq n} \mu_j \\ &= -\sup\left\{t \geq 0 : \textstyle\sum_{j=1}^{\infty} e^{tj}\mu_j < \infty\right\} \\ &= -\sup\left\{t \geq 0 : \textstyle\int e^{tN_j(C,r)}\, d\mu(j) < \infty\right\}.\end{aligned}$$

Thus in this example the upper bound given by Theorem 5.1 for the exponential rate of decay of $P^n(C)$ is in fact the *exact* rate of exponential decay; this coincidence is valid for *any* mixing measure μ_j; and, since μ_j is arbitrary, any nonnegative rate can be exhibited by appropriate choice of μ_j.

Note that since $\lim_{n\to\infty} n^{-1}\log P_j^n(C) = \infty$ for every $j \geq 1$, (1.3) is satisfied by $\Lambda_\theta(C) \equiv \infty$. If, however, one applied the second half of Theorem 1.1 using $\Lambda_\theta(C) \equiv \infty$ (ignoring that Θ is not compact), the resulting bound $\Lambda(C) = \infty$ would in most cases be incorrect (depending on the choice of μ). One might instead attempt to apply Theorem 1.1 by compactifying Θ, and adding a point at infinity: $\bar{\Theta} =: \Theta \cup \{\infty\}$. In order to do this, one would have to define a sequence of probabilities $\{P_\infty^n\}$ for which condition (1.3) were satisfied. In this case one would necessarily have $\Lambda_\infty(C) = \infty$ (since for some μ_j, $\sum \exp(tj)\mu_j = \infty$ for every $t > 0$); the bound resulting from Theorem 1.1 would then be overly conservative if a measure μ were employed such that $\infty \in S(\mu)$ and $\sum \exp(tj)\mu_j < \infty$ for some $t > 0$.

EXAMPLE 5.2 Let P_θ^n be the law of the sequence of sample means of an i.i.d. sequence of random vectors taking values in a separable Banach space E, and let $K \subset E$ be compact. If the rate function λ is jointly lower semicontinuous in θ and x, then it can be shown that 1) $N(K, r)$ is locally bounded; i.e., there exists a neighborhood U_θ for each $\theta \in \Theta$ such that if $0 \leq r < \Lambda_\theta(K)$, then $\sup_{\gamma\in U_\theta} N_\gamma(K, r) < \infty$; and that 2) condition (U2) is satisfied for the compact set K. Thus in this special case the right epi-convergence condition (4.2) is equivalent to the usually stronger condition (U2) for every compact set.

6 Exponential Continuity for Markov Chains

Let S be a Polish space, let $\{\pi_\theta(x, ds) : x \in S,\ \theta \in \Theta\}$ be a family of transition probabilities on S, and let $P(S)$ be the set of probability measures on S endowed with the weak topology, i.e., the relative topology on $P(S)$ when viewed as a closed, convex, metrizable subset of the vector space $E = M(S)$ of finite signed measures on S with the weak topology. We assume throughout that the map $\pi : \Theta \times S \to P(S)$ is weakly continuous. Let $f : S \to P(S)$ be defined by $f(s) = \delta_s$, where δ_s is the point mass at s. (One could also consider, as in de Acosta (1985), a more general functional $f : S \to E$, with E a topological vector space. The results of this section can be easily extended to include this case, although certain additional conditions on the kernel π may be required, depending on the specific nature of the functional.)

For each $\theta \in \Theta$, let $P_{z:\theta}$ be the probability measure on S^∞ with transition probabilities $\{\pi_\theta(x, dy) : x \in S\}$ and initial state $z \in S$. Let Z_i, $i \geq 1$, denote the corresponding coordinate process, a Markov chain. Let p_n denote the empirical probability measure arising from $\{Z_i : 1 \leq i \leq n\}$, so that $p_n = n^{-1}\sum f(Z_i)$; and let $P_{z:\theta}^n = P_{z:\theta} \circ p_n^{-1}$ denote the distribution of this random measure. Let $C(S)$ denote the Banach space of bounded continuous functions on S endowed with the supremum norm; $C(S)$ is the topological dual E^* of E. We assume that π_θ is Feller; in this case we can define an operator T_θ^ξ on $C(S)$ for every $\xi \in E^*$ by

$$T_\theta^\xi(g)(x) = \int_S e^{\xi(f(y))} g(y) \pi_\theta(x, dy).$$

Let $r_\theta(\xi)$ denote the spectral radius of the operator T_θ^ξ, let $\lambda_\theta(v) =: \sup_{\xi \in E^*}\{\xi(v) - \log r_\theta(\xi)\}$, (i.e., the convex conjugate of $\log r_\theta$), let $\Lambda_\theta(A) = \inf_{v \in A}\{\sup_{\xi \in E^*}\{\xi(v) - \log r_\theta(\xi)\}\}$. Under appropriate regularity conditions, it is known that the sequence $\{P_{z:\theta}^n : n \geq 1\}$ satisfies an LDP employing this λ_θ as a rate function; see, for example, Stroock (1984), de Acosta (1985).

The proof of Theorem 6.1 below parallels the i.i.d. proof in Dinwoodie and Zabell (1992, Theorem 3.1), and is based on an idea of de Acosta. The results of Stroock (1984) are used to construct the parallel argument for Markov chains. The following two conditions are required for the lower bound:

(6.1) for every open U and n, $\inf_{z \in S} P_{z:(.)}^n(U) : \Theta \to [0, 1]$ is lower semi-continuous;

(6.2) for each $\theta \in \Theta$, there exist $C_\theta > 0$ and an integer $M_\theta \geq 1$ such that for every $x, y \in S$, $\pi_\theta(x, .) \leq C_\theta \sum_{i=1}^{M_\theta} \pi^i(y, .)$.

THEOREM 6.1 If $\{P_{z:\theta} : z \in S, \theta \in \Theta\}$ satisfies (6.1), $\{\pi_\theta : \theta \in \Theta\}$ satisfies (6.2), and λ_θ is the convex conjugate of $\log r_\theta$, then

$$\liminf_{n \to \infty} \frac{1}{n} \log P_{z:\theta_n}^n(U) \geq -\inf\{\lambda_\theta(v) : v \in U\}.$$

for every $z \in S$, $\theta_n \to \theta \in \Theta$, and open set $U \subset P(S)$.

PROOF We may assume without loss of generality that $\Lambda_\theta(U) < \infty$, and that U is open and convex. It then follows from Theorems 4.1.14 and 4.1.43 of Deuschel and Stroock (1989, pp. 97 and 105) that $\lim_{n \to \infty} n^{-1}\{\inf_{z \in S} \log P_{z:\theta}^n(U)\} \geq -\Lambda_\theta(U)$. It is precisely at this point that assumption (6.2) enters (compare also Stroock, 1984, Theorem 6.2 and Lemma 6.4, Ellis, 1988, Theorem 1.2). Thus for n sufficiently large, say $n \geq N$, we have $\inf_{z \in S} P_{z:\theta}^n(U) > 0$. Let $k \geq 1$, and $n \geq N + 1$. Because U is convex,

$$\inf_{z \in S} P_{z:\theta_n}^n(U) \geq (\inf_{z \in S} P_{z:\theta_n}^k(U))^{[(n-N)/k]} \inf_{z \in S} P_{z:\theta_n}^{N+r_n}(U),$$

where the square brackets denote the greatest integer function, and r_n is the remainder term defined by $n - N = k[(n-N)/k] + r_n$, $0 \leq r_n \leq k-1$.

Let ϵ be such that $0 < \epsilon < \min_{0 \leq i \leq k-1} \inf_{z \in S} P_{z:\theta}^{N+i}(U)$. By condition (6.1), there exists an integer $M \geq 1$ such that if $n \geq M$, then

$$\inf_{z \in S} P_{z:\theta_n}^{N+r_n}(U) > \min_{0 \leq i \leq k-1} \inf_{z \in S} P_{z:\theta}^{N+i}(U) - \epsilon = \delta > 0.$$

Thus we have

$$\begin{aligned}&\liminf_{n\to\infty} \tfrac{1}{n} \log \inf_{z\in S} P^n_{z:\theta_n}(U)\\ &\quad\geq \liminf_{n\to\infty} \tfrac{1}{n} \log(\inf_{z\in S} P^k_{z:\theta_n}(U))^{[(n-N)/k]}\\ &\quad\geq \tfrac{1}{k} \log \inf_{z\in S} P^k_{z:\theta}(U).\end{aligned}$$

This is true for any k, hence taking the limit inferior of the right side we get

$$\liminf_{n\to\infty} \frac{1}{n} \log \inf_{z\in S} P^n_{z:\theta_n}(U) \geq \liminf_{k} \frac{1}{k} \log \inf_{z\in S} P^k_{z:\theta}(U) \geq -\Lambda_\theta(U),$$

again applying Theorems 4.1.14 and 4.1.43 of Deuschel and Stroock (1989). □

REMARK 6.1 If U is open, then the function $P^n_{z:(.)}(U) : \Theta \to [0,1]$ is necessarily lower semicontinuous, because $\pi : \Theta \times S \to P(S)$ is continuous, and hence the measure $P_{z:\theta}$ on S^∞ is (weakly) continuous in θ. If, in addition, $P_{z:\theta}$ is uniformly continuous in Θ, in the sense that for every bounded, continuous function f on S^∞, $\sup_z \left|\int f\,dP_{z:\theta} - \int f\,dP_{z:\gamma}\right| \to 0$ as $\gamma \to \theta$, then condition (6.1) is satisfied.

Let $\rho(T^\xi_\theta)$ be the resolvent set for the operator T^ξ_θ on $C(S)$ and let

$$R(\lambda, T^\xi_\theta) = (\lambda I - T^\xi_\theta)^{-1}$$

be the resolvent operator on $C(S)$ when $\lambda \in \rho(T^\xi_\theta)$. Recall that if $|\lambda| > r_\theta(\xi)$, then $\lambda \in \rho(T^\xi_\theta)$. Lemma 6.1 gives a bound on the probability that p_n lies in a half-space in the vector space E, involving both the spectral radius and the norm of the resolvent operator.

LEMMA 6.1 Let $H_\xi = \{v \in E : \langle v, \xi\rangle > a\}$. If $\lambda > r_\gamma(\xi)$, then

$$P_{z:\gamma}\{p_n \in H_\xi\} \leq e^{-na} e^{\xi(f(z))} \lambda^n \max\{\|R(z, T^\xi_\gamma)\| : |z| = \lambda\}.$$

PROOF First note that for every $z \in S$ and $\theta \in \Theta$,

$$E_{z:\theta}\left[\exp\langle \sum_{i=1}^n f(Z_i), \xi\rangle\right] \leq e^{\xi(f(z))} \left\|[T^\xi_\theta]^{n-1}\right\|;$$

this inequality is contained in the proof of Theorem 4.1 of de Acosta (1985).

From the above inequality, Chebyshev's inequality, and the estimate of Jörgens (1982, p. 70), it is then easily calculated that

$$\begin{aligned}P_{z:\gamma}\left\{\tfrac{1}{n}\textstyle\sum_{i=1}^n f(Z_i) \in H_\xi\right\} &= P_{z:\gamma}\left\{\left\langle \textstyle\sum_{i=1}^n f(Z_i), \xi\right\rangle > na\right\}\\ &\leq e^{-na} E_{z:\gamma} \exp\left\langle \textstyle\sum_{i=1}^n f(Z_i), \xi\right\rangle\\ &\leq e^{-na} e^{\xi(f(z))} \left\|[T^\xi_\gamma]^{n-1}\right\|\\ &\leq e^{-na} e^{\xi(f(z))} \lambda^n \max\left\{\left\|R(z, T^\xi_\gamma)\right\| : |z| = \lambda\right\}. \;\square\end{aligned}$$

The following conditions arise from the bound of Lemma 6.1, and are employed in Theorem 6.2 below.

(6.3) $r_{(.)}(\xi) : \Theta \to [0, \infty)$ is continuous;

(6.4) if $\theta \in \Theta$ and $|C| > r_\theta(\xi)$, then $\|R(z, T^{\xi}_{\theta_n})\| \to \|R(z, T^{\xi}_{\theta})\|$ uniformly on $\{z : |z| = C\}$ as $\theta_n \to \theta$.

Note that $R(z, T^{\xi}_{(.)})$ is defined in a neighborhood of θ when (6.3) is assumed; condition (6.4) is then meaningful. The proof of the large deviation theorem given below employs a technique due to Gärtner (1977); it reduces the vector space problem to a one-dimensional one by applying Chebyshev's inequality to half-spaces.

THEOREM 6.2 If $\{P_{z:\theta} : z \in S,\ \theta \in \Theta\}$ satisfies conditions (6.3) and (6.4), and λ_θ is the convex conjugate of $\log r_\theta$, then

$$\limsup_{n\to\infty} \frac{1}{n} \log P_{z:\theta_n}(p_n \in K) \leq -\inf\{\lambda_\theta(v) : v \in K\}.$$

for every $z \in S$, $\theta_n \to \theta \in \Theta$, and compact set $K \subset P(S)$.

PROOF We can assume that $\Lambda_\theta(K) > 0$, and we will only treat the case where $\Lambda_\theta(K) < \infty$ Let $\epsilon > 0$, and let

$$H_\xi = \{v \in E : \langle v, \xi\rangle - \log r_\theta(\xi) > \Lambda_\theta(K) - \epsilon\}.$$

Then $K \subset \bigcup_\xi H_\xi$, and since K is compact, there exists a finite subcover $K \subset \bigcup_{1\leq k\leq m} H_{\xi_k}$. By only taking the nonempty H_{ξ_k}, we can assume that $r_\theta(\xi_k) < \infty$. Then, letting $a_k = \log r_\theta(\xi_k) + \Lambda_\theta(K) - \epsilon$ and choosing $\lambda_k > \log r_\theta(\xi_k)$ for $1 \leq k \leq m$, it follows that

$$\begin{aligned} P_{z:\theta_n}(p_n \in K) &\leq \textstyle\sum_{k=1}^m P_{z:\theta_n}\left(\frac{1}{n}\sum_{i=1}^n f(Z_i) \in H_{\xi_k}\right) \\ &\leq \textstyle\sum_{k=1}^m e^{-na_k} e^{\xi(f(z))}(\lambda_k)^n \max\left\{\|R(z, T^{\xi_k}_{\theta_n})\| : |z| = \lambda_k\right\}, \end{aligned}$$

by Lemma 6.1, since if n is sufficiently large then $\lambda_k > \log r_{\theta_n}(\xi_k)$ for each $1 \leq k \leq m$ from (6.3). Furthermore, if n is sufficiently large then for each $1 \leq k \leq m$,

$$M_k =: \max\left\{\|R(z, T^{\xi_k}_{\theta_n})\| : |z| = \lambda_k\right\} \leq \max\left\{\|R(z, T^{\xi_k}_{\theta})\| : |z| = \lambda_k\right\} + \epsilon$$

since by (6.4), $\|R(z, T^{\xi_k}_{\theta_n})\| \to \|R(z, T^{\xi_k}_{\theta})\|$ uniformly on $\{z : |z| = \lambda_k\}$. Let $M = \max\{\|R(z, T^{\xi_k}_{\theta})\| : |z| = \lambda_k,\quad 1 \leq k \leq m\}$. Then we have the following bound for n sufficiently large:

$$P_{z:\theta_n}(p_n \in K) \leq \sum_{k=1}^m e^{-na_k} e^{\xi(f(z))}(\lambda_k)^n[M + \epsilon]$$

and it follows that

$$\begin{aligned}\limsup_{n\to\infty} \tfrac{1}{n} \log P_{z:\theta_n}(p_n \in K) &\le \max_{1\le k\le m}\{\log \lambda_k - a_k\} \\ &= \max_{1\le k\le m}\{\log \lambda_k - [\log r_\theta(\xi_k) \\ &\quad +\Lambda_\theta(K) - \epsilon]\} \\ &= -\Lambda_\theta(K) + \epsilon + \max_{1\le k\le m} \\ &\quad [\log \lambda_k - \log r_\theta(\xi_k)].\end{aligned}$$

Since $\epsilon > 0$ and $\lambda_k > \log r_\theta(\xi_k)$ are arbitrary, the theorem follows. □

In order to extend the upper bound to arbitrary closed sets, the following simple technical fact is needed.

LEMMA 6.2 If $\{\pi_\theta(x,.), (\theta, x) \in \Theta \times S\}$ is tight, then for each $a > 0$ there exists a compact set $K_a \subset P(S)$ such that

$$\sup_\theta P_{z:\theta}(p_n \notin K_a) \le e^{-na}.$$

PROOF A simple variant of the proof of Theorem 4.2 in de Acosta (1985). □

THEOREM 6.3 Suppose conditions (6.3) and (6.4) are satisfied, and that the family $\{\pi_\theta(x,.), (\theta, x) \in \Theta \times S\}$ is tight. If $C \subset P(S)$ is any closed set, then

$$\limsup_{n\to\infty} \frac{1}{n} \log P_{z:\theta_n}(p_n \in C) \le -\Lambda_\theta(C)$$

whenever $\theta_n \to \theta$.

PROOF The conventional one using Lemma 6.2: divide C into a compact set K, and a complement $C - K$ with arbitrarily large exponential rate of decay. □

REMARK 6.2 Conditions (6.1), (6.3), and (6.4) are satisfied by a family of finite state-space Markov chains if the transition matrices $\pi_\theta(s_i, s_j)$ are continuous in θ; condition (6.2) is satisfied if the the matrices π_θ are irreducible.

REFERENCES

de Acosta, A. (1985). Upper bounds for large deviations of dependent random vectors. *Z. Wahrscheinlichkeitstheorie verw. Gebiete.* **69**, 551–564.

Attouch, H. (1984). *Variational Convergence for Functions and Operators.* Pitman, Boston.

Bahadur, R. R., Gupta, J. C., and Zabell, S. L. (1980). Large Deviations, Tests, and Estimates. In *Asymptotic Theory of Statistical Tests and Estimation,* ed. I. M. Chakravarti, 33–63. Academic Press, New York.

Deuschel, Jean-Dominique, and Stroock, Daniel W. (1989). *Large Deviations.* Academic Press, San Diego.

Diaconis, Persi, and Freedman, David. (1984). Partial Exchangeability and Sufficiency. In *Statistics, Applications, and New Directions,* eds. J. K. Ghosh and J. Roy, 205–236. Indian Statistical Institute, Calcutta.

Diaconis, Persi, and Freedman, David. (1987). A dozen de Finetti-style results in search of a theory. *Ann. Inst. H. Poincaré Probab. Statist.* **23**, 397–423.

Dinwoodie, I. H., and Zabell, S. L. (1992). Large deviations for exchangeable random vectors. *Annals of Probability,* **20**, 1147–1166.

Gärtner, J. (1977). On large deviations from the invariant measure. *Theory Probab. Appl.* **22**, 24–39.

Jörgens, K. (1982). *Linear Integral Operators.* Pitman Advanced Publishing Program, Boston.

Slaby, M. (1988). On the upper bound for large deviations of sums of i.i.d. random vectors. *Ann. Probab.* **16**, 978–990.

Stroock, D. W. (1984). *An Introduction to the Theory of Large Deviations.* Springer-Verlag, New York.

Vervaat, W. (1981). Une compactification des espaces fonctionnels C et D; une alternative pour la démonstration des théorèmes limites fonctionnels. *Comptes Rendus de l'Académie des Sciences* **292**, 441–444.

Department of Mathematics
Tulane University
New Orleans
LA 70118, USA

Department of Mathematics
Northwestern University
Evanston
IL 60208, USA

J.K. Ghosh, S.K. Mitra, K.R. Parthasarathy and B.L.S. Prakasa Rao (*eds.*),
Statistics and Probability : A Raghu Raj Bahadur Festschrift
Published by Wiley Eastern Limited, Publishers, (1993), 191 – 202

On the Inadmissibility of Some Best Invariant Predictive Distributions: The Translation Parameter Case[1]

MORRIS L. EATON
University of Minnesota, Minneapolis

Abstract

A decision theoretic formulation of a simple prediction problem is described for a translation model. The problem is invariant under translations and a best translation invariant predictive distribution is given. This best invariant decision rule tends to be admissible when the parameter space is $R^p, p = 1, 2$ but tends to be inadmissible when the parameter space is $R^p, p \geq 3$.

1 Introduction

It has long been known that in multidimensional parameter problems, the best invariant estimator of the mean vector tends not to be admissible when the loss is quadratic and the dimension of the translation parameter is larger than two. Stein (1955) demonstrated this for normal samples and Brown (1966) extended his result to other distributions and other types of loss functions. Recent results have shown that, even for samples from the multivariate normal, the dependence of admissibility on dimension can be

[1]This work was supported in part by National Science Foundation Grant DMS-89-22607.

AMS Subject Classifications : Primary 62C05, secondary 62C15
Key words and Phrases : Translation invariant prediction problems, admissibility, inadmissibility, formal Bayes rule.

rather subtle — for example, see Johnstone (1988) for an example where admissibility obtains for dimensions less than five, and inadmissibility prevails in five or more dimensions.

In the present paper, we discuss the relationship between admissibility and dimension in the context of a prediction problem when the underlying probability model depends on an unknown p-dimensional translation parameter. The problem is formulated decision theoretically with a bounded translation invariant loss function. Then, a best translation invariant predictive distribution is exhibited. In the rather elementary setting considered here, it is argued that the above decision rule (i.e. the best translation invariant predictive distribution) is admissible for dimensions 1 and 2, but inadmissible for dimension 3 or more.

Here is a summary of this paper. The emphasis is on understanding the admissibility-inadmissibility dichotomy as a function of dimension, rather than on obtaining general results for the prediction problem. For this reason, only the simplest prediction problems are treated. These are introduced in Section 2 together with a family of translation invariant loss functions. The form of the best translation invariant decision rule is given. In Section 3 we briefly review some results from Eaton (1992) and indicate how these show the best invariant rule is admissible (under mild moment assumptions) for dimensions 1 and 2. Section 4 contains examples of predictive distributions which, when the dimension is 3 or more, dominate the best invariant predictive distributions for a particular type of translation invariant loss function. The inadmissibility results of Brown (1966) for non-convex loss function provide the basic technical tool for establishing the domination.

To put our results into a slightly larger context, we now briefly review some material from Eaton (1992). The best invariant decision rule described in Section 3, is, in fact, a formal Bayes rule obtained from using Lebesgue measure as an improper prior distribution on the parameter space R^p. Thus, the results described above demonstrate that, for the loss functions considered, this formal Bayes rule is admissible for $p = 1, 2$ and inadmissible for $p \geq 3$. More generally, if we are given a parametric model and an improper prior distribution, one can ask for conditions under which a formal Bayes rule (for a given loss function) will be admissible. For the class of quadratically regular decision problems (defined in Eaton (1992, Section 2)), of which the prediction problem in this paper is a special case, sufficient conditions for admissibility of the formal Bayes rules are given. These sufficient conditions are then shown to be equivalent to the recurrence of a natural Markov Chain defined on the parameter space. This suggests the obvious "converse" question:

$$\left.\begin{array}{l}\text{Suppose the natural Markov Chain is transient.}\\ \text{Does there exist a quadratically regular decision problem}\\ \text{for which the formal Bayes rule is inadmissible?}\end{array}\right\} \quad (1.1)$$

At present, very little is known above (1.1), but the results in this paper impinge directly on this question.

The prediction problem considered in this paper is quadratically regular and the improper prior is Lebesgue measure. The natural Markov Chain induced by our translation model and improper prior is a random walk on R^p (See Section 3 for a precise definition of the random walk for the model of this paper). For $p = 1$ and $p = 2$, rather weak moment assumptions guarantee the recurrence of this random walk and hence admissibility. But for $p \geq 3$, there are no non-trivial recurrent random walks on R^p. Therefore, for the rather simple situation considered in this paper, (1.1) has been answered affirmatively for $p \geq 3$, under some regularity conditions. Thus, in this case, there is a fairly complete correspondence between admissibility for quadratically regular problems and recurrence of the Markov Chain defined by the model and the improper prior.

2 The Prediction Problem

In this section we give a decision theoretic formulation of a simple prediction problem based on a translation family model. The problem is invariant under the group of translations since the loss functions considered are invariant. Further, the best invariant decision rule is described.

Consider a random vector X in R^p which is to be observed. It is assumed that X has a distribution $P(\cdot|\theta), \theta \in R^p$ given by

$$P(dx|\theta) = f_1(x - \theta)dx \tag{2.1}$$

Here, the parameter $\theta \in R^p$ is unknown but the density f_1 is known. Also consider a random vector Z in R^p which has a distribution $S(\cdot|\theta)$ given by

$$S(dz|\theta) = f_2(z - \theta)dz. \tag{2.2}$$

Again, the parameter $\theta \in R^p$ is unknown and the density f_2 is known. Given θ, it is assumed that X and Z are independent random variables with the distributions specified in (2.1) and (2.2)

The statistical problem of interest is the following:

Given the observation X, "predict" the random vector Z.

By "predict", we mean that one is to write down a guess (based on X) of the distribution of Z. In decision theoretic terms, this means the action space for the problem is the set M of all probability distributions on R^p (the relevant σ-algebra for M that is generated by the open sets of the w^*-topology for M — see Eaton (1982) for a further discussion). Thus a decision rule is a measurable map Q from R^p to M. The notation $Q(\cdot|x)$ is used for the value of Q at $x \in R^p$, since $Q(\cdot|x)$ is to be thought of as a conditional distribtuion of Z given $X = x$.

It should be mentioned that much of the above formation is motivated by Bayesian considerations. Given a prior distribution π for θ, the "solution" to the prediction problem is dictated by the rules of probability — namely, π and the assumed model determine a joint distribution for X, Z, θ which in turn determines the conditional distribution of Z given $X = x$ — say $Q_\pi(\cdot|x)$. This is obviously the Bayesian solution to the prediction problem and is what has prompted the formulation above.

To complete this decision theoretic formulation of the prediction problem, we now introduce a loss function. Let H be a fixed symmetric distribution on R^p so that the characteristic function

$$q(u) = \int_{R^p} \exp\ [iu't]\ H(dt) \tag{2.3}$$

is real. Let SB denote the real vector space of bounded signed measures defined on R^p. For $\xi_1, \xi_2 \in SB$, consider the bilinear function

$$\langle \xi_1, \xi_2 \rangle = \iint q(u_1 - u_2)\xi_1(du_1)\xi_2(du_2)\,. \tag{2.4}$$

That $\langle \cdot, \cdot \rangle$ is symmetric and non-negative definte is easily checked.

Finally, define the loss function by

$$L(a, \theta)\ = \langle a - S(\cdot|\theta), a - S(\cdot|\theta)\rangle. \tag{2.5}$$

Thus, the loss is a measure of "squared distance" between the action $a \in M$ and the true distribution of Z when the parameter value is θ (see (2.2)). This loss is a fairly natural analog of the quadratic loss functions used for problems of estimating finite dimensional parameter vectors.

We now claim that the decision problem specified by the above model, action space and loss function is invariant under the group $G = R^p$ of translations. The action of the group on R^p is

$$gx \equiv x + g$$

while the action of G on the action space M is

$$(ga)(B)\ =\ a(B - g),\ \ a \in M \tag{2.6}$$

where B is a Borel set in R^p.

PROPOSITION 2.1 *The above decision problem is invariant under G.*

PROOF: The probability model for X is obviously invariant, and a routine calculation shows that

$$L(ga, g\theta)\ =\ L(a, \theta). \tag{2.7}$$

This completes the proof. □

Recall that a decision rule Q is *invariant* if $Q = gQ$ where gQ is defined by

$$(gQ)(B|x) = Q(B - g|x - g). \tag{2.8}$$

Since the decision problem is invariant and the group G acts transitively on the parameter space, all invariant decision rules will have constant risk — that is, the risk function of any decision rule is

$$R(Q,\theta) = \mathcal{E}_\theta L(Q(\cdot|X),\theta), \tag{2.9}$$

and if Q satisfies (2.8), then $R(Q,\theta) = R(Q,0)$ for all θ. We refer the reader to Eaton (1989) for a more complete discussion of group actions and invariant decision problems.

Because the risk function of invariant decision rules is constant in our prediction problem, it makes sense to discuss a best invariant decision rule. To be precise, an invariant decision rule Q_0 is best *invariant* if

$$R(Q_0,0) \leq R(Q,0) \tag{2.10}$$

for all invariant rules Q. In our problem, it is easy to describe a best invariant rule. For each fixed x, the function

$$k(z - x) = \int_{R^p} f_2(z - x + \theta)f_1(\theta)d\theta \tag{2.11}$$

is a probability density on R^p. Define Q_0 by

$$Q_0(B|x) = \int_B k(z - x)dz. \tag{2.12}$$

PROPOSITION 2.2 *The decision rule Q_0 in (2.12) is best invariant.*

PROOF: This is a consequence of a rather general result due to Stein which asserts that, under certain conditions (which are satisfied in our problem), a best invariant decision rule is obtained by using the right Haar measure on G (Lebesgue measure for our problem) as an improper prior, and then calculating the formal Bayes rule for the decision problem. A fairly complete exposition of this result can be found in Eaton (1989). That Q_0 is the formal Bayes rule when the prior is $d\theta$ can easily be checked — see Eaton (1982, 1992) for further details. □

Because Q_0 is best invariant and the group $G = R^p$ is amenable, it follows from Kiefer (1957) that Q_0 is minimax. A discussion of the admissibility of Q_0 is given in the next two sections.

3 Admissibility Results

For dimensions $p = 1$ and $p = 2$, sufficient conditions for the admissibility of Q_0 are given below. This result, essentially given in Eaton (1992), involves the behaviour of a transition function constructed from the assumed model for X and the improper prior $d\theta$ used to construct Q_0. For $\theta, \eta \in R^p$, observe that

$$v(u) = \int f_1(x-u) f_1(x) dx \tag{3.1}$$

satisfies

$$\begin{cases} v(u) = v(-u) \geq 0 \\ \int v(u) du = 1. \end{cases} \tag{3.2}$$

The density v defines a transition function V by

$$V(B|\eta) = \int_B v(\theta - \eta) d\theta \tag{3.3}$$

where $\eta \in R^p$ and $B \subseteq R^p$ is a Borel set.

In turn, the transition function V determines a Markov Chain on R^p, which is in fact a random walk on R^p because V is generated by a translation family. More precisely, let $U_1, U_2, \ldots.$ be *iid* random vectors in R^p whose distribution has density v in (3.1). Let $h \in R^p$ and define the Markov Chain — $Y_0, Y_1, \ldots.$ by

$$\left. \begin{array}{rcl} Y_0 & = & \eta \\ Y_n & = & \eta + \Sigma_{i=1}^n U_i, n \geq 1. \end{array} \right\} \tag{3.4}$$

Obviously, $Y_{n+1} = Y_n + U_n$ for $n \geq 1$ and thus the Markov Chain has the transition function V. Modulo some measure theoretic details, a main result in Eaton (1992) shows that if the random walk $Y = \{Y_0, Y_1, \ldots.\}$ is recurrent, then the decision rule Q_0 is admissible.

We now pause briefly to explain the measure theoretic problems since they are rather important. Let λ denote Lebesgue measure on R^p.

DEFINITION 3.1 *A decision rule Q_1 is almost- λ-admissible if for any decision rule Q_2 which satisfies $R(Q_2, \theta) \leq R(Q_1, \theta)$ for all θ, the set*

$$\{\theta | R(Q_2, \theta) < R(Q_1, \theta)\}$$

has λ measure zero.

DEFINITION 3.2 *The Markov Chain $Y = \{Y_0, Y_1, \ldots.\}$ is locally λ-recurrent if for each set C of positive and finite λ measure, the set*

$$\{\eta \in C | P[Y_i \in C \text{ for some } i \geq 1 | Y_0 = \eta] < 1\}$$

has λ measure zero.

THEOREM 3.1 *If the Markov Chain Y is locally-λ-recurrent, then the decision rule Q_0 is almost- λ-admissible.*

This result is a direct consequence of Theorem 4.3 and 6.1 in Eaton (1992).

Now, recall the well known sufficient conditions for the recurrence (and hence local-λ-recurrence) of Y. If $p = 1$ or $p = 2$ and

$$\int_{R^p} \|\mu\|^p v(u)du < +\infty, \tag{3.5}$$

then Y is recurrent. This yields

THEOREM 3.2 *For $p = 1$ or $p = 2$, if*

$$\int_{R^p} \|x\|^p f_1(x)dx < +\infty, \tag{3.6}$$

the Q_0 is almost-λ-admissible.

PROOF: From the definition of the density v in (3.1), it is easy to see that (3.6) implies (3.5). Thus Y is recurrent and Theorem 3.1 yields the desired conclusion. □

The admissibility of Q_0 for $p = 1, 2$ (When 3.6 holds) now follows easily from

LEMMA 3.3 *For each decision rule Q, the risk function $R(Q, \theta)$ is continuous on R^p.*

PROOF: The risk function, defined at (2.9) is given by

$$\begin{aligned} R(Q,\theta) &= \int \langle Q(\cdot|x) - S(\cdot|\theta), Q(\cdot|x) - S(\cdot|\theta)\rangle \\ &\quad f_1(x-\theta)dx \\ &= \int \langle Q(\cdot|x), Q(\cdot|x)\rangle f_1(x-\theta)dx \\ &\quad -2\int \langle Q(\cdot|x), S(\cdot|\theta)\rangle f_1(x-\theta)dx \\ &\quad +\int \langle S(\cdot|\theta), S(\cdot|\theta)\rangle f_1(x-\theta)dx. \end{aligned} \tag{3.7}$$

We now argue that each of the three terms in the final expression for the risk is continuous. The first term is the convolution of a bounded function of x, namely $\langle Q(\cdot|x), Q(\cdot|x)\rangle$ with the integrable function f_1 and hence is uniformly continuous on R^p (see proposition 33.3 in Choquet (1969, Volume II)). Using the definition of $\langle\cdot,\cdot\rangle$ and making a change of variable, the second term can be written as

$$\begin{aligned} \psi(\theta) &= \\ &-2\iiint q(u_1 - u_2 + \theta)f_2(u_1)Q(du_2|x)d\mu_1 f_1(x-\theta)dx. \end{aligned} \tag{3.8}$$

Suppose $\theta_n \to \theta_0$ in R^p and let $\varepsilon > 0$ be given. Then by the triangle inequality, we have

$$\begin{aligned} \tfrac{1}{2}|\psi(\theta_n) - \psi(\theta_0)| \leq & \\ & \textstyle\int\int\int |q(u_1 - u_2 + \theta_n) - q(u_1 - u_2 + \theta_0)| \\ & f_2(u_1)Q(du_2|x)du_2 f(x - \theta_n)dx \\ & + \left|\textstyle\int r(x)f_1(x - \theta_n)dx - \int r(x)f_1(x - \theta_0)dx\right| \end{aligned} \tag{3.9}$$

where

$$r(x) = \int\int q(u_1 - u_2 + \theta_0)f_2(u_1)Q(du_2|x)du_1.$$

Since the characteristic function q is uniformly continuous, the first expression on the right side of (3.9) can be made less than ϵ by choosing n large enough so that for all u_1, u_2

$$|q(u_1 - u_2 + \theta_n) - q(u_1 - u_2 + \theta_0)| < \epsilon.$$

Since the function r is bounded, the map

$$\theta \to \int r(x)f(x - \theta)dx$$

is uniformly continuous as noted above. Thus, the second expression on the right side of (3.9) can be made smaller than ϵ for all sufficiently large n. This shows the second term on the right side of (3.7) is continuous. A change of variable verifies that the third term on the right side of (3.7) is constant. The proof is complete.

THEOREM 3.4 *Under the assumptions of Theorem 3.2, the decision rule Q_0 is admissible.*

PROOF: If a decision rule Q_1 satisfies $R(Q_1, \theta) \leq R(Q_0, \theta)$ for all θ, then the set

$$B = \{\theta | R(Q_1, \theta) < R(Q_0, \theta)\}$$

has λ-measure zero by Theorem 3.2 and is open by Theorem 3.4. Thus B is empty. □

It should be noticed that the conditions for the admissibility of Q_0 involve only moment assumptions on f_1 and not assumptions on f_2. This is a rather general phenomenon in our formulation of prediction problems (see Eaton (1992, Section 6)).

The argument given above fails completely for $p \geq 3$. The reason for this failure is that there exists no non- trivial recurrent random walks on R^p when $p \geq 3$ (see Guivarc'h, Keane and Roynette (1977)). There are no assumptions on the density function v which will imply that Y is recurrent. In particular the assumptions of the principal result, namely Theorem 3.1, can never be satisfied. This suggests, but certainly does not prove, that Q_0 may not be admissible for $p \geq 3$. It is this question to which we turn in the next section.

4 Inadmissibility for $p \geq 3$

Throughout this section take $p \geq 3$. Under certain assumptions specified below, we construct a predictive distribution Q^* which dominates Q_0 in the sense that $R(Q^*, \theta) < R(Q_0, \theta)$ for all θ in R^p.

In order to simplify the exposition, assume that $f_1 = f_2$ throughout this section — that is, X and Z have the same distribution (see (2.1) and (2.2)). There are three basic assumptions needed in this section. Here are the first two.

$$(\text{A.1}) \begin{cases} \text{The density } f_1 \text{ is symmetric and} \\ \int \|x\|^4 f_1(x)dx < +\infty \end{cases}$$

$$(\text{A.2}) \begin{cases} \text{The symmetric distribution } H \text{ used to define} \\ q \text{ in (2.3) has finite second moments.} \end{cases}$$

Let

$$\psi(v) = \int \exp[iv'x] f_1(x)dx \tag{4.1}$$

be the (real) characteristic function of f_1 and define p on R^p by

$$p(a) = \int \exp[ia't]\psi^3(t)H(dt). \tag{4.2}$$

Because ψ is a characteristic function and H has second moments, p has bounded first and second partial derivatives. Given a constant c (to be specified precisely later in the argument), consider

$$W(a) = c - 2p(a), \quad a \in R^p. \tag{4.3}$$

Then W is bounded and has bounded first and second partial derivatives. Using W, consider the decision theoretic problem of estimating θ with the translation invariant loss function

$$L_1(a, \theta) = W(a - \theta). \tag{4.4}$$

Hence the risk function of any point estimator $t(X)$ of θ is

$$\begin{aligned} R_1(t, \theta) &= \mathcal{E}_\theta W(t(X) - \theta) \\ &= \int W(t(x) - \theta) f_1(x - \theta)dx. \end{aligned} \tag{4.5}$$

To connect the above point estimation problem with our prediction problem, again let $t(X)$ be any point estimator of θ. Define the predictive distribution Q_t by

$$Q_t(B|x) = \int_B k(u - t(x))d\mu \tag{4.6}$$

where k is given by (2.11) when $f_2 = f_1$. Thus Q_t equals Q_0 when $t(X) = X$. Recall that the risk function of the decision rule Q_t is given by (2.9) when the loss is given by (2.5).

THEOREM 4.1 *There exists a choice for the constant c in* (4.3) *such that for all point estimators t*

$$R(Q_t, \theta) = R_1(t, \theta) \quad \text{for} \quad \theta \varepsilon R^p \tag{4.7}$$

PROOF: The proof is routine algebra so we just outline the details.

Since $f_2 = f_1, S(\cdot|\theta) = P(\cdot|\theta)$ is the distribution of X and Z when θ is the parameter value. Using the loss function constructed in Section 2, the risk function of Q_t is

$$\begin{aligned} R(Q_t, \theta) &= \int \langle Q_t(\cdot|x) \\ &\quad -P(\cdot|\theta), Q_t(\cdot|x) - P(\cdot|\theta)\rangle f_1(x-\theta)dx \\ &= \int \langle Q_t(\cdot|x), Q_t(\cdot|x)\rangle f_1(x-\theta)dx \\ &\quad -2\int \langle Q_t(\cdot|x), P(\cdot|\theta)\rangle f_1(x-\theta)dx \\ &\quad +\int \langle P(\cdot|\theta), P(\cdot|\theta)\rangle f_1(x-\theta)dx. \end{aligned} \tag{4.8}$$

Because the loss function defined at (2.7) is translation invariant, the first and third terms in this last expression in (4.8) are constant (independent of θ). Call their sum c. Some tedious algebra and unraveling of notation shows that the second term in the last expression for (4.8) is just

$$-2\int p(t(x) - \theta) f_1(x-\theta)dx.$$

Thus, from the definition of W, we have

$$R(Q_t, \theta) = \int W(t(x) - \theta) f_1(x - \theta)dx$$

Which is just $R_1(t, \theta)$. This completes the proof. □

The utility of Theorem 4.1 is fairly clear. If we can find an estimator $t(X)$ which dominates the estimator X relative to the risk function R_1, then automatically there is a predictive distribution, Q_t which dominates the best invariant rule Q_0. To construct such a t one further assumption is needed.

Given W defined above, let

$$W'(a) = \left(\frac{\partial W}{\partial a_1}(a), ..., \frac{\partial W}{\partial a_p}(a)\right) \tag{4.9}$$

be the row vector of partial derivatives of W.

$$(\text{A.3})\begin{cases}\text{Assume that the } p\times p \text{ matrix}\\ A=\mathcal{E}_0 XW'(X)\\ \text{is non-singular.}\end{cases}$$

Here, as elsewhere, $\mathcal{E}_0$ denotes expectations when $\theta = 0$.

THEOREM 4.2 *Under asssumptions* (A.1), (A.2) *and* (A.3), *there exist positive numbers a and b such that the estimator*

$$t^*(X) = (I_p - \frac{(bA)^{-1}}{a+\|X|^2})X \tag{4.10}$$

dominates the estimator $t_0(X) = X$ *—That, is,*

$$R_1(t^*,\theta) < R_1(t_0,\theta) \text{ forall } \theta.$$

PROOF: When (A.1), (A.2), and (A.3) hold, it is relatively easy to cheeck the assumptions of Theorem 3.3.1 in Brown (1966, p. 1130). This theorem guarantees the existence of positive a and b such that t^* dominates t_0 in the sense that

$$R_1(t^*,\theta) < R_1(t_0,\theta) \quad \text{for} \quad \theta \in R^p. \qquad \square$$

Combining Theorems 4.1 and 4.2 yields

THEOREM 4.3 *Under assumptions (A.1), (A.2) and (A.3), the predictive distribution* Q_{t^*} *defined by* t^* *in (4.10) satisfies*

$$R(Q_t^*,\theta) < R(Q_0,\theta) \quad \text{for} \quad \theta \in R^p. \tag{4.11}$$

Assumptions (A.1) and (A.2) are minor regularity conditions in need of no further comment. However, (A.3) is an assumption of substance and Theorem 4.2 is false without this or some similar type assumption which guarantees that the decision problem defined by W is "not too degenerate". In some cases, (A.3) is not hard to check. For example, if both the distribution H and the density f_1 are invariant under coordinate sign changes and coordinate permutations, then A in (A.3) is a multiple of the identity matrix. In this case one only needs to verify that

$$\mathcal{E}_0 X_1 \frac{\partial W}{\partial a_1}(X) \neq 0. \tag{4.12}$$

This can easily be done if H and f_1 correspond to p- dimensional normal distributions with mean 0 and covariances which are non-zero multiples of the identity (possibly different multiples for H and f_1).

REFERENCES

Brown, L.D. (1966), On the admissibility of invariant estimators of one or more location parameters, *Ann. Math. Statist.* **37**, 1087-1135.

Choquet, G. (1969), *Lectures on Analysis,* Vol. II. In Mathematics Lecture Note Series, W.A. Benjamin, Inc., Reading, Mass.

Eaton, M.L. (1982), A method for evaluating improper prior distributions, *Decision Theory and Related Topics III,* Vol. 1, ed. by Gupta and Berger.

Eaton, M.L. (1989), *Group Invariance Applications in Statistics,* Regional Conference Series in Probability and Statistics, Vol. 1. Published by Institute of Mathematical Statistics and American Statistical Association.

Eaton M.L. (1992), A Statistical Diptych: Admissible inferences — Recurrence of symmetric Markov chains, *Ann. Statist.* **20**, 1147-1179.

Guivarc'h, Y., Keane, M. and Roynette, B. (1977) *Marches aléatoires sur les Groupes de Lie.* Lecture Notes in Math., Vol. 624, Springer-Verlag, New York.

Johnstone, I. (1988), On inadmissibility of some unbiased estimates of loss. In *Statistical Decision Theory and Related Topics IV,* Vol. I. Ed. Gupta and Berger. Springer-Verlag, New York.

Kiefer, J. (1957), Invariance, minimax sequential estimation and continuous time processes. *Ann. Math. Statist.* **28,** 573-601.

Stein, C. (1955), Inadmissibility of the usual estimator for the mean of a multivariate normal distribution. *Proc. Third Berkeley Symp. Math. Statist. Prob. 1,* 197-206. Univ. of California Press, Berkeley.

Department of Theoretical Statistics
University of Minnesota
270, Vincent Hall
206, Church Street S.E.
Minneapolis, A
MN 55455-0488

J.K. Ghosh, S.K. Mitra, K.R. Parthasarathy and B.L.S. Prakasa Rao (*eds.*),
Statistics and Probability : A Raghu Raj Bahadur Festschrift
Published by Wiley Eastern Limited, Publishers, (1993), 203 – 217

Statistical Experiments Admitting and not Admitting "Burkholder Pathologies"

J. FUJII and H. MORIMOTO
Osaka City University, Osaka

Abstract

It is shown that if an undominated experiment $(X, \mathbf{A}, P)$ has a sufficient subfield $\mathbf{B}$ which is "substantially" smaller than $\mathbf{A}$, then there exists a non-sufficient subfield $\mathbf{C}$ including $\mathbf{B}$, under a condition almost trivially satisfied by such experiments. It is suggested that $\mathbf{C}$ rarely takes the form of a statistic.

1 Introduction

In the 1954 paper on sufficiency, Bahadur proved that a subfield including a sufficient subfield is itself sufficient in the dominated case, and asked if it extends to the general undominated case. It was one of the three "problems" posed there, which were to give great impact on the subsequent development of the theory of sufficiency.

The question was later answered in the negative by Burkholder [1961]. The phenomenon thus discovered, that a non-sufficient subfield (or statistic) can include a sufficient one, came to be known as "Burkholder pathology"

AMS Subject Classifications : 62B05, 62B15
Key words : Sufficiency, Burkholder pathology, Burkholder subfield, majorized experiment, truncated experiment.

(e.g. in Basu and Ghosh [1967]). Effort has been directed mainly to removing it from the theory, either by proposing alternative schemes of the theory (LeCam [1964] and others), or by finding out restrictive conditions on the experiments and/or subfields, under which no pathology occurs (Basu and Ghosh [1967], Michaleztky [1985, 1987]).

In this paper we would not particularly try to remove the phenomenon but study some typical though limited cases, as to when it occurs and when it does not. While doing so, we will treat the subfields and statistics separately. It calls for two separate terms: "Burkholder subfield" for a non-sufficient subfield including a sufficient one, and "Burkholder statistic" in case it is induced by a statistic.

It appears to us through this study that a Burkholder subfield is as commonplace a phenomenon as to be found everywhere in the majorized and other cases, while a Burkholder statistic is in fact a rare event. It was already noticed by Basu & Ghosh [1967] that the former exists but not the latter in the discrete experiment. It is this impression that the following Sections are meant to convey.

The Burkholder subfields are treated mainly in Section 2. It turns out that in a majorized experiment $\mathbf{E} = (X, \mathbf{A}, P)$, whenever there is a sufficient subfield $\mathbf{B}$ which is substantially smaller (in the sense specified in due course) than $\mathbf{A}$, one can construct a non-sufficient subfield $\mathbf{C}$ in between $\mathbf{A}$ and $\mathbf{B}$.

This construction can be extended to many other experiments having majorized "subexperiments" and/or majorized "truncated experiments". It should be noticed, however, that the way we construct $\mathbf{C}$ makes it most likely not to be inducible, thus limiting the scope of the result.

On the other hand, the original example of Burkholder [1961] (quoted as Example 2 below) is a non-majorized experiment with an inducible subfield (statistic), i.e., a Burkholder statistic as we call it. In Section 3 we treat one of the typical classes of non-majorized experiments which includes the example.

Assume that every singleton in the sample space X is measurable, i.e., belongs to $\mathbf{A}$ and carries a positive measure wrt. some p in P. A theorem of Michaletzky [1987] with a very general scope can be adapted into this context. It is shown that a Burkholder statistic exists only when the experiment admits a non-measurable set and moreover has essentially the same structure as Burkholder's example.

Section 4 contains final remarks on Burkholder statistics in majorized experiments.

2 Burkholder subfield in majorized experiment

The majorized experiments are one of the well-studied classes of undominated experiments (For details see, e.g., Fujii & Morimoto [1986]). It is defined as an experiment $\mathbf{E} = (X, \mathbf{A}, P)$ with a family of p-measures $P =$

$\{p\}$, such that there exists a measure m on **A** wrt. which each p in P has a density, written dp/dm. Each measure p in P has a support in **A**, the part of X on which $dp/dm > 0$.

A set of countable type (abbreviated as CT) is defined as a set in **A** which is included in a countable union of the supports of measures in P. The subfield of all the CT sets and their complements is denoted by $\mathbf{A}^\circ$. Those sets in **A** which do not belong to $\mathbf{A}^\circ$ are called the sets of uncountable type (UCT).

Every majorized experiment is known to have a "maximal decomposition", an almost disjoint family $\mathbf{D} = \{D\}$ of measurable non-null subsets of supports which almost covers X (Diepenbrock [1971]. "Almost" refers to P. We omit it unless there is a possibility of confusion). It can be taken as a subfamily of the subfield generated by the supports. A set is CT iff it is covered by a countable number of sets in **D**.

Define the experiment **E** *truncated* on A in **A** by $\mathbf{E} \cap A = (A, \mathbf{A} \cap A, P \cap A)$, where $\mathbf{A} \cap A$ means the intersections of all the sets in **A** and A, and $P \cap A$ the normalized restrictions of all the measures in P on A. Then $\mathbf{E} \cap A$ is dominated iff A is a CT.

A sufficient subfield in a majorized experiment has a remarkable property which distinguishes it from a pairwise sufficient subfield.

LEMMA 1 (Theorem 6, Fujii & Morimoto [1986]). Suppose that **B** is a sufficient subfield in a majorized experiment. If a set A belongs to **A** and all the intersections $A \cap D$, $D \in \mathbf{D}$, belong to **B**, then A belongs to **B**.

Hence, in order to have a Burkholder subfield it is enough to construct a subfield which includes a sufficient subfield and does not have this property. That it is easily done by adding $\mathbf{A}^\circ$ to a sufficient subfield will be established as an n&s condition in the following two Theorems.

THEOREM 1. Let **B** be a sufficient subfield in a majorized experiment. Suppose that the subfield **C** generated by **B** and the sets of countable type is smaller than **A**.

Then **C** is a Burkholder subfield.

PROOF. Notice first that there exists a CT set which does not belong to **B** and hence **C** is strictly larger than **B**. Because if otherwise, for any set A in **A** its intersections $A \cap D$ with all D in **D** are in **B**, as they are all CT. Hence A would belong to **B** by Lemma 1, resulting in $\mathbf{B} = \mathbf{A}$, a contradiction.

Next we show that **C** is not sufficient. From the assumption there exists a set A in **A** and not in **C**. Of course A is neither in $\mathbf{A}^\circ$ nor in **B**. Hence A is a UCT not belonging to **B**. The intersection of this set and each D in **D** is a CT and hence is in **C**. If we assume that **C** is sufficient, then

A is forced into $\mathbf{C}$ by Lemma 1 again, leading to a similar contradiction "$\mathbf{A} = \mathbf{C}$".

THEOREM 2. Let $\mathbf{B}$ be a sufficient subfield in a majorized experiment. Suppose that the subfield generated by $\mathbf{B}$ and the sets of countable type is equal to $\mathbf{A}$.

Then every subfield $\mathbf{C}$ including $\mathbf{B}$ is sufficient.

PROOF. Take any set A in $\mathbf{A}$. From the assumption there exists a B in $\mathbf{B}$ such that $A \Delta B$ is a CT, so that it is covered by a set U, a countable union of the supports. U belongs to $\mathbf{B}$, as every sufficient subfield contains the supports by obvious reasons.

Take the truncated experiment $\mathbf{E} \cap U$. Then $\mathbf{B} \cap U$ is sufficient for $\mathbf{E} \cap U$ and so is $\mathbf{C} \cap U$, as $\mathbf{E} \cap U$ is dominated. Hence there exists a $\mathbf{C} \cap U$-measurable function $g(x)$ on U which serves as a version of $P[A \cap U | \mathbf{C} \cap U]$, the conditional probability common to all p in P under $\mathbf{C} \cap U$ (The common conditional probability will henceforth be written with a capital P).

Define a function $f(x)$ by

$$f(x) = \begin{cases} g(x) & \text{on } U, \\ I_B(x) & \text{on } X \setminus U, \end{cases}$$

where $I_B(x)$ means the indicator function of B. Then $f(x)$ serves as $P[A|\mathbf{C}]$, showing that $\mathbf{C}$ is sufficient for $\mathbf{E}$.

The n&s condition thus obtained is hereby written as

CONDITION A. The experiment has a sufficient subfield $\mathbf{B}$ which satisfies $\mathbf{B} \vee \mathbf{A}^\circ \subsetneq \mathbf{A}$.

We submit an argument for this condition to substantiate our claim in the Introduction. A Burkholder pathology obviously prerequires a sufficient subfield $\mathbf{B}$ which is smaller than $\mathbf{A}$. In view of the fact that there is no Burkholder pathology in a dominated experiment (Bahadur [1954]), not only has $\mathbf{B}$ to be smaller but "substantially" smaller than $\mathbf{A}$, in the sense that the part A of X on which $\mathbf{B}$ differs from $\mathbf{A}$ be undominated. This combined with the condition for $\mathbf{E} \cap A$ to be undominated leads to the following precondition on $\mathbf{E}$.

CONDITION P. There exists a sufficient subfield $\mathbf{B}$ which differs from $\mathbf{A}$ on an uncountable number of sets D_i $(i \in I)$ in $\mathbf{D}$. Namely, there exists for each D_i $(i \in I)$ its subset A_i which is in $\mathbf{A}$ but not in $\mathbf{B}$.

Condition A is in fact very close to this obvious precondition. Evidently the latter is prerequisite for the former. Conversely, Condition A would be satisfied if only one of the possibly many families $\{A_i,\ i \in I\}$ postulated by Condition P has its essential supremum A in $\mathbf{A}$. For, in that case the symmetric difference of A and any set in $\mathbf{B}$ is UCT, so that A is not in $\mathbf{B} \vee \mathbf{A}^\circ$. In particular, it is satisfied by the "weakly dominated" experiment, a majorized experiment in which every subfamily of $\mathbf{A}$ has an essential supremum in $\mathbf{A}$.

We can think of some straightforward generalizations of the foregoing result, which apply to many more non-majorized cases than it may appear. Suppose, for example, that in any experiment $\mathbf{E} = (X, \mathbf{A}, P)$, there is a subfamily Q of P such that the "subexperiment" $\mathbf{F} = (X, \mathbf{A}, Q)$ on the same sample space is majorized. A sufficient subfield $\mathbf{B}$ for $\mathbf{E}$ is sufficient for $\mathbf{F}$. If $\mathbf{B}$ satisfies the Condition A in reference to $\mathbf{F}$, then we have a subfield $\mathbf{C}$ including $\mathbf{B}$ modulo Q, which is not sufficient for $\mathbf{F}$. A fortiori $\mathbf{C}$ is not sufficient for $\mathbf{E}$. In case we can take a $\mathbf{C}$ including $\mathbf{B}$ *modulo* P, then it is a Burkholder subfield in $\mathbf{E}$.

A similar argument applies to the case where one of the truncated experiments $\mathbf{E} \cap A$, $A \in \mathbf{A}$, satisfies the Condition A, and, of course, the case where $\mathbf{E} \cap A$ has a majorized "subexperiment". In literature we seldom come across an experiment which does not have any such sub- or truncated experiment or combination of them. Here is an example of the construction of a Burkholder subfield in an apparently complex experiment.

EXAMPLE 1. Let $\mathbf{E} = (X, \mathbf{A}, P)$ where X is an Euclidean plane, $\mathbf{A}$ is the Borel σ-field of X and P is the totality of continuous measures on $\mathbf{A}$ which are symmetric wrt. the horizontal axis. Rao & Yamada [1980] proved that this experiment is not majorized. If we define $\mathbf{B} = \{B \in \mathbf{A};\ B \text{ is symmetric wrt. the horizontal axis}\}$, then $\mathbf{B}$ is sufficient for $\mathbf{E}$. The family P includes a subfamily Q, the totality of the 1-dimensional Normal distributions on the vertical lines in X which have the mean 0. The experiment $\mathbf{F} = (X, \mathbf{A}, Q)$ is majorized and $\mathbf{B}$ is sufficient for $\mathbf{F}$. The vertical lines in X are the supports of the measures in Q and the sets of CT etc. for $\mathbf{F}$ are accordingly defined. It is easy to see that $\mathbf{C} = \mathbf{B} \vee \{CT\}$ strictly includes $\mathbf{B}$, as it contains asymmetric vertical line segments (which are not in $\mathbf{B}$), and is smaller than $\mathbf{A}$, as it does not contain the upper half-plane (which is in $\mathbf{A}$).

Hence $\mathbf{C}$ is a Burkholder subfield for $\mathbf{F}$ by Theorem 1 and that for $\mathbf{E}$ by the foregoing argument.

3 Burkholder Statistic in a Typical Non-majorized Case

Assume throughout this section that the experiment $\mathbf{E} = (X, \mathbf{A}, P)$ satsifies the following conditions.

A1. Every singleton $\{x\}$ $(x \in X)$ belongs to $\mathbf{A}$.

A2. The only P-null set is the empty set, i.e., for each x in X there exists a p in P such that $p(x) > 0$.

The example of Burkholder [1961] satisfies these conditions. The assumption $A2$ implies that the space X is covered by the discrete parts of the measures p in P. It does not rule out continuous measures either, so this type of experiment is generally not majorized. Yet it reveals strong resemblance to the majorized "discrete experiments" of Basu and Ghosh [1967] (See Remark(2) below).

Some of its basic properties are needed later. A *statistic* is defined as a partition, i.e., a family of non-empty subsets of X which collectively cover X. It *induces* the subfield consisting of all those unions of the sets in it which belong to **A**. A subfield of **A** induced in this manner by any partition is called *inducible*. A subfield is inducible iff it is closed under the operations of (possibly uncountable) unions and intersections of its sets provided the resulting unions or intersections are in **A**. This property distinguishes a sufficient subfield from a pairwise sufficient subfield, as it does in the discrete case (Ghosh, Morimoto & Yamada [1981]).

LEMMA 2. Under A1 and A2, any sufficient subfield is inducible.

PROOF. Let B_i, $i \in I$, be sets in a sufficient subfield **B**, denote their union by A and suppose that A is in **A**. There exists a **B**-measurable function $f(x)$ which serves as $P[A|\mathbf{B}]$. It follows that

$$\begin{aligned} p(B_i) &= p(A \cap B_i) = \int_{B_i} I_A \, dp \\ &= \int_{B_i} f \, dp \end{aligned}$$

for all p in P.

Take any x in B_i and q in P such that $q(x) > 0$. It follows that $f(x) = 1$, as otherwise the RHS of the equality shown above for q would be less than the LHS. This means $f(x) = 1$ for all x in B_i and hence on the whole of A.

Next, take any x from $X \backslash A$ and q in P such that $q(x) > 0$. If we assume $f(x) > 0$, then $f(x) = 1$ on A implies that

$$\begin{aligned} q(A) &= \int f \, dq \\ &= \int_A f \, dq + \int_{X \backslash A} f \, dq \geq q(A) + q(x) \, f(x) > q(A), \end{aligned}$$

which is a contradiction. So $f(x) = 0$ on $X \backslash A$ and hence is equal to $I_A(x)$ on the whole of X. Therefore A belongs to **B** and **B** is closed under union of its sets.

The partition which induces a subfield **B** consists of its atoms, which will be denoted by $\{B_i, \ i \in I\}$. Next Lemma shows that if **B** is sufficient, then the atoms are countable sets, so that they are in **A** and hence in **B**, as they are intersections of the sets in **B**.

LEMMA 3. Assume $A1$ and $A2$ and suppose that a subfield $\mathbf{B}$ is sufficient. Then each of its atom is a countable set.

PROOF. Take a point x in X, p in P such that $p(x) > 0$, y in X such that $p(y) = 0$(A p cannot make $p(x) > 0$ for all x in X), and q in P such that $q(y) > 0$.

Then there exists a $\mathbf{B}$-measurable version of $dp/d(p+q)$. Clearly it assumes a positive value at x and 0 at y. Hence there exists a set B in $\mathbf{B}$ which contains x but not y. It means that if $p(y) = 0$ then y does not belong to the $\mathbf{B}$-atom which contains x. The atom can therefore contain only those points for which $p(x) > 0$, and there are only countably many such points.

In order to have a Burkholder subfield it is enough to construct a non-inducible subfield including a sufficient subfield, which is inducible. This is almost a parallelism to Theorems 1 and 2 and their proofs with the maximal decomposition and the sets of CT respectively replaced by the atoms of $\mathbf{B}$ and the countable sets. It is immediate to have the following

THEOREM 3. Under the Assumptions $A1$ and $A2$, the experiment $\mathbf{E}$ has a Burkholder subfield iff it has a sufficient subfield $\mathbf{B}$ such that the subfield $\mathbf{C}$ generated by $\mathbf{B}$ and all the countable sets is smaller than $\mathbf{A}$.

The following discussion is concerned with Burkholder statistic. We are going to assert that, if trivialities and artificialities are properly excluded from the consideration, it occurs essentially only in that experiment which made up the example of Burkholder. The statistic is then constructed in the same way as in the example. We first need to present the example and our discussion starts in some observations on it.

EXAMPLE 2 (Burkholder. Some notations have been altered). Let X be a real line, $\mathbf{A}$ the Borel σ-field and P the totality of the measures on $\mathbf{A}$ which are symmetric w.r.t. 0.

The subfield

$$\mathbf{B} = \{B \in \mathbf{A};\ B = -B\}$$

is sufficient for $\mathbf{E}$ and is induced by the partition

$$\{\{x, -x\},\ 0 < x < \infty\}.$$

Denote $\{x, -x\}$ by B_x. Let S be a symmetric non-Borel set and take $S^* = X \backslash S$ and $A^* = (0, \infty)$. Define

$$\mathbf{C} = \{C;\ C = B \cup A,\ B \subset S^*,\ B \in \mathbf{B},\ A \subset X\backslash S^* \text{ and } A \in \mathbf{A}\}.$$

Then $\mathbf{C}$ is a subfield including $\mathbf{B}$ and is induced by the partition

$$\{B_x;\ B_x \subset S^*\} \cup \{\{x\};\ x \in X\backslash S^*\}.$$

It follows that $\mathbf{C}$ is not sufficient. Because if otherwise, there would exist a $\mathbf{C}$-measurable function $f(x)$ which satisfies $p(A^* \cap C) = \int_c f\, dp$ for any C in $\mathbf{C}$ and p in P.

If we take x in S^* then the set $S^* \cap B_x = \{x, -x\}$ is a $\mathbf{C}$-atom and belongs to $\mathbf{B}$ and hence to $\mathbf{C}$. The function $f(x)$ is a constant, say $c(|x|)$, on it. The integral equality in last paragraph with $C = \{x, -x\}$ and p in P for which $p(x) = p(-x) = 1/2$ gives rise to $p(A^* \cap \{x, -x\}) = p(\{x, -x\})c(|x|)$, which implies $f(x) = c(|x|) = 1/2$.

On the other hand if we take x in $X \setminus S^*$ then $\{x\}$ is a $\mathbf{C}$-atom belonging to $\mathbf{C}$. Now the same equality with $C = \{x\}$ and p in P for which $p(x) > 0$ gives rise to $p(x)I_{A^*}(x) = p(x)f(x)$, hence $f(x) = I_{A^*}(x)$.

It follows that $S^* = [0 < f(x) < 1]$, so that it belongs to $\mathbf{C}$ and hence to $\mathbf{A}$, which is a contradiction.

Remarks (1) Theorem 3 applies here and gives rise to a Burkholder subfield. It is, however, no more than $\mathbf{B} \pm$ all the countable sets, which is not inducible. It is not at all appealing in the face of $\mathbf{C}$ in the Example.

(2) Basu and Ghosh [1967] gave a modification of the same example by removing all the continuous measures from P and showed that $\mathbf{C}$ remains a Burkholder statistic in the example thus modified. But then the power set of X would be a much more reasonable choice for $\mathbf{A}$ than the Borel σ-field. They proposed accordingly the experiment $(X, 2^X, P)$ as a formulation of the problem of sampling from finite populations. Later it came to be known as the "discrete experiment".

(3) The presence of a non-measurable set is essential, whether we remove or retain the continuous measures in P. In fact, if we decide to choose the power set of X as $\mathbf{A}$, that is, to adopt the experiment $(X, 2^X, P)$ with $A1$ and $A2$, then it does not have a Burkholder statistic any more (For the proof see Theorem 4 below. Burkholder subfields still exist). There are alternatives for doing this;

(a) Basu-Ghosh modification of the Example as in Remark (2),

(b) Assume that the cardinality of the continuum is measurable and bring into P the continuous measures thereby accommodated on the real line X (possibly not including the Lebesgue measure) in addition to discrete measures, or

(c) Adopt one of the alternative axioms of the set theory under which all the subsets of X are Lebesgue measurable.

In view of these possibilities we do *not* take the stand that the continuous measures cannot be defined on the power sets. Earlier, when the extension of the Lebesgue measure to the power set was not considered possible, it must have been a compelling reason for the choice of the Borel σ-field as $\mathbf{A}$. We shall avoid only the axiom of choice, which is inconsistent with the alternative axioms, so that the Lebesgue measure be kept within the scope of our results.

The foregoing observations point to a non-measurable set, say S, as a prerequisite for a Burkholder statistic, in addition to a sufficient subfield **B**, which is "substantially smaller" than **A**. This latter in turn requires an uncountable set A in $\mathbf{A} \setminus \mathbf{B}$. The two sets A and S need be placed in the same part of X. Hence emerges the condition: *Uncountably many atoms of* **B** *intersect with both* S *and* A.

This is in fact n&s, and coincides with the condition in a theorem of Michaletzky [1987] when the latter is adapted into the present context. The theorem is quoted below, wherein the word "complete" means that the σ-field contains the essential supremum of any of its subfamilies. That a subfield is "*relatively* complete" (our nomenclature) means that it contains the essential supremum of any of its subfamilies, provided it exists in **A**. That a subfield **C** is "closed" means that a set A in **A** belongs to **C** if for any p and q in P there exists C in **C** such that $p(A\Delta C) = q(A\Delta C) = 0$.

MICHALETZKY'S THEOREM. Let **E** be an experiment and **B** a sufficient subfield. If for any A in **A** the σ-filed $\mathbf{A} \cap \{x; 0 < P[A|\mathbf{B}]\,(x) < 1\}$ is complete, then every relatively complete subfield including **B** is sufficient,

Conversely if every relatively complete and closed sub-field including **B** is sufficient, then for any A in **A** the σ-field $\mathbf{A} \cap \{x;\ 0 < P[A|\mathbf{B}]\,(x) < 1\}$ is complete.

Let us denote the set $\{x;\ 0 < P[A|\mathbf{B}](x) < 1\}$ by $E(A)$. Under $A1$ and $A2$ it is the union of atoms B_i which satisfy $\phi \neq A \cap B_i \subsetneq B_i$. It is not hard to see that under $A1$ and $A2$ an inducible subfield is closed as well as relatively complete. Incidentally the proof of Lemma 2 can be translated almost word by word to the proof of "a sufficient subfield is relatively complete in *any* experiment". Finally, completeness of $\mathbf{A} \cap E(A)$ in the present context means simply that all the subsets of $E(A)$ are measurable, as the singletons are measurable. Hence the following adaptation of the Michaletzky's Theorem, stated in the form of Theorems 4 and 5.

THEOREM 4. Suppose that $\mathbf{E} = (X, \mathbf{A}, P)$ is an experiment satisfying $A1$ and $A2$, and that **B** is a sufficient subfield. Denote by $\{B_i,\ i \in I\}$ the partition of **B** and for each A in **A** let $E(A)$ be the union of B_i's which satisfy $B_i \supsetneq A \cap B_i \neq \phi$. If all the subsets of all $E(A)$, $A \in \mathbf{A}$, are measurable, then any inducible subfield **C** including **B** is sufficient.

THEOREM 5. Suppose that $\mathbf{E} = (X, \mathbf{A}, P)$ is an experiment satisfying $A1$ and $A2$, and that **B** and $E(A)$ are defined as in Theorem 4. If there exists an A in **A** such that $E(A)$ has a non-measurable subset S, then **E** has a Burkholder statistic, constructed as follows:

Find out a non-measurable subset S^* of $E(A)$ and a set A^* in $\mathbf{A}\backslash\mathbf{B}$ such that $S^* \cap B_i \neq A^* \cap B_i \neq \phi$ for each B_i intersecting with S^*. Define a subfield $\mathbf{C}$ by

$$\mathbf{C} = \{C;\ C = K \cup A,\ K = B \cap S^*,\ K \in \mathbf{A},\ B \in \mathbf{B}, A \subset X\backslash S^* \text{ and } A \in \mathbf{A}\}.$$

Then $\mathbf{C}$ is a non-sufficient statistic including $\mathbf{B}$.

We are going to give proofs of these theorems specific to the present case. They are naturally much simpler than those of Michaletzky for the general case, a kind of proofs which exhibit great expertise of the author in places. Also, self-contained proofs would make it easier to see that the axiom of choice is not involved in them, though it is not impossible to check the same with Michaletzky's theorem and the results invoked therein. The way Theorem 5 and its proof are stated shows how the Burkholder statistic is constructed.

PROOF OF THEOREM 4. Take and fix any set A in $\mathbf{A}$ and shorten $E(A)$ to E. We will construct a $\mathbf{C}$-measurable function $f(x)$ which serves as $P[A|\mathbf{C}]$. On the $\mathbf{B}$-measurable sets $[P[A|\mathbf{B}] = 0]$ and $[P[A|\mathbf{B}] = 1]$ we define $f(x)$ to be 0 and 1, respectively. The set $E = [0 < P[A|\mathbf{B}] < 1]$ is a union of the atoms of $\mathbf{C}$ included in it. Denote such an atom by V. Notice that $V \subset [P[V|\mathbf{B}] > 0]$ and define $f(x) = \frac{P[A\cap V|\mathbf{B}]}{P[V|\mathbf{B}]}$ for all x in V. If this is done for all V in E, it amounts to defining $f(x)$ on the whole of E. As $f(x)$ is constant on each atom V and all the subsets of E are measurable, it is $\mathbf{C}$-measurable. Outside E, $f(x)$ is even $\mathbf{B}$-measurable.

In order to prove that $\mathbf{C}$ is sufficient, i.e., $p(A \cap C) = \int_C f\,dp$ for all p in P and C in $\mathbf{C}$, it is enough to prove the same integral equality for each atom of $\mathbf{C}$ in place of C. For V in E, it follows that

$$\begin{aligned}
\int_V f\,dp &= \int_V \frac{P[A\cap V|\mathbf{B}]}{P[V|\mathbf{B}]}\,dp \\
&= \int I_V \frac{P[A\cap V|\mathbf{B}]}{P[V|\mathbf{B}]}\,dp \\
&= \int P[A\cap V|\mathbf{B}]\,dp \\
&= p(A\cap V).
\end{aligned}$$

For an atom outside E, the equality is trivial.

PROOF OF THEOREM 5. First of all, see that $\mathbf{C}$ is a subfield including $\mathbf{B}$ and is induced by the partition

$$\{B_i \cap S^*;\ B_i \cap S^* \neq \phi\} \cup \{\{x\};\ x \in X\backslash S^*\}.$$

We treat two cases separately.

CASE 1. There exists a union of **B**-atoms B_i included in $E(A)$ which does not belong to **A**.

In this case the union is to be taken as S^* and A itself is to be taken as A^*. They satisfy the condition demanded in the Theorem. The definition of **C** and its partition respectively reduce to

$$\mathbf{C} = \{C;\ C = B \cup A,\ B \subset S^*,\ B \in \mathbf{B},\ A \subset X \backslash S^* \text{ and } A \in \mathbf{A}\} \text{ and}$$
$$\{B_i;\ B_i \subset S^*\} \cup \{\{x\};\ x \in X \backslash S^*\}.$$

CASE 2. All the unions of B_i in $E(A)$ belong to **A**.

In this case, either $S \cup (E \cap A)$ or $(E \backslash S) \cup (E \backslash A)$ is non-measurable, as otherwise S would become measurable. In the former case take $(S \cup (E \cap A)) \backslash (\cup\{B_i; S \cap B_i = A \cap B_i\})$ as S^* and take A as A^*. In the latter case take

$$((E \backslash S) \cup (E \backslash A)) \backslash (\cup\{B_i;\ (E \backslash S) \cap B_i = (E \backslash A) \cap B_i\})$$

as S^* and take $E \backslash A$ as A^*.

In either case S^* and A^* together satisfy the same condition as above.

It follows that **C** is not sufficient. Because if otherwise, there would exist a **C**-measurable function $f(x)$ which satisfies $p(A^* \cap C) = \int_C f\, dp$ for any C in **C** and p in P.

If we take x in S^* then the set $S^* \cap B_i$ to which x belongs is a **C**-atom and, being a countable set, belongs to **A** and hence to **C**. The function $f(x)$ is a constant, say $c(i)$, on it. The integral equality in last paragraph with $C = S^* \cap B_i$ and p in P for which $p(x) > 0$ gives rise to

$$p(A^* \cap B_i) = p(S^* \cap B_i)c(i),$$

which implies

$$f(x) = c(i) = p(A^* \cap B_i)/p(S^* \cap B).$$

This is not equal to 0 or 1, because $S^* \cap B_i \supsetneq A^* \cap B_i \neq \phi$.

On the other hand if we take x in $X \backslash S^*$ then $\{x\}$ is a **C**-atom belonging to **C**. Now the same equality with $C = \{x\}$ and p in P for which $p(x) > 0$ gives rise to

$$p(x) I_{A^*}(x) = p(x) f(x),$$

hence $f(x) = I_{A^*}(x)$.

It follows that $S^* = [0 < f(x) < 1]$, so that it belongs to **C** and hence to **A**, which is a contradiction.

Collate this with Example 2 to see that the Case 1 is same as the Example almost to the word. The Case 2 is similar though slightly roundabout. We would rather bypass this type of non-measurability as follows. It is possible to enlarge **A** to a σ-field $\tilde{\mathbf{A}}$ which contains all the subsets of E, while all the measures in P are extended to $\tilde{\mathbf{A}}$ in a very natural way. If so, $\mathbf{A}$ should better be replaced by $\tilde{\mathbf{A}}$, or even by the largest σ-field to which P can be extended. By doing this, all the non-measurable sets in the Case 2 are rendered measurable and the pathologies caused by them disappear. An outline of the enlargement of $\mathbf{A}$ follows.

Define a σ-ideal $\mathbf{W}$ so as to contain all the subsets of $E(A)$ which fall in the Case 2:

$$\mathbf{W} = \{W;\ \text{There exists a set } E \text{ in } \mathbf{B} \text{ such that (i) } E \supset W, \text{ and (ii) all the unions of } \mathbf{B}\text{-atoms included in } E \text{ belong to } \mathbf{B}\}.$$

We will extend P to the σ-field $\tilde{\mathbf{A}}$ generated by $\mathbf{A}$ and $\mathbf{W}$.

Notice first that for each B_i there is a p_i in P such that $p_i(B_i) > 0$. Take W in $\mathbf{W}$ and define a function $f(x)$ by

$$f(x) = p_i(W \cap B_i)/p_i(B_i) \text{ for all } x \text{ in } B_i.$$

The right hand side does not depend on the choice of p_i and is 0 on B_i not intersecting with W, and a fortiori outside E. As all the unions of B_i's in E are in $\mathbf{B}$, this function is $\mathbf{B}$-measurable and does not depend upon the choice of E.

Now define

$$\tilde{p}(W) = \int f\,dp$$

for each W in $\mathbf{W}$. Any set Z in $\tilde{\mathbf{A}}$ is written as a disjoint union of a set A and W in $\mathbf{W}$: $Z = A + W$. Accordingly define $\tilde{p}(Z) = p(A) + \tilde{p}(W)$. This again does not depend upon the choice of A and W. It is easy to check that $\tilde{p}$ is a p-measure on $\tilde{\mathbf{A}}$. Further, it is equal to p on $\mathbf{A}$, as the definition of $f(x)$, if applied to a set A in $\mathbf{A}$, gives rise to $P[A|\mathbf{B}]$.

One can see that the subfield $\mathbf{B}$ remains sufficient for the enlarged experiment $\tilde{\mathbf{E}} = (X, \tilde{\mathbf{A}}, \tilde{P})$ with $\tilde{P} = \{\tilde{p};\ p \in P\}$, and there is no set in $\tilde{\mathbf{A}}$ which falls in the Case 2 in reference to $\mathbf{B}$.

This would justify us to consider the Case 2 as an instance of the "artificialities" to be excluded, and conclude that Burkholder statistic *has to* arise only when:

1. Some measures to be incorporated in P cannot be defined on the power set of X and so some sets are left non-measurable.

2. A sufficient subfield $\mathbf{B}$ is found to exist, such that a union of its atoms is non-measurable, and

3. There exists a measurable set not belonging to $\mathbf{B}$ which intersects all the atoms in the union.

4 Some remarks on Burkholder statistic in majorized experiment

Having worked on the Burkholder statistics in a special class of non-majorized experiments we now comment briefly on the majorized cases. It appears to us that they also do not show sign of contradicting what we have been trying to say, if not they support it.

This is however based upon the rather meagre knowledge available at present, and we, too, cannot add anything tangible to it. Examples are also rarely found in literature. The difficulty in handling statistics (viz. inducible subfields) is caused by the elusive difference between inducibility and relative completeness under the presence of non-empty null sets (unlike in last Section). It prevents use of the knowledge accumulated so far on the relatively complete subfields, especially the result of Michaletzky on the relatively complete Burkholder subfields.

As there is no Burkholder statistic in the dominated and the discrete cases, it is quite likely that the same is true for the whole of weakly dominated cases. In fact it is easily seen for the "decomposable" experiment, defined to be a weakly dominated experiment (See the paragraph following Condition P, Section 2) whose maximal decomposition $\mathbf{D} = \{D\}$ can be taken a disjoint family (For details see, e.g., Luschgy & Mussmann[1985]). We will sketch the proof here.

Suppose that we have a sufficient subfield $\mathbf{B}$ and an inducible subfield $\mathbf{C}$ which includes $\mathbf{B}$. We will prove that $\mathbf{C}$ is sufficient by constructing $P[A|\mathbf{C}]$ for any A in $\mathbf{A}$. Take any D in $\mathbf{D}$. As $\mathbf{B} \cap D$ is sufficient for the truncated experiment $\mathbf{E} \cap D$, which is dominated, $\mathbf{C} \cap D$ is sufficient for $\mathbf{E} \cap D$. Hence there exists on each D a $\mathbf{C} \cap D$-measurable function $f(x, D)$ which serves as $P[A \cap D|\mathbf{C} \cap D]$.

The function $f(x)$ on X, defined by assembling $f(x, D)$ for all D in $\mathbf{D}$, is $\mathbf{C}$-measurable. Because, if we take the set $\{x;\ f(x) < a\}$ for any a and denote it by F, then F is clearly a union of sets in the partition of $\mathbf{C}$. At the same time F is a disjoint union of $F \cap D$, $D \in \mathbf{D}$, all of which belong to $\mathbf{A}$. It is known (Kusama & Yamada [1972]) that in a weakly dominated experiment $\mathbf{A}$ is *weakly complete* as well, that is,

$$(*) \qquad \mathbf{A} = \{A \subset X;\ A \cap D \in \mathbf{A} \text{ for all } D \text{ in } \mathbf{D}\}.$$

Hence F belongs to $\mathbf{A}$. The inducibility of $\mathbf{C}$ implies that F belongs to $\mathbf{C}$. This shows that $f(x)$ is $\mathbf{C}$-measurable. It is easy to check that it acts as $P[A|\mathbf{C}]$.

The weakly dominated experiment without a disjoint maximal decomposition is yet to be studied. In fact we hardly come across such an experiment, let alone a sufficient subfield. And no wonder, if a weakly dominated experiment is decomposable unless the cardinality of X exceeds that of the continuum (Fremlin[1978] and Luschgy & Mussmann[1985]).

On the other hand, if the assumption of weak completeness is dropped while the disjoint maximal decomposition is retained, then the experiment is not weakly dominated any more. This case is represented by the Basu-Ghosh modification of the Burkholder's example (Remark (2), Section 3). There the "pathology" is removed by weak completion, i.e., by replacing **A** with the RHS of the formula (*) above. In the particular example it equals the power set. It is easy to create similar examples and then remove similarly the "pathologies" of one's own making.

Furthermore, if we give up the disjoint maximal decomposition, then again such experiments are not rare. However, we are yet to encounter any example of a Burkholder statistic, after the weak completion of the experiment concerned.

Acknowledgements

The authors express deep thanks to J.K. Ghosh, Alok Goswami, B.V.Rao, K.K. Roy, Haimanti Sarbadhikari, S.M. Srivastava and S. Yamada for their many valuable suggestions. The present work was partly done while the authors visited Indian Statistical Institue, Calcutta. The authors are grateful to the Institute for giving them such an opportunity.

REFERENCES

R.R. Bahadur (1954): Sufficiency and statistical decision functions, *Ann. Math. Stat.* **25**, 423-462.

D. Basu and J.K. Ghosh (1967): Sufficient statistics in sampling from a finite universe, Proc. 36th Session Int. Stat. Inst., 850-859.

D. Burkholder (1961): Sufficiency in the undominated case, *Ann. Math. Stat.* **32**, 1191-1200.

F.R. Diepenbrock (1971): Charakterisierung einer allgemeinen Bedingung als Dominiertheit mit Hilfe von lokalisierbaren Massen. Thesis, Universität von Münster.

D.H. Fremlin (1978): Decomposable measure spaces, Z. Wahrscheinlichkeitstheorie u. Verw. Gebiete **45**, 159-167.

J. Fujii and H. Morimoto (1986): Sufficiency and pairwise sufficiency in majorized experiments, *Sankhyā* **48**, Series A, 315-330.

J.K. Ghosh, H. Morimoto and S. Yamada (1981): Neyman factorization and minimality of pairwise sufficient subfields, *Annals of Statistics* **9**, 514-530.

T. Kusama and S. Yamada (1972): On compactness of statistical structure and sufficiency, *Osaka Journal of Mathematics* **9**, 11-18.

L. LeCam (1964): Sufficiency and approximate sufficiency, *Ann. Math. Stat.* **35**, 1419-1455.

H. Luschgy and D. Mussmann (1985): Equivalent properties and completion of statistical experiments, *Sankhyā* **47**, Series A, 174-195.

G. Michaletzky (1985): A condition to avoid a pathological structure of sufficient σ-fields, *Probability and Mathematical Statistics* **5**, 153-163.

G. Michaletzky (1987): On the σ-fields which are larger than a sufficient one, Mathematical Statistics and Probability Theory, Vol. A, 231-236.

B.V. Rao and S. Yamada (1980): Note on a problem of Diepenbrock, *Osaka Journal of Mathematics* **17**, 325-327.

Department of Mathematics
Osaka City University
Sugimoto, Sumiyoshi-ku
Osaka
JAPAN

J.K. Ghosh, S.K. Mitra, K.R. Parthasarathy and B.L.S. Prakasa Rao (*eds.*),
Statistics and Probability : A Raghu Raj Bahadur Festschrift
Published by Wiley Eastern Limited, Publishers, (1993), 219 – 232

Contiguity in Bahadur Type Representation of a Conditional Quantile and Application in Conditional Quantile Process[1]

ASHIS K. GANGOPADHYAY
Boston University, Boston

and

PRANAB K. SEN
University of North Carolina, Chapel Hill

Abstract

Let $(X_1, Z_1), \ldots, (X_n, Z_n)$ i.i.d. as (X, Z), Z taking values in $\mathbf{R}$. For $0 < p < 1$, define $\xi_p(x_0)$ as the conditional quantile function of Z given $X = x_0$; i.e. $P(Z \le \xi_p(x_0)|X = x_0) = p$. Bhattacharya and Gangopadhyay (1990) proposed a nonparametric nearest neighbor type estimator of $\xi_p(x_0)$, and established a Bahadur type representation of the same using some direct analysis. Here, the notion of contiguity of conditional probability measures developed in Gangopadhyay and Sen (1992) is incorporated in developing a considerably simpler approach to the desired asymptotic representation. This approach also facilitates the study of the estimator under small perturbation of the design variable.

[1] Research partially supported by NSF Grant DMS-92-70285
AMS Subject Classifications: 62G07, 62G20
Keywords and phrases: Bahadur Representation, Conditional Quantile, Nonparametric Curve Estimation

1 Introduction

Let $\{(X_i, Z_i);\ i = 1, 2, \ldots, n\}$ be a sequence of independent and identically distributed random vectors with a distribution function (d.f.) $\pi(x, z)$, $(x, z) \in \mathbf{R}^2$, the real plane. Let $F(x) = \pi(x, \infty)$ be the marginal d.f. of X and let $G(z|x)$ be the conditional d.f. on Z given $X = x$ for $z \in \mathbf{R}$ and $x \in \mathbf{R}$. In recent years, considerable attention has been paid to the problems of nonparametric estimation of conditional functionals. In particular local estimates of conditional expectation (regression) have been explored thoroughly. However, work still remains to be done in the important problem of estimating other types of conditional functionals that are often useful in data analysis.

Suppose we consider the model

$$Z_i = m(X_i) + \epsilon_i,\ i = 1, 2, \ldots, n$$

where ϵ_i's are i.i.d. zero mean random errors. In a regression problem our objective is to estimate the regression function $m(\cdot)$. The usual approach is to consider squared error loss, which is minimized (for $X = x$) at $E(Y|X = x)$. Thus, in nonparametric regression set up, one tries to obtain a local estimate of the expectation using some smoothing technique. One difficulty with this approach is that the estimators obtained by minimizing the squared error loss are generally quite sensitive to the presence of outliers, and so the approach is particularly inappropriate where the error distribution is heavy tailed. Huber and others have proposed the use of loss functions which lead to "robust estimates" that are less sensitive to the presence of outliers. Koenker and Bassett (1978) argued that in many economic studies, the assumption about the normality of error terms are usually invalid, and they strongly advocated the use of more robust techniques of data analysis. As they put it:

> *"The elementary point that there may exist nonlinear, or for that matter-biased estimators superior to least squares for non-Gaussian linear model is a well kept secret in most of the economic literature."*

As an alternative to least square, they suggested the use of loss function (see also Chaudhury (1991))

$$L_p(t) = |t| + (2p - 1)t\ ,\ 0 < p < 1.$$

Thus, for a random variable Y, $E(L_p(Y - b))$ is minimized at $b =$ the pth quantile of Y. An important special case is $p = 1/2$, where the expected loss is minimized at the median of Y. In a regression set up, an analogous minimization problem gives rise to pth conditional quantile and conditional median for $p = 1/2$. Hogg (1975), Koenker and Bassett (1978) and others have considered the problem of estimating a conditional quantile (also called regression quantile and percentile regression) in a parametric set up. Computational aspects of these estimators are discussed in Koenker and D'orey (1987). However, one obvious drawback with this approach is the problem of finding the right parametric model with small number of parameters, which may not be easy in practice. Alternatives are to use nonparametric or semiparametric curve estimation techniques which are developed in the recent past. Gutenbrunner and Jurečková (1992) studied the regression quantile and regression rank score process in linear regression models, which lead to a rank based semiparametric solution under a linear model set up. Although the proceedure does not require any assumption regarding the distribution of the random noise, the approach is still sensitive to the assumption of linear model. This is not the case with nonparametric curve estimation techniques, where unlike parametric modelling, one doesn't require any assumption regarding the shape of the underlying regression function. Thus, the approach is most suitable for exploratory data analysis.

Bhattacharya and Gangopadhyay (1990) and Chaudhury (1991) have proposed nonparametric estimators of conditional quantile function, and established local Bahadur-type representations of those estimators. Gangopadhyay and Sen (1992) considered the problem of estimation of a general class of differentiable functionals, and using an argument involving contiguity of probability measures, they studied asymptotic properties of these estimators. In this paper, we show that this considerably simpler approach is also useful in obtaining a similar Bahadur-type representation of conditional quantile function at a fixed design point $X = x_0$. In sections 2 and 3, we present a unified treatment of these results for fixed design points. This motivates our main asymptotic result related to the conditional quantile process (in the design variable X), which is discussed in section 4, where we explore the asymptotic behavior of the estimator under small perturbation of the design variable.

2 Nearest Neighbor Estimation of Conditional Quantile

Conditional quantile function as a useful tool of data analysis has been discussed by Hogg (1975). Bhattacharya (1963) used a fractile approach to the problem of estimation of conditional quantile. More recently, Bhattacharya and Gangopadhyay (1990) provided a complete account of nearest neighbor approach to the estimation of conditional quantile. For random vector (X, Z), let $G(\cdot|x)$ denote the conditional distribution of Z given $X = x$, with corresponding density by $g(\cdot|x)$; and let $F(\cdot)$, $f(\cdot)$ denote the c.d.f. and p.d.f of X . Suppose $\xi_p(x_0) = \xi$ be the conditional p-quantile of Z given $X = x_0$, i.e. $G(\xi_p(x_0)|x_0) = p$. Based on a suitable $\|\cdot\|$ (e.g. the Eucledian norm), we may define $Y_i = \|X_i - x_0\|$, $i = 1, 2, \ldots, n$; and consider the data transformation $(X_i, Z_i) \to (Y_i, Z_i)$; $i = 1, 2, \ldots, n$. Let $0 < Y_{n_1} < \ldots < Y_{nn}$ be the order statistics corresponding to $(Y_1, \ldots, Y_n)$ and let $Z_{n1}, \ldots, Z_{nn}$ be the induced order statistics (i.e., $Z_{ni} = Z_j$ if $Y_{ni} = Y_j$; for $i, j = 1, \ldots, n$). Now, for every positive integer $k(= k_n)$, the k-nearest neighbor $(k - NN)$ empirical d.f. of Z (with respect to x_0) is defined as:

$$\hat{G}_{nk}(z) = \frac{1}{k} \sum_{i=1}^{k} 1(Z_{ni} \leq z); \; z \in \mathbf{R} \tag{2.1}$$

where $1(A)$ stands for the indicator function of the set A. This suggests the following k-nearest neighbor estimator of ξ_p:

$$\begin{aligned} \hat{\xi}_{nk} &= \text{the } [kp]\text{th order statistics of } \{Z_{n1}, \ldots, Z_{nn}\} \\ &= \inf\{z | \hat{G}_{nk}(z) \geq \frac{[kp]}{k}\}. \end{aligned} \tag{2.2}$$

In this section, and also in section 3, we explore the asymptotic representation of $\hat{\xi}_{nk}$ at a fixed design point $X = x_0$. Based on an argument involving contiguity of conditional densities, we provide a unfied approach leading up to the asymptotic representation of the nearest neighbor estimator $\hat{\xi}_{nk}$. This in turn motivates the study of asymptotic properties of the conditional quantile process (in x) given by

$$W_n(x) = \sqrt{k_n}(\hat{\xi}_{nk}(x) - \xi_p(x)), \; x \in \mathbf{R} \tag{2.3}$$

for fixed $p \in (0, 1)$, which is discussed in section 4.

At this time, we assume that following regularity conditions hold:

[A1] The d.f. F admits an absolutely continuous density function f such that

$$f(x_0) > 0, \tag{2.4}$$

and $f''(x) = (d^2/dx^2)f(x)$ exists in a neighborhood of x_0, and there exists a positive number ϵ, k_0 and α such that $|x - x_0| < \epsilon$ implies that

$$|f''(x) - f''(x_0)| \leq k_0 |x - x_0|^{\alpha}. \tag{2.5}$$

[A2] The d.f. $G(z|x_0)$ has a continuous density $g(z|x_0)$ such that the partial derivatives $g_x(z|x) = (\partial | \partial x) g(z|x)$ and $g_{xx}(z|x) = (\partial^2/\partial x^2) g(z|x)$ exist in a neighborhood of x_0.

Recall that $Y_i = |X_i - x_0|$, $1 \leq i \leq n$, and let $F_Y(y)$, $y \geq 0$ be the marginal d.f. of y_i. Then F_Y admits a density function

$$f_Y(y) = f(x_0 + y) + f(x_0 - y), \; y \geq 0, \tag{2.6}$$

so that using (2.4) and (2.5), we conclude that [A1] holds for f_Y too. $G^*(z|y)$ (and $g^*(z|y)$) the conditional d.f. (and density) of Z given $Y = y$ are given by:

$$G^*(z|y) = \{f(x_0 + y)G(z|x_0 + y) + f(x_0 - y)$$

$$G(z|x_0 - y)\}/f_Y(y) \tag{2.7}$$

$$g^*(z|y) = \{f(x_0 + y)g(z|x_0 + y) + f(x_0 - y)$$
$$g(z|x_0 - y)\}/f_Y(y). \tag{2.8}$$

Note that

$$g^*(z|0) = g(z|x_0) \text{ and } G^*(z|0) = G(z|x_0), \; z \in \mathbf{R}. \tag{2.9}$$

From now on, we will denote $g(z|x_0)$ and $G(z|x_0)$ by $g(z)$ and $G(z)$ respectively.

[A3] There exists $\epsilon > 0$, $\alpha > 0$ and some Lebesgue measurable functions $u_1(z)$ and $u_2(z)$, such that for all y, $y < \epsilon$,

$$\left|(\partial/\partial y^2)g^*(z|y)\right| \leq u_1(z) \text{ a.e. } z \tag{2.10}$$

and

$$\left|\{(\partial/\partial y^2)g^*(z|y)\} - \{(\partial/\partial y^2)g^*(z|y)\Big|_{y=0}\}\right| \tag{2.11}$$
$$\leq u_2(z)y^\alpha.$$

[A4] There exists $\epsilon > 0$ such that

$$I(z|y) = E[\{(\partial/\partial y^2)\log g^*(z|y)\}^2 \Big| Y = y] < \infty \tag{2.12}$$

uniformly in y: $0 \leq y \leq \epsilon$.

REMARK: Note that for [A3] to hold the following conditions suffice: There exist $\epsilon > 0, \alpha > 0$ and Lebesgue measurable functions (on $\mathbf{R}$) $u_1^*(z), u_2^*(z)$ and $u_3^*(z)$ such that for $|x - x_0| < \epsilon$,

$$|g_x(z|x)| \leq u_1^*(z), \; |g_{xx}(z|x)| \leq u_2^*(z), \tag{2.13}$$

$$|g_{xx}(z|x) - g_{xx}(z|x_0)| \leq u_3^*(z)|x - x_0|^\alpha. \tag{2.14}$$

However, expressing in terms of $(\partial/\partial y^2)g^*(z|y)$, we are able to bring the analogy with the classical (location/scale) case treated by Hájek and Sidák (1967).

Bhattacharya and Gangopadhyay (1990) established the following Bahadur type representation for $\hat{\xi}_{nk}$. Bias term show up in their representation because of their choice of $k = k_n = O(n^{4/5})$, which achieves optimum balance between bias and random error. The result is stated in the following theorem:

THEOREM 2.1 *For k lying in*

$$I_n(a, b) = \{k | k_0 = [n^{4/5}a] \leq k \leq [n^{4/5}b] = k_1\} \tag{2.15}$$

$$\hat{\xi}_{nk} - \xi = \beta(\xi)(k/n)^2 + \{kg(\xi)\}^{-1} \sum_{i=1}^{k} [1(Z^*_{ni} > \xi) - (1-p)] + R_{nk}, \tag{2.16}$$

where

$$\beta(\xi) = -[f(x_0)G_{xx}(\xi|x_0) + 2f'(x_0)G_x(\xi|x_0)]/\{24f^3(x_0)g(\xi)\}, \tag{2.17}$$

$$Z^*_{ni} = G^{-1} \circ G^*(Z_{ni}|Y_{ni});\ 1 \le i \le n, \tag{2.18}$$

and

$$\max_{k \in I_n(a,b)} |R_{nk}| = O(n^{-3/5} \log n)\ a.s. \tag{2.19}$$

It should be noted that the rate of convergence of the remainder term corresponds to $O(k^{-3/4} \log k)$ with $k = O(n^{4/5})$. This highlights the similarity of this result to Bahadur's (1966) representation of estimated quantile for one sample data. A proof of the theorem is given in Bhattacharya and Gangopadhyay (1990). The main idea of the proof is to note that the induced order statistics Z_{ni}'s are conditionally independent given $\mathcal{A} = \sigma(Y_1, Y_2, \ldots)$ with Z_{ni} having conditional c.d.f. $G^*(\cdot|Y_{ni})$. Suppose ξ_{nk} be the p-quantile of the random c.d.f. $\overline{G}_{nk}(z) = (1/k)\sum_{i=1}^{k} G^*(\cdot|Y_{ni})$; i.e.,

$$\overline{G}_{nk}(\xi_{nk}) = p, \tag{2.20}$$

then one can think of $\hat{\xi}_{nk}$ as an estimator of ξ_{nk}. The difference between $\overline{G}_{nk}(\cdot)$ and $G(\cdot)$ of which ξ is the p-quantile gives rise to the bias term, as well as random error of small order of magnitude. The difference $\hat{G}_{nk}(\cdot) - \overline{G}_{nk}(\cdot)$ provides the empircal process, which is the main feature of the Bahadur type representation. The order of the remainder term is obtained through a fluctuation result of empirical distribution, which is also quite similar to the one established in Bahadur (1966) paper.

3 Contiguity and Bahadur type representation of Conditional Quantile

We now consider an alternative approach of obtaining a Bahadur type representation for conditional quantile function. This novel and relatively simpler approach involves an application of contiguity of probability measures in a conditional set up.

For every $k(\le n)$; let q_{nk} and Q_{nk} denote the joint conditional density and distribution of $Z_{n1}, \ldots, Z_{nk}$, given $Y_{n1}, \ldots, Y_{nk}$. Using the conditional independence of $(Z_{n1}, \ldots, Z_{nk})$, we have

$$q_{nk}(\cdot) = \prod_{i=1}^{k} g(\cdot|Y_{ni}) = \prod_{i=1}^{k} g_{ni}(\cdot) \tag{3.1}$$

and

$$Q_{nk}(\cdot) = \prod_{i=1}^{k} G(\cdot|Y_{ni}) = \prod_{i=1}^{k} G_{ni}(\cdot). \tag{3.2}$$

Now let $Z_{ni}^0, \ldots, Z_{nk}^0$ be i.i.d. random variables with a density and d.f. $g(z)$ and $G(z)$, and let

$$p_{nk} = \prod_{j=1}^{k} g(\cdot) \text{ and } P_{nk}(\cdot) = \prod_{j=1}^{k} G(\cdot) \tag{3.3}$$

Thus, q_{nk} is the true conditional density of $(Z_{n1}, \ldots, Z_{nk})$ given $(Y_{n1}, Y_{n2}, \ldots, Y_{nk})$, while p_{nk} and P_{nk} stands for the joint density and joint disribution of an i.i.d. model. In this setup, Gangopadhyay and Sen (1992) established the following result:

THEOREM 3.1 *Let $k = [tn^\lambda]$ for some $\lambda (0 < \lambda \le 4/5)$ and for some $t (0 < a < t < b < \infty)$. Then under the regularity conditions [A1]-[A4], the densities q_{nk} are contiguous to the densities p_{nk} a.e. (F).*

Define the likelihood ratio

$$\begin{aligned} L_{nk} &= q_{nk}/p_{nk} \quad \text{if } p_{nk} > 0 \\ &= 1 \qquad\quad \text{if } p_{nk} = q_{nk} = 0 \\ &= \infty \qquad\quad \text{if } p_{nk} = 0 < q_{nk} \end{aligned} \tag{3.4}$$

Gangopadhyay and Sen (1992) established that $\log L_{nk}$ is asymptotically normal $(-(1/2)d^2, d^2)$ for some $d \in \mathbf{R}$. Thus, along the fashion of LeCam's first lemma (see, e.g., Hájek and Sidák (1967), page 203-204), we have the densities q_{nk} are contiguous to the densities p_{nk}.

To apply Theorem 2.2 to develop a Bahadur type representation for conditional quantile, we first establish a modified version of Bahadur's fluctuation result for one sample problem. This result shows that Bahadur's fluctuation result holds uniformly for $k(= k_n)$ in $I_n(a, b)$.

LEMMA 3.1 *Let $W_1, \ldots, W_k$ i.i.d. with distribution H, where $k \in I_n(a, b)$, and let η be the p-quantile of H, i.e., $H(\eta) = p$. Assume that the distribution H is twice differentiable with $H'(\eta) = h(\eta) > 0$. Let a_n be a sequence of constants and that $a_n \sim n^{-2/5} \log n$. Then*

$$V_n^* = \max_{k \in I_n(a,b)} \sup_{|w-\eta| \le a_n} |(\hat{H}_{nk}(w) - \hat{H}_{nk}(\eta)) - (H(w) - H(\eta))|$$

$$= O(n^{-3/5} \log n) \ a.s., \tag{3.5}$$

where $\hat{H}_{nk}(\cdot)$ is the empirical distribution function of $(W_1, \ldots, W_k)$.

PROOF: We will only provide a brief outline of the argument. As in Bahadur (1966), choose $b_n = n^{1/5}$ and divide the interval $[\eta - a_n, \eta + a_n]$ into $2b_n$ equal intervals $J_{nk,r} = [\eta + r(a_n/b_n), \eta + (r-1)(a_n/b_n)] = [\tau_{nk,r}, \tau_{nk,r+1}]$, for $r = -b_n, \ldots, b_n - 1$.

Thus for $w \in J_{nk,r}$,

$$V_n^* = \max_{k \in I_n(a,b)} V_{nk}^* = \max_{k \in I_n(a,b)} \sup_{|w-\eta| \leq a_n} |V_{nk}(w)|$$

$$\leq \max_k \max_{-b_n \leq r \leq b_n} |V_{nk}(\tau_{nk,r})| + \max_k \max_{-b_n \leq r \leq b_n} |\alpha_{nk,r}|,$$

where

$$\alpha_{nk,r} = |H(\tau_{nk,r+1}) - H(\tau_{nk,r})|$$

$$\leq \sup_{|w-\eta| \leq a_n} |H'(w)| n^{-3/5} \log n,$$

for $-b_n \leq r \leq b_n - 1$. Thus

$$\max_k \max_r \alpha_{nk,r} = O(n^{-3/5} \log n).$$

Finally, using Bernstein's inequality,

$$P(\max_{k \in I_n(a,b)} \max_r |V_{nk}(\tau_{nk,r})| > Mn^{-3/5} \log n)$$

$$\leq \sum_k \sum_r P(|V_{nk}(\tau_{nk,r})| > Mn^{-3/5} \log n)$$

$$\leq \sum_k \sum_r \exp[-CM^2 \log n]$$

$$= 2(b-a) n \exp[-CM^2 \log n],$$

for a large constant C. Thus

$$\sum_n P(\max_k \max_r |V_{nk}(\tau_{nk,r})| > Mn^{-3/5} \log n) < \infty,$$

for sufficiently large M, which completes the proof. ■

At this time, we state the following lemmas from Bhattacharya and Gangopadhyay (1990). The proofs are straightforward.

LEMMA 3.2 *Let $S_n = \{Y_{n,[n^{4/5}b]} > n^{-1/5}B\}$. For $B > b/f(x_0)$ and for sufficiently large n,*

$$P[Y_{n,[n^{4/5}b]} > n^{-1/5}B] \leq \exp[-2n^{3/5}(Bf(x_0) - b)^2]$$

LEMMA 3.3 *The following expansions hold for the conditional p.d.f. $g^*(z|y)$ and conditional c.d.f. $G^*(z|y)$:*

$$\begin{aligned} g^*(z|y) &= g(z) + \frac{1}{2}y^2 q(z) + y^3 r(y,z) \\ G^*(z|y) &= G(z) + \frac{1}{2}y^2 Q(z) + y^3 R(y,z) \end{aligned}$$

where

$$\begin{aligned} g(z) &= g(z|x_0) \\ q(z) &= g_{xx}(z|x_0) + 2f'(x_0)g_x(z|x_0)/f(x_0) \\ G(z) &= G(z|x_0) \\ Q(z) &= G_{xx}(z|x_0) + 2f'(x_0)G_x(z|x_0)/f(x_0) \end{aligned}$$

and there exists $\epsilon > 0$ and $M < \infty$ such that $|q(z)|$, $|Q(z)|$, $|r(y,z)|$ and $|R(y,z)|$ are all bounded by M for $0 \le y \le \epsilon$ and $|z-\xi| \le \epsilon$.

LEMMA 3.4 *For every B, on the set S_n, there exist C such that*

$$\max_{k \in I_n(a,b)} |\xi_{nk} - \xi| \le C n^{-2/5}.$$

The following result provides an expression of the bias term in the asymptotic expansion of conditional quantile function.

LEMMA 3.5

$$\max_{k \in I_n(a,b)} |\xi_{nk} - \xi - \beta(\xi)(k/n)^2| = O(n^{-3/5}) \text{ a.s.}$$

where

$$\beta(\xi) = -Q(\xi)\{24 f^2(x_0) g(\xi)\}^{-1}.$$

We now establish the following result:

LEMMA 3.6 *Under regularity conditions [A1]-[A4],*

$$D_n^* = \max_{k \in I_n(a,b)} \sup_{|z-\xi_{nk}| \le a_n} |\{\hat{G}_{nk}(z) - \hat{G}_{nk}(\xi_{nk})\} - \{\overline{G}_{nk}(z) - \overline{G}_{nk}(\xi_{nk})\}|$$

$$= O_p(n^{-3/5}\log n).$$

PROOF: We have shown in lemma 3.1 that under p_{nk} measure, D_n^* is $O(n^{-3/5}\log n)$ a.s. Now, since q_{nk} is contiguous to p_{nk} (Theorem 3.1), under q_{nk} measure, D_n^* is
$O_p(n^{-3/5}\log n)$. ■

REMARK: Note that using this argument involving contiguity, we can obtain the rate $O(n^{-3/5}\log n)$ in probability. However, as we have shown, the result essentially follows from an extension of Bahadur's result for one sample data; and the proof is much simpler than that of Bhattacharya and Gangopadhyay (1990) where they established the rate $O(n^{-3/5}\log n)$ a.s. Also, for most asymptotic analysis $O_p(n^{-3/5}\log n)$ is usually good enough.

LEMMA 3.7

$$P(\max_{k\in I_n(a,b)} |\hat{\xi}_{nk}-\xi_{nk}| > a_n \ i.o.) = 0.$$

PROOF: Proof is given in Bhattacharya and Gangopadhyay (1990). Here we provide an outline of the argument. Note that,

$$\hat{\xi}_{nk} \le \xi_{nk} - a_n \tag{3.6}$$

implies

$$(1/k)\sum_{i=1}^{k}\{1(Z_{ni}\le \xi_{nk}-a_n) - G_{ni}(\xi_{nk}-a_n)\} \ge [kp]/k - \overline{G}_{nk}(\xi_{nk}-a_n)$$

By lemma 3.3

$$\min_{k\in I_n(a,b)} \{[kp]/k - \overline{G}_{nk}(\xi_{nk}-a_n)\} \ge \frac{1}{2}g(\xi)a_n \tag{3.7}$$

on the set S_n, where n large. Thus by an application of Hoeffding's inequality, we get

$$P(\min_{k\in I_n(a,b)} (\hat{\xi}_{nk}-\xi_{nk}) \le -a_n \ i.o.) = 0. \tag{3.8}$$

Similarly,

$$P(\max_{k\in I_n(a,b)} (\hat{\xi}_{nk}-\xi_{nk}) \ge a_n \ i.o.) = 0, \tag{3.9}$$

which completes the proof. ■

Now, lemma 3.6 and lemma 3.7 implies

$$p - \hat{G}_{nk}(\hat{\xi}_{nk}) = \overline{G}_{nk}(\hat{\xi}_{nk}) - \overline{G}_{nk}(\xi_{nk}) + R^*_{nk} \tag{3.10}$$

where

$$\max_{k\in I_n(a,b)} |R^*_{nk}| = O_p(n^{-3/5}\log n). \tag{3.11}$$

But

$$\overline{G}_{nk}(\hat{\xi}_{nk}) - \overline{G}_{nk}(\xi_{nk}) = (\hat{\xi}_{nk}-\xi_{nk})\overline{g}_{nk}(\xi^*_{nk}), \tag{3.12}$$

where ξ^*_{nk} lies between $\hat{\xi}_{nk}$ and ξ_{nk} and $\overline{g}_{nk}(\cdot) = \frac{1}{k}\sum_{i=1}^{k} g(\cdot|Y_{ni})$. Note that

$$\max_{k\in I_n(a,b)} |\xi^*_{nk}-\xi| \le \max_{k\in I_n(a,b)} |\hat{\xi}_{nk}-\xi_{nk}| + \max_{k\in I_n(a,b)} |\xi_{nk}-\xi|, \tag{3.13}$$

and hence by lemma 3.4 and lemma 3.7

$$\max_{k\in I_n(a,b)} |\xi^*_{nk} - \xi| = O(n^{-2/5}\log n), \ a.s. \tag{3.14}$$

Consequently, lemma 3.2 and lemma 3.3 imply

$$\max_{k\in I_n(a,b)} |\overline{g}_{nk}(\xi^*_{nk}) - g(\xi)| = O(n^{-2/5}\log n), \ a.s. \tag{3.15}$$

Also, by Hoeffding's inequality we have

$$p - \hat{G}_{nk}(\xi_{nk}) = O(n^{-2/5}\log n), \ a.s. \tag{3.16}$$

Thus, combining (3.10),(3.11),(3.12),(3.14),(3.15), and (3.16), we have

$$\hat{\xi}_{nk} = \xi_{nk} + \{kg(\xi)\}^{-1}\sum_{i=1}^{k}[1(Z_{ni} > \xi) \tag{3.17}$$

$$-\{1 - G_{ni}(\xi)\}] + R_{nk},$$

where

$$\max_{k\in I_n(a,b)} |R_{nk}| = O_p(n^{-3/5}\log n). \tag{3.18}$$

It is easy to check that the expression (3.17) can be also written as

$$\hat{\xi}_{nk} = \xi_{nk} + \{kg(\xi)\}^{-1}\sum_{i=1}^{k}[1(Z^*_{ni} > \xi) \tag{3.19}$$

$$-(1 - p)] + R_{nk},$$

where $\max_{k\in I_n(a,b)} |R_{nk}| = O_p(n^{-3/5}\log n)$ and $Z^*_{ni} = G^{-1}\circ G_{ni}(Z_{ni})$ and $G(\xi) = p$. Now, (3.19) along with lemma 3.5 completes the proof of Theorem 2.1, where order the remainder term is replaced by

$$\max_{k\in I_n(a,b)} |R_{nk}| = O_p(n^{-3/5}\log n).$$

4 Local Asymptotic Behavior of a Conditional Quantile Process

As in Section 2, let $\xi_p(x)$ be the p-quantile of the d.f. $G(\cdot|x)$, $x\epsilon\mathbf{R}$, $p\epsilon(0,1)$, and let $\hat{\xi}_{nk}(x)$ be defined as in (2.2) where the Z_{ni} are defined with respect to the pivot x. Note that these Z_{ni} are the induced order statistics corresponding to the nearest neighbors of x (i.e., those X_i which are closest to x). As such, if we consider two distinct values, say x_0 and x^0, then the two sets of X_i forming the corresponding nearest neighbors would have at most k_n members in common, and moreover, by the Glivenko-Cantelli Lemma, the cardinality of this intersection subset when divided by n will converge to 0 at a rate faster than $n^{-1}k_n$. As such, the proportion of the two subsets' common elements will be asymptotically $o(n^{-1}k_n)$, and by an appeal to (2.16), we may immediately say that $\hat{\xi}_{nk}(x_0)$ and $\hat{\xi}_{nk}(x^0)$ are asymptotically independent. Keeping (2.16) in mind we set $W_n = \{W_n(x),\ x\epsilon\mathbf{R}\}$, by letting

$$W_n(x) = k_n^{1/2}[\hat{\xi}_{nk_n}(x) - \xi_p(x)],\ x\epsilon\mathbf{R}. \tag{4.1}$$

Then, for any two distinct points, say x_0 and x^0, $W_n(x_0)$ and $W_n(x^0)$ are asymptotically independent and each one having a normal law given by (2.16)-(2.19). Thus the asymptotic behavior of W_n is depicted entirely by the pointwise asymptotic normality and the stochastic dependence pattern only in a shrinking neighborhood of fixed points. Such shrinking neighborhoods are defined by the set of x relative to a given x_0 as $\{x\colon |x - x_0| \le Kn^{-1}k_n\} = I_n(x_0, K)$, say, where K is a finite positive number. For any two points, say x_n and x'_n, belonging to $I_n(x_0, K)$, let $k(x_n, x'_n)$ be the number of elements in common between the two neighbor sets as defined in Section 2 (for x_0). Consider the intersection set $I_n(x_n, x'_n)$:

$$[x_n - (2nf(x_0))^{-1}k_n,\ x_n + (2nf(x_0))^{-1}k_n] \cap [x'_n - (2nf(x_0))^{-1}k_n, x'_n + (2nf(x_0))^{-1}k_n]. \tag{4.2}$$

This set is a null set if the two intervals are not overlapping. Let $m_n(x_n, x'_n)$ be the length of this intersection set $I_n(x_n, x'_n)$. Then, by the Bahadur (1966) representation, we may again claim that

$$n^{-1}k(x_n, x'_n) - f(x_0)m_n(x_n, x'_n) = o(n^{-1}k_n) \quad a.s.,\ \text{as } n \to \infty. \tag{4.3}$$

As such, if we use (2.16) for each of the two points x_n and x'_n, we may conclude that $W_n(x_n)$ and $W_n(x'_n)$ have asymptotically the correlation

$$nf(x_0)m_n(x_n, x'_n)/k_n, \tag{4.4}$$

and it is easy to show that for both x_n and x'_n belonging to $I_n(x_0, K)$, (4.4) has a limit (which vanishes when $m_n(x_n, x'_n) = 0$). In order to justify this asymptotic analysis, there remains a single issue: To extend (2.16) for a suitable convergent sequence $\{x_n\}$ (to x_0), when $x_n\epsilon I_n(x_0, K)$. In this context, we need to replace in (2.16), x_0 by x_n (everywhere) and to show that this holds uniformly in $I_n(x_0, K)$.

To facilitate this, we consider a stochastic process:

$$W_n^0(t) = W_n(x_0 + n^{-1}k_n t),\ t\epsilon\mathbf{R}, \tag{4.5}$$

where $W_n(\cdot)$ is defined in (4.1). Consider a compact interval $C = [-K, K]$ where $K(<\infty)$ is a positive number so chosen that $Kf(x_0) \le 1$. For the convergence of the finite dimensional laws of $W_n^0(C) = \{W_n^0(t),\ t\epsilon C\}$, we may appeal to (2.16) and draw the desired conclusion immediately. Thus, it remains to show only that $W_n^0(C)$ is tight (or relatively compact). The proof of this can easily be worked out by verifying the Billingsley's (1968) type inequality. Alternatively, the original Bahadur (1966) representation of sample quantiles (for the marginal distribution) can easily be extended to our conditional setup. Using the intricate relationship between the order statistics and empirical d.f., the Billingsley type inequality may as well be formulated in terms of the conditional empirical d.f.'s, and as they involve zero-one valued indicator functions, computation of 4th central moment poses no problem. As such, the computations are fairly routine, and therefore, we omit the details. In passing, we may remark that (4.4) possesses a limit when we set $x_n = x_0 + n^{-1}k_n t_1$ and $x'_n = x_0 + n^{-1}k_n t_2$, with $t_1, t_2 \epsilon C$. This is given by

$$[t_1 f(x_0) - 1/2, t_1 f(x_0) + 1/2] \cap [t_2 f(x_0) - 1/2, t_2 f(x_0) + 1/2], \tag{4.6}$$

which, for $t_2 > t_1$ is non-null when $(t_2 - t_1) f(x_0) \leq 1$, and that is the reason we set $K f(x_0) \leq 1$.

Acknowledgment

Some of the results presented here relates to the first author's Ph.D. dissertation at the University of California at Davis under the supervision of Professor P.K. Bhattacharya, whose contributions and guidance are greatfully acknowledged and appreciated.

REFERENCES

Bahadur, R.R. (1966). A note on quantiles in large samples. *Ann. Math. Statist.* **37** 577-580.

Bhattacharya, P.K. (1963). On an analog of regression analysis. *Ann. Math. Statist.* **34** 1459-1473.

Bhattacharya, P.K. (1974). Convergence of sample paths of normalized sums of induced order statistics. *Ann. Statist.* **2** 1034-1039.

Bhattacharya, P.K. and Gangopadhyay, A.K. (1990). Kernel and nearest neighbor estimation of a conditional quantile. *Ann. Statist.* **18** 1400-1415.

Billingsley, P. (1968). *Convergence of probability measures.* Wiley, New York.

Chaudhury, P. (1991). Nonparametric estimates of regression quantile and their local Bahadur representation. *Ann. Statist.* **19** 760-777.

Cheng, K.F. (1983). Nonparametric estimators for percentile regression functions. *Comm. Statist. Theory Methods* **12** 681-692.

Csörgö, M. and Révész, P. (1981). *Strong Approximations in Probability and Statistics.* Academic, New york.

Gangopadhyay, A.K. (1987). Nonparametric estimation of conditional quantile function, Ph.D. dissertation, Univ. California, Davis.

Gangopadhyay, A.K. and Sen, P.K. (1992). Contiguity in nonparametric estimation of conditional functional. *"Nonparametric Statistics and Related Topics"*. North-Holland, Amsterdam.

Gutenbrunner, C. and Jurečkovà, J. (1992). Regression rank scones and regression quantiles. *Ann. Statist.* **20**, 305-330.

Hájek, J. and Sidák, Z. (1967). *Theory of rank tests.* Academic Press, New York.

Hoeffding, W. (1963). Probability inequalities for sums of bounded random variables. *J. Amer. Statist. Assoc.* **58** 13-30.

Hogg, R.V. (1975). Estimates of percentile regression lines using salary data. *J. Amer. Statist. Assoc.* **70** 50-59.

Koenker, R. and Bassett, G. (1978). Regression Quantiles. *Econometrica* **46**, 33-50.

Koenker, R. and D'Orey, V. (1987). Computing regression quantiles. *J. Roy. Statist. Soc. Ser. C* **36**, 383-393.

Krieger, A.M. and Pickands, J., III (1981). Weak convergence and efficient density estimation at a point. *Ann. Statist.* **9** 1066-1078.

Stone, C.J. (1977). Consistent nonparametric regression. *Ann. Statist.* **5** 595-645.

Stute, W. (1986). Conditional empirical processes. *Ann. Statist.* **14** 638-647.

Woodroofe, M. (1970). On choosing a delta-sequence. *Ann. Math. Statist.* **41** 1665-1671.

Boston University
Department of Mathematics
111 Cummington Street
Boston, MA 02215-2411, USA

Department of Biostatistics
University of North Carolina
Chapel Hill, NC 27599-7400,
USA

J.K. Ghosh, S.K. Mitra, K.R. Parthasarathy and B.L.S. Prakasa Rao (*eds.*),
Statistics and Probability : A Raghu Raj Bahadur Festschrift
Published by Wiley Eastern Limited, Publishers, (1993), 233 – 243

Second-order Minimaxity Under Squared Error and Absolute Deviation Loss

J. K. GHOSH
Purdue University, Lafayette

and

RAHUL MUKERJEE
Indian Institute of Management, Calcutta

Abstract

We consider the problem of second-order minimaxity, under squared error and absolute deviation loss functions, with reference to location and scale models, bivariate normal model with unknown correlation parameter and the inverse Gaussian model.

1 Introduction

In the context of second-order efficiency (Ghosh, Sinha and Wieand (1980), Ghosh and Sinha (1981)) or rather third-order efficiency in the terminology of Pfanzagl (1979) or Akahira and Takeuchi (1981), it is well-known (see e.g., Pfanzagl and Wefelmeyer (1978), Ghosh, Sinha and Wieand (1980)) that attention may be confined to the class $\mathcal{C}$ of estimators of the form

$$T_{nc} = \hat{\theta}_n + n^{-1} c(\hat{\theta}_n),$$

AMS Subject Classifications : 62F10, 62F12
Key words and Phrases : Higher Order Minimaxity, Admissibility, Location-Scale Parameters

where $\hat{\theta}_n$ is the maximum likelihood estimate and $c(\cdot)$ is a differentiable function such that the derivative of $c(\cdot)$ fulfils the local Lipschitz condition, the functional form of $c(\cdot)$ being free from n. As indicated in Pfanzagl and Wefelmeyer (1978), Ghosh, Sinha and Wieand (1980) and Ghosh, Sinha and Joshi (1982), under certain regularity conditions, given an estimator T_n, including Bayes estimators with respect to a prior with a smooth and everywhere positive density (Johnson, 1970), one can find an estimator T_{nc} whose risk is less than or equal to that of T_n up to $o(n^{-1})$ for a very wide class of loss functions. Ghosh and Sinha (1981) completely characterized second-order admissibility within a class analogous to $\mathcal{C}$ in the one-parameter case; see Dasgupta and Ghosh (1983) for further results on admissibility in the multiparameter case.

In the present work, we study the question of second order minimaxity within suitable subclasses of $\mathcal{C}$ for various models and under squared error and absolute deviation loss functions. This is in the spirit of an issue left open in Ghosh and Subramanyam (1974). Our definition of second-order minimaxity and some elementary but basic facts have been presented in section 2 where we also consider the estimation of location and scale parameters under squared error loss. In section 3, the problem of estimating the correlation coefficient in a bivariate normal distribution with zero means and unit variances has been treated. The estimation problem of what is known as the location parameter in the inverse Gaussian distribution appears in section 4. In the rest of the paper, we consider the analogous problem when the loss function is based on absolute deviation. Throughout, we attempt to give explicit formulae for risk expansions as well as second-order minimax estimators.

2 Second-order minimaxity under squared error loss

Let $X_1, X_2, \ldots,$ be i.i.d. each with density $f(x,\theta)$ where $\theta \varepsilon \mathbb{R}^1$ or an open subset thereof. We make the assumptions in page 439 of Bhattacharya and Ghosh (1978) taking $s = 4$ in their notation and replacing $f(\cdot,\theta)$ and $g(\cdot,\theta)$ in their notation respectively by log $f(\cdot,\theta)$ and $f(\cdot,\theta)$ in our notation; see also Ghosh, Sinha and Subramanyam (1979). Let $\hat{\theta}_n$ be the maximum likelihood estimate of θ based on $X_1, \ldots, X_n$ (in the sense of Theorem 3 of Bhattacharya and Ghosh (1978)). For $i, j, k, u = 0, 1, 2, \ldots$, let

$$\begin{aligned} L_{ijku} &= E_\theta\{(d \log f(X,\theta)/d\theta)^i (d^2 \log f(X,\theta)/d\theta^2)^j \\ &\quad \times (d^3 \log f(X,\theta)/d\theta^3)^k (d^4 \log f(X,\theta)/d\theta^4)^u\}, \\ L_{ijk} &= L_{ijk0}, L_{ij} = L_{ij0}, L_i = L_{i0}. \end{aligned}$$

Among the expectations defined above, only those which have been useful in the sequel are assumed to exist. These are assumed to be smooth functions of θ. Let $I = I(\theta) = L_2$. Then using an Edgeworth expansion, up to $o(n^{-1})$, for the distribution of $(nI)^{1/2}(T_{nc} - \theta)$ under θ, where $T_{nc} = \hat{\theta}_n +$

$n^{-1}c(\hat{\theta}_n)\varepsilon\mathcal{C}$, it can be seen that

$$E_\theta\{nI(T_{nc}-\theta)^2\} = 1 + n^{-1}a_c(\theta) + o(n^{-1}), \tag{2.1}$$

where

$$a_c(\theta) = I\{\lambda_c(\theta)\}^2 + 2\lambda_c'(\theta) + \psi(\theta), \tag{2.2a}$$

with

$$\lambda_c(\theta) = c(\theta) + I^{-2}(\frac{1}{2}L_{001} + L_{11}),$$

$$\psi(\theta) = I^{-2}L_{02} + I^{-3}(\frac{1}{2}L_{001}^2 + L_{11}^2 + 2L_{001}L_{11}) - 1. \tag{2.2b}$$

Note that in the risk expansion given by (2.2a,b), the expectation operator E_θ can as well be interpreted as the operator E^U for curved exponential families as in Ghosh and Subramanyam (1974) or in the usual sense under the rather strong assumptions of Gusev (1976). Incidentally, under the latter interpretation, the relations (2.2a,b) with $c(\theta) = 0$ correct expression (3.3) in Gusev (1976).

An estimator $T_{nc_0} = \hat{\theta}_n + n^{-1}c_0(\hat{\theta}_n)$ belonging to $\mathcal{C}$ (or a subclass of $\mathcal{C}$) will be said to be second-order minimax in $\mathcal{C}$ (or in the subclass of $\mathcal{C}$) if

$$\sup_\theta a_{c_0}(\theta) \le \sup_\theta a_c(\theta), \tag{2.3}$$

for every $T_{nc} = \hat{\theta}_n + n^{-1}c(\hat{\theta}_n)$ belonging to $\mathcal{C}$ (or the subclass of $\mathcal{C}$), the suprema in (2.3) being over the parametric space for θ.

EXAMPLE 1 *(Location family).* Consider a location model with $f(x,\theta)$ of the form $f(x,\theta) = g(x-\theta), \theta\varepsilon I\!R^1$. Here $I, L_{11}, L_{001}, L_{02}$ are free from θ provided they exist. Let $I = \rho(>0), L_{11} = \ell_{11}, L_{001} = \ell_{001}, L_{02} = \ell_{02}$, where $\rho, \ell_{11}, \ell_{001}, \ell_{02}$ are constants free from θ. Hence by (2.2b), $\psi(\theta) = \psi_0$, say, is a constant. We shall show that for each $T_{nc} = \hat{\theta}_n + n^{-1}c(\hat{\theta}_n)\varepsilon\mathcal{C}$,

$$\sup_\theta a_c(\theta) \ge \psi_0. \tag{2.4}$$

To that effect, consider a univariate normal population with known variance ρ^{-1} and unknown mean $\theta(\varepsilon I\!R^1)$. Observe that for an estimator of θ of the form $\bar{X}_n + n^{-1}\lambda(\bar{X}_n)$, where $\bar{X}_n$ is the mean of n i.i.d. observations and $\lambda(\cdot)$ is a smooth function the functional form of $\lambda(\cdot)$ being free from n,

$$nE_\theta\{\bar{X}_n + n^{-1}\lambda(\bar{X}_n) - \theta\}^2 = \rho^{-1} + n^{-1}[\{\lambda(\theta)\}^2 + 2\rho^{-1}\lambda'(\theta)] + o(n^{-1}). \tag{2.5}$$

Now, $nE_\theta(\bar{X}_n - \theta)^2 = \rho^{-1}$, and by the theorem in Ghosh and Sinha (1981, p. 1335) $\bar{X}_n$ is second-order admissible within the class of estimators considered in (2.5). Hence by (2.5), for every smooth function $\lambda(\cdot)$,

$$\sup_\theta[\{\lambda(\theta)\}^2 + 2\rho^{-1}\lambda'(\theta)] \ge 0,$$

whence by (2.2a,b), the inevitability of (2.4) follows.

Returning now to the general location model, consider the estimator $T_{nc_0} = \hat{\theta}_n + n^{-1}c_0(\hat{\theta}_n)(\varepsilon\mathcal{C})$, where $c_0(\theta) = -\rho^{-2}(\frac{1}{2}\ell_{001} + \ell_{11})$. By (2.2a,b), $a_{c_0}(\theta) = \psi_0$ for each θ and hence by (2.4), T_{nc_0} is second-order minimax in $\mathcal{C}$. It can be seen that, up to $o(n^{-1}), T_{nc_0}$ is risk-equivalent to the Bayes estimator corresponding to the prior $\pi_0(\theta) =$ constant, which happens to be the Jeffreys' prior for the present location model. Also, note that T_{nc_0} is the bias-corrected, up to $o(n^{-1})$, version of the maximum likelihood estimate in the present set-up.

EXAMPLE 2 *(Scale family).* Consider a scale model with $f(x, \theta)$ of the form $f(x,\theta) = \theta^{-1}g(x/\theta), \theta > 0$. Here, assuming that $I, L_{11}, L_{001}, L_{02}$ exist, one obtains $I = \rho\theta^{-2}, L_{11} = \ell_{11}\theta^{-3}, L_{001} = \ell_{001}\theta^{-3}, L_{02} = \ell_{02}\theta^{-4}$, where $\rho, \ell_{11}, \ell_{001}, \ell_{02}$ are constants. We assume that $\rho > 0$. Then by (2.2b), $\psi(\theta) = \psi_0$, say, is free from θ. We shall show that for each $T_{nc} = \hat{\theta}_n + n^{-1}c(\hat{\theta}_n)\varepsilon\mathcal{C}$,

$$\sup_{\theta} a_c(\theta) \geq \psi_0 - \rho^{-1}. \tag{2.6}$$

To that effect, consider a gamma model with unknown scale parameter $\theta(> 0)$ and known shape parameter ρ. With n i.i.d. observations from this model, the maximum likelihood estimate is, say, $\theta_n^* = \rho^{-1}\bar{X}_n, \bar{X}_n$ being the mean of the observations. Observe that for an estimator of the form $\theta_n^* + n^{-1}\lambda(\theta_n^*)$, where $\lambda(\cdot)$ is a smooth function with a functional form free from n,

$$\begin{aligned} &nE_\theta\{\theta_n^* + n^{-1}\lambda(\theta_n^*) - \theta\}^2 \\ &= \rho^{-1}\theta^2 + n^{-1}[\{\lambda(\theta)\}^2 + 2\rho^{-1}\theta^2\lambda'(\theta)] + o(n^{-1}). \end{aligned} \tag{2.7}$$

Following the theorem in Ghosh and Sinha (1981, p. 1335), the estimator $\theta_n^*(1 - n^{-1}\rho^{-1})$ is second-order admissible within the class of estimators considered in (2.7). Furthermore,

$$nE_\theta\{\theta_n^*(1 - n^{-1}\rho^{-1}) - \theta\}^2 = \rho^{-1}\theta^2 - n^{-1}\rho^{-2}\theta^2 + o(n^{-1}).$$

Hence by (2.7) for each smooth function $\lambda(\cdot)$,

$$\sup_{\theta}[\rho\theta^{-2}\{\lambda(\theta)\}^2 + 2\lambda'(\theta)] \geq -\rho^{-1},$$

so that by (2.2a,b), the relation (2.6) follows.

Returning to the general scale model, consider the estimator $T_{nc_o} = \hat{\theta}_n + n^{-1}c_0(\hat{\theta}_n)$, where $c_0(\theta) = -\theta\rho^{-1}\{1 + \rho^{-1}(\frac{1}{2}\ell_{001} + \ell_{11})\}$. By (2.2a,b), $a_{c_0}(\theta) = \psi_0 - \rho^{-1}$. Hence by (2.6) it follows that T_{nc_0} is second-order minimax in $\mathcal{C}$. It can be seen that, up to $o(n^{-1}), T_{nc_0}$ is risk-equivalent to the Bayes estimator corresponding to the prior $\pi_0(\theta) \propto \theta^{-1}$, which is the Jeffreys' prior for the present scale model.

REMARK : Suppose for a given model, the least favourable prior does not depend on n, satisfies the conditions of Johnson (1970) for all θ and, if improper, is such that the posterior distribution of θ given $X_1, ..., X_{n_0}$ is proper for some $n_0(\geq 1)$. Also, let the conditions in Gusev (1976) hold and let the minimax estimator T_M be Bayes with respect to the least favourable prior $\tilde{\pi}(\cdot)$. Then the risk of the minimax estimator T_M has an expansion of the form (2.1) with $a_c(\theta)$ as in (2.2a,b) and

$$c(\theta) = I^{-1}\frac{d}{d\theta}\log\tilde{\pi}(\theta) - I^{-2}(\frac{1}{2}L_{001} + L_{11}) = c_M(\theta) \quad \text{(say)}.$$

If T_{nc_0} be an estimator which is second-order minimax in $\mathcal{C}$ then

$$\sup_\theta a_{c_M}(\theta) \geq \sup_\theta a_{c_0}(\theta).$$

The reverse inequality and hence equality holds if the risk expansions for T_M and T_{nc_0} are uniform, i.e., if the remainder terms are $o(n^{-1})$ uniformly over the parametric space. This holds for a location parameter problem since T_M and T_{nc_0} are both location equivariant provided a location equivariant choice of $\hat{\theta}_n$ is made. A similar remark holds also for the scale parameter problem.

3 Normal correlation model

Let $f(\cdot, \theta)$ represent the bivariate normal model with zero means, unit variances and an unknown correlation coefficient $\theta, -1 < \theta < 1$; see Stuart (1955) for some interesting results on inference about θ. It can be seen that here

$$\left.\begin{array}{lcl} I & = & (1+\theta^2)/(1-\theta^2)^2, L_{11} = -\frac{1}{2}L_{001} = 2\theta(3+\theta^2)/(1-\theta^2)^3, \\ L_{02} & = & (5+26\theta^2+5\theta^4)/(1-\theta^2)^4, \psi(\theta) = 4(1-\theta^2)^2/(1+\theta^2)^3. \end{array}\right\} \tag{3.1}$$

In the spirit of Dasgupta and Ghosh (1983), we consider a subclass, $\mathcal{C}_1$, of $\mathcal{C}$ consisting of estimators $T_{nc} = \hat{\theta}_n + n^{-1}c(\hat{\theta}_n)$, where $c(\cdot)$ is such that either (a) $I\{c(\theta)\}^2 + 2c'(\theta) \to \infty$ along some θ-sequence tending to $1-$ or $-1+$, or (b) $c(\theta)/(1-\theta)$ remains bounded as $\theta \to 1-$ and $c(\theta)/(1+\theta)$ remains bounded as $\theta \to -1+$. It can be seen that the subclass $\mathcal{C}_1$ is very large and contains T_{nc} for almost all natural choices of $c(\cdot)$.

Here $\frac{1}{2}L_{001} + L_{11} = 0$. Hence for $T_{nc}\varepsilon\mathcal{C}_1$ with $c(\cdot)$ satisfying (a), by (2.2a,b), (3.1),$\sup_\theta a_c(\theta) = \infty$. Consider now $T_{nc}\varepsilon\mathcal{C}_1$ with $c(\cdot)$ satisfying (b). For $p > 0$ and $j = 2, 3, ...$, let

$$\pi_{1pj}(\theta) = z_{1pj}^{-1}(1-\theta^2)^{p-1}, \ \ |\theta| \leq 1 - j^{-1}, \tag{3.2}$$

where $z_{1pj} = \int(1-\theta^2)^{p-1}d\theta$, the integral being over the range from $-(1-j^{-1})$ to $(1-j^{-1})$. Using integration by parts, from (2.2a,b), (3.1), it follows after some simplification that for $p > 0$ and $j = 2, 3, ...$,

$$\int_{-(1-j^{-1})}^{1-j^{-1}} a_c(\theta)\pi_{1pj}(\theta)d\theta$$
$$\geq \int_{-(1-j^{-1})}^{1-j^{-1}} \frac{4(1-\theta^2)^2}{(1+\theta^2)^3}\pi_{1pj}(\theta)d\theta - \int_{-(1-j^{-1})}^{1-j^{-1}} \frac{\{\pi'_{1pj}(\theta)\}^2}{I(\theta)\pi_{1pj}(\theta)}d\theta$$
$$+ 2[c(\theta)\pi_{1pj}(\theta)]_{-(1-j^{-1})}^{1-j^{-1}}. \tag{3.3}$$

Since $c(\cdot)$ satisfies (b), by (3.1), (3.2), it can be seen that for fixed $p(>0)$, the right-hand member of (3.3) tends to J_{1p} as $j \to \infty$, where

$$J_{1p} = 4z_{1p}^{-1}\int_{-1}^{1}\{\frac{(1-\theta^2)^{1+p}}{(1+\theta^2)^3} - (p-1)^2\frac{\theta^2(1-\theta^2)^{p-1}}{(1+\theta^2)}\}d\theta,$$

with

$$z_{1p} = \int_{-1}^{1}(1-\theta^2)^{p-1}d\theta.$$

As a consequence, for $T_{nc}\varepsilon\mathcal{C}_1$ with $c(\cdot)$ satisfying (b),

$$\sup_\theta a_c(\theta) \geq \sup_{p>0} J_{1p}. \tag{3.4}$$

Numerical studies show that, up to two places of decimals, the right-hand member of (3.4) equals 1.73. Therefore, for each $T_{nc}\varepsilon\mathcal{C}_1$,

$$\sup_\theta a_c(\theta) \geq 1.73. \tag{3.5}$$

We now consider estimators $T_{nc} = \hat{\theta}_n + n^{-1}c(\hat{\theta}_n)(\varepsilon\mathcal{C}_1)$ with $c(\theta)$ of the form

$$c(\theta) = (1-\theta^2)(\zeta_1\theta + \zeta_3\theta^3 + \zeta_5\theta^5 + ...)$$

where $\zeta_1, \zeta_3, \zeta_5, ...$ are constants such that $c(\theta)$ is positive for $-1 < \theta < 0$ and negative for $0 < \theta < 1$. Thus, in a sense, such estimators may be looked upon as shrinkage estimators. By a computer search, we attempt to choose $\zeta_1, \zeta_3, \zeta_5, ...$ so as to make $\sup_\theta a_c(\theta)$ as small as possible. This leads to the estimator $T_{nc_0} = \hat{\theta}_n + n^{-1}c_0(\hat{\theta}_n)(\varepsilon\mathcal{C}_1)$ where

$$c_0(\theta) = (1-\theta^2)(-1.13\theta + 1.85\theta^3 - 1.95\theta^5 + 1.05\theta^7 - 0.15\theta^9). \tag{3.6}$$

From (2.2a,b), (3.1), (3.6), one can check that up to two places of decimals, $\sup_\theta a_{c_0}(\theta) = 1.75$. This indicates that the lower bound (3.5) is very sharp and that the estimator T_{nc_0} is nearly optimal in $\mathcal{C}_1$ with regard to second-order minimaxity. A more extensive search is likely to yield an even better estimator. Anyway, T_{nc_0} itself is considerably better than the maximum likelihood estimate for which $c(\theta) = 0$ and $\sup_\theta a_c(\theta)$ is as large as 4.

In the present problem, the bound (3.5) was obtained numerically. It could, therefore, be intuitively anticipated that an analytical solution to the problem of second-order minimaxity would be intractable. Interestingly, if the parametric space for θ be slightly modified by excluding the point 0 then, as shown below, an analytical solution is available. Let the parametric space be $\{\theta : 0 < |\theta| < 1\}$. Then, as before, it can be seen that for each $T_{nc}\varepsilon\mathcal{C}_1$,

$$\sup_\theta a_c(\theta) \geq -2. \tag{3.7}$$

For example, if $c(\cdot)$ satisfies (b) then (3.7) can be proved by integrating $a_c(\theta)$ with respect to $\pi_{2j}(\theta)(j = 1, 2, ...)$, where

$$\begin{aligned} \pi_{2j}(\theta) &= z_{2j}^{-1}(1+\theta)^{j^{-1}-1} \quad \text{if } -1+\exp(-j^2) \leq \theta \leq -\tfrac{1}{2}, \\ &= 0 \quad \text{otherwise,} \end{aligned}$$

and $z_{2j} = j\{2^{-1/j} - \exp(-j)\}$. The lower bound (3.7) is attained by the estimator $T_{nc_0} = \hat{\theta}_n + n^{-1}c_0(\hat{\theta}_n)(\varepsilon\mathcal{C}_1)$, where $c_0(\theta) = \theta^{-1}(1-\theta^2)$.

4 Inverse Gaussian model

We now consider the estimation problem of the 'location' parameter θ of the inverse Gaussian model specified by

$$f(x;\theta) = (2\pi x^3)^{-\frac{1}{2}}\exp\{-\frac{1}{2}(\theta^2 x)^{-1}(x-\theta)^2\}, x > 0,$$

where $0 < \theta < \infty$. Note that θ is not a location parameter in the usual sense of the term (cf. Example 1). Here

$$I = \theta^{-3}, L_{11} = -\frac{1}{2}L_{001} = -3\theta^{-4}, L_{02} = 9\theta^{-5} + \theta^{-6}, \psi(\theta) = 0. \tag{4.1}$$

Again in the spirit of Dasgupta and Ghosh (1983), we consider a subclass $\mathcal{C}_1$ of $\mathcal{C}$ consisting of estimators $T_{nc} = \hat{\theta}_n + n^{-1}c(\hat{\theta}_n)$, where $c(\cdot)$ is such that either (a) $I\{c(\theta)\}^2 + 2c'(\theta) \to \infty$ along some θ-sequence tending to 0+, or (b) $c(\theta)/\theta$ is bounded as $\theta \to 0+$. It can be seen that this subclass is very large.

Here $\frac{1}{2}L_{001} + L_{11} = 0$. Hence for $T_{nc}\varepsilon\mathcal{C}_1$ with $c(\cdot)$ satisfying (a), by (2.2a,b), (4.1), $\sup_\theta a_c(\theta) = \infty$. Consider now $T_{nc}\varepsilon\mathcal{C}_1$ with $c(\cdot)$ satisfying (b). For $j = 2, 3, ...$, let

$$\pi_{3j}(\theta) = (\log j)^{-1}\theta^{-1}, j^{-1} \leq \theta \leq 1. \tag{4.2}$$

Analogously to (3.3), by (2.2a,b), (4.1), for $j = 2, 3, ...$,

$$\begin{aligned} &\int_{j^{-1}}^{1} a_c(\theta)\pi_{3j}(\theta)d\theta \\ &\geq -\int_{j^{-1}}^{1} \theta^3\{\pi_{3j}'(\theta)\}^2\{\pi_{3j}(\theta)\}^{-1}d\theta + 2[c(\theta)\pi_{3j}(\theta)]_{j^{-1}}^{1}. \end{aligned} \tag{4.3}$$

Since $c(\cdot)$ satisfies (b), by (4.2), one can check that the right-hand member of (4.3) tends to 0 as $j \to \infty$. It follows that for each $T_{nc} \varepsilon \mathcal{C}_1, \sup_\theta a_c(\theta) \geq 0$. By (2.2a,b), (4.1), it can be seen that if $T_{nc} = \hat{\theta}_n + n^{-1}c(\hat{\theta}_n)(\varepsilon \mathcal{C}_1)$ be such that $c(\theta) = \zeta\theta^2$, where ζ is a constant, then

$$a_c(\theta) = (\zeta^2 + 4\zeta)\theta. \tag{4.4}$$

The right hand side of (4.4) is minimized, uniformly in $\theta(> 0)$, when $\zeta = -2$ and for this choice of $\zeta, \sup_\theta a_c(\theta) = 0$. Thus $T_{nc_0} = \hat{\theta}_n + n^{-1}c_0(\hat{\theta}_n)$, where $c_0(\theta) = -2\theta^2$, is second-order minimax in $\mathcal{C}_1$.

5 Second-order minimaxity under absolute deviation loss

We make assumptions as in section 2. Then for an estimator $T_{nc} = \hat{\theta}_n + n^{-1}c(\hat{\theta}_n)(\varepsilon\mathcal{C})$ one can calculate the approximate cumulants of $(nI)^{\frac{1}{2}}(T_{nc} - \theta)$ under θ and hence, using an Edgeworth expansion, up to $o(n^{-1})$, for the distribution of $(nI)^{\frac{1}{2}}(T_{nc} - \theta)$ under θ, show that

$$E_\theta\{(\frac{1}{2}\pi nI)^{1/2}|T_{nc} - \theta|\} = 1 + n^{-1}b_c(\theta) + o(n^{-1}), \tag{5.1}$$

where

$$b_c(\theta) = \frac{1}{2}I\{\mu_c(\theta)\}^2 + \mu_c'(\theta) + \phi(\theta), \tag{5.2a}$$

with

$$\mu_c(\theta) = c(\theta) - \frac{1}{6}I^{-2}L_3, \tag{5.2b}$$

$$\phi(\theta) = \frac{1}{24}I^{-2}(5L_{0001} + 12L_{101} + 15L_{02} + 6L_{21})$$
$$+ \frac{1}{36}I^{-3}(19L_{001}^2 + 27L_{11}^2 + 54L_{001}L_{11}) - \frac{3}{8}. \tag{5.2c}$$

The detailed derivation of (5.1), (5.2a-c) is omitted here to save space. An estimator $T_{nc*} = \hat{\theta}_n + n^{-1}c^*(\hat{\theta}_n)$ belonging to $\mathcal{C}$ (or a subclass of $\mathcal{C}$) will be said to be second-order minimax in $\mathcal{C}$ (or in the subclass of $\mathcal{C}$) under absolute deviation loss if

$$\sup_\theta b_{c*}(\theta) \leq \sup_\theta b_c(\theta), \tag{5.3}$$

for every $T_{nc} = \hat{\theta}_n + n^{-1}c(\hat{\theta}_n)$ belonging to $\mathcal{C}$ (or the subclass of $\mathcal{C}$), the suprema in (5.3) being over the parametric space for θ.

For the location model considered in Example 1, it can be seen that I, L_3 and $\phi(\theta)$ are free from θ. Let $I = \rho, L_3 = \ell_3, \phi(\theta) = \phi_0$, where $\rho(> 0), \ell_3, \phi_0$ are constants. Then, as in Example 1, from (5.2a-c), it can be seen that for each $T_{nc}(\varepsilon\mathcal{C}), \sup_\theta b_c(\theta) \geq \phi_0$ and that this lower bound is attained by $T_{nc*} = \hat{\theta}_n + n^{-1}c^*(\hat{\theta}_n)(\varepsilon\mathcal{C})$, where $c^*(\theta) = \frac{1}{6}\rho^{-2}\ell_3$. It can be seen that $P_\theta(T_{nc^*} \leq \theta) = \frac{1}{2} + o(n^{-\frac{1}{2}})$ for each θ.

Consider next the scale model of Example 2. Here $I = \rho\theta^{-2}, L_3 = \ell_3\theta^{-3}, \phi(\theta) = \phi_0$, where ρ, ℓ_3, ϕ_0 are constants and ρ is assumed to be positive. As in Example 2, from (5.2a-c), one can check that for each $T_{nc}(\varepsilon\mathcal{C}), \sup_\theta b_c(\theta) \geq \phi_0 - \dfrac{1}{2}\rho^{-1}$ and that this lower bound is attained by $T_{nc^*} = \hat{\theta}_n + n^{-1}c^*(\hat{\theta}_n)(\varepsilon\mathcal{C})$, where $c^*(\theta) = -\theta\rho^{-1}(1 - \frac{1}{6}\rho^{-1}\ell_3)$. Note that $c^*(\theta) = -q_1\theta$ while, as noted in Example 2, $c_0(\theta) = -q_2\theta$, where $q_1 = \rho^{-1}(1-\frac{1}{6}\rho^{-1}\ell_3), q_2 = \rho^{-1}\{1+\rho^{-1}(\frac{1}{2}\ell_{001}+\ell_{11})\}$. From regularity conditions like $I'(\theta) = -(L_{11}+L_{001})$ and $L_3+3L_{11}+L_{001} = 0$, it is not hard to see that $q_1 = \frac{2}{3}q_2$. Thus, under a scale model, one obtains a simple rule for finding a second-order minimax estimator under absolute deviation loss from an estimator which is second-order minimax under squared error loss and this illustrates an application of the explicit formulae considered here.

We now consider the normal correlation model of section 3 with the parameteric space for θ given by $\{\theta : -1 < \theta < 1\}$. Here

$$\left.\begin{array}{lcl} L_3 & = & -2\theta(3+\theta^2)/(1-\theta^2)^3, \\ L_{0001} & = & -18(1+6\theta^2+\theta^4)/(1-\theta^2)^4 = -3L_{101} \\ L_{21} & = & -(5+26\theta^2+5\theta^4)/(1-\theta^2)^4, \\ \phi(\theta) & = & \frac{1}{36}(27+9\theta^2+69\theta^4+7\theta^6)/(1+\theta^2)^3. \end{array}\right\}. \tag{5.4}$$

As in section 3, we restrict to a large subclass, $\mathcal{C}_1^*$, of $\mathcal{C}$ consisting of estimators $T_{nc} = \hat{\theta}_n + n^{-1}c(\hat{\theta}_n)$, where $c(\cdot)$ is such that either (a) $\frac{1}{2}I\{\mu_c(\theta)\}^2 + \mu_c'(\theta) \to \infty$ along some θ-sequence tending to $1-$ or $-1+$, or (b) $\mu_c(\theta)/(1-\theta)$ remains bounded as $\theta \to 1-$ and $\mu_c(\theta)/(1+\theta)$ remains bounded as $\theta \to -1+$. By (5.2a), (5.4), for $T_{nc}\varepsilon\mathcal{C}_1^*$ with $c(\cdot)$ satisfying (a), $\sup_\theta b_c(\theta) = \infty$. On the other hand, as in section 3, for $T_{nc}\varepsilon\mathcal{C}_1^*$ with $c(\cdot)$ satisfying (b), integrating $b_c(\theta)$ with respect to $\pi_{1pj}(\theta)$, using (3.1), (3.2), (5.2a, c), (5.4), and allowing j to tend to infinity,

$$\sup_\theta b_c(\theta) \geq \sup_{p>0} J_{1p}^*, \tag{5.5}$$

where for $p > 0$,

$$J_{1p}^* = z_{1p}^{-1} \int_{-1}^{1} \{\phi(\theta) - 2(p-1)^2\theta^2(1+\theta^2)^{-1}\}(1-\theta^2)^{p-1}\, d\theta,$$

$\phi(\cdot)$ being given by (5.4). By numerical methods, up to two places of decimals, the right-hand member of (5.5) is seen to equal 0.53. Hence for each $T_{nc}\varepsilon\mathcal{C}_1^*, \sup_\theta b_c(\theta) \geq 0.53$. Considering now the estimator $T_{nc*} = \hat{\theta}_n + n^{-1}c^*(\hat{\theta}_n)(\varepsilon\mathcal{C}_1^*)$, with

$$c^*(\theta) = (1-\theta^2)(-0.21\theta + 0.26\theta^3 - 0.12\theta^5)$$

$$-\frac{1}{3}\theta(1-\theta^2)(3+\theta^2)(1+\theta^2)^{-2}, \tag{5.6}$$

from (3.1), (5.2a-c),(5.4), one can check that $\sup_\theta b_{c*}(\theta) = 0.54$. Thus T_{nc*}, obtained by numerical methods, is nearly optimal in $\mathcal{C}_1^*$ in terms of second-order minimaxity under absolute deviation loss. We also note that T_{nc*} is considerably better than the maximum likelihood estimate for which $\sup_\theta b_c(\theta)$ equals 1.75.

Under the normal correlation model and with absolute deviation loss, an exact solution to the problem of second-order minimaxity is available as soon as one modifies the parametric space for θ as $\{\theta : 0 < |\theta| < 1\}$. Then, as in section 3, one can show that for each $T_{nc} \varepsilon \mathcal{C}_1^*, \sup_\theta b_c(\theta) \geq -\frac{11}{18}$ and that this lower bound is attained by the estimator $T_{nc*} = \hat{\theta}_n + n^{-1}c^*(\hat{\theta}_n)$, where

$$c^*(\theta) = \theta^{-1}(1-\theta^2) - \frac{1}{3}\theta(1-\theta^2)(3+\theta^2)(1+\theta^2)^{-2} \tag{5.7}$$

The expressions for $c^*(\theta)$, as in (5.6), (5.7), can be motivated along the line of section 3. The only change is that now one has to work initially with $\mu_c(\theta)$ and then employ (5.2b) to find $c(\theta)$ appropriately.

Finally, we consider the inverse Gaussian model of section 4. Here

$$L_3 = 3\theta^{-4}, L_{0001} = -3L_{101} = -36\theta^{-5}, L_{21} = -9\theta^{-5} - \theta^{-6}, \phi(\theta) = \frac{5}{8}\theta. \tag{5.8}$$

We restrict to a large subclass $\mathcal{C}_1^*$ of $\mathcal{C}$ consisting of estimators $T_{nc} = \hat{\theta}_n + n^{-1}c(\hat{\theta}_n)$, where $c(\cdot)$ is such that either (a) $\frac{1}{2}I\{\mu_c(\theta)\}^2 + \mu_c'(\theta) \to \infty$ along some θ-sequence tending to 0+, or (b) $\mu_c(\theta)/\theta$ is bounded as $\theta \to 0+$. Then by (4.1), (5.2b,c), (5.8), proceeding as in section 4, one can check that for each $T_{nc}\varepsilon\mathcal{C}_1^*, \sup_\theta b_c(\theta) \geq 0$, and that this lower bound is attained by $T_{nc*} = \hat{\theta}_n + n^{-1}c^*(\hat{\theta}_n)(\varepsilon\mathcal{C}_1^*)$, where $c^*(\theta) = -\frac{3}{2}\theta^2$. This again can be motivated exactly as in section 4.

Acknowledgement

The authors are thankful to a referee for very constructive suggestions. The work of RM was supported by a grant from the Centre for Management and Development Studies, Indian Institute of Management Calcutta.

REFERENCES

Akahira, M. and Takeuchi, K. (1981). *Asymptotic Efficiency of Statistical Estimators.* Springer-Verlag, New York.

Bhattacharya, R.N. and Ghosh, J.K. (1978). On the validity of formal Edgeworth expansions. *Ann. Statist.* **6**, 434-451.

Dasgupta, A. and Ghosh, J.K. (1983). Some remarks on second-order admissibility in the multiparameter case. *Sankhya* A **45**, 181-190.

Ghosh, J.K. and Sinha, B.K. (1981). A necessary and sufficient condition for second-order admissibility with applications to Berkson's bioassay problem. *Ann. Statist.* **9**, 1334-1338.

Ghosh, J.K., Sinha, B.K. and Joshi, S.N. (1982). Expansions for posterior density and integrated Bayes risk. In: *Statistical Decision Theory and Related Topics III*, vol. 1 (eds: S.S. Gupta and J.O. Berger), Academic Press, New York, pp. 403-456.

Ghosh, J.K., Sinha, B.K. and Subramanyam, K. (1979). Edgeworth expansions for Fisher-consistent estimators and second-order efficiency. *Calcutta Statist. Assoc. Bull.* **28**, 1-18.

Ghosh, J.K., Sinha, B.K. and Wieand, H.S. (1980). Second-order efficiency of the maximum likelihood estimate with respect to any bounded bowl-shaped loss function. *Ann. Statist.* **8**, 506-521.

Ghosh, J.K. and Subramanyam, K. (1974). Second-order efficiency of maximum likelihood estimators. *Sankhya* A **36**, 325-358.

Gusev, S.I. (1976). Asymptotic expansions associated with some statistical estimators in the smooth case II. Expansions of moments and distributions. *Theory Probab. and Appl.* **21**, 14-32.

Johnson, R.A. (1970). Asymptotic expansions associated with posterior distributions. *Ann. Math. Statist.* **41**, 851-864.

Pfanzagl, J. (1979). Asymptotic expansions in parametric statistical theory. In: *Developments in Statistics*, vol. 3 (ed: P.R. Krishnaiah), Academic Press, New York, pp. 1-97.

Pfanzagl, J. and Wefelmeyer, W. (1978). A third-order optimum property of the maximum likelihood estimator. *J. Multivariate Anal.* **8**, 1-29.

Stuart, A. (1955). A paradox in statistical estimation. *Biometrika* **42**, 527-529.

Department of Statistics
Mathematical Sciences Building
Purdue University
W. Lafayette, IN - 47907, U.S.A.
and
Indian Statistical Institute
203, B.T. Road, Calcutta - 700 035
INDIA

Indian Institute of Management
Joka, Diamond Harbour Road
Post Box No. 16757
Alipore Post Office
Calcutta - 700 027
INDIA

J.K. Ghosh, S.K. Mitra, K.R. Parthasarathy and B.L.S. Prakasa Rao (*eds.*),
Statistics and Probability : A Raghu Raj Bahadur Festschrift
Published by Wiley Eastern Limited, Publishers, (1993), 245 - 264

A Decision Theoretic Approach to Point and Interval Estimation of the Variance

MALAY GHOSH[1]
University of Florida, Gainesville

and

SUDEEP KUNDU[2]
Merck Research Laboratories, Rahway

Abstract

This paper considers the point and interval estimation of the error variance, say σ^2, in the fixed-effects balanced one-way ANOVA model. In the point estimation case, the general problem of the estimation of $\sigma^{2\alpha}$ is considered. Two general classes of hierarchical Bayes (HB) estimators are proposed which dominate the best multiple estimator S^α (S is the error sum of squares) under the entropy loss, $L(a, \sigma^{2\alpha}) = a/\sigma^{2\alpha} - ln(a/\sigma^{2\alpha}) - 1$. The dominance of our proposed estimators over the best multiple (also the best equivariant) estimator is proved by using a technique of Kubokawa (1991). For $\alpha = 1$, numerical studies of the percentage risk improvement of the proposed estimators over the best multiple estimator are presented for

[1]Research partially supported by NSF Grant Numbers DMS-8901334 and SES-9201210.

[2]This work was done when the second author was a graduate student in the Department of Statistics at the University of Florida.

AMS Subject Classifications : 62C15, 62F10, 62F15, 62F25
Key Words and Phrases: Fixed effects, balanced, ANOVA, error variance, entropy loss, hierarchical Bayes, risk dominance, highest posterior density, minimum length, shortest unbiased, coverage probability.

some special cases. These calculations indicate that the risk improvement can often be quite substantial. Then we find two classes of confidence intervals based on the two classes of priors for σ^2 using all the cell means and the error sum of squares which dominate (in the sense that the former have smaller length and greater frequentist coverage probability) the usual confidence intervals based on the error sum of squares only. Also, under an alternative formulation due to Cohen (1972) and Shorrock (1982, 1990), we have provided two classes of HPD credible intervals which have the same length but higher coverage probability than the usual intervals. One of these classes includes as its member a confidence interval earlier proposed by Brewster and Zidek (1974) and Shorrock (1990). Sarkar (1989) found non-smooth confidence intervals which dominated the shortest invariant confidence interval for the generalized variance. In the same problem, it seems possible to extend the present approach in finding intervals that have smaller length and greater coverage probability than the shortest invariant confidence intervals.

1 Introduction

Consider a fixed-effects balanced one-way analysis of variance (ANOVA) model with homoscedastic errors. This paper takes a decision-theoretic approach towards point and interval estimation of $\sigma^{2\alpha}$ ($\alpha > 0$, known), where σ^2 denotes the error variance. First, consider the point estimation problem. We denote by S, k and n the error sum of squares, the number of cells and the number of observations per cell respectively. It is well-known that under the group of location-scale transformations, the best equivariant estimator of $\sigma^{2\alpha}$ under the loss $L(a, \sigma^{2\alpha}) = \sigma^{-4\alpha}(a - \sigma^{2\alpha})^2$ is given by $c_\alpha S^\alpha$ where $c_\alpha = \Gamma(\frac{1}{2}((n-1)k + 2\alpha)) \Big/ \Gamma(\frac{1}{2}((n-1)k + 4\alpha))$. Under the above loss, this is also the constant risk minimax estimator of $\sigma^{2\alpha}$. In addition, c_α is the best multiple of S^α under any arbitrary quadratic loss

$$L_Q(a, \sigma^{2\alpha}) = Q(\sigma^{2\alpha})(a - \sigma^{2\alpha})^2, \tag{1.1}$$

where $Q(\sigma^{2\alpha}) > 0$.

However, within the bigger class of estimators equivariant only under scale transformations, $c_\alpha S^\alpha$ is an inadmissible estimator of $\sigma^{2\alpha}$ under every single L_Q loss given in (1.1). This illuminating fact was first pointed out in the seminal work of Stein (1964) when $\alpha = 1$. Stein-type estimators dominating $c_\alpha S^\alpha$ for a general α (> 0) were produced by Gelfand and Dey (1988).

Ghosh (1992b) derived a class of hierarchical Bayes (HB) estimators of $\sigma^{2\alpha}$ ($\alpha > 0$) under the loss $L(a, \sigma^{2\alpha}) = \sigma^{-4\alpha}(a - \sigma^{2\alpha})^2$. Such estimators, unlike the Stein-type estimators, are smooth. Also, Ghosh found a subclass of these HB estimators which dominated $c_\alpha S^\alpha$ under the loss (1.1). An estimator proposed earlier by Brewster and Zidek (1974) when $\alpha = 1$ is a member of this subclass of estimators. The numerical findings of Ghosh (when $\alpha = 1$) indicated that often these risk improvements could be quite substantial.

However, the quadratic loss given in (1.1) is often inadequate for estimation of $\sigma^{2\alpha}$. This was pointed out very forcibly in Brown (1968). An alternate loss, usually referred to as the entropy loss, is given by

$$L(a, \sigma^{2\alpha}) = \frac{a}{\sigma^{2\alpha}} - log\left(\frac{a}{\sigma^{2\alpha}}\right) - 1 \tag{1.2}$$

This loss was proposed by Stein (1964) and, later on, was studied very extensively in Brown (1968). Under the loss, the best equivariant estimator of $\sigma^{2\alpha}$ is given by its UMVUE, $d_\alpha S^\alpha$ where

$$d_\alpha = \frac{\Gamma(\frac{1}{2}(n-1)k)}{\Gamma(\frac{1}{2}((n-1)k+2\alpha))} \tag{1.3}$$

The present paper aims at finding a class of HB estimators of $\sigma^{2\alpha}$ under the loss (1.2) which dominates $d_\alpha S^\alpha$. The derivation of such a class of HB estimators is given in Section 2. The dominance of such estimators over $d_\alpha S^\alpha$ is proved by appealing to a theorem of Kubokawa (1991). Also, some numerical calculations are provided to indicate the extent of risk improvement of such estimators over $d_\alpha S^\alpha$.

In Section 3, we find the Highest Posterior Density (HPD) credible intervals for $\sigma^{2\alpha}$ (or σ^2) under two classes of hierarchical priors. We then demonstrate that such intervals have nice frequentist properties as well in the sense that they have smaller length and greater coverage probability (in the frequentist sense) than either the minimum length or the shortest unbiased confidence intervals based only on S, the error sum of squares. Also, we consider classes of HPD credible intervals, one of which includes as its members the ones considered earlier by Shorrock (1982, 1990). These latter intervals have the same length but higher coverage probability than the usual intervals based on S only. For an interesting review of interval estimation of σ^2, see Maatta and Casella (1990).

Sarkar (1989) improved on the shortest invariant confidence interval for the generalized variance by using certain non-smooth confidence intervals along the lines of Cohen (1972). It will be interesting to investigate whether there exists HPD credible intervals for this problem having similar frequentist properties as Sarkar's intervals.

2 The HB Model

Consider the following balanced fixed effects one-way ANOVA model:

$$Y_{ij} = \theta_i + e_{ij} \qquad (j = 1, \ldots, n; \quad i = 1, \ldots, k)$$

where the θ_i's are unknown fixed effects, and the e_{ij}'s are i.i.d. $N(0, \sigma^2)$. The minimal sufficient statistic is $(\bar{Y}_1, \ldots, \bar{Y}_k, S)$, where $\bar{Y}_i = n^{-1}\Sigma_{j=1}^n Y_{ij}$ $(i = 1, \ldots, k)$, and $S = \Sigma_{i=1}^k \Sigma_{j=1}^n (Y_{ij} - \bar{Y}_i)^2$. We use the notation Y_i for $\bar{Y}_i$ $(i =$

$1,\ldots,k$). Also, let $\boldsymbol{Y} = (Y_1,\ldots,Y_k)^T$, $\boldsymbol{\theta} = (\theta_1,\ldots,\theta_k)^T$. We consider the estimation of $\sigma^{2\alpha}$ under the entropy loss given in (1.2).

We first consider a hierarchical Bayes model which is constructed by using independent normal priors with known means, say zeroes, and a common unknown variance at the first stage for the k cell means. In the second stage, diffuse priors are assigned to the first stage prior variance and the error variance, σ^2. The models considered in Brewster and Zidek (1974) and Strawderman (1971, 1974) are members of this class. Such priors are used for simultaneous estimation of means by Morris (1983) and Ghosh (1992a). The hierarchical model is given in (I) – (III) below, where $R = \sigma^{-2}$.

(I) Conditional on $\boldsymbol{\Theta} = \boldsymbol{\theta}$, $R = r$ and $\Lambda = \lambda$, $\boldsymbol{Y}$ and S are mutually independently with $\boldsymbol{Y} \sim N(\boldsymbol{\theta},\ (nr)^{-1}\boldsymbol{I}_k)$ and $S \sim r^{-1}\chi^2_{(n-1)k}$, $(n \geq 2)$;

(II) Conditional on $R = r$ and $\Lambda = \lambda$, $\boldsymbol{\Theta} \sim N(\boldsymbol{0},\ (\lambda r)^{-1}\boldsymbol{I}_k)$;

(III) R and Λ are marginally independently distributed, where R has pdf $f_R(r) \propto r^{-\frac{1}{2}a}$ $(0 \leq a < nk+2)$ while Λ has pdf $f_\Lambda(\lambda) \propto \lambda^{-\frac{1}{2}b}(n+\lambda)^{-\frac{1}{2}(4-b)}$, $(b < k+2)$.

Writing $U = \frac{\Lambda}{n+\Lambda}$, it is shown in Ghosh (1992b) that a version of Jeffreys' prior based on (I) and (II) only is a member of the above class of priors. From (I) – (III) one has

(i) conditional on $\boldsymbol{Y} = \boldsymbol{y}$, $S = s$, $R = r$ and $U = u$,

$$\boldsymbol{\Theta} \sim N((1-u)\boldsymbol{y},\ (rn)^{-1}\boldsymbol{I}_k)\,;$$

(ii) conditional on $\boldsymbol{Y} = \boldsymbol{y}$, $S = s$ and $U = u$,

$$R \sim Gamma\left(\frac{1}{2}(s+ut),\ \frac{1}{2}(nk-a+2)\right),$$

where $t = n\Sigma_{i=1}^k y_i^2$. [We say that $Z \sim Gamma(\alpha, \beta)$ if it has pdf $f_Z(z) \propto exp(-\alpha z)z^{\beta-1}I_{(0,\infty)}(z)$, $\alpha > 0$, $\beta > 0$ and I is the usual indicator function.]

(iii) conditional on $\boldsymbol{Y} = \boldsymbol{y}$ and $S = s$, the pdf of U is given by

$$f_U(u|\boldsymbol{y}, s) \propto u^{\frac{1}{2}(k-b)}(1+uw)^{-\frac{1}{2}(nk-a+2)}I_{(0,1)}(u), \tag{2.1}$$

where $w = t/s$.

For the loss (1.2), the Bayes estimator of $\sigma^{2\alpha} = R^{-\alpha}$ is given by $[E(R^\alpha|\boldsymbol{Y}, S)]^{-1}$.

Now, from (i) – (iii), we have,

$$E_a(R^\alpha \mid u, \boldsymbol{y}, s) = \frac{\Gamma(\frac{1}{2}(nk-a+2+2\alpha))}{\Gamma(\frac{1}{2}(nk-a+2))}\left(\frac{s+ut}{2}\right)^{-\alpha}$$

$$= \frac{\Gamma(\frac{1}{2}(nk-a+2+2\alpha))}{\Gamma(\frac{1}{2}(nk-a+2))}\left(\frac{s}{2}\right)^{-\alpha}(1+uw)^{-\alpha}.$$

Therefore, under the above loss, the Hierarchical Bayes estimator of $\sigma^{2\alpha}$ $(\alpha > 0)$ is given by,

$$e^{HB}_{a,b,\alpha}(\boldsymbol{Y}, S) = \frac{1}{E(R^\alpha|\boldsymbol{Y}, S)}$$

$$= \left(\frac{s}{2}\right)^{\alpha}\frac{\Gamma\left(\frac{1}{2}(nk-a+2)\right)}{\Gamma\left(\frac{1}{2}(nk-a+2+2\alpha)\right)}$$

$$\times\left\{E[(1+UW)^{-\alpha}\,|\boldsymbol{Y}, S]\right\}^{-1}$$

$$= S^\alpha\ \phi_{a,b,\alpha}(w), \quad \text{say} \tag{2.2}$$

We appeal to the following version of Theorem 2.1 of Kubokawa (1991) which provides sufficient conditions under which $e^{HB}_{a,b,\alpha}(\boldsymbol{Y}, S)$ dominates the best equivariant estimator $e_{0,\alpha}(S) = d_\alpha S^\alpha$ of $\sigma^{2\alpha}$ under the loss (1.2).

THEOREM 1 *Suppose*

(a) $\phi_{a,b,\alpha}(w)$ *is* $\uparrow$ *in* w, *and* $\lim_{w\to\infty}\phi_{a,b,\alpha}(w) = d_\alpha$*;*

(b) $\phi_{a,b,\alpha}(w) \ \geq\ \phi_{0,\alpha}(w)$, *where*

$$\phi_{0,\alpha}(w) = \frac{\int_0^\infty exp(-\frac{s}{2})s^{\frac{(n-1)k}{2}-1}(\int_0^{ws} e^{-\frac{y}{2}}y^{\frac{k}{2}-1}dy)ds}{\int_0^\infty s^\alpha exp(-\frac{s}{2})s^{\frac{(n-1)k}{2}-1}(\int_0^{ws} e^{-\frac{y}{2}}y^{\frac{k}{2}-1}dy)ds}$$

Then $e^{HB}_{a,b,\alpha}(\boldsymbol{Y}, S)$ *dominates* $e_{0,\alpha}(S)$ *under the loss (1.2), i.e.,*

$$E_{\boldsymbol{\theta},\sigma^2}\left[\frac{e^{HB}_{a,b,\alpha}(\boldsymbol{Y}, S)}{\sigma^{2\alpha}} - ln\left(\frac{e^{HB}_{a,b,\alpha}(\boldsymbol{Y}, S)}{\sigma^{2\alpha}}\right) - \frac{e_{0,\alpha}(S)}{\sigma^{2\alpha}} + ln\left(\frac{e_{0,\alpha}(S)}{\sigma^{2\alpha}}\right)\right]$$

$$< \quad 0$$

for all $\boldsymbol{\theta}, \sigma^2$.

Note that $\phi_{0,\alpha}(w)$ can be simplified further as

$$\phi_{0,\alpha}(w) = \frac{\int_0^\infty\int_0^w exp(-\frac{1}{2}s(1+x))\,x^{\frac{1}{2}(k-2)}\,s^{\frac{1}{2}nk-1}\,dxds}{\int_0^\infty\int_0^w exp(-\frac{1}{2}s(1+x))\,x^{\frac{1}{2}(k-2)}\,s^{\frac{1}{2}nk+\alpha-1}\,dxds}$$

$$= 2^{-\alpha}\,\frac{\Gamma(\frac{nk}{2})}{\Gamma(\frac{nk+2\alpha}{2})}\,\frac{\int_0^w x^{\frac{1}{2}(k-2)}(1+x)^{-\frac{nk}{2}}dx}{\int_0^w x^{\frac{1}{2}(k-2)}(1+x)^{-\frac{nk+2\alpha}{2}}dx}$$

$$= \phi_{2,2,\alpha}(w), \quad \text{using (2.1) and (2.2).} \tag{2.3}$$

We now confine our attention to the subclass of HB estimators $e^{HB}_{a,a,\alpha}(\mathbf{Y}, S)$ of $\sigma^{2\alpha}$. The next theorem provides conditions for a under which $\phi_{a,a,\alpha}(w)$ satisfies the conditions of Theorem 1.

THEOREM 2 *Let $2 \le a < k+2$. Then, $\phi_{a,a,\alpha}(w)$ satisfies conditions (a) and (b) of Theorem 1.*

PROOF. We first prove condition (b). For every $0 < a < k+2$, using (2.1), we simplify $\phi_{a,a,\alpha}(w)$ given in (2.2) as

$$\phi_{a,a,\alpha}(w) = 2^{-\alpha}\frac{\Gamma(\frac{(n-1)k}{2})}{\Gamma(\frac{(n-1)k+2\alpha}{2})}\frac{P\left[Beta(\frac{k-a+2}{2}, \frac{(n-1)k}{2}) \le v\right]}{P\left[Beta(\frac{k-a+2}{2}, \frac{(n-1)k+2\alpha}{2}) \le v\right]} \tag{2.4}$$

where $Beta(m,p)$ stands for a beta-variable with parameters m and p.

Since $\phi_{0,\alpha}(w) \equiv \phi_{2,2,\alpha}(w)$, we have

$$\begin{aligned}\frac{\phi_{a,a,\alpha}(w)}{\phi_{0,\alpha}(w)} &= \frac{P\left[Beta(\frac{1}{2}(k-a+2), \frac{1}{2}(n-1)k) \le v\right]}{P\left[Beta(\frac{1}{2}(k-a+2), \frac{1}{2}((n-1)k+2\alpha)) \le v\right]} \\ &\quad \div \frac{P\left[Beta(\frac{1}{2}k, \frac{1}{2}(n-1)k) \le v\right]}{P\left[Beta(\frac{1}{2}k, \frac{1}{2}((n-1)k+2\alpha)) \le v\right]}.\end{aligned} \tag{2.5}$$

The following lemma, proved in Ghosh (1992b), shows that the above expression $\phi_{a,a,\alpha}(w)$ satisfies (b) of Theorem 1 for $2 \le a < k+2$.

LEMMA 1 *If $0 < p_1 < p_2$, then for every $0 < v < 1$,*

$$\frac{P\left[Beta(m,p_1) \le v\right]}{P\left[Beta(m,p_2) \le v\right]} \text{ is } \downarrow \text{ in } m.$$

To verify condition (a) of Theorem 1, note that $v = \frac{w}{1+w}$, so, as $w \to \infty$, $v \to 1$. From (2.4), we have

$$\lim_{w\to\infty} \phi_{a,a,\alpha}(w) = 2^{-\alpha}\frac{\Gamma(\frac{1}{2}(n-1)k)}{\Gamma(\frac{1}{2}(n-1)k+2\alpha)} = d_\alpha \tag{2.6}$$

To show that $\phi_{a,a,\alpha}(w)$ is $\uparrow$ in w, that is, $\uparrow$ in v, define a new pdf,

$$f_v(z) \propto z^{\frac{1}{2}(k-1)}\,(1-z)^{\frac{1}{2}((n-1)k-2)}I_{(0,v)}(z), \tag{2.7}$$

Then, for $0 < v < v'$, $\frac{f_{v'}(z)}{f_v(z)} \propto \frac{I_{(0,v')}(z)}{I_{(0,v)}(z)} \uparrow$ in z. Since $\phi_{a,a,\alpha}(w) \propto E(1-Z)^{-\alpha}$ (expectation being taken here with respect to the pdf given in (2.7)), it is $\uparrow$ in v, that is $\uparrow$ in w.

We now consider the second hierarchical model. In this case, $k \ge 2$. The basic difference between this model and the previous one is that, unlike the previous case, the first stage prior mean is taken as M, and a diffuse prior is assigned to M as well in the second stage. The model is described below.

(I)′ Conditional on $\boldsymbol{\Theta} = \boldsymbol{\theta}$, $M = m$, $R = r$ and $\Lambda = \lambda$, $\boldsymbol{Y}$ and S are mutually independently with $\boldsymbol{Y} \sim N(\boldsymbol{\theta}, (nr)^{-1}\boldsymbol{I}_k)$ and $S \sim r^{-1}\chi^2_{(n-1)k}, (n \geq 2)$;

(II)′ Conditional on $M = m$, $R = r$ and $\Lambda = \lambda$,

$$\boldsymbol{\Theta} \sim N(m\mathbf{1}, (\lambda r)^{-1}\boldsymbol{I}_k);$$

(III)′ M, R and Λ are marginally independent, with $M \sim Uniform(-\infty, \infty)$, $f_R(r) \propto r^{-\frac{1}{2}a}$ $(0 \leq a < nk+1)$ and $f_\Lambda(\lambda) \propto \lambda^{-\frac{1}{2}b}(n+\lambda)^{-\frac{1}{2}(4-b)}$, $(b < k+1)$.

Again, writing $U = \frac{\Lambda}{n+\Lambda}$, we have from (I)′ – (III)′,

(i)′ conditional on $\boldsymbol{Y} = \boldsymbol{y}$, $S = s$, $R = r$ and $U = u$,

$$\boldsymbol{\Theta} \sim N\left((1-u)\boldsymbol{y} + u\bar{y}\mathbf{1}_k, (nr)^{-1}((1-u)\boldsymbol{I}_k + uk^{-1}\mathbf{1}_k\mathbf{1}_k^T)\right);$$

(ii)′ conditional on $\boldsymbol{Y} = \boldsymbol{y}$, $S = s$ and $U = u$,

$$R \sim Gamma\left(\frac{1}{2}(s+ut), \frac{1}{2}(nk-a+1)\right);$$

(iii)′ conditional on $\boldsymbol{Y} = \boldsymbol{y}$ and $S = s$, the pdf of U is given by

$$f_U(u|\boldsymbol{y}, s) \sim u^{\frac{1}{2}(k-1-b)}(1+uw^*)^{-\frac{1}{2}(nk-a+1)} I_{(0,1)}(u), \tag{2.8}$$

where $w^* = n\Sigma_{i=1}^k (y_i - \bar{y})^2/s$ and $\bar{y} = k^{-1}\Sigma_{i=1}^k y_i$.

The HB estimator of $\sigma^{2\alpha}$ based on (i)′ – (iii)′ is given by,

$$\begin{aligned} e^*_{a,b,\alpha}(\boldsymbol{Y}, S) &= \frac{1}{E(R^\alpha|\boldsymbol{Y}, S)} \\ &= \left(\frac{s}{2}\right)^\alpha \frac{\Gamma\left(\frac{1}{2}(nk-a+1)\right)}{\Gamma\left(\frac{1}{2}(nk-a+1+2\alpha)\right)} \\ &\quad \times E[(1+UW^*)^{-\alpha}\,|\boldsymbol{Y}, S]^{-1} \\ &= S^\alpha\ \phi^*_{a,b,\alpha}(w), \text{ say} \end{aligned} \tag{2.9}$$

Using arguments similar to those in Theorems 1 and 2, we now have the following theorem, which we present here without the proof.

THEOREM 3 *Under the loss (1.2),* $e^*_{a,a,\alpha}(\boldsymbol{Y}, S)$ *dominates* $e_{0,\alpha}(S)$ *for every* $a \in [2, k+1)$.

Risk Improvement when $\alpha = 1$

Now we do some numerical calculations to find the risk improvement of $e^{HB}_{a,a,1}(\boldsymbol{Y}, S)$ for $0 \le a < k+2$ and $e^{*}_{a,a,1}(\boldsymbol{Y}, S)$ over $e_{0,1}(S)$ under the loss (1.2) for different values of n, k and a.

Note that for $\alpha = 1$, $e^{HB}_{a,a,1}(\boldsymbol{Y}, S)$ simplifies to

$$\begin{aligned} e^{HB}_{a,a,1}(\boldsymbol{Y}, S) &= \frac{S}{(n-1)k}[1 - \phi_a(V)] \\ \text{where } \phi_a(V) &= \frac{2}{nk - a + 2} \frac{v^{\frac{k-a+2}{2}}(1-v)^{\frac{(n-1)k}{2}}}{\int_0^v x^{\frac{k-a}{2}}(1-x)^{\frac{(n-1)k}{2}}dx} \end{aligned} \quad (2.10)$$

The next theorem provides an expression for the risk difference of $e_{0,1}(S)$ and $e^{HB}_{a,a,1}(\boldsymbol{Y}, S)$ under the loss (1.2).

THEOREM 4 *Consider the model under which* $\boldsymbol{Y}$ *and* S *are mutually independent with* $\boldsymbol{Y} \sim N(\Theta, n^{-1}\sigma^2 \boldsymbol{I}_k)$ *and* $S \sim \sigma^2\chi^2_{(n-1)k}$. *Then for* $0 \le a < k+2$,

$$E_{\boldsymbol{\theta},\sigma^2}[L(e_{0,1}(S), \sigma^2) - L(e^{HB}_{a,a,1}(\boldsymbol{Y}, S), \sigma^2)]$$

$$= E[(nk + 2L)A_{L,a}] \quad (2.11)$$

where

$$A_{L,a} = E\left[\left\{\frac{(1-V)\phi_a(V)}{(n-1)k} + \frac{log(1-\phi_a(V))}{nk+2L}\right\}\Big| L\right] \quad (2.12)$$

where $L \sim Poisson\left(\frac{1}{2\sigma^2}\sum_{i=1}^k \theta_i^2\right)$ *and* $V|L \sim Beta\left(\frac{k+2L}{2}, \frac{(n-1)k}{2}\right)$.

PROOF. Let $T = n\sum_{i=1}^k Y_i^2$ and $G = T + S$ so that

$$V = \frac{W}{1+W} = \frac{T/S}{1+T/S} = \frac{T}{G}.$$

Then $e_{0,1}(S) = G\psi_*(V)$ and $e^{HB}_{a,a,1}(\boldsymbol{Y}, S) = G\psi_a(V)$ where $\psi_*(V) = \frac{(1-V)}{(n-1)k}$ and $\psi_a(V) = \frac{(1-V)(1-\phi_a(V))}{(n-1)k}$. For an estimator $G\psi(V)$ of σ^2, one has

$$\begin{aligned} E_{\boldsymbol{\theta},\sigma^2}[L(G\psi(V), \sigma^2)] &= E_{\boldsymbol{\theta},\sigma^2}\left[\frac{G\psi(V)}{\sigma^2} - log\left(\frac{G\psi(V)}{\sigma^2}\right) - 1\right] \\ &= E_{\boldsymbol{\theta}/\sigma,1}[G\psi(V) - log\, G - log\, \psi(V) - 1] \end{aligned} \quad (2.13)$$

If we write $\eta = (2\sigma^2)^{-1}\sum_{i=1}^k \theta_i^2$, then, under the reparametrization $(\boldsymbol{\theta}/\sigma, 1)$, S and T are independently distributed with $S \sim \chi^2_{(n-1)k}$ and $T \sim \chi^2_k(\eta)$. Introducing the dummy variable $L \sim Poisson(\eta)$, V and G are independent conditional on L with $V|L \sim Beta(\frac{1}{2}(k+2L), \frac{1}{2}(n-1)k)$ and $G|L \sim \chi^2_{nk+2L}$. Then,

$$E_{\boldsymbol{\theta},\sigma^2}[L(e_{0,1}(S),\sigma^2) - L(e^{HB}_{a,a,1}(\boldsymbol{Y},S),\sigma^2)]$$

$$= E_{\boldsymbol{\theta}/\sigma,1}\, E\left[\{\frac{G(1-V)\phi_a(V)}{(n-1)k} + log(1-\phi_a(V))\}\Big|L\right]$$

$$= E_{\boldsymbol{\theta}/\sigma,1}\left[(nk+2L)\, E\left[\{\frac{(1-V)\phi_a(V)}{(n-1)k} + \frac{log(1-\phi_a(V))}{nk+2L}\}\Big|L\right]\right]$$

The next theorem provides an expression for the risk difference of $e_{0,1}(S)$ and $e^*_{a,a,1}(\boldsymbol{Y},S)$ under the loss (1.2).

THEOREM 5 *For $0 \leq a < k+1$ and $k \geq 2$,*

$$E_{\boldsymbol{\theta},\sigma^2}[L(e_{0,1}(S),\sigma^2) - L(e^*_{a,a,1}(\boldsymbol{Y},S),\sigma^2)]$$

$$= E[(nk-1+2L^*)A^*_{L^*,a}] \qquad (2.14)$$

where

$$A^*_{L^*,a} \;=\; E\left[\left\{\frac{(1-V)\phi^*_a(V)}{(n-1)k} + \frac{log(1-\phi^*_a(V))}{nk-1+2L^*}\right\}\Big|L^*\right], \qquad (2.15)$$

$L^* \sim Poisson\left(\frac{1}{2\sigma^2}\sum_{i=1}^k(\theta_i-\bar{\theta})^2\right)$, $V|L^* \sim Beta\left(\frac{k-1+2L^*}{2}, \frac{(n-1)k}{2}\right)$ *and*

$$\phi^*_a(v) = \frac{2}{nk-a+1}\;\frac{v^{\frac{k-a+1}{2}}(1-v)^{\frac{(n-1)k}{2}}}{\int_0^v x^{\frac{k-a-1}{2}}(1-x)^{\frac{(n-1)k}{2}}dx}.$$

The proof is omitted because of its similarity to the proof of the previous theorem.

Using the expressions obtained in Theorem 4, we now present two graphs which demonstrate the extent of the risk improvement of the proposed hierarchical Bayes estimator $e^{HB}_{a,a,1}(\boldsymbol{Y},S)$ over the best multiple estimator $e_{0,1}(S) = S\big/(n-1)k$ of σ^2. In Fig. 1, the percent risk improvement of $e^{HB}_{a,a,1}(\boldsymbol{Y},S)$ over $S\big/(n-1)k$ is plotted against $\eta = (2\sigma^2)^{-1}\sum_{i=1}^k \theta_i^2$ for $k=5$ and $n=2$. The three curves represent the percent risk improvement of $e^{HB}_{a,a,1}(\boldsymbol{Y},S)$ over $e_{0,1}(S)$ for $a = 1,2,3$ respectively. Although the HB estimator $e^{HB}_{a,a,1}(\boldsymbol{Y},S)$ when $a=1$ is not minimax, it provides maximum risk improvement for higher values of η. Unlike this estimator, for $a=2$ or $a=3$ the estimator is always better than the best multiple estimator, the improvement being quite substantial for smaller values of η. It seems that the improvement is proportional to a when η is close to zero, and inversely proportional to a when η is larger than a certain value.

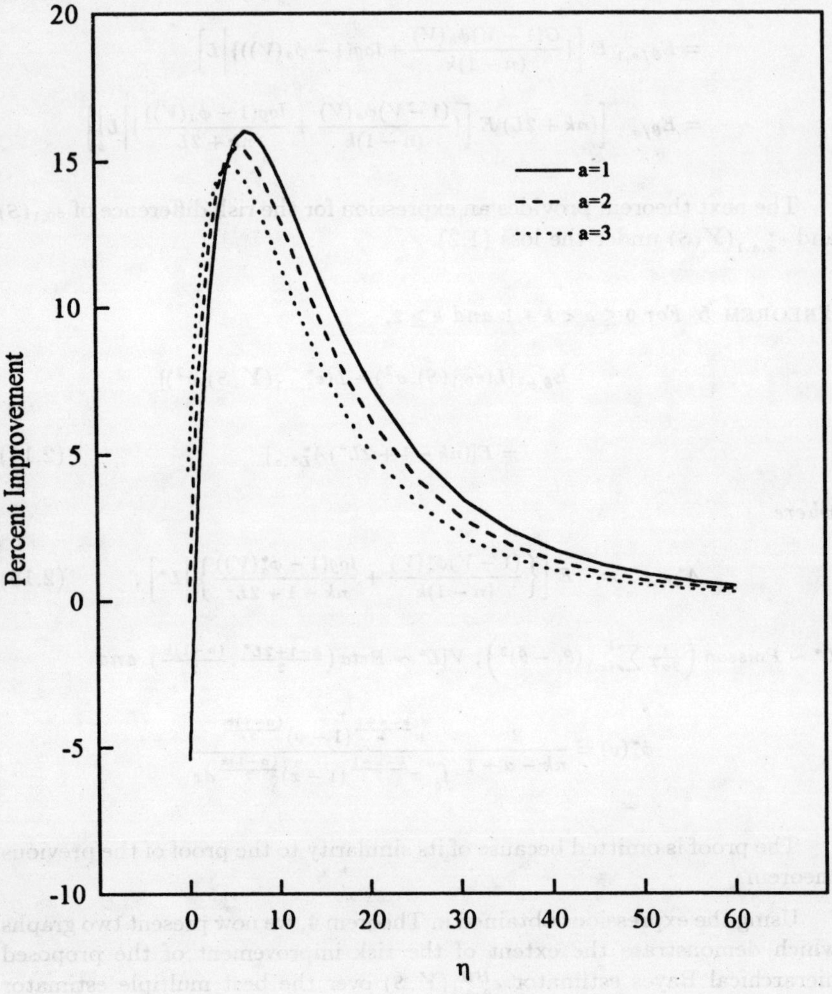

Figure 1. Percent Risk Improvement of $e^{HB}_{a,a,1}(\mathbf{Y}, S)$ over $S\big/(n-1)k$ for $k = 5$

A very similar figure is obtained for the same plot with $k = 10$ which is given in Figure 2. The maximum improvement for each estimator is more than the maximum for the corresponding one with $k = 5$, whereas the improvement tends to zero at a faster rate for $k = 10$ than for $k = 5$ as η

becomes large.

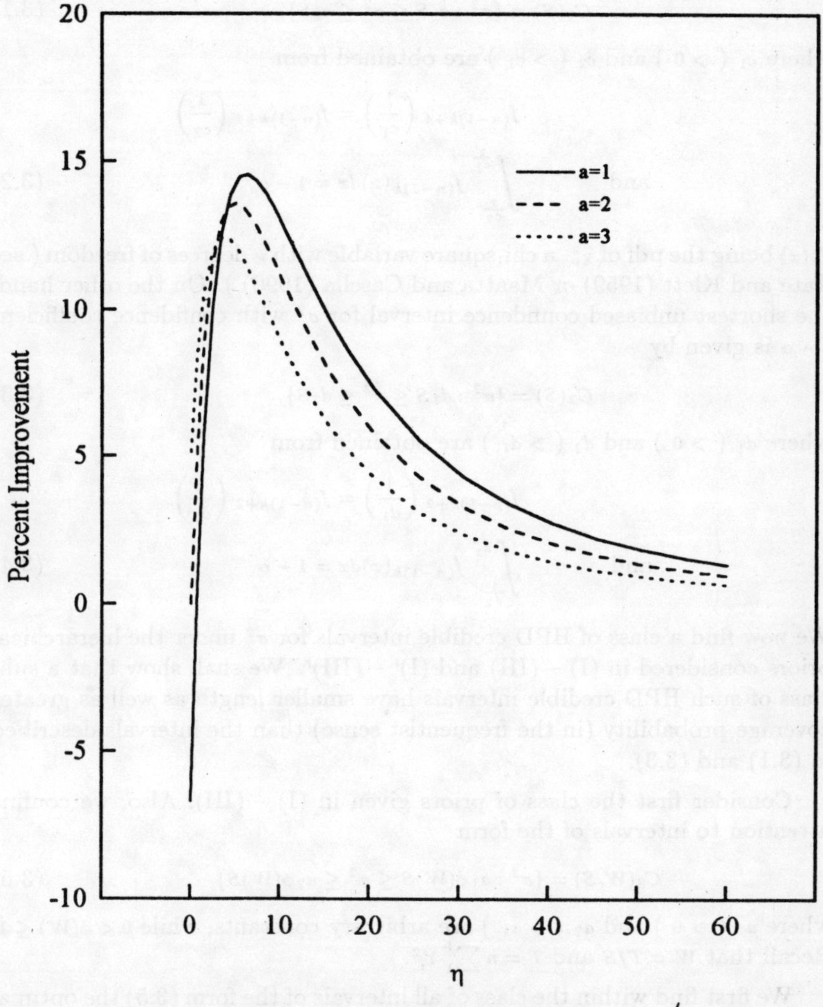

Figure 2. Percent Risk Improvement of $e^{HB}_{a,a,1}(\boldsymbol{Y}, S)$ over $S \big/ (n-1)k$ for $k = 10$

Using the expression obtained in Theorem 5, one can similarly construct graphs providing percent risk improvement of $e^{*}_{a,a,1}(\boldsymbol{Y}, S)$ over $S \big/ (n-1)k$ of σ^2 for different values of a. It is to be noted that in this case $\eta = \left(2\sigma^2\right)^{-1} \sum_{i=1}^{k} \left(\theta_i - \bar{\theta}\right)^2$. For brevity, we have not provided these figures.

3 Interval Estimation of σ^2

This section is devoted to the comparison of different confidence intervals for σ^2. Using S alone, the minimum length confidence interval for σ^2 with confidence coefficient $1-\alpha$ is given by

$$C_1(S) = \{\sigma^2 : c_1 S \le \sigma^2 \le c_2 S\}, \tag{3.1}$$

where c_1 (>0) and c_2 ($>c_1$) are obtained from

$$f_{(n-1)k+4}\left(\frac{1}{c_1}\right) = f_{(n-1)k+4}\left(\frac{1}{c_2}\right)$$

$$\text{and} \qquad \int_{\frac{1}{c_2}}^{\frac{1}{c_1}} f_{(n-1)k}(x)dx = 1-\alpha, \tag{3.2}$$

$f_\nu(x)$ being the pdf of χ^2_ν, a chi-square variable with ν degrees of freedom (see Tate and Klett (1959) or Maatta and Casella (1990)). On the other hand, the shortest unbiased confidence interval for σ^2 with confidence coefficient $1-\alpha$ is given by

$$C_2(S) = \{\sigma^2 : d_1 S \le \sigma^2 \le d_2 S\}, \tag{3.3}$$

where d_1 (>0) and d_2 ($>d_1$) are obtained from

$$f_{(n-1)k+2}\left(\frac{1}{d_1}\right) = f_{(n-1)k+2}\left(\frac{1}{d_2}\right)$$

$$\text{and} \qquad \int_{\frac{1}{d_2}}^{\frac{1}{d_1}} f_{(n-1)k}(x)dx = 1-\alpha. \tag{3.4}$$

We now find a class of HPD credible intervals for σ^2 under the hierarchical priors considered in (I) – (III) and (I)′ – (III)′. We shall show that a subclass of such HPD credible intervals have smaller length as well as greater coverage probability (in the frequentist sense) than the intervals described in (3.1) and (3.3).

Consider first the class of priors given in (I) – (III). Also, we confine attention to intervals of the form

$$C_3(W,S) = \{\sigma^2 : a_1\phi(W)S \le \sigma^2 \le a_2\phi(W)S\}, \tag{3.5}$$

where a_1 (>0) and a_2 ($>a_1$) are arbitrary constants, while $0<\phi(W)\le 1$. Recall that $W = T/S$ and $T = n\sum_1^k Y_i^2$.

We first find within the class of all intervals of the form (3.5) the optimal $\phi_0(W)$ which leads to the HPD credible interval based on the prior given in (I) – (III). Specifically, the following theorem is proved.

THEOREM 6 *Consider the class of priors given in (I) – (III) with $(n-1)k+b-a>2$ and $b<k+2$. Then, within the class of intervals of the form (3.5), the optimal $\phi(w)$ is given by $\phi_0(w)$, where $\phi_0(w)$ is a solution of*

$$\begin{aligned} & f_{(n-1)k+b-a+4}\left(\frac{1}{\phi_0(w)a_1}\right) F_{k-b+2}\left(\frac{w}{\phi_0(w)a_1}\right) \\ = \; & f_{(n-1)k+b-a+4}\left(\frac{1}{\phi_0(w)a_2}\right) F_{k-b+2}\left(\frac{w}{\phi_0(w)a_2}\right), \end{aligned} \tag{3.6}$$

where $f_\nu(x)$ denotes the pdf of χ^2_ν, and $F_\nu(x) = P(\chi^2_\nu \le x)$.

PROOF. As before, write $R = (\sigma^2)^{-1}$. Then from (I) – (III), the joint posterior pdf of R and $U = \Lambda/(n+\Lambda)$ given $\boldsymbol{Y} = \boldsymbol{y}$ and $S = s$ is

$$f(r, u|\boldsymbol{y}, s) \propto exp[-\frac{1}{2}r(s+ut)]r^{\frac{1}{2}(nk-a)}u^{\frac{1}{2}(k-b)}. \tag{3.7}$$

Now integrating with respect to u in (3.7), one gets

$$\begin{aligned} f(r|\boldsymbol{y}, s) &\propto exp(-\frac{1}{2}rs)r^{\frac{1}{2}((n-1)k+b-a)-1}P(\chi^2_{k-b+2} \leq tr) \\ &= exp(-\frac{1}{2}rs)r^{\frac{1}{2}((n-1)k+b-a)-1}P(\chi^2_{k-b+2} \leq wrs). \end{aligned} \tag{3.8}$$

Next we show that $f(r|\boldsymbol{y}, s)$ is log-concave. Since the expression $exp(-\frac{1}{2}rs)$ $r^{\frac{1}{2}((n-1)k+b-a)-1}$ is log-concave, it suffices to show that $P(\chi^2_{k-b+2} \leq tr)$ is log-concave. This fact follows immediately by noting that $F_\nu(x) = P(\chi^2_\nu \leq x)$, $\nu > 0$, $x > 0$ is log-concave.

Since the posterior pdf of R given $\boldsymbol{Y} = \boldsymbol{y}$ and $S = s$ is log-concave, using (3.8) the HPD credible interval for Rs within the class of all intervals of the form (3.5) is given by $\left(\frac{1}{a_2\phi_0(w)}, \frac{1}{a_1\phi_0(w)}\right)$ where $\phi_0(w)$ is obtained by maximizing

$$\int_{\frac{1}{a_2\phi(w)}}^{\frac{1}{a_1\phi(w)}} exp(-\frac{1}{2}z)z^{\frac{1}{2}((n-1)k+b-a)-1}P(\chi^2_{k-b+2} \leq wz)dz \tag{3.9}$$

with respect to $\phi(w)$. This leads to the solution

$$\begin{aligned} &\frac{1}{(a_1\phi_0(w))^2}f_{(n-1)k+b-a}\left(\frac{1}{a_1\phi_0(w)}\right)F_{k-b+2}\left(\frac{w}{a_1\phi_0(w)}\right) \\ &= \frac{1}{(a_2\phi_0(w))^2}f_{(n-1)k+b-a}\left(\frac{1}{a_2\phi_0(w)}\right)F_{k-b+2}\left(\frac{w}{a_2\phi_0(w)}\right) \end{aligned} \tag{3.10}$$

which is equivalent to (3.6) since $f_\nu(x)$ denotes the pdf of χ^2_ν. This proves the theorem.

Next we improve on confidence intervals of the form (3.1) and (3.3) in a frequentist sense using a subclass of HPD credible intervals described in (3.6). Denote the interval given in (3.1) and (3.3) as I_{ML}, and let I_ϕ denote an interval of the form

$$\{\sigma^2 : c_1\phi(w)s \leq \sigma^2 \leq c_2\phi(w)s\}, \tag{3.11}$$

where c_1 and c_2 are determined from (3.2). Also, we denote by $L(I)$ the length of an interval I.

We consider the special case $b = a$ ($< k+2$). If we now set $a_1 = c_1$ and $a_2 = c_2$, from Theorem 6, we have found a class of HPD credible intervals of the form

$$\{\sigma^2 : c_1\phi_0(w)s \leq \sigma^2 \leq c_2\phi_0(w)s\} \tag{3.12}$$

for σ^2 where $\phi_0(w)$ satisfies

$$\begin{aligned} & f_{(n-1)k+4}\left(\frac{1}{\phi_0(w)c_1}\right) F_{k-a+2}\left(\frac{w}{\phi_0(w)c_1}\right) \\ = \; & f_{(n-1)k+4}\left(\frac{1}{\phi_0(w)c_2}\right) F_{k-a+2}\left(\frac{w}{\phi_0(w)c_2}\right) \end{aligned} \tag{3.13}$$

We now prove the following theorem.

THEOREM 7 *For an interval* I_{ϕ_0} *given in (3.12) and (3.13) with* $0 < \phi_0(w) < 1$, $L(I_{\phi_0}) < L(I_{ML})$ *and* $P_{\sigma^2}(\sigma^2 \in I_{\phi_0}) \geq P_{\sigma^2}(\sigma^2 \in I_{ML})$ *for all* σ^2.

PROOF. We appeal to Theorem 3.1 of Kubokawa (1991) in the special case of chi-squared distribution. By definition $L(I_{\phi_0}) < L(I_{ML})$. To prove that I_{ϕ_0} has at least as large a confidence coefficient as I_{ML}, following Kubokawa (1991), we need to verify

(i) $\phi_0(w)$ is ↑ in w and $\lim_{w\to\infty} \phi_0(w) = 1$,

(ii) $f_{(n-1)k+4}\left(\frac{1}{a_1\phi_0}\right) F_k\left(\frac{w}{a_1\phi_0}\right) \geq f_{(n-1)k+4}\left(\frac{1}{a_2\phi_0}\right) F_k\left(\frac{w}{a_2\phi_0}\right)$, where $\phi_0 \equiv \phi_0(w)$.

First we prove (i). Making $w \to \infty$, we get from (3.13)

$$\begin{aligned} & f_{(n-1)k+4}\left(\frac{1}{(\lim_{w\to\infty}\phi_0(w))c_1}\right) \\ = \; & f_{(n-1)k+4}\left(\frac{1}{(\lim_{w\to\infty}\phi_0(w))c_2}\right) \end{aligned} \tag{3.14}$$

Since $f_{(n-1)k+4}(\frac{1}{c_1}) = f_{(n-1)k+4}(\frac{1}{c_2})$, and f is a chi-squared pdf, we must have $\lim_{w\to\infty} \phi_0(w) = 1$.

Next to show that $\phi_0(w)$ is ↑ in w, differentiate both sides of (3.13) with respect to w, and get

$$\phi_0'(w)(A + wB) = \phi_0(w)B, \tag{3.15}$$

where

$$\begin{aligned} A \equiv A(w) \;=\; & \frac{1}{c_2} f'_{(n-1)k+4}\left(\frac{1}{\phi_0 c_2}\right) F_{k-a+2}\left(\frac{1}{\phi_0 c_2}\right) \\ & -\frac{1}{c_1} f'_{(n-1)k+4}\left(\frac{1}{\phi_0 c_1}\right) F_{k-a+2}\left(\frac{1}{\phi_0 c_1}\right), \end{aligned} \tag{3.16}$$

$$\begin{aligned} B \equiv B(w) \;=\; & \frac{1}{c_2} f_{(n-1)k+4}\left(\frac{1}{\phi_0 c_2}\right) f_{k-a+2}\left(\frac{w}{\phi_0 c_2}\right) \\ & -\frac{1}{c_1} f_{(n-1)k+4}\left(\frac{1}{\phi_0 c_1}\right) f_{k-a+2}\left(\frac{w}{\phi_0 c_1}\right). \end{aligned} \tag{3.17}$$

To prove that $B \geq 0$, we require the following lemma.

LEMMA 2 *Let $f_\nu(x)$ denote the pdf of a χ^2_ν random variable and $F_\nu(x) = P(\chi^2_\nu \le x)$. Then $xf_\nu(x)\big/F_\nu(x)$ is $\downarrow$ in x.*

PROOF. Let $h(x) = xf_\nu(x)\big/F_\nu(x)$. Then,

$$\begin{aligned} h(x) &= \frac{1}{F_\nu(x)}\left[\frac{xe^{-x/2}x^{\nu/2-1}}{2^{\nu/2}\Gamma(\nu/2)}\right] \\ &= \frac{\nu f_{\nu+2}(x)}{F_\nu(x)} \\ &= \frac{\nu}{2}\left[1-\frac{F_{\nu+2}(x)}{F_\nu(x)}\right] \end{aligned}$$

using the fact that $F_{\nu+2}(x) = F_\nu(x) - 2f_{\nu+2}(x)$, which is obtained by integrating by parts. Since, $F_{\nu+2}(x)\big/F_\nu(x)$ is $\uparrow$ in x, the result follows.

From (3.17) and (3.13),

$$\begin{aligned} B &= f_{(n-1)k+4}\left(\frac{1}{\phi_0 c_1}\right)F_{k-a+2}\left(\frac{w}{\phi_0 c_1}\right) \\ &\quad \times\left(\frac{\phi_0}{w}\right)\left[\frac{w}{\phi_0 c_2}\frac{f_{k-a+2}(\frac{w}{\phi_0 c_2})}{F_{k-a+2}(\frac{w}{\phi_0 c_2})} - \frac{w}{\phi_0 c_1}\frac{f_{k-a+2}(\frac{w}{\phi_0 c_1})}{F_{k-a+2}(\frac{w}{\phi_0 c_1})}\right] \\ &\geq 0 \end{aligned} \tag{3.18}$$

since $xf_\nu(x)/F_\nu(x)$ is $\downarrow$ in x for every ν, and $0 < \frac{1}{c_2} < \frac{1}{c_1}$. Now, using (3.16) and (3.13),

$$\begin{aligned} A &= f_{(n-1)k+4}\left(\frac{1}{\phi_0 c_1}\right)F_{k-a+2}\left(\frac{w}{\phi_0 c_1}\right) \\ &\quad \times\phi_0\left[\frac{1}{\phi_0 c_2}\frac{f'_{(n-1)k+4}(\frac{1}{\phi_0 c_2})}{f_{(n-1)k+4}(\frac{1}{\phi_0 c_2})} - \frac{1}{\phi_0 c_1}\frac{f'_{(n-1)k+4}(\frac{1}{\phi_0 c_1})}{f_{(n-1)k+4}(\frac{1}{\phi_0 c_1})}\right] \\ &\geq 0 \end{aligned} \tag{3.19}$$

since $xf'_\nu(x)/f_\nu(x) = \frac{1}{2}[\nu - 2 - x]$ is $\downarrow$ in x for every ν, and $0 < \frac{1}{c_2} < \frac{1}{c_1}$. It follows from (3.15) – (3.19) that $\phi_0'(w) \geq 0$, i.e. $\phi_0(w)$ is $\uparrow$ in w.

Next to prove (ii), use (3.13) to get

$$\begin{aligned} & f_{(n-1)k+4}\left(\frac{1}{a_1\phi_0}\right)F_k\left(\frac{w}{a_1\phi_0}\right) - f_{(n-1)k+4}\left(\frac{1}{a_2\phi_0}\right)F_k\left(\frac{w}{a_2\phi_0}\right) \\ &= f_{(n-1)k+4}\left(\frac{1}{a_1\phi_0}\right)\left[F_k\left(\frac{w}{a_1\phi_0}\right) - \frac{F_{k-a+2}(\frac{w}{a_1\phi_0})}{F_{k-a+2}(\frac{w}{a_1\phi_0})}F_k\left(\frac{w}{a_2\phi_0}\right)\right] \\ &= \frac{f_{(n-1)k+4}\left(\frac{1}{a_1\phi_0}\right)}{F_{k-a+2}\left(\frac{w}{a_1\phi_0}\right)} \times \\ &\quad \left[F_k\left(\frac{w}{a_1\phi_0}\right)F_{k-a+2}\left(\frac{w}{a_2\phi_0}\right) - F_k\left(\frac{w}{a_2\phi_0}\right)F_{k-a+2}\left(\frac{w}{a_1\phi_0}\right)\right] \\ &\geq 0, \end{aligned}$$

using the monotone likelihood property of chi-squares.

This completes the proof of Theorem 7.

To improve on intervals given in (3.3) and (3.4), put $a_i = d_i$ $(i = 1, 2)$ and choose $b = a - 2$ and $a < k + 4$ in (3.6) to get ϕ_0 from

$$
\begin{aligned}
& f_{(n-1)k+2}\left(\frac{1}{\phi_0 d_1}\right) F_{k-a+4}\left(\frac{w}{\phi_0 d_1}\right) \\
= \quad & f_{(n-1)k+2}\left(\frac{1}{\phi_0 d_2}\right) F_{k-a+4}\left(\frac{w}{\phi_0 d_2}\right)
\end{aligned}
\tag{3.20}
$$

Then, following the line of the proof of Theorem 7, one can show that $L(I_{\phi_0}) < L(I_{SU})$, where I_{SU} denotes the shortest unbiased confidence interval given in (3.3) and (3.4), and $P_{\sigma^2}(\sigma^2 \in I_{\phi_0}) \geq P_{\sigma^2}(\sigma^2 \in I_{SU})$ for all σ^2, where ϕ_0 is defined in (3.20).

For the class of priors defined in (I)′ – (III)′, first analogous to Theorem 6, define a class of confidence intervals of the form (3.5), where $\phi_0(w)$ is a solution of

$$
\begin{aligned}
& f_{(n-1)k+b-a+4}\left(\frac{1}{\phi_0(w)a_1}\right) F_{k-b+1}\left(\frac{w}{\phi_0(w)a_1}\right) \\
= \quad & f_{(n-1)k+b-a+4}\left(\frac{1}{\phi_0(w)a_2}\right) F_{k-b+1}\left(\frac{w}{\phi_0(w)a_2}\right)
\end{aligned}
\tag{3.21}
$$

Such intervals are HPD within the class of intervals given in (3.5) under the prior (I)′ – (III)′. Now putting $b = a - 2$ and $a < k + 3$, $a_i = c_i$ $(i = 1, 2)$, (3.21) reduces to

$$
\begin{aligned}
& f_{(n-1)k+2}\left(\frac{1}{\phi_0 c_1}\right) F_{k-a+3}\left(\frac{w}{\phi_0 c_1}\right) \\
= \quad & f_{(n-1)k+2}\left(\frac{1}{\phi_0 c_2}\right) F_{k-a+3}\left(\frac{w}{\phi_0 c_2}\right)
\end{aligned}
\tag{3.22}
$$

The resulting interval I_{ϕ_0} satisfies $L(I_{\phi_0}) < L(I_{SU})$ and $P_{\sigma^2}(\sigma^2 \in I_{\phi_0}) \geq P_{\sigma^2}(\sigma^2 \in I_{SU})$ for all σ^2. Thus, we have produced different classes of HPD intervals which dominate the minimum length or shortest unbiased confidence intervals. Unlike the intervals of Cohen (1972) and Shorrock (1990) which have the same length, our intervals have smaller length and greater coverage probability. The intervals proposed by Goutis and Casella (1991) are different from ours although they achieve the same objective as ours.

However, it is also possible to develop HPD intervals using the formulation of Cohen (1972) and Shorrock (1982, 1990). The minimum length confidence interval under their formulation for σ^2 with confidence coefficient $1 - \alpha$ is given by

$$
C_4(S) = \{\sigma^2 : a_1 S \leq \sigma^2 \leq (a_1 + c_1)S\},
\tag{3.23}
$$

where a_1 (>0) and c_1 (>0) are obtained from

$$f_{(n-1)k+4}\left(\frac{1}{a_1}\right) = f_{(n-1)k+4}\left(\frac{1}{a_1+c_1}\right)$$

and
$$\int_{\frac{1}{(a_1+c_1)}}^{\frac{1}{a_1}} f_{(n-1)k}(x)dx = 1-\alpha, \tag{3.24}$$

$f_\nu(x)$ being the pdf of χ^2_ν, a chi-square variable with ν degrees of freedom (see Maatta and Casella (1990)). On the other hand, the shortest unbiased confidence interval for σ^2 with confidence coefficient $1-\alpha$ is given by

$$C_5(S) = \{\sigma^2 : a_2 S \le \sigma^2 \le (a_2+c_2)S\}, \tag{3.25}$$

where a_2 (>0) and c_2 (>0) are obtained from

$$f_{(n-1)k+2}\left(\frac{1}{a_2}\right) = f_{(n-1)k+2}\left(\frac{1}{a_2+c_2}\right)$$

and
$$\int_{\frac{1}{(a_2+c_2)}}^{\frac{1}{c_2}} f_{(n-1)k}(x)dx = 1-\alpha. \tag{3.26}$$

We now find a class of HPD credible intervals for σ^2 under the hierarchical priors considered in (I) – (III) and (I)′ – (III)′. A subclass of such HPD credible intervals have greater coverage probability (in the frequentist sense) than the intervals described in (3.23) and (3.25).

Consider first the class of priors given in (I) – (III). Also, we confine attention to intervals of the form

$$C_6(W,S) = \{\sigma^2 : \phi(W)S \le \sigma^2 \le (\phi(W)+c)S\}, \tag{3.27}$$

where c (>0) is an arbitrary constant, while $0<\phi(W)\le 1$. Recall that $W=T/S$ and $T = n\sum_1^k Y_i^2$.

The optimal $\phi_0(W)$ within the class of all intervals of the form (3.27) which leads to the HPD credible interval based on the prior given in (I) – (III) is obtained from the following theorem.

THEOREM 8 *Consider the class of priors given in (I) – (III) with $(n-1)k+b-a>2$ and $b<k+2$. Then, within the class of intervals of the form (3.27), the optimal $\phi(w)$ is given by $\phi_0(w)$, where $\phi_0(w)$ is a solution of*

$$\begin{aligned} & f_{(n-1)k+b-a+4}\left(\frac{1}{\phi_0(w)}\right) F_{k-b+2}\left(\frac{w}{\phi_0(w)}\right) \\ = \; & f_{(n-1)k+b-a+4}\left(\frac{1}{\phi_0(w)+c}\right) F_{k-b+2}\left(\frac{w}{\phi_0(w)+c}\right), \end{aligned} \tag{3.28}$$

where $f_\nu(x)$ denotes the pdf of χ^2_ν, and $F_\nu(x) = P(\chi^2_\nu \le x)$.

The proof of this theorem is omitted because of its similarity to the proof of Theorem 6.

For the special case $b = a$ ($< k+2$), we have, from Theorem 8, a class of HPD credible intervals of the form

$$I(\phi_0) = \{\sigma^2 : \phi_0(w)s \leq \sigma^2 \leq (\phi_0(w) + c_1)s\} \tag{3.29}$$

for σ^2 where $\phi_0(w)$ satisfies

$$f_{(n-1)k+4}\left(\frac{1}{\phi_0(w)}\right) F_{k-a+2}\left(\frac{w}{\phi_0(w)}\right) = f_{(n-1)k+4}\left(\frac{1}{\phi_0(w)+c_1}\right) F_{k-a+2}\left(\frac{w}{\phi_0(w)+c_1}\right) \tag{3.30}$$

THEOREM 9 *For an interval I_{ϕ_0} given in (3.29) with $0 < \phi_0(w) < a_1$, $L(I_{\phi_0}) = L(I_{ML})$ and $P_{\sigma^2}(\sigma^2 \in I_{\phi_0}) \geq P_{\sigma^2}(\sigma^2 \in I_{ML})$ for all σ^2.*

The proof of this theorem is similar to the proof of Theorem 7. The main difference in the proof is that in proving the monotonicity of $\phi_0(w)$, we appeal to Lemma 3.1 of Shorrock (1990) instead of (3.15) – (3.19). Also, we may note that the class of intervals given in (3.29) includes as its members the ones considered by Brewster and Zidek (1974) and Shorrock (1990) for the special case $a = b = 2$.

To improve on intervals given in (3.25) and (3.26), choose $b = a - 2$ and $a < k+4$ in (3.28) to get ϕ_0 from

$$f_{(n-1)k+2}\left(\frac{1}{\phi_0}\right) F_{k-a+4}\left(\frac{w}{\phi_0}\right) = f_{(n-1)k+2}\left(\frac{1}{\phi_0+c_2}\right) F_{k-a+4}\left(\frac{w}{\phi_0+c_2}\right) \tag{3.31}$$

Then, one can show that $P_{\sigma^2}(\sigma^2 \in I_{\phi_0}) \geq P_{\sigma^2}(\sigma^2 \in I_{SU})$ for all σ^2, where ϕ_0 is defined in (3.31) and I_{SU} denotes the shortest unbiased confidence interval given in (3.25) and (3.26).

For the class of priors defined in (I)′ – (III)′, the class of HPD intervals is given by $(\phi_0 s, (\phi_0 + c_1)s)$ where $\phi_0(w)$ is a solution of

$$f_{(n-1)k+b-a+4}\left(\frac{1}{\phi_0(w)}\right) F_{k-b+1}\left(\frac{w}{\phi_0(w)}\right) = f_{(n-1)k+b-a+4}\left(\frac{1}{\phi_0(w)+c_1}\right) F_{k-b+1}\left(\frac{w}{\phi_0(w)+c_1}\right) \tag{3.32}$$

Now putting $b = a - 2$ and $a < k+3$, (3.32) reduces to

$$f_{(n-1)k+2}\left(\frac{1}{\phi_0}\right) F_{k-a+3}\left(\frac{w}{\phi_0}\right) = f_{(n-1)k+2}\left(\frac{1}{\phi_0+c_1}\right) F_{k-a+3}\left(\frac{w}{\phi_0+c_1}\right)$$

The resulting interval I_{ϕ_0} satisfies $P_{\sigma^2}(\sigma^2 \in I_{\phi_0}) \geq P_{\sigma^2}(\sigma^2 \in I_{SU})$ for all σ^2.

Acknowledgements

The authors would like to thank Prof. S. K. Sarkar, Dr. H. H. Nguyen and an anonymous referee for their helpful comments.

REFERENCES

Brewster, J. F. and Zidek, J. V. (1974). Improving on equivariant estimators. *Ann. Statist.*, **2**, 21-38.

Brown, L. (1968). Inadmissibility of the usual estimators of scale parameters in problems with unknown location and scale parameters. *Ann. Math. Statist.*, **39**, 29-48.

Cohen, A. (1972). Improved confidence intervals for the variance of a normal distribution. *J. Amer. Statist. Assoc.*, **67**, 382-387.

Gelfand, A. E. and Dey, D. K. (1988). On the estimation of a variance ratio. *J. Stat. Plan. Inf.*, **19**, 121-131.

Ghosh, M. (1992a). Hierarchical and empirical Bayes multivariate estimation. *Current Issues in Statistical Inference: Essays in Honor of D. Basu*, Eds. M. Ghosh and P. K. Pathak. IMS Lecture Notes Monograph Series **17**, 151-177.

Ghosh, M. (1992b). On some Bayesian solutions of the Neyman-Scott problem. Tech. Report No. 407, Dept. of Statistics, University of Florida.

Goutis, C. and Casella, G. (1991). Improved invariant confidence intervals for a normal variance. *Ann. Statist.*, **19**, 2015-2031.

Kubokawa, T. (1991). A unified approach to improving equivariant estimators. To appear in *Ann. Statist.*

Maatta, J. M. and Casella, G. (1990). Developments in decision-theoretic variance estimation (with discussion). *Statistical Science*, **5**, 90-120.

Morris, C. N. (1983). Parametric empirical Bayes confidence intervals. *Scientific Inference, Data Analysis and Robustness.* Eds. G. E. P. Box, T. Leonard and C. F. J. Wu. Academic Press, New York, 25-50.

Sarkar, S. K. (1989). On improving the shortest length confidence interval for the generalized variance. *J. Mult. Anal.*, **31**, 136-147.

Shorrock, G. (1982). A minimax generalized Bayes confidence interval for a normal variance. Ph.D. dissertation, Dept. of Statistics, Rutgers University.

Shorrock, G. (1990). Improved confidence intervals for a normal variance. *Ann. Statist.*, **18**, 972-980.

Stein, C. (1964). Inadmissibility of the usual estimator for the variance of a normal distribution with unknown mean. *Ann. Inst. Statist. Math.*, **16**, 155-160.

Strawderman, W. E. (1971). Proper Bayes minimax estimators of the multivariate normal mean. *Ann. Math. Statist.*, **42**, 385-388.

Strawderman, W. E. (1974). Minimax estimation of powers of the variance of a normal population under squared error loss. *Ann. Statist.*, **2**, 190-198.

Tate, R. F. and Klett, G. W. (1959). Optimal confidence intervals for the variance of a normal distribution. *J. Amer. Statist. Assoc.*, **54**, 674-682.

Department of Statistics
University of Florida
Gainesville, FL 32611
U.S.A.

Clinical Biostatistics Dept.
Merck Research Laboratories
Rahway, NJ 07065
U.S.A.

J.K. Ghosh, S.K. Mitra, K.R. Parthasarathy and B.L.S. Prakasa Rao (*eds.*),
Statistics and Probability: A Raghu Raj Bahadur Festschrift
Published by Wiley Eastern Limited, Publishers, (1993), 265 – 278

Bayes and Empirical Bayes Rules for Selecting Fair Multinomial Populations

SHANTI S. GUPTA
Purdue University

and

TACHEN LIANG
Wayne State University

Abstract

This paper deals with the problem of selecting fair multinomial populations compared with a standard. Two selection procedures are investigated: the natural selection procedure of Gupta and Leu (1990) and an empirical Bayes simultaneous selection procedure. It is proved that the natural selection procedure is a Bayes procedure relative to a symmetric Dirichlet prior, and therefore it is admissible. For the empirical Bayes simultaneous selection procedure, the associated asymptotic optimality is investigated. It is shown that the proposed procedure is asymptotically optimal relative to a class of symmetric Dirichlet priors, with a rate of convergence of order $O(\exp(-\tau k + \ln k))$ for some positive constant τ, where k is the number of populations involved in the selection problem. Also, presented are results of a simulation study of the small sample performance of the empirical Bayes procedure.

1 Introduction

The concept of diversity within a population is of considerable importance in statistical theory and applications. The problem of measuring diversity

AMS Subject Classifications : Primary 62F07; secondary 62C12
Key words and Phrases: Fair Multinomial Population, Entropy Function, Gini-Simpson Index, Natural Selection Procedure, Empirical Bayes Selection Procedure, Asymptotically Optimal, Rate of Convergence.

arises in a variety of studies in ecology, sociology, econometrics, genetics and many other sciences. For a multinomial population with m cells, the index of diversity is a function of the corresponding probability parameter vector $\underline{p} = (p_1, \ldots, p_m)$. In practice, a Schur-convex or Schur-concave function of $\underline{p}$ may be appropriate. There are two measures of diversity of a multinomial population which have been commonly used. These are Shannon's entropy function and the Gini-Simpson index. The notion of the entropy function was introduced by Shannon (1948). The Gini-Simpson index was introduced by Gini (1912) and Simpson (1949). Both are Schur-concave function of $\underline{p}$.

In the literature, selection procedures using indices of diversity as selection criteria have been studied by many authors. Gupta and Huang (1976) studied the problem of selecting the population with the largest entropy function for binomial distributions. Gupta and Wong (1975) considered the problem of selecting a subset containing the population with the largest entropy for multinomial distributions. Dudewicz and Van der Meulen (1981) investigated a selection procedure based on a generalized entropy function. Alam, Mitra, Rizvi and Saxena (1986) studied selection procedures based on Shannon's entropy function and the Gini-Simpson index using the indifference zone approach. Rizvi, Alam and Saxena (1987) also considered a subset selection procedure based on certain other diversity indexes. Recently, Gupta and Leu (1990) and Liang and Panchapakesan (1991) have studied certain selection procedures based on the Gini-Simpson index.

In this paper, we are dealing with the problem of selecting fair multinomial populations compared with a standard level. Consider k independent multinomial populations $\pi_1, \ldots, \pi_k$. For each $i = 1, \ldots, k$, population π_i has m cells, and is characterized by the corresponding probability parameter vector $\underline{p}_i = (p_{i1}, \ldots, p_{im})$, where $0 \leq p_{ij} \leq 1$, $j = 1, \ldots, m$, and $\sum_{j=1}^{m} p_{ij} = 1$ for each $i = 1, \ldots, k$. Define

$$\theta_i = \Psi(\underline{p}_i) = \sum_{j=1}^{m} (p_{ij} - \frac{1}{m})^2. \tag{1.1}$$

We use θ_i as a measure of diversity (or uniformity) of population π_i. Note that since $\theta_i = \sum_{j=1}^{m} p_{ij}^2 - \frac{1}{m}$, it is essentially equivalent to the Gini-Simpson index. Also note that $0 \leq \theta_i \leq 1 - \frac{1}{m}$. For a given constant θ_0, $0 < \theta_0 < 1 - \frac{1}{m}$, population π_i is said to be a fair population if $\theta_i \leq \theta_0$ and a bad population, otherwise. Our goal is to derive statistical selection procedures for selecting all fair populations while excluding all bad populations. Since a fair multinomial would imply equal cell probabilities analogous to a fair coin, the problem at hand is equivalent to selecting multinomial populations that are fairer (more homogeneous) than the standard. It should be noted that the problem of selecting fair multinomial populations has been considered by Gupta and Leu (1990) through a classical approach.

Let $\Omega = \{\underline{\underline{p}} = (\underline{p}_1, \ldots, \underline{p}_k) | \underline{p}_i = (p_{i1}, \ldots, p_{im}), 0 \le p_{ij} \le 1,\ j = 1, \ldots, m,\ \sum_{j=1}^{m} p_{ij} = 1$ for each $i = 1, \ldots, k\}$ be the parameter space and let $\mathcal{A} = \{s | s \subset \{1, \ldots, k\}\}$ be the action space. When action s is taken, it means that population π_i is selected as a fair population if $i \in s$ and excluded as a bad population if $i \notin s$. For $\underline{\underline{p}} \in \Omega$ and action $s \in \mathcal{A}$, the loss function $L(\underline{\underline{p}}, s)$ is defined to be:

$$L(\underline{\underline{p}}, s) = \sum_{i \in s} (\theta_i - \theta_0) I_{(\theta_0, 1 - \frac{1}{m}]}(\theta_i) + \sum_{i \notin s} (\theta_0 - \theta_i) I_{[0, \theta_0]}(\theta_i). \tag{1.2}$$

In (1.2), the first summation is the loss due to selecting certain bad populations and the second summation is the loss due to not selecting certain fair populations.

The content of this paper consists of two parts. In Section 2, we investigate some optimal properties of the natural selection procedure of Gupta and Leu (1990). It is shown that, for the loss function $L(\underline{\underline{p}}, s)$ of (1.2), the natural selection procedure is Bayes relative to some symmetric Dirichlet prior, and therefore, it is admissible. Section 3 deals with this selection problem through a parametric empirical Bayes approach. An empirical Bayes selection procedure is proposed and the corresponding asymptotic optimality is investigated. It is shown that the proposed empirical Bayes selection procedure is asymptotically optimal relative to a class of symmetric Dirichlet priors. The rate of convergence of the proposed empirical Bayes selection procedure is also established, and shown to be of order $O(\exp(-\tau k + \ln k))$ for some positive constant τ, where k is the number of populations involved in the selection problem. Finally, a simulation study is carried out to investigate the small sample performance of the empirical Bayes selection rule. The results of this study are presented in the last section of the paper.

2 Optimality of Natural Selection Procedure

For each $i = 1, \ldots, k$, let $\underline{X}_i = (X_{i1}, \ldots, X_{im})$ be the random observation associated with population π_i, where X_{ij}, $1 \le j \le m$, are nonnegative integer random variables such that $0 \le X_{ij} \le N$ and $\sum_{j=1}^{m} X_{ij} = N$. Then, $\underline{X}_i$ has the probability function

$$f(\underline{x}_i | \underline{p}_i) = \frac{N!}{\prod_{j=1}^{m} (x_{ij}!)} \prod_{j=1}^{m} p_{ij}^{x_{ij}} \tag{2.1}$$

at point $\underline{x}_i = (x_{i1}, \ldots, x_{im})$ for which $0 \le x_{ij} \le N$, $1 \le j \le m$ and $\sum_{j=1}^{m} x_{ij} = N$. Let $\mathcal{X}_i$ be the sample space generated by $\underline{X}_i$. Let $\underline{\underline{X}} = (\underline{X}_1, \ldots, \underline{X}_k)$

and denote the corresponding observed value by $\underline{x} = (\underline{x}_1, \ldots, \underline{x}_k)$. Also, let $\mathcal{X} = \mathcal{X}_1 \times \ldots \times \mathcal{X}_k$ denote the sample space of $\underline{X}$.

A selection procedure $d = (d_1, \ldots, d_k)$ is defined to be a mapping from the sample space $\mathcal{X}$ into the product space $[0,1]^k$. That is, for each $\underline{x} = (\underline{x}_1, \ldots, \underline{x}_k) \in \mathcal{X}$, $d(\underline{x}) = (d_1(\underline{x}), \ldots, d_k(\underline{x}))$, where $d_i(\underline{x})$ is the probability of selecting population π_i as a fair population given $\underline{X} = \underline{x}$ is observed. We let D denote the class of all selection procedures defined in the above way.

For each $i = 1, \ldots, k$, $(\frac{X_{i1}}{N}, \ldots, \frac{X_{im}}{N})$ is the maximum likelihood estimator of $\underline{p}_i = (p_{i1}, \ldots, p_{im})$. From (1.1), it is natural and reasonable to estimate θ_i by $\hat{\theta}_i = \sum_{j=1}^{m} (\frac{X_{ij}}{N})^2 - \frac{1}{m}$. Gupta and Leu (1990) proposed a natural selection procedure $d^N = (d_1^N, \ldots, d_k^N)$ based on $\hat{\theta}_i$, $i = 1, \ldots, k$, which is equivalent to the following: For each $i = 1, \ldots, k$,

$$d_i^N(\underline{X}) = \begin{cases} 1 & \text{if } \hat{\theta}_i \leq \delta, \\ 0 & \text{otherwise,} \end{cases} \tag{2.2}$$

where δ is a prespecified positive constant such that $0 < \delta < 1 - \frac{1}{m}$. Since this natural selection procedure is heavily dependent on the constant δ, we denote this procedure by $d^{N(\delta)} = (d_1^{N(\delta)}, \ldots, d_k^{N(\delta)})$.

In the following, it is assumed that for each $i = 1, \ldots, k$, the parameter vector $\underline{p}_i = (p_{i1}, \ldots, p_{im})$ is a realization of the random vector $\underline{P}_i = (P_{i1}, \ldots, P_{im})$. It is also assumed that $\underline{P}_1, \ldots, \underline{P}_k$ are iid with a common prior distribution G_α belonging to a class of symmetric Dirichlet distributions $\mathcal{C}$, where

$$\begin{aligned} \mathcal{C} = \{g_\alpha | g_\alpha(\underline{p}_i) &= \frac{\Gamma(m\alpha)}{[\Gamma(\alpha)]^m} \prod_{j=1}^{m} p_{ij}^{\alpha-1}, \quad 0 \leq p_{ij} \leq 1, \\ & \qquad j = 1, \ldots, m, \quad \sum_{j=1}^{m} p_{ij} = 1\}. \end{aligned} \tag{2.3}$$

For a prior distribution $G_\alpha \in \mathcal{C}$ and a selection procedure $d = (d_1, \ldots, d_k) \in D$, we denote the corresponding Bayes risk by $r(G_\alpha, d)$. From (1.2) and the statistical model described previously,

$$r(G_\alpha, d) = \sum_{i=1}^{k} r_i(G_\alpha, d_i) \tag{2.4}$$

where

$$\begin{aligned} r_i(G_\alpha, d_i) &= \sum_{\underline{x} \in \mathcal{X}} d_i(\underline{x}) \left[\sum_{j=1}^{m} E[P_{ij}^2 | \underline{x}_i] - \frac{1}{m} - \theta_0 \right] \prod_{j=1}^{k} f(\underline{x}_j) \\ &\quad + \int_{\Omega_i(\theta_0)} (\theta_0 - \theta_i) g_\alpha(\underline{p}_i) d\underline{p}_i \end{aligned} \tag{2.5}$$

$$f(\underline{x}_i) = \int f(\underline{x}_i|\underline{p}_i)g_\alpha(\underline{p}_i)d\underline{p}_i = \frac{N!}{\prod_{j=1}^{m}(x_{ij}!)} \frac{\Gamma(m\alpha)}{[\Gamma(\alpha)]^m} \frac{\prod_{j=1}^{m}\Gamma(\alpha + x_{ij})}{\Gamma(m\alpha + N)},$$

and $\Omega_i(\theta_0) = \{\underline{p}_i = (p_{i1}, \ldots, p_{im})|\theta_i \leq \theta_0\}$.

Since the second term of (2.5) does not depend on the selection procedure d, a Bayes selection procedure $d^{G_\alpha} = (d_1^{G_\alpha}, \ldots, d_k^{G_\alpha})$ can be obtained as follows: For each $i = 1, \ldots, k$,

$$d_i^{G_\alpha}(\underline{\underline{x}}) = \begin{cases} 1 & \text{if } \sum_{j=1}^{m} E[P_{ij}^2|\underline{x}_i] - \frac{1}{m} \leq \theta_0, \\ 0 & \text{otherwise.} \end{cases} \tag{2.6}$$

Then, we have the following theorem.

THEOREM 2.1. For each positive constant δ such that $0 < \delta < 1 - \frac{1}{m}$ and $\frac{m\delta N^2 + mN - N}{mN(N+1)} > \theta_0$, for the loss function $L(\underline{\underline{p}}, s)$, the natural selection procedure $d^{N(\delta)}$ given in (2.2) is a Bayes procedure relative to some symmetric prior distribution G_α.

PROOF: First, straightforward computation yields for each $i = 1, \ldots, k$ and $j = 1, \ldots, m$,

$$E[P_{ij}^2|\underline{X}_i] = \frac{(X_{ij} + \alpha + 1)(X_{ij} + \alpha)}{(m\alpha + N + 1)(m\alpha + N)}$$

and therefore

$$\sum_{j=1}^{m} E[P_{ij}^2|\underline{X}_i] - \frac{1}{m} = \frac{(m^2 - m)\alpha + mN + m\sum_{j=1}^{m} X_{ij}^2 - N^2 - N}{m(m\alpha + N + 1)(m\alpha + N)}. \tag{2.7}$$

Note that

$$\hat{\theta}_i \leq \delta \Longleftrightarrow \sum_{j=1}^{m} E[P_{ij}^2|\underline{X}_i] - \frac{1}{m} \leq H(\alpha)$$

where

$$H(\alpha) = \frac{(\delta + \frac{1}{m})mN^2 + (m^2 - m)\alpha + mN - N^2 - N}{m(m\alpha + N + 1)(m\alpha + N)}.$$

Thus, it suffices to prove that for given θ_0, $0 < \theta_0 < 1 - \frac{1}{m}$, there exists a positive α such that $H(\alpha) = \theta_0$. Since $H(\alpha)$ is decreasing in α, $H(0) = \dfrac{m\delta N^2 + mN - N}{mN(N+1)} > \theta_0$ by the assumption, $\lim_{\alpha\to\infty} H(\alpha) = 0 < \theta_0$ and $H(\alpha)$ is a continuous function of α on $[0, \infty)$, there exists a unique

$\alpha \equiv \alpha(\theta_0) > 0$ such that $H(\alpha) = \theta_0$. This implies that the natural selection procedure $d^{N(\delta)}$ is the Bayes procedure relative to the symmetric Dirichlet prior $G_{\alpha(\theta_0)}$. The proof is now complete. □

The following corollary is a direct consequent of Theorem 2.1.

COROLLARY 2.1. For each positive constant δ such that $0 < \delta < 1 - \frac{1}{m}$, $\frac{m\delta N^2 + mN - N}{mN(N+1)} > \theta_0$, the natural selection procedure is admissible for the loss function $L(\underline{p}, s)$.

3 An Empirical Bayes Selection Procedure

We assume that the hyperparameter α of the symmetric Dirichlet prior G_α is unknown. In this situation, it is not possible to apply the Bayes selection procedure d^{G_α} for the problem at hand. Thus, the empirical Bayes approach is employed here.

For each $i = 1, \ldots, k$, let $W_i = \sum_{j=1}^{m} X_{ij}^2$ and let w_i denote the observed value of W_i. Under the preceding statistical model, $W_1, \ldots, W_k$ are iid random variables such that $\frac{N^2}{m} \leq W_i \leq N^2$. It follows from straightforward computations that

$$\mu_2 \equiv E[W_1] = N[1 + \frac{(N-1)(\alpha+1)}{m\alpha+1}], \tag{3.1}$$

and therefore,

$$\alpha = \frac{N^2 - \mu_2}{m\mu_2 - N(m+N-1)}. \tag{3.2}$$

From (3.2) and (2.7), for each $i = 1, \ldots, k$,

$$\begin{aligned} &\textstyle\sum_{j=1}^m E[P_{ij}^2|\underline{x}_i] - \frac{1}{m} \\ &= \{(m^2-m)(N^2-\mu_2) + [m\mu_2 - N(m+N-1)] \\ &\quad [mN + mw_i - N^2 - N]\} \\ &\times \frac{(m\mu_2 - mN - N^2 + N)}{[m(N-1)(m\mu_2 - N^2)N(m\mu_2 - N^2 - m + 1)]} \\ &\equiv Q_i(\mu_2|w_i), \end{aligned} \tag{3.3}$$

where $w_i = \sum_{j=1}^{m} x_{ij}^2$.

Note that for w_i(or $\underline{x}_i$) being kept fixed, $Q_i(\mu_2|w_i)$ is increasing in μ_2.

Define, $\hat{\mu}_2 = \frac{1}{k}\sum_{i=1}^{k} W_i$. We will use $\hat{\mu}_2$ to estimate μ_2 and use $Q_i(\hat{\mu}_2|w_i)$ to estimate $Q_i(\mu_2|w_i)$. However, by the definition of μ_2 (see (3.1)), $N + \frac{N(N-1)}{m} < \mu_2 < N^2$ and $Q_i(\mu_2|w_i)$ tends to zero as μ_2 tends to $N + \frac{N(N-1)}{m}$.

Also, it is possible that $\hat{\mu}_2 \leq N + \frac{N(N-1)}{m}$. Thus we define

$$\varphi_i^*(w_i) = \begin{cases} Q_i(\hat{\mu}_2|w_i) & \text{if } \hat{\mu}_2 > N + \frac{N(N-1)}{m}, \\ 0 & \text{otherwise.} \end{cases} \tag{3.4}$$

We now propose an empirical Bayes selection procedure $d^* = (d_1^*, \ldots, d_k^*)$ as follows: For each $i = 1, \ldots, k$,

$$d_i^*(\underline{x}) = \begin{cases} 1 & \text{if } \varphi_i^*(w_i) \leq \theta_0, \\ 0 & \text{otherwise.} \end{cases} \tag{3.5}$$

The following example describes the empirical Bayes selection procedure.

Example: Suppose that $k = 10$, $m = 5$, $N = 20$ and $\theta_0 = 0.1$. Let the data $(X_{i1}, X_{i2}, X_{i3}, X_{i4}, X_{i5})$, $i = 1, \ldots, 10$, from each of the 10 populations, be as listed below.

i	X_{i1}	X_{i2}	X_{i3}	X_{i4}	X_{i5}
1	3	5	4	7	1
2	4	1	2	8	5
3	10	4	3	2	1
4	5	0	0	10	5
5	6	0	5	6	3
6	8	8	2	1	1
7	13	1	1	1	4
8	7	8	2	3	0
9	3	2	2	9	4
10	12	2	3	1	2

Using $W_i = \sum_{j=1}^{m} X_{ij}^2$ and (3.4), the values of the empirical Bayes estimators $\varphi_i^*(W_i)$ are as follows:

i	1	2	3	4	5
W_i	100	110	130	150	106
$\varphi_i^*(W_i)$	0.04610	0.06566	0.09085	0.11603	0.06063

i	6	7	8	9	10
W_i	134	188	126	114	162
$\varphi_i^*(W_i)$	0.09588	0.16388	0.08581	0.07070	0.13114

Comparing the values of $\varphi_i^*(W_i)$ with $\theta_0 = 0.1$, the empirical Bayes rule selects populations π_1, π_2, π_3, π_5, π_6, π_8 and π_9 as fair populations. According to this rule π_4, π_7, and π_{10} are not selected.

In the following, we will investigate the asymptotic performance of the empirical Bayes selection procedure d^* for the case where k, the number of populations involved in the selection problem under study, is sufficiently large.

Since $d^{G_\alpha} = (d_1^{G_\alpha}, \ldots, d_k^{G_\alpha})$ is the Bayes selection procedure, for the empirical Bayes selection procedure $d^* = (d_1^*, \ldots, d_k^*)$, $r_i(G_\alpha, d_i^*) - r_i(G_\alpha, d_i^{G_\alpha}) \geq 0$ for each $i = 1, \ldots k$ and therefore, $r(G_\alpha, d^*) - r(G_\alpha, d^{G_\alpha}) = \sum_{i=1}^{k} [r_i(G_\alpha, d_i^*) - r_i(G_\alpha, d_i^{G_\alpha})] \geq 0$. This nonnegative regret value $r(G_\alpha, d^*) - r(G_\alpha, d^{G_\alpha})$ will be used as a measure of performance of the empirical Bayes selection procedure d^*.

DEFINITION 3.1. A selection procedure $d = (d_1, \ldots, d_k) \in D$ is said to be asymptotically optimal of order $\{\beta_k\}$ relative to the prior distribution G if

$$r(G, d) - r(G, d^G) = O(\beta_k)$$

where $\{\beta_k\}$ is a sequence of positive numbers such that $\lim_{k \to \infty} \beta_k = 0$.

For each $i = 1, \ldots, k$ and for the fixed μ_2, $Q_i(\mu_2 | w_i)$, which is defined in (3.3), can be viewed as a function of w_i. It is clear that $Q_i(\mu_2 | w_i)$ is increasing in w_i. Let

$$A_i = \{w_i | w_i = w_i(\underline{x}_i) = \sum_{j=1}^{m} x_{ij}^2,\ \underline{x}_i \in \mathcal{X}_i,\ Q_i(\mu_2 | w_i) < \theta_0\}$$

$$B_i = \{w_i | w_i = w_i(\underline{x}_i) = \sum_{j=1}^{m} x_{ij}^2,\ \underline{x}_i \in \mathcal{X}_i,\ Q_i(\mu_2 | w_i) > \theta_0\}$$

From the statistical model under consideration, $Q_i(\mu_2 | w) = Q_j(\mu_2 | w) = Q(\mu_2 | w)$ (say) for all $i, j = 1, \ldots, k$. Thus, $A_1 = \cdots = A_k = A$ (say) and $B_1 = \cdots = B_k = B$ (say).

Let $h(w)$ denote the common marginal probability function of the iid random variables $W_i = \sum_{j=1}^{m} X_{ij}^2$, $i = 1, \ldots, k$.

From (2.5), (2.6) and (3.5), straightforward computation yields

$$\begin{aligned} 0 &\leq r_i(G_\alpha, d_i^*) - r_i(G_\alpha, d_i^{G_\alpha}) \\ &\leq \sum_{w_i \in A} P_i\{\varphi_i^*(w_i) > \theta_0 | W_i = w_i\} h(w_i) \\ &\quad + \sum_{w_i \in B} P_i\{\varphi_i^*(w_i) \leq \theta_0 | W_i = w_i\} h(w_i) \end{aligned} \tag{3.6}$$

where the probability measure P_i is computed with respect to $(W_1, \ldots, W_{i-1}, W_{i+1}, \ldots, W_k)$.

Thus, it suffices to investigate the asymptotic behavior of $P_i\{\varphi_i^*(w_i) \leq \theta_0 | W_i = w_i\}$ for $w_i \in B$ and $P_i\{\varphi_i^*(w_i) > \theta_0 | W_i = w_i\}$ for $w_i \in A$.

LEMMA 3.1. For each $c > 0$ and for sufficiently large k,

(a) $P_i\{\hat{\mu}_2 - \mu_2 < -c|W_i = w_i\} \leq O(\exp(-kc^2N^{-4}(1-\frac{1}{m})^{-2}/2))$.

(b) $P_i\{\hat{\mu}_2 - \mu_2 > c|W_i = w_i\} \leq O(\exp(-kc^2N^{-4}(1-\frac{1}{m})^{-2}/2))$.

(c) $P_i\{\hat{\mu}_2 \leq N+\frac{N(N-1)}{m}|W_i = w_i\} \leq O(\exp(-k(\mu_2 - N-(N-1)N/m)^2N^{-4}(1-\frac{1}{m})^{-2}/2))$.

Note that the above upper bounds are independent of the value w_i.

PROOF: (a). Let $\hat{\mu}_2(i) = \frac{1}{k-1}\sum_{\substack{j=1\\j\neq i}}^{k} W_j$. Then,

$$P_i\{\hat{\mu}_2 - \mu_2 < -c|W_i = w_i\} = P_i\{\hat{\mu}_2(i) - \mu_2 < -\frac{kc}{k-1} + \frac{\mu_2 - w_i}{k-1}\}.$$

Note that $\frac{N^2}{m} \leq w_i \leq N^2$ for all $w_i = w_i(\underline{x}_i) = \sum_{j=1}^{m} x_{ij}^2$. Thus for k sufficiently large $-\frac{kc}{k-1} + \frac{\mu_2 - w_i}{k-1} \leq -\frac{c}{2}$. Hence, we obtain, for k sufficiently large, that

$$\begin{aligned} &P_i\{\hat{\mu}_2 - \mu_2 < -c|W_i = w_i\} \\ &\leq P_i\{\hat{\mu}_2(i) - \mu_2 < -\tfrac{c}{2}\} \\ &\leq \exp\{-kc^2N^{-4}(1-\tfrac{1}{m})^{-2}/2\}, \end{aligned}$$

where the last inequality follows from Theorem 2 of Hoeffding (1963). Note that the upper bound is independent of w_i.

The proof of part (b) is similar to that of part (a). By letting $c = \mu_2 - [N + \frac{N(N-1)}{m}]$, then $c > 0$. Thus, the result of part (c) follows directly from part (a). □

LEMMA 3.2. For $w_i \in A$,

$$P_i\{\varphi_i^*(w_i) > \theta_0|W_i = w_i\} \leq O(\exp(-k(Q_i^{-1}(\theta_0|w_i) - \mu_2)^2N^{-4}(1-\frac{1}{m})^{-2}/2))$$

where $Q_i^{-1}(\cdot|w_i)$ is the inverse function of $Q_i(\cdot|w_i)$.

PROOF: From (3.4) and the fact that $\theta_0 > 0$,

$$P_i\{\varphi_i^*(w_i) > \theta_0|W_i = w_i\} = P_i\{Q_i(\hat{\mu}_2|w_i) > \theta_0|W_i = w_i\}. \quad (3.7)$$

Now, for each fixed $w_i \in A$, $Q_i(\mu|w_i)$ is strictly increasing in μ for $N + \frac{N(N-1)}{m} < \mu < N^2$ and $Q_i(\mu_2|w_i) < \theta_0$. Thus $\mu_2 < Q_i^{-1}(\theta_0|w_i)$. Then,

$$\begin{aligned} Q_i(\hat{\mu}_2|w_i) > \theta_0 &\iff \hat{\mu}_2 > Q_i^{-1}(\theta_0|w_i) \\ &\iff \hat{\mu}_2 - \mu_2 > Q_i^{-1}(\theta_0|w_i) - \mu_2 > 0. \end{aligned} \tag{3.8}$$

Combining (3.7) and (3.8), by Lemma 3.1 (b), we obtain, for $w_i \in A$,

$$\begin{aligned} P_i\{\varphi_i^*(w_i) > \theta_0|W_i = w_i\} &= P_i\{\hat{\mu}_2 - \mu_2 > Q_i^{-1}(\theta_0|w_i) - \mu_2|W_i = w_i\} \\ &\leq O(\exp(-kb)) \end{aligned}$$

where $b = (Q_i^{-1}(\theta_0|w_i) - \mu_2)^2 N^{-4}(1 - \frac{1}{m})^{-2}/2))$.

Thus, the proof of this lemma is complete. □

LEMMA 3.3. For $w_i \in B$,

$$\begin{aligned} &P_i\{\varphi_i^*(w_i) \leq \theta_0|W_i = w_i\} \\ &\leq O(\exp(-k(Q_i^{-1}(\theta_0|w_i) - \mu_2)^2 N^{-4}(1 - \tfrac{1}{m})^{-2}/2)) \\ &+O(\exp(-k(\mu_2 - N - (N-1)N/m)^2 N^{-4}(1 - \tfrac{1}{m})^{-2}/2)). \end{aligned}$$

PROOF:

$$\begin{aligned} &P_i\{\varphi_i^*(w_i) \leq \theta_0|W_i = w_i\} \\ &= P_i\{\varphi_i^*(w_i) \leq \theta_0,\ \hat{\mu}_2 \leq N + \tfrac{N(N-1)}{m}|W_i = w_i\} \\ &\quad +P_i\{\varphi_i^*(w_i) \leq \theta_0,\ \hat{\mu}_2 > N + \tfrac{N(N-1)}{m}|W_i = w_i\}. \end{aligned} \tag{3.9}$$

From Lemma 3.1 (c),

$$\begin{aligned} &P_i\{\varphi_i^*(w_i) \leq \theta_0,\ \hat{\mu}_2 \leq N + \tfrac{N(N-1)}{m}|W_i = w_i\} \\ &\leq O(\exp(-k(\mu_2 - N - (N-1)N/m)^2 N^{-4}(1 - \tfrac{1}{m})^{-2}/2)). \end{aligned} \tag{3.10}$$

From (3.4) and an argument analogous to that given in the proof of Lemma 3.2, we have

$$\begin{aligned} &P_i\{\varphi_i^*(w_i) \leq \theta_0,\ \hat{\mu}_2 > N + \tfrac{N(N-1)}{m}|W_i = w_i\} \\ &= P_i\{Q_i(\hat{\mu}_2|w_i) \leq \theta_0,\ \hat{\mu}_2 > N + \tfrac{N(N-1)}{m}|W_i = w_i\} \\ &\leq P_i\{\hat{\mu}_2 - \mu_2 \leq Q_i^{-1}(\theta_0|w_i) - \mu_2|W_i = w_i\} \\ &= O(\exp(-k(Q_i^{-1}(\theta_0|w_i) - \mu_2)^2 N^{-4}(1 - \tfrac{1}{m})^{-2}/2)) \end{aligned} \tag{3.11}$$

where $Q_i^{-1}(\theta_0|w_i) - \mu_2 < 0$ since $w_i \in B$. Thus, the lemma follows from (3.9)–(3.11) □

Let $\tau_1 = \min\{(Q_i^{-1}(\theta_0|w_i) - \mu_2)^2 N^{-4}(1 - \frac{1}{m})^{-2}/2|w_i \in A_i \cup B_i\}$. Then, $\tau_1 > 0$ since $Q_i^{-1}(\theta_0|w_i) - \mu_2 \neq 0$ for all $w_i \in A_i \cup B_i$ and $A_i \cup B_i$ is a finite set. Then $\tau \equiv \min(\tau_1, (\mu_2 - N - (N-1)N/m)^2 N^{-4}(1 - \frac{1}{m})^{-2}/2)) > 0$.

The following theorem describes the asymptotic optimality property of the empirical Bayes selection procedure $d^* = (d_1^*, \ldots, d_k^*)$.

THEOREM 3.1. Let $d^* = (d_1^*, \ldots, d_k^*)$ be the empirical Bayes selection procedure defined through (3.4) and (3.5). Suppose that the prior is a member of the class $\mathcal{C}$ of symmetric Dirichlet distributions given in (2.3). Then

(a) For each $i = 1, \ldots, k$, $r_i(G_\alpha, d_i^*) - r_i(G_\alpha, d_i^{G_\alpha}) \le O(\exp(-\tau k))$, and

(b) $r(G_\alpha, d^*) - r(G_\alpha, d^{G_\alpha}) \le O(\exp(-\tau k + \ln k))$

where τ is the positive constant defined previously.

PROOF: Part (b) is a direct result of part (a). Thus, we need to prove part (a) only. From (3.6), and Lemmas 3.2 and 3.3,

$$\begin{aligned} & r_i(G_\alpha, d_i^*) - r_i(G_\alpha, d_i^{G_\alpha}) \\ & \le O(\exp(-\tau k)) \textstyle\sum_{w_i \in A \cup B} h(w_i) \\ & = O(\exp(-\tau k)). \end{aligned}$$

The theorem now follows. □

4 Small Sample Performance: A Simulation Study

A Monte Carlo study was designed to investigate the small sample performance of the proposed empirical Bayes selection rule d^*. The simulation scheme used is as follows:

(1) For given α and m, generate $k-1$ random vectors $\underline{p}_i = (p_{i1}, \ldots, p_{im})$, $i = 1, \ldots, k-1$, according to the symmetric Dirichlet distribution

$$g_\alpha(\underline{p}) = \frac{\Gamma(m\alpha)}{[\Gamma(\alpha)]^m} \prod_{i=1}^{m} p_i^{\alpha-1}$$

where $\underline{p} = (p_1, \ldots, p_m)$, $0 \le p_i \le 1$ and $\sum_{i=1}^{m} p_i = 1$.

(2) For each N and each $i = 1, \ldots, k-1$, generate random vector $\underline{X}_i = (X_{i1}, \ldots, X_{im})$ according to the multinomial distribution

$$f(\underline{x}|\underline{p}_i) = \frac{N!}{\prod_{j=1}^{m} x_j!} p_{ij}^{x_j},$$

where $0 \le x_j \le N$ and $\sum_{j=1}^{m} x_j = N$. Note that under the given statistical model,

$$\begin{aligned} r_1(G_\alpha, d_1^*) = r_2(G_\alpha, d_2^*) &= \ldots = r_k(G_\alpha, d_k^*), \quad \text{and} \\ r_1(G_\alpha, d_1^{G_\alpha}) = r_2(G_\alpha, d_2^{G_\alpha}) &= \ldots = r_k(G_\alpha, d_k^{G_\alpha}). \end{aligned}$$

Therefore,

$$r(G_\alpha, d^*) - r(G_\alpha, d^{G_\alpha}) = k[r_k(G_\alpha, d_k^*) - r_k(G_\alpha, d_k^{G_\alpha})].$$

Hence, it suffices to generate $r_k(G_\alpha, d_k^*) - r_k(G_\alpha, d_k^{G_\alpha})$ only.

(3) Let $\underline{x}_k$ be an observed vector obtained from a population with probability function $f(\underline{x}) = \int f(\underline{x}|\underline{p})g_\alpha(\underline{p})d\underline{p}$. Use $\underline{X}_1, \ldots, \underline{X}_{k-1}$ and $\underline{x}_k$ to compute $\varphi_k^*(w_k)$ and $d_k^*(\underline{X}_1, \ldots, \underline{X}_{k-1}, \underline{x}_k)$ according to (3.3), (3.4) and (3.5). Then compute

$$\begin{aligned} & r_k(G_\alpha, d_k^*|\underline{X}_1, \ldots, \underline{X}_{k-1}) - r_k(G_\alpha, d_k^{G_\alpha}) \\ \equiv\ & \sum_{\underline{x}_k} d_k^*(\underline{X}_1, \ldots, \underline{X}_{k-1}, \underline{x}_k)[Q_k(\mu_2|w_k) - \theta_0]f(\underline{x}_k) \\ & -\sum_{\underline{x}_k} d_k^{G_\alpha}(\underline{X}_1, \ldots, \underline{X}_{k-1}, \underline{x}_k)[Q_k(\mu_2|w_k) - \theta_0]f(\underline{x}_k) \\ \equiv\ & D_k(\underline{X}_1, \ldots, \underline{X}_{k-1}). \end{aligned}$$

The quantity $D_k(\underline{X}_1, \ldots, \underline{X}_{k-1})$ can be viewed as the difference between the conditional Bayes risk of the empirical Bayes selection rule d_k^* given $(\underline{X}_1, \ldots, \underline{X}_{k-1})$ and the minimum Bayes risk $r_k(G_\alpha, d_k^{G_\alpha})$.

(4) The process was repeated 500 times. The average of the differences based on the 500 repetitions denoted by $\overline{D}_k$ is used as an estimator of the difference $r_k(G_\alpha, d_k^*) - r_k(G_\alpha, d_k^{G_\alpha})$. We then use $k\overline{D}_k$ as an estimator of the overall difference $r(G_\alpha, d^*) - r(G_\alpha, d^{G_\alpha})$.

Tables 1–3 list simulation results on the small sample performance of the empirical Bayes selection rule d^* for selected values of the parameters k, m, α and N. $SE(\overline{D}_k)$ denotes the estimated standard error $\overline{D}_k$.

In each of the three tables, the results indicate that the estimated overall difference $k\overline{D}_k$ decreases as k becomes large. However, the rate of convergence to zero depends on the value of the parameter α. For $\alpha = 4$ we see that for $k \geq 50$, $\overline{D}_k$ is very close to zero. This indicates that the asymptotic optimality of the empirical Bayes selection rule d^* may also hold for moderate sample sizes.

Table 1. *Small Sample Performance of d^**
$\alpha = 2,\ m = 5,\ N = 10$ *and* $\theta_0 = 0.3$ [1]

k	$\overline{D}_k$	$SE(\overline{D}_k)$	$k\overline{D}_k$
10	3.6898(−5)	5.883(−6)	36.898(−5)
20	0.7199(−5)	1.254(−6)	14.398(−5)
30	0.4678(−5)	0.595(−6)	14.034(−5)
40	0.2892(−5)	0.479(−6)	11.568(−5)
50	0.2297(−5)	0.430(−6)	11.485(−5)
60	0.1871(−5)	0.390(−6)	11.226(−5)
70	0.0595(−5)	0.223(−6)	4.165(−5)
80	0.0851(−5)	0.266(−6)	6.808(−5)
90	0.0595(−5)	0.223(−6)	5.355(−5)
100	0.0425(−5)	0.189(−6)	4.25 (−5)

Table 2. *Small Sample Performance of d^**
$\alpha = 3,\ m = 5,\ N = 15$ *and* $\theta_0 = 0.3$

k	D_k	$SE(\overline{D}_k)$	$k\overline{D}_k$
10	0.1980(−5)	0.483(−6)	1.980(−5)
20	0.0715(−5)	0.250(−6)	1.430(−5)
30	0.0043(−5)	0.008(−6)	0.129(−5)
40	0.0070(−5)	0.031(−6)	0.280(−5)
50	0.0014(−5)	0.004(−6)	0.070(−5)
60	0.0011(−5)	0.004(−6)	0.066(−5)
70	0.0007(−5)	0.003(−6)	0.049(−5)
80	0.0004(−5)	0.002(−6)	0.032(−5)
90	0.0004(−5)	0.002(−6)	0.036(−5)
100	0.0001(−5)	0.001(−6)	0.010(−5)

Table 3. *Small Sample Performance of d^**
$\alpha = 4,\ m = 5,\ N = 15$ *and* $\theta_0 = 0.3$

k	$\overline{D}_k$	$SE(\overline{D}_k)$	$k\overline{D}_k$
10	0.0235(−5)	0.090(−6)	0.235(−5)
20	0.0036(−5)	0.017(−6)	0.072(−5)
30	0.0003(−5)	0.001(−6)	0.009(−5)
40	0.0010(−5)	0.010(−6)	0.040(−5)
50	0.	0.	0.
60	0.	0.	0.
70	0.	0.	0.
80	0.	0.	0.
90	0.	0.	0.
100	0.	0.	0.

[1] An entry such as 3.6898 (-5) stands for 3.6898×10^{-5}.

Acknowledgments

This research was supported in part by the office of Naval Research Contract N00014-88-K-0170 and NSF Grants DMS-8702620, DMS-8923071 and DMS-8717799 at Purdue University. The authors are grateful to a referee for helpful suggestions to improve the presentation of the article.

REFERENCES

Alam, K., Mitra, A., Rizvi, M. H. and Saxena, K. M. L. (1986). Selection of the most diverse multinomial population. *American Journal of Mathematical and Management Sciences*, Vol. **6**, 65–86.

Dudewicz, E. J. and Van der Meulen, E. C. (1981). Selection procedures for the best binomial population with generalized entropy goodness. *Tamkang Journal of Mathematics*, **12**, 206–208.

Gini, C. (1912). Variabilita e Mutabilitá. *Studi Economico-aguridici della facotta di Giurisprudegza dell*, Universite di Cagliari, III, Part II.

Gupta, S. S. and Huang, D. Y. (1976). On subset selection procedures for the entropy function associated with the binomial populations. *Sankhyā*, **38A**, 153–173.

Gupta, S. S. and Leu, L. Y. (1990). Selecting the fairest of $k(\geq 2)$ m-sided dice. *Communications in Statistics – Theory and Methods*, **A(19)**, 2159–2177.

Gupta, S. S. and Wong, W. Y. (1975). Subset selection procedures for finite schemes in information theory. *Colloquia Mathematica Societatis János Bolyai*, **16**, 279–291.

Hoeffding, W. (1963). Probability inequalities for sums of bounded random variables. *Journal of American Statistical Association*, **58** 13–30.

Liang, T. and Panchapakesan, S. (1991). An empirical Bayes procedure for selecting the most homogeneous multinomial population according to the Gini-Simpson index. *Proceedings of the 1990 Taipei Symposium in Statistics June 28–30, 1990*, ed. by M. T. Chao and P. E. Cheng, Institute of Statistical Science, Academia Sinica, 447–460.

Rizvi, M. H., Alam, K. and Saxena, K. M. L. (1987). Selection procedure for multinomial populations with respect to diversity indices. *Contributions to the Theory and Application of Statistics*, A Volume in Honor of Herbert Solomon, ed. by Alan E. Gelfand, Academic Press, Inc. 485–509.

Shannon, C. E. (1948). A mathematical theory of communication. *Bell System Technical Journal*, **27**, 379–423, 623–656.

Simpson, E. H. (1949). Measurement of diversity. *Nature*, **163**, 688.

Department of Statistics
Purdue University
West Lafayette, IN 47907
U.S.A.

Department of Mathematics
Wayne State University
Detroit, MI 48202
U.S.A.

J.K. Ghosh, S.K. Mitra, K.R. Parthasarathy and B.L.S. Prakasa Rao (*eds.*),
Statistics and Probability : A Raghu Raj Bahadur Festschrift
Published by Wiley Eastern Limited, Publishers, (1993), 279 – 292

Asymptotic Equivalence of Regression Rank Scores Estimators and R-estimators in Linear Models

JANA JUREČKOVÁ
Charles University, Prague

and

PRANAB KUMAR SEN
University of North Carolina at Chapel Hill

Abstract

The classical R-estimators in linear models are based on (nonlinear) estimating equations involving a vector of aligned regression rank statistics, while the regression rank scores estimators are based on appropriate linear programming algorithms and are regression invariant. Computationally, the R-estimators may generally be more cumbersome than the other ones (particularly, for larger number of regressors). Under appropriate regularity conditions, both the methods are shown to be asymptotically equivalent. Also, a coordinate-wise modification of the regression rank scores estimators is considered.

AMS Subject Classifications : 62G05, 62J05

Keywords and Phrases : Coordinate-wise estimation; estimating equation; linear programming; linear rank statistics; multi-sample model; regression quantiles; uniform asymptotic linearity; weak Bahadur representation.

1 Introduction

Consider the usual linear (regression) model:

$$\mathbf{Y} = (Y_1, \cdots, Y_n)' = \mathbf{X}^0_{(n)}\beta^0 + \mathbf{e}_n; \ \mathbf{e}_n = (e_1, \cdots, e_n)', \tag{1.1}$$

where

$$\mathbf{X}^0_n = (\mathbf{1}_n, \mathbf{X}_n) = (\mathbf{1}_n, \mathbf{X}_{n1}, \mathbf{X}_{n2}); \mathbf{1}_n = (1, \cdots, 1)', \tag{1.2}$$

$\mathbf{X}_{n1}$ and $\mathbf{X}_{n2}$ are $n \times p$ and $n \times q$ matrices of known regression constants,

$$\boldsymbol{\beta}^0 = (\beta_0, \boldsymbol{\beta}')' = (\beta_0, \boldsymbol{\beta}_1', \boldsymbol{\beta}_2')' \tag{1.3}$$

is a vector of unknown parameters, and the e_i are independent and identically distributed (i.i.d.) random variables (r.v.) having a continuous distribution function (d.f.) F, defined on IR_1. The model in (1.1) may also be expressed as

$$\mathbf{Y} = \beta_0\mathbf{1}_n + \mathbf{X}_n\boldsymbol{\beta} + \mathbf{e}_n = \beta_0\mathbf{1}_n + \mathbf{X}_{n1}\boldsymbol{\beta}_1 + \mathbf{X}_{n2}\boldsymbol{\beta}_2 + \mathbf{e}_n. \tag{1.4}$$

In this setup, treating β_0 and $\boldsymbol{\beta}_1$ as nuisance parameters, we intent to construct rank-based estimators of $\boldsymbol{\beta}_2$.

Motivated by the work of Adichie (1967) (for $p = 0$, $q = 1$), Jurečková (1969, 1971a,b) and Koul (1970) constructed different versions of rank (R-) estimators of $\boldsymbol{\beta}$, while Jaeckel (1972) proposed another version of R-estimators minimizing a specific measure of rank dispersion; all these methods are asymptotically equivalent. Heiler and Willers (1988), using a convexity argument, eliminated one of the regularity conditions on X_n. Note that the ranks are translation–invariant, and hence, to estimate β_0, one may need to use signed rank statistics [Jurečková (1971b)]. Moreover, in this approach, to estimate $\boldsymbol{\beta}_2$, one needs to estimate the full vector $\boldsymbol{\beta}$, so that from the estimation point of view, there is no tactical gain in treating $\boldsymbol{\beta}$ as a nuisance parameter. Simultaneous estimation of $\boldsymbol{\beta}_1$, particularly, for $p+q$ not so small, generally entails serious computational complexities, and this is one of the reasons why the R-estimators inspite of having global robustness properties, have not met the light of applications.

Based on the concept of regression rank scores developed in Gutenbrunner (1986) and Gutenbrunner and Jurečková (1992), an estimator of $\boldsymbol{\beta}_2$ has been proposed by Jurečková (1991). The regression rank scores are extensions of ordinary rank scores to linear models, and compared to the translation–invariance of the usual ranks, they are regression-invariant. This makes the corresponding estimator computationally simpler than the usual R-estimators.

Our primary task is to establish an asymptotic equivalence of regression rank scores estimators and R-estimators when they are based on a common score function. This enables us to examine their relative computational complexities without a compromise on their asymptotic efficiency and related properties. This also leads us to an idea of estimating $\boldsymbol{\beta}_2$ component–wise (based on regression rank scores) without distorting the mentioned before asymptotic equivalence results.

Besides the computational aspects, we have a diagnostic point of view too. Avoiding direct estimation of $\boldsymbol{\beta}_1$ may also avoid eventual leverage on influential points in $\mathbf{X}_{n1}$. This feature is shared by the subhypothesis testing problem (H_0 : $\boldsymbol{\beta}_2 = 0$ vs. $\boldsymbol{\beta}_2 \neq 0$), as has been treated in Grutenbrunner et al. (1992).

Jaeckel's estimator, regression rank scores estimators and Jurečková estimators are all introduced in Section 2. Their asymptotic equivalence, as further extended to a coordinate–wise estimator, is established in Section 3. The k-sample model is treated as an illustration in Section 4.

2 Alternative Rank Based Estimators

We write $\mathbf{X}_n = (\mathbf{x}_1, \cdots, \mathbf{x}_n)'$. For $\mathbf{b} \in I\!R_{p+q}$, let $R_{ni}(\mathbf{b})$ be the rank of $Y_i - \mathbf{x}_i'\mathbf{b}$ among $Y_1 - \mathbf{x}_1'\mathbf{b}, \cdots, Y_n - \mathbf{x}_n'\mathbf{b}$, for $i = 1, \cdots, n$. Let $a_n(i)$, $i = 1, \cdots, n$ be monotone scores derived from a nondecreasing, square-integrable score function $\varphi : [0,1] \to I\!R_1$, in either of the following manner:

$$a_n(i) = E\varphi(U_{n:i}) \text{ or } \varphi\left[\frac{i}{n+1}\right] \text{ or } n\int_{(i-1)/n}^{i/n} \varphi(t)dt, \; i = 1, \cdots, n, \tag{2.1}$$

where $U_{n:1} \leq \cdots \leq U_{n:n'}$ are the order statistics of a sample size n from the uniform (0,1)d.f. Let $\mathbf{S}_n(\mathbf{b}) = (S_{n1}(\mathbf{b}), \cdots, S_{n,p+q}(\mathbf{b}))'$ be a vector of linear rank statistics of the form

$$\mathbf{S}_n(\mathbf{b}) = \sum_{i=1}^{n} (\mathbf{x}_i - \bar{\mathbf{x}}_n) a_n(R_{ni}(\mathbf{b})), \; \mathbf{b} \in I\!R_{p+q}, \tag{2.2}$$

where $\bar{\mathbf{x}}_n = n^{-1} \sum_{i=1}^{n} \mathbf{x}_i = (\bar{x}_{n1}, \cdots, \bar{x}_{n,p+q})'$. Jurečková (1969, 1971a) established the 'uniform asymptotic linearity of $\mathbf{S}_n(\mathbf{b})$ in $\mathbf{b}$ in an $n^{-\frac{1}{2}}$–neighbourhood of $\boldsymbol{\beta}$, and incorporated the same in the estimation of $\boldsymbol{\beta}$ based on a L_1-norm on $\mathbf{S}_n(\mathbf{b})$. Specifically, the estimator is

$$\hat{\boldsymbol{\beta}}_n = \text{Arg. min} \left\{ \sum_{j=1}^{p+q} |S_{nj}(\mathbf{b})| : \mathbf{b} \in I\!R_{p+q} \right\}. \tag{2.3}$$

The related asymptotic distribution theory was developed under appropriate regularity conditions on F, φ and $\mathbf{X}_n$. Heiler and Willers (1988) weakened a concordance-discordance condition on $\mathbf{X}_n$ and derived the same asymptotic linearity results via a convexity argument, so that the asymptotic multinormality result for $\hat{\boldsymbol{\beta}}_n$ holds even under this weakened condition. Jaeckel (1972) proposed to estimate $\boldsymbol{\beta}$ by minimizing the dispersion function:

$$D_n(\mathbf{b}) = \sum_{i=1}^{n} \{a_n(R_{ni}(\mathbf{b})) - \overline{a}_n\} (Y_i - \mathbf{x}_i'\mathbf{b}), \ \mathbf{b} \in I\!R_{p+q}, \tag{2.4}$$

which is continuous, piecewise linear and convex in $\mathbf{b}$; here $\overline{a}_n = n^{-1} \sum_{i=1}^{n} a_n(i)$. The vector of derivatives of $D_n(\mathbf{b})$ with respect to the components of $\mathbf{b}$ coincides with $(-1)\mathbf{S}_n(\mathbf{b})$ whenever it exists. Hence, under the regularity conditions of Heiler and Willers (1988), the minimization of (2.3) and (2.4) are equivalent. As such, denoting $\hat{\boldsymbol{\beta}}_n' = (\hat{\boldsymbol{\beta}}_{n1}', \hat{\boldsymbol{\beta}}_{n2}')$, we may use the component $\hat{\boldsymbol{\beta}}_2$ for estimating $\boldsymbol{\beta}_2$. However, this involves the simultaneous estimation of $\boldsymbol{\beta}_1$ and $\boldsymbol{\beta}_2$, and as no algebraic expression is generally available for $\hat{\boldsymbol{\beta}}$, an iterative procedure may have to be employed. The computational scheme becomes quite cumbersome, especially when $p+q$ is not small.

Jurečková (1991) proposed another estimator of $\boldsymbol{\beta}_2$ based on regression rank scores. To pose this estimator, we consider the full model $\mathbf{Y} = \mathbf{X}_n^0 \boldsymbol{\beta}^0 + \mathbf{e}$. For $\alpha : 0 < \alpha < 1$, let $\hat{\boldsymbol{\beta}}_n^0(\alpha)$ denote the α-regression quantile, proposed by Koenker and Bassett (1978), which can be expressed as a solution of the minimization of

$$\sum_{i=1}^{n} \rho_\alpha(Y_i - \mathbf{x}_i^{0'}\mathbf{b}), \ \mathbf{b} \in I\!R_{p+q+1}, \tag{2.5}$$

where $\mathbf{x}_i^{0'} = (1, \mathbf{x}_i')$, $i = 1, \cdots, n$, and

$$\rho_\alpha(x) = |x| \{\alpha I(x > 0) + (1-\alpha) I(x < 0)\}, \ x \in I\!R_1. \tag{2.6}$$

Koenker and Bassett also characterized $\hat{\boldsymbol{\beta}}_n^0(\boldsymbol{\alpha})$ as the component of $\boldsymbol{\beta}^0$ in the optimal solution $(\boldsymbol{\beta}^0, \mathbf{r}^+, \mathbf{r}^-)$ of the linear programme

$$\begin{aligned} \alpha \mathbf{1}_n'\mathbf{r}^+ + (1-\alpha)\mathbf{1}_n'\mathbf{r}^- &:= \min, \mathbf{X}_n^0\boldsymbol{\beta}^0 + \mathbf{r}^+ - \mathbf{r}^- = \mathbf{Y} \\ (\boldsymbol{\beta}^0, \mathbf{r}^+, \mathbf{r}^-) &\in I\!R_{p+q+1} \times I\!R_n^+ \times I\!R_n^+, \ 0 < \alpha < 1. \end{aligned} \tag{2.7}$$

Such regression quantiles provide a basis for L-estimation in linear models and are themselves extensions of sample quantiles to linear models. We may refer to the work of Ruppert and Carroll (1980), Jurečková (1984), Koenker and Portnoy (1987), Gutenbrunner and Jurečková (1992), among others. The dual programme to (2.7) can be written in the following manner:

$$\begin{aligned} \mathbf{Y}'\hat{\mathbf{a}}(\alpha) &:= \max, \ \mathbf{X}_n^{0'}\hat{\mathbf{a}}(\alpha) = (1-\alpha)\mathbf{X}_n^{0'}\mathbf{1}_n; \\ \hat{\mathbf{a}}(\alpha) &\in [0,1]^n, \ 0 < \alpha < 1. \end{aligned} \tag{2.8}$$

In the location model [i.e. $Y_i = \beta_0 + e_i$, $i \geq 1$], the optimal solution $\hat{\mathbf{a}}_n(\alpha) = (\hat{a}_{n1}(\alpha), \cdots, \hat{a}_{nn}(\alpha))'$ coincides with the vector of rank scores $(a_n^*(R_i, \alpha),$

$i = 1, \cdots, n$), where

$$a_n^*(R_i, \alpha) = \begin{cases} 1, & \text{if } 0 \le \alpha < (R_i - 1)/n, \\ a_n^*(R_i, \alpha) = R_i - n\alpha, & \text{if } (R_i - 1)/n \le \alpha \le R_i/n, \\ 0, & \text{if } R_i/n < \alpha \le 1, \end{cases} \tag{2.9}$$

and R_i is the rank of Y_i among $Y_1, \cdots, Y_n$, for $i = 1, \cdots, n$ [viz., Hájek and Šidák (1967, Ch. V)]. For this reason, for the linear model too, we find it convenient to call the $\hat{a}_{ni}(\alpha)$, $1 \le i \le n$, for $0 < \alpha < 1$, the regression rank scores. We may refer to Gutenbrunner and Jurečková (1992) for a detailed study of various properties of such regression rank scores and other related references.

An important property of regression rank scores is their invariance with respect to the $\mathbf{X}_n^0$-regression, i.e.,

$$\hat{\mathbf{a}}_n(\alpha, \mathbf{Y} + \mathbf{X}_n^0 \mathbf{b}) = \hat{\mathbf{a}}_n(\alpha, \mathbf{Y}), \ \forall\, \mathbf{b} \in \mathbb{R}_{p+q+1}, \tag{2.10}$$

and this plays a basic role in the developments to follow. Moreover, $\hat{a}_{ni}(\alpha)$ is a continuous function of bounded variation and is piecewise linear to $[0, 1]$, for every i $(= 1, \cdots, n)$. Relations of regression quantiles and regression rank scores, following from the duality of linear programmes, extend the duality of order statistics and ranks from the location to the linear model. As such, statistical inference based on regression rank scores is typically based on *linear regression rank scores statistics*, which may be defined as follows: Take a nondecreasing and square integrable score generating function $\varphi : (0, 1) \to \mathbb{R}_1$ and define the *scores* as

$$\hat{b}_{ni} = -\int_0^1 \varphi(\alpha) d\hat{a}_{ni}(\alpha), \ i = 1, \cdots, n, \tag{2.11}$$

and the linear regreesion rank scores statistics as

$$Z_n = \sum_{i=1}^{n} c_{ni} \hat{b}_{ni}, \tag{2.12}$$

where $c_{n1}, \cdots, c_{nn}$ are given coefficients.

Gutenbrunner et al. (1992) constructed a class of tests of the hypothesis $H_0 : \boldsymbol{\beta}_2 = 0$ (vs. $\boldsymbol{\beta}_2 \ne 0$) based on quadratic forms of a vector of statistics of the type (2.12). Such tests do not explicitly involve the estimation of the nuisance parameters β_0, $\boldsymbol{\beta}_1$, whereas the usual aligned rank tests, studied in detail by Sen and Puri (1977) and Adichie (1978, 1984), among others, rest on the elimination of nuisance parameters through R–estimation.

For $\mathbf{t} \in \mathbb{R}_q$, consider the pseudo-observation $Y_i - \mathbf{x}_{i2}'\mathbf{t}$, $1 \le i < n$, where $\mathbf{x}_{i2}'$ is the ith row of $\mathbf{X}_{n2}$, $1 \le i < n$. Let $\hat{a}_{ni}$ $(\alpha, \mathbf{Y} - \mathbf{X}_{n2}\mathbf{t})$, $i = 1, \cdots, n$, be the regression rank scores calculated from (2.8) with the Y_i replaced by $Y_i - \mathbf{x}_{i2}'\mathbf{t}$ and $\mathbf{X}_n^0$ replaced by $\mathbf{X}_{n1}^0 = (\mathbf{1}_n, \mathbf{X}_{n1})$ (so that

$\boldsymbol{\beta}_1^0 = (\beta_0, \boldsymbol{\beta}_1')' \in \mathbb{R}_{p+1}$. Jurečková (1991) proposed to estimate $\boldsymbol{\beta}_2$ by minimizing $D_n(\mathbf{t})$ with respect to $\mathbf{t} \in \mathbb{R}_q$, where

$$D_n(\mathbf{t}) = \sum_{i=1}^{n}(Y_i - \mathbf{t}'\mathbf{x}_{i2})\left[\hat{b}_{ni}(\mathbf{Y} - \mathbf{X}_{n2}\mathbf{t}) - \overline{\varphi}\right], \tag{2.13}$$

where

$$\overline{\varphi} = \int_0^1 \varphi(\alpha)d\alpha. \tag{2.14}$$

Note that by the second part of equation (2.8), $n^{-1}\sum_{i=1}^{n}\hat{a}_{ni}(\alpha) = 1 - \alpha$, so that $n^{-1}\sum_{i=1}^{n}\hat{b}_{ni} = \overline{\varphi}$, $\forall n \geq 1$. Thus, we have

$$\overline{\boldsymbol{\beta}}_n^{(2)} = \text{Arg. min}\left\{D_n(\mathbf{t}) : \mathbf{t} \in \mathbb{R}_q\right\}. \tag{2.15}$$

The function $D_n(\mathbf{t})$ corresponds to Jaeckel's function in (2.4) (albeit with possibly different score functions), and hence, is continuous, piece-wise linear and convex in $\mathbf{t} \in \mathbb{R}_q$. It is differentiable (with respect to $\mathbf{t}$) a.e., and the vector of its derivatives coincides, whenever it exists, with a special vector of linear regression rank scores statistics. Jurečková (1992a) showed that the latter statistics are 'uniformly asymptotically linear in $\mathbf{t}'$ and this, in turn, enables us to derive the asymptotic distribution of $n^{\frac{1}{2}}\left(\overline{\boldsymbol{\beta}}_n^{(2)} - \boldsymbol{\beta}^{(2)}\right)$.

It may be remarked that the linear regression rank scores statistics in (2.12) involve the scores $\hat{b}_{ni}$, which, in turn, depend on φ and the $\hat{a}_{ni}(\alpha)$. On the other hand, for every $\alpha\,(0 < \alpha < 1)$, the scores $\hat{a}_{ni}(\alpha)$ as obtained from (2.8) involves (i) $\mathbf{Y} - \mathbf{X}_{n2}\mathbf{t}$ instead of $\mathbf{Y}$ and (ii) $\mathbf{X}_{n1}^0$ instead of $\mathbf{X}_n^0$. Since (i) involves a linear function of $\mathbf{t}$ and (ii) involves a subset of $\mathbf{X}_n^0$, computational complications are less than in (2.8) (when $q > 1$). Note also that $D_n(\mathbf{t})$ and hence, $\hat{\boldsymbol{\beta}}_{n2}$ are invariant to the $\mathbf{X}_{n1}^0$-regression. Further, if in (2.13) we work with $\mathbf{X}_{n2} = \mathbf{X}_n$ (i.e., $p = 0$), then we have the Jaeckel estimation of $\boldsymbol{\beta}$, so that it involves comparatively more involved computations. But in either case, we treat β_0 as a nuisance parameter, and its estimation can be accomplished by the use of signed-ranks, as has been done in Jurečková (1971b).

Our main contention is to show that under suitable regularity conditions, $\hat{\boldsymbol{\beta}}_{n2}$ and $\overline{\boldsymbol{\beta}}_{n2}$ are asymptotically (efficiency-) equivalent, and hence, our choice between them can be based mainly on computational and diagnostic aspects. Moreover, one may also estimate $\boldsymbol{\beta}_2$ coordinatewise, applying Jurečková's (1991) procedure successively for single components of $\boldsymbol{\beta}_2$ (so that there are $p + q$ nuisance parameter at each stage). Thus, this will involve q linear programmes, each followed by a minimization of a function $D_n(t)$ but with respect to a single argument, so that computationally this will be simpler. On the other hand, the linear programming algorithm will involve $p + q$ instead of $p + 1$ arguments, and may be somewhat more complicated. These opposing views need to be reconciled in choosing such a coordinatewise approach, which will be shown to be asymptotically equivalent to the procedure outlinead in (2.13)-(2.15).

3 Comparison of Estimations

Consider the model in the form (1.4), in which we want to estimate $\boldsymbol{\beta}_2$. Because all estimator under consideration are invariant to the translation, they will involve matrices $(\mathbf{I}_n - \mathbf{H})\mathbf{X}_{n1}$, $(\mathbf{I}_n - \mathbf{H})\mathbf{X}_{n2}$ where $\mathbf{H} = n^{-1}(\mathbf{1}_n, \cdots, \mathbf{1}_n)$ is the $(n \times n)$ projection matrix corresponding to $\mathbf{1}_n$. Hence, for simplicity of notation, we shall assume that

$$\mathbf{1}'_n\mathbf{X}_{n1} = \mathbf{0}',\ \mathbf{1}_n\mathbf{X}_{n2} = 0' \tag{3.1}$$

If we want to compare the two types of estimators, we need to assume that our model satisfies a combined set of regularity conditions imposed in Jaeckel (1972), Heiler and Willers (1988) and Jurečková (1991). However, based on some very recent results of Jurečková (1992b), we are able to relax these regularity conditions a bit more. As such, we impose the following regularity conditions:

(F.1) The d.f. F has an absolutely continuous density f which is positive and bounded for $A < x < B$ and monotonically decreasing as $x \to A^+$ or $x \to B^-$, and f' is bounded on (A, B), where $A = \sup\{x : F(x) = 0\}$ and $B = \inf\{x : F(x) = 1\}$.

(F.2) For some $a > 0$,

$$\sup_{0<\alpha<1} \left\{\alpha(1-\alpha)\left|f'(F^{-1}(\alpha))\right| / f^2(F^{-1}(\alpha))\right\} \leq 1 + a.$$

Also, let

$$\alpha_n^* = n^{-1/(1+2b)} \text{ and } \sigma_\alpha = \{\alpha(1-\alpha)\}^{\frac{1}{2}} / f(F^{-1}(\alpha)),$$

where b is a fixed number satisfying $0 < \eta < b - a < a + \eta$, for some $\eta > 0$. Note that α_n^* converges to 0 as $n \to \infty$, and

$$\left|f'(F^{-1}(\alpha))\right| \Big/ f(F^{-1}(\alpha)) \leq (1+a)\{\alpha(a-\alpha)\}^{-\frac{1}{2}} \Big/ \sigma_\alpha, \forall\alpha \in (0,1),$$

so that the connection with finite Fisher information may easily be detected by imposing suitable integrability condition on σ_α^{-1}. However, we do not need this explicitly, and hence, we shall refrain from such a condition.

(X.1) $\lim_{n\to\infty} \mathbf{Q}_n = \mathbf{Q}$ where $\mathbf{Q}_n = n^{-1}\mathbf{X}'_n\mathbf{X}_n$ and $\mathbf{Q}$ is a positive definite (p.d.) $(p+q) \times (p+q)$ matrix.

(X.2) $\max_{1\leq i\leq n} \|\mathbf{x}_i\| = 0(n^{\tau(a,b,\eta)})$ where $\tau(a, b, \eta) = (2(b-a) - \eta)/(1+4b)$, $\eta > 0$ and $b > 0$ is such that $0 < b - a < \epsilon/2$, $\epsilon > 0$ (hence $0 < b < \frac{1}{4} - \frac{\epsilon}{2}$). Moreover,

$$n^{-1}\sum_{i=1}^{n} \|\mathbf{x}_i\|^4 = O(1).$$

(X.3) $\lim_{n\to\infty} \mathbf{Q}_n^{(2)} = \mathbf{Q}^{(2)}$ where $\mathbf{Q}^{(2)}$ is p.d. and

$$\mathbf{Q}_n^{(2)} = n^{-1}\Big[\mathbf{X}_{n2} - \hat{\mathbf{X}}_{n2}\Big]'\Big[\mathbf{X}_{n2} - \hat{\mathbf{X}}_{n2}\Big]$$

and where

$$\hat{\mathbf{X}}_{n2} = \mathbf{X}_n^0(\mathbf{X}_n^{0'}\mathbf{X}_n^0)^{-1}\mathbf{x}_n^{0'}\mathbf{X}_{n2}$$

is the projection of $\mathbf{X}_{n2}$ on the space spanned over the columns of $\mathbf{X}_{n0}$.

(φ.1) $\varphi : (0,1) \to I\!R_1$ is a nondecreasing, square-integrable function which has a derivative $\varphi'(\alpha)$ for $0 < \alpha < \alpha_0$ and $1 - \alpha_0 < \alpha < 1$ satisfying

$$|\varphi'(\alpha)| \le c(\alpha(1-\alpha))^{-1-\eta^*} \quad \text{with } 0 < \eta^* < \eta. \tag{3.2}$$

For any $\mathbf{b} \in I\!R_{p+q}$, let $R_{ni}(\mathbf{b})$ denote the rank of pseudo observation $\mathbf{Y}_i - \mathbf{b}'\mathbf{x}_i$, $i = 1, \cdots, n$. Calculate the scores generated by φ in either of the ways in (2.1) and the Jaeckel estimator $\hat{\boldsymbol{\beta}}_n$ of $\boldsymbol{\beta}$, consider its part $\hat{\boldsymbol{\beta}}_{n2}$ as an estimator of $\boldsymbol{\beta}_2$.

Moreover, for $\mathbf{t} \in I\!R_q$, calculate the regression rank scores $\hat{a}_n(\alpha, \mathbf{Y} - \mathbf{X}_{n2}\mathbf{t})$ from (2.8) with $\mathbf{Y}$ replaced by $\mathbf{Y} - \mathbf{X}_{n2}\mathbf{t}$ and then calculate the scores

$$\hat{b}_{ni}(\mathbf{Y} - \mathbf{X}_{n2}\mathbf{t}) = -\int_0^1 \varphi(\alpha)d\hat{a}_{ni}(\alpha\mathbf{Y} - \mathbf{X}_{2n}\mathbf{t}) \tag{3.3}$$

$i = 1, \cdots, n$. The Jurečková estimator $\hat{\boldsymbol{\beta}}_2$ then minimizes $D_n(\mathbf{t})$ in (2.13) in $\mathbf{t} \in I\!R_q$.

The following theorem gives the Bahadur-type representations for both $\hat{\boldsymbol{\beta}}_{n2}$ and $\overline{\boldsymbol{\beta}}_{n2}$ which yield their asymptotic equivalence and distributions.

THEOREM 3.1. *Assume that F, $\mathbf{X}$ and φ satisfy the above conditions and that*

$$0 < \gamma = -\int_0^1 \varphi(\alpha)df(F^{-1}(\alpha)) < \infty. \tag{3.4}$$

Then, as $n \to \infty$,

$$n^{\frac{1}{2}}(\hat{\boldsymbol{\beta}}_{n2} - \boldsymbol{\beta}_2) = n^{-\frac{1}{2}}\gamma^{-1}\left(\mathbf{Q}^{(2)}\right)^{-1}\sum_{i=1}^{n}(\mathbf{x}_{i2} - \hat{x}_{i2})\varphi(F(e_i)) + o_p(1) \tag{3.5}$$

and

$$n^{\frac{1}{2}}(\tilde{\boldsymbol{\beta}}_{n2} - \boldsymbol{\beta}_2) = n^{-\frac{1}{2}}\gamma^{-1}(\mathbf{Q}^{(2)})^{-1}\sum_{i=1}^{n}(\mathbf{x}_{i2} - \hat{\mathbf{x}}_{i2})\varphi(F(e_i)) + o_p(1) \tag{3.6}$$

with $\mathbf{x}_{i2}$, *and* $\hat{\mathbf{x}}_{i2}$, *being the i-th row of* $\mathbf{X}_{n2}$ *and* $\hat{\mathbf{X}}_{n2}$, *respectively,* $i = 1, \cdots, n$. *Consequently,*

$$n^{\frac{1}{2}} \left\| \hat{\boldsymbol{\beta}}_{n2} - \tilde{\boldsymbol{\beta}}_{n2} \right\| \xrightarrow{P} 0, \text{ as } n \to \infty \tag{3.7}$$

and both $n^{\frac{1}{2}}(\hat{\boldsymbol{\beta}}_{n2} - \boldsymbol{\beta}_2)$ *and* $n^{\frac{1}{2}}(\tilde{\boldsymbol{\beta}}_{n2} - \boldsymbol{\beta}_2)$ *have the same asymptotic normal distribution*

$$N_q(\mathbf{0}, (\mathbf{Q}^{(2)})^{-1} A^2(\varphi, F)\gamma^{-2}) \tag{3.8}$$

where $A^2(\varphi, F) = \int_0^1 \varphi^2(\alpha)d\alpha - \varphi^{-2}$..

PROOF. Note that by (F.2) and (3.2), $\gamma < \infty$, even without requiring that f has a finite Fisher information. Actually, it suffices to assume [instead of (3.4)] that $\int_0^1 \left(|\varphi(\alpha)| / \left\{ \sqrt{\alpha(1-\alpha)}\sigma_\alpha \right\} \right) d\alpha < \infty$. Moreover, using the convexity argument in Heiler and Willers (1988), one can directly modify the basic result of Jurečková (1971, 1977), and obtain that under the assumed regularity conditions,

$$n^{\frac{1}{2}}(\hat{\boldsymbol{\beta}}_n - \boldsymbol{\beta}) = \gamma^{-1}\mathbf{Q}_n^{-1}\mathbf{S}_n(\boldsymbol{\beta}) + o_p(1) \tag{3.9}$$

where

$$\mathbf{S}_n(\boldsymbol{\beta}) = n^{-\frac{1}{2}} \sum_{i=1}^{n} \mathbf{x}_i a_n(R_{ni}(\boldsymbol{\beta})). \tag{3.10}$$

Writing

$$\mathbf{X}_n'\mathbf{X}_n = \begin{bmatrix} \mathbf{V}_{11} & \mathbf{V}_{12} \\ \mathbf{V}_{21} & \mathbf{V}_{22} \end{bmatrix} \tag{3.11}$$

where $\mathbf{V}_{ij} = \mathbf{X}_{ni}'\mathbf{X}_{nj}$, $i, j = 1, 2$, we obtain (for $n \geq n_0$)

$$(\mathbf{X}_n'\mathbf{X})^{-1} = \begin{bmatrix} \mathbf{V}^{11} & \mathbf{V}^{12} \\ \mathbf{V}^{21} & \mathbf{V}^{22} \end{bmatrix} \tag{3.12}$$

where

$$\mathbf{V}^{21} = -\mathbf{V}_0^{-1}\mathbf{V}_{21}\mathbf{V}_{11}^{-1}, \ \mathbf{V}^{22} = \mathbf{V}_0^{-1} \tag{3.13}$$

and

$$\begin{aligned} \mathbf{V}_0 &= \mathbf{V}_{22} - \mathbf{V}_{21}\mathbf{V}_{11}^{-1}\mathbf{V}_{12} \\ &= \left[\mathbf{X}_{n2} - \hat{\mathbf{X}}_{n2}\right]' \left[\mathbf{X}_{n2} - \hat{\mathbf{X}}_{n2}\right] \end{aligned} \tag{3.14}$$

Writing (3.10) as

$$\mathbf{S}_n(\tilde{\boldsymbol{\beta}}) = \begin{bmatrix} S_n^{(1)}(\tilde{\boldsymbol{\beta}}) \\ S_n^{(2)}(\tilde{\boldsymbol{\beta}}) \end{bmatrix} = \begin{bmatrix} n^{-\frac{1}{2}} \sum_{i=1}^{n} \mathbf{x}_{i1} a_n(R_{ni}(\tilde{\boldsymbol{\beta}})) \\ n^{-\frac{1}{2}} \sum_{i=1}^{n} \mathbf{x}_{i2} a_n(R_{ni}(\tilde{\boldsymbol{\beta}})) \end{bmatrix} \tag{3.15}$$

we obtain, in view of (3.9), (3.13) and (3.14),

$$\begin{aligned} n^{\frac{1}{2}}\gamma(\hat{\boldsymbol{\beta}}_{n2} - \boldsymbol{\beta}_2) &= n^{-\frac{1}{2}} \sum_{i=1}^{n} (\mathbf{V}^{21}\mathbf{x}_{i1} + \mathbf{V}^{22}\mathbf{x}_{i2}) a_n(R_{ni}(\boldsymbol{\beta})) + o_p(1) \\ &= n^{-\frac{1}{2}} (Q_n^{(2)})^{-1} \sum_{i=1}^{n} (\mathbf{x}_{i2} - \hat{x}_{i2}) a_n(R_{ni}(\boldsymbol{\beta})) + o_p(1) \end{aligned}$$

and this leads to (3.5) with the aid of results of Hájek and Sidák (1967), Chapter V.

For the estimator $\tilde{\boldsymbol{\beta}}_n^{(2)}$, Jurečková (1991) proved

$$n^{\frac{1}{2}}\gamma(\tilde{\boldsymbol{\beta}}_{n2} - \boldsymbol{\beta}_{n2}) = n^{-\frac{1}{2}} (\mathbf{Q}_n^{(2)})^{-1} \sum_{i=1}^{n} (\mathbf{x}_{i2} - \hat{\mathbf{x}}_{i2}) \hat{b}_{ni}(\mathbf{Y} - \mathbf{X}_n\boldsymbol{\beta}) + o_p(1)$$

and this, by Theorem 4.1 in Gutenbrunner et al. (1992) implies

$$n^{\frac{1}{2}}\gamma(\tilde{\boldsymbol{\beta}}_{n2} - \boldsymbol{\beta}_2) = n^{-\frac{1}{2}} (\mathbf{Q}_n^{(2)})^{-1} \sum_{i=1}^{n} (\mathbf{x}_{i2} - \hat{\mathbf{x}}_{i2}) \varphi(\mathbf{F}(e_i)) + o_p(1)$$

and hence (3.6) follows. The remaining propositions easily follow from (3.5) and (3.6).

If we apply the Jurečková procedure separately to every component of $\boldsymbol{\beta}^{(2)}$, we obtain the coordinate estimator of $\boldsymbol{\beta}_2$ which we shall denote $^{\circ}\boldsymbol{\beta}_{n2}$. More precisely, put $\boldsymbol{\beta}_2 = \boldsymbol{\beta}_{p+j}$ (the $(p+j)^{\text{th}}$ component of $\boldsymbol{\beta}$) and $\mathbf{X}_{n2} = \mathbf{x}_{np+j}$ (the $(p+j)^{\text{th}}$ column of $\mathbf{X}_n$), $j = 1, \cdots, q$. Calculate the regression rank scores corresponding to the matrix $\mathbf{X}_n^0 \backslash \mathbf{x}_{np+j}$ and to the pseudo observations $Y_i - t\mathbf{x}_{i,p+q}$ and the scores generated by $\varphi(j)$ by (3.2), which we shall denote by $\hat{b}_{ni}^{(j)}(\mathbf{Y} - t\mathbf{x}_{np+j})$, $i = 1, \cdots, n$; $j = 1, \cdots, q$; $t \in I\!R_1$. The estimator $^{\circ}\boldsymbol{\beta}_{n,p+j}$ of $\boldsymbol{\beta}_{p+j}$ is then a solution of the minimization

$$D_n^{(j)}(t) = \sum_{i=1}^{n} (Y_i - tx_{i,p+j})(\hat{b}_{ni}^{(j)}(\mathbf{Y} - t\mathbf{x}_{np+j}) - \overline{\varphi}) := \min,\ t \in I\!R_1. \quad (3.16)$$

The function $D_n^{(j)}(t)$ is convex and piecewise linear; its right-hand derivative

$$Z_n^{(j)}(t) = -\sum_{i=1}^{n} x_{i,j+p} \left[\hat{b}_{ni}^{(j)}(\mathbf{Y} - t\mathbf{x}_{np+j}) - \overline{\varphi}\right] \quad (3.17)$$

is nondecreasing in t. Hence, $^{\circ}\beta_{n,p+j}$ could be characterized in the following way

$$\begin{aligned} ^{\circ}\beta_{n,p+j} &= \frac{1}{2}(t_j^- + t_j^+) \\ t_j^- &= \sup\left\{t : Z_n^{(j)}(t) > 0\right\} \\ t_j^+ &= \inf\left\{t : Z_n^{(j)}(t) < 0\right\}. \end{aligned} \quad (3.18)$$

We propose $^\circ\boldsymbol{\beta}_{n2} = (^\circ\beta_{n,p+1}, \cdots, ^\circ\beta_{n,p+q})'$, as an estimator of $\boldsymbol{\beta}_2$.

Surprisingly, even $^\circ\boldsymbol{\beta}_{n2}$ is asymptotically equivalent to $\hat{\boldsymbol{\beta}}_{n2}$ and $\tilde{\boldsymbol{\beta}}_{n2}$, as it is formulated in the following theorem. Analogously, we could estimate $\boldsymbol{\beta}_2$ not only componentwise but also in arbitrary groups of components. This provides us with a variety of alternative possibilities.

THEOREM 3.2. *Consider the model (1.4) with the design matrix satisfying conditions (X.1) and (X.2) and with every column of* $\mathbf{X}_{n2}$ *satisfying (X.3). Then, under (F.1)-(F.3),* (φ.1) *and (3.3), as* $n \to \infty$,

$$n^{\frac{1}{2}} \left\| ^\circ\boldsymbol{\beta}_{n2} - \hat{\boldsymbol{\beta}}_{n2} \right\| = o_p(1),\ n^{\frac{1}{2}} \left\| \tilde{\boldsymbol{\beta}}_{n2} - \hat{\boldsymbol{\beta}}_{n2} \right\| = o_p(1) \tag{3.19}$$

and $^\circ\boldsymbol{\beta}_{n2}$ *admits the asymptotic representation (3.4) and is asymptotically normally distributed according to (3.7).*

PROOF. It follows from Theorem 3.1 that

$$n^{\frac{1}{2}}(^\circ\beta_{n,p+j} - \hat{\beta}_{n,p+j}) = o_p(1),\ j = 1, \cdots, q,$$

hence $n^{\frac{1}{2}} \left\| ^\circ\boldsymbol{\beta}_{n2} - \hat{\boldsymbol{\beta}}_{n2} \right\| = o_p(1)$

and this further implies the remaining propositions. □

4 Estimation in k-sample model

As an illustration, apply the above results to the special model

$$Y_{ij} = \beta_0 + \beta_i + e_{ij},\ j = 1, \cdots, n_i;\ i = 1, \cdots, k \tag{4.1}$$

with $\beta_1 = 0$. Consider the problem of estimating β_k.

The regression rank scores corresponding to the submodel (4.1) under $\beta_1 = \ldots \beta_k = 0$ satisfy

$$\sum_{i=1}^{k} \sum_{j=1}^{n_i} Y_{ij}\hat{a}_{ij}(\alpha) := \max$$

$$\sum_{i=1}^{n_1} \hat{a}_{ij}(\alpha) = n_i(1-\alpha),\ i = 2, 3, \cdots, k-1 \tag{4.2}$$

$$\sum_{j=1}^{n_i} \hat{a}_{1j}(\alpha) + \sum_{j=1}^{n_k} \hat{a}_{kj}(\alpha) = (n_1 + n_k)(1-\alpha),\ 0 < \alpha < 1.$$

this implies that $(\hat{a}_{i1}(\alpha), \cdots, \hat{a}_{in_i}(\alpha))$ generate the ranks of the ith sample $Y_{i1}, \cdots, Y_{in_i}$, $i = 2, \cdots, k-1$, in the same manner as in (2.9). On the other hand, $(\hat{a}_{11}(\alpha), \cdots, \hat{a}_{1n_1}(\alpha), \hat{a}_{k1}(\alpha), \cdots, \hat{a}_{kn_k}(\alpha))$ analogously generate the ranks in the combined 1st and k^{th} samples.

Now, let us calculate regression rank scores analogously, with the only difference that the kth sample in (4.2) is replaced by pseudo observations $Y_{k1} - t_k, \cdots, Y_{kn_k} - t_k$. Calculate the Wilcoxon scores corresponding to $\varphi(\alpha) = \alpha - \frac{1}{2}$, $0 \leq \alpha \leq 1$, i.e.

$$\hat{b}_{ij}(t_k) = -\int_0^1 (\alpha - \frac{1}{2}) d\hat{a}_{ij}(\alpha, t_k) \tag{4.3}$$

$j = 1, \cdots, n_i$; $i = 1, \cdots, k$. The estimator of β_k then minimizes the function

$$D_n(t_k) = \sum_{i=1}^{k-1} \sum_{j=1}^{n_j} \hat{b}_{ij}(t_k) Y_{ij} + \sum_{j=1}^{n_k} \hat{b}_{kj}(t_k)(Y_{kj} - t_k) \tag{4.4}$$

with respect to $t_k \in I\!R_1$. the right-hand derivative of (4.4) in t_k is equal to $-\sum_{j=1}^{n_k} \hat{b}_{kj}(t_k)$, while the above remark shows that $(n\hat{b}_{11}(t_k), \cdots, n\hat{b}_{1n_1}(t_k), n\hat{b}_{k1}(t_k), \cdots, n\hat{b}_{kn_k}(t_k))$ are the ranks of $(Y_{11}, \cdots, Y_{1n_1}, Y_{k1} - t_k, \cdots, Y_{kn_k} - t_k)$ centered by $\frac{1}{2}(n_1 + n_k + 1)$. Hence, we immediately obtain that the estimator $^\circ\beta_k$ and β_k based on regression rank scores coincide with the Wilcoxon scores estimator of the shift between the kth and the 1st samples, i.e.

$$^\circ\beta_k = \text{med}\,\{Y_{kr} - Y_{1s} : 1 \leq r \leq u_k,\ 1 \leq s \leq n_1\}, \tag{4.5}$$

considered by Hodges and Lehmann (1963) and Sen (1963). When we estimate $\beta_2, \cdots, \beta_k$ analogously we obtain the coordinatewise estimator $(^\circ\beta_2, \cdots, ^\circ\beta_k)$ of $(\beta_2, \cdots, \beta_k)$. The jth component $^\circ\beta_j$ is the Wilcoxon scores estimator between the jth and the 1st samples. It follows from Theorem 3.2 that $(^\circ\beta_2, \cdots, ^\circ\beta_k)$ is asymptotically equivalent to Jaeckel's estimator of $(\beta_2, \cdots, \beta_k)$, corresponding to the Wilcoxon scores.

REFERENCES

Adichie, J.N. (1967). Estimating of regression coefficients based on rank tests. *Ann. Math. Statist.* **38**, 894-904.

Adichie, J.N. (1978). Rank tests for sub-hypotheses in the general linear regression. *Ann. Statist.* **6**, 1012-1026.

Adichie, J.N. (1984). Rank tests in linear models. *Handbook of Statistics, Vol. 4, Nonparametric Methods.* (P.R. Krishnaiah and P.K. Sen, eds.), pp. 229-257. Elsevier, New York.

Gutenbrunner, C. (1986). Zur Asymptotik von Rgression quantileprocessen und daraus abgeleiten Statistiken. *Ph. D. Dissertation, Universitat Freiburg.*

Gutenbrunner, C. and Jurečková, J. (1992). Regression rank scores and regression quantiles. *Ann. Statist.* **20**, 305-330.

Gutenbrunner, C., Jurečková, J., Koenker, R. and Portnoy, S. (1992). Tests of linear hypotheses based on regression rank scores. *Working paper 92-0169, Bus adm., Univ. Illinois* (to appear in *J. Nonpara. Statist.* **3**, 1993).

Hajek, J. and Šidak, Z. (1967). *Theory of Rank Tests.* Academia, Prague.

Heiler, S. and Willers, R. (1988). Asymptotic normality of R-estimates in the linear model. *Statistics*, **19**, 173-184.

Hodges, J.L. and Lehmann, E.L. (1963). Estimates of location based on rank tests. *Ann. Math. Statist.* **34**, 598-611.

Jaeckel, L.A. (1972). Estimating regression coefficients by minimizing the dispersion of residulas. *Ann. Math. Statist.* **43**, 1449-1458.

Jurečková, J. (1969). Asymptotic linearity of a rank statistic in regression parameter. *Ann. Math. Statist.* **40**, 1889-1900.

Jurečková, J. (1971a). Nonparametric estimate of regression coefficients. *Ann. Math. Statist.* **42**, 1328-1338.

Jurečková, J. (1971b). Asymptotic independence of rank test statistic for testing symmetry on regression. *Sankhyā A*, **33**, 1-18.

Jurečková, J. (1977). Asymptotic relations of M-estimates and R-estimates in linear regression model. *Ann. Statist.* **5**, 664-672.

Jurečková, J. (1984). Regression quantiles and trimmed least squares estimator under a general design. *Kybernetika* **20**, 345-357.

Jurečková, J. (1991). Estimation in a linear model based on regression rank scores. *Nonparametric Statist.* **1**, 197-203.

Jurečková, J. (1992a). Uniform asymptotic linearity of regression rank scores process. *Proc. Intern. Symp. on Nonparametric Statistic and Related Topics* (A.K. Md. E. Saleh, ed.), pp. 217-228, Elsevier, New York.

Jurečková, J. (1992b). Regression rank scores applied to heavy-tailed distributions. Preprint 088-92, *Math. Sc. Res. Inst.*, Berkeley.

Koenker, R. and Bassett, G. (1978), Regression quantiles. *Econometrica* **46**, 33-50.

Koenker, R. and d'Orey, V. (1987). Computing regression quantiles. *Applied Statist.* **36**, 383-393.

Koenker, R. and Portnoy, S. (1987). L-estimation for linear models. *J. Amer. Statist. Assoc.* **82**, 851-857.

Koul, H.L. (1970). A class of ADF tests for subhypothesis in the multiple linear regression. *Ann. Math. Statist.* **41**, 1273-1281.

Puri, M.L. and Sen, P.K. (1985). *Nonparametric Methods in General Linear Models*. J. Wiley, New York.

Ruppert, D. and Carroll, R.J. (1980). Trimmed least squares estimation in the linear model. *J. Amer. Statist Assoc.* **75**, 256-272.

Sen, P.K. (1963). On the estimation of relative potency in dilution (direct) assays by distribution-free methods. *Biometrics* **19**, 532-552.

Sen, P.K. (1968). Estimates of the regression coefficient based on Kendall's tau. *Jour. Amer. Statist. Assoc.* **63**, 1379-1389.

Sen, P.K. and Puri, M.L. (1977). Asymptotically distribution-free aligned rank order tests for composite hypotheses for general linear models. *Zeit. Warsch. Verw. Geb.* **39**, 175-186.

Department of Statistics & Probability
Charles University
Sokolovska ul 83
186 00 Prague 8, Czechoslovakia

Department of Statistics
University of North Carolina
Chapel Hill, NC 27599- 3260
USA

J.K. Ghosh, S.K. Mitra, K.R. Parthasarathy and B.L.S. Prakasa Rao (*eds.*),
Statistics and Probability : A Raghu Raj Bahadur Festschrift
Published by Wiley Eastern Limited, Publishers, (1993), 293 – 302

A Test for Lorenz Ordering between Two Distributions

B.K. KALE
University of Poona, Pune

and

HARSHINDER SINGH
Panjab University, Chandigarh

Abstract

An asymptotic distribution-free and consistent test is developed for comparing Lorenz curves of two distributions on a region of interest given by $[p_1, p_2]$, $0 < p_1 < p_2 < 1$. The test will be useful for making comparisons of income inequality between different income distributions.

1 Introduction

Lorenz (1905) introduced Lorenz curve for measuring income inequality of a distribution. The Lorenz curve corresponding to a particular population of individuals is a function, say $L(u)$, defined on the interval $[0, 1]$, such that for each $u \in [0, 1]$, $L(u)$ represents the proportion of the total income of the population which accrues to the poorest $100u$ % of the population. Apart from representing the detailed structure of inequality, the Lorenz curve

AMS Subject Classifications: 62G10 *Key words and Phrases* : Lorenz Ordering, Gaussian Process

helps identify those regions of a distribution where significant inequality differences occur. This curve is the basis for several measures of wealth inequality such as the Gini co-efficient, Goldie index, Pietra index etc. [ref. Arnold (1987) Csörgö *et al* (1986)]. Outside economics, Goldie (1977) mentioned applications of Lorenz curves to the distribution of scientific grants and fisheries. Chandra and Singpurwalla [(1978), (1981)] and Klefsjö (1984) discussed the relevance of Lorenz curves in the area of Reliability. We refer to a recent paper by Moothathu (1990-91) for a more detailed review.

Goldie (1977) provided a thorough convergence theory for empirical Lorenz curves. Sendler (1979) considered the problem of estimating the theoretical Lorenz curve from the data. Beach and Davidson (1983) advanced the results of Sendler (1979) and others by deriving the asymptotic joint variance-covariance structure for fixed finite number of the ordinates of the empirical Lorenz curve showing that the entire structure can be expressed in distribution-free terms and provided consistent estimates of the variances and covariances. In particular, they used these results to construct an asymptotically distribution-free test for testing the equality of the two Lorenz curves at finite number of specified points p_i. This test has the limitation that it can not be used to compare two Lorenz curves in the whole interval, say $[p_1, p_2]$, of interest.

In this paper we have constructed an asymptotically distribution-free and consistent test for comparing Lorenz curves of two income distributions in the entire region of interest, say, $[p_1, p_2]$, where $0 < p_1 < p_2 < 1$.

2 Testing for Lorenz Ordering between Two Distributions

Let F and G denote the cumulative distribution functions (cdf's) of two income distributions having continuous densities f and g respectively, where all incomes are assumed to be nonnegative. Let μ_F and μ_G denote the means of the distributions with cdf's F and G respectively. Let σ_F^2 and σ_G^2 denote the corresponding variances. We shall assume throughout that F^{-1} and G^{-1} are continuous functions and the distributions F and G have finite support. This assumption of finite supports is not too restrictive as far as practical situations are concerned.

The Lorenz curves corresponding to the distributions with cdf's F and G respectively are defined as

$$L_1(p) = \frac{1}{\mu_F} \int_0^{F^{-1}(p)} x\,dF(x), \qquad 0 < p < 1. \tag{2.1}$$

and

$$L_2(p) = \frac{1}{\mu_G} \int_0^{G^{-1}(p)} x dG(x), \quad 0 < p < 1. \tag{2.2}$$

Let $[p_1, p_2]$ be a closed interval with $0 < p_1 < p_2 < 1$. We consider testing of the hypotheses

$$H_0 \; : \; L_1(p) \leq L_2(p) \text{ for some } p \in [p_1, p_2]$$

against the alternative

$$H_1 \; : \; L_1(p) > L_2(p) \text{ for all } p \in [p_1, p_2].$$

Let

$$\xi_p \; = \; F^{-1}(p), \quad \eta_p = G^{-1}(p) \tag{2.3}$$

$$\alpha_p \; = \; \frac{1}{p} \int_0^{\xi_p} u dF(u), \quad \beta_p = \frac{1}{p} \int_0^{\eta_p} u dG(u) \tag{2.4}$$

$$\lambda_p^2 \; = \; \frac{1}{p} \int_0^{\xi_p} u^2 dF(u) - \alpha_p^2 \tag{2.5}$$

$$\delta_p^2 \; = \; \frac{1}{p} \int_0^{\eta_p} u^2 dG(u) - \beta_p^2 \tag{2.6}$$

Let $F_m(x)$ and $G_n(x)$ denote the empirical cdf's. Then using order statistics of the samples consistent estimators of the above parameters are obtained by putting

$$\hat{\xi}_p \; = \; X_{([mp]+1)} \text{ and } \hat{\eta}_p = Y_{([np]+1)} \tag{2.7}$$

$$\hat{\alpha}_p \; = \; \frac{1}{mp} \sum_{i=1}^{[mp]} X_{(i)} \tag{2.8}$$

$$\hat{\beta}_p \; = \; \frac{1}{np} \sum_{i=1}^{[np]} Y_{(i)} \tag{2.9}$$

$$\hat{\lambda}_p^2 \; = \; \frac{1}{mp} \sum_{i=1}^{[mp]} X_{(i)}^2 - (\hat{\alpha}_p)^2 \tag{2.10}$$

$$\hat{\delta}_p^2 \; = \; \frac{1}{np} \sum_{i=1}^{[np]} Y_{(i)}^2 - (\hat{\beta}_p)^2 \tag{2.11}$$

$$\hat{L}_1(p) \; = \; p\,\hat{\alpha}_p/\overline{X} \tag{2.12}$$

$$\hat{L}_2(p) \; = \; p\,\hat{\beta}_p/\overline{Y} \tag{2.13}$$

where $\overline{X}$ and $\overline{Y}$ denote the respective sample means and $[x]$ denotes the greatest integer less than or equal to x.

From Theorem 2 of Beach and Davidson (1983), it follows that for each $p_1 \leq p \leq p_2$, $\sqrt{m}\,[\hat{L}_1(p) - L_1(p)]$ has limiting normal distribution with mean 0 and variance

$$\begin{aligned} V_p^2 &= \frac{1}{\mu_F^2}[p(\lambda_p^2 + \alpha_p^2) + p(1-p)(\xi_p^2 - (p\alpha_p)^2 - 2(p\alpha_p)\xi_p] + \left[\frac{p\alpha_p}{\mu_F^2}\right]^2 \sigma_F^2 \\ &\quad - \frac{2}{\mu_F^3}[(p\alpha_p)p(\lambda_p^2 + \alpha_p^2) - \mu_F(p\alpha_p)^2 - \xi_p(p\alpha_p)^2] \end{aligned} \tag{2.14}$$

and $\sqrt{m}\,(\hat{L}_2(p) - L_2(p))$ has asymptotic normal distribution with mean zero and variance W_p^2 where W_p^2 is defined in an analogous way.

Let $\hat{V}_p^2$ denote an estimate of V_p^2 obtained by replacing μ_F, λ_p, α_p, ξ_p, σ_F^2 in (2.14) by their consistent estimators $\overline{X}$, $\hat{\lambda}_p$, $\hat{\alpha}_p$, $\hat{\xi}_p$, and $(1/m)\sum_{i=1}^{m}(X_i - \overline{X})^2$ respectively. Let $\hat{W}_p^2$ denote a similar estimate of W_p^2.

Now for any fixed $p \in [p_1, p_2]$, we will first define an asymptotic level α and consistent test for the simple problem.

$$H_{op} : L_1(p) \leq L_2(p) \text{ against } H_{1p} : L_1(p) > L_2(p)$$

and then reject H_0 in favour of H_1 if H_{op} is rejected at level α for all $p \in [p_1, p_2]$.

An obvious test statistic for testing H_{op} against H_{1p} is

$$\begin{aligned} U_{m,n}(p) &= (\hat{L}_1(p) - \hat{L}_2(p))/[(\hat{V}_p^2/m) + (\hat{W}_p^2/n)]^{1/2} \\ &= \frac{A_{m,n}(p)}{D_{m,n}(p)} + \frac{B_{m,n}(p)}{D_{m,n}(p)} \end{aligned} \tag{2.15}$$

where

$$\begin{aligned} A_{m,n}(p) &= \sqrt{m}(\hat{L}_1(p) - L_1(p)) - (\sqrt{m}/\sqrt{n})\,\sqrt{n}\,(\hat{L}_2(p) \\ &\quad - L_2(p)), \end{aligned} \tag{2.16}$$

$$B_{m,n}(p) = \sqrt{m}(L_1(p) - L_2(p)) \tag{2.17}$$

and

$$D_{m,n}^2(p) = \hat{V}_p^2 + (m/n)\,\hat{W}_p^2 \tag{2.18}$$

THEOREM 2.1 : Let $m, n \to \infty$ in such a way that $(m/(m+n)) \to \lambda$, where $0 \leq \lambda < 1$ then,

$$P[U_{m,n}(p) > Z_\alpha] \to \begin{cases} 0 & \text{if } L_1(p) < L_2(p), \\ \alpha & \text{if } L_1(p) = L_2(p), \\ 1 & \text{if } L_1(p) > L_2(p) \end{cases}$$

where Z_α is the $(1-\alpha)$-th percentile of the standard normal distribution.

PROOF. It follows from Theorem 2 of Beach and Davidson (1983) that

$$A_{m,n}(p) \xrightarrow{\text{d}} N(0, D^2(p)), \tag{2.19}$$

where

$$D^2(p) = V_p^2 + (\lambda/(1-\lambda))W_p^2. \tag{2.20}$$

Further from the consistency of $\hat{V}_p^2$ and $\hat{W}_p^2$ it follows that

$$D_{m,n}^2(p) \overset{\text{a.s.}}{\to} D^2(p), \tag{2.21}$$

$$\frac{A_{m,n}(p)}{D_{m,n}(p)} \xrightarrow{\text{d}} Z \tag{2.22}$$

where Z is a standard normal variable. Furthermore (2.17) implies

$$B_{m,n}(p) \to \begin{cases} \infty & \text{if } L_1(p) > L_2(p), \\ 0 & \text{if } L_1(p) = L_2(p), \\ -\infty & \text{if } L_1(p) < L_2(p). \end{cases} \tag{2.23}$$

The Theorem now follows immediately from (2.15), (2.22), and (2.23).

Now for large sample sizes m and n, we propose a test that rejects H_0 in favour of H_1 if and only if

$$\inf_{p_1 \le p \le p_2} U_{m,n}(p) > Z_\alpha. \tag{2.24}$$

In the following theorem, we show that the proposed test (2.24) for testing H_0 versus H_1 has the prescribed asymptotic size α.

THEOREM 2.2 : Let m, $n \to \infty$ in such a way that $(m/(m+n)) \to \lambda$, $0 \le \lambda < 1$, then for F, G in H_0

$$\overline{\lim}\, P[\inf_{p_1 \le p \le p_2} U_{m,n}(p) > Z_\alpha] \le \alpha.$$

PROOF: Since F, G are in $\cup_{p\in[p_1,p_2]} H_{op}$ there exists a $p_0 \in [p_1, p_2]$ such that $L_1(p_0) - L_2(p_0) \le 0$ and we have

$$P[\inf_{p_1 \le p \le p_2} U_{m,n}(p) > Z_\alpha] \le P[U_{m,n}(p_0) > Z_\alpha]$$

and hence

$$\overline{\lim}\, P[\inf_{p_1 \le p \le p_2} U_{m,n}(p) > Z_\alpha] \le \overline{\lim}\, P[U_{m,n}(p_0) > Z_\alpha].$$

The result now follows from Theorem 2.1.

In the following Theorem we show that the suggested test (2.24), for testing H_0 against H_1, is consistent.

THEOREM 2.3 : For, $F, G \in H_1$,

$$P[\inf_{p_1 \leq p \leq p_2} U_{m,n}(p) > Z_\alpha] \to 1 \text{ as } m, n \to \infty.$$

For proving the above theorem, we shall need the following two Lemmas.

LEMMA 2.1 : Let $m, n \to \infty$ such that $(m/(m+n)) \to \lambda$, $0 \leq \lambda < 1$, then

$$\sup_{p_1 \leq p \leq p_2} |A_{m,n}(p) - A(p)| \overset{\text{a.s.}}{\to} 0,$$

where $\{A(p), p_1 \leq p \leq p_2\}$ is a mean zero Gaussian process with continuous paths almost surely.

This follows from Goldie (1977) and page 9 of Cszorgo *et al* (1986).

LEMMA 2.2 : Let $m, n \to \infty$ in such a way that $(m/(m+n)) \to \lambda$, $0 \leq \lambda < 1$, then

$$\sup_{p_1 \leq p \leq p_2} |D_{m,n}(p) - D(p)| \overset{\text{a.s.}}{\to} 0.$$

PROOF : In order to establish the Lemma we first prove that as $m \to \infty$

$$\sup_{p_1 \leq p \leq p_2} |p\hat{\alpha}_p - p\alpha_p| \overset{\text{a.s.}}{\to} 0, \tag{2.25}$$

$$\sup_{p_1 \leq p \leq p_2} |p(\hat{\lambda}_p^2 + \hat{\alpha}_p^2) - p(\lambda_p^2 + \alpha_p^2)| \overset{\text{a.s.}}{\to} 0, \tag{2.26}$$

and

$$\sup_{p_1 \leq p \leq p_2} |\hat{\xi}_p - \xi_p| \overset{\text{a.s.}}{\to} 0. \tag{2.27}$$

We can express

$$\begin{aligned} p\hat{\alpha}_p - p\alpha_p &= F_m^{-1}(p)\,[p - F\,F_m^{-1}(p)] \\ &\quad - \int_0^{F_m^{-1}(p)} [F_m(u) - F(u)]\,du + \int_{F^{-1}(p)}^{F_m^{-1}(p)} [udF(u)]. \end{aligned}$$

Thus

$$\begin{aligned} \sup_{p_1 \leq p \leq p_2} |p\hat{\alpha}_p - p\alpha_p| &\leq F_m^{-1}(p_2) \sup_{p_1 \leq p \leq p_2} |F\,F_m^{-1}(p) - p| \\ &\quad + F_m^{-1}(p_2) \sup_{0 \leq u < \infty} |F_m(u) - F(u)| \\ &\quad + \max(F^{-1}(p_2), F_m^{-1}(p_2)) \sup_{p_1 \leq p \leq p_2} |F\,F_m^{-1}(p) - p|. \end{aligned}$$

Let [0, A] denote the support of the distribution having cdf F. Then we have

$$\sup_{p_1 \le p \le p_2} |p(\hat{\alpha}_p - \alpha_p)| \le 2A \sup_{p_1 \le p \le p_2} |FF_m^{-1}(p) - p| + A \sup_{0 \le x < \infty} |F_m(x) - F(x)|.$$

By using arguments similar to those in Chung (1974), p.133 for proving the Glivenko-Cantelli theorem, it follows easily that $\sup_{p_1 \le p \le p_2} |FF_m^{-1}(p) - p| \stackrel{\text{a.s.}}{\to} 0$ and (2.25) is established by observing that $\sup |F_m(x) - F(x)| \stackrel{\text{a.s.}}{\to} 0$.

Proofs of (2.26) and (2.27) being similar are omitted.

Since the support of F is finite, the sequences $p\hat{\alpha}_p$, $p(\hat{\lambda}_p^2 + \hat{\alpha}_p^2)$ and $\hat{\xi}_p$ are bounded. Also by the strong law of large numbers $\overline{X} \stackrel{\text{a.s.}}{\to} \mu_F$ and $(1/m) \sum_{i=1}^{m} (X_i - \overline{X})^2 \stackrel{\text{a.s.}}{\to} \sigma_F^2$. Now from (2.25) – (2.27) and the definitions of $\hat{V}_p^2$ and V_p^2 given by (2.14) one can similarly show that

$$\sup_{p_1 \le p \le p_2} |\hat{V}_p^2 - V_p^2| \stackrel{\text{a.s.}}{\to} 0 \text{ as } m \to \infty, \tag{2.28}$$

$$\sup_{p_1 \le p \le p_2} |\hat{W}_p^2 - W_p^2| \stackrel{\text{a.s.}}{\to} 0 \text{ as } n \to \infty, \tag{2.29}$$

Lemma 2.2 now follows from (2.28) and (2.29).

PROOF OF THEOREM 2.3 : Let F, G be in H_1 and let

$$\Delta = \inf_{p_1 \le p \le p_2} [L_1(p) - L_2(p)].$$

Since $L_1(p)$ and $L_2(p)$ is continuous and positive on the compact interval $[p_1, p_2]$ it follows that $\Delta > 0$ implying that

$$\inf_{p_1 \le p \le p_2} B_{m,n}(p) = \sqrt{m}\,\Delta > 0. \tag{2.30}$$

From Lemma 2.2,

$$\sup_{p_1 \le p \le p_2} |D_{m,n}(p) - D(p)| \stackrel{\text{a.s.}}{\to} 0. \tag{2.31}$$

and

$$\sup_{p_1 \le p \le p_2} D(p) = r > 0 \tag{2.32}$$

because $D(p)$ is continuous and positive on the compact interval $[p_1, p_2]$. Therefore from (2.3) – (2.32), we have for large m, n

$$\inf_{p_1 \le p \le p_2} \frac{B_{m,n}(p)}{D_{m,n}(p)} > \frac{\sqrt{m}\,\Delta}{2r} \text{ a.s.} \tag{2.33}$$

Thus

$$P\left[\inf_{p_1\le p\le p_2} U_{m,n}(p)\le Z_\alpha\right]$$
$$\le P\left[\inf_{p_1\le p\le p_2}\frac{A_{m,n}(p)}{D_{m,n}(p)}\le Z_\alpha - m^{1/4}\right]+P\left[\inf_{p_1\le p\le p_2}\frac{B_{m,n}(p)}{D_{m,n}(p)}\le m^{1/4}\right]$$

Therefore from (2.33), it follows that

$$\overline{\lim} P\left[\inf_{p_1\le p\le p_2} U_{m,n}(p)\le Z_\alpha\right]$$
$$\le \quad \overline{\lim} P\left[\inf_{p_1\le p\le p_2}\frac{A_{m,n}(p)}{D_{m,n}(p)}\le Z_\alpha - m^{1/4}\right] \tag{2.34}$$

Now from Lemmas 2.1, 2.2 it follows that

$$\inf_{p_1\le p\le p_2}\left|\frac{A_{m,n}(p)}{D_{m,n}(p)}-\frac{A(p)}{D(p)}\right|\overset{\text{a.s.}}{\to} 0 \text{ as } m,n\to\infty$$

implying that $\frac{A_{m,n}(p)}{D_{m,n}(p)}\to\frac{A(p)}{D(p)}$ weakly in $D[p_1, p_2]$, the Skorohod space on $[p_1, p_2]$ (refer Billingsley (1968)). Since $\inf\limits_{[p_1,p_2]}(.)$ is a continuous functional on $D[p_1,p_2]$, it follows that

$$\inf_{p_1\le p\le p_2}\frac{A_{m,n}(p)}{D_{m,n}(p)}\overset{\text{d}}{\to}\inf_{p_1\le p\le p_2}\frac{A(p)}{D(p)}.$$

Observe that $\left\{\frac{A(p)}{D(p)}, p_1\le p\le p_2\right\}$ is a mean zero Gaussian process with continuous sample paths. Thus from (2.34), we have

$$\overline{\lim}\, P\left[\inf_{p_1\le p\le p_2} U_{m,n}(p)\le Z_\alpha\right]=0$$

implying

$$\lim P\left[\inf_{p_1\le p\le p_2} U_{m,n}(p)> Z_\alpha\right]=1.$$

For the evaluation of the test statistic $\inf U_{m,n}(p)$, it is easily seen from the definitions of $\hat{L}_1(p)$, $\hat{L}_2(p)$, $\hat{V}_p^2$ and $\hat{W}_p^2$ that inf will be attained at one of the points p_1, p_2, i/m, $i=1,2,3,\cdots, m-1$, j/n, $j=1,2,\cdots,n-1$ which lie in the inerval (p_1,p_2) and at $p=1/2$ if it lies in (p_1,p_2). Thus for computing $\inf\limits_{p_1\le p\le p_2} U_{m,n}(p)$, we need to compute $U_{m,n}(p)$ only at the points p_1,p_2, all those i/m, $i=1,2,\cdots,m-1$ and all those j/n, $j=1,\cdots,n-1$ which lie in (p_1, p_2) and at $p=1/2$ if it lies in (p_1,p_2). All these points are finite in number. The minimum value of all such values of $U_{m,n}(p)$ will be the same as $\inf\limits_{p_1\le p\le p_2} U_{m,n}(p)$. The expressions (2.7) - (2.13) are helpful in computing the values of $U_{m,n}(p)$ at the above mentioned points and the procedure can be easily carried out on a computer. To see the feasib y of this we worked out a case with $m = 50$ and $n = 40$ and we were able to compute the statistics $\inf_{p_1 \le p \le p_2} U_{m,n}(p)$ at a PC using a program written in BASIC. The major part of CPU time was used for obtaining the order statistics of the two samples.

REMARKS

The test proposed in this paper requires specification of the interval of interest $[p_1, p_2]$ where $0 < p_1 < p_2 < 1$ but otherwise no restriction on p is assumed. The test proposed by Beach and Davidson (1983) can be used only for testing the equality of two Lorenz curves at a specified finite number of points but it can not be used to test the one sided hypotheses considered in this paper.

By letting say $n \to \infty$ and noting that $\hat{L}_2(p) \overset{\text{a.s.}}{\to} L_2(p)$ and $\hat{W}_p^2/n \overset{\text{p}}{\to} 0$, we observe that

$$p - \lim_{n \to \infty} U_{m,n}(p) = \left[(\hat{L}_1(p) - L_2(p)) / \sqrt{\hat{V}_p^2/m} \right]$$

and we can use the test statistic $\inf_{p_1 \le p \le p_2} \left[\hat{L}_1(p) - L_2(p) \right] / \sqrt{\hat{V}_p^2/m}$ to test $H_0 : L_1(p) \le L_2(p)$ for some $p \in [p_1, p_2]$, where $L_2(p)$ is a specified Lorenz curve.

It may be pointed out that analogous to the Kolmogorov–Smirnov one sided two sample test one could use the test statistic $K_{mn}^- = \inf_{p_1 \le p \le p_2} \left[\hat{L}_1(p) - \hat{L}_2(p) \right]$ which is simpler to compute than the test statistics proposed in this paper. However, the critical points of K_{mn}^- are not easy to obtain and the asymptotic distribution theory of K_{mn}^- is rather difficult due to the complicated variance covariance structure of the empirical Lorenz process. The general problem of developing the theory of Kolmogorov–Smirnov type statistics, for one sample and two sample problems based on estimated Lorenz curves, though not a simple one, is interesting and would be attempted later.

For practical applications we need test statistics based on grouped data rather than the raw data, since in most cases involving income distributions, the sample sizes m and n are usually very large and the data is grouped in income intervals. For estimation of Lorenz curve from grouped data we refer to a paper by Kakawani and Podder (1973). The problem of testing hypotheses of Lorenz ordering based on grouped data has raised some non-trivial problems and would be pursued later.

REFERENCES

Arnold, B.C. (1987) : *Majorization and the Lorenz Order : A Brief Introduction*, Lecture Notes in Statistics, No. **43**, Ed. D. Brillinger *et al* Springer-Verlag.

Beach, C.M. and Davidson, R. (1983). Distribution free Statistical Inference with Lorenz curves and income shares. *Review of Economic Studies* L, 723-735.

Billingsley, P. (1968). *Convergence of Probability Measures*, John Wiley, New York.

Chung, K.L. (1974). *A Course in Probability Theory*, Second Edition, Academic Press, New York.

Chandra, M. and Singpurwalla, N.D. (1978). The Gini Index, the Lorenz Curve and the Total Time on Test Transform, George Washington University Serial T-368.

Chandra, M. and Singpurwalla, N.D. (1981). Relationships between some notions which are common to Reliability Theory and Economics, *Maths. Oper. Res.* **6**, 113-121.

Cszorgo, M., Cszorgo S. and Harvath, L. (1986). *An Asymptotic Theory for Empirical Reliability and Concentration Processes*, Lecture Notes in Statistics, No. **33**, Ed. D. Brillinger *et al.* Springer-Verlag.

Goldie, C.M. (1977). Convergence theorems for empirical Lorenz curves and their inverses. *Advances in App. Prob.* **9**, 765-791.

Kakwani, N.C. and Podder, N. (1973). On the estimation of Lorenz Curves from grouped observations. *Int. Economic Rev.* **14**, 278-292.

Klefsjö, B. (1984). Reliability interpretations of some concepts from Economics. *Nav. Res. Logist. Quart.*, **31**, 301-308.

Lorenz, M.O. (1905). Method of measuring the concentration of wealth, *Publication of Amer. Stat. Assn.*, **9**, 209-219.

Moothathu, T.S. K. (1990-91). Lorenz Curve and Gini Index, *Calcutta Statist. Assoc. Bull.* **40**, 307-324.

Sendler, W. (1979). On Statistical Inference in Concentration Measurement, *Metrika* **26**, 109-122.

Department of Statistics
University of Poona
Pune 411007, India

Department of Statistics
Panjab University
Chandigarh 160014, India

J.K. Ghosh, S.K. Mitra, K.R. Parthasarathy and B.L.S. Prakasa Rao (*eds.*),
Statistics and Probability : A Raghu Raj Bahadur Festschrift
Published by Wiley Eastern Limited, Publishers, (1993), 303 – 324

Comparison of Tests, and Local Families

WILBERT C.M. KALLENBERG
University of Twente, Enschede

and

FEIKE C. DROST[1]
University of Tilburg, Tilburg

Abstract

Several aspects of comparison of tests using efficiency concepts are treated in this paper. It is shown that in exponential families Bahadur optimality over most of the parameter space is essentially restricted to the likelihood ratio test. Further the limiting efficiency is investigated as approximation tool for the finite sample relative efficiency. The likelihood ratio test, Rao's test and Wald's test are compared in a typical testing problem, both in a local way and nonlocally. The theoretical results are compared with numerical results. Local efficiency and local limits of nonlocal efficiency are often applied in practice. The role of the way along one goes from alternative to hypothesis is hardly investigated so far. For accurate approximation the choice of an appropriate local family is important. Using global approximation such families are derived for the Cressie-Read statistics for testing goodness-of-fit. Although often dismissed, approximate Bahadur efficiency turns out to work well here as approximation to the finite sample relative efficiency.

[1] The second author is research fellow of the Royal Netherlands Academy of Arts and Sciences (KNAW).

AMS Subject Classifications : 62 F 05, 60 F 10, 62 E 20.

Keywords: exact Bahadur efficiency, approximate Bahadur efficiency, likelihood ratio test, exponential families, Rao's test, Wald's test, Pitman efficiency, Pitman deficiency, local families, global approximation, contamination family, Cressie- Read statistic, goodness-of-fit.

1 Introduction

Traditionally the performance of statistical tests is measured by the behaviour of the power functions. In a few testing problems an optimal test exists in the sense that it has maximal power against all alternatives of interest. Usually no optimal test exists and we want to compare several tests as candidates for application. Even in the case where an optimal test exists, we might be interested in comparison, for instance to solve the question how much we loose when applying a more simple test.

Many criteria for comparison of tests have been proposed. A unified description of a lot of them is obtained by using the number of observations $N(\alpha, \beta, \theta)$ needed by a level$-\alpha$ test to achieve power β at the alternative θ. If we have two sequences of tests $\{\phi_n\}$ and $\{\psi_n\}$ (n referring to the number of observations) the relative efficiency of $\{\phi_n\}$ w.r.t. $\{\psi_n\}$ at (α, β, θ) is defined as

$$e_{\phi,\psi}(\alpha, \beta, \theta) = \frac{N^{\psi}(\alpha, \beta, \theta)}{N^{\phi}(\alpha, \beta, \theta)}, \tag{1}$$

where N^{ϕ} and N^{ψ} are of course the previously defined numbers of observations corresponding to ϕ and ψ, respectively. A value of the relative efficiency (1) larger than one indicates that, for the given $(\alpha, \beta, \theta), \{\phi_n\}$ is to be preferred to $\{\psi_n\}$, since with $\{\phi_n\}$ fewer observations are needed to obtain the same performance.

Application of (1) is as a rule very difficult. Since (1) depends upon three arguments we cannot expect in general that one test dominates another for all choices of (α, β, θ). Moreover, the numbers N^{ϕ} and N^{ψ} are often hard to compute. To make things more simple one or another asymptotic approach is considered, for instance by keeping β and θ fixed and sending α to 0 (Bahadur efficiency), or keeping α and β fixed and sending θ to the null set (Pitman efficiency). For a more detailed discussion see e.g. Serfling (1980, chapter 10), Kallenberg (1983).

The limiting values then may be viewed as approximations to (1) for some particular triple (α, β, θ). In Groeneboom and Oosterhoff (1981) the accuracy of such an approximation is investigated numerically in a number of examples. A basic theorem to compute Bahadur efficiency is Theorem 7.2 in Bahadur (1971). It is seen that under the alternative only convergence to a constant is needed, for instance a law of large numbers, and nothing more refined, like a central limit theorem. From an approximation point of view this means that the distribution of the test statistic under the alternative is replaced by a constant, and this is in general not a very accurate approximation. Under the null hypothesis the exponential rate of convergence to 0 determines the efficiency, and thus, for example, all chi- square distributions are treated as equal, irrespective of the number of degrees of freedom.

Two topics arise from the preceding discussion First, the suggestion that due to the rough approximations many tests perform equally well w.r.t. Bahadur efficiency and hence the criterion should be not very discriminating, many tests being optimal. Surprisingly, the contrary seems to be the case! In Hoeffding (1965) it is shown that in a multinomial setting tests which are "sufficiently different" from a likelihood ratio test are inferior at "most" points in the set of alternatives w.r.t. a criterion also based on small levels of significance. In section 2 we obtain related results: for being optimal w.r.t. Bahadur efficiency the test statistic has to be close to a Kullback-Leibler "distance". (In multinomial distributions the likelihood ratio test is based on the Kullback-Leibler information.)

So, in a sense Bahadur-optimal tests are scarce. Note that this is in contrast with another efficiency concept, the so-called Hodges-Lehmann efficiency, w.r.t. which in fact many tests are asymptotically optimal, cf. Nikitin (1986,1987), Kallenberg and Kourouklis (1992).

On the other hand, although by means of Bahadur efficiency we might get an answer on the question which of, for example, three competing tests is the best one, the theoretical answer may disagree with the finite sample behaviour. In section 3 this is discussed by comparing the likelihood ratio test, Rao's and Wald's test in a typical example. This leads to the second topic, where the limiting efficiency is investigated as an approximation tool. Here we concentrate on an aspect in local efficiency-approaches and local limits of nonlocal efficiencies, which is hardly investigated so far, although it is important in determining efficiencies: the way along which one goes from alternative to hypothesis. Pitman efficiency and local limits of (approximate) Bahadur efficiency are often applied in comparison of tests. In both concepts the answer also depends on the bridge between alternative and hypothesis.

In section 4 it is shown that several local families are of interest. Global approximation plays an important role here. Goodness-of-fit tests in a multinomial setting are discussed. In particular, the Cressie-Read class is investigated. It is well-known that the Cressie-Read statistics are equivalent in the sense of Pitman efficiency. However, when evaluating Pitman efficiencies along appropriate families a second order difference shows up. This second order term makes it possible to discriminate between the different members in the class. The results correspond perfectly to simulation studies of Cressie and Read (1984) and Drost et al. (1989). Moreover, there is a relation with approximate Bahadur efficiency, implying that this concept works here very well.

2 Bahadur optimality

Let $X_1, X_2, \ldots$ be i.i.d. r.v.'s each distributed according to a k-parameter exponential family

$$dP_\theta(x) = \exp\{\theta^T x - \psi(\theta)\} d\mu(x), \; \theta \in \Theta \subset \mathbb{R}^k, x \in \mathbb{R}^k, \tag{2}$$

where μ is a σ-finite nondegenerate measure, Θ denotes the natural parameter space, i.e. $\Theta = \{\theta \in \mathbb{R}^k : \int \exp(\theta^T x) d\mu(x) < \infty\}$, and $\psi(\theta) = \log \int \exp(\theta^T x) d\mu(x)$, $\theta \in \Theta$. By θ^T we denote the transpose of θ. Without loss of generality assume that Θ has a non-empty interior and that μ is not supported on a flat. Let $\Theta^* = \{\theta \in \Theta : E_\theta \|X_1\| < \infty\}$, where $\|\cdot\|$ denotes the Euclidean norm.
Note that int $\Theta \subset \Theta^* \subset \Theta$. For $\theta \in \Theta^*$ define $\lambda(\theta) = E_\theta X_1$. The mapping λ is $1-1$ on Θ^*. Defining $\Lambda = \lambda(\Theta^*) = \{\lambda(\theta) : \theta \in \Theta^*\}$, the inverse mapping λ^{-1} exists on Λ. Note that $\lambda(\theta) =$ grad $\psi(\theta)$ if $\theta \in$ int Θ. Moreover, for $\theta \in$ int Θ, the covariance matrix Σ_θ of X_1 is the Hessian of ψ. The distribution of $\bar{X}_n = n^{-1} \Sigma_{i=1}^n X_i$ is denoted by $\bar{P}_\theta^n$, the distribution of $S = (X_1, X_2, \ldots)$ by $\mathbb{P}_\theta$.
Consider the testing problem $H_0 : \theta \in \Theta_0$ against $H_1 : \theta \in \Theta_1 = \Theta - \Theta_0$. Let $\{\phi_n(\alpha) : n \in \mathbf{N},\ 0 < \alpha < 1\}$ be a family of (randomized) level$-\alpha$ tests of H_0, $\phi_n(\alpha)$ being based on $X_1, \ldots, X_n$. The dependence on $s = (x_1, x_2, \ldots)$ is sometimes explicitly written as $\phi_n(\alpha; s)$. Further define

$$N^\phi(\alpha, \beta, \theta) = \min\{n : E_\theta \phi_m(\alpha) \geq \beta \text{ for all } m \geq n\} \tag{3}$$

and $N^+(\alpha, \beta, \theta) = \inf_\phi N^\phi(\alpha, \beta, \theta)$, where $\phi = \{\phi_n(\alpha)\}$ runs through all families of tests of H_0. The family $\{\phi_n(\alpha)\}$ is called *efficient* in the sense of Bahadur at θ as $\alpha \downarrow 0$ if for all $0 < \beta < 1$

$$\lim_{\alpha \downarrow 0} N^\phi(\alpha, \beta, \theta)/N^+(\alpha, \beta, \theta) = 1. \tag{4}$$

For a discussion of the relation with exact slopes see Kallenberg (1981) and references therein. By $\phi_n^{LR}(\alpha)$ we denote the likelihood ratio (LR) level-α test. If $\bar{x}_n \in \Lambda$ it is given by

$$\phi_n^{LR}(\alpha; \bar{x}_n) = \begin{cases} 1 & > \\ \delta_n \text{ if } K(\lambda^{-1}(\bar{x}_n), \Theta_0) & = \; d_n, \\ 0 & < \end{cases} \tag{5}$$

where $K(\theta, \Theta_0) = \inf\{K(\theta, \theta_0) : \theta_0 \in \Theta_0\}$ and

$$\begin{aligned} K(\theta, \vartheta) &= E_\theta \log(dP_\theta / dP_\vartheta) \\ &= \psi(\vartheta) - \psi(\theta) + (\theta - \vartheta)^T \lambda(\theta),\ \theta \in \Theta^*,\ \vartheta \in \Theta, \end{aligned} \tag{6}$$

the Kullback-Leibler information number. It is well known that for each $\theta \in$ int Θ,

$$K(\lambda^{-1}(\bar{X}_n), \Theta_0) \to K(\theta, \Theta_0) \text{ w.p.1 } [\mathbb{P}_\theta] \text{ as } n \to \infty. \tag{7}$$

The critical region of the LR test may be interpreted as the region outside a Kullback-Leibler "sphere", although Kullback-Leibler information numbers do not define a "distance" in the technical sense. The following lemma expresses this in another way, saying that if at a certain point the asymptotic power is smaller than 1, then for every point closer to Θ_0 in Kullback-Leibler "distance" the asymptotic power equals 0.

LEMMA 1 *Let $\{\alpha_n\}$ be a sequence of levels such that*

$$\limsup_{n\to\infty} E_\theta \phi_n^{LR}(\alpha_n) < 1 \tag{8}$$

for some $\theta \in$ int Θ_1, then

$$\lim_{n\to\infty} E_\vartheta \phi_n^{LR}(\alpha_n) = 0 \tag{9}$$

for all $\vartheta \in$ int Θ_1 with $K(\vartheta, \Theta_0) < K(\theta, \Theta_0)$.

PROOF. In view of (7) and (8) the critical value d_n of $\phi_n^{LR}(\alpha_n)$ in (5) satisfies

$$\liminf_{n\to\infty} d_n \geq K(\theta, \Theta_0)$$

and hence, again by (7), we get (9). □

If Θ_0 is a Borel set and the LR test satisfies the following very weak similarity-condition: for all n

$$\sup_{\theta_0 \in \Theta_0 \cap C} E_{\theta_0} \phi_n^{LR}(\alpha) \geq \epsilon\alpha \tag{10}$$

for some compact subset C of int Θ and some $\epsilon > 0$, then the LR test is efficient in the sense of Bahadur at θ for those points $\theta \in$ int Θ_1, satisfying $K(\theta, \Theta_0) < K(\Theta_0 \cap C)$, where $K(T)$ is a sort of Kullback-Leibler "distance" from the boundary of Θ to a set $T \subset$ int Θ, defined by

$$K(T) = \sup\{a \in \mathbb{R} : \{\theta : K(\theta, T) \leq a\} \subset K_a \subset \text{int } \Theta, \text{ where } K_a \text{ is compact}\},$$

cf. Kallenberg (1981, Theorem 2.1). (Under the preceding conditions we even get Bahadur deficiency of order $O(\log N)$, which agrees with the "first order efficiency implies second order efficiency" property, cf. Bahadur, Chandra and Lambert (1984, Proposition 2.2) and Kallenberg (1983, Theorem 3.1).)

It is the purpose of this section to show that tests, which are sufficiently different from the LR test are not optimal in the sense of Bahadur. This extends in a way the results of Hoeffding (1965) for the multinomial distribution, and shows that Bahadur optimality over "most of the parameter space" is essentially restricted to the LR test. The property "sufficiently different from the LR test" means that the critical region of the test is not the complement of a Kullback-Leibler sphere in the sense of Lemma 1.

THEOREM 2 *Let $\{\phi_n(\alpha)\}$ be a family of tests. Suppose that there exists a sequence of levels $\{\alpha_n\}$ such that*

$$\limsup_{n\to\infty} E_\theta \phi_n(\alpha_n) < 1 \tag{11}$$

for some $\theta \in int\ \Theta_1$ *and*

$$\limsup_{n\to\infty} E_\vartheta \phi_n(\alpha_n) > 0 \tag{12}$$

for some $\vartheta \in\ int\ \Theta_1$ *with* $K(\vartheta,\Theta_0) < K(\theta,\Theta_0)$. *Then* $\{\phi_n(\alpha)\}$ *is not efficient in the sense of Bahadur at* θ.

Before proving Theorem 2 we prove the following lemma, which is essentially Stein's lemma (cf. Bahadur et al. (1980), Theorem 2.1 and Kester (1985), Corollary 1.2 on p.21).

LEMMA 3 *If (12) holds then*

$$\liminf_{n\to\infty} -n^{-1}\log\alpha_n \le K(\vartheta,\Theta_0). \tag{13}$$

PROOF. Let $\theta_0 \in \Theta_0$ satisfy $K(\vartheta,\theta_0) \le K(\vartheta,\Theta_0)+1$. Writing $r_n(s) = (\theta_0-\vartheta)^T(\sum_{i=1}^n x_i) - n\psi(\theta_0) + n\psi(\vartheta)$ we get for any $\epsilon > 0$

$$E_{\theta_0}\phi(\alpha_n) = \int \phi_n(\alpha_n;s)d\mathbf{P}_{\theta_0}(s) = \int \phi_n(\alpha_n;s)\exp\{r_n(s)\}d\mathbf{P}_\vartheta(s)$$

$$\ge \exp\{-n[K(\vartheta,\theta_0)+\epsilon]\}\int_{r_n(s)\ge -n[K(\vartheta,\theta_0)+\epsilon]} \phi_n(\alpha_n;s)d\mathbf{P}_\vartheta(s)$$

$$\ge \exp\{-n[K(\vartheta,\theta_0)+\epsilon]\}\{E_\vartheta\phi_n(\alpha_n) - \mathbf{P}_\vartheta(r_n(S) < -n[K(\vartheta,\theta_0)+\epsilon])\}.$$

Since $K(\vartheta,\theta_0)$ is bounded, it follows by Lemma 3.1(b) of Kallenberg (1981) that also $\|\vartheta-\theta_0\|$ is bounded and hence

$$n^{-1}r_n(S) \xrightarrow{\mathbf{P}_\theta} -K(\vartheta,\theta_0)$$

uniformly on the set $A = \{\theta_0 \in \Theta_0 : K(\vartheta,\theta_0) \le K(\vartheta,\Theta_0)+1\}$. Therefore

$$\liminf_{n\to\infty} -n^{-1}\log\alpha_n \le \liminf_{n\to\infty} -n^{-1}\log[\sup\{E_{\theta_0}\phi_n(\alpha_n);\theta_0\in A\}]$$

$$\le K(\vartheta,\Theta_0)+\epsilon+\liminf_{n\to\infty} -n^{-1}\log[E_\vartheta\phi_n(\alpha_n) -$$

$$\sup_{\theta_0\in A}\mathbf{P}_\vartheta(n^{-1}r_n(S) < -K(\vartheta,\theta_0)-\epsilon)] = K(\vartheta,\Theta_0)+\epsilon,$$

and since $\epsilon > 0$ was arbitrary chosen, (13) follows. □

The next lemma describes the behaviour of $N^+(\alpha,\beta,\theta)$.

LEMMA 4 *For each $\theta \in$ int Θ_1 we have*

$$\liminf_{\alpha \to 0} \frac{-\log \alpha}{N^+(\alpha, \beta, \theta)} \geq K(\theta, \Theta_0). \tag{14}$$

PROOF. Consider the tests which reject H_0 if $\|\bar{X}_n - \lambda(\theta)\|\sqrt{n} \geq c_1$, where the constant c_1 is so large that the powers of the tests at θ exceed β. For each $\theta_0 \in \Theta_0$ we get for some $c_2 > 0$ (not depending on θ_0)

$$\mathbb{P}_{\theta_0}(\|\bar{X}_n - \lambda(\theta)\|\sqrt{n} \geq c_1) =$$

$$\int_{\|\bar{x} - \lambda(\theta)\|\sqrt{n} \geq c_1} \exp\{n(\theta_0 - \theta)^T \bar{x} - n\psi(\theta_0) + n\psi(\theta)\} d\bar{P}_\theta^n(\bar{x})$$

$$\leq \exp\{-nK(\theta, \theta_0) + n^{1/2}\|\theta_0 - \theta\| c_1\} \leq \exp\{-nK(\theta, \theta_0) + n^{1/2} c_2 K(\theta, \theta_0)\}$$

$$= \exp[-nK(\theta, \theta_0)\{1 - c_2 n^{-1/2}\}],$$

where the last inequality follows from the convexity of $K(\theta, \cdot)$ and the fact that $\theta \in$ int Θ_1. Therefore, for all $n > c_2^2$,

$$\sup_{\theta_0 \in \Theta_0} \mathbb{P}_{\theta_0}(\|\bar{X}_n - \lambda(\theta)\|\sqrt{n} \geq c_1) \leq \exp[-nK(\theta, \Theta_0)\{1 - c_2 n^{-1/2}\}]. \tag{15}$$

For a given α let m be the smallest integer such that the right-hand side of (15) is less than or equal to α. Then we get a level-α test with power at least β at θ and hence $N^+(\alpha, \beta, \theta) \leq m$. Since $\lim_{\alpha \to 0} -m^{-1} \log \alpha = K(\theta, \Theta_0)$ the result is obtained. □

PROOF OF THEOREM 2. Take $1 > \beta > \limsup_{n \to \infty} E_\theta \phi_n(\alpha_n)$, then $N^\phi(\alpha_n, \beta, \theta) > n$ for sufficiently large n. On the other hand it follows from Lemma 4. that

$$\liminf_{\alpha \to 0} -\frac{\log \alpha}{N^+(\alpha, \beta, \theta)} \geq K(\theta, \Theta_0).$$

In view of lemma 3 we therefore get

$$\limsup_{n \to \infty} \frac{N^\phi(\alpha_n, \beta, \theta)}{N^+(\alpha_n, \beta, \theta)} \geq \limsup_{n \to \infty} \frac{n}{(-\log \alpha_n)} \liminf_{n \to \infty} \frac{(-\log \alpha_n)}{N^+(\alpha_n, \beta, \theta)}$$

$$\geq \frac{K(\theta, \Theta_0)}{K(\vartheta, \Theta_0)} > 1$$

and thus (4) does not hold, which completes the proof. □

Many test statistics are measuring the "distance" between the observations and the null hypothesis in one or another way. To be more explicit consider test statistics of the form

$$T_n(X_1, \ldots, X_n) = d(\lambda^{-1}(\bar{X}_n), \Theta_0) \text{ if } \bar{X}_n \in \Lambda. \tag{16}$$

(Note that $\bar{X}_n$ is sufficient for θ.) If d is different from the Kullback- Leibler "distance" then the family of tests based on T_n corresponding to d is not efficient, as is seen from the following corollary.

COROLLARY 5 *Let* $\{\phi_n(\alpha)\}$ *be a family of tests which reject the hypothesis for large values of* T_n, *given by (16). Suppose that there exist* $\theta, \vartheta \in$ *int* Θ_1 *such that*

$$d(\vartheta, \Theta_0) > d(\theta, \Theta_0), \text{ while } K(\vartheta, \Theta_0) < K(\theta, \Theta_0). \tag{17}$$

Further assume that $d(\cdot, \Theta_0)$ *is continuous at* θ *and* ϑ. *Then* $\{\phi_n(\alpha)\}$ *is not efficient in the sense of Bahadur at* θ.

PROOF. Let a satisfy $d(\theta, \Theta_0) < a < d(\vartheta, \Theta_0)$. Let $\alpha_n = \sup\{\mathbb{P}_{\theta_0}(T_n > a) : \theta_0 \in \Theta_0\}$. Then

$$E_\theta \phi_n(\alpha_n) = \mathbb{P}_\theta(T_n > a) \leq \mathbb{P}_\theta(d(\lambda^{-1}(\bar{X}_n), \Theta_0) > a, \bar{X}_n \in \Lambda) + \mathbb{P}_\theta(\bar{X}_n \notin \Lambda)$$

and hence $\lim_{n\to\infty} E_\theta \phi_n(\alpha_n) = 0$. Similarly, $\lim_{n\to\infty} E_\vartheta \phi_n(\alpha_n) = 1$ and thus (11) and (12) hold. Application of Theorem 2 yields the result. □

REMARK 1. Although in many applications d will behave like a distance, formally this is not necessary in Corollary 5. Condition (17) and the continuity of d are sufficient to obtain the results. □

When taking the multinomial distribution, a simple hypothesis $\Theta_0 = \{\theta_0\}$, and d corresponding to the chi-square distance, Corollary 5 shows that the chi-square test is not efficient on "most" of the parameter space, thus obtaining the results on p.31,32 of Bahadur (1971). The same result holds true for tests based upon the Cressie-Read (1984) test statistic $R^\lambda (\lambda \neq 0)$. As another example we compare the LR test, Rao's test and Wald's test in a typical example.

EXAMPLE 1. Let $X_1, X_2, \ldots$ be i.i.d. r.v.'s each with a normal $N(0, \sigma^2)$-distribution. Consider the testing problem $H_0 : \sigma^2 = 1$ against $H_1 : \sigma^2 \neq 1$. The $N(0, \sigma^2)$-distribution may be written in the form (2) with μ the distribution of X_1^2, when X_1 has a $N(0,1)$-distribution, yielding $\psi(\theta) = -\frac{1}{2}\log(1-2\theta)$. The null hypothesis corresponds to $\theta = 0$ and H_1 to $\theta \neq 0$. Writing $\bar{Y}_n = n^{-1} \sum_{i=1}^n X_i^2$ the LR test statistic can be taken as

$$2K(\lambda^{-1}(\bar{Y}_n), 0) \text{ with } K(\theta, 0) = \tfrac{1}{2}\log(1-2\theta) + \theta(1-2\theta)^{-1} \text{ and}$$

$$\lambda^{-1}(\bar{Y}_n) = \tfrac{1}{2}(1 - \bar{Y}_n^{-1}).$$

The Wald test statistic (in the original parametrization in terms of σ^2) equals

$$d^W(\lambda^{-1}(\bar{Y}_n), 0) \text{ with } d^W(\theta, 0) = 2\theta^2,$$

while the score test statistic proposed by Rao is given by

$$d^R(\lambda^{-1}(\bar{Y}_n),0) \text{ with } d^R(\theta,0) = 2\theta^2(1-2\theta)^{-2}.$$

Since for each $\theta \in \Theta$ with $\theta > 0$ we have $K(\theta,0) > K(-\theta,0)$ and both $d^W(\cdot,0)$ and $K(\cdot,0)$ are continuous, it is easily seen that for each $\theta \in \Theta$ with $\theta > 0$ the conditions of Corollary 5 with $d = d^W$ are satisfied and hence Wald's test is not efficient in the sense of Bahadur at θ for each $\theta \in \Theta$ with $\theta > 0$. Similarly, if $\theta < 0$ and $d^R(\theta_1,0) = d^R(\theta,0)$ with $\theta_1 > 0$ we get $K(\theta_1,0) < K(\theta,0)$ and hence again by continuity of $d^R(\cdot,0)$ and $K(\cdot,0)$ the conditions of Corollary 5 are satisfied, implying that Rao's test is not efficient in the sense of Bahadur at θ for each $\theta < 0$.

In fact the following ordering holds with respect to Bahadur efficiency

$$\begin{aligned} W = LR > R \text{ if } \sigma^2 < 1(\text{ or } \theta < 0) \\ R = LR > W \text{ if } \sigma^2 > 1(\text{ or } \theta > 0) \end{aligned} \tag{18}$$

In the next section this example will be discussed further on, considering also other efficiency concepts and comparing them with numerical results.□

3 Efficiency concepts as approximation tools

Exact power comparison of tests is usually impossible. Therefore one relies on simplifications by taking limits in one or another way. By the simplifications we get in many cases simple answers indicating for instance preference of one test above the others. Whether the same preference still is appropriate in the finite sample case, is not clear. This depends on whether the invoked limiting results give good approximations. In this section we start the discussion of efficiency concepts as approximation tools with considering the LR test, Wald's and Rao's test in the simple, but rather typical example of testing the variance of normal r.v.'s, as presented in section 2. This example is also discussed in Madansky (1989), while Peers (1971) and Chandra and Joshi (1983) discuss the problem in a more general context. Since the main features of the comparison of the three tests can be illustrated in this (simple) example and finite-sample power can be obtained, we restrict ourselves here to this example. So, let $X_1, X_2, \ldots$ be i.i.d. r.v.'s each with a normal $N(0,\sigma^2)$-distribution. Consider the testing problem $H_0 : \sigma^2 = 1$ against $H_1 : \sigma^2 \neq 1$. If the level of significance equals α, the acceptance regions of the three tests are

LR test:

$$c_{n\alpha}^{LR} < \Sigma X_i^2 < d_{n\alpha}^{LR} \text{ with } P_{\sigma^2=1}(c_{n\alpha}^{LR} < \Sigma X_i^2 < d_{n\alpha}^{LR}) = 1-\alpha \tag{19}$$

$$\text{and } n\log(d_{n\alpha}^{LR}/c_{n\alpha}^{LR}) = d_{n\alpha}^{LR} - c_{n\alpha}^{LR}$$

Wald's test:

$$c_{n\alpha}^W < \Sigma X_i^2 < d_{n\alpha}^W \text{ with } P_{\sigma^2=1}(c_{n\alpha}^W < \Sigma X_i^2 < d_{n\alpha}^W) = 1-\alpha \qquad (20)$$

$$\text{and } d_{n\alpha}^W = \begin{cases} \infty & \geq \\ & \text{if } P_{\sigma^2=1}(\Sigma X_i^2 < \tfrac{1}{2}n) \quad \alpha \\ nc_{n\alpha}^W(2c_{n\alpha}^W - n)^{-1} & \leq \end{cases}$$

Rao's test:

$$c_{n\alpha}^R < \Sigma X_i^2 < d_{n\alpha}^R \text{ with } P_{\sigma^2=1}(c_{n\alpha}^R < \Sigma X_i^2 < d_{n\alpha}^R) = 1-\alpha \qquad (21)$$

$$\text{and } c_{n\alpha}^R = \begin{cases} 0 & \geq \\ & \text{if } P_{\sigma^2=1}(\Sigma X_i^2 > 2n) \quad \alpha \\ 2n - d_{n\alpha}^R & \leq \end{cases}$$

Although all three tests are "two-sided chi-square tests", none of them coincides with the commonly used equal tailed two-sided chi-square test (cf. for instance Lehmann (1959), p.130). Chandra and Joshi (1983) superposed on the tests the condition of equal tails, thus obtaining one common test, different from the three tests under consideration. This explains the difference between Peers' and Chandra and Joshi's results, cf. also Madansky (1989).

In Example 1 the nonlocal behaviour was discussed. Here we supply it with an investigation of the local behaviour and a discussion of both approaches by comparing them with numerical results for finite sample size. For the local approach we consider alternatives σ^2 tending to 1. Define $N^{LR}(\alpha,\beta,\sigma^2)$ as the number of observations necessary to obtain power $\beta \in (\alpha,1)$ at the alternative $\sigma^2(\neq 1)$ with the level-α LR test; similarly we define $N^W(\alpha,\beta,\sigma^2)$ and $N^R(\alpha,\beta,\sigma^2)$. By straightforward calculus one obtains the following local expansions as $\sigma^2 \to 1$

$$N^{LR}(\alpha,\beta,\sigma^2) = 2t^2(1-\sigma^2)^{-2} + f(t)(1-\sigma^2)^{-1} + o((1-\sigma^2)^{-1}), \qquad (22)$$

$$N^W(\alpha,\beta,\sigma^2) = 2t^2(1-\sigma^2)^{-2} + \{f(t) - \tfrac{8}{3}u_{\alpha/2}^2\}(1-\sigma^2)^{-1} + o((1-\sigma^2)^{-1}), \quad (23)$$

$$N^R(\alpha,\beta,\sigma^2) = 2t^2(1-\sigma^2)^{-2} + \{f(t) + \tfrac{4}{3}u_{\alpha/2}^2\}(1-\sigma^2)^{-1} + o((1-\sigma^2)^{-1}), \quad (24)$$

where $\Phi(u_{\alpha/2}) = \alpha/2$ with Φ the standard normal distribution function, where $t = t(\alpha,\beta) > 0$ is determined by $\Phi(u_{\alpha/2}+t) + \Phi(u_{\alpha/2}-t) = \beta$, and where

$$f(t) = \tfrac{2}{3}\left[-1 - 2t^2 - tu_{\alpha/2}\left\{\frac{\phi(u_{\alpha/2}+t) + \phi(u_{\alpha/2}-t)}{\phi(u_{\alpha/2}+t) - \phi(u_{\alpha/2}-t)}\right\}\right]$$

with ϕ the standard normal density.

These results show that locally the three tests are equivalent at first order, i.e. their asymptotic relative Pitman efficiencies with respect to each other are equal to one. In such a case it is of interest to consider the concept of Pitman deficiency (cf. Hodges and Lehmann (1970)). Writing $g(\sigma^2) \sim h(\sigma^2)$ for $\lim_{\sigma^2 \to 1} g(\sigma^2)/h(\sigma^2) = 1$, we obtain (as $\sigma^2 \to 1$)

$$N^{LR}(\alpha, \beta, \sigma^2) - N^{W}(\alpha, \beta, \sigma^2) \sim \tfrac{8}{3} u^2_{\alpha/2}(1-\sigma^2)^{-1} \tag{25}$$

implying that the LR test is better than Wald's test if $\sigma^2 > 1$ and reversed if $\sigma^2 < 1$. Similarly we have (as $\sigma^2 \to 1$)

$$N^{LR}(\alpha, \beta, \sigma^2) - N^{R}(\alpha, \beta, \sigma^2) \sim -\tfrac{4}{3} u^2_{\alpha/2}(1-\sigma^2)^{-1} \tag{26}$$

and

$$N^{R}(\alpha, \beta, \sigma^2) - N^{W}(\alpha, \beta, \sigma^2) \sim 4 u^2_{\alpha/2}(1-\sigma^2)^{-1} \tag{27}$$

So we have the ordering

$$\begin{aligned} W > LR > R \text{ if } \sigma^2 < 1 \\ R > LR > W \text{ if } \sigma^2 > 1. \end{aligned} \tag{28}$$

This ordering agrees with Peer's (1971) result in example (a), p.586. Note that $(1-\sigma^2)^{-1}$ is of order $N^{1/2}$ (cf.(22) - (24)) and hence the deficiency is of the rather unfamiliar order $N^{1/2}$. This does *not* contradict the well-known phenomenon of "first order efficiency implies second order efficiency" (cf. for instance Bickel, Chibisov and Van Zwet (1981), Kallenberg (1983)), because none of the three tests is Pitman-efficient, i.e. none of the three tests has Pitman efficiency greater than or equal to one with respect to *each* other test.

What can be concluded from the investigations in section 2 and 3?

1. Rao's conjecture that the test proposed by him is "locally" more powerful than the others does not come true.
2. None of the three tests is "locally" more powerful than the others for all (local) alternatives simultaneously.
3. The LR test is never worst, never best in a local sense.
4. For (very) small levels of significance the LR test is optimal.

To examine how well the asymptotic results reflect the performance of the tests for small and moderate sample sizes some numerical results for finite sample sizes are presented. Note that the Pitman efficiency $e^P_{LR,R}$ and the Bahadur efficiency $e^B_{LR,R}$ of the LR test with respect to Rao's test are defined by

$$e^P_{LR,R} = \lim_{\sigma^2 \to 1} \frac{N^R(\alpha, \beta, \sigma^2)}{N^{LR}(\alpha, \beta, \sigma^2)} \text{ and } e^B_{LR,R} = \lim_{\alpha \downarrow 0} \frac{N^R(\alpha, \beta, \sigma^2)}{N^{LR}(\alpha, \beta, \sigma^2)},$$

respectively, and similarly for the other tests. Further note that all Pitman efficiencies are equal to one. The quantities d^P are defined by

$$d^P_{LR,R} = -\tfrac{4}{3}u^2_{\alpha/2}(1-\sigma^2)^{-1}\; d^P_{R,W} = 4u^2_{\alpha/2}(1-\sigma^2)^{-1},$$

which are the first order terms in the expansions of the deficiencies (cf. (26), (27)).

Inspection of Table 1 and 2 leads to the following remarks:

1. the ordering of the tests as given in (18) and (28) is affirmed, just as the above conclusions;
2. Bahadur efficiency is a slightly better predictor than Pitman efficiency;
3. Pitman deficiency is a much better indicator (see especially $N^{LR}-N^R$ compared with $d^P_{LR,R}$);
4. Rao's test is not locally optimal: if $\sigma^2 = 0.9$ and $\alpha = 0.05$ one needs 4.2 times as much observations for Rao's test as for Wald's tests to obtain power 0.1!

Although the qualitative statements (18) and (28) derived from the efficiency calculations are in reasonable agreement with the numerical results, quantitatively the Pitman and Bahadur efficiencies may sometimes be rather different from the (finite-sample) relative efficiencies. So it is questionable whether the limiting efficiencies are suitable as *approximation* tools. A promising new approach in this area is the so-called global approximation, where it is tried to write the limiting results in such a form that the same smooth formula can be used in local and nonlocal approaches. This is obtained by avoiding to insert local properties of the parameters as long as possible, thus getting results with a global character. In the next section global approximations for the Cressie-Read statistics in goodness-of-fit are investigated.

4 Local families

Pitman efficiency is often interpreted as an approximation to the relative efficiency (1). As is illustrated in section 3 sometimes such an approximation may be rather poor. Technically, Pitman efficiency is calculated as the limit of (1) as $\theta \to \Theta_0$. An aspect, which has not attained much attention so far, is the way along which one goes from alternative to hypothesis. Especially in goodness-of-fit testing problems one meets this question, although in all testing problems the same point can and should be made. Consider, for instance, the testing problem of an exponential null hypothesis with unknown scale, and let G be any given alternative distribution

Table 1. $\alpha = 0.05$

β	σ^2	N^{LR}	N^W	N^R	N^R/N^{LR}	N^R/N^W	$e^B_{LR,R}$	$e^B_{W,R}$	$N^{LR} - N^R$	$N^R - N^W$	$d^P_{LR,R}$	$d^P_{R,W}$
0.5	0.5	19	15	29	1.5	1.9	2.0	2.0	-10	14	-10	31
	0.8	161	127	186	1.2	1.5	1.3	1.3	-25	59	-26	77
	1.2	226	293	199	0.9	0.7	1	0.6	27	-94	26	-31
	1.5	45	82	34	0.8	0.4	1	0.4	11	-48	10	-77
0.9	0.5	45	38	55	1.2	1.4	2.0	2.0	-10	17	-10	31
	1.5	129	162	118	0.9	0.7	1	0.4	11	-44	10	-31
0.1	0.9	81	32	133	1.6	4.2	1.1	1.1	-52	101	-51	154
	1.1	91	213	40	0.4	0.2	1	0.8	51	-173	51	-154

Table 2. $\alpha = 0.01$

β	σ^2	N^{LR}	N^W	N^R	N^R/N^{LR}	N^R/N^W	$e^B_{LR,R}$	$e^B_{W,R}$	$N^{LR} - N^R$	$N^R - N^W$	$d^P_{LR,R}$	$d^P_{R,W}$
0.5	0.5	33	29	55	1.7	1.9	2.0	2.0	-22	26	-18	53
	0.8	282	238	329	1.2	1.4	1.3	1.3	-47	91	-44	133
	1.2	384	516	342	0.9	0.7	1	0.6	42	-174	44	-133
	1.5	74	152	60	0.8	0.4	1	0.4	14	-92	18	-53
0.9	0.5	66	59	86	1.3	1.5	2.0	2.0	-20	27	-18	53
	1.5	178	247	161	0.9	0.7	1	0.4	17	-86	18	-53
0.1	0.9	322	227	417	1.3	1.8	1.1	1.1	-95	190	-88	265
	1.1	349	585	268	0.8	0.5	1	0.8	81	-317	88	-265

function. In local theory one often considers the contamination family of scale distributions

$$G_\theta(x) = (1-\theta)\{1-\exp(-x/\sigma)\} + \theta G(x/\sigma). \tag{29}$$

The idea is that the behaviour at G is reflected in the asymptotic behaviour for $n \to \infty$ and $\theta \to 0$. In the sense of approximation one derives limiting formulae and inserts $\theta = 1$ in these formulae for the approximation. Taking G in (29) equal to the exponential distribution $1 - \exp(-x/4)$, it is clear that G belongs to the null set, but the members of the contamination family G_θ for $0 < \theta < 1$ not! It can be shown that generalized chi-square tests give appreciable asymptotic power at the local alternatives $G_\theta(\theta \to 0)$, leading also to Pitman efficiencies different from 1 for several members of the class of generalized chi-square tests. It is clear that here the power behaviour at G is not reflected in the asymptotic power behaviour for local alternatives G_θ in the contamination family (29). (For more details see Drost (1989) p. 1294 and Drost (1988) p. 25).

The same problem of how to connect alternative and null set arises when one considers local limits of (exact or approximate) Bahadur efficiency. In this section the role of the local family will be discussed.

Let $Y_1, \ldots, Y_n$ be i.i.d. observations with density θ^Y and consider the testing problem $H_0 : \theta^Y = \theta_0$. Let T_n be some general goodness-of-fit statistic for H_0. At some interesting alternative θ_1 we want to calculate the minimal sample size to achieve a power of (say) 0.8. Often exact power computations are troublesome and an asymptotic approach is adopted. However, if the sample size increases, the power at θ_1 of any reasonable test tends to one. Hence we suppose that θ_1 is embedded in some "nice" family $\mathcal{F}(\theta_1) = \{\theta_t : 0 \le t \le 1\}$ between θ_1 and θ_0. Suppose that T_n converges in distribution under H_0 and under a sequence of contiguous alternatives $\{\theta_t\}$. Typically these limit distributions depend on $\lim_{n\to\infty} nt^2, \theta_1$ and the family $\mathcal{F}(\theta_1)$. We denote them by $\mathcal{L}(\lim_{n\to\infty} nt^2, \theta_1; \mathcal{F}(\theta_1))$. Since the family $\mathcal{F}(\theta_1)$ forms a bridge between the hypothesis θ_0 and the alternative of interest θ_1, this result is often applied as an approximation theorem in the following way. Substituting $t = 1$ the power of T_n at θ_1 is calculated via an approximation of the true distribution by $\mathcal{L}(n, \theta_1; \mathcal{F}(\theta_1))$. Generally the approximation depends on the family $\mathcal{F}(\theta_1)$. Since this family serves merely as a bridge between θ_0 and θ_1 the choice of an appropriate family is crucial. Often the contamination family is used. Pitman efficiencies are also evaluated along this family or along some "natural" parametric family between θ_0 and θ_1. We want to show that other families are of interest too. We restrict attention to a discretized sampling scheme (with k cells). Densities θ are represented as k-dimensional probability vectors. For each Cressie-Read (1984) test statistic we derive an appropriate family $\mathcal{F}(\theta)$ such that approximations are accurate along this family. It is well-known that the Cressie-Read statistics are equivalent in the sense of Pitman efficiency.

However, when evaluating Pitman efficiencies along appropriate families a second order difference shows up.

Let N_{in} be the number of observations in the i-th cell ($i = 1, \ldots, k$), put $N_n = (N_{1n}, \ldots, N_{kn})^T$, T denoting the transpose. It is assumed that N_n follows a multinomial (n, θ) distribution for some k-dimensional probability vector θ. Consider the class of Cressie-Read (1984) test statistics

$$R^\lambda = 2n\, I^\lambda(N_n/n, \theta_0), \tag{30}$$

where I^λ denotes the directed divergence measure (cf. Rényi (1961))

$$I^\lambda(q,p) = \{\lambda(\lambda+1)\}^{-1}\, \Sigma\, q\{(q/p)^\lambda - 1\}.$$

(The summation Σ of a vector denotes the sum of the components; products, ratios and other operations between vectors are defined componentwise.) Define I^λ by continuity when $\lambda = 1, 0$. The class (30) contains several well-known test statistics, e.g. Pearson's χ^2-statistic X^2 ($\lambda = 1$), the likelihood ratio statistic G^2 ($\lambda = 0$), Freeman-Tukey's statistic FT ($\lambda = -\frac{1}{2}$), the modified likelihood ratio statistic $\tilde{G}^2$ ($\lambda = -1$) and Neyman's modified χ^2-statistic $\tilde{X}^2$ ($\lambda = -2$). Cressie and Read (1984) advise $\lambda = \frac{2}{3}$. See Moore (1984) for an interpretation of I^λ as a measure of lack of fit. It is well-known that, under H_0, R^λ converges to a χ^2-distribution with $k-1$ degrees of freedom. Under a sequence of contiguous alternatives $\{\theta_t\}$ R^λ converges to a noncentral χ^2_{k-1}-distribution with noncentrality parameter

$$\lim_{n\to\infty} \Sigma(\theta_t - \theta_0)^2/\theta_0 = \lim_{n\to\infty} nt^2 \Sigma\{\partial\theta_t/\partial t|_{t=0}\}^2/\theta_0 \equiv \lim_{n\to\infty} nt^2 \Delta_{\mathcal{F}}(\theta_1). \tag{31}$$

Substituting $t = 1$ the power of R^λ is calculated via an approximation of the true distribution by a noncentral χ^2_{k-1}-distribution with noncentrality parameter $n\Delta_{\mathcal{F}}(\theta_1)$. Here we investigate the choice of the bridge $\mathcal{F}(\theta)$. Generally the contamination family (CF)

$$\mathcal{F}_c(\theta) = \{\theta_t = (1-t)\theta_0 + t\theta : 0 \le t \le 1\}$$

is considered (cf. e.g. Gvanceladze and Chibisov (1979), Quine and Robinson (1985)). Kallenberg et al. (1985) and Kallenberg (1990) consider also the exponential family (EF)

$$\mathcal{F}_e(\theta) = \{\theta_t = c_\theta(t)\theta_0^{1-t}\theta^t : 0 \le t \le 1\},$$

where $c_\theta(t)$ is a normalizing constant. Observe that these families yield different power approximations because the parameters

$$\Delta_c(\theta) \equiv \Delta_{\mathcal{F}_c}(\theta) = \Sigma(\theta - \theta_0)^2/\theta_0 \text{ and}$$

$$\Delta_e(\theta) \equiv \Delta_{\mathcal{F}_e}(\theta) = \Sigma\theta_0 \log^2(\theta/\theta_0) - \{\Sigma\theta_0 \log(\theta/\theta_0)\}^2$$

are in general different. Simulations show that the EF is not very succesful in approximating the power of X^2 and G^2. The CF gives only a partial solution: for X^2 the approximations are very accurate but for G^2 errors of ten per cent or more are quite common (cf. Kallenberg et al.(1985)). For X^2 Kallenberg (1990) gives a theoretical explanation: he shows that the true distribution of X^2 is closer to the approximate distribution along the CF than to the approximate distribution along the EF. E.g. when testing discretized $N(0,1)$ versus discretized $N(1,1)$ the nonnormal CF $(1-t)N(0,1)+tN(1,1)$ yields better approximations than the normal EF $N(t,1)$ (cf. Kallenberg (1990)).

In the following theorem we derive a sufficient condition upon the family $\mathcal{F}(\theta)$ to obtain accurate power approximations for R^λ (cf. Theorem 6). As a consequence the CF is appropriate iff $\lambda = 1$, i.e. for X^2. Second order efficiency considerations lead to an appropriate choice of λ. In Example 2 we derive a class of families $\mathcal{F}_\lambda(\theta)$ such that approximations are also accurate for R^λ. It turns out that the EF corresponds to Neyman's modified chi-square test.

Denoting by $\|\cdot\|$ the Euclidean norm and by $\chi^2_k(\delta)$ an r.v. with a noncentral chi-square distribution with k degrees of freedom and noncentrality parameter δ, we have the following result.

THEOREM 6 *Let $\mathcal{F}(\theta)$ be some family between θ and θ_0 and assume*

$$\Delta_{\mathcal{F}}(\theta) = \Delta_c(\theta) + \tfrac{1}{3}(\lambda-1)\Sigma(\theta-\theta_0)^3/\theta_0^2 + O(\|\theta-\theta_0\|^4) \text{ as } \theta \to \theta_0. \tag{32}$$

Let $\{s_n\}, s_n > 0$ be a nondecreasing sequence such that $s_n^2/\sqrt{n} = O(1)$ as $n \to \infty$ and put $\Theta(s_n) = \{\theta : \|\theta - \theta_0\| \le s_n/\sqrt{n}\}$. Then as $n \to \infty$

$$\sup_{\theta\in\Theta(s_n)} \sup_{c>0} |P_\theta(R^\lambda > c) - \Pr(\chi^2_{k-1}(n\Delta_{\mathcal{F}}(\theta)) > c)| = O(s_n/\sqrt{n}) \tag{33}$$

and

$$\sup_{\theta\in\Theta(s_n)} \sup_{c>0} |P_\theta(R^\mu > c) - \Pr(\chi^2_{k-1}(n\Delta_{\mathcal{F}}(\theta)) > c)| = O(s_n^2/\sqrt{n}), \tag{34}$$

when $\mu \ne \lambda$. The second bound is sharp as $s_n \to \infty$.

PROOF. Similar to the proof of Theorem 2.1 of Kallenberg (1990), see also Drost et al. (1989). □

Note that (32) can be realized by taking the Rényi-type families as presented further on in Example 2.

REMARK 2 The conditions of Theorem 6 can be relaxed. Suppose $s_n^2/\sqrt{n} \to \infty$, then the error bound in (34) is of order $O(1)$ and this bound is sharp if $s_n/\sqrt{n} \to 0$; moreover, the error bound in (33) is in that case of order $O(s_n^3/n)$. Even for these distant alternatives the maximum approximation error along $\mathcal{F}(\theta)$ tends to zero for R^λ if $s_n^3 n^{-1} \to 0$, while the approximation may be useless for R^μ when $\mu \neq \lambda$. Similar extensions are possible if the remainder term in (32) is replaced by a term of order $o(\|\theta - \theta_0\|^3)$. □

REMARK 3 Suppose that the family $\mathcal{F}$ is nested, i.e. $\mathcal{F}(\theta_u) \subset \mathcal{F}(\theta)$ and $(\theta_u)_t = \theta_{ut}$ for $0 \leq u, t \leq 1$. Then $\Delta_{\mathcal{F}}(\theta) = \Delta_c(\theta) + o(\|\theta - \theta_0\|^2)$ as $\theta \to \theta_0$, since

$$\lim_{\theta \to \theta_0} [\Delta_{\mathcal{F}}(\theta) - \Delta_c(\theta)]/\|\theta - \theta_0\|^2 = \lim_{u \to 0} [\Delta_{\mathcal{F}}(\theta_u) - \Delta_c(\theta_u)]/\|\theta_u - \theta_0\|^2$$

$$= \lim_{u \to 0} \lim_{t \to 0} [\Sigma\{(\theta_{ut} - \theta_0)/(ut)\}^2/\theta_0 - \Sigma\{(\theta_u - \theta_0)/u\}^2/\theta_0]/\Sigma\{(\theta_u - \theta_0)/u\}^2 = 0.$$

Hence second order cross-terms are absent in nested families. Unfortunately this is not in general true for third order cross-terms. Condition (32) is not automatically satisfied for nice families. □

Observe that the error bounds are equivalent for contiguous alternatives (i.e. s_n remains bounded). For noncontiguous alternatives the family $\mathcal{F}(\theta)$ yields better approximations for R^λ than for R^μ ($\mu \neq \lambda$). It is well-known that the Cressie-Read statistics are equivalent in the sense of Pitman efficiency. On the other hand the LR test G^2 turns out to be optimal in the sense of Bahadur slope. Simulations, however, show a variety of power differences between the R^λ tests. To decide which test statistic to use the efficiency considerations seem to be not very helpful. Both approaches are rather extreme: we are not interested in alternatives very close to the hypothesis nor in the case of very small significance levels. This is also seen in the choice of alternatives and levels in simulation studies. Theorem 6 provides a more illuminating view. Using Pitman efficiency of $R^{\lambda(1)}$ and $R^{\lambda(2)}$ as an approximation to the relative efficiency (1) at θ, the powers are compared along a sequence of contiguous alternatives $\{\theta_t\}$. If, however, we want to predict power or to explain simulated power or to approximate (1) at θ the sequence $\{\theta_t\}$ itself is of little interest. In the best case it provides a guideline. An accurate power approximation seems to be more valuable. Theorem 6 implies that at most one of the powers is accurately estimated under the sequence $\{\theta_t\}$ when $nt^2 \to \infty$. It is better to approximate the powers of $R^{\lambda(1)}$ and $R^{\lambda(2)}$ along different families $\mathcal{F}_{\lambda(1)}(\theta)$ and $\mathcal{F}_{\lambda(2)}(\theta)$ such that (32) is satisfied for $\lambda = \lambda(1)$ respectively $\lambda = \lambda(2)$ (see e.g. formula (36) in Example 2 with $a = \{\lambda(1) + 2\}/3$ respectively $a = \{\lambda(2) + 2\}/3$. Since the null-hypothesis distributions of $R^{\lambda(1)}$ and $R^{\lambda(2)}$ are the same up to high accuracy ($O(n^{-1/2})$), comparison of the powers of $R^{\lambda(1)}$ and

$R^{\lambda(2)}$, using the accurate power approximations provided by Theorem 6 with families $\mathcal{F}_{\lambda(1)}(\theta)$ and $\mathcal{F}_{\lambda(2)}(\theta)$, respectively, yields eventually comparison of noncentral chi-square distributions with the same degrees of freedom $(k-1)$ and noncentralities

$$n\Delta_{\mathcal{F}_{\lambda(1)}}(\theta) \text{ and } n\Delta_{\mathcal{F}_{\lambda(2)}}(\theta) \text{ respectively.}$$

Therefore to get the same approximate power for $R^{\lambda(1)}$ with n_1 observations and $R^{\lambda(2)}$ with n_2 observations we need

$$n_1\Delta_{\mathcal{F}_{\lambda(1)}}(\theta) = n_2\Delta_{\mathcal{F}_{\lambda(2)}}(\theta).$$

This leads to the following efficiency concept (cf. (1)):

$$e(R^{\lambda(1)}, R^{\lambda(2)}) = \Delta_{\mathcal{F}_{\lambda(1)}}(\theta)/\Delta_{\mathcal{F}_{\lambda(2)}}(\theta) \tag{35}$$

$$= 1 + \tfrac{1}{3}\{\lambda(1) - \lambda(2)\}\Delta_c(\theta)^{-1}\Sigma(\frac{\theta}{\theta_0} - 1)(\theta - \theta_0)^2/\theta_0 + O(\|\theta - \theta_0\|^2)$$

as $\theta \to \theta_0$. The first term on the right-hand-side shows that the Cressie-Read statistics are first order equivalent. The second term shows that there is a second order effect. Large values of λ are promising if there are high peaks in the vector θ/θ_0 while small values of λ are interesting for deep dips (cf. also sections 2.4 and 3.3 in Drost et al. (1989)). This corresponds perfectly to simulation studies (cf. Cressie and Read (1984), Drost et al. (1989)). If there are neither high peaks nor deep dips small values of λ seem to be more promising. Since the limiting null-distribution is difficult to control for extreme λ values we propose to use Pearson's chi-square test ($\lambda = 1$) if one is mainly interested in alternatives with high peaks and the likelihood ratio statistic ($\lambda = 0$) in all other cases.

REMARK 4 Approximate Bahadur efficiency is often qualified as a concept of little value, since monotone transformations of test statistics, yielding the same tests, may lead to entirely different approximate Bahadur efficiencies, cf. e.g. Bahadur (1960,pp.283,284), Gleser (1964, pp.1543, 1544). Note however also the remarks of Gleser on p.1544 about "its intuitive appeal as an approximate extension of the exact efficiency" and the responses of "that oracle" (electronic computer), indicating that it is just possible in certain circumstances that approximate Bahadur efficiency has more relevance than exact Bahadur efficiency (Bahadur (1967, p.312)). The efficiency (35) is in fact equal to the approximate Bahadur efficiency of $R^{\lambda(1)}$ w.r.t. $R^{\lambda(2)}$. Although the derivation of (35) using Theorem 6 with different local families is formally quite different from the concept of approximate Bahadur efficiency, in a sense Theorem 6 and its efficiency-implication (35) show that approximate Bahadur efficiency is useful in these circumstances. □

From the previous discussion it is seen that for approximation tools it is important to choose for each test statistic a suitable local family, guaranting the connection between the local limit and the alternative under consideration. The next example shows that (32) can be established.

EXAMPLE 2 *Rényi-type families.* Consider the family $\mathcal{F}_a(\theta)$ with θ_t given by

$$\theta_t = c_\theta(t)\{(1-t)\theta_0^a + t\theta^a\}^{1/a}, 0 \le t \le 1, \tag{36}$$

with $c_\theta(t)$ a normalizing constant. Straightforward calculations show

$$\Delta_{\mathcal{F}_a}(\theta) = \Sigma\theta_0 \left\{\frac{(\theta/\theta_0)^a - 1}{a}\right\}^2 - \left\{\Sigma\theta_0 \frac{(\theta/\theta_0)^a - 1}{a}\right\}^2$$

$$= \Delta_c(\theta) + (a-1)\Sigma(\theta-\theta_0)^3/\theta_0^2 + O(\|\theta-\theta_0\|^4) \text{ as } \theta \to \theta_0.$$

The Rényi family with parameter a does not correspond to the statistic R^a but to the Cressie-Read statistic with parameter $\lambda = 3a - 2$. E.g. the CF ($a = 1$) corresponds to the chi-square statistic, the EF ($a = 0$) corresponds to the modified chi-square statistic and the Hellinger type family $\mathcal{F}_{1/2}$ with

$$\theta_t = c_\theta(t)\{(1-t)\theta_0^{1/2} + t\theta_1^{1/2}\}^2$$

corresponds to the Freeman-Tukey statistic. Observe that the family $\mathcal{F}_a(\theta)$ is not nested when $a \neq 0, 1$.

To illustrate the strength of the efficiency criterion (35) and the power approximations suggested by Theorem 6 we re-examine various testing problems discussed in Kallenberg et al. (1985) and Drost et al. (1989). Here we use exact powers produced by randomized tests. A large variety of alternatives is considered (216 alternatives in case of 50 observations and 204 alternatives if $n = 100$). This yields the following table of power differences between the LR-test based upon G_n^2 and Pearson's χ^2-test based on X_n^2.

Table 3. Number of absolute power differences between G_n^2 and X_n^2 for $n = 50, 100$. Testing problems are taken from Kallenberg et al. (1985) and Drost et al. (1989).

absolute power difference	< 0.02	0.02 – 0.05	> 0.05
n=50	61%	32%	7%
n=100	59%	39%	3%

The efficiency criterion (35) predicts the best test in 86% (91%) of the cases if $n = 50$ ($n = 100$). Another 9% of the cases is accounted for by power differences less than 2%. Hence, in most of the cases leading to a false prediction of the best test the power difference is very small.

To obtain good power approximations for G_n^2 ($\lambda = 0$) and X_n^2 ($\lambda = 1$) the results above suggest to use the family with $a = \frac{2}{3}$ respectively $a = 1$. This suggestion is confirmed by Table 4.

Table 4. Number of approximation errors for G_n^2 and X_n^2 and $n = 50, 100$. Testing problems are taken from Kallenberg et al. (1985) and Drost et al. (1989).

approximation		error	< 0.02	$0.02 - 0.05$	> 0.05
G_n^2	$n = 50$	$a = 0$	45%	22%	33%
		$a = \frac{1}{2}$	75%	22%	3%
		$a = \frac{2}{3}$	77%	22%	1%
		$a = 1$	57%	22%	21%
	$n = 100$	$a = 0$	40%	21%	39%
		$a = \frac{1}{2}$	82%	18%	0%
		$a = \frac{2}{3}$	85%	15%	0%
		$a = 1$	48%	24%	29%
X_n^2	$n = 50$	$a = 0$	40%	19%	42%
		$a = \frac{1}{2}$	53%	28%	19%
		$a = 1$	82%	16%	2%
	$n = 100$	$a = 0$	32%	21%	47%
		$a = \frac{1}{2}$	52%	35%	13%
		$a = 1$	63%	30%	7%

□

Acknowledgement. The authors are much obliged to B.F.Schriever, who performed the numerical computations in section 3.

REFERENCES

Bahadur, R. R. (1960) Stochastic comparison of tests, *Ann. Math. Statist.* **31**, 276-295.

Bahadur, R. R. (1967) Rates of convergence of estimates and test statistics, *Ann. Math. Statist.* **38**, 303-324.

Bahadur, R. R. (1971) *Some Limit Theorems in Statistics*, SIAM, Philadelphia.

Bahadur, R. R., Gupta, J. C. and Zabell, S. L. (1980) Large deviations, tests and estimates, In *Asymptotic Theory of Statistical Tests and Estimation* (I. M. Chakravarti, ed.) 33-64, Academic, New York.

Bahadur, R. R., Chandra, T. K. and Lambert, D. (1984) Some further properties of likelihood ratios on general sample spaces, In *Statistics: Applications and New Directions* (J. K. Ghosh, J. Roy, eds.) 1-19, Indian Statist. Inst. Calcutta.

Bickel, P. J., Chibisov, D. M. and van Zwet, W. R. (1981) On efficiency of first and second order, *Int. Statist. Rev.* **49**, 169-175.

Chandra, T. K. and Joshi, S. N. (1983) Comparison of likelihood ratio, Rao's and Wald's tests and a conjecture of C. R. Rao, *Sankhyā Ser. A* **45**, 226-246.

Cressie, N. and Read, T. R. C. (1984) Multinomial goodness-of-fit tests, *J. Roy. Statist. Soc. Ser. B* **46**, 440-464.

Drost, F. C. (1988) *Asymptotics for Generalized Chi-square Goodness-of-fit Tests*, CWI Tract 48, Amsterdam.

Drost, F. C. (1989) Generalized chi-square goodness-of-fit tests for location-scale models when the number of classes tends to infinity, *Ann. Statist.* **17**, 1285-1300.

Drost, F. C., Kallenberg, W. C. M., Moore, D. S. and Oosterhoff, J. (1989) Power approximations to multinomial tests of fit, *J. Amer. Statist. Assoc.* **84**, 130-141.

Gleser, L. J. (1964) On a measure of test efficiency proposed by R. R. Bahadur, *Ann. Math. Statist.* **35**, 1537-1544.

Groeneboom, P. and Oosterhoff, J. (1981) Bahadur efficiency and small sample efficiency, *Int. Statist. Rev.* **49**, 127-141.

Gvanceladze, L. G. and Chibisov, D. M. (1979) On tests of fit based on grouped data, In *Contributions to Statistics, J. Hájek Memorial Volume* (J. Jurečková, ed.) 79-89, Academia, Prague.

Hodges, J. L. and Lehmann, E. L. (1970) Deficiency, *Ann. Math. Statist.* **41**, 783-801.

Hoeffding, W. (1965) Asymptotically optimal tests for multinomial distributions, *Ann. Math. Statist.* **36**, 369-401.

Kallenberg, W. C. M. (1981) Bahadur deficiency of likelihood ratio tests in exponential families, *J. Multivariate Anal.* **11**, 506-531.

Kallenberg, W. C. M. (1983) Asymptotic efficiency and deficiency of tests, *Proceedings of the 44th Session of the International Statistical Institute*, Bulletin of the ISI Vol. L., Bk. 2, 1173-1189.

Kallenberg, W. C. M. (1990) Inequalities for noncentral chi-square distributions, *Statist. Probab. Lett.* **9**, 273-278.

Kallenberg, W. C. M., Kourouklis, S. (1992) Hodges-Lehmann optimality of tests, *Statist. Probab. Lett.* **14**, 31-38.

Kallenberg, W. C. M., Oosterhoff, J. and Schriever, B. F. (1985) The number of classes in chi-square goodness-of-fit tests, *J. Amer. Statist. Assoc.* **80**, 959-968.

Kester, A. D. M. (1985) *Some Large Deviation Results in Statistics*, CWI Tract 18, Amsterdam.

Lehmann, E. L. (1959) *Testing Statistical Hypotheses*, Wiley, New York.

Madansky, A. (1989) A comparison of the likelihood ratio, Wald and Rao tests. In *Contributions to Probability and Statistics, Essays in Honor of Ingram Olkin* (L. J. Gleser, M. D. Perlman, S. J. Press, A. R. Sampson, eds.) 465-471, Springer-Verlag, New York.

Moore, D. S. (1984) Measures of lack of fit from tests of chi-square type, *J. Statist. Planning and Inf.* **10**, 151-166.

Nikitin, Ya. Yu. (1986) Hodges-Lehmann efficiency of nonparametric tests, *Proc. of the IV Vilnius Conference on Probab. Theory and Math. Statist.* Utrecht: VNU Science Press, vol. II, 391-408.

Nikitin, Ya. Yu. (1987) On the Hodges-Lehmann asymptotic efficiency of nonparametric tests of goodness-of-fit and homogeneity, *Theory Probab. Appl.* **32**, 77-85.

Peers, H. W. (1971) Likelihood ratio and associated test criteria, *Biometrika* **58**, 577-587.

Quine, M. P. and Robinson, J. (1985) Efficiencies of chi-square and likelihood ratio goodness-of-fit tests, *Ann. Statist.* **13**, 727-742.

Rényi, A. (1961) On measures of entropy and information, *Proc. Fourth Berkeley Symp. Math. Stat. Prob.* **1**, 547-561.

Serfling, R. J. (1980) *Approximation Theorems of Mathematical Statistics*, Wiley, New York.

Faculty of Applied Mathematics
University of Twente
P. O. Box 217,
7500 AE Enschede
The Netherlands

Econometric Institute
University of Tilburg
P. O. Box 90153
Tilburg
The Netherlands

J.K. Ghosh, S.K. Mitra, K.R. Parthasarathy and B.L.S. Prakasa Rao (*eds.*),
Statistics and Probability : A Raghu Raj Bahadur Festschrift
Published by Wiley Eastern Limited, Publishers, (1993), 325 – 347

Estimation of Hilbert Space Valued Parameters by the Method of Sieves[1]

G. KALLIANPUR
University of North Carolina at Chapel Hill

and

R.S. SELUKAR
University of South Alabama

Abstract

The approach developed by Ibragimov, Hasminski and others for investigating the consistency of the m.l.e. of a finite dimensional parameter is extended to the case of a Hilbert space valued parameter via the method of sieves. The generalization also yields a large deviation inequality for the sieve estimator. Examples are discussed, among them, the estimation of the drift function in a linear stochastic differential equation and the estimation of the intensity of a nonstationary Poisson process.

[1]Research supported by the Air Force Office of Scientific Research Contract No. 91-0030.

AMS Subject Classifications: Primary 62G05, secondary 62M99.

Key words : Hilbert space, sieve estimation, drift function, intensity estimation, large deviation.

1 Introduction

Many nonparametric estimation problems can be regarded as problems of estimation of infinite dimensional parameters. However, there are situations where a direct application of maximum likelihood (m.l.) method or its variants fails. Several possible estimation procedures have been considered in this case, the more popular ones being kernel estimation, penalized likelihood estimation and sieve estimation. Among these techniques the method of sieves, first systematically developed by Grenander in his 1981 book, appears to lend itself most naturally to an extension of the finite dimensional theory.

The idea of sieves is to restrict the parameter space to a subset S_α (α is a positive index called the mesh size) chosen such that the m.l. estimate exists. It is also required that as the mesh size tends to zero, the family S_α, called a sieve, eventually includes the entire parameter space and is sufficiently rich to allow the m.l. estimate to converge to the true parameter value [8].

When the parameter space is finite dimensional, the theory of local asymptotic normality, developed by Le Cam and presented in the book of Ibragimov and Hasminski has proved to be a powerful approach to asymptotic estimation (see [9]). The present paper is an attempt to extend the work of the latter authors to the case of infinite dimensional (Hilbert space valued) parameters. Such an attempt is natural and perhaps overdue, in view of the great success that this approach has enjoyed in asymptotic statistical theory.

There are several papers in the recent literature devoted to sieve estimation. We briefly mention some of those having a bearing on the subject of this paper. Under fairly general conditions Geman and Hwang [7] have established the strong consistency of the sieve estimators. In [17] Nguyen and Pham prove weak consistency of the sieve estimator of the drift function of a linear stochastic differential equation (SDE). There they also prove the asymptotic normality of certain functionals of this estimator. In [16] McKeague deals with a general semimartingale regression model and obtains strong consistency of certain sieve estimators. Nonparametric sieve estimation problems in point processes have been studied by Karr [10], Leskow [14], and by Leskow and Rozanski [13]. For density estimation by the method of sieves see Walter and Blum [5]. In [3, 4] Beder has considered nonparametric sieve estimation of the mean and covariance of a Gaussian process.

Among papers dealing with estimation of Hilbert space valued parameters we mention the work of Antoniadis [2] and Karr and Zucker [11]. In the latter paper the penalized partial likelihood method is used for the intensity estimation in a survival analysis problem. The parameter space is a Sobolev space and weak consistency is obtained in the uniform norm. The rate of convergence and pointwise asymptotic normality are also discussed. Another paper of Antoniadis [1] obtains a large deviation inequality for a sieve estimator of the mean of a Banach space valued Gaussian process.

The present paper is arranged in the following manner. Section 2 is devoted to the derivation of a basic large deviation inequality (Theorem 2.1) and its consequences. The latter include the strong consistency of the sieve estimator. Theorem 2.1 is a generalization of Ibragimov and Hasminski's result for the finite dimensional problem. In Section 3 the drift function estimation problem is re-examined. Under fairly mild assumptions on the parameter set, the results of Nguyen and Pham are strengthened. Poisson intensity estimation is briefly discussed in Section 4. Finally in Section 5, one of Antoniadis's results in [1] is derived as a consequence of Theorem 2.1 of this paper.

2 Large deviation inequality

Let $(\Omega^n, \Im^n, \mathbf{P}^{\mathbf{n}}_\theta)$, $n \geq 1$, be a sequence of probability spaces and let $\theta \in \Theta$. Points (observations) from Ω^n will be denoted by X^n. Assume that the parameter space Θ is a subset of some fixed infinite dimensional separable Hilbert space $\mathbf{H}$. Also assume that, for every n, there exists a σ-finite measure ν^n such that $\mathbf{P}^{\mathbf{n}}_\theta$ is absolutely continuous with respect to ν^n for every $\theta \in \Theta$. Let $f^n(X^n, \theta)$ be the corresponding density.

DEFINITION 1 **(sieve)** *A sequence $(\mathbf{S_k})$ of subsets of Θ is called a sieve if $\cup \mathbf{S_k}$ is a dense set in Θ and for each n and k the maximum of $f^n(X^n, \theta)$ over $\mathbf{S_k}$ is attained in $\mathbf{S_k}$.*

The above definition of a sieve is slightly narrow and a more general definition would admit, besides m.l.e.s, least squares estimates, partial m.l.e.s etc. The results of the present paper have obvious analogues in these other cases, however, in order to keep the treatment as direct as possible and the notation relatively simple, we will confine ourselves to the m.l.e.s only.

DEFINITION 2 **(sieve estimator)** *An estimator $\hat{\theta}^n_k(X^n)$ is called a sieve estimator corresponding to the sieve $\mathbf{S_k}$ if it is a point in $\mathbf{S_k}$ such that*

$$f^n(X^n, \hat{\theta}^n_k) = \sup_{t \in \mathbf{S_k}} f^n(X^n, t).$$

Our aim is to study the asymptotic behavior of the sieve estimator $\hat{\theta}^n_k$ corresponding to a sieve $(\mathbf{S_k})$, $k \geq 1$, of the following type :

i) for each $k \geq 1$, $\mathbf{S_k}$ is a compact, convex set in the span of $(g_j,\ 1 \leq j \leq k)$ where (g_j) is some fixed set of independent vectors from $\mathbf{H}$.

ii) $\mathbf{S_k} \subset \mathbf{S_{k+1}}$ and $\cup \mathbf{S_k}$ is dense in Θ.

REMARK 2.1 : It is assumed that for the problem in question such a sieve is known.

For any $\theta \in \Theta$, let θ_k denote its projection in $\mathbf{S_k}$ i.e.

$$||\theta - \theta_k|| = \inf_{t \in \mathbf{S_k}} ||\theta - t||.$$

Since $\mathbf{S_k}$ is compact and convex, θ_k always exists and is unique. Furthermore, $||\theta - \theta_k|| \to 0$ *as* $k \to \infty$.

Let $\theta \in \Theta$ be the true parameter and let $\mathbf{U_k^n}$ be defined as follows:

$$\mathbf{U_k^n} = (\phi_{n,k})(\mathbf{S_k} - \theta_k)$$

where $\phi_{n,k}$ are positive constants which will be required to satisfy suitable conditions later on. Suppose that Z_k^n are random functions such that, for $u \in \mathbf{U_k^n}$,

$$Z_k^n(u) = l_n(\theta_k + (\phi_{n,k})^{-1}u) - l_n(\theta_k)$$

where, for $t \in \Theta$, $l(X^n, t) = l_n(t)$ is the log likelihood function of the data at t. Then, since $\hat{\theta}_k^n$ maximizes $l_n(.)$ in $\mathbf{S_k}$ and $Z_k^n(0) = 0$ it is easy to verify that $\{ ||(\phi_{n,k})(\hat{\theta}_k^n - \theta_k)|| \geq h \} \subset \{ \sup_{||u||>h} Z_k^n(u) \geq 0 \}$. Therefore

$$\mathbf{P}_\theta^\mathbf{n}\{||(\phi_{n,k})(\hat{\theta}_k^n - \theta_k)|| \geq h\} \leq \mathbf{P}_\theta^\mathbf{n}\{ \sup_{||u||>h} Z_k^n(u) \geq 0\}. \tag{2.1}$$

The required large deviation inequality will be obtained by bounding the probability on the RHS above. In getting this bound we use arguments analogous to those in [9].

Let $\mathbf{G}$ be a class of real valued functions g defined on $[0, \infty)$ with the following two properties:

i) There exists x_0 such that g is positive and strictly increasing on $[x_0, \infty)$

ii) For every $N > 0$,

$$\int_0^\infty y^N \exp(-g(y))\, dy < \infty.$$

Throughout this article C, with or without a subscript, will denote a positive constant independent of n and k (it need not always be the same). Also, $[x]$ will denote the largest integer less than or equal to x.

PROPOSITION 2.1 *Suppose that for all integers* k, n, $Z_k^n(u)$ *are measurable and separable processes satisfying the following conditions:*

C1) There exist numbers $\alpha = \alpha(k) > k$, $m = m(k) \geq \alpha$ *and positive constants* B_k^n *such that for all* u, v,

$$E_\theta^n |Z_k^n(u)|^m \leq B_k^n \text{ and,}$$

$$E^n_\theta |Z^n_k(u) - Z^n_k(v)|^m \leq B^n_k \ ||u - v||^\alpha.$$

C2) There exists $g^n_k \in \mathbf{G}$ *such that for some* $\eta > 0$ *(independent of* n, k*),*

$$E^n{}_\theta \ \exp(\eta Z^n_k(u)) \ \leq \ \exp(-g^n_k(||u||)), \quad \forall \, u \in \mathbf{U^n_k}.$$

Then, for $h > x_0$,

$$\mathbf{P^n_\theta}\{||(\phi_{n,k})(\hat{\theta}^n_k - \theta_k)|| \ \geq h\} \ \leq \ B_1 \sum_{r=0}^{\infty} (1 + h + r)^{B_2} g^n_{kr} \ \exp(-b_1 g^n_{kr})$$

where
i) $x_0 \ = \ x_0(n, k)$ *is a point such that* g^n_k *is positive and increasing on* $[x_0, \ \infty)$;
ii) $g^n_{kr} \ = \ g^n_k(h + r)$,
iii) $b_1 = (\alpha - k)/(\alpha - k + mk) > 0$, $B_1 = C(B^n_k)^{1/m} D$ *where* $D = 64^k((1 - 2^{-(\alpha-k)/m})^{-1} + (2^{(m-\alpha)/m} - 1)^{-1})$.
iv) $B_2 = \alpha/m$.

Sufficient conditions for the existence of a consistent sieve estimator can be deduced from the above proposition.

THEOREM 2.1 *Assume the conditions of Proposition 2.1 and the following additional conditions.*
T1) The constants $m = m(k)$ *and* $\alpha = \alpha(k)$ *in condition* C1 *are constant multiples of* k*, the multiplying constant being independent of* k.
T2) $B^n_k \leq C^k \, n^{Ck} \, k^{Ck}$ *for some* C.
T3) $g^n_k(x) = n^\delta k^{-\mu} |x - a(n, k)|^\nu - t(n, k)$ *where* δ, μ *and* ν *are fixed positive numbers and* $t(n, k)$ *and* $a(n, k)$ *are non-negative constants which may depend on* n *and* k.
T4) There exists a positive number s*, less than* $\delta(\mu + 2)^{-1}$*, such that* $a(n, n^s) \to 0$, $t(n, n^s)/n^s \to 0$ *and* $|\phi_{n,[n^s]}| \to \infty$ *as* $n \to \infty$.

Then,
i) $\hat{\theta}^n_{[n^s]}$ *is a strongly consistent estimator of* θ.
ii) For any $h > 0$*, there exists a positive integer* $N = N(\theta, h)$ *such that, for* $n > N$,

$$\mathbf{P^n_\theta}\{ \, || \, \hat{\theta}^n_{[n^s]} - \theta \, || > h \, \} \ \leq \ \exp(-Cn^b \, h^\nu) \tag{2.2}$$

where the positive constant C *depends on* θ *alone and* $b = \delta - s\mu - s$.

REMARK 2.2 In the above setting if, in addition, we also know that $(\phi_{n,[n^s]})(\theta - \theta_{[n^s]})$ tends to zero then the following stronger version of Theorem 2.1 holds
1) $(\phi_{n,[n^s]})(\, \hat{\theta}^n_{[n^s]} - \theta \,) \ \to \ 0$ a.s. and
2) for any $h > 0$, there exists a positive integer $N = N(\theta, h)$ such that, for $n > N$,

$$\mathbf{P^n_\theta}\{ \, || \, (\phi_{n,[n^s]})(\, \hat{\theta}^n_{[n^s]} - \theta \,) \, || > h \, \} \leq \exp(-Cn^b \, h^\nu) \tag{2.3}$$

where $b = \delta - s\mu - s$ and the positive constant C depends on θ alone.

PROOFS: First, we need a result from [9] (Theorem 19, Appendix).

LEMMA 2.1 *Let $\xi(t)$ be a real valued random function defined on F, a closed set in R^k. We shall assume that the process $\xi(t)$ is measurable and separable. Also assume that the following conditions are fulfilled :*
I) There exist $m \geq \alpha > k$ such that for all $x \in F$,

$$E|\xi(x)|^m \leq H(x)$$

II) For $y \in R^k$ such that $(x+y) \in F$,

$$E|\xi(x+y) - \xi(x)|^m \leq H(x)||y||^\alpha$$

where $H(x)$ is some positive continuous function defined on F.
Then with probability one the realizations of $\xi(t)$ are continuous functions. Moreover, if $\omega(h, \xi, L) = \sup\{|\xi(x) - \xi(y)|\}$ where the sup is taken over $x, y \in F$ with $||x - y|| < h$, $||x||, ||y|| < L$ then

$$E\omega(h, \xi, L) \leq D(L^k \sup_{|x| \leq L} H(x))^{1/m} h^{(\alpha-k)/m} \log(h^{-1})$$

where the constant $D = D(k, \alpha, m)$ is as in Proposition 2.1 (iii).

This constant D is not explicitly given in [9] but can be obtained by carefully going over their proof; when $m = \alpha$, the bound given there appears to be incorrect, an additional log factor as is given above is necessary. In this paper the lemma is applied when the domain F is not just closed but is also compact, then the function $H(.)$ can be taken to be identically equal to a constant.

PROOF OF PROPOSITION 2.1: For $r = 0, 1, 2, ...$, let

$$\gamma_r = \{u \in \mathbf{U}_{\mathbf{k}}^{\mathbf{n}} : h + r < ||u|| \leq h + r + 1\}.$$

We will show that

$$\mathbf{P}_\theta^{\mathbf{n}}[\sup_{u \in \gamma_r} Z_k^n(u) \geq 0] \leq B_1(1 + h + r)^{B_2} g_{kr}^n \exp(-b_1 g_{kr}^n) \tag{2.4}$$

where b_1, B_1 and B_2 are as in Proposition 2.1.

Since the set $\{u \in \mathbf{U}_{\mathbf{k}}^{\mathbf{n}} : ||u|| > h\}$ is the union of γ_r's it is easy to see that (2.1) together with (2.4) will prove the assertion of the proposition. Now we proceed to prove (2.4).

We subdivide the set γ_r into N regions, each with diameter at most $2(1 + h + r)\delta$, δ to be specified later. This subdivision can be done such that, $N \leq 2^k(\delta)^{-k}$. Denote the above subdivision by $\gamma_{r1}, \gamma_{r2} \cdots \gamma_{rN}$ and let u_j be any fixed point in γ_{rj}. Then

$$\mathbf{P}_\theta^{\mathbf{n}}\{ \sup_{u \in \gamma_r} Z_k^n(u) \geq 0 \} \leq \sum_{j=1}^{N} \mathbf{P}_\theta^{\mathbf{n}}\{Z_k^n(u_j) \geq -1/2\} + \mathbf{P}_\theta^{\mathbf{n}}\{ \sup_{\substack{\|u-v\|<2\delta^* \\ \|u\|,\|v\|<1+h+r}} |Z_k^n(u) - Z_k^n(v)| \geq 1/2 \} \quad (2.5)$$

where $\delta^* = (1+h+r)\delta$.
The first term in the RHS above can be bounded using condition C2

$$\sum_{j=1}^{N} \mathbf{P}_\theta^{\mathbf{n}}\{ Z_k^n(u_j) \geq -1/2 \} \leq N e^{\eta/2} \exp(-g_{kr}^n). \quad (2.6)$$

Now, taking the function $H(.)$ in Lemma 2.1 to be identically equal to B_k^n, the condition C1 implies that

$$\begin{aligned} & E_\theta^n[\omega(2\delta^*, Z_k^n, (1+h+r))] \\ \leq\ & D[B_k^n(1+h+r)^\alpha]^{1/m}(1+h+r)^{k/m}(2\delta^*)^{(\alpha-k)/m} \log(2\delta^*)^{-1} \\ \leq\ & D[B_k^n(1+h+r)^\alpha]^{1/m}[2\delta]^{(\alpha-k)/m} \log \delta^{-1}. \end{aligned} \quad (2.7)$$

Using (2.7) the second term in the RHS of (2.5) can be bounded as follows

$$\begin{aligned} & \mathbf{P}_\theta^{\mathbf{n}}\{ \sup_{\substack{\|u-v\|<2\delta^* \\ \|u\|,\|v\|<1+h+r}} |Z_k^n(u) - Z_k^n(v)| \geq 1/2\} \\ \leq\ & 4D[B_k^n(1+h+r)^\alpha]^{1/m}\delta^{(\alpha-k)/m} \log \delta^{-1}. \end{aligned} \quad (2.8)$$

Thus from (2.5), (2.6) and (2.8) we get

$$\begin{aligned} & \mathbf{P}_\theta^{\mathbf{n}}\{ \sup_{u \in \gamma_r} Z_k^n(u) \geq 0\} \\ & \leq N e^{\eta/2} \exp(-g_{kr}^n) \\ & +4D[B_k^n(1+h+r)^\alpha]^{1/m}\delta^{(\alpha-k)/m} \log \delta^{-1}. \end{aligned} \quad (2.9)$$

So far we have not chosen δ. We will choose δ such that $N \exp(-g_{kr}^n)$ and $\delta^{(\alpha-k)/m}$ are of the same order of magnitude.
Let $\delta = \exp(-[m/(\alpha - k + mk)]g_{kr}^n)$.
Then $\delta^{(\alpha-k)/m} = \exp(-[(\alpha-k)/(\alpha-k+mk)]g_{kr}^n)$ and, since $N \leq 2^k\delta^{-k}$,

$$\begin{aligned} N \exp(-g_{kr}^n) &\leq 2^k\delta^{-k} \exp(-g_{kr}^n) \\ &= 2^k \exp(-[(\alpha-k)/((\alpha-k)+mk)]g_{kr}^n). \end{aligned} \quad (2.10)$$

From (2.9), (2.10) and $(\alpha - k)/(\alpha - k + mk) < 1$, we get

$$\begin{aligned} & \mathbf{P}_\theta^{\mathbf{n}}[\sup_{u \in \gamma_r} Z_k^n(u) \geq 0] \\ & \leq \exp(-[(\alpha - k)/(\alpha - k + mk)] g_{kr}^n) \\ & \times [2^k e^{\eta/2} + 4D[B_k^n(1 + h + r)^\alpha]^{1/m} g_{kr}^n]. \end{aligned}$$

Hence

$$\mathbf{P}_\theta^{\mathbf{n}}\{\sup_{u \in \gamma_r} Z_k^n(u) \geq 0\} \leq B_1(1 + h + r)^{B_2} \exp(-b_1 g_{kr}^n)\, g_{kr}^n$$

which completes the proof. □

PROOF OF THEOREM 2.1 : From the expression of functions g_k^n in the statement of the theorem it is clear that they belong to **G** for each n and k. Let $g_n(x) = g_{[n^s]}^n(x)$. Then

$$g_n(x) = n^{\delta - \mu s}|x - a(n)|^\nu - t(n) \tag{2.11}$$

where $a(n) = a(n, [n^s])$ and $t(n) = t(n, [n^s])$. From (2.11) it is easy to see that g_n is positive and increasing on $[x_0(n), \infty)$ where

$$x_0(n) = a(n) + \{n^{-(\delta - \mu s - s)} \cdot \frac{t(n)}{[n^s]}\}^{1/\nu}.$$

In view of assumption (T4) $\delta - \mu s - s > s > 0$ and, $a(n)$ and $t(n)$ are such that

$$a(n) \to 0 \quad and \quad t(n)/[n^s] \to 0.$$

Therefore $x_0(n) \to 0$ as $n \to \infty$. Hence for every $h > 0$, $x_0(n) < h$ for all $n > n_0(h)$ and the following inequality is then a consequence of Proposition 2.1.

$$\begin{aligned} & \mathbf{P}_\theta^{\mathbf{n}}\{||(\phi_{n,[n^s]}(\hat{\theta}_{[n^s]}^n - \theta_{[n^s]})|| > h\} \\ & \leq B_1 \sum_{r=0}^{\infty} (1 + h + r)^{B_2} \exp(-b_1 g_n(h + r))\, g_n(h + r) \end{aligned} \tag{2.12}$$

where, because of assumptions (T1) and (T2) of the theorem,

$$\begin{aligned} b_1 &= b_1(k) = (\alpha - k)/(\alpha - k + mk) \sim C/k = C/[n^s], \\ B_2 &= \alpha/m \sim C, \\ B_1 &= C(B_k^n)^{1/m} D(k, \alpha, m) \\ &\leq C(24)^{C[n^s]} n^{C(1+s)} = C(24)^{C[n^s]} n^C \text{ and} \\ |g_n(h + r)| &\leq n^C k^C (h + r)^C. \end{aligned}$$

(In all of the above expressions the constant C denotes a positive number independent of n but it could be different in each of them.)

This means that the expression in (2.12) can be written as

$$\begin{aligned} &\mathbf{P}_\theta^{\mathbf{n}}\{||(\phi_{n,[n^s]}(\hat{\theta}^n_{[n^s]} - \theta_{[n^s]})|| > h\} \\ &\leq Cn^C(24)^{C[n^s]} \sum_{r=0}^{\infty} (2+h+r)^{B_2} \exp(-\frac{C}{[n^s]} g_n(h+r)). \end{aligned}$$

For fixed $h > 0$, our choice of the constants δ, μ and ν is such that, for large n and some $C > 0$,

$$Cn^C(24)^{C[n^s]} \exp(-\frac{C}{2[n^s]} g_n(h+r)) \leq \exp(-Cn^{(\delta-\mu s-s)}(h+r)^\nu).$$

To see this, note that for large enough n

$$a(n) < (h+r)/2 \quad and \quad t(n)/[n^s] < 1/2.$$

Hence

$$\begin{aligned} C[n^s]^{-1} g_n(h+r) &= [n^s]^{-1} Cn^{\delta-\mu s}|h+r-a(n)|^\nu - Ct(n) \\ &\geq Cn^{(\delta-\mu s-s)}(h+r)^\nu - C/2. \end{aligned}$$

Therefore, for large n,

$$\begin{aligned} & Cn^C(24)^{C[n^s]} \exp(-C[n^s]^{-1} g_n(h+r)) \\ \leq\ & Cn^C(24)^{C[n^s]} \exp(-Cn^{(\delta-\mu s-s)}(h+r)^\nu + C/2) \\ \leq\ & \exp(-Cn^{(\delta-\mu s-s)}(h+r)^\nu) \end{aligned}$$

since $h > 0$ and $\delta - \mu s - s > s$.

Hence for large n,

$$\begin{aligned} & \mathbf{P}_\theta^{\mathbf{n}}\{||(\phi_{n,[n^s]}(\hat{\theta}^n_{[n^s]} - \theta_{[n^s]})|| > h\} \\ \leq\ & \sum_{r=0}^{\infty} (1+h+r)^{B_2} \exp(-Cn^{(\delta-\mu s-s)}(h+r)^\nu) \\ \leq\ & \exp(-C/2\, n^{(\delta-\mu s-s)} h^\nu). \end{aligned} \tag{2.13}$$

The assertion in (2.13) and the Borel-Cantelli lemma essentially prove the theorem since, $|\phi_{n,[n^s]}| \to \infty$ as $n \to \infty$ implies that

$$||\hat{\theta}^n_{[n^s]} - \theta_{[n^s]}|| \to 0 \ as \ n \to \infty \ a.s.$$

and then we only need to observe that the deterministic difference $||\theta - \theta_{[n^s]}||$ always $\to 0$ *as* $n \to \infty$. □

3 Drift function estimation in a linear SDE

Consider the model

$$dx_t = \theta_t\, x_t dt + dW_t, \quad x_0 = X_0, \quad 0 \le t \le 1; \qquad (3.1)$$

where (W_t) is the standard Wiener process and the initial value X_0 is a zero mean Gaussian random variable with variance $\sigma_0^2 > 0$. Furthermore X_0 is assumed to be independent of (W_t).

The goal is to estimate $\theta \in \Theta = L^2([0,1], dt)$, an unknown deterministic function, using a sample of n independent trajectories of (x_t).

The above problem was first discussed by Nguyen and Pham [17]. They proposed a sieve estimator and showed its *weak* L^2 consistency; no regularity assumptions on θ were made. The rate of convergence of this estimator was not discussed; however, under certain conditions the asymptotic normality of some functionals of the estimator was established. Later, Kutoyants [12] proposed a kernel estimator and proved its *weak pointwise* consistency under mild differentiability assumptions on θ. He also discusses the rate of convergence of this estimator. In McKeague [16] this problem is treated as a special example of semimartingale regression and consistency of the sieve estimator is proved.

Here we consider a sieve estimator which is based on the method suggested in the last section. It turns out that our estimator is a smoothed version of the estimator proposed by Nguyen and Pham. It will be seen that unless some kind of regularity assumption on θ is made the method in Section 2 does not yield the consistency of the proposed estimator. However, under mild regularity conditions, the conditions of Theorem 2.1 can be verified to obtain a large deviation inequality. This enables us to establish the strong L^2 consistency of the estimator and in addition to obtain lower bounds on the rates of convergence. If the CONS used in defining the sieves is chosen to be the one consisting of trigonometric polynomials then one can use standard results from Fourier theory which connect the L^2 norm of a differentiable function to that of its uniform norm. These results imply the strong uniform-norm consistency of our estimators.

It is easy to verify that (3.1) has a unique strong solution (x_t) which is a continuous, mean zero Gaussian process. If $\sigma_\theta^2(t) = Ex_t^2$ then it can be shown that $\sigma_\theta^2(t) = \sigma_0^2 \exp(2\int_0^t \theta_u du) + \int_0^t \exp(2\int_s^t \theta_u du) ds$. Obviously $\sigma_\theta^2(t)$ is a continuous function on $[0,1]$. Let

$$M_\theta = \sup_{0 \le t \le 1} \sigma_\theta^2(t) \quad \text{and} \quad m_\theta = \inf_{0 \le t \le 1} \sigma_\theta^2(t).$$

Then, since $[0, 1]$ is compact and $\sigma_\theta^2(t)$ continuous, $M_\theta < \infty$. Also, $m_\theta > 0$ since $\sigma_0^2 > 0$. This means that, for $f \in \Theta$,

$$m_\theta ||f||^2 \le ||f||_\theta^2 \le M_\theta\, ||f||^2 \qquad (3.2)$$

where

$$||f||_\theta^2 = \int_0^1 f_t^2 \sigma_\theta^2(t)dt \quad \text{and} \quad ||f||^2 = \int_0^1 f_t^2 dt.$$

Thus for fixed θ the two norms are equivalent. For $\theta \in \Theta$, let P_θ denote the measure induced by the paths of (x_t) on $(C, \mathbf{C})$ where C is the space of continuous functions on $[0,1]$ endowed with the supremum norm topology and $\mathbf{C}$ is the Borel σ-field of subsets of C. Then it is well known that (see, for example, Liptser and Shiryayev [15], vol. I, Theorem 7.19, pp 277) P_θ and P_0 are mutually absolutely continuous (P_0 denoting the measure P_θ when $\theta \equiv 0$) and,

$$\frac{dP_\theta}{dP_0}(\omega) = \exp\left(\int_0^1 \theta_t \omega_t d\omega_t - 1/2 \int_0^1 \theta_t^2 \omega_t^2 dt\right). \tag{3.3}$$

Let C^n be the n-fold product of C, $\mathbf{C^n}$ the n- fold product of $\mathbf{C}$ and P_θ^n the n-fold product of P_θ. Also, ω^n will denote a typical point in C^n.

The sieve S_k to be used is defined as follows:

Let $\{f_j\}$ be an arbitrary but fixed complete orthonormal system (CONS) in $\Theta = L^2([0,1], dt)$. For $k \geq 1$,

$$S_k = \{ \theta \in \Theta : \theta = \sum_{j=1}^{k} \theta_j f_j \, ; \, \sum_{j=1}^{k} \theta_j^2 \leq k^{2q} \}; \tag{3.4}$$

$q > 0$ will be determined later. Let $\hat{\theta}_k^n = \hat{\theta}_k^n(\omega^n)$ be the sieve estimator corresponding to this sieve.

For $\theta \in \Theta$, let $\Pi_k \theta$ and $\underline{\theta}_k$ be the projections of θ on the linear span of $\{f_j, 1 \leq j \leq k\}$ and S_k respectively (clearly, for fixed θ and large k, these two projections will be the same). For $0 < \epsilon < 1/2$ let

$$\Theta_\epsilon = \{ \theta \in \Theta : n^{1-\epsilon} \, ||\theta - \Pi_{[n^\epsilon]}\theta\,||^2 \to 0 \text{ as } n \to \infty \}. \tag{3.5}$$

That is, if θ_j's are the Fourier coefficients of θ w.r.t. $\{f_j\}$, $\theta \in \Theta_\epsilon$ if and only if

$$n^{1-\epsilon} \sum_{j=[n^\epsilon]}^{\infty} \theta_j^2 \to 0 \text{ as } n \to \infty.$$

From the definition it is easy to see that, $\epsilon_1 < \epsilon_2 \Rightarrow \Theta_{\epsilon_1} \subset \Theta_{\epsilon_2}$; that is, smaller ϵ implies higher smoothness. Now we can state the main result of this section.

THEOREM 3.1 *Assume that the true parameter θ belongs to Θ_ϵ where $0 < \epsilon < 1/2$. Choose $\beta \geq 0$ and $q > 0$ such that*

$$b = 1 - 2\beta - \epsilon - 2\epsilon q > \epsilon > 0.$$

Then

1) $n^{\beta}(\hat{\theta}^{n}_{[n\epsilon]} - \theta) \to 0$ *a.s. and*

2) *for all* $h > 0$ *there exists an integer* $n_0 = n_0(h, \theta)$ *such that, for* $n > n_0$,

$$P^{n}_{\theta}\{ n^{\beta}\|\hat{\theta}^{n}_{[n\epsilon]} - \theta\| > h \} \le \exp(-C n^{b} h^{2})$$

where $C > 0$, *depends on* θ *alone.*

It is clear that the type of functions which belong to Θ_ϵ depends on the CONS used in defining the sieve S_k and for an arbitrary CONS it is difficult to ascertain whether a given function is a member of Θ_ϵ or not. However, for some special choices of CONS's it is possible to identify large classes of functions which belong to Θ_ϵ. One such choice of CONS $\{f_j\}$ is illustrated below:

Set $f_1 \equiv 1$ and for $j \ge 2$, let

$$\begin{aligned} f_j(t) &= \sqrt{2}\cos(\pi j t) \text{ for even } j\text{'s and} \\ f_j(t) &= \sqrt{2}\sin(\pi(j+1)t) \text{ for odd } j\text{'s}; \quad t \in [0,1]. \end{aligned} \tag{3.6}$$

For $m \ge 0$ and $0 \le \alpha \le 1$ let $\Theta_{m,\alpha}$ be the class of *m-times differentiable* functions with *m-th derivative Lipschitz* continuous of order α. More precisely,

$\Theta_{m,\alpha} = \{ \theta \in \Theta : \theta$ is m times continuously differentiable with m-th derivative Lipschitz continuous of order α. Furthermore, for $0 \le j \le m, \theta^{(j)}(0) = \theta^{(j)}(1)$ where $\theta^{(j)}(t) = \frac{d^j}{dt^j}\theta(t)\}$.

If the CONS in (3.6) is used in defining the sieve S_k then it can be shown that (see (3.18)),

$$\Theta_{m,\alpha} \subset \Theta_\epsilon \quad \text{when} \quad 1 + 2(m+\alpha) > \frac{1}{\epsilon}.$$

For Theorems 3.2, 3.3 and Corollary 3.1, it is assumed that the CONS $\{f_j\}$ used in defining the sieve S_k is the one given in (3.6).

THEOREM 3.2 *Let* $\theta \in \Theta_{m,\alpha}$, $m + \alpha > 1/2$. *Then for every* β, $\beta < \frac{(m+\alpha)-1/2}{2(m+\alpha)+1}$, *there exist* $\epsilon > 0$ *and* $q > 0$ *(which depend on* β*) such that,*

$$n^{\beta}\, \|\hat{\theta}^{n}_{[n\epsilon]} - \theta\| \to 0 \text{ a.s. as } n \to \infty.$$

PROOF : When $m + \alpha > 1/2$, for every β, $\beta < \frac{(m+\alpha)-1/2}{2(m+\alpha)+1}$, there exist $\epsilon > 0$ and $q > 0$ such that $\Theta_{m,\alpha} \subset \Theta_\epsilon$ and $1 - 2\beta - \epsilon - 2\epsilon q > \epsilon > 0$. Therefore the result follows from Theorem 3.1. □

We now turn to the problem of consistency in *uniform norm*. It is well known that (see (3.17)) if $\theta \in \Theta_{m,\alpha}$ and $m + \alpha \geq 1/2$, the Fourier series of θ converges absolutely and uniformly. Moreover, it can be shown that, for large n,

$$\sup_{0 \leq t \leq 1} |\hat{\theta}^n_{[n^\epsilon]}(t) - \underline{\theta}_{[n^\epsilon]}(t)| \leq n^{\epsilon/2} \, \|\hat{\theta}^n_{[n^\epsilon]} - \underline{\theta}_{[n^\epsilon]}\|.$$

Therefore, in order to show the uniform norm consistency of $\hat{\theta}^n_{[n^\epsilon]}$ at the rate n^β, it suffices to obtain L^2-consistency at the rate $n^{\beta+\epsilon/2}$.

THEOREM 3.3 *Let $\theta \in \Theta_{m,\alpha}$, $m + \alpha > 1$. Then for every β, $\beta < \frac{(m+\alpha)-1}{2(m+\alpha)+1}$, there exist $\epsilon > 0$ and $q > 0$ (which depend on β) such that*

$$n^\beta \sup_{0 \leq t \leq 1} |\hat{\theta}^n_{[n^\epsilon]}(t) - \theta(t)| \to 0 \text{ a.s., as } n \to \infty.$$

When it is known that the true parameter θ is a C^∞ function the above result yields the following corollary.

COROLLARY 3.1 *Let θ be a C^∞ function. Then for every $\beta < 1/2$ there exist $\epsilon > 0$ and $q > 0$ such that*

$$n^\beta \sup_{0 \leq t \leq 1} |\hat{\theta}^n_{[n^\epsilon]}(t) - \theta(t)| \to 0 \text{ a.s., as } n \to \infty.$$

Now we begin proving the results given in this section.

PROOF OF THEOREM 3.1 : This is accomplished by verifying the conditions of Theorem 2.1. For this we need several lemmas. The first lemma contains some facts, relevant to our purpose, about the L^2 norm of a square integrable Gaussian process.

LEMMA 3.1 *Let $(Y_t), 0 \leq t \leq T$ be a zero mean Gaussian process such that*

$$E\int_0^T Y_t^2\,dt = \text{tr}(R) < \infty. \text{ Then, } \forall\, k \geq 1,$$

$$E\left(\int_0^T Y_t^2\,dt\right)^k \leq k^k \left[E\left(\int_0^T Y_t^2\,dt\right)\right]^k. \tag{3.7}$$

Furthermore, if $0 \leq \alpha < (2\text{tr}(R))^{-1}$,

$$E\exp\left(\alpha \int_0^T Y_t^2\,dt\right) \leq \exp\left(\frac{\alpha\text{tr}(R)}{1 - 2\alpha\text{tr}(R)}\right). \tag{3.8}$$

Also,

$$E\exp(-\int_0^T Y_t^2 dt) \le \exp(\frac{-\mathrm{tr}(R)}{1+2\mathrm{tr}(R)}). \qquad (3.9)$$

PROOF : If $R(t,s) = EY_tY_s$ then R, the corresponding covariance operator defined on $L^2[0,T]$, is a self adjoint, non-negative definite, trace class operator. In fact

$$\mathrm{tr}(R) := \mathrm{Trace}(R) = \int_0^T R(t,t)\,dt = E\int_0^T Y_t^2\,dt.$$

Let $\{f_j\}$ be the system of eigenfunctions of R and λ_j's be the corresponding eigenvalues then, using the Karhunen-Loeve expansion of (Y_t) we get

$$\int_0^T Y_t^2 dt = \sum_{j=1}^{\infty}\lambda_j X_j^2 \text{ and } E\int_0^T Y_t^2 dt = \sum_{j=1}^{\infty}\lambda_j = \lambda^* \textit{ say}. \qquad (3.10)$$

The coefficients (X_j) are independent standard normal variables. Therefore

$$E(\int_0^T Y_t^2 dt)^k = E(\sum_{j=1}^{\infty}\lambda_j X_j^2)^k = (\lambda^*)^k\, E[(\lambda^*)^{-1}\sum_{j=1}^{\infty}\lambda_j X_j^2]^k.$$

Then the application of Jensen's inequality in the square bracket and the i.i.d. nature of (X_j)s imply

$$E(\int_0^T Y_t^2 dt)^k \le (\lambda^*)^{k-1}E(\sum_{j=1}^{\infty}\lambda_j X_j^{2k}) = (\lambda^*)^k EX_1^{2k}.$$

Moreover because X_1 is $N(0,1)$, using Stirling's approximation (see Rao (1973) pp 59),

$$E(\int_0^T Y_t^2 dt)^k \le (\lambda^*)^k\frac{(2k)!}{2^k\,k!} \le k^k(\lambda^*)^k.$$

This completes the proof of the first inequality (see (3.10)). The last two inequalities are easy to deduce from (3.10) and the fact that if X is a $N(0,1)$ variable then, for $\beta < 1/2$, $E\exp(\beta X^2) = (1-2\beta)^{-1/2}$. □

Now, for $n,k \ge 1$ and $\phi_{n,k} = n^\beta$, $\beta \ge 0$, define random functions $Z_k^n(u)$ with domains U_k^n as in Section 2 using the sieve S_k defined in (3.4). From (3.3) and the fact that P_θ^n is a product measure, the following expression for the loglikelihood function is obtained.

$$l_n(\theta,\omega^n) = \sum_{i=1}^{n}\int_0^1 \theta_t\,\omega_t^i\,d\omega_t^i - 1/2\int_0^1 \theta_t^2\,(\omega_t^i)^2\,dt. \qquad (3.11)$$

The derivation of the separability and the measurability of $Z_k^n(u,\omega^n)$ is easy to see from the above expression. A brief sketch of this derivation is given below.

LEMMA 3.2 *The random function $Z_k^n(u,\omega^n)$ is measurable and separable.*

PROOF : From (3.11),

$$\begin{aligned} Z_k^n(u,\omega^n) &= l_n(\underline{\theta}_k + n^{-\beta}u) - l_n(\underline{\theta}_k) \\ &= \sum_{i=1}^n \{ n^{-\beta} \int_0^1 u_t\, \omega_t^i\, d\omega_t^i \\ &\quad -1/2 \int_0^1 [(\underline{\theta}_k + n^{-\beta}u)_t^2 - (\underline{\theta}_k)_t^2\,]\,(\omega_t^i)^2\, dt\}. \end{aligned}$$

If we set $X_{i,j}(\omega^n) = \int_0^1 f_j(t)\, \omega_t^i\, d\omega_t^i$ where f_j is a function from the CONS used in defining the sieve S_k then

$$\sum_{i=1}^n \int_0^1 u_t\, \omega_t^i\, d\omega_t^i = \sum_{i=1}^n \sum_{j=1}^k u_j X_{i,j}(\omega^n)$$

where u_j are the Fourier coefficients of u_t with respect to the CONS $\{f_j\}$. Note that the terms in the summation on the right hand side are of products of purely deterministic and purely random terms. Therefore it is easy to see that the right hand side above is measurable (in (u,ω^n)) and, continuous in $(u_1, u_2, ..., u_k)$ for *every* fixed (ω^n). Therefore, in particular, it is measurable and separable. This means that the first term in the expression of $Z_k^n(u,\omega^n)$ is measursble and separable. It is equally easy to show that the second term is also measurable and separable (in fact both the terms are continuous in $(u_1, u_2, ..., u_k)$). Therefore $Z_k^n(u,\omega^n)$ is measursble and separable.

Condition C1 of Theorem 2.1 is verified in the next lemma.

LEMMA 3.3 *For $u, v \in U_k^n$,*

$$E|Z_k^n(u)|^{2k} \le B_k^n \quad \text{and}$$

$$E|Z_k^n(u) - Z_k^n(v)|^{2k} \le B_k^n\, ||u-v||^{2k}$$

where

$$B_k^n = n^{(2k+1)}\, k^{k(2q+3)}\, ||\theta||_\theta^{2k}\, (32M_\theta)^{2k}.$$

PROOF : We will prove the second assertion of the lemma with a smaller constant, say C_k^n, in place of B_k^n. From this the first assertion can be deduced because, note that

a) $Z_k^n(0) = 0$ and,

b) For $u \in U_k^n$, $||u|| \leq 2n^\beta k^q$ (see (3.4)).

Now the first assertion follows by taking $v = 0$ and $B_k^n = (2n^\beta k^q)^{2k} C_k^n$, which is the constant given in the statement of the lemma.

Let us consider the second assertion. From (3.11),

$$\begin{aligned} Z_k^n(u) - Z_k^n(v) &= l_n(\underline{\theta}_k + n^{-\beta} u) - l_n(\underline{\theta}_k + n^{-\beta} v) \\ &= \sum_{i=1}^n \{ n^{-\beta} \int_0^1 (u_t - v_t)\, \omega_t^i \, d\omega_t^i \\ &\quad -1/2 \int_0^1 [(\underline{\theta}_k + n^{-\beta} u)_t^2 - (\underline{\theta}_k + n^{-\beta} v)_t^2]\, (\omega_t^i)^2 \, dt\}. \end{aligned}$$

Hence, using the elementary inequality, $(\sum_1^m a_i)^k \leq m^k (\sum_1^m a_i^k)$, for positive numbers a_i repeatedly, and the fact that (ω_t^i) are i.i.d., we get

$$E|Z_k^n(u) - Z_k^n(v)|^{2k} \leq 2^{2k} n^{2k(1-\beta)+1} E(\int_0^1 (u_t - v_t)\omega_t d\omega_t)^{2k}$$

$$+ n^{2k+1} E(\int_0^1 [(\underline{\theta}_k + n^{-\beta} u)_t^2 - (\underline{\theta}_k + n^{-\beta} v)_t^2] \omega_t^2 dt)^{2k}. \tag{3.12}$$

We now bound each term in (3.12) separately. Since

$$\int_0^1 (u_t - v_t)\omega_t d\omega_t = \int_0^1 (u_t - v_t)\theta_t \omega_t^2 dt + \int_0^1 (u_t - v_t)\omega_t dW_t(\omega),$$

$W_t(\omega)$ being the standard Wiener process under P_θ,

$$\begin{aligned} E(\int_0^1 (u_t - v_t)\omega_t d\omega_t)^{2k} &\leq 2^{2k} E(\int_0^1 (u_t - v_t)\theta_t \omega_t^2 dt)^{2k} \\ &+ E(\int_0^1 (u_t - v_t)\omega_t dW_t(\omega))^{2k}. \end{aligned} \tag{3.13}$$

Then by repeated application of the Cauchy-Schwarz inequality and (3.7) of Lemma 3.1 one gets

$$E(\int_0^1 (u_t - v_t)\theta_t \omega_t^2 dt)^{2k} \leq k^k ||u - v||_\theta{}^{2k} \, ||\theta||_\theta{}^{2k}. \tag{3.14}$$

Next, by the martingale inequality of Burkholder, Davis and Gundy (see Dellacherie and Meyer (1982), pp 287) and by Lemma 3.1,

$$E(\int_0^1 (u_t - v_t)\omega_t dW_t(\omega))^{2k} \leq 8^{2k} k^{3k} \, ||u - v||_\theta^{2k}. \tag{3.15}$$

Hence, in view of (3.13), (3.14) and (3.15), the first term in (3.12) has the bound

$$||u - v||_{\theta}^{2k}\ n^{2k(1-\beta)+1}\ k^{3k}\ 8^{4k}\ ||\theta||_{\theta}^{2k}.$$

The second term in (3.12) can be bounded in the same fashion and the lemma now follows upon combining the two bounds and using the inequality in (3.2). □

Lemma 3.5 shows that condition C2 of Theorem 2.1 is also satisfied. We first prove Lemma 3.4 as a preparatory result.

LEMMA 3.4

$$E\exp(n^{-\beta}/4\int_0^1 u_t\omega_t d\omega_t - 1/8\int_0^1 [(\underline{\theta}_k + n^{-\beta}u)_t^2 - (\underline{\theta}_k)_t^2]\omega_t^2 dt)$$

$$\leq\ (E\exp(-1/4\int_0^1 [n^{-\beta}u - 2(\theta - \underline{\theta}_k)]_t^2\omega_t^2 dt))^{1/4}$$

$$\times(E\exp(\int_0^1 (\theta - \underline{\theta}_k)_t^2\omega_t^2 dt))^{1/4}.$$

PROOF : Writing $\theta_t = (\theta - \underline{\theta}_k)_t + (\underline{\theta}_k)_t$ and $d\omega_t = \theta_t\omega_t dt + dW_t(\omega)$ we get

$$n^{-\beta}/4\int_0^1 u_t\omega_t d\omega_t - 1/8\int_0^1 [(\underline{\theta}_k + n^{-\beta}u)_t^2 - (\underline{\theta}_k)_t^2]\omega_t^2 dt$$

$$= n^{-\beta}/4\int_0^1 u_t\omega_t dW_t(\omega) + n^{-\beta}/4\int_0^1 u_t(\theta - \underline{\theta}_k)_t\omega_t^2 dt$$

$$-n^{-2\beta}/8\int_0^1 u_t^2{\omega_t}^2 dt.$$

Moreover,

$$E_\theta\exp(n^{-\beta}/2\int_0^1 u_t\omega_t dW_t(\omega) - n^{-2\beta}/8\int_0^1 u_t^2\omega_t^2 dt) = 1$$

since the expression in E_θ is the Radon-Nikodym derivative of $P_{\theta+n^{-\beta/2}u}$ with respect to P_θ. The conclusion of the lemma follows by repeated application of the Cauchy-Schwarz inequality. □

LEMMA 3.5 *For* $u\in U_k^n$ *and* k *large so that*

$$1 - 2||\theta - \underline{\theta}_k||_{\theta}^2 > 0,$$

$$E\exp(1/4\ Z_k^n(u))\ <\ \exp(\frac{-nC}{k^{2q}}|\, n^{-\beta}||u|| - 2||\theta - \underline{\theta}_k||\,|^2 - 2n||\theta - \underline{\theta}_k||^2).$$

PROOF : Note that Z_k^n can be written as

$$Z_k^n(u) = \sum_{i=1}^n n^{-\beta} \int_0^1 u_t \omega_t^i d\omega_t^i - 1/2 \int_0^1 ((\underline{\theta}_k + n^{-\beta} u)_t^2 - (\underline{\theta}_k)_t^2)\, (\omega_t^i)^2 dt$$

where $(\omega_t^i), 0 \le t \le 1$, $i = 1, 2 \ldots n$ are i.i.d processes. From this and Lemma 3.4 we get,

$$\begin{aligned} & E \exp(1/4 Z_k^n(u)) \\ \le \;& (E \exp(-1/4 \int_0^1 [n^{-\beta} u_t - 2(\theta - \underline{\theta}_k)_t]^2 \omega_t^2 dt))^{n/4} \\ & \times (E \exp(\int_0^1 (\theta - \underline{\theta}_k)_t^2 \omega_t^2 dt))^{n/4}. \end{aligned}$$

The application of Lemma 3.1 to the first and the last term of the product in the RHS above yields us the following:

$$\begin{aligned} E \exp(1/4\, Z_k^n(u)) \;\; & \le \;\; \exp(\frac{-n}{16} \frac{\|n^{-\beta} u - 2(\theta - \underline{\theta}_k)\|_\theta^2}{1 + \|n^{-\beta} u - 2(\theta - \underline{\theta}_k)\|_\theta^2}) \\ & \times \;\; \exp(\frac{n\|(\theta - \underline{\theta}_k)\|_\theta^2}{4(1 - 2\|\theta - \underline{\theta}_k\|_\theta^2)}). \end{aligned}$$

The conclusion of the lemma is obtained by simplifying the above bound using the following facts:

(i) $| \, \|x\|_\theta - \|y\|_\theta \, | \le \|x - y\|_\theta \le M_\theta \|x - y\|$

(ii) $u \in U_k^n = n^\beta (S_k - \underline{\theta}_k)$ implies that (see (3.2), (3.4))

$$\|u\|_\theta \le 2\sqrt{M_\theta}\, k^q n^\beta.$$

(iii) For large k, $1 + \|n^{-\beta} u - 2(\theta - \underline{\theta}_k)\|_\theta^2 \le C\, k^{2q}$. □

Condition C2 now holds with $g_k^n(x) = C n^\delta k^{-\mu} |x - a(n,k)|^\nu - t(n,k)$ where $\delta = 1 - 2\beta$, $\mu = 2q$, $\nu = 2$, $a(n,k) = 2n^\beta \|\theta - \underline{\theta}_k\|$ and $t(n,k) = 2n\|\theta - \underline{\theta}_k\|^2$ (see (T3) of Theorem 2.1).

Note that here $m = \alpha = 2k$ and $B_k^n \le n^{(2k+1)} k^{k(3+2q)} C_\theta^{2k}$.

It is easy to check the remaining condition (T4) of Theorem 2.1. For this we take the number s appearing in the condition equal to ϵ where ϵ is less than min $(\delta/(\mu+2), 1 - 2\beta)$. We then have

$a(n, [n^\epsilon]) = 2n^\beta \|\theta - \underline{\theta}_{[n^\epsilon]}\| \to 0$ since, by assumption, $\theta \in \Theta_\epsilon$. A similar argument shows that

$$\frac{t(n, [n^\epsilon])}{[n^\epsilon]} = 4n^{1-\epsilon} \|\theta - \underline{\theta}_{[n^\epsilon]}\|^2 \to 0$$

as $n \to \infty$. All the conditions of Theorem 2.1 hold good in the present case and thus Theorem 3.1 follows. □

The proof of Theorem 3.2 appears following its statement. Now we show that Theorem 3.3 follows from Theorem 3.2. First some standard facts from Fourier theory are reviewed. Let $\{f_j\}$ denote the CONS defined in (3.6) i.e. the one consisting of the trigonometric functions.

For $\theta \in \Theta_{m,\alpha}^{-}$, there exists $C = C(m, \alpha, \theta)$ such that

$$||\theta - \Pi_n\theta||^2 = \int_0^1 [\theta_t - (\Pi_n\theta)_t]^2 dt \leq Cn^{-2(m+\alpha)}. \tag{3.16}$$

If, in addition, $m + \alpha > 1/2$ then

$$\sup_{0 \leq t \leq 1} |\theta(t)| < \sqrt{2} \sum_{1}^{\infty} |\theta_j| < \infty \text{ and } |\theta_j| \approx O(j^{-m-\alpha}) \tag{3.17}$$

where θ_j are the Fourier coefficients of θ.

Inequality (3.16) is a simple consequence of a result called Jackson's inequality (see [21], Theorem 13.6, pp 115, vol 1) and the fact that $\Pi_n\theta$ is the projection of θ on the linear span of $\{f_j\}$, $1 \leq j \leq n$. For (3.17) see [21] (pp 240 vol 2, pp 71, Theorem 5, vol 1).

From (3.16) and (3.17) we get the following important relation

$$\Theta_{m,\alpha} \subset \Theta_\epsilon \text{ if } (1 + 2(m + \alpha)) > 1/\epsilon. \tag{3.18}$$

PROOF OF THEOREM 3.3 In view of the first part of (3.17)

$$n^\beta \sup_{0 \leq t \leq 1} |\hat{\theta}^n_{[n^\epsilon]}(t) - \theta(t)| \leq n^\beta \sqrt{2} \{ \sum_{j=1}^{[n^\epsilon]} |\hat{\theta}_j - \theta_j| + \sum_{[n^\epsilon]+1}^{\infty} |\theta_j| \}.$$

Hence the result is proved if

(1) $n^\beta \sum_{j=1}^{[n^\epsilon]} |\hat{\theta}^n_j - \theta_j| \to 0$ and (2) $n^\beta \sum_{[n^\epsilon]+1}^{\infty} |\theta_j| \to 0$ as $n \to \infty$.

In view of the Cauchy – Schwarz inequality (1) is proved if

$$n^{\beta+\epsilon/2} ||\hat{\theta}^n_{[n^\epsilon]} - \underline{\theta}_{[n^\epsilon]}|| \to 0 \text{ a.s. as } n \to \infty.$$

But this follows from Theorem 3.2 because, for $\beta < \frac{(m+\alpha)-1}{2(m+\alpha)+1}$, we can find an ϵ and q such that $\Theta_{m,\alpha} \subset \Theta_\epsilon$ and $1 - 2(\beta + \epsilon/2) - 2\epsilon - 2\epsilon q > 0$. Next, in view of the second part of (3.17), there exists a C such that for large n,

$$n^\beta \sum_{[n^\epsilon]+1}^{\infty} |\theta_j| \leq Cn^\beta (n^\epsilon)^{1-m-\alpha} \leq Cn^{\beta+\epsilon(1-m-\alpha)}.$$

For $\beta < \frac{(m+\alpha)-1}{2(m+\alpha)+1}$ one can find an ϵ such that $\Theta_{m,\alpha} \subset \Theta_\epsilon$ and $\beta + \epsilon(1-m-\alpha) < 0$; thus (2) is proved. □

4 Intensity estimation of a Poisson process

Let $(N_t), 0 \leq t \leq 1$ be a Poisson process with intensity function $\theta = (\theta_t), 0 \leq t \leq 1$. The deterministic function θ is assumed to be a point in Θ, the space of functions which are *positive, bounded and bounded away from zero.* The aim is to estimate θ using a sample of n independent trajectories of (N_t). This problem has been treated as a special example in many papers, see e.g., McKeague [16], Karr [10] and Leskow and Rozanski [13]. In [10] a histogram sieve estimator is shown to be strongly consistent in L^1 norm and its rate of convergence is obtained. In [13] it is shown that the histogram sieve estimator is weakly pointwise consistent and that its finite dimensional distributions are asymptotically normal. Here, as in Section 3, we can verify the conditions of Theorem 2.1 and obtain results identical to Theorems 3.1 - 3.3 and Corollary 3.1 for the sieve estimators corresponding to the sieves given below:

(I) For $k \geq 1$, $q > 0$ and $f_j(t) = 1_{[j-1/k, j/k)}(t)$, $1 \leq j \leq k$,

$$S_k = \{\theta \in \Theta \ : \ \theta = \sum_{j=1}^{k} \theta_j f_j, \ \ k^{-q} \leq \theta_t \leq k^q \text{ for all t } \}$$

(This type of sieve is sometimes referred to as the *Histogram sieve.*)

(II)

$$S_k = \{ \ \theta \in \Theta \ : \ \theta = \sum_{j=1}^{k} \ \theta_j f_j, \ \ k^{-q} \leq \theta_t \leq k^q \text{ for all t } \}$$

where for $j \geq 1$, f_j's are as in (3.6).

For more details we refer the reader to ([19], Chapter 4).

5 Mean estimation of an abstract Wiener process

In [1] Antoniadis has derived a general result on the strong consistency of a sieve estimator of the mean θ of an abstract Wiener process based on i.i.d. observations. The parameter space Θ is a closed subset of the reproducing kernel Hilbert space (RKHS) H of the process. The covariance K of the Gaussian measure μ is assumed to be known. In this section we shall indicate how an analogue of Theorem 4.1 of [1] can be deduced from Theorem 2.1 of the present paper. The sieve estimators used in the two cases, however, are different. We omit the proof and other details for the sake of brevity.

Let **B** be a real separable Banach space and μ a mean zero Gaussian measure on its Borel subsets with (known) covariance K. Let H be the corresponding RKHS. It is known that H can be realized as a subset of **B** and is isometric to the closure of $\mathbf{B}^*$ (the dual of **B**) in $L^2(\mathbf{B},\mu)$. Let this isometry be denoted by i.

Assume the setting of Theorem 4.1 of [1], i.e., let the parameter space Θ be a closed infinite dimensional linear subspace of H for which there is a sequence $\{e_j^*, j \geq 1\}$ of $\mathbf{B}^*$ such that $\{f_j := i(e_j^*), j \geq 1\}$ is a CONS in Θ. Suppose that for $\theta \in \Theta$, μ_θ denotes the translation of μ by θ. Then, under μ_θ, $\{e_j^*\}$ is a sequence of independent normal r.v.s with mean θ_j and unit variance; θ_j being the Fourier coefficients of θ with respect to $\{f_j\}$.

Define k-dimensional, compact, convex subsets S_k of Θ as follows:

$$S_k = \{\eta \in \Theta : \eta = \sum_1^k \eta_j f_j; \sum_1^k \eta_j^2 \leq k^{2q}\}, \quad (q > 0).$$

As in the earlier sections it is easy to see that $\{S_k, k \geq 1\}$ is a sieve. Let $\hat{\theta}_k^n$ be the corresponding sieve estimator based on n i.i.d. observations with the common distribution μ_θ. Let μ_θ^n be the joint distribution of the sample.

In the above setting it is easy to verify the conditions of Theorem 2.1 and the following results hold:

THEOREM 5.1 *For every ϵ, $0 < \epsilon < 1/2$, $\hat{\theta}_{[n^\epsilon]}^n$ is a strongly consistent estimator of θ. Moreover, for every $h > 0$, there exists an integer $n_0 = n_0(h, \theta)$ such that, for $n > n_0$,*

$$\mu_\theta^n\{\|\hat{\theta}_{[n^\epsilon]}^n - \theta\| > h\} \leq \exp(-Cn^{1-\epsilon}h^2),$$

C being a positive constant depending only on θ.

Suppose the parameter θ is such that in its Fourier expansion $\theta = \sum_1^\infty \theta_j f_j$ the coefficients satisfy the condition

$$n^\beta \sum_{[n^\epsilon]}^\infty \theta_j^2 \to 0 \text{ as } n \to \infty$$

where β is a positive number less than $1/2 - \epsilon$ (for $\beta = 0$ this holds for all θ in Θ). Then we have the following:

THEOREM 5.2 *(1) $n^\beta \|\hat{\theta}_{[n^\epsilon]}^n - \theta\| \to 0$ as $n \to \infty$ a.s..*

(2) For every $h > 0$, there exists an integer $n_1 = n_1(h, \theta)$ such that, for $n \geq n_1$,

$$\mu_\theta^n\{n^\beta \|\hat{\theta}_{[n^\epsilon]}^n - \theta\| > h\} \leq \exp(-Cn^b h^2),$$

where $b = 1 - 2\beta - 2\epsilon$ and C is a positive constant depending only on θ.

In conclusion, we thank the referees for providing us with many of the references and for several helpful suggestions.

REFERENCES

[1] Antoniadis, A., (1988), "Parametric estimation for the mean of a Gaussian process by the method of sieves," *J. Multivariate Anal.*, **26** : 1–15.

[2] Antoniadis, A., (1989), "Penalty method for nonparametric estimation of the intensity of a counting process," *Ann. Inst. Statist. Math.*, **41** : 781–807.

[3] Beder, J.H., (1987), "A sieve estimator for the mean of a Gaussian process," *Ann. Statist.*, **15,** 59–78.

[4] Beder, J.H., (1988), "A sieve estimator for the covariance of a Gaussian process," *Ann. Statist.*, **16,** 648–660.

[5] Blum, J.R. and Walter, G.G., (1984), "A simple solution to a nonparametric maximum likelihood estimation problem," *Ann. Statist.*, **12** : 372–379.

[6] Dellacherie, C. and Meyer, P., (1982), *Probability and Potential B,* North-Holland.

[7] Geman, S, and Hwang, C.R., (1982) "Nonparametric maximum likelihood estimation by the method of sieves," *Ann. Statist.*, **10:** 401–414.

[8] Grenander, U., (1981), *Abstract inference,* John Wiley & Sons, New York.

[9] Ibragimov, I.A. and Hasminski, R.Z., (1981), *Statistical estimation: Asymptotic theory,* Springer-Verlag, New York. (This book contains references to Le Cam's work.)

[10] Karr, A.F., (1987), "Maximum likelihood estimation in multiplicative intensity model," *Ann. Stat.* **15:** 473–490.

[11] Karr, A.F. and Zucker, D.M., (1990), "Nonparametric survival analysis with time dependent covariate effects: a penalized partial likelihood approach," *Ann. Statist.*, **18** : 329–353.

[12] Kutoyants, Yu.A., (1984), "On estimation of trend coefficients in a diffusion process", *Statistics and control of stochastic processes, Steklov Seminar,* edited by Krylov, N.V. et al, *Optimization Software Inc., Publications Div., N.Y.* 231–250.

[13] Leskow, J. and Rozanski, R., (1989), "Histogram maximum likelihood estimator in multiplicative intensity model," *Stochastic Processes and their Applications,* **31** : 151–159.

[14] Leskow, J., (1989), "A note on kernel smoothing of an estimator of a periodic function in multiplicative intensity model." *Statistics and Probability Letters,* **7** : 395–400.

[15] Liptser, R.S. and Shiryayev, A.N., (1977), *Statistics of random processes I, II.*, Springer-Verlag, Berlin.

[16] McKeague, I.W., (1986), "Estimation for a semimartingale regression model using the method of sieves," *Ann. Statist.,* **14** : 579–589.

[17] Nguyen, H.T., and Pham, D.P., (1982), "Indentification of the Nonstationary Diffusion Model by the method of sieves," *SIAM J. Control and Optimization,* **20(5)**.

[18] Rao, C.R., (1973), *Linear statistical inference and its applications,* John Wiley & Sons, New York.

[19] Selukar, R.S., (1989) *On estimation of Hilbert space valued parameters,* Ph.D. thesis, University of North Carolina at Chapel Hill.

[20] Wald, A., (1949), "Note on the consistency of the maximum likelihood estimate," *Ann. Math. Statist.,* **20:** 595–601.

[21] Zygmund, A., (1968), *Trigonometric Series, Vol. I & II,* University Press, Cambridge.

University of South Alabama
Dept of Math and Stat.
Mobile, AL 36688
U.S.A.

Department of Statistics
The University of North
Carolina at Chapel Hill
CB# 3260 Phillips Hall
Chapel Hill, NC 27599-3260
U.S.A.

J.K. Ghosh, S.K. Mitra, K.R. Parthasarathy and B.L.S. Prakasa Rao (*eds.*),
Statistics and Probability : A Raghu Raj Bahadur Festschrift
Published by Wiley Eastern Limited, Publishers, (1993), 349 – 364

Bahadur Type Representations for Some Minimum Distance Estimators in Linear Regression

HIRA L. KOUL and ZHIWEI ZHU
Michigan State University, East Lansing

Abstract

Certain classes of minimum distance estimators are shown to be strongly consistent for the regression parameters in multiple linear regression models. A Bahadur type representation for these estimators is also obtained. In the process, an almost sure linearity result about weighted residual empirical processes of Ghosh and Sen (1972) for a simple linear regression model is extended to the multiple linear regression model.

1 Introduction

Let $p \geq 1$ be an integer, F be a distribution function (d.f.) which has density f and $\{\varepsilon_i\}$, $1 \leq i \leq n$, be independent and identically distributed (i.i.d.) F random variables (r.v.'s). Consider the linear regression model

$$Y_{ni} = \mathbf{x}'_{ni}\boldsymbol{\beta} + \varepsilon_i, \qquad 1 \leq i \leq n, \tag{1}$$

where $\mathbf{x}'_{ni}$ is the ith row of the known $n \times p$ design matrix $\mathbf{X}$, $1 \leq i \leq n$, and $\boldsymbol{\beta}$ is the regression parameter vector of interest of dimension $p \times 1$.

AMS Subject Classifications : Primary 62G05, 62G20; secondary 62J05.

Key words and Phrases: L_2-distance, minimum distance estimators, strong consistency, multiple linear regression, weighted residual empiricals.

This paper is concerned with the strong consistency and the Bahadur representation of certain minimum distance estimators (m.d.e.'s) of $\boldsymbol{\beta}$. The m.d. estimation method, where one obtains an estimator of a parameter by minimizing some distance between the empirical d.f. and the modeled d.f., was elevated to a general method of estimation by Wolfowitz (1953, 1954, 1957). In these papers he demonstrated that compared to the maximum likelihood estimation method, the m.d. estimation method yielded consistent estimators rather cheaply in several problems of varied levels of difficulty.

This methodology saw increasing research activity from the mid seventies when several authors demonstrated various robustness properties of certain m.d.e.'s. See, e.g., Beran (1977, 1978), Parr and Schucany (1979), Millar (1981, 1984), Donoho and Liu (1988a,b), and the bibliography by Parr (1981). All of these authors focus on either the one sample setup or the two sample location-scale models.

Little was known till 1979 about the extension of m.d. estimation methodology to linear regression models. Koul (1979), Williamson (1979), Koul and De Wet (1983) and Koul (1985a,b) gave extensions of this methodology to linear regression models. These estimators are known to be locally asymptotically minimax and qualitatively robust against certain gross error models.

On the other hand the Bahadur type representation of an estimator is a fundamental tool in using the given estimator in sequential analysis and in obtaining a deeper understanding of its large sample behavior. It is thus only natural to seek this representation of the above m.d.e.'s.

We now describe the estimators that will be considered in this paper. To that effect we need some notation. For an $m \times l$ matrix $\mathbf{B}$, b_{ij} will denote its (i,j)th entry, $\mathbf{b}_{i\cdot}$ its ith row, $1 \leq i \leq m$, and $\mathbf{b}_{\cdot j}$ the jth column, $1 \leq j \leq l$, where m, l are positive integers. For any $m \times 1$ vector $\mathbf{t}$, $\mathbf{t}'$ will denote its transpose, $|\mathbf{t}| = \max_{1 \leq i \leq m} |t_i|$, and $\| \mathbf{t} \|^2 = \sum_{i=1}^{m} t_i^2$.

Let $\mathbf{S} = \mathbf{X}'\mathbf{X}$ and assume that $\mathbf{S}^{-1}$ exists. Let $\mathbf{C} = \mathbf{X}\mathbf{S}^{-1/2}$. Then the model (1) is equivalent to

$$Y_i = \mathbf{c}_{i\cdot}\boldsymbol{\Delta} + \varepsilon_i, \quad 1 \leq i \leq n, \tag{2}$$

where $\boldsymbol{\Delta} = \mathbf{S}^{1/2}\boldsymbol{\beta}$. Here and in the sequel, the dependence of $\{Y_i\}$, $\{\mathbf{c}_{i\cdot}\}$ etc. on n is not exhibted for the sake of convenience.

For $\mathbf{b} \in R^p$ and $y \in R$, let

$$\mathbf{v}(y, \mathbf{C}\,\mathbf{b}) = \Big(I(Y_1 \leq y + \mathbf{c}_{1\cdot}\mathbf{b}), \cdots, I(Y_n \leq y + \mathbf{c}_{n\cdot}\mathbf{b})\Big)',$$

$$\mathbf{u}(y, \mathbf{C}\,\mathbf{b}) = \Big(F(y + \mathbf{c}_{1\cdot}\mathbf{b}), \cdots, F(y + \mathbf{c}_{n\cdot}\mathbf{b})\Big)',$$

$$\mathbf{V}_I(y, \mathbf{C}\,\mathbf{b}) = \mathbf{C}'\mathbf{v}(y, \mathbf{C}\,\mathbf{b}), \quad \mathbf{V}_F(y, \mathbf{C}\,\mathbf{b}) = \mathbf{C}'\mathbf{u}(y, \mathbf{C}\,\mathbf{b}),$$

$$\mathbf{U}(y, \mathbf{C\,b}) = \mathbf{V}_I(y, \mathbf{C\,b}) - \mathbf{V}_F(y, \mathbf{0}),$$
$$\mathbf{W}(y, \mathbf{C\,b}) = \mathbf{V}_I(y, \mathbf{C\,b}) - \mathbf{V}_F(y, \mathbf{C\,b}).$$

Further, for a nondecreasing right continuous function H from R to R, let

$$M(\mathbf{b}) = \int \| \mathbf{U}(y, \mathbf{C\,b}) \|^2 \, dH(y), \qquad \mathbf{b} \in R^p.$$

Observe that

$$M(\mathbf{b}) = \int \| \sum_{i=1}^{n} \mathbf{c}_{i\cdot} \{ I(Y_i \le y + \mathbf{c}_{i\cdot}\mathbf{b}) - F(y) \} \|^2 \, dH(y)$$

and that under (2), $E\,\mathbf{U}(y, \mathbf{C}\boldsymbol{\Delta}) = \mathbf{0}$. This motivates one to define m.d.e. $\tilde{\boldsymbol{\Delta}}$ of $\boldsymbol{\Delta}$, in the case F is known, as a minimizer of M:

$$\tilde{\boldsymbol{\Delta}} = \text{argmin}_{\mathbf{b}} \; M(\mathbf{b}).$$

Similarly, in the case F is unknown but symmetric around 0, m.d.e. $\boldsymbol{\Delta}^+$ of $\boldsymbol{\Delta}$ is defined as a minimizer of M^+:

$$\boldsymbol{\Delta}^+ = \text{argmin}_{\mathbf{b}} \; M^+(\mathbf{b}),$$

where, for $\mathbf{b} \in R^p$,

$$\begin{aligned} M^+(\mathbf{b}) &= \int \| \mathbf{V}_I(y, \mathbf{C\,b}) - \sum_{i=1}^{n} \mathbf{c}'_{i\cdot} + \mathbf{V}_I(-y, \mathbf{C\,b}) \|^2 \, dH(y) \\ &= \int \| \sum_{i=1}^{n} \mathbf{c}_{i\cdot} \{ I(Y_i \le y + \mathbf{c}_{i\cdot}\mathbf{b}) - I(-Y_i < y - \mathbf{c}_{i\cdot}\mathbf{b}) \} \|^2 \, dH(y). \end{aligned}$$

See Koul and De Wet (1983) and Koul (1985b) for more motivation and some other small and large sample properties of $\tilde{\boldsymbol{\Delta}}$ and $\boldsymbol{\Delta}^+$, respectively.

In the sequel, $\| \cdot \|_H$ denotes the L_2 norm with respect to H and

$$|\mathbf{C}| = \max_{1 \le i \le n} \| \mathbf{c}_{i\cdot} \|, \quad |\mathbf{c}_{\cdot j}| = \max_{1 \le i \le n} |c_{ij}|, \; 1 \le j \le p.$$

Throughout this paper, we shall be using various subsets of the following assumptions.

(A) H is a bounded nondecreasing right continuous real function which induces a finite measure on $(R, \mathcal{B})$, the Borel line.

(B) F has a bounded density f such that $\int f \, dH > 0$ and

$$\sup_{|y| < \infty} |f(y + h) - f(y)| = O(|h|), \text{ as } h \to 0.$$

(C) $\mathbf{S}^{-1} = (\mathbf{X}'\mathbf{X})^{-1}$ exists for all $n \geq p$.

(D) $|\mathbf{C}| = o\Big((\ln n)^{-a}\Big)$ for some $a \geq \frac{1}{2}$.

Without loss of generality, we shall assume that H is a d.f. whenever **(A)** is assumed to hold.

In what follows all limits are taken as $n \to \infty$ and *wp1n* means with probability 1 as $n \to \infty$.

THEOREM 1 . *In addition to (2), assume that* **(A)–(D)** *hold. Then, there exists a constant* $K_1 > 0$ *such that*

$$\| \tilde{\boldsymbol{\Delta}} - \boldsymbol{\Delta} \| < (K_1 \ln n)^{\frac{1}{2}}, \qquad wp1n,$$

and

$$\| \boldsymbol{\Delta}^+ - \boldsymbol{\Delta} \| < (K_1 \ln n)^{\frac{1}{2}}, \qquad wp1n. \qquad \square$$

Now recall that $\tilde{\boldsymbol{\beta}} = \mathbf{S}^{-1/2}\tilde{\boldsymbol{\Delta}}, \ \ \boldsymbol{\beta}^+ = \mathbf{S}^{-1/2}\boldsymbol{\Delta}^+$.

COROLLARY 1 . *Under assumptions* **(A)–(D)**, *for model (1),*

$$\| \tilde{\boldsymbol{\beta}} - \boldsymbol{\beta} \| < (K_1 \| \mathbf{S}^{-1/2} \|^2 \ln n)^{1/2}, \qquad wp1n,$$

$$\| \boldsymbol{\beta}^+ - \boldsymbol{\beta} \| < (K_1 \| \mathbf{S}^{-1/2} \|^2 \ln n)^{1/2}, \qquad wp1n. \qquad \square$$

REMARK 1 . *The corollary implies that if the design matrix* $\mathbf{X}$ *is such that*

$$\| \mathbf{S}^{-1/2} \|^2 = o((\ln n)^{-1}),$$

then the sequences of estimators $\{\tilde{\boldsymbol{\beta}}\}$ *and* $\{\boldsymbol{\beta}^+\}$ *are strongly consistent for* $\boldsymbol{\beta}$ *in (1). This condition is satisfied by a large class of designs including polynomials.*

Before stating the next theorem we need to define

$$\psi(x) = \int_{-\infty}^{x} f(t)\, dH(t), \qquad \varphi(x) = \psi(-x) - \psi(x), \quad x \in R.$$

THEOREM 2 . *In addition to (2), assume that* **(A)–(D)** *hold. Then*

$$\| f \|_H^2 (\tilde{\Delta} - \Delta)' = \sum_{i=1}^{n} \mathbf{c}_{i\cdot} [\psi(\varepsilon_i) - E\psi(\varepsilon_i)] + \sigma_n,$$

$$\| f \|_H^2 (\Delta^+ - \Delta)' = \sum_{i=1}^{n} \mathbf{c}_{i\cdot} \varphi(\varepsilon_i) + \tau_n,$$

where

$$\sigma_n = O(|\mathbf{C}|^{1/2} (\ln n)^{3/4} + |\mathbf{C}| \ln n + |\mathbf{C}|^{1/4} (\ln n)^{5/8}), \qquad wp1n,$$

and

$$\tau_n = O(\sigma_n), \qquad wp1n. \qquad \square$$

COROLLARY 2 . *In addition to (1), assume that* **(A)–(D)** *hold. Then*

$$\tilde{\beta} - \beta = (\| f \|_H^2 \mathbf{X'X})^{-1} \sum_{i=1}^{n} \mathbf{x}_{ni} [\psi(\varepsilon_i) - E\psi(\varepsilon_i)] + \tilde{R}_n$$

$$\beta^+ - \beta = (\| f \|_H^2 \mathbf{X'X})^{-1} \sum_{i=1}^{n} \mathbf{x}_{ni} \varphi(\varepsilon_i) + \hat{R}_n,$$

where

$$\tilde{R}_n = O(|\mathbf{X}(\mathbf{X'X})^{-1}|^{1/2} (\ln n)^{3/4} + |\mathbf{X}(\mathbf{X'X})^{-1}| \ln n + |\mathbf{X}(\mathbf{X'X})^{-1}|^{1/4} (\ln n)^{5/8}), \qquad wp1n,$$

and

$$\hat{R}_n = O(\tilde{R}_n), \qquad wp1n. \qquad \square$$

2 The Proofs

Because the estimators $\tilde{\Delta}$, Δ^+ are translation invariant the true parameter Δ will be assumed to be $\mathbf{0}$ throughout this section. We also need to define some approximating r.v.'s and processes as follows.

$$\mathbf{T} = -\int \mathbf{U}(y, \mathbf{0}) f(y)\, dH(y),$$

$$\hat{M}(\mathbf{b}) = \int \| \mathbf{U}(y, \mathbf{0}) + \mathbf{b} f(y) \|^2 \, dH(y),$$

$$\hat{M}^{+}(\mathbf{b}) = \int \| \mathbf{V}_I(y,\mathbf{0}) - \sum_{i=1}^{n} \mathbf{c}'_{i\cdot} + \mathbf{V}_I(-y,\mathbf{0}) + 2\mathbf{b}f(y) \|^2 \, dH(y), \quad \mathbf{b} \in R^p,$$

$$\hat{\mathbf{\Delta}} = \text{argmin } \hat{M}(\mathbf{b}), \quad \hat{\mathbf{\Delta}}^{+} = \text{argmin } \hat{M}^{+}(\mathbf{b}).$$

The following facts will be used repeatedly in the proofs:

1. $\mathbf{C}'\mathbf{C} = \mathbf{I}_{p\times p}$. 2. $\sum_{i=1}^{n} \| \mathbf{c}_{i\cdot} \|^2 = \text{trace } \mathbf{C}'\mathbf{C} = p$.
3. $|c_{ij}| \le |\mathbf{c}_{\cdot j}| \le \max_{1\le i\le n}(\| \mathbf{c}_{i\cdot} \|) = |\mathbf{C}|$.
4. $|\mathbf{C}| \ge (\frac{p}{n})^{1/2} \ge n^{-1/2}$. 5. $\| f \|_H^2 \, \hat{\mathbf{\Delta}} = \mathbf{T}$.

To begin with, the assumptions **(B)** and **(D)** imply, in a straight forward fashion, that for any $K > 0$,

$$(\mathbf{D}') \qquad \sup_{\|\mathbf{b}\|\le (K\ln n)^{1/2}} \int \| \sum_{i=1}^{n} \mathbf{c}_{i\cdot}\Big(F(y+\mathbf{c}_i.\mathbf{b}) - F(y) - \mathbf{c}_i.\mathbf{b}f(y)\Big) \|^2 \, dH(y)$$
$$= O((|\mathbf{C}|\ln n)^2).$$

We divide the proofs of the two theorems into several steps. Each step consists of a lemma.

LEMMA 1 . *If f is continuous and if for a σ-finite measure μ on $(R, \mathcal{B})$, $0 < \int f \, d\mu < \infty$, then there exists a $\delta > 0$ such that*

$$\int \inf_{|a|\le\delta} f(y+a)\, d\mu(y) = \eta > 0.$$

PROOF. By the continuity of f,

$$\lim_{\delta\to 0} \inf_{|a|\le\delta} f(y+a) = f(y), \qquad y \in R.$$

Also, $\inf_{|a|\le\delta} f(y+a) \le f(y)$. The claim now follows from the Dominated Convergence Theorem. □

For the sake of completeness, we present an exponential inequality from Marcus and Zinn (1984) as

LEMMA 2 . *Let $\{X_{ni}\}_{i=1}^{n}$ be a triangular array of independent real valued r.v.'s and $\{a_{ni}\}_{i=1}^{n}$ be an array of real numbers such that $\sum_{i=1}^{n} a_{ni}^2 = 1$. Then for all $\lambda \ge 0$ and all $n \ge 1$,*

$$P\Big(\sup_{|t|<\infty} \Big| \sum_{i=1}^{n} a_{ni}\Big(I(X_{ni} \le t) - P(X_{ni} \le t)\Big)\Big| \ge \lambda\Big)$$
$$\le (1 + 2\sqrt{2\pi}\lambda)\exp(-\frac{\lambda^2}{8}).$$
□

Actually Marcus and Zinn (1984) proved this inequality for non- triangular sequences and nonnegative $\{X_i\}$. A close scrutiny of their proof shows that it is valid for the above setup.

The following two lemmas are extensions of Theorem 3.1 of Ghosh and Sen (1972). The idea of the proof of the first lemma is similar to that of Ghosh and Sen but the details are involved and hence not given here. They can be found in Zhu (1992). To state these lemmas, we need to define, for $y, u \in R,\ \mathbf{s}, \mathbf{t} \in R^n$,

$$W_{\mathbf{t}}(y, u\,\mathbf{s}) = \sum_{i=1}^{n} t_i\Big(I(Y_i \le y + u\,s_i) - F(y + u\,s_i)\Big). \tag{3}$$

LEMMA 3 . *Let F satisfy* **(B)** *and $Y_1, \cdots, Y_n$ be i.i.d. F. Let $\mathbf{s}_n$, $\mathbf{t}_n$ be sequences of vectors in R^n satisfying the following:*

- $\| \mathbf{s}_n \| \le 1,\ \| \mathbf{t}_n \| \le 1,$
- *there is a positive number sequence $\{a_n\}_{n=1}^{\infty}$, $a_n = o((\ln n)^{-k_0})$ for some $k_0 \ge \frac{1}{2}$, such that $|\mathbf{s}_n| \le a_n$ and $|\mathbf{t}_n| \le a_n$.*

Then for any $K_1 > 0$ and $k > 0$, there exists $K_{11} > 0$ and n^* (both of which depend on K_1, k, and $\{a_n\}$) such that for all $n \ge n^*$,

$$P\Big(\sup_{|y|<\infty, |u|\le b_n} |W_{\mathbf{t}_n}(y, u\,\mathbf{s}_n) - W_{\mathbf{t}_n}(y, \mathbf{0})| > K_{11} a_n^{1/2} (\ln n)^{\frac{3}{2}k_0}\Big) \le 2\,n^{-k},$$

where $b_n = K_1(\ln n)^{k_0}$. Consequently,

$$\sup_{|y|<\infty, |u|\le b_n} \big|W_{\mathbf{t}_n}(y, u\,\mathbf{s}_n) - W_{\mathbf{t}_n}(y, \mathbf{0})\big| \le K_{11}\,|a_n|^{1/2}\,(\ln n)^{\frac{3}{2}k_0}, \quad wp1n.$$

□

To state the next lemma, we need to define

$$\begin{aligned} \mathcal{E} &= \{\mathbf{e} \in R^p : \| \mathbf{e} \| = 1\}, \\ \mathcal{D} &= \{\mathbf{d} \in R^n : \mathbf{d} = \mathbf{C}\mathbf{e},\ \mathbf{e} \in \mathcal{E}\}. \end{aligned}$$

LEMMA 4 . *Let $\{Y_i\}$ obey the model (2) with the true $\mathbf{\Delta} = \mathbf{0}$. In addition, assume that* **(B)–(D)** *hold. Then, there exists a constant $K_2 > 0$ such that for any $\mathbf{t} \in R^n$, $|\mathbf{t}| \le |\mathbf{C}|$ and $\| \mathbf{t} \| \le 1$,*

$$\sup_{|y|<\infty, 0\le b\le b_n, \mathbf{d}\in\mathcal{D}} |W_{\mathbf{t}}(y, b\,\mathbf{d}) - W_{\mathbf{t}}(y, \mathbf{0})| \le K_2\,|\mathbf{C}|^{1/2}\,(\ln n)^{3/4}, \quad wp1n. \tag{4}$$

PROOF. Let $\gamma_n = K_2 |\mathbf{C}|^{1/2} (\ln n)^{3/4}$. We will assume that all components of $\mathbf{t}$ are nonnegative. Because of the triangle inequality, there is no loss of generality in doing so.

For each n, we can split $\mathcal{D}$ into m_n parts: $\mathcal{D}_1, \cdots, \mathcal{D}_{m_n}$ such that

1. the diameter of each $\mathcal{D}_k$ is no larger than n^{-1}, $1 \le k \le m_n$,
2. $m_n \le (pn)^p$.

Fix a point $\mathbf{d}^k$ in each $\mathcal{D}_k$. Let $\boldsymbol{\delta}_n = (n^{-1}, \cdots, n^{-1}) \in R^n$ and let

$$\mathcal{D}_0 = \{\mathbf{d}^k \pm \boldsymbol{\delta}_n, k = 1, \cdots, m_n\}.$$

Then $\mathcal{D}_0$ contains $2m_n$ points.

Now, for any $\mathbf{d} \in \mathcal{D}$, $\mathbf{d} \in \mathcal{D}_k$ for some k. Use this and the fact that the ith summand in $W_{\mathbf{t}}(y, b\mathbf{d})$ is a difference of two nondecreasing functions of $b\,d_i$, $1 \le i \le n$, to obtain, for every $y \in R$,

$$\begin{aligned} W_{\mathbf{t}}(y, b\mathbf{d}) &= \sum_{i=1}^{n} t_i \Big(I(Y_i \le y + b\,d_i) - F(y + b\,d_i) \Big) \\ &\le \sum_{i=1}^{n} t_i \Big(I(Y_i \le y + b(d_i^k + n^{-1})) - F(y + b(d_i^k - n^{-1})) \Big). \end{aligned}$$

Therefore,

$$\begin{aligned} & W_{\mathbf{t}}(y, b\mathbf{d}) - W_{\mathbf{t}}(y, \mathbf{0}) \\ & \qquad \le W_{\mathbf{t}}(y, b(\mathbf{d}^k + \boldsymbol{\delta}_n)) - W_{\mathbf{t}}(y, \mathbf{0}) \\ & \qquad\quad + \sum_{i=1}^{n} t_i \Big(F(y + b(d_i^k + n^{-1}) - F(y + b(d_i^k - n^{-1})) \Big). \end{aligned}$$

Similarly,

$$\begin{aligned} & W_{\mathbf{t}}(y, b\mathbf{d}) - W_{\mathbf{t}}(y, \mathbf{0}) \\ & \qquad \ge W_{\mathbf{t}}(y, b(\mathbf{d}^k - \boldsymbol{\delta}_n)) - W_{\mathbf{t}}(y, \mathbf{0}) \\ & \qquad\quad - \sum_{i=1}^{n} t_i \Big(F(y + b(d_i^k + n^{-1})) - F(y + b(d_i^k - n^{-1})) \Big). \end{aligned}$$

Hence, by the assumption **(B)**, uniformly in $y \in R$,

$$\begin{aligned} & \sup_{\mathbf{d} \in \mathcal{D}} |W_{\mathbf{t}}(y, b\mathbf{d}) - W_{\mathbf{t}}(y, \mathbf{0})| \\ & \quad \le \max_{\mathbf{d} \in \mathcal{D}_0} |W_{\mathbf{t}}(y, b\mathbf{d}) - W_{\mathbf{t}}(y, \mathbf{0})| \\ & \qquad\quad + \max_{1 \le k \le m_n} \sum_{i=1}^{n} t_i \Big(F(y + b(d_i^k + n^{-1})) - F(y + b(d_i^k - n^{-1})) \Big) \\ & \quad \le \max_{\mathbf{d} \in \mathcal{D}_0} |W_{\mathbf{t}}(y, b\mathbf{d}) - W_{\mathbf{t}}(y, \mathbf{0})| + O(bn^{-1/2}). \end{aligned}$$

The second term of the right hand side does not depend on $\mathbf{t}$ and $\mathbf{d}^k$, and for $0 \le b \le b_n$, it becomes $O((n \ln n)^{-1/2}) = o(\gamma_n)$, by the fact 4. To complete the proof of (4), it suffices to show that

$$\sup_{|y|<\infty} \sup_{0 \le b \le b_n} \sup_{\mathbf{d} \in \mathcal{D}_0} |W_{\mathbf{t}}(y, b\mathbf{d}) - W_{\mathbf{t}}(y, \mathbf{0})| \le \gamma_n, \qquad wpln.$$

Apply Lemma 3 with $k_0 = \frac{1}{2}$, $a_n = |\mathbf{C}|$, and $k = p + 2$, to obtain

$$\begin{aligned} & P\Big(\sup_{|y|<\infty} \sup_{0 \le b \le b_n} \sup_{\mathbf{d} \in \mathcal{D}_0} |W_{\mathbf{t}}(y, b\mathbf{d}) - W_{\mathbf{t}}(y, \mathbf{0})| \ge \gamma_n\Big) \\ & \le \sum_{\mathbf{d} \in \mathcal{D}_0} P\Big(\sup_{|y|<\infty} \sup_{0 \le b \le b_n} |W_{\mathbf{t}}(y, b\mathbf{d}) - W_{\mathbf{t}}(y, \mathbf{0})| \ge \gamma_n\Big) \\ & \le 2m_n 2n^{-k} \\ & = O(n^{-2}). \end{aligned}$$

The claim now follows from the Borel-Cantelli Lemma. □

LEMMA 5 . *In addition to (2) with true* $\mathbf{\Delta} = \mathbf{0}$, *assume that* **(A)** *holds. Then there exists a* $0 < K_0 < \infty$ *such that*

$$M(\mathbf{0}) < K_0 \ln n, \qquad wpln. \tag{5}$$

PROOF.

$$\begin{aligned} M(\mathbf{0}) &= \int \| \mathbf{U}(y, \mathbf{0}) \|^2 \, dH(y) \\ &= \sum_{j=1}^{p} \int \Big\{ \sum_{i=1}^{n} c_{ij} \Big(I(Y_i \le y) - F(y) \Big) \Big\}^2 dH(y) \\ &\le p \max_{1 \le j \le p} \sup_{|y|<\infty} \Big| \sum_{i=1}^{n} c_{ij} \Big(I(Y_i \le y) - F(y) \Big) \Big|^2 . \end{aligned}$$

Because $\mathbf{C}'\mathbf{C} = \mathbf{I}_{p \times p}$, $\sum_{i=1}^n c_{ij}^2 = 1$, $1 \le j \le p$. Now apply Lemma 2 with $a_{ni} = c_{ij}$, $\lambda_n = (p^{-1} K_0 \ln n)^{1/2}$, to conclude that for all $1 \le j \le p$,

$$P(\sup_{|y|<\infty} \Big| \sum_{i=1}^{n} c_{ij} \Big(I(Y_i \le y) - F(y) \Big) \Big| \ge \lambda_n) \le \Big(1 + 2\sqrt{2\pi p^{-1} K_0 \ln n} \Big) n^{-\frac{K_0}{8p}} . \tag{6}$$

The claim now follows by the Borel-Cantelli Lemma upon choosing $K_0 \ge 9p$. □

REMARK 2 . *Because of (5) and the fact that* $\| \mathbf{T} \|^2 \le M(\mathbf{0}) \| f \|_H^2$, *it follows from fact 5 that*

$$\| \hat{\boldsymbol{\Delta}} \| < (K \ln n)^{1/2}, \qquad wp1n,$$

where $K = K_0 \| f \|_H^{-2}$.

LEMMA 6 . *In addition to model (2) with true* $\boldsymbol{\Delta} = \mathbf{0}$, *assume that* **(B)–(D)** *hold and that* H *induces a sigma finite measure on* $(R, \mathcal{B})$. *Then there exists a constant* $0 < K_1 < \infty$, *such that*

$$\inf_{\|\mathbf{b}\| \ge (K_1 \ln n)^{1/2}} M(\mathbf{b}) > K_0 \ln n, \qquad wp1n \qquad (7)$$

with K_0 *as in Lemma 5.*

REMARK 3 . *In this lemma, the function* H *is not required to be bounded.*

PROOF OF LEMMA 6. Let $b_n = (K_1 \ln n)^{1/2}$, $\mathcal{E}$ and $\mathcal{D}$ be as in Lemma 4 and let g be a nonnegative function on R such that $0 < \int g\, dH < \infty$, $\int g^2\, dH = 1$. Define

$$\mu(x) = \int_{-\infty}^{x} g(y)\, dH(y), \qquad x \in R; \quad \mu_0 = \mu(\infty).$$

Note that μ is a bounded nondecreasing function with $\sup_x \mu(x) \le \mu_0$.

For any $\mathbf{b} \in R^p, \mathbf{b} \ne \mathbf{0}$, there is unique $\mathbf{e} \in \mathcal{E}$ such that $\mathbf{b} = \| \mathbf{b} \| \mathbf{e} = b\mathbf{e}$, $b \in R^+$. Therefore (7) is equivalent to

$$\inf_{b \ge b_n, \mathbf{e} \in \mathcal{E}} M(b\mathbf{e}) > K_0 \ln n, \qquad wp1n. \qquad (8)$$

By the Cauchy-Schwarz inequality, with $\mathbf{d} = \mathbf{Ce}$,

$$\begin{aligned} M(\mathbf{b}) &= M(b\mathbf{e}) \\ &= \int \| \mathbf{e} \|^2 \| \mathbf{C}' \Big(\mathbf{v}(y, b\mathbf{Ce}) - \mathbf{u}(y, \mathbf{0}) \Big) \|^2 \; dH(y) \\ &\ge \int \Big[\mathbf{d}' \Big(\mathbf{v}(y, b\mathbf{Ce}) - \mathbf{u}(y, \mathbf{0}) \Big) \Big]^2 \; dH(y) \\ &\ge \Big[\int \mathbf{d}' \Big(\mathbf{v}(y, b\mathbf{Ce}) - \mathbf{u}(y, \mathbf{0}) \Big) d\mu(y) \Big]^2 \\ &= \Big[\int U_d^*(y, b\mathbf{d})\, d\mu(y) \Big]^2 \end{aligned} \qquad (9)$$

where

$$U_d^*(y, b\,\mathbf{d}) := \sum_{i=1}^{n} d_i\Big(I(Y_i \le y + bd_i) - F(y)\Big).$$

Observe that for all y, $U_d^*(y, b\,\mathbf{d})$ is an nondecreasing function of b so that when $b \ge b_n$,

$$U_d^*(y, b\,\mathbf{d}) \ge U_d^*(y, b_n\mathbf{d}), \qquad \forall\ \mathbf{d} \in \mathcal{D},\ y \in R.$$

Therefore, to prove (8), it suffices to show that

$$\sup_{\mathbf{d}\in\mathcal{D}} \Big(-\int U_d^*(y, b_n\mathbf{d})\, d\mu(y)\Big) < -(K_0 \ln n)^{1/2}, \qquad wp1n. \quad (10)$$

Now, let m_n and $\mathcal{D}_1, \cdots, \mathcal{D}_{m_n}$ be as in the proof of Lemma 4. Recall the definition of η from Lemma 1. Let K_1 be such that $\eta\sqrt{K_1} - 2\sqrt{K_0} > 0$. We first prove that for each $\mathbf{d}^k \in \mathcal{D}_k$,

$$\begin{aligned} & P\Big(-\int U_{d^k}^*(y, b_n\mathbf{d}^k)\, d\mu(y) \ge -2(K_0 \ln n)^{1/2}\Big) \\ \le\ & \exp\Big\{-\frac{1}{2\mu_0^2}\Big(\eta\sqrt{K_1} - 2\sqrt{K_0}\Big)^2 \ln n\Big\}. \end{aligned} \qquad (11)$$

Lemma 1 and the assumption **(D)** imply that for sufficiently large n,

$$\begin{aligned} E\int U_{d^k}^*(y, b_n\mathbf{d}^k)\, d\mu(y) &= \int \sum_{i=1}^{n} d_i^k (F(y + b_n d_i^k) - F(y))\, d\mu(y) \\ &= \sum_{i=1}^{n} (d_i^k)^2 b_n \int f(y + \theta_i)\, d\mu(y), \quad |\theta_i| \le |\, d_i^k b_n \,| \\ &\ge \eta(K_1 \ln n)^{1/2}. \end{aligned} \qquad (12)$$

Now observe that $\int U_{d^k}^*(y, b_n\mathbf{d}^k)\, d\mu(y)$ is a sum of n independent bounded r.v.'s

$$X_{ni} = d_i^k \int \Big(I(Y_i \le y + d_i^k b_n) - F(y)\Big)\, d\mu(y)$$

with the bounds given as

$$-\left|d_i^k \mu_0\right| \le X_{ni} \le \left|d_i^k \mu_0\right|, \quad 1 \le i \le n.$$

Therefore, by an exponential inequality of Hoeffding (Theorem 2, 1963) and (12), for n sufficiently large,

$$P\Big(-\int U_{d^k}^*(y, b_n\mathbf{d}^k)\, d\mu(y) \ge -2(K_0 \ln n)^{1/2}\Big)$$

$$\le P\Big(-\int U^*_{d^k}(y,b_n\mathbf{d}^k)\,d\mu(y) + E\int U^*_{d^k}(y,b_n\mathbf{d}^k)\,d\mu(y) \ge \big(\eta\sqrt{K_1}-2\sqrt{K_0}\big)(\ln n)^{1/2}\Big)$$

$$\le \exp\Big(-\frac{1}{2\mu_0^2}\big(\eta\sqrt{K_1}-2\sqrt{K_0}\big)^2\ln n\Big).$$

This proves (11). Note that the upper bound in (11) does not depends on k. Hence, for n sufficiently large,

$$\begin{aligned} P\Big(\max_{1\le k\le m_n}\Big\{ &- \int U^*_{d^k}(y,b_n\mathbf{d}^k)\,d\mu(y)\Big\} \ge -2(K_0\ln n)^{1/2}\Big) \\ &\le (pn)^p \exp\Big\{-\frac{1}{2\mu_0^2}\big(\eta\sqrt{K_1}-2\sqrt{K_0}\big)^2\ln n\Big\} \\ &\le n^{-\frac{1}{2\mu_0^2}(\eta\sqrt{K_1}-2\sqrt{K_0})^2+p+1} \end{aligned}$$

Clearly there exists a positive constant K_1 so that this upper bound is summable in n. Hence, the Borel-Cantelli lemma gives

$$\max_{1\le k\le m_n}\Big\{-\int U^*_{d^k}(y,b_n\mathbf{d}^k)\,d\mu(y)\Big\} < -2(K_0\ln n)^{1/2}, \qquad wp1n. \tag{13}$$

Next, for any $\mathbf{d}\in\mathcal{D}$, $\mathbf{d}\in\mathcal{D}_k$ for some k,

$$\begin{aligned} &\Big|\int\Big[U^*_d(y,b_n\mathbf{d}) - U^*_{d^k}(y,b_n\mathbf{d}^k)\Big]\,d\mu(y)\Big| \\ &= \Big|\int\Big[U^*_d(y,b_n\mathbf{d}) - U^*_{d^k}(y,b_n\mathbf{d})\Big]\,d\mu(y) \\ &\qquad + \int\Big[U^*_{d^k}(y,b_n\mathbf{d}) - U^*_{d^k}(y,b_n\mathbf{d}^k)\Big]\,d\mu(y)\Big| \\ &\le \Big(\sup_{|y|<\infty}\Big|\sum_{i=1}^n (d_i^k - d_i)\Big[I(Y_i\le y+b_nd_i) - F(y)\Big]\Big| \\ &\qquad + p\sup_{|y|<\infty}\Big|\sum_{i=1}^n d_i^k\Big[I(Y_i\le y+b_nd_i^k) - I(Y_i\le y+b_nd_i)\Big]\Big|\Big)\mu_0 \\ &:= (I_1+I_2)\mu_0, \qquad say. \end{aligned} \tag{14}$$

Recall the definition of $W_{\mathbf{t}}$ from (3). By Lemma 4 and the consequence (D′),

$$\begin{aligned} I_1 &= \sup_{|y|<\infty}\Big|\sum_{i=1}^n (d_i^k - d_i)\Big[I(Y_i\le y+b_nd_i) - F(y)\Big]\Big| \\ &\le \|\,\mathbf{d}^k - \mathbf{d}\,\|\; n^{1/2} \end{aligned}$$

$$= \quad o((\ln n)^{-1/2}), \qquad wpln, \tag{15}$$

$$\begin{aligned} I_2 &= \sup_{|y|<\infty} \left| \sum_{i=1}^{n} d_i^k \Big[I(Y_i \le y + b_n d_i^k) - I(Y_i \le y + b_n d_i) \Big] \right| \\ &\le \sup_{|y|<\infty} \Big\{ \left| W_{\mathbf{d}^k}(y, b_n \mathbf{d}^k) - W_{\mathbf{d}^k}(y, \mathbf{0}) \right| + \left| W_{\mathbf{d}^k}(y, b_n \mathbf{d}) - W_{\mathbf{d}^k}(y, \mathbf{0}) \right| \\ &\quad + \left| \sum_{i=1}^{n} d_i^k \Big(F(y + b_n d_i^k) - F(y + b_n d_i) \Big) \right| \Big\}. \\ &\le K_2 \, |\mathbf{C}|^{1/2} (\ln n)^{3/4}, \qquad wpln. \end{aligned} \tag{16}$$

Combine (14) – (16) with the assumption **(D)** to obtain

$$\sup_{\mathbf{d} \in \mathcal{D}} \left| \int \Big[U_d^*(y, b_n \mathbf{d}) - U_{d^k}^*(y, b_n \mathbf{d}^k) \Big] \, d\mu(y) \right| = o((\ln n)^{1/2}), \qquad wpln. \tag{17}$$

Finally, by (13) and (17),

$$\begin{aligned} &\sup_{\mathbf{d} \in D} \Big(- \int U_d^*(y, b_n \mathbf{d}) \, d\mu(y) \Big) \\ &= \sup_{\mathbf{d} \in D} \Big(- \int U_{d^k}^*(y, b_n \mathbf{d}^k) \, d\mu(y) + \int \Big[U_{d^k}^*(y, b_n \mathbf{d}^k) - U_d^*(y, b_n \mathbf{d}) \Big] \, d\mu(y) \Big) \\ &\le \max_{1 \le k \le m_n} \Big\{ - \int U_{d^k}^*(y, b_n \mathbf{d}^k) \, d\mu(y) \Big\} + o((\ln n)^{1/2}) \\ &< -(K_0 \ln n)^{1/2}, \qquad wpln, \end{aligned}$$

thereby proving (10) and also the lemma. □

PROOF OF THEOREM 1. We only give the proof for $\tilde{\boldsymbol{\Delta}}$. The proof for $\boldsymbol{\Delta}^+$ is similar. By the definition of $\tilde{\boldsymbol{\Delta}}$,

$$M(\tilde{\boldsymbol{\Delta}}) \le M(\mathbf{0}).$$

On the other hand, by Lemmas 5 and 6,

$$\inf_{\|\mathbf{b}\| \ge b_n} M(\mathbf{b}) > K_0 \ln n > M(\mathbf{0}), \qquad wpln,$$

where $b_n = (K_1 \ln n)^{1/2}$. Therefore

$$\| \tilde{\boldsymbol{\Delta}} \| < b_n = (K_1 \ln n)^{1/2}, \qquad wpln.$$

This completes the proof of Theorem 1. □

PROOF OF THEOREM 2. Let b_n be as before. By using the fact 5, we obtain

$$\hat{M}(\tilde{\boldsymbol{\Delta}}) - \hat{M}(\hat{\boldsymbol{\Delta}})$$

$$\begin{aligned}
&= \int \Big(\| \mathbf{U}(y,\mathbf{0}) + f(y)\tilde{\boldsymbol{\Delta}} \|^2 - \| \mathbf{U}(y,\mathbf{0}) + f(y)\hat{\boldsymbol{\Delta}} \|^2 \Big)\, dH(y) \\
&= \int (\tilde{\boldsymbol{\Delta}}' - \hat{\boldsymbol{\Delta}}')2\Big[\mathbf{U}(y,\mathbf{0}) + f(y)(\tilde{\boldsymbol{\Delta}} + \hat{\boldsymbol{\Delta}})\Big] f(y)\, dH(y) \\
&= \| f \|_H^2 \, \| \tilde{\boldsymbol{\Delta}} - \hat{\boldsymbol{\Delta}} \|^2. \qquad (18)
\end{aligned}$$

By Theorem 1 and Remark 2,

$$\| \tilde{\boldsymbol{\Delta}} \| < b_n \quad and \quad \| \hat{\boldsymbol{\Delta}} \| < b_n, \qquad wp1n.$$

Hence,

$$\begin{aligned}
\hat{M}(\tilde{\boldsymbol{\Delta}}) - \hat{M}(\hat{\boldsymbol{\Delta}}) &\le \Big| \hat{M}(\tilde{\boldsymbol{\Delta}}) - M(\tilde{\boldsymbol{\Delta}}) \Big| + \Big| M(\hat{\boldsymbol{\Delta}}) - \hat{M}(\hat{\boldsymbol{\Delta}}) \Big| \\
&\le 2 \sup_{\|\mathbf{b}\| \le b_n} \Big| M(\mathbf{b}) - \hat{M}(\mathbf{b}) \Big|, \qquad wp1n. \qquad (19)
\end{aligned}$$

For a fixed $\mathbf{b} \in R^p$, we have

$$\begin{aligned}
\Big| M(\mathbf{b}) - \hat{M}(\mathbf{b}) \Big| &= \Big| \int \Big(\| \mathbf{U}(y, \mathbf{C}\,\mathbf{b}) \|^2 - \| \mathbf{U}(y,\mathbf{0}) + f(y)\,\mathbf{b} \|^2 \Big)\, dH(y) \Big| \\
&\le \int \| I \|^2 \, dH + \int \| III \|^2 \, dH \\
&\quad + 2 \int |II'(I + III)| \, dH + 2 \int |I' III| \, dH \qquad (20)
\end{aligned}$$

where

$$\begin{aligned}
I &= \mathbf{W}(y, \mathbf{C}\,\mathbf{b}) - \mathbf{W}(y, \mathbf{0}) \\
II &= \mathbf{U}(y, \mathbf{0}) + f(y)\,\mathbf{b} \\
III &= \mathbf{C}'\Big(\mathbf{u}(y, \mathbf{C}\,\mathbf{b}) - \mathbf{u}(y, \mathbf{0})\Big) - f(y)\mathbf{b}.
\end{aligned}$$

But

$$\| III \|^2 = \sum_{j=1}^{p} \Big\{ \sum_{i=1}^{n} c_{ij} \Big(F(y + \mathbf{c}_i.\mathbf{b}) - F(y) - \mathbf{c}_i.\mathbf{b}\, f(y) \Big) \Big\}^2 .$$

Therefore, by $(\mathbf{D}')$,

$$\sup_{\|\mathbf{b}\| \le b_n} \| III \|_H = O(|\mathbf{C}|\, b_n^2) = O(|\mathbf{C}| \ln n). \qquad (21)$$

Next, from the definition of b_n, Lemma 5 and the fact that the $\| f \|_H < \infty$, it readily follows that

$$\sup_{\|\mathbf{b}\| \le b_n} \| II \|_H = O((\ln n)^{1/2}), \qquad wp1n. \qquad (22)$$

Moreover, Lemma 4 and assumption (**A**) imply that

$$\sup_{\|\mathbf{b}\| \le b_n} \| I \|_H = O(|\mathbf{C}|^{1/2} (\ln n)^{3/4}), \quad wpln. \tag{23}$$

Combining (21), (22), and (23), we obtain

$$\sup_{\|\mathbf{b}\| \le b_n} |M(\mathbf{b}) - \hat{M}(\mathbf{b})| = O\Big(|\mathbf{C}|^{1/2} (\ln n)^{5/4} + |\mathbf{C}| (\ln n)^{3/2} + (|\mathbf{C}| \ln n)^2\Big). \tag{24}$$

Finally, by using the fact $(a+b+c)^{1/2} \le \sqrt{a} + \sqrt{b} + \sqrt{c}$ for a, b, and c positive, the claim follows from (18), (19), (24) and the fact that

$$\| f \|_H^2 \hat{\mathbf{\Delta}}' = \sum_{i=1}^{n} \mathbf{c}_i. [\psi(\varepsilon_i) - E\psi(\varepsilon_i)].$$

This proves the theorem for $\tilde{\Delta}$. The proof for Δ^+ is similar and uses the approximating r.v.'s $\hat{M}^+$ and $\hat{\mathbf{\Delta}}^+$. The details are left out for the sake of brevity. □

Acknowledgment

Authors would like to thank the referee for helpful comments that resulted in an improved presentation. □

REFERENCES

Beran, R. J. (1977). Minimum Hellinger distance estimates for parametric models. *Ann. Statist.* **5**, 445–463.

Beran, R. J. (1978). An efficient and robust adaptive estimator of location. *Ann. Statist.* **6**, 292–313.

Donoho, D. L. and Liu, R. C. (1988a). The "automatic" robustness of minimum distance functionals. *Ann. Statist.* **16**, 552–586.

Donoho, D. L. and Liu, R. C. (1988b). Pathologies of some minimum distance estimators. *Ann. Statist.* **16**, 587–608.

Ghosh, M. and Sen, P. K. (1972). On bounded length confidence interval for the regression coefficient based on a class of rank statistics. *Sankhya, Ser. A.* **34**, 33–52.

Hoeffding, W. (1963). Probability inequalities for sums of bounded random variables. *J. A. S. A.* **58**, 13-30.

Koul, H. L. (1979). Weighted empirical processes and the regression model. *J. Ind. Statist. Assoc.* **17**, 83–91.

Koul, H. L. (1985a). Minimum distance estimation in multiple linear regression. *Sankhya, Ser. A.* **47**, 57–74.

Koul, H. L. (1985b). Minimum distance estimation in linear regression with unknown error distributions. *Statist. & Prob. Letters.* **3**, 1-8.

Koul, H. L. and DeWet, T. (1983). Minimum distance estimation in a linear regression model. *Ann. Statist.* **11**, 921–932.

Marcus, M. B. and Zinn, J. (1984). The bounded law of the iterated logarithm for the weighted empirical distribution process in the non-i.i.d. case. *Ann. Probab.* **12**, 335–360.

Millar, P. W. (1981). Robust estimation via minimum distance methods. *Z. Wahrsch. Verw. Gebiete.* **55**, 73–89.

Millar, P. W. (1984). A general approach to the optimality of minimum distance estimators. *Trans. Amer. Math. Soc.* **286**, 377-418.

Parr, W. (1981). Minimum distance estimation: a bibliography. *Comm. Statist. Theor. Meth. A10* **12**, 1205–1224.

Parr, W. C. and Schucany, W. R. (1979). Minimum distance and robust estimation. *J.A.S.A.* **75**, 616–624.

Williamson, M. A. (1979). Weighted empirical-type estimators of regression parameter. Ph. D. Thesis, MSU.

Wolfowitz, J. (1953). Estimation by minimum distance method. *Ann. Inst. Statist. Math.* **5**, 9–23.

Wolfowitz, J. (1954). Estimation of the components of stochastic structures. *Proc. Nat. Acad. Sci.* **40**, 602–606.

Wolfowitz, J. (1957). Minimum distance estimation method. *Ann. Math. Statist.* **28**, 75–88.

Zhu, Z. (1992). Some almost sure results on weighted residual empirical processes. To appear in *Sankhya, Ser. A.*

Department of Statistics and Probability
Michigan State University
East Lansing, Michigan 48824-1027
U.S.A.

J.K. Ghosh, S.K. Mitra, K.R. Parthasarathy and B.L.S. Prakasa Rao (*eds.*),
Statistics and Probability : A Raghu Raj Bahadur Festschrift
Published by Wiley Eastern Limited, Publishers, (1993), 365 – 370

On the Optimality of Bayes Test Statistics and Fisher's Combination Procedure

JAMES A. KOZIOL
The Scripps Research Institute, La Jolla

Abstract

The optimality of Bayes tests in the sense of exact slopes is described and applied to Fisher's combination procedure.

1 Introduction

Let (S, A) be a sample space of infinitely many independent and identically distributed observations $s = (x_1, x_2, \cdots)$ on a random variable x, the distribution of which is determined by a parameter θ taking values in a set Θ. For Θ_0 a given subset of Θ, we wish to test the null hypothesis that some θ in Θ_0 obtains. For each n, let $T_n(s)$ be a real valued A-measurable function depending on s through $x_1, x_2, \cdots, x_n$ only, such that in testing the null hypothesis, large values of T_n are significant. For any θ and t, let

$$F_n(t, \theta) = P_\theta(T_n(s) < t)$$

and

$$G_n(t) = \inf \{F_n(t, \theta) : \theta \in \Theta_0\}.$$

Then the level attained by T_n is defined as

$$L_n(s) = 1 - G_n(T_n(s)).$$

AMS Subject Classifications : 62F15 *Key words and Phrases* : Likelihood ratio, Kullback-Leibler information, exact slope.

The rate at which L_n tends to zero when a non-null θ obtains is considered by Bahadur (1960, 1967, 1971) as a measure of the asymptotic efficiency of the sequence of test statistics $\{T_n\}$ against that θ. If there exists a function $c(\theta)$ defined for $\theta \in \Theta - \Theta_0$ such that

$$\lim_{n\to\infty} n^{-1} \log L_n = -\frac{1}{2} c(\theta) [P_\theta],$$

the sequence of test statistics $\{T_n\}$ is said to have exact slope $c(\theta)$ when θ obtains.

Under certain regularity conditions, the likelihood ratio statistic is an optimal statistic in the sense of exact slopes. Suppose that for each θ the distribution of the single observation x has a density $f(x, \theta)$ with respect to a σ-finite measure μ. For any θ and θ_0 in Θ, define the Kullback-Liebler information number $K(\theta, \theta_0)$ by

$$K(\theta, \theta_0) = E_\theta \log [f(x, \theta)/ f(x,\theta_0)],$$

and let

$$J(\theta) = \inf_{\theta_0\in\Theta_0} K(\theta, \theta_0).$$

Denote by λ_n the likelihood ratio statistic for testing $H_0 : \theta \in \Theta_0$ based on $(x_1, \cdots, x_n)$, and let $\hat{T}_n(x) = -n^{-1} \log \lambda_n(s)$. Then, Bahadur (1965) proved:

(i) If c is the exact slope of a sequence of test statistics $\{T_n\}$, then $c(\theta) \leq 2J(\theta)$ for each non-null θ;

(ii) The exact slope of $\{\hat{T}_n\}$ exists and equals $2J(\theta)$ for each non-null θ.

Raghavachari (1970) and Bahadur and Raghavachari (1971) have generalised and refined this result in several directions. In particular, they formulated the ***likelihood ratio*** method of finding the exact slope of a given sequence, which does not require explicit estimation of large deviation probabilities. Koziol (1976) further described this technique, and gave readily verifiable conditions under which it may be invoked.

The purpose of this note is to demonstrate that Bayes tests are also optimal in the sense of exact slopes. This is formally stated in the next section; the proof is closely related to the Bahadur-Raghavachari results for likelihood ratio statistics. In the concluding section it is shown that Fisher's combination procedure is a Bayes test in a particular hypothesis testing situation, hence is optimal.

2 The Main Result

Let $X = \{x\}$ be the sample space of a single observation. Assume $dP_\theta = f(x, \theta)\, d\mu$ on X, where $\theta \in \Theta$, and Θ is a metric space. Consider the simple hypothesis $H_0 : \theta = \theta_0$, where θ_0 is given. Let λ be a prior probability on Θ such that $\lambda(\{\theta_0\}) = \alpha$ satisfies $0 < \alpha < 1$ and such that for each $\theta_1 \in \Theta - \{\theta_0\}$ every spherical neighbourhood of θ_1 has positive λ-measure. Assume that $\{f(x, \theta)\}$ is well-behaved in the sense that, with

$$h(x, \theta_1, \epsilon) = \inf \{[f(x, \theta)/ f(x, \theta_0)] : \theta \in \Theta, d(\theta, \theta_1) < \epsilon\},$$
$$E_{\theta_1} (\log h(x, \theta_1, \epsilon)) \to K(\theta_1, \theta_0)$$

as $\epsilon \to 0$ for each $\theta_1 \in \Theta - \{\theta_0\} = \Theta_1$.

Under this scenario, the following theorem establishes the optimality of Bayes tests in the sense of exact slopes.

THEOREM. Let $x_1, \cdots, x_n$ be i.i.d. observations on x, and let T_n^* be the posterior odds in favour of θ_1 when the sample is $(x_1, \cdots, x_n)$, i.e.,

$$T_n^* = \frac{\int_{\Theta_1} \{\Pi_1^n f(x_i, \theta)\} \, \lambda(d\theta)}{\alpha \, \Pi_1^n f(x_i, \theta_0)}.$$

Then $\{T_n^*\}$ has exact slope $K(\theta_1, \theta_0)$ against each non-null θ_1.

PROOF. Choose and fix θ_1. Let $\epsilon > 0$ be so small that

$$B = \{\theta : \theta \in \Theta, d(\theta, \theta_1) < \epsilon\} \subset \Theta_1.$$

Then

$$\begin{aligned} T_n^* &\geq \alpha^{-1} \int_B \{\Pi_1^n f(x_i, \theta)/ f(x_i, \theta_0)\} \, \lambda(d\theta) \\ &\geq \alpha^{-1} \, [\Pi_1^n h(x_i, \theta_1, \epsilon)] \int_B \lambda(d\theta). \end{aligned}$$

Hence,

$$n^{-1} \log T_n^* \geq n^{-1} \log \left[\alpha^{-1} \lambda(B)\right] + \frac{1}{n} \sum_{i=1}^{n} h(x_i, \theta_1, \epsilon),$$

and so

$$\liminf_{n\to\infty} n^{-1} \log T_n^* \geq E_{\theta_1} (h(x, \theta_1, \epsilon)) \, [P_{\theta_1}].$$

Since ϵ is arbitrary,

$$\liminf_{n\to\infty} n^{-1} \log T_n^* \geq K(\theta_1, \theta_0) \, [P_{\theta_1}]. \tag{2.1}$$

Next, for any real t,

$$\begin{aligned} P_{\theta_0} \left(\frac{1}{n} \log T_n^* \geq t \right) &= P_{\theta_0} \left(\int_{\Theta_1} f_\theta^{(n)} \, \lambda(d\theta) \geq e^{nt} \, \alpha \, f_{\theta_0}^{(n)} \right) \\ &= \int_C f_{\theta_0}^{(n)} \mu^{(n)}(dx^{(n)}) \\ &\leq \alpha^{-1} \, e^{-nt} \int_C \left\{ \int_{\Theta_1} f_\theta^{(n)} \lambda(d\theta) \right\} \mu^{(n)}(dx^{(n)}) \\ &\leq \alpha^{-1} \, e^{-nt} \int_{X^{(n)}} \left\{ \int_\Theta f_\theta^{(n)} \lambda(d\theta) \right\} \mu^{(n)}(dx^{(n)}) \\ &= \alpha^{-1} e^{-nt} \end{aligned} \tag{2.2}$$

Here, the superscript n denotes the n-fold Cartesian product (with the argument suppressed for notational convenience), and

$$C = \left\{ x^{(n)} : \int_{\Theta_1} f_\theta^{(n)} \lambda(d\theta) \geq e^{nt} \alpha f_{\theta_0}^{(n)} \right\}$$

Let L_n^* be the level attained by T_n^*; this is also the level attained by $1/n \log T_n^* = U_n^*$ say. It follows from (2.2) that

$$L_n^* \leq \alpha^{-1} e^{-n} U_n^*.$$

Hence, by (2.1)

$$\liminf_{n\to\infty} \frac{1}{n} \log L_n^* \leq -K(\theta_1, \theta_0)\,[P_{\theta_1}];$$

thus

$$\lim \frac{1}{n} \log L_n^* \leq -K(\theta_1, \theta_0)\,[P_{\theta_1}]$$

by Raghavachari's (1970) theorem. This concludes the proof.

3 Combining Independent Tests of Significance

A fundamental problem of meta-analysis is the combination of independent tests of significance. A simple formulation of this problem is as follows: Let θ be an unknown parameter in some metric space Θ and let T_i, $i = 1, \cdots, n$ be independent statistics for testing a given null hypothesis, e.g., $H_0 : \theta \in \Theta_0$, concerning the value of θ with large values of T_i leading to rejection of H_0. Let L_i denote the level attained by T_i, $i = 1, \cdots, n$. L_i is sometimes called the p-value associated with T_i. It is well-known that in the null case, L_i is always stochastically larger than a uniform random variable over $(0, 1)$, that is,

$$P_\theta(L_i \leq \alpha) \leq \alpha,\ \theta \in \Theta_0,\ 0 < \alpha < 1. \tag{3.1}$$

The inequality (3.1) can be taken as the starting point for determining slopes; see Bahadur (1965, 1967, 1971) and Bahadur and Raghavachari (1972) for further discussion.

A widely used procedure for combining independent p-values is that of Fisher (1950). Suppose the distribution of T_i is fully specified and continuous under the null hypothesis; then, invoking the probability integral transformation, one may take L_i to be uniformly distributed under H_0. Fisher's procedure uses the product $\Pi_1^n L_i$ to combine the p-values. In particular, one would reject H_0 at level α with Fisher's method if

$$F_n = -2 \sum_1^n \log L_i \geq c_\alpha,$$

where the critical value c_α is the upper $(1 - \alpha)$ percentage point of the chi-square distribution with $2n$ degrees of freedom.

Littell and Folks (1973) established that Fisher's method is asymptotically optimal in the sense of exact slopes among all "reasonable" methods of combining independent tests. An alternative demonstration of the optimality of Fisher's method is afforded from the theorem of Section 2, upon showing that it is a Bayes test.

To this end, let $x_1, \cdots, x_n$ be i.i.d. observations on $(0,1)$ with

$$f(x,\theta) = (1-\theta)\, x^{-\theta},\ \theta \in \Theta = [0,1). \tag{3.2}$$

Under H_0, $\theta \in \Theta_0 = \{0\}$, and under the alternative, $\theta \in \Theta_1 = (0,1)$. The class of Lehmann alternatives to uniformity in (3.2) is chosen in light of (3.1); see also Lipták (1958) for further discussion. [Note, nevertheless, that this alternative is less general than is typically postulated under other formulations of the combination problem, as in Koziol and Perlman (1978).]

Consider the family of beta priors

$$g(\theta;\, a, b) = \frac{1}{B(a,b)}\theta^{a-1}(1-\theta)^{b-1},\ a > 0,\ b > 0, \tag{3.3}$$

for $\theta \in \Theta_1$. Now,

$$\begin{aligned}&\int_{\Theta_1} \Pi_{i=1}^{n} f(x_i,\theta)\, g(\theta;\, a, b)\, d\theta = \\ &\int_0^1 \frac{1}{B(a,b)}\, \theta^{a-1}\,(1-\theta)^{b+n-1} \exp\,(\theta \cdot T(x_1,\cdots,x_n))\, d\theta = \\ &\frac{1}{B(a,b)} B(a, b+n)\, M(a, a+b+n, T(x_1,\cdots,x_n));\end{aligned} \tag{3.4}$$

here,

$$T(x_1,\cdots,x_n) = -\sum_{i=1}^{n} \log x_i$$

and $M(a,b,x)$ is Kummer's function, a confluent hypergeometric function with series expansion

$$M(a,b,x) = 1 + \frac{ax}{b} + \frac{(a)_2}{(b)_2}\frac{x^2}{2!} + \cdots + \frac{(a)_n}{(b)_n}\frac{x^n}{n!} + \cdots$$

where

$$(a)_n = a(a+1)\cdots(a+n-1),\ (a)_0 = 1$$

and similarly for $(b)_n$ (Abramowitz and Stegun (1964), eqn. 13.1.2).

In particular, one may adopt a noninformative prior for θ by selecting $a = 1$, $b = 1$ in (3.3); then, it is clear from the series expansion of $M(1, n+2, x)$ that the integral in (3.4) is a monotone function of $T(x_1, \cdots, x_n)$. It follows that the Bayes test of H_0 versus the alternative from (3.2) will be based on $T(x_1, \cdots, x_n)$, which is equivalent to Fisher's procedure.

Acknowledgement

The theorem in Section 2 is not new: it was described to me by Professor Bahadur at the University of Chicago many years ago. I am deeply grateful to Professor Bahadur for his consideration, encouragement, and friendship.

REFERENCES

Abramowitz, M. and Stegun, I.A., eds. (1964). *Handbook of Mathematical Functions*, U.S. Government Printing Office, Washington, D.C.

Bahadur, R.R. (1960). Stochastic comparison of tests. *Ann. Math. Statist.* **31**, 276-295.

Bahadur, R.R. (1965). An optimal property of the likelihood ratio statistic. *Proc. Fifth Berkeley Symp. Math. Statist. Prob.* **1**, 13-26. Univ. of California Press, Berkeley.

Bahadur R.R. (1967). Rates of convergence of estimates and test statistics. *Ann. Math. Statist.* **38**, 303-324.

Bahadur, R.R. (1971). Some limit theorems in statistics. *Regional Conference Series in Applied Mathematics* **4**, SIAM, Philadelphia.

Bahadur, R.R. and Raghavachari, M. (1972). Some asymptotic properties of likelihood ratios on general sample spaces. *Proc. Sixth Berkeley Symp. Math. Statist. Prob.* **1**, 129-152. Univ. of California Press, Berkeley.

Fisher, R.A. (1950). *Statistical Methods for Research Workers*, 11th Edition. Oliver and Boyd, London, pp. 99-101.

Koziol, J.A. (1978). Exact slopes of certain multivariate tests of hypotheses. *Ann. Statist.* **6**, 546-558.

Koziol, J.A. and Perlman M.D. (1978). Combining independent chi-squared tests. *J. Amer. Statist. Assoc.* **73**, 753-763.

Lipták, T. (1958). On the combination of independent tests. *Magyar. Tud. Akad. Mat. Kutato Int. Közl.* **3**, 171-197.

Littell, R.C. and Folks, J.L. (1973). Asymptotic optimality of Fisher's method of combining independent tests II. *J. Amer. Statist. Assoc.* **68**, 193-194.

Raghavachari, M. (1970). On a theorem of Bahadur on the rate of convergence of test statistics. *Ann. Math. Statist.* **41**, 1695-1699.

Department of Molecular and
Experimental Medicine
The Scripps Research Institute
10666 North Torrey Pines Road
La Jolla, California
U. S. A.

J.K. Ghosh, S.K. Mitra, K.R. Parthasarathy and B.L.S. Prakasa Rao (*eds.*),
Statistics and Probability : A Raghu Raj Bahadur Festschrift
Published by Wiley Eastern Limited, Publishers, (1993), 371 – 380

The Bertrand-Borel Debate and the Origins of the Neyman-Pearson Theory[1]

E.L. Lehmann
University of California, Berkeley

Abstract

The views of the mathematicians Bertrand and Borel on hypothesis testing and the usefulness of probability models are reviewed as background to the Neyman-Pearson theory. There is a discussion of Neyman's claim that these views had an influence on his work with Pearson, and of Pearson's denial of this claim.

1 Introduction

Throughout the 19th Century, hypothesis testing was carried out in various different areas of application along the lines laid down in the "Théorie Analytique des Probabilités" by Laplace (1812, 1814, 1820) and in Poisson's "Recherches sur la Probabilité des Jugements" (1837). Nearly all of this work was asymptotic, based on De Moivre's normal approximation to the

[1] Research supported by NSF Grant No. DMS-8908670.
AMS Subject Classifications : Primary 62-03; secondary 62 F 04
Key words and Phrases: Hypothesis Testing, Neyman-Pearson Theory, Probability Model.

binomial and Laplace's central limit theorem. Toward the end of the century, the approach was summarized by Edgeworth (1895) and extended in an important new direction by Karl Pearson (1900).

A new phase, small-sample theory, was ushered in by Student (1908) and was developed by Fisher in a series of papers culminating in his book "Statistical Methods for Research Workers" (1925). In all the work up to this point the test was taken as given (on intuitive grounds) and the emphasis was on the calculation of the p-value or, when a Bayesian approach was used, on the probability of the hypothesis being true.

The final step, asking how to choose the test, was taken by Neyman and E.S. Pearson (1928, 1933) who formulated the problem and to a large extent solved it. These authors give somewhat different accounts of the main influences that inspired them in this work. Pearson (1939, 1966) attributes the suggestion that the appropriate test must depend on the alternatives against which the hypothesis is to be tested to a letter of Student while Neyman cites the influence of a debate between Bertrand and Borel.

The Bertrand-Borel debate[2] is concerned not only with the problem of how to choose a test, but also with the more fundamental question of whether calculations of p-values are appropriate at all since they are based on probability models that at best are rough approximations of a much more complex real situation. In presenting the views of Bertrand and Borel in sections 2 and 3, we shall be concerned with both these aspects of the debate. Sections 4 and 5 discuss its relation to the Neyman-Pearson theory.

2 Bertrand's views

Neyman's source for Bertrand's views was Bertrand's book "Calcul des Probabilités (1889, 1897, 1907). The author explains the purpose of his book in a short preface. Probability theory is one of the most attractive branches of the mathematical sciences, he says, and yet one of the most neglected. He suggests that one of the reasons for this state of affairs may be Laplace's book (the Théorie Analytique), which is indispensable for the study of probability theory, but requires a great deal of mathematical preparation. Bertrand's book is to remedy this situation by providing an account of the most useful and best known results with the simplest possible proofs.

Emulating Laplace, Bertrand begins his book with a nontechnical essay, "Les lois du hasard" [The laws of chance], which gives an introductory overview.[3] Besides the essay, the relevant chapter for the present purpose is chapter 7 on the probability of causes. Much of this chapter is devoted

[2] This seems the best way to describe the situation although it was not a debate in the sense of a formal discussion. Bertrand stated certain views in his probability text of 1889. He was no longer alive when Borel criticized them rather sharply in a popular book on probability in 1914.

[3] However, in contrast to Laplace's ebullience and optimism, Bertrand's attitude toward applications is rather negative.

to an attack on hypothesis testing on the grounds that the plausibility of a hypothesis depends on the prior probability of it being true, which is rarely known.

Then, speaking about Michell's famous problem[4] of whether the closeness to each other of the six (visible to the naked eye) stars forming the Pleiades can be attributed to chance, he changes the direction of attack. He again first refers to the unavailability of prior probabilities, but then asks: "How can we decide on the unusual results that chance is incapable of producing?" and continues (p.166)[5]

"The Pleiades appear closer to each other than one would naturally expect. This statement deserves thinking about; but when one wants to translate the phenomenon into numbers, the necessary ingredients are lacking. In order to make the vague idea of closeness more precise, should we look for the smallest circle that contains the group? the largest of the angular distances? the sum of squares of all the distances? the area of the spherical polygon of which some of the stars are the vertices and which contains the others in its interior? Each of these quantities is smaller for the group of the Pleiades than seems plausible. Which of them should provide the measure of implausibility? If three of the stars form an equilateral triangle, do we have to add this circumstance, which is certainly very unlikely apriori, to those that point to a cause?"

What Bertrand seems to consider here is the choice between a number of possible test statistics, small values of which are significant. However, in the last sentence he addresses an additional issue, namely what features of the available material to include in the observations to be analysed.

Beyond these particular questions, Bertrand here also raises doubts about the scientific validity of such a probabilistic approach to testing. In the introductory essay he goes much further. After discussing Michell's own analysis, he states (p.xviii): "The application of such calculations to questions of this kind is a delusion and an abuse."

Quite generally, Bertrand opposes probability models, particularly in the social sciences, as quite unrealistic,[6] and this aspect of the debate continues today. His statement concerning Laplace's law of succession: "If time is to confirm and increase [the conviction that the sun will rise tomorrow], it will be by the discovery of the laws of astronomy and not by the repeated success of an analogous game of chance", is echoed in recent writing of David Freedman (for example, Freedman (1985) and (1987)).

---

[4] For recent reviews of some of the extensive literature on this problem see Gower (1982) and Dale (1991).

[5] Since copies of the book by Bertrand (and that of Borel) cited here are easily available, I have translated the passages to be quoted for easier reading. The page references are to the edition of 1907.

[6] For further discussion of his views see for example Stigler (1986) and Daston (1988).

3 Borel's reply

A very different position is taken by Borel in his book "Le Hasard" (1914); the passages given here in translation are from the revised edition of (1948). His treatment of these issues occurs in chapter 4 which, like Bertrand's corresponding chapter, is entitled "probability of causes". After discussing Bayes' theorem, Borel mentions the example of the frequency of male births and restates it as an urn problem. After some discussion he points out that "this example shows the particular form that problems of causes often take on. This form is the following: Is such and such a result due to chance or does it have a cause? It has often been observed how much this statement lacks in precision. Bertrand has strongly emphasized this point.[7] But whatever objection one can raise from a logical point of view cannot prevent the preceding question from arising in many situations: the theory of probability cannot refuse to examine it and to give an answer; the precision of the response will naturally be limited by the lack of precision in the question; but to refuse to answer under the pretext that the answer cannot be absolutely precise, is to place oneself on purely abstract grounds and to misunderstand the essential nature of the application of mathematics. True, the calculations provide an exact answer to every exact question; but in practice the questions are never exact: the given experimental facts necessarily permit a certain amount of play; the same lack of precision then affects the result of the calculations and the would-be absolute theoretical precision of these calculations is pure delusion."

Borel next turns to problems in astronomy and after some introductory remarks quotes the passage about the Pleiades from Bertrand that was given in the preceding section. Referring to Bertrand's equilateral triangle, he writes: "If one has observed a number such as that quoted above [1.342517] as evaluation of an angle in tenths of seconds (in the decimal system), one would not think of asking to know the probability that this angle is exactly $13^G\ 42'\ 51''\ 7$; because one would never have asked that precise question before having measured the angle...

"The question is whether one has the same reservations in the case in which one states that one of the angles of the triangle formed by three stars has "une valeur *remarquable*" [a striking or noteworthy value], and is for example equal to the angle of the equilateral triangle ($66^G\ 66'\ 66''$),[8] or to a half right angle: 50 decimal degrees, or 5.000.000 tenths of a second. Here is what one can say on this subject:

"One should carefully guard against the tendency to consider as striking an event that one has not specified *beforehand*, because the number of

[7] Borel here refers to Bertrand's discussion of the different interpretations that can be given to the idea of randomness, as illustrated by his well-known treatment of Buffon's needle problem; cf Bertrand (1889, ch.1).

[8] The old-fashioned french angular measure "grade" is a decimal measure; a right angle is 100 grades.

such events that may appear striking, from different points of view, is very substantial.

"In contrast to this, the question of grouping [i.e. clustering] in space is one that it is natural to ask apriori; the theory of probability does not enable one to resolve it with certainty but makes precise the conclusions one can draw from observations.

"This precision is quite useful, because it often effectively replaces inferences at which one arrives too quickly by 'common sense' ... Between the two extreme situations [those in which the results clearly are or are not significant] there are many others, namely those in which the probability lies between for example one hundredth and one hundred-thousandth: the value [the probability] has between these limits is not at all useless and what one can learn from it may be very valuable, no matter what Bertrand thinks".

In summary, Bertrand raises two questions: (i) on what grounds to prefer one test statistic to another; (ii) whether the calculation of a p-value, once the test statistic has been chosen, is at all useful in view of the shaky assumptions underlying the typical probability model.

To the first point Borel replies with the idea that the test statistic should represent some striking feature of the situation; on the second he asserts his unshaken faith in the value of probabilistic calculations even though they are only approximations to the true values.

It should be noted that neither of these authors offers any direct suggestion that the test should be based on more plausible alternatives although both Paul Lévy and Neyman were later to interpret Borel's comments in this way.

4 Circonstances remarquables

The first mention of the views of Bertrand and Borel in Neyman's papers occur in Neyman (1929) where he refers to Bertrand's critique of the Bayesian approach to hypothesis testing and then mentions:[9]

"M. Borel who also discusses analogous questions, observes that the correct judgment concerning a statistical hypothesis depends on the choice of the aspect of the observed fact $f(E)$ which in some way must be 'remarquable'. This idea of a remarkable aspect which is a bit vague can be made precise if one agrees to consider also alternative hypotheses."

The debate between the two french mathematicians is discussed at greater length in Neyman and Pearson (1933), and Neyman returns to it again in Neyman (1935) where he writes:[9]

"... the origin of the general theory of testing hypotheses is linked to most interesting remarks of french mathematicians, particularly of J.

[9] My translation of the original french.

Bertrand (1889) and M.E. Borel (1932). They considered the process of hypothesis testing which is basically the one employed by the english school...

"It is known that Bertrand was skeptical about the scientific value of the results of such a process. On the other hand, M. Borel insisted that such a result can be valid provided the feature $f(E)$ of the observed facts that serve as basis for testing N, be 'en quelque sorte remarquable'.

"This last remark preceded a series of works [by Neyman Pearson, and others] which basically strove to give a precise meaning to Borel's somewhat vague words..."

How much importance Neyman attached to the influence of Borel's remark can be seen from the fact that late in life he returned to it again in Neyman (1977), an account of his views on statistics. He mentions that the methodology for testing hypotheses "was the subject of a lively discussion by Borel and others", and goes on to say that "Borel was optimistic but insisted that: (a) the criterion to test a hypothesis ... must be selected *not after the examination of the results of observation*, but before, and (b) this criterion should be a function of the observations 'en quelque sorte remarquable'. It is these remarks of Borel that served as an inspiration to Egon S. Pearson and myself in our effort to build a frequentist theory of testing hypotheses".

Neyman sent a preprint of this paper to E.S. Pearson who expressed himself[10] delighted with it ("so delighted that I have started devouring it, although only received five hours ago!" he wrote) and who attached to his reply to Neyman a large number of comments. In one of the first of these he takes exception to Neyman's attribution of inspiration cited above. He writes:

"I remember that you produced this quotation when we began to get our [1933] paper into shape and write it up. This must have been during the period 1930/31, or even as early as my summer visit to Poland of 1929. I had not heard of the Borel quotation until you started to draft the introductory section of our paper. The words may have for a year or two been acting as an inspiration to you, but while I at once felt the words relevant and important, my approach was a different one, and was not I think *altered* by inspiration from the words."

Pearson goes on to say that in 1925 or 1926 he had begun to question the basis of the t-test and to ask: Why use that particular statistic? and that this led to the question: 'How choose the rejection region?' He then continues:

"The above stages led up to Borel's requirement of finding ... a criterion which was 'a function of the observations 'en quelque sorte remarquable'.

"Now my point is that you and I (perhaps my first leading) had our-

[10] In an unpublished letter of September 28, 1976, which is contained in the Neyman Collection of the Bancroft Library, University of California, Berkeley.

selves reached the Borel requirement independently of Borel, because we were serious humane thinkers; Borel's expression neatly capped our own requirement."

As mentioned in the introduction, a strong outside influence had also affected Pearson's thinking. This was a letter[11] by Student written in an answer to a question by Pearson. This letter contains the crucial suggestion: "If there is an alternative hypothesis which will explain the occurrence of the sample with a more reasonable probability ... you will be very much more inclined to consider that the original hypothesis is not true."

Pearson must very soon have communicated this idea to Neyman because a letter of Neyman's to Pearson, dated December 9, 1926 begins: "I think that to have the possibility of testing it is necessary to adopt such a principle as Student's, ..." (Reid (1982), p.70). On the other hand, there appears to be no mention of Borel in this or other letters of Neyman's to Pearson of this period. This combination of circumstances lends some support to Pearson's recollections.[12]

It is interesting to note that Neyman was not the only one on whom Borel's discussion of tests of randomness made a deep impression. Paul Lévy in his "Calcul des Probabilités (1925) treats this topic in a section which he acknowledges to be largely borrowed from Borel. He there also mentions the role played by "propriétés remarquables" and goes on to analyse the meaning of this phrase. And he comes to the conclusion which — by whatever common or separate routes — Neyman and Pearson were also to reach. "Une propriété remarquable", he writes, "est une propriété facile à expliquer par une cause autre que le hasard."[13] (p.93). [A striking property is a property that is easy to explain by a cause other than pure chance.]

5 The Bertrand-Borel debate and the Neyman-Pearson theory

The basic concepts of the Neyman-Pearson theory: rejection region, type 1 error, alternatives to the hypothesis, and power, have became so embedded in our statistical language and thought that it is difficult for us to see the situation as it was in the first quarter of the century. The debate of the French probabilists of that period provides a window into their thinking.

What strikes the modern reader is the vagueness of the discussion which is of course inevitable when no conceptual framework and suitable terminology are available. As a partial explanation of "the obscurity which envelops the theoretical bases of statistical methods" Fisher in his fundamental paper, "On the mathematical foundations of theoretical statistics" (1922) suggests that "it appears to be widely thought, or rather felt, that in

---

[11] This letter of May 11, 1926 was published in part in Pearson (1939).

[12] The Neyman-Pearson story is told from Pearson's point of view in Pearson (1966) and from Neyman's in Reid (1982).

[13] Neyman's only reference to Lévy in this general context is in Neyman (1932) where be attributes the concept of "remarquable" characters to "E. Borel and P. Lévy" in the books quoted here.

a subject in which all results are liable to greater or smaller errors, precise definition of ideas or concepts is, if not impossible, at least not a practical necessity."

In this paper Fisher did in fact create the needed theoretical framework in the context of estimation. He formulated his theory in terms of a parametric family of distributions (of the observations), posed the problem of choosing a suitable estimator, defined a measure of accuracy, and introduced both relative and absolute efficiency. However, Fisher did not extend the theory to hypothesis testing, which — like his predecessors — he discussed only in the context of specific examples or classes of examples.

It remained for Neyman and Pearson to formulate a general theory of hypothesis testing. Fisher's parametric families provided a way to specify both hypothesis and alternatives. Instead of speaking, as Bertrand and Borel had done, of choosing among vaguely indicated test criteria, the class of available tests is now clearly defined as the class of all possible level α rejection regions. Furthermore the concept of power made possible a comparison of the advantages and disadvantages of different tests and led to maximal power as a principle for chosing a "best" test.

There is a world of difference between the vague musings of Bertrand and Borel, and the clearly defined structure of the Neyman-Pearson theory. In particular, the latter of course permits quantitative evaluations and comparisons.

It is just this clarity and precision, at the time only an unrealized idea, that is at the heart of the debate between Bertrand and Borel. Bertrand would undoubtedly object to many of the uses to which tests are put (and he would not be alone in this). It is true, in particular, that the models from which the analysis is derived often have a rather flimsy basis and even at best are only approximations. The theory has tried to meet these and related objections by studying the robustness of existing tests, developing new more robust alternative tests, by providing simultaneous test procedures to deal with the problem of multiplicity, and so on, all within the general Neyman-Pearson framework.

To the many objections that remain, Borel has an answer in the passage quoted in Section 3 which can be summarized as follows. "The theory of probability cannot refuse to examine" the situation and to come up with an answer. Of course, the real situation is much more complex than the model allows for; the precise answers that the calculations provide for hypothetically precise questions are only rough approximate solutions of the much less precise real problems. We should therefore not take the analysis too literally, put too much stock in the precise numerical answers, but — we should not give up on the enterprise. It provides invaluable guidance, a check on undisciplined speculation, on "inferences at which one arrives too quickly by 'common sense'."

REFERENCES

Bertrand, J. (1889, 1897, 1907) Calcul des Probabilités Gauthier-Villars, Paris.

Borel, E. (1914, 1920, 1932; New Revised Edition 1948) Le Hasard. Presses Universaires de France, Paris.

Dale, A.I. (1991) A History of Inverse Probability: From Thomas Bayes to Karl Pearson, Springer Verlag, New York.

Daston, L. (1988) Classical Probability in the Enlightenment, Princeton Univ. Press.

Edgeworth, F.Y. (1895) Methods of Statistics, Jubilee Volume of the Statist. Soc., 181-217.

Fisher, R.A. (1925) *Statistical Methods for Research Workers*, Oliver and Boyd, Edinburgh.

Gower, B. (1982) Astronomy and probability. Forbes versus Michell on the distribution of the stars. *Annals of Science*, **39**, 145-160.

Freedman, D. (1985) Statistics and the scientific method, Cohort Anal., 343-359.

Freedman, D. (1987) As others see us: A case study in path analysis, *J. Educ. Statist.*, **12**, 101-128.

Hochberg, Y. and Tamhane, A. (1987) *Multiple Comparison Procedures*, Wiley, New York.

Laplace, P.S. (1812, 1814, 1820) *Théorie Analytique des Probabilités*, Courcier, Paris.

Lévy, P. (1925) *Calcul des Probabilités*, Gauthier-Villars. 1925.

Neyman, J. (1929) Méthodes nonvelles de vérification des hypothèses. Comptes Rendus du 1 Congrès de Mathématiciens des Pays Slaves. Warsaw, 355-366.

Neyman, J. (1932) On methods of testing hypotheses, *Atti Congr. Internat. Matematici*, **6**, 35-41.

Neyman, J. (1935) Sur la vérification des hypothèses statistiques composées, *Bull. Soc. Math. de France*, **63**, 346-366.

Neyman, J. and Pearson, E.S. (1928) On the use and interpretation of certain test criteria, *Biometrika*, **20A**, 175-240 and 263-294.

Neyman, J. and Pearson, E.S. (1933) On the problem of the most efficient tests of statistical hypotheses, *Phil. Trans. Roy. Soc. (A)*, **231**, 289-337.

Pearson, E.S. (1939) "Student" as a statistician, *Biometrika*, **30**, 205-250.

Pearson, E.S. (1966) *The Neyman-Pearson story, 1926-34*, In Research Papers in Statistics (Festschrift for J. Neyman), F.N. David (Ed). Wiley. New York.

Pearson, K. (1900) On the criterion that a given system of deviations from the probable in the case of a correlated system of variables is such that it can be reasonably supposed to have arisen from random sampling, *Phil. Mag.* *5*, **Vol. 50**, 157-175.

Poisson, S.D. (1837) Recherches sur la Probabilité des Jugements, Bachelier, Paris.

Reid, C. (1982) *Neyman from Life*, Springer-Verlag, New York.

Shaffer, J. (1988) *Simultaneous testing*, In Encycl. Statist. Sci. Vol. 8, 484-490. Wiley, New York.

Stigler, S. (1986) *The History of Statistics*, Belknap Press, Cambridge.

Student (1908) On the probable error of a mean, *Biometrika*, **6**, 1-25.

Department of Statistics
367 Evans Hall
University of California
Berkeley, California 94720
U. S. A.

J.K. Ghosh, S.K. Mitra, K.R. Parthasarathy and B.L.S. Prakasa Rao (*eds.*),
Statistics and Probability : A Raghu Raj Bahadur Festschrift
Published by Wiley Eastern Limited, Publishers, (1993), 381 – 394

Strong Representation for Kernel Estimates of Conditional Medians and Applications [1]

Z. J. LIU
Mississippi State University, Mississippi State

C. RADHAKRISHNA RAO and D. S. TU
Pennsylvania State University, University Park

Abstract

Let (X,Y) be a pair of random variables which are, respectively, d and one dimensional, and let the conditional distribution of Y given $X=x$ be $F(y|x)$. The kernel estimate $\theta_h(x)$ of the conditional median $\theta(x) = F^{-1}(2^{-1}|x)$ is considered in this paper, where $h=h_n$ is a sequence of window widths. The Bahadur type representation is obtained for $\theta_h(x)$ under some mild conditions on $F(y|x)$ and h. As applications of this representation, the law of iterated logarithm and the asymptotic normality of $\theta_h(x)$ are established.

1 Introduction

Let the conditional distribution function of Y given $X=x$ be $F(y|x)$ and $\theta(x) = F^{-1}(2^{-1}|x)$ be the median of the conditional distribution $F(y|x)$,

[1] This research is sponsored by the Air Force Office of Scientific Research under Grant AFOSR-89-0279. The United States Government is authorized to reproduce and distribute reprints for governmental purposes notwithstanding any copyright notation hereon.

AMS Subject Classifications: 62G05, 60F15.

Key words and Phrases: Bahadur representation, Conditional median, Kernel methods, L_1-Norm, Nonparametric regression.

where the definition of the inverse function of $F(y|x)$ is given by $F^{-1}(p|x) = \inf\{y : F(y|x) > p\}$. Further let $(X_1, Y_1), \ldots, (X_n, Y_n)$ be an iid sample from the joint distribution of (X, Y). We consider an estimator of $\theta(x)$ of the form

$$\theta_n(x) = F_n^{-1}(2^{-1}|x) \tag{1.1}$$

with $F_n(y|x)$ defined as

$$F_n(y|x) \triangleq \sum_{i=1}^{n} W_{ni}(x) I(Y_i \le y) \tag{1.2}$$

where the weights $W_{ni}(x)$ depend on $X_1, X_2, \ldots, X_n$ only.

The estimation of conditional medians is not a new topic in nonparametric statistics. In the seminal work on nonparametric regression, Stone (1977) considered conditional medians and obtained a consistency theorem of the nearest neighbor estimate of conditional medians. Härdle and Luckhaus (1984) obtained the L^∞-norm convergence rate of a class of robust nonparametric estimators including an estimator of conditional medians. Zheng (1985) and Stute (1986) established the asymptotic normality of a class of nearest neighbor estimates of conditional medians. Liu and Tu (1987) established strong consistency, asymptotic normality and bootstrap approximation for a class of kernel estimates, and recently Truong (1989) considered the convergence rate for a class of kernel estimates, and Bhattacharya and Gangopadhyay (1990) obtained a strong representation of k-nearest neighborhood estimate and kernel estimate of conditional median when the kernel is uniform and the predictor is univariate and smooth. In this paper we concentrate our attention on the estimator proposed by Liu and Tu (1987), where the weight functions in (1.2) are given by using the kernel method. The exact definition of this estimator is given in Section 2. We establish a Bahadur type strong representation of the conditional median estimator with a general compact supported kernel function. Note that the strong representation in Section 2 still holds when the predictor X is multivariate and may not be absolutely continuous.

If X and Y are independent, then $\theta_h(x)$ is just the usual (unconditional) sample median of $Y_1, \ldots, Y_n$ (by taking uniform weights), which has been extensively studied during the last few decades (see Serfling (1980)). For this estimate, Bahadur (1966) has shown that under some very mild conditions on the distribution of the sample, it can be approximated by the mean of functions of sample observations with the order of $O(n^{-3/4}(\log n)^{3/4})$. This approximation, called the Bahadur representation, is very useful in establishing the asymptotic properties of the sample median. For example, the law of the iterated logarithm and the asymptotic normality of the sample median can be easily obtained from this representation.

The results of Bahadur (1966) have been extended by Cheng (1984) and Babu (1988) to the sample median regression function of fixed-design nonparametric regression models and the least absolute deviation estimate of linear regression models, respectively. The purpose of this paper is to establish a Bahadur type representation for the kernel estimates of the conditional median given by Liu and Tu (1987). From this expression we will derive the law of iterated logarithm and asymptotic normality for the kernel estimates.

The rest of the paper is organized in the following way. The main results on Bahadur's representation of the kernel estimates of conditional median and their applications are presented in Section 2. Section 3 contains the proofs of the main results.

2 The Main Results

Let U_t be a nonempty bounded neighborhood of t in R^d. The norm of $x = (x_1, x_2, \ldots, x_d)$ is defined to be $\| x \| = (x_1^2 + \ldots + x_d^2)^{1/2}$ and $S_{x,r}$ is defined by $\{u : \| u - x \| \leq r\}$. The joint distribution of (X, Y), the conditional distribution of Y given $X = x$, and the marginal density of X are denoted by $F(x,y)$, $F(y|x)$ and $F(x)$, respectively. For convenience, we use $F(\bullet)$ to denote the distribution function as well as the probability measure of X.

Suppose that $(X_1, Y_1), \ldots, (X_n, Y_n)$ is a random sample from the distribution of (X, Y). Taking positive numbers $h = h_n$ that tend to 0 as n tends to ∞, the kernel estimate of the conditional median $\theta(x_0) = F^{-1}(p|x_0)$ is taken to be

$$\theta_h(x_0) \triangleq \theta_{h_n}(x_0) = F_h^{-1}(2^{-1}|x_0) \tag{2.1}$$

where

$$F_h(y|x) \triangleq F_{h_n}(y|x) \triangleq \sum_{i=1}^{n} W_{n,i}(x) I(Y_i \leq y), \tag{2.2}$$

$I(\bullet)$ is the indicator function and $W_{n,i}(x)$, $i = 1, \ldots, n$ are weight functions defined to be

$$W_{n,i}(x) = \begin{cases} K\left(\frac{X_i - x}{h}\right) / \sum_{j=1}^{n} K\left(\frac{X_j - x}{h}\right) & \text{if } \sum_{j=1}^{n} K\left(\frac{X_j - x}{h}\right) \neq 0 \\ 0 & \text{otherwise.} \end{cases} \tag{2.3}$$

where $K(\bullet)$ is a positive and integrable function defined on R^d. Liu and Tu (1987) proposed the estimator (2.1) of the conditional median and investigated its strong consistency, asymptotic normality and bootstrap approximation. Truong (1989) obtained a convergence rate of a special case of (2.1) with uniform kernel function. Bhattacharya and Gangopadhyay (1990) considered the strong representation of (2.1) for univariate predictor X and uniform kernel function.

The smoothness conditions on the conditional distribution of Y given $X = x$ are given below.

CONDITION 1 *$F(y|x)$, as a bivariate function of (y, x), is twice differentiable with respect to y in a closed neighborhood $B(x_0) = U_{\theta(x_0)} \times U_{x_0}$ of $(\theta(x_0), x_0)$ with*

$$\left.\frac{\partial F(y|x_0)}{\partial y}\right|_{y=\theta(x_0)} = f(\theta(x_0)|x_0) > 0$$

$$\sup_{(y,x)\in B(x_0)} \left|\frac{\partial^2 F(y|x)}{\partial y^2}\right| \leq M_1 \tag{2.4}$$

where M_1 is a positive constant.

CONDITION 2 *$F(y|x)$ satisfies the following condition in $B(x_0) = U_{\theta(x_0)} \times U_{x_0}$.*

$$\sup_y \left|\int_{S_{x_0,r}} [F(y|x) - F(y|x_0)]\, dF(x)\right| \leq M_2 r^\alpha F(S_{x_0,r}),$$

$$\sup_y \left|\int_{S_{x_0,r}} [f(y|x) - f(y|x_0)]\, dF(x)\right| \leq M_3 r^\alpha F(S_{x_0,r}). \tag{2.5}$$

for some $\alpha > 0$, where M_2 and M_3 are positive constants.

CONDITION 3 *We only consider points x_0 such that*

$$\liminf_{n\to\infty} (S_{x_0,r})/r^d > 0. \tag{2.6}$$

The assumption on the kernel function K is given below:

CONDITION 4 *$K(x)$ is a positive and integrable function on R^d. There exist positive constants $0 < C_1 \leq C_2 < \infty$ and ρ such that*

$$C_1 I(\| x \| \leq \rho) \leq K(x) \leq C_2 I(\| x \| \leq \rho). \tag{2.7}$$

For h_n we suppose that they satisfy the following condition.

CONDITION 5 *The window width satisfies*

$$\lim_{n\to\infty} (S_{x_0,\rho h})\, n/\log n = \infty, \limsup_{n\to\infty} F(S_{x_0,\rho h})\, h^{2\alpha} n/\log n < \infty. \tag{2.8}$$

REMARK 2.1 *Our conditions on the distribution of (X, Y) do not require the existence of the marginal density of X. The results based on Conditions 1 and 2 can be applied when the predictor X is multivariate and has a distribution which is not absolutely continuous. If $F(x)$, $F(y|x)$ and $f(y|x)$ have bounded second derivatives in $B(x_0)$ with respect to x, then $\alpha = 2$. Also note that if $F(x)$ is finite discrete, then (2.5) is automatically satisfied and α can be any number. Condition on the kernel function provides us more choices for the kernel function. Even though from a large sample point of view it makes no difference whatever kernel function is chosen, but a simulation study by Härdle (1991) showed that a nonparametric estimator with uniform kernel is not smooth if the sample is not large.*

THEOREM *Suppose that Conditions 1-5 hold. Then with probability 1*

$$\theta_h(x_0) - \theta(x_0) = \frac{1/2 - F_h(\theta(x_0)|x_0)}{f(\theta(x_0)|x_0)} + O\left[(F(S_{x_0,\rho h})n)^{-3/4}(\log n)^{3/4}\right]. \tag{2.9}$$

REMARK 2.2 *If X has a discrete distribution and $P\{X = x_0\} > 0$, then the remainder in (2.9) becomes $O\left[(n/\log n)^{-3/4}\right]$.*

The proof of the theorem will be given in the next section. In the following we give some applications of the theorem.

COROLLARY 1 (*The law of iterated logarithm of $\theta_h(x_0)$*) *Let $d = 1$ and suppose that the conditions of the theorem hold. We also assume that the marginal density $f(x)$ of X exists and is differentiable, and h satisfies*

$$nh^5/\log n \to 0, \sum_{n=1}^{\infty} n^{-2}h^{-1} < \infty$$

and

$$\lim_{\varepsilon\to\infty}\lim_{n\to\infty}\sup_{\{m:|m-n|>\varepsilon n\}}\left|\frac{h_m}{h_n} - 1\right| = 0. \tag{2.10}$$

Then we have

$$\limsup_{n\to\infty}[\theta_h(x_0) - \theta(x_0)][nh/(2\log\log n)]^{1/2} = \frac{\int K^2(u)du}{[4f(\theta(x_0))f^2(\theta(x_0)|x_0)]^{1/2}}, \quad a.s. \tag{2.11}$$

and

$$\liminf_{n\to\infty} f[\theta_h(x_0) - \theta(x_0)][nh/(2\log\log n)]^{1/2} = -\frac{\int K^2(u)du}{[4f(\theta(x_0))f^2(\theta(x_0)|x_0)]^{1/2}}, \quad a.s. \tag{2.12}$$

The proof of this corollary can be obtained by combining the conclusions of the above theorem and that of Theorem 4 of Härdle (1984).

We need some notations for Corollary 2. Denote by D the operator $(\partial/\partial x_{(1)} \dots \partial/\partial x_{(d)})'$ and by D^2 the operator $(\partial^2/\partial x_{(i)}\partial x_{(j)}), i, j = 1, \dots, d$. Let

$$\Delta = DF(\theta(x_0)|x_0)D'f(x_0) + 2^{-1}D^2F(\theta(x_0)|x_0)f(x_0).$$

$V(k)$ denotes the volume of the unit ball in R^k, $B(\bullet, \bullet)$, the Beta-function and dx the product $dx_{(1)} \dots dx_{(d)}$.

COROLLARY 2 (*The asymptotic normality of* $\theta_h(x_0)$) *Besides the Conditions* 1 − 5, *if we assume that* $f(x)$ *is differentiable, then for* $nh^{d+4} \to 0$ *we have*

$$(nh^d)^{1/2}[\theta_h(x_0) - \theta(x_0)] \xrightarrow{L} N(0, \sigma^2), \tag{2.13}$$

where $\sigma^2 = \int K^2(u)du[4f^2(\theta(x_0)|x_0)f(\theta(x_0))]^{-1}$. *If in addition we assume that* $nh^{d+4} \to C > 0$, *and the third derivative of* $F(\theta(x_0)|x)$ *with respect to* x *is bounded in a neighborhood of* x_0 *and* $f(x)$ *is twice differentiable in a neighborhood of* x_0, *then we have*

$$(nh^d)^{1/2}[\theta_h(x_0) - \theta(x_0)] \xrightarrow{L} N(\mu, \sigma^2) \quad a.e.\ x_0 \tag{2.14}$$

where

$$\mu = -\rho^2 C^{1/2} \frac{V(d-1)}{V(d)} B\left(\frac{3}{2}, \frac{d-1}{2}\right) tr(\Delta)/f(\theta(x_0)|x_0). \tag{2.15}$$

PROOF: Note that the first part of the corollary was proved by Liu and Tu (1987) under weaker conditions. We only prove the second part of the corollary. By the theorem, we need only to consider the asymptotic normality of $2^{-1} - F_h(\theta(x_0)|x_0)$. Using the notations in Lemma 2 we have

$$\frac{1}{2} - F_h(\theta(x_0)|x_0)$$

$$= \frac{1}{2} - E^*F_h(\theta(x_0)|x_0) + E^*F_h(\theta(x_0)|x_0) - F_h(\theta(x_0)|x_0). \tag{2.16}$$

It is seen that

$$\frac{(nh^d)^{1/2}}{F(\theta(x_0)|x_0)}[E^*F_h(\theta(x_0)|x_0) - F_h(\theta(x_0)|x_0)] \xrightarrow{L} N(0, \sigma^2). \tag{2.17}$$

On the other hand, we have

$$E[1/2 - E^*F_h(\theta(x_0)|x_0)]$$

$$= \sum_{i=1}^{n} W_{n,i}(x_0) \int_{S_{x_0,\rho h}} [F(\theta(x_0)|x_0) - F(\theta(x_0)|x)]f(x)dx/F(S_{x_0,\rho h}).$$

Using the conditions on $F(y|x)$ and $f(x)$, we have

$$\begin{aligned} F(\theta(x_0)|x_0) - F(\theta(x_0)|x) &= -(x - x_0)'DF(\theta(x_0)|x_0) - \\ &\quad 2^{-1}(x - x_0)'D^2F(\theta(x_0)|x_0)(x - x_0) + \\ &\quad o(\| x - x_0 \|^2) \end{aligned} \tag{2.18}$$

and

$$f(x) = f(x_0) + (x - x_0)'Df(x_0) + o(\| x - x_0 \|). \tag{2.19}$$

Note that $\int_{S_{x_0,\rho h}}(x-x_0)'DF(\theta(x_0)|x_0)dx = 0$. Therefore

$$\begin{aligned} &E[1/2 - E^* F_h(\theta(x_0)|x_0)] \\ &= -\int_{S_{x_0,\rho h}} (x-x_0)'\Delta(x-x_0)dx/F(S_{x_0,\rho h}) + o(h^2). \end{aligned} \tag{2.20}$$

It is easy to show

$$(2.20) = -\frac{V(d-1)}{V(d)}B\left(\frac{3}{2},\frac{d-1}{2}\right)tr(\Delta)(\rho h)^2 + o(h^2).$$

If $nh^{d+4} \to 0$, then $(nh^d)^{1/2}E[1/2 - E^* F_h(\theta(x_0)|x_0)] \to 0$. If $nh^{d+4} \to K$, then

$$\begin{aligned} &(nh^d)^{1/2}E[1/2 - E^* F_h(\theta(x_0)|x_0)] \\ &\to -\frac{V(d-1)}{V(d)}B\left(\frac{3}{2},\frac{d-1}{2}\right)tr(\Delta)\rho^2 K^{1/2}. \end{aligned} \tag{2.21}$$

In addition, it is seen that

$$Var\{(nh^d)^{1/2}E[1/2 - E^* F_h(\theta(x_0)|x_0)]\} \to 0. \tag{2.22}$$

Therefore the corollary is proved by combining (2.21) and (2.22).

In order to use the above corollary to get the confidence interval for $\theta(x_0)$, we should have consistent estimates of $f(\theta(x_0)|x_0)$ and $f(\theta(x_0))$. For the estimation of $f(\theta(x_0))$, we can use the usual kernel density estimate proposed by Parzen (1962). A simple estimate of $f(\theta(x_0)|x_0)$ can be

$$f_h(\theta(x_0)|x_0) = [F_h(\theta(x_0)+\lambda|x_0) - F_h(\theta(x_0)-\lambda|x_0)]/(2\lambda) \tag{2.23}$$

where $\lambda = \lambda_n$ is a sequence of positive numbers tending to zero as n tends to ∞. This estimate is a special case of the general estimates proposed by Rosenblatt (1969) and Zhao and Liu (1985). The strong consistency was proved in the later paper. As an application of Lemma 5 of Section 3, we can obtain the strong convergence rate of $f_h(\theta(x_0)|x_0)$.

COROLLARY 3 *Under the conditions of the theorem, if we take* $\lambda = [\log n/(nh^d)]^{1/2}$, *then we have*

$$f_h(\theta(x_0)|x_0) - f(\theta(x_0)|x_0) = O[(nh^d)^{-1/4}(\log n)^{1/4}] \quad a.s. \tag{2.24}$$

REMARK 2.3 *The results of this paper can be extended to the kernel estimates of conditional p-quantiles which are defined by*

$$\theta_h(x_0,p) = F_h^{-1}(p|x_0) \tag{2.25}$$

for $0 < p < 1$ *without any difficulty. We note that these estimates can also be obtained by minimizing*

$$\sum_{i=1}^n W_{n,i}(x_0)[p|Y_i - \theta_h(x_0,p)| + (Y_i - \theta_h(x_0,p))I(Y_i - \theta_h(x_0,p) < 0)]. \tag{2.26}$$

REMARK 2.4 *By assuming only that $F(y|x_0)$ is differentiable at $\theta(x_0)$ with derivative $f(\theta(x_0)|x_0)$ and $\sup|F(y|u) - F(y|x_0)| \leq M \parallel u - x_0 \parallel$ in a neighborhood of $(\theta(x_0), x_0)$, we can obtain a Bahadur type representation with the order of $o_p[(nh^d)^{-1/2}]$. This has been done in Tu (1988) by following the proof of Ghosh (1971). The asymptotic normality of the conditional quantile can be established under somewhat weaker conditions.*

3 The Proof of the Main Theorem

The proof of the theorem is divided into several lemmas. We let C denote a positive constant independent of n and ε. It may take different values in different formulae.

LEMMA 1 *Let $N = \#\{\parallel X_i - x_0 \parallel \leq r, i = 1, \ldots, n\}$. Then*

$$P\{|N/n - F(S_{x_0,r})| > \varepsilon\} \leq 2\exp\{-\frac{\varepsilon^2 n}{f(S_{x_0,r}) + \varepsilon}\} \tag{3.1}$$

holds for every $n, \varepsilon > 0$ and $r > 0$.

PROOF: Note that N is a summation of n iid Bernoulli random variables with expectation $F(S_{x_0,r})$. Then (3.1) follows by using Bernstein's inequality (Serfling, 1980, P.95).

LEMMA 2 *Denote $\delta = min\{F[\theta(x_0) + \varepsilon|x_0] - 2^{-1}, 2^{-1} - F[\theta(x_0) + \varepsilon|x_0]\}$. If $\tau = 2^{-1}\delta - M_2(\rho h)^\alpha > 0$, where α is as given in Condition 3, then for every n*

$$P\{|\theta_h(x_0) - \theta(x_0)| > \varepsilon\} \leq 4\exp\left(-\frac{C\delta^2 F(S_{x_0,\rho h})\, n}{1+\delta}\right)$$

$$+ 4\exp\left[-\frac{\tau^2 F(S_{x_0,\rho h})^n}{1+\tau}\right] + 16\exp\{-CF(S_{x_0,\rho h})\, n\}. \tag{3.2}$$

PROOF: Denote by $E^*(\bullet)$ and $P^*(\bullet)$ the conditional expectation and conditional probability given $\{X_1, \ldots, X_n\}$, respectively. Consider $P\{\theta_h(x_0) > \theta(x_0) + \varepsilon\}$, and note that

$$P\{\theta_h(x_0) > \theta(x_0) + \varepsilon\} \leq P\{\frac{1}{2} > F_h[\theta(x_0) + \varepsilon|x_0]\}$$

$$\leq P\{L_1 + L_2 < -\delta\} \leq P\{L_1 < -\frac{1}{2}\delta\} + P\{L_2 < -\frac{1}{2}\delta\} \tag{3.3}$$

where

$$L_1 = F_h[\theta(x_0) + \varepsilon|x_0] - E^* F_h[\theta(x_0) + \varepsilon|x_0],$$

$$L_2 = E^* F_h[\theta(x_0) + \varepsilon | x_0] - F[\theta(x_0) + \varepsilon | x_0].$$

Further note that given $\{X_1, \ldots, X_n\}$, L_1 is a summation of independent and bounded random variables with mean value 0. By Bernstein's inequality we have

$$\begin{aligned} P\{L_1 < -\frac{1}{2}\delta\} &= E[P^*\{L_1 < -\frac{1}{2}\delta\} I(N \neq 0)] + P\{N = 0\} \\ &\leq 2E\left[\exp\left(-\frac{C\delta^2 N}{1+\delta}\right)\right] + P\{N = 0\}. \end{aligned} \tag{3.4}$$

By Lemma 1

$$P\{N < \frac{1}{2} n F(S_{x_0,\rho h})\} \leq 2\exp\{-CF(S_{x_0,\rho h})n\}. \tag{3.5}$$

Combining (3.4) and (3.5), we have

$$\begin{aligned} P\{L_1 < -\frac{1}{2}\delta\} \ \leq \ & 2\exp\left(-\frac{C\delta^2 F(S_{x_0,\rho h})n}{1+\delta}\right) + \\ & 4\exp\{-CF(S_{x_0,\rho h})\, n\}. \end{aligned} \tag{3.6}$$

We now consider $P\{L_2 < -2^{-1}\delta\}$. Let $A_i = \{X_i \in S_{x_0,\rho h}\}, i = 1, \ldots, n$. Note that for given $B = \{A_1, \ldots, A_n\}, X_1, \ldots, X_n$ are iid random variables with distribution function

$$\hat{F}(\bullet) = F(\bullet \cap S_{x_0,\rho h}) / F(S_{x_0,\rho h}).$$

Denote by $\hat{E}(\bullet)$ the conditional expectation given B. Consider decomposition of L_2

$$L_2 = L_3 + L_4 \tag{3.7}$$

where

$$\begin{aligned} L_3 &= E^* F_h[\theta(x_0) + \varepsilon | x_0] - \hat{E} F_h[\theta(x_0) + \varepsilon | x_0], \\ L_4 &= \hat{E} F_h[\theta(x_0) + \varepsilon | x_0] - F[\theta(x_0) + \varepsilon | x_0]. \end{aligned}$$

Then by Condition 2

$$|L_4| = \left| \int_{S_{x_0,\rho h}} \frac{F(\theta(x_0) + \varepsilon | x) - F(\theta(x_0) + \varepsilon | x_0)}{F(S_{x_0,\rho h})} dF(x) \right| \leq M_2(\rho h)^\alpha. \tag{3.8}$$

Therefore

$$P\{L_2 < -\frac{1}{2}\delta\} \leq P\{L_3 \leq -\frac{1}{2}\delta + M_2(\rho h)^\alpha, N \neq 0\} + P\{N = 0\}.$$

When $\tau = \frac{1}{2}\delta - M_2(\rho h)^\alpha > 0$, by the same argument as in (3.6), we have

$$\begin{aligned} P\{L_2 < -1/2\delta\} \ \leq \ & 2\exp\left[-\frac{\tau^2 F(S_{x_0,\rho h})n}{1+\tau}\right] + \\ & 4\exp\{-CF(S_{x_0,\rho h})n\}. \end{aligned} \tag{3.9}$$

Therefore as long as $\tau > 0$, we have

$$P\{\theta_h(x_0) > \theta(x_0) + \varepsilon\} \leq 2\exp\left(-\frac{C\delta^2 F(S_{x_0,\rho h})\, n}{1+\delta}\right)$$

$$+ 2\exp\left[-\frac{\tau^2 F(S_{x_0,\rho h})n}{1+\tau}\right] + 8\exp\{-CF(S_{x_0,\rho h})n\}. \qquad (3.10)$$

Similarly, so long as $\tau > 0$, we have

$$P\{\theta_h(x_0) < \theta(x_0) - \varepsilon\} \leq 2\exp\left(-\frac{C\delta^2 F(S_{x_0,\rho h})\, n}{1+\delta}\right)$$

$$+ 2\exp\left[-\frac{\tau^2 F(S_{x_0,\rho h})n}{1+\tau}\right] + 8\exp\{-CF(S_{x_0,\rho h})n\}. \qquad (3.11)$$

This lemma is proved by combining (3.10) and (3.11).

REMARK 3.1 *From the inequality (3.2) we can easily get the weak convergence rate of $\theta_h(x_0)$, which has been derived by Truong (1989). We note that our conditions of smoothness are imposed on the conditional distribution of Y given $X = x$ rather than on $\theta(x)$ itself.*

LEMMA 3 *Suppose that the Conditions 1-5 are satisfied. Then with probability 1*

$$|\theta_h(x_0) - \theta(x_0)| = O[(F(S_{x_0,\rho h})n)^{-1/2}(\log n)^{1/2}]. \qquad (3.12)$$

PROOF: Put $\varepsilon_n = 2[f(\theta(x_0)|x_0)]^{-1}\{b[F(S_{x_0,\rho h})n]^{-1/2}(\log n)^{1/2} + M_2(\rho h)^\alpha\}$, where b is a positive constant. When n is sufficiently large, we have

$$\tau = \frac{1}{2}\delta - M_2(\rho h)^\alpha = f(\theta(x_0)|x_0)\varepsilon_n + O(\varepsilon_n^2) - M_2(\rho h)^\alpha$$

$$= b[F(S_{x_0,\rho h})n]^{-1/2}(\log n)^{1/2} + O(\varepsilon_n^2) > 0. \qquad (3.13)$$

Therefore, from Lemma 2 and the Condition 5 we have

$$P\{|\theta_h(x_0) - \theta(x_0)| > \varepsilon_n\} \leq Cn^{-Cb}. \qquad (3.14)$$

We choose b such that $Cb > 1$. The lemma is proved by using the Borel-Cantelli lemma.

REMARK 3.2 *Assume that the marginal density of X exists. Then $F(S_{x_0,\rho h}) \sim h^d$. In this case, we can choose $h = (n/\log n)^{-1/(d+2\alpha)}$ which balances the bias and the random variation, and (3.12) becomes*

$$|\theta_h(x_0) - \theta(x_0)| = O[(n/\log n)^{-(d+2\alpha-1)/(2d+4\alpha)}].$$

For $d = 1$ and $\alpha = 2$, the window width is of order $O(n/\log n)^{-1/5}$ and the order of the remainder term in (2.9) is $O[(n/\log n)^{-3/5}]$. We note that Bhattacharya and Gangopadhyay's (1990) claim that the optimum balance between bias and random error is achieved by choosing the window width of order $O(n^{-1/5})$ is true only for $\alpha = 2$ i.e., the $F(y|x)$ has bounded second derivative. If $F(y|x)$ has only first bounded derivative, then $\alpha = 1$ and a good choice of the window width is of the order of $O[(n/\log n)^{-1/3}]$ just as Cheng (1984) and Stute (1986) suggested. If X has finite discrete distribution, then the remainder term in (2.9) is of the order of $O[(n/\log n)^{-3/4}]$.

LEMMA 4 *Let $a_n = O[(F(S_{x_0,\rho h})n)^{-1/2}(\log n)^{1/2}]$ and*

$$H_n = \sup_{|y|\le a_n} |[F_h(\theta(x_0)+y|x_0) - F_h(\theta(x_0)|\,x_0)] - [E^*F_h(\theta(x_0)+y|x_0) - E^*F_h(\theta(x_0)|x_0)]|, \tag{3.15}$$

and further let the Conditions 1-5 be satisfied. Then

$$H_n = O[(F(S_{x_0,\rho h})n)^{-3/4}(\log n)^{3/4}] \quad a.s. \tag{3.16}$$

PROOF: Take integers $d_n = O[(F(S_{x_0,\rho h})n)^{1/4}(\log n)^{1/2}]$. For $r = -d_n, \ldots, d_n$, define

$$\eta_{r,n} = \theta(x_0) + a_n b_n^{-1} r,$$

$$\alpha_{r,n} = E^*F_h(\eta_{r+1,n}|x_0) - E^*F_h(\eta_{r,n}|x_0),$$

$$G_{r,n} = |[F_h(\eta_{r,n}|x_0) - F_h(\theta(x_0|x_0)] - [E^*F_h(\eta_{r,n}|x_0) - E^*F_h(\theta(x_0|x_0)]|.$$

It is easy to see from the monotonicity of $F_h(y|x_0)$ and $E^*F_h(y|x_0)$ that

$$H_n \le K_n + \beta_n \tag{3.17}$$

where $K_n = \max\{G_{r,n} : -d_n \le r \le d_n\}$ and $\beta_n = \max\{\alpha_{r,n} : -d_n \le r \le d_n\}$. Since $\eta_{r+1,n} - \eta_{r,n} = a_n d_n^{-1} = O[(F(S_{x_0,\rho h})n)^{-3/4}]$, when n is large enough, we have

$$\alpha_{r,n} \le C \sum_{i=1}^{n} \frac{I(A_i)}{N}\,[F(\eta_{r+1,n}|X_i) - F(\eta_{r,n}|X_i)]$$

$$= O\left[(F(S_{x_0,\rho h})n)^{-3/4}\right].$$

The last inequality above is from the Condition 1. Hence

$$\beta_n = O\left[(F(S_{x_0,\rho h})n)^{-3/4}\right]. \tag{3.18}$$

Next we prove that there exists a $t > 0$ such that

$$\sum_{n=1}^{\infty} P\{K_n \ge \gamma_n t\} < \infty \tag{3.19}$$

where $\gamma_n = [F(S_{x_0,\rho h})n]^{-3/4}(\log n)^{3/4}$. In fact we have

$$P\{K_n \geq \gamma_n t\} \leq \sum_{r=-d_n}^{d_n} P\{G_{r,n} \geq \gamma_n t\}.$$

From the definition of $G_{r,n}$ we have

$$G_{r,n} = \left|\sum_{i=1}^{n} W_{n,i}(x_0)[I\{Y_i \in (\theta(x_0), \eta_{r,n}]\} - p_i]\right|$$

where $p_i = F(\eta_{r,n}|X_i) - F(\theta(x_0)|X_i)$. By the same argument as in Lemma 2

$$P\{G_{r,n} \geq \gamma_n t\} \leq 2E\exp\left[-\frac{C(\gamma_n t)^2 N}{a_n + \gamma_n t}\right].$$

Note that $\gamma_n/a_n \to 0$ and therefore, for large n,

$$\begin{aligned} P\{G_{r,n} \geq \gamma_n t\} &\leq 2\exp\{-C(\gamma_n t)^2 a_n^{-1} F(S_{x_0,\rho h})n\} + 4\exp\{-CF(S_{x_0,\rho h})\,n\} \\ &\leq 2\exp\{-Ct^2\log n\} + 4\exp\{-CF(S_{x_0,\rho h})\,n\}. \qquad (3.20) \end{aligned}$$

We can choose t large enough such that $Ct^2 > 2$. Then

$$P\{K_n \geq \gamma_n t\} \leq C d_n n^{-2} \leq C n^{-3/2}. \qquad (3.21)$$

By the Borel-Cantelli lemma, we get

$$K_n = O[(F(S_{x_0,\rho h})n)^{-3/4}(\log n)^{3/4}]. \qquad (3.22)$$

Therefore the lemma is proved by combining (3.18) and (3.22).

PROOF OF THE THEOREM: From the conditions (2.4) and (2.5), we have

$$\begin{aligned} &E^*F_h(\theta(x_0) + y|x_0) - E^*F_h\,(\theta(x_0)|x_0)] \\ &= (\theta_h(x_0) - \theta(x_0)) \sum_{i=1}^{n} W_{n,i}(x_0) f(\theta(x_0)|X_i) + O[(\theta_h(x_0) - \theta(x_0))^2]. \end{aligned}$$

By the same reason as in the proofs of Lemma 2 and Lemma 3, we have

$$\sum_{i=1}^{n} W_{n,i}(x_0) f(\theta(x_0)|X_i) = f(\theta(x_0)|x_0) + O[(F(S_{x_0,\rho h})n)^{-1/2}(\log n)^{1/2}] \ a.s.$$

Then from Lemma 3, we have

$$\begin{aligned} &E^*F_h(\theta(x_0) + y|x_0) - E^*F_h(\theta(x_0)|x_0)] \\ &= (\theta_h(x_0) - \theta(x_0))\, f(\theta(x_0)|x_0) + O[(F(S_{x_0,\rho h})n)^{-1}\log n] \quad a.s. \qquad (3.23) \end{aligned}$$

Therefore by Lemma 4, we have

$$\begin{aligned} F_h(\theta_h(x_0)|x_0) - F_h(\theta(x_0)|x_0) &= (\theta_h(x_0) - \theta(x_0))\, f(\theta(x_0)|x_0) \\ &\quad + O[(F(S_{x_0,\rho h})n)^{-3/4}(\log n)^{3/4}] \quad a.s. \end{aligned} \tag{3.24}$$

Finally, it is seen that

$$F_h(\theta_h(x_0)|x_0) - 1/2 = O[(F(S_{x_0,\rho h})n)^{-3/4}(\log n)^{3/4}] \quad a.s. \tag{3.25}$$

and the theorem is proved.

REFERENCES

Babu, G. J. (1989). Strong representations for LAD estimators in linear models, *Probability Theory and Related Fields*, 83, 547-558.

Bahadur, R. R. (1966). A note on quantiles in large samples, *Ann. Math. Statist.*, 37, 577-580.

Cheng, K. F. (1984). Nonparametric estimation of regression function using linear combinations of sample quantile regression functions, *Sankhya* Ser. A, 46, 287-302.

Collomb, G. (1980). Estimation non parametrique de probabilities conditionnelles, *C. R. Acad. Sc. Paris*, Series A, 291, 427-430.

Devroye, L. (1981). On the almost everywhere convergence of nonparametric regression function estimates, *Ann. Statist.*, 9, 1310-1319.

Dodge, Y. (Eds) (1987). *"Statistical Data Analysis Based on the L_1-norm and Related Methods"*, North-Holland, Amsterdam.

Ghosh, J. K. (1971). A new proof of the Bahadur representation of quantiles and an application, *Ann. Math. Statist.*, 42, 1957-1961.

Härdle, W. (1991). *"Smoothing Techniques"*, Springer-Verlag, New York.

Härdle, W. (1984). A law of iterated logarithm for nonparametric regression function estimators, *Ann. Statist.*, 12, 624-635.

Härdle, W. and Luckhaus, S. (1984). Uniform consistency of a class of regression function estimators, *Ann. Statist.*, 12, 624-635.

Liu, Z. K. and Tu, D. S. (1987). Kernel methods on conditional median estimation, *Kexue Tongbao (Chinese Bulletin of Science)*, 32, 642-643.

Parzen, E. (1962). On estimation of a probability density function, *Ann. Statist.*, 33, 1065-1076.

Rao, C. R. (1989). Methodology based on the L_1-norm in statistical inference, *Sankhya A*, 50, 289-313.

Rosenblatt, M. (1969). Conditional probability density and regression estimators, in *"Multivariate Analysis II"*, (Krishnaih, eds), 25-31.

Serfling, R. J. (1980). *"Approximation Theorems of Mathematical Statistics"*, John Wiley and Sons, New York.

Stone, C. J. (1977). Consistent nonparametric regression (with discussion), *Ann. Statist.*, 17, 606-617.

Stute, W. (1986). Conditional empirical processes, *Ann. Statist.*, 14, 638-647.

Truong, Y. K. (1989). Asymptotic properties of kernel estimators based on local medians, *Ann. Statist.*, 17, 606-617.

Tu, D. S. (1988). The kernel estimates of the conditional L-functional and their bootstrapping statistics, *Acta Math. Appl. Sinica*, 11, 53-68.

Wheeden, R. L. and Zygmund, A. (1977). *"Measure and Integral"*, Marcel Dekker, New York.

Zhao, L. C. and Liu, Z. J. (1985). Strong consistency of the kernel estimates of conditional density function, *Acta Math. Appl. Sinica, New Series*, 1, 314-318.

Zheng, Z. G. (1985). On the asymptotic behavior of nearest neighbor estimator of conditional median, *Acta Math. Appl. Sinica, New Series*, 1, 206-212.

Department of Mathematics
and Statistics
Mississippi State University
Mississippi State
MS 39762, USA

Center for Multivariate Analysis
Department of Statistics
Pennsylvania State University
University Park
Pennsylvania 16802, USA

J.K. Ghosh, S.K. Mitra, K.R. Parthasarathy and B.L.S. Prakasa Rao (*eds.*),
Statistics and Probability : A Raghu Raj Bahadur Festschrift
Published by Wiley Eastern Limited, Publishers, (1993), 395 – 410

Skorokhod Integral and Differentiation for Gaussian Processes

V. MANDREKAR[1]
Michigan State University

and

SIXIANG ZHANG[2]
Gannon University

Abstract

Let $\{X_t, t \in T\}$ be a Gaussian process with covariance C and reproducing kernel Hilbert space $K(C)$. Let $L_2(\Omega, K(C))$ be the space of square-integrable, Bochner measurable functions. We define the Skorokhod Integral as a map I^s from $L_2(\Omega, K(C))$ to $L_2(\Omega, \sigma(X_t, t \in T), \mathcal{R})$ and prescribe its domain. As a consequence, we derive, in special cases, the known results. To relate our definition to that of Hitsuda, we define the differentiation operator D from $L_2(\Omega, \sigma(X_t, t \in T), \mathcal{R})$ to $L_2(\Omega, K(C))$ and show that $I^s = D^*$.

[1] Partially supported by ONR Grant: N000-14-91-J-1084.
[2] Partially supported by ONR Grant: N000-14-85-K-0150.
AMS Subject Classifications : 60H05, 60H07.
Key words and Phrases : Multiple Wiener Integral, Homogeneous chaos expansion, Malliavin differentiation.

1 Introduction

Motivated by problems in filtering and stochastic partial differential equations, there is increasing interest in the definition of stochastic integral for a non-anticipative functional. The first definition of such a stochastic integral was proposed by Hitsuda [4], using the concept of stochastic partial derivative in the white noise case. Skorokhod [12], in his fundamental paper, gave the definition in the context of random linear functional. His technique is to give the definition of multiple integrals through induction motivated by a property of multiple integrals [5] and the Wiener-Ito Chaos expansion [6]. A relation of Skorokhod integral to gradient differential operator of Malliavin is given in [2].

The purpose of this paper is to complete the program, started in [8], to define stochastic integrals with respect to all Gaussian processes $\{X_t, t \in T\}$. Using ideas of Skorokhod and Ito we define multiple stochastic integrals with respect to any Gaussian process indexed by any set. This definition and a suitable generalization of Wiener-Ito chaos decomposition [1] are used to define Skorokhod integrals for any Gaussian process. We define also Skorokhod differentiation as an operator on $L^2(\Omega, P) \to L^2(\Omega, K(C))$ and it is shown that its adjoint is the Skorokhod Integral with $K(C)$ being the reproducing kernel Hilbert space of $\{X_t, t \in T\}$.

As a consequence we derive the results of [12], [4], [2] and recent examples in [9], [10], and [11].

2 The Structure of the Stochastic Integral

Let T be any set and C be a real function on $T \times T$. Then C is called a covariance on T if

1. $C(s,t) = C(t,s)$.
2. $\sum_{t,s \in i} a_t a_s C(t,s) \geq 0$ for all finite subsets i of T and $\{a_s, s \in i\}$ of R.

Let T be any set and C be a real valued covariance on T. Then there exists a unique Hilbert space $K(C)$ of functions on T satisfying:

$$\begin{array}{ll} C(\cdot,t) \in K(C) & \text{for each } t \in T \text{ and} \\ (f(\cdot), C(\cdot,t))_{K(C)} = f(t) & \text{for each } t \in T \text{ and } f \in K(C). \end{array}$$

Here for each t, $C(\cdot,t)$ denotes the function of the first variable and $(\cdot,\cdot)_{K(C)}$ means the inner product in $K(C)$. $K(C)$ is called the reproducing kernel Hilbert space (RKHS) of C.

Definition 2.1. Let T be a set and $(\Omega, \mathcal{F}, P)$ be a complete probability space. Then a family $\{X_t, t \in T\}$ of random variables is called a centered Gaussian process if every linear combination of finite elements of

$\{X_t, t \in T\}$ is a Gaussian random variable with mean zero.

If $\{X_t, t \in T\}$ is a centered Gaussian process on a probability space $(\Omega, \mathcal{F}, P)$, then $C_X(s,t) = E(X_s X_t)$ is a covariance on T. C_X is also called the covariance of $\{X_t, t \in T\}$ and $K(C_X) = \{f : f(t) = E(X_t Y_f)$ for a unique Y_f in $H(X)\}$, where E is the expectation under P and $H(X)$ is the closed linear subspace of $L_2(\Omega, \mathcal{F}, P)$ spanned by $\{X_t, t \in T\}$.

LEMMA 2.1. ([1]) Let C be a covariance on T, then there exists a Gaussian process $\{X_t, t \in T\}$ defined on a complete probability space $(\Omega, \mathcal{F}, P)$ such that $E(X_s X_t) = C(s,t)$, where $\mathcal{F}$ is the σ-field generated by $\{X_t, t \in T\}$. Let C_i be a covariance on $T_i (i = 1, 2, ..., p)$ and $K(C_i)$ be the RKHS of C_i for each i. Then $K(C_1) \otimes K(C_2) \otimes \ldots \otimes K(C_p) = K(C_1 \otimes C_2 \otimes \ldots \otimes C_p)$, consists of functions f of p variables of the form

$$f(t_1, \ldots, t_p) = \sum_{\alpha_1 \ldots \alpha_p} a_{\alpha_1 \ldots \alpha_p} e^{(1)}_{\alpha_1}(t_1) e^{(2)}_{\alpha_2}(t_2) \ldots e^{(p)}_{\alpha_p}(t_p)$$

with $\sum_{\alpha_1 \ldots \alpha_p} a^2_{\alpha_1 \ldots \alpha_p}$ is finite. Here $\left\{ e^{(j)}_{\alpha_j}, \alpha_j \in J_j \right\}$ is an orthonormal base of $K(C_j)$ $(j = 1, 2, \ldots, p)$.

We note that

$$(f, f')_{K(C_1) \otimes \ldots \otimes K(C_p)} = \sum_{\alpha_1 \ldots \alpha_p} a_{\alpha_1 \ldots \alpha_p} a'_{\alpha_1 \ldots \alpha_p}.$$

The proof of the following lemma is simple and hence is omitted.

LEMMA 2.2. i) If $f \in K(C_1) \otimes \ldots \otimes K(C_p)$ and π is a permutation of $\{1, 2, \ldots p\}$, then

$$f^{(\pi)}(t_1, \ldots, t_p) = f(t_{\pi(1)}, \ldots, t_{\pi(p)}) \in K(C_1) \otimes \ldots \otimes K(C_p)$$

ii) If $f \in K(C_1) \otimes \ldots \otimes K(C_p)$ and $g \in K(C_{p+1}) \otimes \ldots \otimes K(C_{p+q})$ then $(fg)(t_1, \ldots, t_{p+q}) = f(t_1, \ldots, t_p) g(t_{p+1}, \ldots, t_{p+q}) \in K(C_1) \otimes \ldots \otimes K(C_{p+q})$.

When $C_1 = C_2 = \ldots = C_p = C$, we denote $K(C) \otimes \ldots \otimes K(C)$ by $K(C)^{\otimes p}$ and call it the p-th tensor product of $K(C)$. If there is no confusion, we denote it by $K^{\otimes p}$. For $f \in K^{\otimes p}$ its symmetrization is defined by $\tilde{f}(t_1, \ldots, t_p) = \frac{1}{p!} \Sigma'_\pi f(t_{\pi(1)}, \ldots, t_{\pi(p)})$ where the sum is over all permutations. We denote by $K^{\odot p} \subseteq K^{\otimes p}$, the functions symmetric in $(t_1, \ldots, t_p)$. We note that if $f \in K^{\otimes p}$ then by Lemma 2.2 $\tilde{f} \in K^{\odot p}$. For $p = 0$, $K^{\otimes 0} = K^{\odot 0} = \mathcal{R}$.

Given a covariance C on T, let $\{X_t, t \in T\}$ be a centered Gaussian process defined on $(\Omega, \mathcal{F}, P)$ as in Lemma 2.1. We denote by $L^2(\Omega, \mathcal{F}, P)$ the space of square integrable functions on $(\Omega, \mathcal{F}, P)$. Let $f \in K(C)$ then $f(t) = E(X_t Y_f)$, we define $I_1(f) = Y_f$. Then I_1 is an isometry of $K(C)$ into

$L^2(\Omega,\mathcal{F},P)$. We call I_1 the stochastic integral of order one. Of course, we define $I_0(r) = r$ for $r \in K^{\otimes 0} = \mathcal{R}$. We now define the multiple integral I_{p+1} as a map on $K^{\otimes p+1}$ to $L_2(\Omega,\mathcal{F},P)$. Consider $\{e_\alpha, \alpha \in J\}$ an orthonormal basis in K. Because of the form of elements of $K^{\otimes p+1}$ it is enough to define I_{p+1} for functions of the form

$$f(t_1,\ldots,t_p,t_{p+1}) = e_{\alpha_1}(t_1)\ldots e_{\alpha_{p+1}}(t_{p+1}) \tag{1}$$

where among $\alpha_1,\ldots,\alpha_{p+1}$ only n are different with repeats $p_1,p_2,\ldots p_n$ $(p_1+\ldots+p_n = p+1)$. Let us denote the corresponding n different e_{α_i}'s as $u_1,u_2,\ldots,u_n \in K$. Define

$$I_{p+1}(f) = \Pi_{i=1}^n H_{p_i}(I_1(u_i))$$

We note that $I_{p+1}(f) = I_{p+1}(f^\pi)$. Extend I_{p+1} by linearity to the linear manifold generated by elements of type (1).

LEMMA 2.3. Let φ, $\psi \in K^{\otimes p+1}$ be of the form (1). Then

(a) $I_{p+1}(\varphi) = I_{p+1}(\tilde{\varphi})$.

(b) $E[I_{p+1}(\varphi)] = 0$ and $E[I_{p+1}(\varphi)I_{p+1}(\psi)] = (p+1)!(\tilde{\varphi},\tilde{\psi})_{K^{\otimes p+1}}$ which is zero if $\tilde{\varphi} \neq \tilde{\psi}$).

PROOF: a) follows as $I_{p+1}(f) = I_{p+1}(f^\pi)$. To prove b) we first note that if $\tilde{\varphi} \neq \tilde{\psi}$ then $I_{p+1}(\varphi)$ and $I_{p+1}(\psi)$ involve two different order polynomials giving LHS $= 0$ and since in (1) at least one p_i corresponding to e_{α_i} is different we get that RHS $= 0$. So without loss of generality we can assume that

$$\varphi(\underline{t}) = \psi(\underline{t}) = e_1(t_1)\ldots e_1(t_{p_1})e_2(t_{p_1+1})\ldots e_2(t_{p_1+p_2})\ldots e_n(t_{p_1+\ldots p_n}).$$

where $\{e_1\ldots e_n\} \subset \{e_\alpha, \alpha \in J\}$. Then

$$\begin{aligned}\|\tilde{\varphi}\|^2 = (\tilde{\varphi},\tilde{\varphi})_{K^{\otimes p+1}} &= \left(\tfrac{1}{(p+1)!}\right)^2 \textstyle\sum_{\pi,\pi'} \left(\varphi^\pi,\varphi^{\pi'}\right)_{K^{\otimes p+1}} \\ &= \left(\tfrac{1}{(p+1)!}\right)^2 \textstyle\sum_{\pi,\pi'} (1)\end{aligned}$$

where the summation contains $(p+1)!\ p_1!\ldots p_n!$ elements. This follows as there are $(p+1)!$ choices of π and for each fixed π, there are $p_1!\ldots p_n!$ choices of π'. Thus

$$(p+1)!\,\|\tilde{\varphi}\|^2_{K^{\otimes p+1}} = (p+1)!\left(\frac{1}{(p+1)!}\right)^2 (p+1)!p_1!\ldots p_n!$$

giving

$$(p+1)!\,\|\tilde{\varphi}\|_{K^{\otimes p+1}} = p_1!\ldots p_n! = E\left\{\prod_{i=1}^n H_{p_i}(I_1(u_i))\right\}^2.$$

From Lemma 2.3, for the basis elements $\tilde{\varphi}, \tilde{\psi}$ of $K^{\odot p+1}$, the map $U_{p+1} = ((p+1)!)^{-\frac{1}{2}} I_{p+1}$ satisfies $(U_{p+1}\tilde{\varphi}, U_{p+1}\tilde{\psi})_{L^2(\Omega,\mathcal{F},P)} = (\tilde{\varphi}, \tilde{\psi})_{K^{\otimes p+1}}$. Hence, U_{p+1} extends to a unitary operator $\tilde{U}_{p+1}$ on $K^{\odot p+1}$ into $L^2(\Omega, \mathcal{F}, P)$. We define

$$I_{p+1} = ((p+1)!)^{\frac{1}{2}} \tilde{U}_{p+1}$$

We call I_p the multiple integral of order p with respect to X. We note that $I_p(f) = I_p(\tilde{f})$ defines I_p on $K^{\otimes p}$.

THEOREM 2.1. Let I_p be defined on $K^{\otimes p}$ as above for each $p \geq 0$. Then for $f, g \in K^{\otimes p}$,

(a) $E(I_p(f)) = 0$ and $E[I_p(f)I_p(g)] = p!\,(\tilde{f}, \tilde{g})_{K^{\otimes p}}$ for $p > 0$.

(b) $I_p(K^{\otimes p}) \perp I_k(K^{\otimes k})$ for all $k \neq p$.

PROOF: We note that $E(I_p(f)) = 0$ for f of the form (1) by the orthogonality of $\mathrm{H}_{p_i}(I(u_i))$ and $H_o(I_1(u_i)) = 1$ for all $p_i > 0$. For general g we get $I_p(g)$ as a limit in $L^2(\Omega, \mathcal{F}, P)$ and hence $L_1(\Omega, \mathcal{F}, P)$ of $I_p(g_n)$ with g_n finite linear combinations of elements of the form (1) giving first part of (a). The second part of (a) follows by the fact that $(p!)^{-\frac{1}{2}} I_p$ is an isometry. To prove (b), consider f_p and f_k of the form (1) with $\sum p_i = p$ and $\sum p_i' = k$ respectively. Then at least one $p_i \neq p_i'$ for some i and hence by the orthogonality of Hermite polynomials we get the orthogonality of $I_p(f_p)$ and $I_k(f_k)$. Thus for every element f_p of the basis of $K^{\otimes p}$, $I_p(f_p) \perp I_k(f_k)$ where f_k is the basis of element of $K^{\otimes k}$. This gives (b).

Let us consider for f of the form (1). Let $\omega_i = I_1(u_i)$ and assume $e_{\alpha_{p+1}}(t) = u_j(t)$ for some j. Then using the fact that

$$H_n(t) = tH_{n-1}(t) - (n-1)H_{n-2}(t),$$

we get,

$$\begin{aligned} I_{p+1}(f) &= \Pi_{i=1}^n H_{p_i}(\omega_i) \\ &= H_{p_1}(\omega_1)\ldots.H_{p_{j-1}}(\omega_{j-1})\{H_{p_{j-1}}(\omega_j)\omega_j - (p_{j-1})H_{p_j-2}(\omega_j)\} \\ &\quad H_{p_{j+1}}(\omega_{j+1})\ldots.H_{p_n}(\omega_n) \\ &= I_p(g)I_1(h) - \sum_{k=1}^{\prime} I_{p-1}(g \underset{k}{\otimes} h). \end{aligned}$$

Here $g(t_1\ldots.t_p) = e_{\alpha_1}(t_1)\ldots.e_{\alpha_p}(t_p)$, $h(t) = e_{\alpha_{p+1}}(t)$ and

$$(g \underset{k}{\otimes} h)(t_1, \ldots, t_{k-1}, t_{k+1}, \ldots, t_p) = (g(t_1, \ldots, t_{k-1}, \cdot, t_{k+1}, \ldots t_p), h(\cdot))_K.$$

Given $g \in K^{\otimes p}, h \in K$ of the form

$$g(t_1, \ldots t_p) = \sum_{\alpha_1,\ldots\alpha_p} a_{\alpha_1,\ldots\alpha_p} e_{\alpha_1}(t_1)\ldots.e_{\alpha_p}(t_p)$$

and

$$h(t) = \sum_{\beta} b_\beta e_\beta(t)$$

we have

$$(g\ h)(t_1, \ldots, t_{p+1}) = \sum_{\alpha_1, \ldots \alpha_p, \beta} a_{\alpha_1, \ldots, \alpha_p} b_\beta e_{\alpha_1}(t_1), \ldots, e_{\alpha_p}(t_p) e_{\alpha_{p+1}}(t_{p+1}).$$

Using the above property for the elements of the form (1) one gets

$$I_{p+1}(g\ h) = I_p(g) I_1(h) - \sum_{k=1}^{p} I_{p-1}(g \underset{k}{\otimes} h) \tag{2}$$

We remark that it can be proved from property (2) that, for f as in (1), $I_p(f) = \Pi_{i=1}^n H_{p_i}(\omega_i)$ and hence I_p defined by us coincides with the definition in [12].

$\{I_1(e_\alpha), \alpha \in J\}$ are the i.i.d. standard normal random variables corresponding to $\{e_\alpha, \alpha \in J\}$ which is a complete orthonormal basis of $K(C)$. According to our construction, the linear space spanned by all random variables of type $\Pi_{i=1}^n H_{p_i}(I_1(e_{\alpha_i}))$ is the same as $I_p(K^{\odot p})$, where H are Hermite polynomials and p_i's are nonnegative integers whose sum is p. From ([1], p. 191) we have

LEMMA 2.4. $L^2(\Omega, \mathcal{F}, P) = \sum_{p=0}^{\infty} \oplus I_p(K^{\odot p})$.

3 Applications

a. Skorokhod integral. Let $\{g_t, t \in T\}$ be a measurable stochastic process from Ω to $K(C)$ such that $g_.(\omega)$ can be considered as a $K(C)$-valued random variable in Bochner sense and assume that $g_.(\omega) \in L^2(\Omega, K(C))$.
Now

$$|g_t(\omega)|^2 = |< g_.(\omega), C(\cdot, t) >|^2 \leq \|g_.(\omega)\|^2_{K(C)} C(t,\ t)$$

implies that $E[g_t^2(\omega)] < \infty$ for each t. Hence, by Lemma 2.4, for each t, there exists unique $f_p^t(\cdot) \in K^{\odot p}$, $p = 0, 1, 2, \ldots$, such that

$$g_t(\omega) = \sum_{p=0}^{\infty} I_p(f_p^t) \tag{3}$$

LEMMA 3.1. If $g_.(\omega)$ has the decomposition as in (3), then

(a) $f_p^t(t_1,\ t_2, \ldots, t_p) \equiv f_p(t_1,\ t_2, \ldots, t_p,\ t)$ can be chosen such that $f_p(t_1,\ t_2, \ldots, t_p,\ t) \in K^{\otimes p+1}$ and f_p is a symmetric function in the first p variables.

(b) if $g_t(\omega) = \sum_{p=0}^{\infty} I_p(f_p(\cdot)) = \sum_{p=0}^{\infty} I_p(f_p'(\cdot))$, where both f_p and f_p' satisfy the requirement in (a), then $f_p = f_p'$ for all p.

PROOF: If $g_t = \sum_{k=1}^{n} F_k(\omega) g_k(t)$, where $F_k(\omega) \in L^2$ and $g_k \in K$, then

$$\begin{aligned} g_t = \sum_{k=1}^{n}\left[\sum_{p=0}^{\infty} I_p\left[f_{k,p}(\cdot)\right]\right] g_k(t) &= \sum_{p=0}^{\infty} \sum_{k=1}^{n} I_p\left[f_{k,p}(\cdot)\right] g_k(t) \\ &= \sum_{p=0}^{\infty} I_p\left\{\sum_{k=1}^{n} f_{k,p}(\cdot) g_k(t)\right\}. \end{aligned}$$

In general, g_t can be approximated by $g_t^n(\omega) = \sum_{p=0}^{\infty} I_p\left\{f_{n,p}(\cdot,t)\right\}$, where $f_{n,p}(\cdot,\cdot) \in K^{\otimes p+1}$ and for fixed t, $f_{n,p}(\cdot,t) \in K^{\odot p}$, and the approximation is in the sense that $E\left\|g_t(\omega) - g_t^n(\omega)\right\|_K^2 \to 0$ as $n \to \infty$. Clearly $E\left\|g_t^n(\omega) - g_t^m(\omega)\right\|_K^2 \to 0$ as $m, n \to \infty$.

Let $f_{n,p}(t_1, t_2, \ldots, t_p, t) - f_{m,p}(t_1, t_2, \ldots t_p, t)$

$$= \sum_{\alpha_1 \ldots \alpha_{p+1}} a(n,m,p)_{\alpha_1 \ldots \alpha_{p+1}} e_{\alpha_1}(t_1) \cdots e_{\alpha_p}(t_p)\, e_{\alpha_{p+1}}(t)$$

$$= \sum_{\alpha_{p+1}} \left\{ \sum_{\alpha_1 \cdots \alpha_p} a(n,m,p)_{\alpha_1 \cdots \alpha_{p+1}} e_{\alpha_1}(t_1) \cdots e_{\alpha_p}(t_p) \right\} e_{\alpha_{p+1}}(t).$$

Then

$$E\left\|g_t^n(\omega) - g_t^m(\omega)\right\|_K^2 = E\left\|\sum_{p=0}^{\infty} I_p\left\{f_{n,p}(\cdot,t) - f_{m,p}(\cdot,t)\right\}\right\|_K^2$$

$$= \mathrm{E}\left(\sum_{p,p'=0}^{\infty} (I_p[f_{n,p}(\cdot,t) - f_{m,p}(\cdot,t)], I_{p'}[f_{n,p'}(\cdot,t) - f_{m,p'}(\cdot,t)]\right)_K$$

$$= \sum_{p=0}^{\infty} E\left\|I_p[f_{n,p}(\cdot,t) - f_{m,p}(\cdot,t)]\right\|_K^2.$$

But

$$E\left\|I_p[f_{n,p}(\cdot,t) - f_{m,p}(\cdot,t)]\right\|_K^2$$

$$= E \sum_{\alpha_{p+1}} (I_p[f_{n,p}(\cdot,t) - f_{m,p}(\cdot,t)], e_{\alpha_{p+1}}(t))_K^2$$

$$= \sum_{\alpha_{p+1}} E\left(\sum_{\alpha_1 \cdots \alpha_p} a(n,m,p)_{\alpha_1 \cdots \alpha_p \alpha_{p+1}} I_p[e_{\alpha_1}(t_1) \cdots e_{\alpha_p}(t_p)] \right)^2$$

$$= \sum_{\alpha_{p+1}} p! \sum_{\alpha_1 \cdots \alpha_p} [a(n,m,p)_{\alpha_1 \cdots \alpha_p \alpha_{p+1}}]^2$$

$$= p! \sum_{\alpha_1 \cdots \alpha_{p+1}} [a(n,m,p)_{\alpha_1 \cdots \alpha_p \alpha_{p+1}}]^2$$

$$= p! \left\|f_{n,p}(\cdot,t) - f_{m,p}(\cdot,t)\right\|_{K^{\otimes p+1}}^2.$$

Hence

$$E\left\|g_t^n(\omega) - g_t^m(\omega)\right\|_K^2 = \sum_{p=0}^{\infty} p! \left\|f_{n,p} - f_{m,p}\right\|_{K^{\otimes p+1}}^2 \to 0 \text{ as } m, n \to \infty,$$

which implies that

$$\|f_{n,p} - f_{m,p}\|^2_{K^{\otimes p+1}} \to 0 \text{ as } m, n \to \infty \text{ for each } p.$$

So $\{f_{n,p}\}$ as a sequence in $K^{\otimes p+1}$ is Cauchy, therefore $f_{n,p} \to f_p$ as $n \to \infty$, for some $f_p \in K^{\otimes p+1}$. We can easily show that: $\sum_{p=0}^{\infty} p!\, \|f_{n,p} - f_p\|^2_{K^{\otimes p+1}} \to 0$ as $n \to \infty$. Let $h_t(\omega) = \sum_{p=0}^{\infty} I_p(f_p(\cdot, t))$, then

$$\begin{aligned} &E\,\|g_t^n(\omega) - h_t(\omega)\|^2_K \\ &= E\left\|\sum_{p=0}^{\infty} I_p\{f_{n,p}(\cdot, t) - f_p(\cdot, t)\}\right\|^2_K \\ &= \sum_{p=0}^{\infty} E\,\|I_p[f_{n,p}(\cdot, t) - f_p(\cdot, t)]\|^2_K \\ &= \sum_{p=0}^{\infty} p!\,\|f_n - f_p\|^2_{K^{\otimes p+1}} \to 0 \text{ as } n \to \infty, \text{ this implies } g_t(\omega) = h_t(\omega). \end{aligned}$$

Since

$$\begin{aligned} &E\left\|\sum_{p=0}^{\infty} I_p(f_p(\cdot, t) - \sum_{p=0}^{\infty} I_p(f_p'(\cdot, t))\right\|^2_K \\ &= \sum_{p=0}^{\infty} E\,\|I_p(f_p(\cdot, t) - f_p'(\cdot, t))\|^2_K \\ &= \sum_{p=0}^{\infty} p!\,\|f_p(\cdot) - f_p'(\cdot)\|^2_{K^{\otimes p+1}} = 0. \end{aligned}$$

This gives $f_p(\cdot) = f_p'(\cdot)$ and hence (b).□

In the proof of the previous lemmas we also showed that if $g_t(\omega) \in L^2(\Omega, K)$ and

$$g_t(\omega) = \sum_{p=0}^{\infty} I_p(f_p(\cdot, t)),$$

where $f_p(\cdot,\cdot) \in K^{\otimes p+1}$, and for fixed t, $f_p(\cdot, t) \in K^{\odot p}$, then

$$E\,\|g.(\omega)\|^2_K = \sum_{p=0}^{\infty} p!\,\|f_p(\cdot)\|^2_{K^{\otimes p+1}},$$

In fact, the proof gives the following lemma.

LEMMA 3.2. If $f(\cdot,\cdot) \in K^{\otimes p+1}$ and for each fixed t, $f(\cdot, t) \in K^{\odot p}$, and $g(\cdot,\cdot) \in K^{\otimes q+1}$, and for fixed t, $g(\cdot, t) \in K^{\odot q}$, let $u_t(\omega) = I_p(f(\cdot, t))$ and $v_t(\omega) = I_q(g(\cdot, t))$. Then

$$E(u.(\omega),\ v.(\omega))_K = \begin{cases} 0 & \text{if } p \neq q \\ p!(f, g)_{K^{\otimes p+1}} & \text{if } p = q \end{cases}$$

Let $u \in L^2(\Omega, K(C))$, then $u(\omega)(t) = \sum_{p=0}^{\infty} I_p(f_p^t(\cdot))$, where f_p^t satisfies (a) of Lemma 3.1, and hence f_p^t is unique. Let $f_p(\cdot, t) = f_p^t(\cdot)$.

If $\sum_{p=1}^{\infty} I_{p+1}(f_p) = \sum_{p=1}^{\infty} I_{p+1}(\tilde{f}_p)$ converges in L^2, then this sum is defined as the Skorokhod integral of u w.r.t. the Gaussian process X and is denoted by $I^s(u)$. With the above notation we have:

LEMMA 3.3. $I^s(u) \in L^2(\Omega)$ iff

$$\sum_{p=1}^{\infty}(p+1)! \left\|\tilde{f}_p\right\|^2_{K^{\odot p+1}} < \infty. \tag{4}$$

and in this case L^2-norm of $I^s(u)$ coincides with the sum on the left side in (4). Furthermore the domain of I^s consists of all u for which (4) holds.

b. Stochastic differentiation. Let $\{X_t, t \in T\}$ be a Gaussian process defined on $(\Omega, \mathcal{F}, P)$ with covariance $C(s,t)$. Assume that its reproducing kernel Hilbert space $K(C)$ is separable. Let us denote by $L^2(\Omega, P)$ and $L^2(\Omega, K(C))$ the spaces of P-square integrable functions and P-square integrable $K(C)$-valued Bochner measurable functions with respect to $\mathcal{F} = \sigma\{X_t,\ t \in T\}$ respectively. By Lemma 3.2 we have for $F = \sum_{m=0}^{\infty} I_m(f_m(\cdot))$,

$$E\left\|\sum_{m=0}^{\infty} m I_{m-1}(f_m(\cdot,t))\right\|_K^2 = \sum_{m=0}^{\infty} m^2 \cdot (m-1)!\, \|f_m\|^2_{K^{\odot m}}$$

$$= \sum_{m=0}^{\infty} m \cdot m!\, \|f_m\|^2_{K^{\odot m}}. \tag{5}$$

Let $D_t F = \sum_{m=1}^{\infty} m I_{m-1}(f_m(\cdot,t))$.

LEMMA 3.4. Let $F = \sum_{m=1}^{\infty} I_m(f_m(\cdot)) \in L^2(\Omega, P)$ with $f_m(\cdot) \in K^{\odot m}$, then $D_t F \in L^2(\Omega, K(C))$ if

$$\sum_{m=1}^{\infty} m \cdot m!\, \|f_m\|^2_{K^{\odot m}} < \infty. \tag{6}$$

and

$$E\,\|D.F\|^2_{K(C)} = \sum_{m=1}^{\infty} m\, m!\, \|f_m\|^2_{K^{\odot m}}.$$

We call $D_t F$ the Skorokhod derivative of F. Also $D.$ can be considered as an operator from $L^2(\Omega, P) \to L^2(\Omega, K(C))$ and the domain of D consists of all

F that satisfy (6). Given $u_t(\omega) \in L^2(\Omega, K)$ which is Skorokhod integrable, let

$$u_t(\omega) = \sum_{p=0}^{\infty} I_p(f_p(\cdot, t)).$$

Let us consider $D_s(u_t(\omega)) = g(s,t)$. Assume that for each fixed t, $D_s(u_t)$ is well defined and $E\|g(s,t)\|^2_{K^{\otimes 2}} < \infty$. Suppose

$$f_p(t_1, \ldots, t_p, t) = \sum_{\alpha_1, \alpha_{p+1}} a(p)_{\alpha_1 \cdots \alpha_{p+1}} e_{\alpha_1}(t_1) \cdots e_{\alpha_p}(t_p) e_{\alpha_{p+1}}(t),$$

then

$$g(s,t) = D_s(u_t) = \sum_{p=0}^{\infty} p I_{p-1}(f_p(\cdot, s, t)),$$

with

$$I_{p-1}(f_p(\cdot, s, t)) = \sum_{\alpha_1 \cdots \alpha_{p+1}} a(p)_{\alpha_1 \cdots \alpha_{p+1}} I_{p-1}[e_{\alpha_1}(t_1) \cdots e_{\alpha_{p-1}}(t_{p-1})] e_{\alpha_p}(s) e_{\alpha_{p+1}}(t)$$

$$= \sum_{\alpha_p, \alpha_{p+1}} \left\{ I_{p-1} \left\{ \sum_{\alpha_1 \cdots \alpha_{p-1}} a(p)_{\alpha_1 \cdots \alpha_{p+1}} e_{\alpha_1}(t_1) \cdots e_{\alpha_{p-1}}(t_{p-1}) \right\} \right\} e_{\alpha_p}(s) e_{\alpha_{p+1}}(t)$$

Since

$$E \sum_{\alpha_p, \alpha_{p+1}} \left\{ I_{p-1} \left\{ \sum_{\alpha_1 \cdots \alpha_{p-1}} a(p)_{\alpha_1 \cdots \alpha_{p+1}} e_{\alpha_1}(t_1) \cdots e_{\alpha_{p-1}}(t_{p-1}) \right\} \right\}^2$$

$$= \sum_{\alpha_p, \alpha_{p+1}} (p-1)! \sum_{\alpha_1 \cdots \alpha_{p-1}} [a(p)_{\alpha_1 \cdots \alpha_{p+1}}]^2$$

$$= (p-1)! \sum_{\alpha_1, \alpha_2 \cdots \alpha_{p+1}} [a(p)_{\alpha_1 \alpha_2 \cdots \alpha_p \alpha_{p+1}}]^2 = (p-1)! \|f_p\|^2_{K^{\otimes p+1}} < \infty$$

we get a.e. P,

$$\sum_{\alpha_p, \alpha_{p+1}} \left\{ I_{p-1} \left\{ \sum_{\alpha_1 \cdots \alpha_{p-1}} a(p)_{\alpha_1 \cdots \alpha_{p+1}} e_{\alpha_1}(t_1) \cdots e_{\alpha_{p-1}}(t_{p-1}) \right\} \right\}^2 < \infty.$$

So $I_{p-1}(f_p(\cdot, s, t)$ for fixed ω is an element in $K^{\otimes 2}$. One can also show that

$$E(I_{p-1}(f_p(\cdot, s, t), \, I_{q-1}(f_q(\cdot, s, t))_{K^{\otimes 2}} = \begin{cases} 0 & \text{if } p \neq q \\ (p-1)!(f_p, f_p)_{K^{\otimes p+1}} & \text{if } p = q \end{cases}$$

This implies

$$E\|g(s,t)\|^2_{K^{\otimes 2}} = E\left\{ \sum_{p=0}^{\infty} p I_{p-1}(f_p(\cdot, s, t) \right\}^2 = \sum_{p=0}^{\infty} p^2 E\|I_{p-1}(f_p(\cdot, s, t)\|^2_{K^{\otimes 2}}$$

$$= \sum_{p=0}^{\infty} p^2 (p-1)! \|f_p\|^2_{K^{\otimes p+1}} = \sum_{p=0}^{\infty} p p! \|f_p\|^2_{K^{\otimes p+1}}.$$

Now let us compute

$$E(g(s,t), \, g(t,s))_{K^{\otimes 2}}$$

$$= E\left(\sum_{p=0}^{\infty} pI_{p-1}(f_p(\cdot,s,t)), \sum_{p=0}^{\infty} pI_{p-1}(f_p(\cdot,t,s))\right)_{K^{\otimes 2}}$$

$$= \sum_{p=0}^{\infty} p^2 E\left(I_{p-1}(f_p(\cdot,s,t)),\ I_{p-1}(f_p(\cdot,t,s))\right)_{K^{\otimes 2}}.$$

But

$$E\left(I_{p-1}(f_p(\cdot,s,t)),\ I_{p-1}(f_p(\cdot,t,s))\right)_{K^{\otimes 2}}$$

$$= \sum_{\alpha_p=\alpha_{p+1}} (p-1)! \sum_{\alpha_1,\alpha_2\cdots\alpha_{p-1}} [a(p)_{\alpha_1\alpha_2\cdots\alpha_p\alpha_{p+1}}]^2$$

$$= (p-1)! \sum_{\alpha_p=\alpha_{p+1}} \sum_{\alpha_1,\alpha_2\cdots\alpha_{p-1}} [a(p)_{\alpha_1\alpha_2\cdots\alpha_p\alpha_{p+1}}]^2,$$

giving

$$E(g(s,t),\ g(t,s))_{K^{\otimes 2}} = \sum_{p=0}^{\infty} p^2(p-1)! \sum_{\alpha_p=\alpha_{p+1}} \sum_{\alpha_1\cdots\alpha_{p-1}} [a(p)_{\alpha_1\cdots\alpha_p\alpha_{p+1}}]^2.$$

Note that, $\left\|\tilde{f}_p\right\|^2_{K^{\otimes p+1}} = \left\|\frac{1}{p+1}\sum_{\pi} f_p(\pi\underline{t})\right\|^2_{K^{\otimes p+1}}$, where $\underline{t} = (t_1,\ldots,t_{p+1})$, π are all permutations which only permute t_{p+1} with one of the t_i's.Hence

$$\begin{aligned}\left\|\tilde{f}_p\right\|^2_{K^{\otimes p+1}} &= \frac{1}{(p+1)^2}\sum_{\pi,\pi'} (f_p(\pi\underline{t}),\ f_p(\pi'\underline{t}))_{K^{\otimes p+1}}\\ &= \frac{1}{(p+1)^2}\sum_{\pi} (f_p(\pi\underline{t}),\ f_p(\pi\underline{t}))_{K^{\otimes p+1}}\\ &+ \frac{1}{(p+1)^2}\sum_{\pi\neq\pi'} (f_p(\pi\underline{t}),\ f_p(\pi'\underline{t}))_{K^{\otimes p+1}}.\end{aligned}$$

The first sum has $p+1$ terms which are the same as $\|f_p\|^2_{K^{\otimes p+1}}$, second sum has $(p+1)p$ terms, which are of the form

$$\sum_{\alpha_i=\alpha_{p+1}} \sum_{\alpha_1,\cdots\alpha_{i-1},\alpha_{i+1}\cdots\alpha_p} [a(p)_{\alpha_1\alpha_2\cdots\alpha_p\alpha_{p+1}}]^2,\ (i < p+1).$$

But all these terms are the same because f_p is symmetric in the first p variables, hence the second sum is

$$(p+1)p \sum_{\alpha_p=\alpha_{p+1}} \sum_{\alpha_1,\alpha_2\cdots\alpha_{p-1}} [a(p)_{\alpha_1\alpha_2\cdots\alpha_p\alpha_{p+1}}]^2.$$

Therefore,

$$\left\|\tilde{f}_p\right\|^2_{K^{\otimes p+1}} = \frac{1}{p+1}\|f_p\|^2_{K^{\otimes p+1}} + \frac{p}{(p+1)} \sum_{\alpha_p=\alpha_{p+1}} \sum_{\alpha_1\cdots\alpha_{p-1}} [a(p)_{\alpha_1\cdots\alpha_{p+1}}]^2.$$

Also, $E[I^s(u)]^2 = \sum_{p=0}^{\infty} (p+1)! \left\|\tilde{f}_p\right\|^2_{K^{\otimes p+1}}$

$$= \sum_{p=0}^{\infty} (p+1)! \left\{\frac{1}{p+1}\|f_p\|^2_{K^{\otimes p+1}} + \frac{p}{(p+1)} \sum_{\alpha_p=\alpha_{p+1}} \sum_{\alpha_1\cdots\alpha_{p-1}} [a(p)_{\alpha_1\cdots\alpha_{p+1}}]^2\right\}$$

$=\sum_{p=0}^{\infty} p! \|f_p\|^2_{K^{\otimes p+1}} + \sum_{p=0}^{\infty} pp! \sum_{\alpha_p=\alpha_{p+1}\alpha_1,\alpha_2\cdots\alpha_{p-1}} \sum [a(p)_{\alpha_1\alpha_2\cdots\alpha_p\alpha_{p+1}}]^2$
$= E\|u.(\omega)\|^2_K + E(g(s,t),\ g(t,s))_{K^{\otimes 2}}$
$= E\|u.(\omega)\|^2_K + E(D_s(u_t),\ D_t(u_s))_{K^{\otimes 2}}.$
We summarize the above work in the following theorem.

THEOREM 3.1. Let $u_t(\omega) \in L^2(\Omega, K)$ be Skorokhod integrable. Assume that for each fixed t, $D_s(u_t)$ is well defined and $E\|D_s u_t\|^2_{K^{\otimes 2}} < \infty$. Then

$$E[I^s(u)]^2 = E\|u.(\omega)\|^2_K + E(D_s(u_t),\ D_t(u_s))_{K^{\otimes 2}}.$$

c. Examples.

1. *Brownian Motion.*

Let us consider $C(t,s) = \min(t,s)$, $0 \le s,t \le 1$, the covariance of Brownian Motion W. Let us define $V : L^2[0,1] \to K = K(C)$ given by $V(f)=\int_0^{\cdot} f(s)ds$ as $K = \left\{\int_0^{\cdot} f(s)ds,\ f \in L^2[0,1]\right\}$([1]). We note that as in [8], we can define $\prod[V(f)] = \int_0^1 f(s)dW(s)$, where the integral is in the sense of Doob. The multiple integral $I_p(g_p)$ can now be defined for $g_p \in K^{\odot p}$ as in section 2, so that $I_p(e_1^{p_1} e_2^{p_2} \cdots e_n^{p_n}) = \prod_{i=1}^{n} H_{p_i}(\prod(e_i))$, where $e_1^{p_1} e_2^{p_2} \cdots e_n^{p_n} = e_1(t_1)\cdots e_1(t_{p_1})e_2(t_{p_1+1})\cdots e_2(t_{p_1+p_2})\cdots e_n(t_{p_1+\cdots+p_{n-1}}+1)\cdots e_n(t_{p1+\cdots+p_n})$ $(p_1+\cdots+p_n = p)$, $e_1, e_2, \cdots e_n$ are orthonormal in K, and if $e_i(\cdot) = \int_0^{\cdot} f_i(s)ds$, then f_i's are orthonormal in $L^2[0,1]$. Hence if we define $I_p^i(f_1^{p_1} f_2^{p_2} \cdots f_n^{p_n}) = \prod_{i=1}^{n} H_{p_i}(\prod(V(f_i)))$, then by [5] we get that I_p^i can be uniquely extended to $L^2([0,1])^{\odot p} = L^2([0,1]^p)^\wedge$, the symmetric functions in $L^2([0,1]^p)$. It is easy to check that for elementary functions the definition coincides with the definition in [5]. Since $I_p(K^{\odot p}) = I_p^i[L^2([0,1]^p)^\wedge]$, we have $L^2(\Omega, \sigma(W_t,\ t \in [0,1]),\ P) = \sum_{p=0}^{\infty} \oplus I_p^i[L^2([0,1]^p)^\wedge]$. Thus the definition of the Skorokhod integral given by us coincides with the one given in [9]. As the measure $P \circ W^{-1}$ sits on $C[0,1]$ with support $\overline{K(C)}$ (closure in $\|\ \|_\infty$ norm on $C[0,1]$), our D operator coincides with that of [9]. The results on Skorokhod integral and relation between the integral and Malliavin derivative in [10] can be easily derived in a similar manner by noting that for a Gaussian random measure W with covariance $C(A,A') = \mu(A \cap A')(A, A' \in \mathcal{B}_0(\mu) = \{B,\ \mu(B) < \infty\})$ gives $K(C) = \{\int_A f d\mu,\ f \in L^2\}$. Here $V(f) = \int f d\mu$.

2. *Random Linear Functional on a Hilbert Space H ([12]).*

Let us consider a random linear functional ξ on a separable Hilbert space H into a Gaussian space ([1]) continuous in the norm of H such that $< \xi(h),\ \xi(h') >= (h, h')_H$. We consider the Gaussian process $\{\xi(h),\ h \in H\}$ with covariance $C(h, h') = (h, h')_H$. In view of Riesz Representation Theorem, we can identify elements of $K(C) = K$ with H. Hence our definition of Skorokhod Integral is for an H-valued random variable η with $E\,\|\eta\|_H^2 < \infty$. This is precisely the class considered by Skorokhod ([12]). To compare our definition to that of Skorokhod, $K^{\otimes p} = K(C)^{\otimes p}$ has reproducing kernel $C(x_1, y_1)\cdots C(x_p, y_p) = (x_1, y_1)_H \cdots (x_p, y_p)_H$ and hence $f \in K(C)^{\otimes p}$ implies:
$f(x_1, \ldots x_p) = (f,\ C(x_1, \cdot)\ldots C(x_p, \cdot))_{K^{\otimes p}}$. This shows that elements of $f \in K^{\odot p}$ are symmetric multilinear forms and

$$f(x_1, \ldots x_p) = \sum_{\alpha_1, \cdots \alpha_p} f(e_{\alpha_1}, \ldots, e_{\alpha_p}) e_{\alpha_1}(x_1) \ldots e_{\alpha_p}(x_p)$$

with $\{e_\alpha, \alpha \in J\}$ an orthonormal basis in $K(C) = H$. Now $f \in K^{\odot p}$ implies that $f(e_{\alpha_1}, \ldots e_{\alpha_p})$ is symmetric in $(\alpha_1, \ldots, \alpha_p)$ and $\sum\limits_{\alpha_1, \ldots \alpha_p} |f(e_{\alpha_1}, \ldots, e_{\alpha_p})|^2 < \infty$. Thus I_p is defined on the class of p-linear operators A on H^p with $\operatorname{tr}(A * A)^p < \infty$ in the sense of [12]. As remarked earlier $I_p(e_1^{p_1} e_2^{p_2} \ldots e_n^{p_n}) = \prod\limits_{i=1}^{n} H_{p_i}((\xi, e_i))$ and since $I_1(x) = (\xi, x)$ we get that the definition of I_p in [12] coincides with I_p above. Hence our work gives results in [12].

3. *Multidimensional Brownian Motion [10].*

Let us consider the Gaussian process $\{W_i(t)\}$ on $\{1, 2, \ldots, d\} \times [0, 1]$ given by covariance $C((i, t), (j, s)) = \delta_{ij} \min(t, s)$. Then, clearly,
$K(C) = \{g_i(t) = \int\limits_0^t f_i(s) ds\ :\ f_i \in L^2[0, 1],\ i \in \{1, 2, \ldots, d\}\}$. We can identify $K(C)$ with $K(C') \otimes R^d$ where $C'(t, s) = \min(t, s)$, with the above notation. Hence we get that there exists an isometry map
$V : L^2([0, 1]), R^d) \to K(C)$. Now we note that, as in [8], we can see that $\prod[V(f)] = \sum\limits_{i=1}^{d} \int\limits_0^1 f_i(s) dW_i(s)$. We can adopt the arguments in Example 1 to show that the corresponding multiple integrals coincide with those defined in [10] and hence Skorokhod integral is the same. The details are left to the reader.

One can also consider cylindrical Brownian motion $W(t, h)$ with covariance on $H \times [0, 1]$ given by $C((t, h), (t', h')) = \min(t, t')(h, h')_H$. In this case $K(C) = H \otimes K(C')$ and one can again consider
$I_1(f) = \sum\limits_{i=1}^{\infty} (h, e_i) \int\limits_0^1 (f(s), e_i) dW(e_i, s)$. The details of this case will be

given elsewhere.

4. *Hitsuda Integral with respect to White Noise [4].*
Let us consider $T = \mathcal{S}(R)$ and $C(\varphi, \phi) = \int \varphi\phi dt$. Then $K \equiv K(C) = \{F : F \in \mathcal{S}^*(R),\ F(\varphi) = \int f\varphi dt,\ f \in L^2(R)\}$ with $(F,G)_K = (f,g)_{L^2(R)}$. Thus we can identify through an isometry $V : L^2(R) \to K(C)$ given by $V(f) = F$. Furthermore, $I_1^i(f) = \prod \circ V(f)$ gives as in Example 1 a map on $L^2(R) \to \overline{sp}\{X(\varphi) : (\varphi) \in \mathcal{S}(R)\}$ with X being a white noise process. Now just as in Example 1 we can define I_p^i on $L^2(R^p)^\wedge$ (symmetric functions) and get a representation $L^2(\Omega, \sigma(X(\varphi), \varphi \in \mathcal{S}(R)), P) = \sum_{p=0}^{\infty} \oplus I_p^i[L^2(R^p)^\wedge] = \sum_{p=0}^{\infty} \oplus I_p(K^{\odot p})$. Suppose $\int_R E(u_t^2)dt < \infty$, then with $F(\varphi) = \int_R u_t\varphi dt$, F is K-valued and $E\,\|F\|_K^2 < \infty$. We recall that $I^s(F) = \sum_{p=0}^{\infty} I_{p+1}(\tilde{f}_p)$. We now need the following general lemma:

LEMMA 3.5. Let $u \in L^2(\Omega, K)$, then u is Skorokhod integrable if there exists $M > 0$ such that

$$|E(u., D.F)_K| \leq M\,\|F\|_{L^2}$$

for any F which has Skorokhod derivative. In this case

$$E((u., D.F)_K = E(I^s(u)F).$$

PROOF: In view of Lemma 2.3 and linearity, it suffices to prove the result for $F = I_p(g)$ with $g \in K^{\odot p}$, $u_t = \sum_{p=0}^{\infty} I_p(f_p(\cdot, t))$ with $f_p \in K^{\otimes p+1}$ and for fixed t, $f_p(\cdot, t) \in K^{\odot p}$. Note $D_tF = pI_{p-1}(g(\cdot, t))$. Now assume $\{e_\alpha,\ \alpha \in J\}$ is orthonormal basis of K. Denote by

$$(I_q(f_q(\cdot, *)),\ pI_{p-1}(g(\cdot, *)))_K = \sum_\alpha Z_{q,p}^\alpha$$

where

$$Z_{q,p}^\alpha = p(I_q(f_q(\cdot, *), e_\alpha(*))_K(I_{p-1}(g(\cdot, *), e_\alpha(*))_K,$$

$$\begin{aligned} E((u., D.F)_K = \sum_{\alpha,q} E(Z_{q,p}^\alpha) &= p(p-1)!(f_{p-1}, g_{p-1})_{K^{\otimes p}} \\ &= p!(\tilde{f}_{p-1}, g_{p-1})_{K^{\odot p}} \end{aligned}$$

the last equality holds because g_{p-1} is symmetric. Hence for $F \in L^2(\Omega)$, $E((u., D.F)_K = E(I^s(u)F)$ if u is Skorokhod integrable. Conversely by the inequality for $u \in L^2(\Omega, K)$ we get using for any $n \in N$, $\left| E\left(\sum_{q=1}^{n} I_{q-1}(\tilde{f}_q)F\right) \right| \leq$

$M\|F\|_{L^2}$ giving for any n,
$\left|E\left(\sum_{q=1}^{n} I_{q-1}(\tilde{f}_q)\right)\right|^2 \leq M$. Hence $\sum_{q=1}^{\infty} q!\left\|\tilde{f}_{q-1}\right\|^2_{K^{\odot q}} < \infty$
completing the proof. □

The above lemma shows that Skorokhod integral is the dual of the differential operator $D. : L^2(\Omega) \to L^2(\Omega, K)$ giving the result of [2] and identifying the Skorokhod integral given here with the one given by Hitsuda [4] using $C(\varphi, \phi) = \int \varphi\phi dt$.

REMARK. Let $\{X_t, t \in T\}$ be a Gaussian process defined on $(\Omega, \mathcal{F}, P)$ such that $E(X_t X_s) = C(s,t)$. Let us now consider the functional representation of $\{X_t, t \in T\}$, i.e. $\Omega = R^T$, $\mathcal{F} = \sigma$-field generated by cylinder sets and P the Kolmogorov measure corresponding to $\{X_t, t \in T\}$ and $X_t(\omega) = \omega(t), t \in T$. Clearly $K(C) \subseteq R^T$. Let us assume that P is Radon. Denote by $\mathcal{X}$ the support of P in R^T. Then $K(C)$ is separable. Since R^T is a locally convex topological space, we get that $\mathcal{X}$ is a locally convex topological vector space under induced topology from R^T and $\mathcal{X} = \overline{K(C)}$, where closure is in the induced (locally convex) topology of R^T. So $\mathcal{X}$ is a locally convex topological space. Let $\{e_i\}_{i=1}^{\infty} \subseteq \mathcal{X}^*$ be dense in $K(C)$, then $e_i(t) = E\{X_t e_i(x)\}$ by Karhünen representation. Then $I_1(f)(x) = \sum_{i \in A} < f, e_i > e_i(x)$ a.e. P and we get L^2-Frechet derivative of $I_1(f)$ is $f(t)$. One can show a similar result for Hermite polynomials also essentially following [2], we can prove that Skorokhod derivative coincides with Malliavin derivative. This structure will be needed to study relation between Ramer integral and Skorokhod integral. This will be done in our later work.

REFERENCES

[1] Chatterji, S.D. and Mandrekar, V., Equivalence and singularity of gaussian measure and applications in *Probabilistic Analysis and Related Topics* (A.T. Barucha-Reid ed.) Vol. 1, Academic Press, New York, 1978, 169-197.

[2] Gaveau, B. and Tauber, P. (1982). *L'*integral stochastique comme opèrateur de divergence dans l'espace fonctional. *J. Functional Analysis* **46** 230-238.

[3] Gross, L. (1967). Potential theory on Hilbert spaces. *J. Functional Analysis* **1** 123-181.

[4] Hitsuda, M. (1972). Formula for Brownian partial derivatives. *Proc. 2nd Japan-USSR Symp. in Prob. Theory* (Kyoto) **2** 111-114.

[5] Ito, K. (1951). Multiple Wiener integral. *J. Math. Soc. Japan.* **3** 157-169.

[6] Kakutani, S. (1961). Spectral analysis of stationary Gaussian processes. *Proc. 4th Berkeley Symp.* **2** 239-247.

[7] Kuo, H. (1975). Gaussian measures in Banach spaces. *Lect. Notes in Math.* **463** Springer-Verlag, New York.

[8] Mandrekar, V. (1984) Stochastic integral with respect to Gaussian processes. *Lect. Notes in Math.* **1089** Springer-Verlag, New York, 288-293.

[9] Nualart, D. (1988). Noncausal stochastic integrals and calculus. *Lect. Notes in Math.* **1316** Springer-Verlag, New York.

[10] Nualart, D. and Pardoux, E. (1988). Stochastic calculus with anticipating integrands. *Prob. Th. and Rel. Fields.* **78** 535-581.

[11] Nualart, D. and Zakai, M. (1986). Generalized stochastic integrals and Mallianvin calculus. *Prob. Th. and Rel. Fields* **73** 255-280.

[12] Skorokhod, A. V. (1975). On a generalization of stochastic integral. *Theory of Prob. and its Applications.* **20** 219-233.

Department of Statistics and Probability
Michigan State University
East Lansing, MI 48824
U.S.A.

Department of Mathematics
Gannon University
Erie, PA 16541
U.S.A.

J.K. Ghosh, S.K. Mitra, K.R. Parthasarathy and B.L.S. Prakasa Rao (*eds.*),
Statistics and Probability : A Raghu Raj Bahadur Festschrift
Published by Wiley Eastern Limited, Publishers, (1993), 411 – 444

Nonnegative Estimation of Individual Variance Components in Some Mixed Linear Models: A Survey[1]

THOMAS MATHEW[1] and BIMAL KUMAR SINHA
University of Maryland Baltimore County, Baltimore

Abstract

The object of this paper is to provide a survey of a variety of procedures for nonnegative estimation of individual variance components in a general mixed linear model, with special emphasis on improving over the ANOVA estimators in models with two variance components. Due to the well known fact that the standard unbiased estimators of these variance components, obtained via the ANOVA table, are not guaranteed to be nonnegative and are inadmissible, a host of techniques to improve over them has emerged during the last thirty years. This paper summarizes most of the available results for balanced mixed models, provides some new results for unbalanced mixed models, and ends with a list of a few open problems in this area.

1 Introduction

In the huge literature on variance components spanning over the last three decades, a problem that has received considerable attention is that of nonnegative estimation of variance components. This is primarily due to the

[1]Research sponsored by Grant *AFOSR* 89-0237.
AMS Subject Classifications : Primary 62J10; secondary 62F10.
Key words and Phrases: Admissibility, Balanced models, Invariant quadratic estimator (IQE), Maximum likelihood estimator, Minimum norm quadratic unbiased estimator (MINQUE), Restricted maximum likelihood estimator, One-way classification, Testimator, Unbalanced models.

well known fact that, in the context of ANOVA models, the ANOVA estimator of a random effect variance component, which is the standard unbiased estimator obtained via the ANOVA table, can assume negative values, thus making it unacceptable in practice. The object of this paper is to provide a comprehensive survey of a variety of procedures which emerged to take care of the negativity of the ANOVA estimators. Even for the estimation of the error variance component, for which the usual ANOVA estimator is nonnegative, techniques to devise nonnegative estimators with smaller mean squared errors (MSEs) will be surveyed. General balanced mixed linear models which admit a very simple canonical form and for which explicit results are available will be considered first in Sections 2, 3 and 4. Unbalanced mixed linear models are addressed in Section 5. Some open problems are listed in Section 6.

In the context of estimation of variance components, it is worth recalling a general result of LaMotte (1973) that, for estimating a random effect variance component, nonnegative quadratic unbiased estimators do not exist. (However, see Kleffe and Rao (1986), where the existence of asymptotically unbiased nonnegative quadratic estimators of variance components is investigated). It is then clear that any modification of the ANOVA estimator of a random effect variance component to yield a nonnegative estimator must give up either unbiasedness or quadratic nature or both. In fact, all the nonnegative estimators proposed so far and reviewed below, are biased and mostly nonquadratic.

Since the ANOVA estimator of a variance component is its uniformly minimum variance unbiased estimator (UMVUE) in the balanced case, it is natural and desirable that in the balanced case, the proposed estimators are not merely nonnegative, but offer some improvement in terms of MSE over the ANOVA estimators. It is well known that any unbiased quadratic estimator of a variance component, in particular, the ANOVA estimator, can be uniformly improved by multiplication with a positive scalar (see LaMotte (1979) and Kleffe (1984) for a proof of this and Perlman (1972) for a more general result). However, the improved estimator so obtained can obviously be negative. Motivated by this, Mathew, Sinha and Sutradhar (1992a) came up with nonnegative quadratic estimators, and Kelly and Mathew (1992) and Kubokawa (1991) with nonnegative nonquadratic estimators which offer substantial MSE improvement compared to the ANOVA estimators. However, in the unbalanced case, UMVUE's do not exist for estimating the variance components. Hence, even though the ANOVA estimator is a standard estimator, a nonnegative estimator that provides MSE improvement only over the ANOVA estimator may not be satisfactory. In view of this observation, Mathew, Sinha and Sutradhar (1992c) investigated the existence of a nonnegative quadratic estimator that is uniformly better than every quadratic unbiased estimator for estimating the random effect variance component in unbalanced models that involve only two variance components. Throughout this paper, we have used the MSE criterion to compare rival estimators of variance components. While this may not be a direct measure of closeness of an estimator to its true value, an estimator of a variance component with a small MSE is likely to have a high probability of concentration around the true value, as noted in Conerly and Webster (1987, Section 2). We hope this justifies our exclusive use of the MSE to some extent.

We shall now briefly discuss a general mixed linear model, both balanced and unbalanced, and indicate its canonical form in the balanced case. Such a model is given by

$$\begin{aligned}\mathbf{y} &= X_1\alpha_1 + \cdots + X_k\alpha_k + Z_1\mathbf{u}_1 + \cdots + Z_{c-1}\mathbf{u}_{c-1} + \mathbf{u}_c \\ &= X\alpha + Z_1\mathbf{u}_1 + \cdots + Z_{c-1}\mathbf{u}_{c-1} + \mathbf{u}_c, \qquad (1.1)\end{aligned}$$

where $\mathbf{y}$ is the $N \times 1$ vector of observations, $X_1 = \mathbf{1}_N$ (the $N \times 1$ vector of ones), X_i's and Z_j's are fixed design matrices of appropriate dimensions, $X = (X_1, ..., X_k)$, α_1 is a scalar representing the general mean, each α_i is an $r_i \times 1$ vector of fixed effects (i = 2, 3, ..., k), $\alpha = (\alpha_1', \alpha_2', ..., \alpha_k')'$, each $\mathbf{u}_j$ is an $s_j \times 1$ vector of random effects (j = 1, 2, ..., c-1) of the various factors (crossed or nested) and their interactions, and $\mathbf{u}_c$ is the vector of experimental errors. We also assume that the $\mathbf{u}_j$'s are independent random variables with $\mathbf{u}_j \sim N(\mathbf{0}, \sigma_j^2 I_{s_j})$ for j = 1, 2, ..., c-1, and $\mathbf{u}_c \sim N(0, \sigma^2 I_N)$. The parameters $\sigma_j^2 \geq 0$ for j = 1, 2, ..., c-1, and $\sigma^2 > 0$ are known as the variance components. The problem we shall address is, without loss of generality, the nonnegative estimation of the random effect variance component σ_1^2, and estimation of the error variance σ^2.

We first discuss a canonical form of the model (1.1) in the balanced case. Note that in the balanced case, each X_i and each Z_j is a Kronecker product of identity matrices and the vectors $\mathbf{1}$ of appropriate dimensions. Let U_i (i = 1, 2, ..., k) and V_j (j = 1, 2, .., c) respectively denote the sum of squares due to α_i and $\mathbf{u}_j$ (as in the fixed effects model). Note that V_c is the error sum of squares. Furthermore, the model being balanced $\frac{1}{\beta_i}U_i$ has a noncentral chi-squared distribution with noncentrality parameter say, λ_i (where $\lambda_i = \frac{\gamma_i'\gamma_i}{\beta_i}$, γ_i being a linear combination of the α_i's) and degrees of freedom say, p_i, and $\frac{1}{\delta_j}V_j$ has a central chi-squared distribution with degrees of freedom say, q_j, where β_i and δ_j are nonnegative linear combinations of σ_i^2's and σ^2. As is well known, the ANOVA estimator of any variance component is a linear combination of the V_j's and is also its uniformly minimum variance unbiased estimator. (For further details regarding the above, we refer to Mathew and Sinha (1988)). It is clear that the ANOVA estimator of σ_1^2, say $\hat{\sigma}_{1A}^2$ is given by

$$\hat{\sigma}_{1A}^2 = \frac{1}{\lambda_1}\left\{\frac{1}{q_1}V_1 + \sum_{j=2}^{c}\frac{\epsilon_j}{q_j}V_j\right\}, \qquad (1.2)$$

where each ϵ_j is +1 or -1 for j = 2, ..., c. Moreover, the ANOVA estimator of σ^2, say $\hat{\sigma}_A^2$, is given by

$$\hat{\sigma}_A^2 = \frac{V_c}{q_c}. \tag{1.3}$$

The analysis in the balanced case is simplified by the fact that the V_j's form a set of complete sufficient statistics for the variance components. Incidentally, estimators of σ_i^2's based on just the V_j's are known as invariant estimators. However, such an explicit representation of the complete sufficient statistic is not possible in the unbalanced case, except when $c = 2$ in (1.1), *i.e*, we have exactly one random effect, namely, $\mathbf{u}_1$. Because of the above difficulty, the nonnegative estimation of σ_1^2 and σ^2 in the unbalanced case will be addressed in this paper only for the special case $c = 2$. Thus the model we shall consider in the unbalanced case is

$$\mathbf{y} = X\alpha + Z_1\mathbf{u}_1 + \mathbf{u}_2, \tag{1.4}$$

where $\mathbf{u}_2$ is the vector of experimental errors. Thus $\mathrm{E}(\mathbf{y}) = X\alpha$ and $\mathrm{Cov}(\mathbf{y}) = \sigma_1^2 Z_1 Z_1' + \sigma^2 I_N$. Let $r = \mathrm{rank}(X)$, $t = N - r$, and let H be an $N \times t$ matrix satisfying $H'X = 0$ and $H'H = I_t$. Define $\mathbf{u} = H'\mathbf{y}$. Then

$$\mathrm{E}(\mathbf{u}) = 0 \text{ and } \mathrm{Cov}(\mathbf{u}) = \sigma_1^2 W + \sigma^2 I_t, \tag{1.5}$$

where $W = H'Z_1Z_1'H$. The estimators of σ_1^2 and σ^2 that we shall consider will be functions of $\mathbf{u}$, since only such estimators are invariant under the transformation $\mathbf{y} \longrightarrow \mathbf{y} + X\theta$, where θ is an arbitrary vector. This invariance property is certainly a natural requirement. Let $s = \mathrm{rank}(W)$ and let $\lambda_1 > \lambda_2 > \cdots > \lambda_g > 0$ be the g distinct non-zero eigenvalues of W with respective multiplicities $m_1, m_2, \ldots, m_g$. Clearly $\sum_{i=1}^g m_i = s$. We shall assume that $t - s > 0$. Consider the spectral decomposition

$$W = \sum_{i=1}^{g} \lambda_i E_i \tag{1.6}$$

where E_i is an idempotent matrix of rank m_i and $E_iE_j = 0$ ($i \neq j$; $i, j = 1, 2, \ldots, g$). Write $E_{g+1} = I - \sum_{i=1}^g E_i$ and $m_{g+1} = \mathrm{rank}(E_{g+1}) = t - s$. Then it is known that the quadratic forms $\mathbf{u}'E_i\mathbf{u}$ ($i = 1, 2, \ldots, g+1$) form a set of minimal sufficient statistic for the normal family of distributions of $\mathbf{u}$ in (1.5). The balanced case considered in Section 2 corresponds to $g = 1$, in which case the minimal sufficient statistic $(\mathbf{u}'E_1\mathbf{u}, \mathbf{u}'E_2\mathbf{u})$ is also complete. When $g > 1$, due to the lack of completeness of the minimal sufficient statistic, unbiased estimators of σ_1^2 and σ^2 are not unique. In particular, the ANOVA estimator $\hat{\sigma}_{1A}^2$ of σ_1^2 ($\hat{\sigma}_A^2$ of σ^2) and the minimum norm quadratic unbiased estimator (MINQUE; see Rao and Kleffe, 1988) $\hat{\sigma}_{1M}^2$ of σ_1^2 ($\hat{\sigma}_M^2$ of σ^2) are distinct. These are given by

$$\hat{\sigma}_{1A}^2 = \frac{1}{\mathrm{tr}W}\left\{\sum_{i=1}^{g} \mathbf{u}'E_i\mathbf{u} - \frac{s}{t-s}\mathbf{u}'E_{g+1}\mathbf{u}\right\}, \tag{1.7}$$

$$\hat{\sigma}_A^2 = \frac{1}{t-s}\mathbf{u}'E_{g+1}\mathbf{u}, \tag{1.8}$$

$$\hat{\sigma}_{1M}^2 = \{t\ \mathrm{tr}W^2 - (\mathrm{tr}W)^2\}^{-1}\{t\ \mathbf{u}'W\mathbf{u} - (\mathrm{tr}W)\mathbf{u}'\mathbf{u}\}, \tag{1.9}$$

and

$$\hat{\sigma}_M^2 = \{t\ \mathrm{tr}W^2 - (\mathrm{tr}W)^2\}^{-1}\{(\mathrm{tr}W^2)\mathbf{u}'\mathbf{u} - (\mathrm{tr}W)\mathbf{u}'W\mathbf{u}\}. \tag{1.10}$$

It may be noted that a major source of the literature emerged due to the fact that $\hat{\sigma}_{1A}^2$ defined in (1.2) for the balanced model, and in (1.7) for the unbalanced model, can assume negative values. Various authors have suggested nonnegative estimators obtained by sacrificing unbiasedness and suitably modifying $\hat{\sigma}_{1A}^2$ (see Rao and Chaubey 1978, Chaubey 1983, Conerly and Webster 1987). It may be noted that these estimators provide mean squared error (MSE) improvement over $\hat{\sigma}_{1A}^2$ only in a certain region of the parameter space but not uniformly. However, in the absence of any a priori information about the variance components, it is natural to ask for uniform, substantial MSE improvement over $\hat{\sigma}_{1A}^2$ while still satisfying the nonnegativity requirement. Bayes estimators having good MSE properties have been recommended by Portnoy (1971). However, it is not clear if such estimators, which are nonquadratic in nature, offer uniform MSE improvement compared to the ANOVA estimator. Obviously, as noted in Herbach (1959) and Thompson (1962), a simple truncation of $\hat{\sigma}_{1A}^2$ at zero yields uniform improvement over $\hat{\sigma}_{1A}^2$. However, the amount of MSE improvement of the resultant nonquadratic estimator is marginal. A natural question then arises if one can obtain nonnegative *quadratic* or *nonquadratic* estimators of σ_1^2 with uniform, substantial, MSE improvement over $\hat{\sigma}_{1A}^2$. This is precisely what Mathew, Sinha and Sutradhar (1992a), Kelly and Mathew (1992), and Kubokawa (1991) recently accomplished.

In Section 2, due to several reasons, we have first considered a special case of general balanced models involving only two variance components, namely σ_1^2 and σ^2. This special case admits a simple canonical representation and the complexity of the underlying problem is well understood in this situation. Moreover, most of the modifications of $\hat{\sigma}_{1A}^2$ to generate nonnegative estimators of σ_1^2 are centered around this case. It may be noted that all the modified estimators are invariant in the sense that they depend only on V_1 and V_2. In this section, besides presenting all the available earlier results in this case, we have reviewed necessary and sufficient conditions due to Mathew, Sinha and Sutradhar (1992a) for the existence of a nonnegative invariant quadratic estimator (IQE) of σ_1^2 uniformly better than the ANOVA estimator $\hat{\sigma}_{1A}^2$. It is interesting to note that LaMotte (1985) indicated the possibility that nonnegative IQE's uniformly better that $\hat{\sigma}_{1A}^2$ may exist for the balanced one-way random model. Our analysis in this case shows, however, that such estimators exist only when the number of treatments is 9 or less. We have also presented some nonnegative nonquadratic estimators of σ_1^2 due to Kelly and Mathew (1992) and Kubokawa (1991) which are uniformly better than the ANOVA estimator.

In Section 3, the results for the case of two variance components are generalized to the general balanced model (1.1). It is pointed out that in many situations, this problem reduces to the same problem in a model involving only one random effect (i.e., two variance components) and hence the earlier results can be applied to a general balanced model as well, although such a reduction may not always be possible for every variance component. Nevertheless, it is shown that it is quite possible to derive the conditions for the existence of a nonnegative IQE uniformly better than the ANOVA estimator of the variance component under consideration in any balanced mixed model.

In Section 4, we address the problem of improved estimation of the error variance component σ^2 in a general balanced mixed model. We first show that the best multiple of the ANOVA estimator $\hat{\sigma}_A^2$ of σ^2 can be uniformly improved using a nonsmooth estimator, based on the ideas in Stein (1964). Our result is a generalization of a corresponding result due to Klotz, Milton and Zacks (1969), proved for the balanced one-way model. Our second result in Section 4 provides a class of smooth estimators uniformly better than $\hat{\sigma}_A^2$, following the ideas in Strawderman (1974).

Section 5 deals with the unbalanced set up of the model in Section 2, *i.e.*, an unbalanced model involving only two variance components, displayed in (1.4) and (1.5). Sufficient conditions are derived under which there will exist a nonnegative IQE uniformly better than *every* unbiased IQE for estimating the random effect variance component σ_1^2. For the one-way unbalanced mixed model, sufficient conditions are also obtained for a multiple of the usual treatment sum of squares to uniformly dominate the ANOVA estimator of σ_1^2. For estimating the error variance component σ^2, it is shown that the best multiple of the ANOVA estimator can be uniformly improved using nonquadratic estimators similar to those in Section 4. We would like to point out that it has not been possible to derive similar results for unbalanced models involving more than two variance components.

The problems addressed in this paper deal exclusively with the estimation of the individual variance components. Our results on the improved nonnegative estimation of the individual variance components are clearly of practical interest in applications where point estimates of the individual variance components are desired. A more general problem is that of estimating a nonnegative linear combination of the variance components. Unlike the individual random effects variance components, a nonnegative linear combination of the variance components will admit an unbiased nonnegative estimator under appropriate conditions (see Pukelsheim (1981a,b), Mathew (1984)). For estimating a nonnegative linear combination of variance components in models involving only two variance components, namely the models considered in Section 5, a class of admissible nonnegative IQE's has been obtained by Gnot, Kleffe and Zmyslony (1985). A review of these results is given in Chapter 6 of Rao and Kleffe (1988) and they will not be presented in this paper.

2 Nonnegative Estimation of σ_1^2 in Balanced Mixed Models With Two Variance Components

In this section, we consider the balanced model (1.1) with $c = 2$, and discuss various procedures to deal with the negativity of the ANOVA estimator of σ_1^2. Note that since we are dealing with a balanced model involving exactly one random effect, namely $\mathbf{u}_1$, there are only two variance components, namely σ_1^2 and σ^2. In this case V_1 is the sum of squares due to $\mathbf{u}_1$ and V_2 is the sum of squares due to error (where the V_j's are defined in Section 1). Clearly, $V_1 \sim (\sigma^2 + \lambda_1\sigma_1^2)\chi^2_{q_1}$ and $V_2 \sim \sigma^2\chi^2_{q_2}$, where λ_1 is a positive scalar. The ANOVA estimator of σ_1^2, namely $\hat{\sigma}^2_{1A}$, then simplifies to

$$\hat{\sigma}^2_{1A} = \frac{1}{\lambda_1}\left\{\frac{1}{q_1}V_1 - \frac{1}{q_2}V_2\right\}. \tag{2.1}$$

As is well known, $\hat{\sigma}^2_{1A}$ is an invariant quadratic estimator (IQE), i.e., a quadratic form in $\mathbf{y}$ which is invariant under the transformation $\mathbf{y} \longrightarrow \mathbf{y} + X\theta$, where θ is a vector. Furthermore, as noted before, $\hat{\sigma}^2_{1A}$ is also the UMVUE of σ_1^2.

The seriousness of negativity of $\hat{\sigma}^2_{1A}$ was realized long time back, and a simple truncation at 0 was proposed. However, Herbach (1959) was the first to derive the maximum likelihood estimator (MLE) of σ_1^2, say $\hat{\sigma}^2_{1H}$, by maximizing the entire likelihood of $\mathbf{y}$. This estimator is given by

$$\hat{\sigma}^2_{1H} = \max\{0, \frac{1}{\lambda_1}[\frac{1}{q_1+1}V_1 - \frac{1}{q_2}V_2]\}. \tag{2.2}$$

Later, by maximizing the likelihood function based on V_1 and V_2 (the restricted maximum likelihood (REML)), Thompson (1962) came up with the REML estimator of σ_1^2, which is simply $\hat{\sigma}^2_{1A}$ truncated at zero.

For some reason, research in this area to create substantially better nonnegative estimators of σ_1^2 seems to have stopped for quite some time until in the eighties. Motivated by some what different principles, Hartung (1981) and later Chaubey (1983) proposed nonnegative quadratic estimators of σ_1^2, given respectively by $\frac{\lambda_1}{q_1(\lambda_1^2+1)}V_1$ and $\frac{1}{\lambda_1 q_1}V_1$. It turns out that from a decision theoretic point of view, under squared error loss function, Chaubey's estimator is always inadmissible and Hartung's estimator is admissible only under a rather strong condition (the details are given later in this section). Recently, a nonnegative nonquadratic estimator with good MSE properties has been suggested by Chow and Shao (1988) for the one-way balanced model with k treatments and n observations per treatment. This estimator (their equation (2.2), page 351), say $\hat{\sigma}^2_{1CS}$, is given by

$$\hat{\sigma}^2_{1CS} = \begin{cases} \frac{k-1}{nk(k+1)}MSS_{\mathrm{Tr}} & \text{if } MSS_{\mathrm{Tr}} \le \frac{k+1}{k-1}MSS_E \\ \frac{1}{n}\left[\frac{k-1}{k}MSS_{\mathrm{Tr}} - MSS_E\right], & \text{otherwise,} \end{cases} \tag{2.3}$$

where MSS_{Tr} and MSS_E respectively denote the mean sum of squares due to treatments and error. Our numerical results indicate that this estimator does not offer *uniform* MSE improvement over $\hat{\sigma}^2_{1A}$, although it provides a substantial MSE improvement in a certain region of the parameter space.

Motivation for the most recent works of Mathew, Sinha and Sutradhar (1992a), Kelly and Mathew (1992) and Kubokawa (1991) came from an attempt to derive nonnegative estimators of σ_1^2 with uniform and substantial MSE improvement compared to the ANOVA estimator. For improving $\hat{\sigma}^2_{1A}$ using a nonnegative IQE, Mathew, Sinha and Sutradhar (1992a) consider an estimator of the form

$$\delta = a_1 V_1 + a_2 V_2, \tag{2.4}$$

with $a_1 \geq 0$, $a_2 \geq 0$, and by a direct comparison of the MSEs of δ and $\hat{\sigma}^2_{1A}$ show that the inequalities (2.5), (2.6) and (2.7) below are sufficient conditions for $\text{MSE}(\delta) \leq \text{MSE}(\hat{\sigma}^2_{1A})$, for all σ_1^2, σ^2.

$$2\left\{a_1^2 q_1 + a_2^2 q_2\right\} + \left\{a_1 q_1 + a_2 q_2\right\}^2 \leq \frac{2(q_1+q_2)}{\lambda_1^2 q_1 q_2} \tag{2.5}$$

$$2a_1^2 q_1 \lambda_1^2 + (a_1 q_1 \lambda_1 - 1)^2 \leq \frac{2}{q_1} \tag{2.6}$$

$$2a_1^2 q_1 \lambda_1 + \{a_1 q_1 + a_2 q_2\}(a_1 q_1 \lambda_1 - 1) \leq \frac{2}{\lambda_1 q_1}. \tag{2.7}$$

Indeed, (2.5) and (2.6) are necessary as well for $\text{MSE}(\delta) \leq \text{MSE}(\hat{\sigma}^2_{1A})$, for all σ_1^2, σ^2. A lot of simplification results from the important observation that if (2.5) holds for some $a_1, a_2 \geq 0$, then it also holds with $a_2 = 0$, for the same a_1. This coupled with a further analysis of the above conditions leads to the following theorem (Mathew, Sinha and Sutradhar (1992a)).

THEOREM 2.1 *Consider the model (1.1) with $c = 2$. (i) For $q_1 = 1$ or 2, the estimator $a_1 V_1$ has a uniformly smaller MSE than $\hat{\sigma}^2_{1A}$ whenever $0 \leq a_1 \leq \min\{\frac{1}{\lambda_1 q_1}, \frac{1}{\lambda_1 q_1}[\frac{2(q_1+q_2)}{q_2(q_1+2)}]^{\frac{1}{2}}\}$. (ii) For $q_1 \geq 3$, there exists a nonnegative IQE with uniformly smaller MSE than $\hat{\sigma}^2_{1A}$ if and only if $(q_1-2)^2 q_2 \leq 2(q_1+q_2)(q_1+2)$. When this condition holds, $a_1 V_1$ is one such estimator where $\frac{q_1-2}{q_1(q_1+2)\lambda_1} \leq a_1 \leq \min\{\frac{1}{\lambda_1 q_1}, \frac{1}{\lambda_1 q_1}\left[\frac{2(q_1+q_2)}{q_2(q_1+2)}\right]^{\frac{1}{2}}\}$.*

REMARK 2.1. Theorem 2.1 suggests that there may exist a range of values of a_1 for which $a_1 V_1$ is uniformly better than $\hat{\sigma}^2_{1A}$. Which value of a_1 to select clearly depends on the magnitude of σ_1^2 relative to σ^2. For example, if σ_1^2 is very small compared to σ^2, the leading term in MSE of $a_1 V_1$ will be the term involving σ^4. We should therefore choose the smallest possible value of a_1 which, from Theorem 2.1, is $a_1 = \frac{q_1-2}{q_1(q_1+2)\lambda_1}$ for $q_1 \geq 3$ and

$a_1 = 0$ for $q_1 = 1$ or 2. On the other hand, if σ_1^2 is large compared to σ^2, we should choose $a_1 = \frac{1}{\lambda_1(q_1+2)}$ to minimize the coefficient of σ_1^4 in the MSE.

Mathew, Sinha and Sutradhar (1992a) also studied the admissibility of the estimator $a_1 V_1$ for estimating σ_1^2 under squared error loss in the class of nonnegative IQE's, and observed that $a_1 V_1$ is admissible if and only if $a_1 \leq \frac{1}{\lambda_1(q_1+2)}$. From this, the inadmissibility of Chaubey's (1983) estimator, namely $\frac{1}{\lambda_1 q_1} V_1$, is immediate. It also follows that Hartung's (1981) estimator, namely $\frac{\lambda_1}{q_1(\lambda_1^2+1)} V_1$, is admissible if and only if $2\lambda_1^2 \leq q_1$. This condition appears to be too severe a restriction. Application to the one-way model (discussed below) shows that the condition $2\lambda_1^2 \leq q_1$ holds only when the number of treatments is very large compared to the number of replications per treatment. The inadmissibility of Hartung's estimator in some situations (for the one-way model) has also been observed by LaMotte (1985).

As an application of the above theorems, we briefly discuss the balanced one-way random model which provided inspirations to many researchers in this area. Let there be k random treatment effects each treatment replicated n times. Then $N = nk$, $q_1 = k - 1$, $q_2 = k(n-1)$ and $\lambda_1 = n$. Furthermore, V_1 = treatment sum of squares = SS_{Tr} (say), and V_2 = error sum of squares = SS_{E}. Recall that $\hat{\sigma}_{1A}^2 = \frac{1}{n}\left\{\frac{1}{k-1}SS_{\text{Tr}} - \frac{1}{k(n-1)}SS_{\text{E}}\right\}$ and $\hat{\sigma}^2 = \frac{1}{k(n-1)}SS_{\text{E}}$. The preceding discussion and the results in Theorem 2.1 can then be summarized as follows.

COROLLARY 2.1 *(i) For $k = 2$ or 3, the estimator $a_1 SS_{\text{Tr}}$ has a uniformly smaller MSE than $\hat{\sigma}_{1A}^2$ provided $0 \leq a_1 \leq \frac{1}{n(k-1)} \min\{1, [\frac{2(nk-1)}{k(n^2-1)}]^{\frac{1}{2}}\}$.*
(ii) For $k \geq 4$, there exists a nonnegative IQE with a uniformly smaller MSE than $\hat{\sigma}_{1A}^2$ if and only if $k(k-3)^2(n-1) \leq 2(nk-1)(k+1)$. When this condition holds, $a_1 SS_{\text{Tr}}$ is one such estimator whenever $\frac{k-3}{n(k^2-1)} \leq a_1 \leq \frac{1}{n(k-1)} \min\{1, [\frac{2(nk-1)}{k(n^2-1)}]^{\frac{1}{2}}\}$.
(iii) The estimator $a_1 SS_{\text{Tr}}$ (for $a_1 \geq 0$) of σ_1^2 is admissible among nonnegative IQE's if and only if $a_1 \leq \frac{1}{n(k+1)}$.
(iv) Hartung's estimator of σ_1^2, given by $\frac{n}{(n^2+1)(k-1)} SS_{\text{Tr}}$, is admissible among nonnegative IQE 's if and only if $2n^2 \leq k - 1$.

REMARK 2.2. The condition in Corollary 2.1 *(ii)* holds for all n if $k \leq 7$. For $k = 8$, it holds for $n = 2$ and 3, for $k = 9$ it holds for $n = 2$ and for $k \geq 10$, it never holds.

Motivated by the above fact that nonnegative IQE's uniformly better than $\hat{\sigma}_{1A}^2$ does not always exist, Kelly and Mathew (1992) suggested a different approach to derive a satisfactory nonnegative estimator uniformly better than $\hat{\sigma}_{1A}^2$. The procedure is to first derive an IQE (which may take negative values) that has a uniformly smaller MSE compared to $\hat{\sigma}_{1A}^2$ and then truncate this IQE at zero. The desired IQE is derived by expressing the estimator in the form

$$\hat{\sigma}^2_{1(c,\epsilon)} = \frac{c}{\lambda_1}\left[\frac{V_1}{q_1} - \epsilon\frac{V_2}{q_2}\right],$$

and then minimizing the coefficient of σ_1^4 with respect to c and the coefficient of σ^4 with respect to ϵ in the MSE of $\hat{\sigma}^2_{1(c,\epsilon)}$. The resulting estimator, say $\tilde{\sigma}_1^2$, is obtained by direct computations and is given by

$$\tilde{\sigma}_1^2 = \frac{1}{\lambda_1(q_1+2)}\left[\frac{V_1}{q_1} - \frac{V_2}{q_2+2}\right]. \tag{2.8}$$

It can be shown that $\text{MSE}(\tilde{\sigma}_1^2) \leq \text{MSE}(\hat{\sigma}_{1A}^2)$, uniformly in σ^2 and σ_1^2. In fact, the numerical results in Kelly and Mathew (1992) indicate that the reduction in MSE over $\hat{\sigma}_{1A}^2$ achievable through the use of $\tilde{\sigma}_1^2$ turns out to be quite substantial irrespective of the magnitude of the ratio σ_1^2/σ^2. Furthermore, the bias of the estimator $\tilde{\sigma}_1^2$ is rather insignificant. However, note that $\tilde{\sigma}_1^2$ can be negative. Truncation of $\tilde{\sigma}_1^2$ at zero provides a satisfactory nonnegative estimator of σ_1^2. It should also be noted that $\tilde{\sigma}_1^2$ has a smaller probability of being negative compared to $\hat{\sigma}_{1A}^2$. In Kelly and Mathew (1992), a class of estimators having uniformly smaller MSE and uniformly smaller probability of negativity compared to $\hat{\sigma}_{1A}^2$ is derived.

For the one-way balanced model, recently a class of nonquadratic estimators uniformly better than $\hat{\sigma}_{1A}^2$ has been constructed by Kubokawa (1991), who also identified some known nonnegative estimators in the class. By direct comparison of the MSE's, it is easy to verify that the estimator $\hat{\sigma}_{10}^2$ given below has a uniformly smaller MSE compared to $\hat{\sigma}_{1A}^2$.

$$\hat{\sigma}_{10}^2 = \frac{1}{n}\left[\frac{V_1}{k-1} - \frac{V_2}{k(n-1)+2}\right], \tag{2.9}$$

where, as noted before, V_1 and V_2 are respectively the treatment and error sums of squares. The estimators constructed by Kubokawa (1991) are in fact uniformly better than $\hat{\sigma}_{10}^2$ given above. To describe his class of estimators, consider

$$\hat{\sigma}_{1\phi}^2 = \frac{1}{n}\left[\frac{V_1}{k-1} - V_2\phi(\frac{V_1}{V_2})\right]. \tag{2.10}$$

Theorem 2.2 given below provides sufficient conditions on the function ϕ under which $\hat{\sigma}_{1\phi}^2$ is uniformly better than $\hat{\sigma}_{10}^2$. For a proof of Theorem 2.2, which is based on the properties of a definite integral, we refer to Kubokawa (1991).

THEOREM 2.2 *Consider the class of estimators given by (2.10) where ϕ satisfies the following conditions:*

(i) ϕ is nondecreasing,
(ii) $lim_{w\to\infty}\phi(w) = \frac{1}{k(n-1)+2}$, and
(iii) $\phi_0(w) \le \phi(w) \le \frac{1}{k(n-1)+2}$, where

$$\phi_0(w) = \frac{1}{(k-1)} \frac{\int_0^w z^{(k-1)/2}(1+z)^{-(nk+3)/2}dz}{\int_0^w z^{(k-3)/2}(1+z)^{-(nk+3)/2}dz}.$$

Then $\hat{\sigma}^2_{1\phi}$ is uniformly better than $\hat{\sigma}^2_{10}$.

It can be verified that the choices ϕ_1 and ϕ_2 given below satisfy the conditions (i) - (iii) in Theorem 2.2.

$$\begin{aligned} \phi_1(w) &= \min\{\frac{1}{n+2}, \frac{w}{k-1}\} \\ \phi_2(w) &= \min\{\frac{1}{n+2}, \frac{w}{k+1}\}. \end{aligned}$$

The corresponding estimators of σ_1^2 are

$$\hat{\sigma}^2_{1\phi_1} = \max\{\hat{\sigma}^2_{10}, 0\} \tag{2.11}$$

$$\hat{\sigma}^2_{1\phi_2} = \max\{\hat{\sigma}^2_{10}, \frac{2V_1}{n(k-1)(k+1)}\}. \tag{2.12}$$

Clearly $\hat{\sigma}^2_{1\phi_1}$ and $\hat{\sigma}^2_{1\phi_2}$ are both nonnegative and are also uniformly better than $\hat{\sigma}^2_{10}$ according to Theorem 2.2. It can also be verified that with ϕ_0 given in Theorem 2.2, the estimator

$$\hat{\sigma}^2_{1\phi_0} = \frac{1}{n}\left[\frac{V_1}{k-1} - V_2\phi_0(\frac{V_1}{V_2})\right] \tag{2.13}$$

is also nonnegative (see Kubokawa (1991)). The estimators $\hat{\sigma}^2_{1\phi_1}$ and $\hat{\sigma}^2_{1\phi_2}$ are clearly nonsmooth. However, $\hat{\sigma}^2_{1\phi_0}$ is a smooth nonnegative estimator that is also uniformly better than $\hat{\sigma}^2_{10}$.

We now turn our attention to another natural approach to the problem of nonnegative estimation of σ_1^2, namely, the Bayesian approach (see Hill, 1965; Klotz, Milton and Zacks, 1969; Portnoy, 1971; Box and Tiao, 1973; Broemeling, 1985). An exact Bayes estimator of σ_1^2 using a noninformative prior has been discussed by Klotz, Milton and Zacks (1969) and Box and Tiao (1973). Broemeling (1985) derived an approximate Bayes estimator of σ_1^2 for an inverse gamma prior. However, the main drawback of the Bayes solution is the poor performance in terms of MSE of the Bayes estimator of σ_1^2 using a non-informative prior (see Klotz Milton and Zacks, 1969; Box and Tiao, 1973). A second criticism against using a Bayes estimator, which probably is of less significance these days, is its analytical intractability and lack of suitable frequentist properties.

Our object in the rest of this section is to discuss empirical Bayes estimators of the two variance components in a one-way balanced model.

Numerical computations reveal the superiority of the proposed empirical Bayes estimators over most of the previously known estimators in terms of MSE. To derive the empirical Bayes estimators, we have worked with conjugate priors involving unknown hyperparameters, which are subsequently estimated from the data. It turns out that a balanced model involving more than two variance components can be dealt with using results for the model with two variance components, thus providing an easy solution even in the general case. As expected, the computation of the empirical Bayes estimators involves incomplete Beta functions, as do the exact Bayes estimators (see Klotz, Milton and Zacks, 1969; Box and Tiao, 1973). Following Tierney and Kadane (1986), one can derive suitable approximations of the empirical Bayes estimators, which are readily computable. We refer to Niyogi (1992) for details.

Recall that for the balanced one-way random effects model,

$$V_1 \sim (\sigma^2 + n\sigma_1^2)\chi^2_{k-1}, \quad V_2 \sim (\sigma^2)\chi^2_{k(n-1)} \tag{2.14}$$

and V_1 and V_2 are mutually independent. In view of (2.14), the likelihood function of V_1 and V_2 can be written as

$$L(\sigma^2, \sigma_1^2 | V_1, V_2) \propto (\sigma^2)^{-\frac{k(n-1)}{2}} (\sigma^2 + n\sigma_1^2)^{-\frac{k-1}{2}} \exp\left[-\frac{V_1}{2(\sigma^2 + n\sigma_1^2)} - \frac{V_2}{2\sigma^2}\right]. \tag{2.15}$$

We assume the following informative prior for σ^2 and σ_1^2 :

$$\pi(\sigma^2, \sigma_1^2) \propto \frac{\exp[-\frac{\delta}{2(\sigma^2+n\sigma_1^2)} - \frac{\delta}{2\sigma^2}]}{(\sigma^2)^\nu (\sigma^2 + n\sigma_1^2)^\nu} \tag{2.16}$$

where $\delta > 0, \nu > 2$ are hyperparameters.

From (2.15) and (2.16), the joint posterior density of σ^2 and σ_1^2 is readily obtained as

$$\begin{aligned} p(\sigma^2, \sigma_1^2 | V_1, V_2) &\propto (\sigma^2)^{-\frac{k(n-1)}{2}+\nu} (\sigma^2 + n\sigma_1^2)^{-\frac{k-1}{2}+\nu} \\ &\times \exp\left[-\frac{V_1+\delta}{2(\sigma^2 + n\sigma_1^2)} - \frac{V_2+\delta}{2\sigma^2}\right]. \end{aligned}$$

Hence the posterior mean of σ_1^2, as its Bayes estimator, is given by

$$\begin{aligned} \hat{\sigma}^2_{1B} = & \left\{ \int_0^\infty \int_0^\infty \sigma_1^2 (\sigma^2)^{-\frac{k(n-1)}{2}+\nu} (\sigma^2 + n\sigma_1^2)^{-\frac{k-1}{2}+\nu} \right. \\ & \left. \times \exp\left[-\frac{V_1+\delta}{2(\sigma^2 + n\sigma_1^2)} - \frac{V_2+\delta}{2\sigma^2}\right] d\sigma^2 d\sigma_1^2 \right\} \div \\ & \left\{ \int_0^\infty \int_0^\infty (\sigma^2)^{-\frac{k(n-1)}{2}+\nu} (\sigma^2 + n\sigma_1^2)^{-\frac{k-1}{2}+\nu} \right. \\ & \left. \times \exp\left[-\frac{V_1+\delta}{2(\sigma^2 + n\sigma_1^2)} - \frac{V_2+\delta}{2\sigma^2}\right] d\sigma^2 d\sigma_1^2 \right\} \end{aligned}$$

$$= \frac{1}{n}\Big[\frac{(V_1+\delta)I_z(\frac{k-1}{2}+\nu-2, \frac{k(n-1)}{2}+\nu-1)}{(k+2\nu-5)I_z(\frac{k-1}{2}+\nu-1, \frac{k(n-1)}{2}+\nu-1)} - \frac{(V_2+\delta)I_z(\frac{k-1}{2}+\nu-1, \frac{k(n-1)}{2}+\nu-2)}{(kn-k+2\nu-4)I_z(\frac{k-1}{2}+\nu-1, \frac{k(n-1)}{2}+\nu-1)}\Big], \tag{2.17}$$

where $z = \frac{V_1+\delta}{V_1+V_2+2\delta}$ and $I_x(a,b) = \int_0^x u^{a-1}(1-u)^{b-1}du / \int_0^1 u^{a-1}(1-u)^{b-1}du$ is an incomplete Beta function. The above integrals are evaluated by the formula; see Box and Tiao (1973, equation (A 5.2.1)): for $p_1 > 0, p_2 > 0, a_1 > 0, a_2 > 0$ and $c > 0$,

$$\int_0^\infty \int_0^\infty x^{-(p_1+1)}(x+cy)^{-(p_2+1)} \exp[-(\frac{a_1}{x} + \frac{a_2}{x+cy})]dxdy = \frac{\Gamma(p_1)\Gamma(p_2)}{ca_1^{p_1}a_2^{p_2}} I_{a_2/(a_1+a_2)}(p_2, p_1). \tag{2.18}$$

Having derived the Bayes estimators, empirical Bayes estimators are obtained by estimating the unknown hyperparameters ν and δ. It turns out that the maximum likelihood approach for estimating ν and δ is computationally difficult and one must employ numerical solution techniques for computing them. However, using the method of moments approach, we can easily obtain explicit analytical estimators for ν and δ. Noting that for fixed σ^2 and σ_1^2,

$$\begin{aligned} E(V_1|\sigma^2, \sigma_1^2) &= (k-1)(\sigma^2 + n\sigma_1^2) \\ E(V_2|\sigma^2, \sigma_1^2) &= k(n-1)\sigma^2, \end{aligned}$$

under a prior distribution of σ^2 and σ_1^2, we can write,

$$\begin{aligned} E(V_1) &= (k-1)E_\pi(\sigma^2 + n\sigma_1^2) \\ E(V_2) &= k(n-1)E_\pi(\sigma^2). \end{aligned}$$

We now observe that

$$\begin{aligned} &E_\pi(\sigma^2 + n\sigma_1^2) \\ &= \frac{\int\int(\sigma^2)^{-\nu}(\sigma^2+n\sigma_1^2)^{-(\nu-1)} \exp[-\frac{\delta}{2(\sigma^2+n\sigma_1^2)} - \frac{\delta}{2\sigma^2}]\, d\sigma^2 d\sigma_1^2}{\int\int(\sigma^2)^{-\nu}(\sigma^2+n\sigma_1^2)^{-\nu} \exp[-\frac{\delta}{2(\sigma^2+n\sigma_1^2)} - \frac{\delta}{2\sigma^2}]\, d\sigma^2 d\sigma_1^2} \\ &= (\frac{\delta}{\nu-2})I_{\frac{1}{2}}(\nu-2, \nu-1) \end{aligned}$$

and

$$\begin{aligned} E_\pi(\sigma^2) &= \frac{\int\int(\sigma^2)^{-(\nu-1)}(\sigma^2+n\sigma_1^2)^{-\nu} \exp[-\frac{\delta}{2(\sigma^2+n\sigma_1^2)} - \frac{\delta}{2\sigma^2}]\, d\sigma^2 d\sigma_1^2}{\int\int(\sigma^2)^{-\nu}(\sigma^2+n\sigma_1^2)^{-\nu} \exp[-\frac{\delta}{2(\sigma^2+n\sigma_1^2)} - \frac{\delta}{2\sigma^2}]\, d\sigma^2 d\sigma_1^2} \\ &= (\frac{\delta}{\nu-2})I_{\frac{1}{2}}(\nu-1, \nu-2), \end{aligned}$$

using (2.18) mentioned earlier. This readily leads to the following two equations for estimating δ and ν.

$$\frac{V_1}{k-1} = \frac{\delta}{\nu-2} I_{\frac{1}{2}}(\nu-2, \nu-1) \quad (2.19)$$

$$\frac{V_2}{k(n-1)} = \frac{\delta}{\nu-2} I_{\frac{1}{2}}(\nu-1, \nu-2). \quad (2.20)$$

Note that, $I_{\frac{1}{2}}(\nu-2, \nu-1) = 1 - I_{\frac{1}{2}}(\nu-1, \nu-2)$. From (2.19) and (2.20), we get,

$$I_{\frac{1}{2}}(\nu-2, \nu-1) = \frac{k(n-1)V_1}{(k-1)V_2 + k(n-1)V_1}.$$

Computationally, it turns out that ν, which must be greater than 2, tends to ∞ if the right hand side quantity of the above equation is close to 0.5, thus causing a difficulty. Hence it becomes necessary to assume ν is known. Detailed numerical computations (see Niyogi (1992)) reveal that $\nu = 5$ will be a reasonable choice. For known ν, δ can be estimated as

$$\hat{\delta} = (\nu-2)\left[\frac{(k-1)V_2 + k(n-1)V_1}{k(k-1)(n-1)}\right]. \quad (2.21)$$

Hence, the empirical Bayes estimator of σ_1^2, say $\hat{\sigma}_{1EB}^2$, can be obtained from (2.17) with $\nu = 5$ and δ estimated as a function of ν, k, n, V_1 and V_2, given by (2.21). The estimator is given by

$$\hat{\sigma}_{1EB}^2 = \frac{1}{n}\left[\frac{(V_1+\hat{\delta})I_z(\frac{k-1}{2}+3, \frac{k(n-1)}{2}+4)}{(k+5)I_z(\frac{k-1}{2}+4, \frac{k(n-1)}{2}+4)} - \frac{(V_2+\hat{\delta})I_z(\frac{k-1}{2}+4, \frac{k(n-1)}{2}+3)}{(kn-k+6)I_z(\frac{k-1}{2}+4, \frac{k(n-1)}{2}+4)}\right], \quad (2.22)$$

Analogously, we can estimate σ^2 empirically as

$$\hat{\sigma}_{EB}^2 = \frac{(V_2+\hat{\delta})I_z(\frac{k-1}{2}+\nu-1, \frac{k(n-1)}{2}+\nu-2)}{(nk-k+2\nu-4)I_z(\frac{k-1}{2}+\nu-1, \frac{k(n-1)}{2}+\nu-1)} \quad (2.23)$$

with $\hat{\delta} = (\nu-2)[\frac{(k-1)V_2+k(n-1)V_1}{k(k-1)(n-1)}]$ and $\nu = 5$.

In Mathew, Sinha and Sutradhar (1992a), Kelly and Mathew (1992) and Niyogi (1992), numerical results are reported for comparing several nonnegative estimators of σ_1^2 in the balanced one-way model having k treatments and n observations per treatment. In Table 2.1, we have reported the MSE's and squared biases of several estimators of σ_1^2 for the following values of k and n:

$$2(i)\ k=6, n=4;\ 2(ii)\ k=8, n=4 \text{ and } 2(iii)\ k=6, n=10. \quad (2.24)$$

Table 2.1. MSE's and squared biases of $\hat{\sigma}^2_{1A}$, $\hat{\sigma}^2_{1H}$, $\hat{\sigma}^2_{CS}$, $\hat{\sigma}^2_{1P_1}$, $\hat{\sigma}^2_{1P_2}$, $\tilde{\sigma}^2_{1+}$ and $\hat{\sigma}^2_{1EB}$ for the one-way balanced models 2(*i*), 2(*ii*) and 2(*iii*) in (2.24) with $\sigma^2 = 1$ (the squared biases are given within brackets).

	σ_1^2	0.1	0.3	0.5	1.0	4.0	9.0
2(*i*)	$\hat{\sigma}^2_{1A}$	0.055	0.128	0.221	0.628	7.633	35.742
	$\hat{\sigma}^2_{1H}$	0.022	0.075	0.176	0.487	5.084	26.507
		(0)	(0.004)	(0.012)	(0.041)	(0.493)	(2.334)
	$\hat{\sigma}^2_{1CS}$	0.020	0.073	0.150	0.473	5.684	27.008
		(0)	(0.003)	(0.011)	(0.040)	(0.493)	(2.334)
	$\hat{\sigma}^2_{1P_1}$	0.012	0.027	0.071	0.338	5.984	31.601
		(0.003)	(0.004)	(0.032)	(0.215)	(4.731)	(25.283)
	$\hat{\sigma}^2_{1P_2}$	0.048	0.070	0.116	0.330	4.616	23.188
		(0.023)	(0.009)	(0.001)	(0.011)	(0.919)	(5.666)
	$\tilde{\sigma}^2_{1+}$	0.020	0.061	0.125	0.386	4.907	23.760
		(0)	(0.003)	(0.014)	(0.070)	(1.254)	(6.459)
	$\hat{\sigma}^2_{1EB}$	0.015	0.035	0.087	0.342	5.199	25.657
		(0.004)	(0.002)	(0.018)	(0.127)	(2.472)	(12.639)
2(*ii*)	$\hat{\sigma}^2_{1A}$	0.040	0.091	0.159	0.423	5.190	25.714
	$\hat{\sigma}^2_{1H}$	0.021	0.060	0.120	0.396	4.491	21.470
		(0)	(0.003)	(0.007)	(0.023)	(0.268)	(1.269)
	$\hat{\sigma}^2_{1CS}$	0.017	0.060	0.121	0.349	4.225	20.940
		(0)	(0.002)	(0.007)	(0.022)	(0.268)	(1.269)
	$\hat{\sigma}^2_{1P_1}$	0.019	0.026	0.056	0.229	4.221	22.661
		(0.009)	(0)	(0.007)	(0.092)	(2.657)	(14.761)
	$\hat{\sigma}^2_{1P_2}$	0.051	0.069	0.104	0.271	3.604	18.049
		(0.030)	(0.017)	(0.007)	(0.001)	(0.465)	(3.164)
	$\tilde{\sigma}^2_{1+}$	0.018	0.052	0.105	0.313	3.867	18.628
		(0)	(0.002)	(0.008)	(0.041)	(0.743)	(3.836)
	$\hat{\sigma}^2_{1EB}$	0.012	0.030	0.075	0.284	4.074	19.917
		(0.003)	(0.002)	(0.015)	(0.094)	(1.709)	(8.681)
2(*iii*)	$\hat{\sigma}^2_{1A}$	0.017	0.063	0.143	0.456	6.914	34.787
	$\hat{\sigma}^2_{1H}$	0.009	0.051	0.107	0.359	5.012	23.693
		(0)	(0.004)	(0.010)	(0.033)	(0.456)	(2.246)
	$\hat{\sigma}^2_{1CS}$	0.009	0.046	0.106	0.355	5.199	25.955
		(0)	(0.004)	(0.009)	(0.033)	(0.456)	(2.246)
	$\hat{\sigma}^2_{1P_1}$	0.003	0.029	0.084	0.371	6.183	31.532
		(0)	(0.016)	(0.057)	(0.278)	(5.012)	(25.920)
	$\hat{\sigma}^2_{1P_2}$	0.010	0.033	0.079	0.293	4.579	23.150
		(0.002)	(0)	(0.005)	(0.045)	(1.133)	(6.173)
	$\tilde{\sigma}^2_{1+}$	0.008	0.038	0.092	0.323	4.676	23.225
		(0)	(0.007)	(0.019)	(0.079)	(1.285)	(6.521)
	$\hat{\sigma}^2_{1EB}$	0.006	0.038	0.095	0.348	5.126	25.523
		(0)	(0.0071)	(0.024)	(0.102)	(1.652)	(8.367)

The estimators considered are the following:

(i) the ANOVA estimator $\hat{\sigma}^2_{1A}$,
(ii) the maximum likelihood estimator $\hat{\sigma}^2_H$, due to Herbach (1959), given in (2.2),
(iii) the estimator $\hat{\sigma}^2_{1CS}$, due to Chow and Shao (1988), given in (2.3),
(iv) the estimator $\hat{\sigma}^2_{1P_1} = \frac{k-3}{n(k+1)(k-1)} SS_{\text{Tr}}$ (for $k \geq 4$),
(v) the estimator $\hat{\sigma}^2_{1P_2} = \frac{1}{n(k+1)} SS_{\text{Tr}}$,
(vi) the estimator $\tilde{\sigma}^2_{1+} = \max\{\tilde{\sigma}^2_1, 0\}$ where $\tilde{\sigma}^2_1$ is given in (2.8),
(vii) the estimator $\hat{\sigma}^2_{1EB}$ given in (2.22).

The restricted maximum likelihood estimator due to Thompson (1962) was not included in the numerical study since this estimator is known to have a larger MSE compared to $\hat{\sigma}^2_{1H}$ (see Corbeil and Searle (1976) and Swallow and Monahan (1984)). The MSE computations of $\hat{\sigma}^2_{1A}$, $\hat{\sigma}^2_{1P_1}$ and $\hat{\sigma}^2_{1P_2}$ were direct and those of $\hat{\sigma}^2_H$, $\hat{\sigma}^2_{1CS}$, $\tilde{\sigma}^2_{1+}$ and $\hat{\sigma}^2_{1EB}$ were based on 20,000 simulations. The numerical results indicate that the nonnegative estimators $\tilde{\sigma}^2_{1+}$ and $\hat{\sigma}^2_{1EB}$ should be preferred in the absence of any information regarding the magnitude of the ratio σ^2_1/σ^2. $\tilde{\sigma}^2_{1+}$ appears to have a slight advantage over $\hat{\sigma}^2_{1EB}$ for large values of σ^2_1 and the converse is true for small values of σ^2_1. Furthermore, $\tilde{\sigma}^2_{1+}$ appears to have a smaller bias compared to $\hat{\sigma}^2_{1EB}$ in most cases. It should also be noted that the estimator $\tilde{\sigma}^2_{1+}$ has an explicit analytic expression and can be easily computed, whereas $\hat{\sigma}^2_{1EB}$ has to be obtained numerically. In the absence of any information regarding the magnitude of σ^2_1/σ^2, the estimator that we recommend is $\tilde{\sigma}^2_{1+}$. If the ratio σ^2_1/σ^2 is known to be large or small, it is possible to recommend other nonnegative estimators. The estimator $\hat{\sigma}^2_{1P_2}$ turned out to have a definite edge over the others when the ratio σ^2_1/σ^2 is large. For small values of this ratio, $\hat{\sigma}^2_{1P_1}$ had a smaller MSE compared to the other estimators.

3 Nonnegative Estimation of σ^2_1 in General Balanced Mixed Models

Following Mathew, Sinha and Sutradhar (1992a) we now discuss nonnegative quadratic estimation of σ^2_1 in a general balanced mixed model involving any number of variance components. Suppose the set of quadratic forms V_j defined in Section 1 contains two quadratic forms, say V_1 and V_2, satisfying the following:

$$\frac{1}{q_1}\mathrm{E}(V_1) = \lambda_1\sigma^2_1 + \sum_{j=2}^{c-1} \lambda_j\sigma^2_j + \sigma^2, \tag{3.1}$$

and

$$\frac{1}{q_2}\mathrm{E}(V_2) = \sum_{j=2}^{c-1} \lambda_j\sigma^2_j + \sigma^2. \tag{3.2}$$

It should be noted that this situation obtains quite often in balanced set ups, even though this is not always the case. Accordingly, we consider two cases. First, when (3.1) and (3.2) hold, in which case $\hat{\sigma}^2_{1A}$ is given by (2.1). (3.1) and (3.2) suggest that for improving $\hat{\sigma}^2_{1A}$ using a nonnegative IQE in the model (1.1), it is enough to consider the same problem in a model based on V_1 and V_2 involving only two variance components, namely σ_1^2 and $\sum_{j=2}^{c-1}\lambda_j\sigma_j^2 + \sigma^2$. Mathew, Sinha and Sutradhar (1992a) demonstrate that this is indeed the case. Hence, from (3.1) and (3.2), and applying Theorem 2.1, we immediately get the following result.

THEOREM 3.1 *Consider the model (1.1) and let V_1, V_2, q_1, q_2 and λ_1 be as defined in Section 1. If (3.1) and (3.2) hold, then for the existence of a nonnegative IQE uniformly better than the ANOVA estimator of σ_1^2, the conclusions in Theorem 2.1 (i) and (ii) are valid.*

We next consider the case when the assumptions leading to (3.1) and (3.2) do not hold, and demonstrate that it is still possible to obtain necessary and sufficient conditions for the existence of a nonnegative IQE uniformly better than the ANOVA estimator of *any* variance component.

Without loss of generality, assume that we are interested in estimating σ_1^2 and the quadratic form V_1 satisfies (3.1) and furthermore, there does not exist a quadratic form V_2 satisfying (3.2). When (3.2) does not hold, the ANOVA estimator of σ_1^2 involves V_1 along with several other quadratic forms, say V_2, ..., V_t, and the estimator is given by (1.2), where each ϵ_j is +1 or -1 for $j = 2, ..., t$. It should be noted that for $j = 2, ..., t$, $V_j \sim \delta_j\chi^2_{q_j}$, where δ_j is a linear combination of $\sigma_2^2, ..., \sigma_c^2$, and δ_j does not involve σ_1^2. It can be shown that for improving $\hat{\sigma}^2_{1A}$ using a nonnegative IQE, it is enough to consider a_1V_1, for $a_1 \geq 0$. Thus we need to obtain necessary and sufficient conditions for the existence of $a_1 \geq 0$ so that a_1V_1 has a uniformly smaller MSE than $\hat{\sigma}^2_{1A}$. Using the observations $\mathrm{E}(V_j) = q_j\delta_j$ and $\mathrm{Var}(V_j) = 2q_j\delta_j^2$, it is readily verified that a_1V_1 has a uniformly smaller MSE than $\hat{\sigma}^2_{1A}$ if and only if the inequality

$$2a_1^2q_1\left\{\lambda_1\sigma_1^2 + \sum_{l=2}^{c}\lambda_l\sigma_l^2\right\}^2 + \left\{a_1q_1(\lambda_1\sigma_1^2 + \sum_{l=2}^{c}\lambda_l\sigma_l^2) - \sigma_1^2\right\}^2$$

$$\leq \frac{2}{\lambda_1^2}\left\{\frac{1}{q_1}(\lambda_1\sigma_1^2 + \sum_{l=2}^{c}\lambda_l\sigma_l^2)^2 + \sum_{j=2}^{t}\frac{\delta_j^2}{q_j}\right\} \tag{3.3}$$

holds for all $\sigma_l^2 \geq 0$, $l = 1, ..., c$. (Recall that $\sigma_c^2 = \sigma^2$, the error variance component, and $\lambda_c = 1$). For the inequality (3.3) to hold, it is necessary that the coefficient of σ_1^4 on the left hand side (LHS) of (3.3) is less than or equal to the corresponding coefficient on the right hand side (RHS) of (3.3). Similarly, for every $l \neq 1$, the coefficient of $\lambda_l^2\sigma_l^4$ on the LHS of (3.3) must be less than or equal to the corresponding coefficient on the RHS of

(3.3). Comparing the coefficients of σ_1^4 on both sides of the inequality (3.3), we get

$$2a_1^2 q_1 \lambda_1^2 + (a_1 q_1 \lambda_1 - 1)^2 \le \frac{2}{q_1}. \tag{3.4}$$

Note that since the δ_j's ($j = 2, ..., t$) are linear combinations of the σ_l^2's ($l = 2, ..., c$), there may be several terms involving $\lambda_l^2 \sigma_l^4$ (for $l \ne 1$) on the RHS of (3.3). Also note that for any $l \ne 1$, the coefficient of $\lambda_l^2 \sigma_l^4$ on the LHS of (3.3) is the same quantity, namely, $2a_1^2 q_1 + a_1^2 q_1^2$. Thus for comparing the coefficients of $\lambda_l^2 \sigma_l^4$ ($l \ne 1$) on both sides of the inequality (3.3), it is enough to compare the coefficients of those $\lambda_l^2 \sigma_l^4$'s that occur the least number of times on the RHS of (3.3). Without loss of generality, assume that for some $g \ge 2$, each of the quantities $\lambda_2^2 \sigma_2^4, ..., \lambda_g^2 \sigma_g^4$ occurs exactly h times on the RHS of (3.3) and any other $\lambda_l^2 \sigma_l^4$ occurs more than h times. Also assume that in the linear combination of the variance components that represent δ_j, there exist g_j variance components, say $\sigma_{j_i}^2$ ($i = 1, ..., g_j$), from among the set ($\sigma_2^2, ..., \sigma_g^2$). Comparing the terms that involve only $\lambda_2^2 \sigma_2^4, ..., \lambda_g^2 \sigma_g^4$ on both sides of (3.3), we get the second inequality

$$(2a_1^2 q_1 + a_1^2 q_1^2)(\sum_{l=2}^{g} \lambda_l \sigma_l^2)^2 \le \frac{2}{\lambda_1^2 q_1}(\sum_{l=2}^{g} \lambda_l \sigma_l^2)^2 + \frac{2}{\lambda_1^2} \sum_{j=2}^{t} \frac{1}{q_j} (\sum_{i=1}^{g_j} \lambda_{j_i} \sigma_{j_i}^2)^2. \tag{3.5}$$

Additionally, comparing the coefficients of $2\sigma_1^2 \sigma_c^2$ on both sides of (3.3), we get the third inequality

$$2a_1^2 q_1 \lambda_1 + a_1 q_1 (f_1 q_1 \lambda_1 - 1) \le \frac{2}{\lambda_1 q_1}. \tag{3.6}$$

It may be verified that the inequalities (3.4) and (3.5) are necessary conditions for (3.3) to hold, while (3.4), (3.5) and (3.6) are sufficient. Note that (3.4) is equivalent to

$$\max\left\{0, \frac{(q_1 - 2)}{q_1(q_1 + 2)\lambda_1}\right\} \le a_1 \le \frac{1}{q_1 \lambda_1} \tag{3.7}$$

and furthermore, (3.5) implies (3.6). Thus (3.4) and (3.5) are necessary and sufficient conditions for the inequality (3.3) to hold.

In order to simplify the inequality (3.5), note that (3.5) is equivalent to

$$a_1^2 q_1 (q_1 + 2) \le \frac{2}{\lambda_1^2 q_1} \left\{ 1 + q_1 \sum_{j=2}^{t} \frac{1}{q_j} (\sum_{i=1}^{g_j} \lambda_{j_i} \sigma_{j_i}^2)^2 / (\sum_{l=2}^{g} \lambda_l \sigma_l^2)^2 \right\}. \tag{3.8}$$

Let

$$m_0 = \min_{\sigma_2^2 \ge 0, ..., \sigma_g^2 \ge 0} \left[\{\sum_{j=2}^{t} \frac{1}{q_j} (\sum_{i=1}^{g_j} \lambda_{j_i} \sigma_{j_i}^2)^2\} / (\sum_{l=2}^{g} \lambda_l \sigma_l^2)^2 \right]. \tag{3.9}$$

Computation of the above minimum is rather straightforward in any given situation. Using (3.9), it follows that (3.5) holds for all $\sigma_l^2 \geq 0$, $l = 2, ..., g$, if and only if

$$a_1 \leq \frac{1}{q_1\lambda_1} m_*, \tag{3.10}$$

where

$$m_* = \left\{\frac{2}{q_1+2}(1+q_1 m_0)\right\}^{\frac{1}{2}}. \tag{3.11}$$

Thus $a_1 V_1$ has a uniformly smaller MSE than $\hat{\sigma}_{1A}^2$ if and only if a_1 satisfies (3.7) and (3.10). We summarize our conclusions in the following theorem.

THEOREM 3.2 *Consider the balanced model (1.1) and let the V_j's be as defined in Section 1. Suppose V_1 satisfies (3.1) and let m_* be given by (3.11).*
(i) If $q_1 = 1$ or 2, the nonnegative IQE $a_1 V_1$ has a uniformly smaller MSE than $\hat{\sigma}_{1A}^2$ for $0 \leq a_1 \leq \min\{\frac{1}{\lambda_1 q_1}, \frac{1}{\lambda_1 q_1} m_*\}$.
(ii) For $q_1 \geq 3$, there exists a nonnegative IQE with a uniformly smaller MSE than $\hat{\sigma}_{1A}^2$ if and only if $(q_1 - 2) \leq (q_1+2)m_*$, in which case $a_1 V_1$ is one such estimator for $\frac{q_1-2}{q_1(q_1+2)\lambda_1} \leq a_1 \leq \min\{\frac{1}{\lambda_1 q_1}, \frac{1}{\lambda_1 q_1} m_*\}$.

REMARK 3.1. Bayes or empirical Bayes estimators of variance components for the general model (1.1) are not available in the literature. We feel that while their derivation is possible as in the case of balanced models with two variance components, analytical study of their frequentist properties would be impossible.

4 Improved Estimation of the Error Variance in General Balanced Mixed Models

We now address the problem of improved estimation of the error variance σ^2 in the general balanced mixed effects model (1.1), and provide two classes of improved estimators, which can be considered as generalizations of Klotz, Milton and Zacks (1969) and Strawderman (1974). Our first result, which is a two-way generalization of a result of Klotz, Milton and Zacks (1969), is based on the ideas of Stein (1964). The results in this section are based on Mathew, Sinha and Sutradhar (1992b) and Kubokawa (1991). Recall that the ANOVA estimator of σ^2, namely $\hat{\sigma}_A^2$, is given by V_c/q_c and its best multiple is given by $\delta_0 = \frac{V_c}{q_c+2}$.

THEOREM 4.1 *The testimator defined by*

$$\tilde{\sigma^2} = \min\left[V_c/(q_c+2), (\sum_{j=1}^{c} V_j + \sum_{i=1}^{k} U_i)/(\sum_{j=1}^{c} q_j + \sum_{i=1}^{k} p_i + 2)\right] \tag{4.1}$$

has a uniformly smaller MSE than δ_0.

The main idea of the proof is to consider, following Stein (1964) and Sinha (1976), a class of estimators of σ^2 given by $\sigma^2* = (V_c/(q_c+2))\phi$ $(V_1/V_c, \cdots, V_{c-1}/V_c, U_1/V_c, \cdots, U_k/V_c)$ where ϕ is a positive real-valued function. It is then easy to verify that the conditional MSE of σ^2*, conditionally given $V_j/V_c = \zeta_j$ (say), $j = 1, \cdots, c-1$, and $U_i/V_c = \xi_i$ (say), $i = 1, \cdots, k$, is a minimum when the function $\phi(.)$ is chosen as $\phi_0(.|\delta, \beta, \lambda)$ given by

$$\phi_0(\zeta, \xi|\delta, \beta, \lambda) = \sigma^2(q_c+2)E_{\delta,\beta,\lambda}[V_c|\zeta,\xi]/E_{\delta,\beta,\lambda}[V_c^2|\zeta,\xi] \quad (4.2)$$

where $\zeta = (\zeta_1, \cdots, \zeta_{c-1})$, $\xi = (\xi_1, \cdots, \xi_k)$, $\delta = (\delta_1, \cdots, \delta_c)$, $\beta = (\beta_1, \cdots, \beta_k)$, $\lambda = (\lambda_1, \cdots, \lambda_k)$. Mathew, Sinha and Sutradhar (1992b) then show that

$$\begin{aligned}\phi_0(\zeta, \xi|\delta, \beta, \lambda) &\leq \sigma^2(q_c+2)(1+\sum_{j=1}^{c-1}\zeta_j + \sum_{i=1}^{k}\xi_i)/(\sum_{j=1}^{c} q_j + \sum_{i=1}^{k} p_i + 2) \\ &= \tilde{\phi}(\zeta, \xi) \text{ (say), whatever } \delta, \beta, \lambda. \quad (4.3)\end{aligned}$$

Using the convexity of the MSE criterion, it then follows that given any estimator of σ^2 of the form

$$\sigma^2* = (V_c/(q_c+2))\phi(\zeta, \xi), \quad (4.4)$$

the estimator given by

$$\tilde{\sigma^2} = (V_c/(q_c+2))\min[\phi(\zeta,\xi), \tilde{\phi}(\zeta,\xi)] \quad (4.5)$$

is uniformly better. Taking $\phi(\zeta, \xi) = 1$ proves the theorem.

REMARK 4.1. It is rather surprising that the components U_i's can be used to construct a testimator though a similar test for $H_0 : \lambda_i = 0$ *vs* $H_1 : \lambda_i > 0$ does not exist.

REMARK 4.2. It is also clear from the above proof that a class of testimators each better than the usual estimator $V_c/(q_c+2)$ can be constructed by choosing any subset of $V_1, \cdots, V_{c-1}, U_1, \cdots, U_k$.

We next proceed to state another class of improved estimators of the error variance σ^2, which are smooth in contrast to the improved nonsmooth estimators given in Theorem 4.1 above. Following Strawderman (1974), we consider two classes of estimators of σ^2, say $\tilde{\sigma}^2_{(j)}$ and $\tilde{\sigma}^2_{[i]}$ given by

$$\tilde{\sigma}^2_{(j)} = (V_c/(q_c+2))[1 - \epsilon_j V_c/(V_c+V_j)] \text{ for } j = 1, \cdots, c-1, \quad (4.6)$$

and

$$\tilde{\sigma}^2_{[i]} = (V_c/(q_c+2))[1 - \rho_i V_c/(V_c+U_i)] \text{ for } i = 1, \cdots, k, \quad (4.7)$$

where ϵ_j and ρ_i are positive scalars. The following theorem gives conditions on ϵ_j and ρ_i so that $\tilde{\sigma}^2_{(j)}$ and $\tilde{\sigma}^2_{[i]}$ have a uniformly smaller MSE than δ_0.

THEOREM 4.2 *The estimators* $\tilde{\sigma}^2_{(j)}$ *for* $j = 1, \cdots, c-1$, *and* $\tilde{\sigma}^2_{[i]}$ *for* $i = 1, \cdots, k$ *given by (4.6) and (4.7) respectively, are all uniformly better than the usual estimator* $V_c/(q_c + 2)$ *whenever*

$$0 < \epsilon_j < 4q_j/[(q_c + q_j + 2)(q_c + 4)], \ \ j = 1, \cdots, c-1 \qquad (4.8)$$

and

$$0 < \rho_i < 4p_i/[(q_c + p_i + 2)(q_c + 4)], \ \ i = 1, \cdots, k. \qquad (4.9)$$

Moreover, $\epsilon_{j\text{``}opt\text{''}} = 2q_j/[(q_c + q_j + 2)(q_c + 4)]$.

REMARK 4.3. It would be interesting to construct a smooth improved estimator of the error variance based on all the components $V_1, \cdots, V_c, U_1, \cdots, U_k$. However, so far such an estimator is not available.

Mathew, Sinha and Sutradhar (1992b) have tabulated the MSE's of the following three competing estimators of σ^2 in a balanced one-way model having k treatments and n observations per treatment for the values of k and n given in (2.24).

(*i*) the ANOVA estimator $\hat{\sigma}^2_A$
(*ii*) the nonquadratic nonnegative estimator from Theorem 4.1, which was also derived by Klotz, Milton and Zacks (1969) given by $\delta_1 = \min\{\frac{V_2}{k(n-1)+2}, \frac{V_1+V_2}{nk+1}\}$,
(*iii*) the estimator, say δ_2, obtained from Theorem 4.2 using the value $\epsilon_{\text{``opt''}}$ for ϵ.

The MSE's are given in Table 4.1. The MSE of $\hat{\sigma}^2_A$ was computed directly and those of δ_1 and δ_2 were based on simulations. The numerical results show that the MSE of δ_1, as well as that of δ_2, is significantly less than that of $\hat{\sigma}^2_A$. It is also clear that the MSE of δ_2 is smaller or larger than that of δ_1 depending on whether σ_1^2/σ^2 is small or large.

Similar to Theorem 2.2 in Section 2, we now exhibit a different class of estimators of σ^2 uniformly better than $\delta_0 = \frac{V_2}{k(n-1)+2}$ for the one-way balanced model. For a proof, we refer to Kubokawa (1991).

THEOREM 4.3 *Consider the class of estimators* $\hat{\sigma}^2_\phi = V_2\phi(\frac{V_2}{V_1})$, *where* ϕ *satisfies the following conditions:*
(i) ϕ *is nondecreasing*
(ii) $\lim_{w\to\infty}\phi(w) = \frac{1}{k(n-1)+2}$, *and*
(iii) $\phi_0(w) \leq \phi(w) \leq \frac{1}{k(n-1)+2}$, *where*

$$\phi_0(w) = \frac{1}{nk+1}\frac{\int_0^w z^{(k-3)/2}(1+z)^{-(nk+1)/2}dz}{\int_0^w z^{(k-3)/2}(1+z)^{-(nk+3)/2}dz}.$$

Then $\hat{\sigma}^2_\phi$ is uniformly better than δ_0.

For the one-way balanced model, the estimators in Theorems 4.1 and 4.2 can be shown to belong to the class of estimators $\hat{\sigma}^2_\phi$ in Theorem 4.3 for appropriate choices of the function ϕ satisfying the conditions in Theorem 4.3. It is also interesting to observe the similarity between the empirical Bayes estimator $\hat{\sigma}^2_{EB}$ of σ^2, given in (2.21), and $\hat{\sigma}^2_{\phi_0} = V_2\phi_0(V_2/V_1)$.

Table 4.1. MSE's of $\hat{\sigma}^2_A$, δ_1 and δ_2 for the one-way balanced models 2(*i*), 2(*ii*) and 2(*iii*) given in (2.24) with $\sigma^2 = 1$.

	σ_1^2	0.1	0.3	0.5	1.0	4.0	9.0
2(*i*)	$\hat{\sigma}^2_A$	0.1667	0.1667	0.1667	0.1667	0.1667	0.1667
	δ_1	0.0984	0.1008	0.1015	0.1048	0.1034	0.1026
	δ_2	0.1021	0.1034	0.1026	0.1045	0.1031	0.1023
2(*ii*)	$\hat{\sigma}^2_A$	0.1250	0.1250	0.1250	0.1250	0.1250	0.1250
	δ_1	0.0741	0.0762	0.0721	0.0793	0.0751	0.0725
	δ_2	0.0779	0.0780	0.0732	0.0790	0.0748	0.0723
2(*iii*)	$\hat{\sigma}^2_A$	0.0370	0.0370	0.0370	0.0370	0.0370	0.0370
	δ_1	0.0360	0.0366	0.0365	0.0368	0.0349	0.0345
	δ_2	0.0364	0.0368	0.0366	0.0368	0.0349	0.0344

5 Nonnegative Estimation of Individual Variance Components in Unbalanced Mixed Models With Two Variance Components

In this section we provide a survey of the results on the nonnegative estimation of the two variance components in the unbalanced model (1.5). For the estimation of σ_1^2, Mathew, Sinha and Sutradhar (1992c) obtained sufficient conditions for the existence of a nonnegative invariant quadratic estimator (IQE), uniformly better than *every* unbiased IQE. This is obviously quite a powerful result. The nonnegative IQE they have considered is a scalar multiple of $\mathbf{u}'W^+\mathbf{u}$ where, for W given in (1.5), W^+ denotes its Moore-Penrose generalized inverse, i.e., $W^+ = \sum_{i=1}^{g} \frac{1}{\lambda_i} E_i$. This choice of the nonnegative IQE follows quite naturally in this context when σ^2 is small (because $\mathbf{u}'W^+\mathbf{u}$ happens to be a complete sufficient statistic for σ_1^2 if indeed $\sigma^2 = 0$). Among such multiples, an admissible characterization is also available. It turns out that for the special case of an unbalanced one-way model, $\mathbf{u}'W^+\mathbf{u}$ is not a scalar multiple of the usual sum of squares due to treatments, although such multiples are quite natural at least in the balanced case, as observed earlier. For this special model, conditions are also stated on the number of treatments (k) and the treatment replication numbers ($\mathbf{n} = (n_1, ..., n_k)'$) under which $\hat{\sigma}^2_{1A}$ can be uniformly improved by a multiple of the treatment sum of squares.

To describe improved nonnegative quadratic estimators, let $\mathbf{u}'A\mathbf{u}$ be an IQE of σ_1^2 where we write $A = \sum_{i=1}^{g+1} a_i E_i$. Then the MSE of $\mathbf{u}'A\mathbf{u}$ is given by

$$\begin{aligned}\mathrm{MSE}(\mathbf{u}'A\mathbf{u}) &= \sigma^4\left\{2\sum_{i=1}^{g+1} m_i a_i^2 + \left(\sum_{i=1}^{g+1} m_i a_i^2\right)^2\right\} \\ &\quad +\sigma_1^4\left\{2\sum_{i=1}^{g} m_i\lambda_i^2 a_i^2 + \left(\sum_{i=1}^{g} m_i\lambda_i a_i - 1\right)^2\right\} \\ &\quad +2\sigma^2\sigma_1^2\left\{2\sum_{i=1}^{g} m_i\lambda_i a_i^2\right. \\ &\quad \left. + \left(\sum_{i=1}^{g+1} m_i a_i\right)\left(\sum_{i=1}^{g} m_i\lambda_i a_i - 1\right)\right\}. \qquad (5.1)\end{aligned}$$

If $B = \sum_{i=1}^{g+1} b_i E_i$, then it is obvious that $\mathbf{u}'B\mathbf{u}$ is an unbiased estimator of σ_1^2 if and only if $\sum_{i=1}^{g+1} m_i b_i = 0$ and $\sum_{i=1}^{g} m_i b_i \lambda_i = 1$. Also, the coefficients of σ^4, σ_1^4 and $2\sigma^2\sigma_1^2$ in the variance of $\mathbf{u}'B\mathbf{u}$ are respectively $2\sum_{i=1}^{g+1} m_i b_i^2$, $2\sum_{i=1}^{g} m_i\lambda_i^2 b_i^2$ and $2\sum_{i=1}^{g} m_i\lambda_i b_i^2$. It can be verified that the minimum values of these expressions (subject to the unbiasedness conditions) are respectively $2t/\{t\ \mathrm{tr}W^2 - (\mathrm{tr}W)^2\}$, $2/s$ and $2/\mathrm{tr}W$. Suppose $a_i \geq 0$, $\forall i$, so that $\mathbf{u}'A\mathbf{u}$ is nonnegative. It then follows from the above discussion that (5.2) and (5.3) below are necessary conditions for $\mathbf{u}'A\mathbf{u}$ to have a uniformly smaller MSE than every unbiased IQE of σ_1^2. Furthermore, (5.2), (5.3) and (5.4) are sufficient.

$$2\sum_{i=1}^{g+1} m_i a_i^2 + \left(\sum_{i=1}^{g+1} m_i a_i\right)^2 \leq \frac{2t}{t\ \mathrm{tr}(W^2) - (\mathrm{tr}W)^2} \qquad (5.2)$$

$$2\sum_{i=1}^{g} m_i\lambda_i^2 a_i^2 + \left(\sum_{i=1}^{g} m_i\lambda_i a_i - 1\right)^2 \leq \frac{2}{s} \qquad (5.3)$$

$$2\sum_{i=1}^{g} m_i\lambda_i a_i^2 + \left(\sum_{i=1}^{g+1} m_i a_i\right)\left(\sum_{i=1}^{g} m_i\lambda_i a_i - 1\right) \leq \frac{2}{\mathrm{tr}W}. \qquad (5.4)$$

To simplify the above conditions, Mathew, Sinha and Sutradhar (1992c) first show that for satisfying the inequality (5.3), it is enough to consider estimators which are scalar multiples of $\mathbf{u}'W^+\mathbf{u}$. For such an estimator of the type $\epsilon\mathbf{u}'W^+\mathbf{u}$ for some $\epsilon > 0$, (5.3) is equivalent to $2s\epsilon^2 + (\epsilon s - 1)^2 \leq \frac{2}{s}$ or, equivalently, to $\max\left\{0, \frac{s-2}{s(s+2)}\right\} \leq \epsilon \leq \frac{1}{s}$. Moreover, note that when ϵ is in the range

$$\max\left\{0, \frac{s-2}{s(s+2)}\right\} \leq \epsilon \leq \frac{1}{s+2}, \qquad (5.5)$$

the estimator $\epsilon\mathbf{u}'W^+\mathbf{u}$ also satisfies (5.4). Therefore, it only remains to derive the condition on ϵ so that (5.2) holds. For the estimator $\epsilon\mathbf{u}'W^+\mathbf{u}$, noting that $\text{tr}W^+ = \sum_{i=1}^{g} \frac{m_i}{\lambda_i}$ and $\text{tr}\,(W^+)^2 = \sum_{i=1}^{g} \frac{m_i}{\lambda_i^2}$, (5.2) simplifies to

$$\epsilon^2 \left\{2\text{tr}\left(W^{+^2}\right) + \left(\text{tr}W^+\right)^2\right\} \leq \frac{2t}{t\ \text{tr}(W^2) - (\text{tr}W)^2}. \tag{5.6}$$

Thus if ϵ satisfies (5.5) and (5.6), it also satisfies (5.2), (5.3) and (5.4). Let

$$\epsilon_* = \left[\frac{2t}{\{t\ \text{tr}(W^2) - (\text{tr}W)^2\}\{2\text{tr}(W^{+^2}) + (\text{tr}W^+)^2\}}\right]^{\frac{1}{2}}. \tag{5.7}$$

The above considerations lead to the following theorem.

THEOREM 5.1 *Consider the model (1.5) and let ϵ_* be given by (5.7). (i) If $s = 1$ or 2, the nonnegative IQE $\epsilon\mathbf{u}'W^+\mathbf{u}$ has a uniformly smaller MSE than every unbiased IQE of σ_1^2 whenever $0 \leq \epsilon \leq \min\{\epsilon_*, \frac{1}{s+2}\}$. (ii) For $s \geq 3$, the nonnegative IQE $\epsilon\mathbf{u}'W^+\mathbf{u}$ has a uniformly smaller MSE than every unbiased IQE of σ_1^2 if $\frac{(s-2)}{s(s+2)} \leq \epsilon_*$. When this condition holds, the estimator $\epsilon\mathbf{u}'W^+\mathbf{u}$ has the stated property whenever $\frac{(s-2)}{s(s+2)} \leq \epsilon \leq \min\{\epsilon_*, \frac{1}{s+2}\}$.*

REMARK 5.1. We have not been able to conclude if, in the context of Theorem 5.1, it is indeed enough to consider IQE's that are scalar multiples of $\mathbf{u}'W^+\mathbf{u}$. In other words, it is not clear if there will exist a nonnegative IQE different from a scalar multiple of $\mathbf{u}'W^+\mathbf{u}$ uniformly better than every unbiased IQE of σ_1^2 when a nonnegative scalar multiple of $\mathbf{u}'W^+\mathbf{u}$ fails to satisfy this requirement, although, as noted before, such scalar multiples are natural candidates in this context.

Mathew, Sinha and Sutradhar (1992c) also proved the following admissibility result in a somewhat restricted context.

THEOREM 5.2 *If $\epsilon \geq 0$, the estimator $\epsilon\mathbf{u}'W^+\mathbf{u}$ (for estimating σ_1^2) is admissible among all estimators that are scalar multiples of $\mathbf{u}'W^+\mathbf{u}$ if and only if $\epsilon \leq \frac{1}{s+2}$.*

We now state some specific results for the unbalanced one-way random effects model. It is readily verified that the treatment sum of squares, say SS_{tr}, is given by $SS_{tr} = \sum_{i=1}^{g} \mathbf{u}'E_i\mathbf{u}$. Thus the estimator $\epsilon\mathbf{u}'W^+\mathbf{u}$ in Theorem 5.1 is not a scalar multiple of SS_{tr}. Therefore it makes sense to explore the possibility of improving the ANOVA estimator $\hat{\sigma}_{1A}^2$ of σ_1^2 using scalar multiples of SS_{tr}.

Let there be k treatments and let n_i denote the replication number for the ith treatment. Write $\mathbf{n} = (n_1, \ldots, n_k)'$, $\sum_{i=1}^{k} n_i = N$ and $J_m = \mathbf{1}_m\mathbf{1}_m'$, where $\mathbf{1}_m$ is the $m \times 1$ vector of ones. If $\mathbf{y}$ is the $N \times 1$ vector of observations, then $E(\mathbf{y}) = \mu \mathbf{1}_N$, where μ is the general mean and $Cov(\mathbf{y}) = \sigma^2 I_N + \sigma_1^2 \text{diag}(J_{n_1}, \ldots, J_{n_k})$. Furthermore,

$$SS_{tr} = \mathbf{y}' \left[\text{diag}(\frac{1}{n_1} J_{n_1}, \ldots, \frac{1}{n_k} J_{n_k}) - \frac{1}{N} J_N \right] \mathbf{y}.$$

To derive the conditions under which an estimator aSS_{tr} (for $a \geq 0$) will be uniformly better than the ANOVA estimator of σ_1^2, straightforward algebra gives

$$\begin{aligned} & E(aSS_{tr} - \sigma_1^2)^2 = \sigma^4 a^2 (k-1)(k+1) \\ & +\sigma_1^4 \left\{ 2\frac{a^2}{N^2} (N^2 \Sigma n_i^2 + (\Sigma n_i^2)^2 - 2N \Sigma n_i^3) + \left[\frac{a}{N}(N^2 - \Sigma n_i^2) - 1 \right]^2 \right\} \\ & +2\sigma^2 \sigma_1^2 \left\{ 2\frac{a^2}{N}(N^2 - \Sigma n_i^2) + a(k-1) \left[\frac{a}{N}(N^2 - \Sigma n_i^2) - 1 \right] \right\}. \end{aligned} \tag{5.8}$$

On the other hand, the variance of the ANOVA estimator of σ_1^2 is given by (see Searle (1971), p. 474)

$$\begin{aligned} \text{Var}(\hat{\sigma}_{1A}^2) &= \sigma^4 \frac{2N^2(N-1)(k-1)}{(N-k)(N^2 - \Sigma n_i^2)^2} \\ & +\sigma_1^4 \frac{2\left\{ N^2 \Sigma n_i^2 + (\Sigma n_i^2)^2 - 2N \Sigma n_i^3 \right\}}{(N^2 - \Sigma n_i^2)^2} \\ & +4\sigma^2 \sigma_1^2 \frac{N}{N^2 - \Sigma n_i^2}. \end{aligned} \tag{5.9}$$

For comparing (5.8) with (5.9), consider the following inequalities similar to (5.2), (5.3) and (5.4).

$$a^2(k-1)(k+1) \leq \frac{2N^2(N-1)(k-1)}{(N-k)\epsilon_1^2} \tag{5.10}$$

$$\frac{2a^2}{N^2}\epsilon_2 + \left(\frac{a}{N}\epsilon_1 - 1\right)^2 \leq \frac{2\epsilon_2}{\epsilon_1^2} \tag{5.11}$$

$$\frac{2a^2}{N}\epsilon_1 + a(k-1)\left(\frac{a}{N}\epsilon_1 - 1\right) \leq \frac{2N}{\epsilon_1} \tag{5.12}$$

where

$$\epsilon_1 = N^2 - \Sigma n_i^2 \tag{5.13}$$

and

$$\epsilon_2 = N^2 \Sigma n_i^2 + (\Sigma n_i^2)^2 - 2N \Sigma n_i^3. \tag{5.14}$$

Obviously (5.10) and (5.11) are necessary conditions for aSS_{tr} to be uniformly better than $\hat{\sigma}_{1A}^2$. (5.10), (5.11) and (5.12) are, of course, sufficient. It is not difficult to show that (5.11) implies (5.12) so that (5.10) and (5.11) are necessary and sufficient, and hence it is enough to characterize scalars $a \geq 0$ satisfying (5.10) and (5.11). Let

$$a_* = \left\{ \frac{2N^2(N-1)}{(N-k)(k+1)\epsilon_1^2} \right\}^{\frac{1}{2}}. \tag{5.15}$$

The following result is proved in Mathew, Sinha and Sutradhar (1992c).

THEOREM 5.3 *Consider the unbalanced one-way random effects model and let* ϵ_1, ϵ_2 *and* a_* *be as given in (5.13), (5.14) and (5.15) respectively. (i) If* $\epsilon_1^2 \leq 2\epsilon_2$, *the nonnegative IQE* aSS_{tr} *has a uniformly smaller MSE than* $\hat{\sigma}_{1A}^2$ *whenever* $0 \leq a \leq \min\left\{a_*, \frac{N}{\epsilon_1}\right\}$. *(ii) If* $\epsilon_1^2 > 2\epsilon_2$, *there exists* $a > 0$ *such that the estimator* aSS_{tr} *has a uniformly smaller MSE than* $\hat{\sigma}_{1A}^2$ *if and only if* $\frac{N(\epsilon_1^2-2\epsilon_2)}{\epsilon_1(\epsilon_1^2+2\epsilon_2)} \leq a_*$. *When this condition holds,* aSS_{tr} *has the stated property whenever* $\frac{N(\epsilon_1^2-2\epsilon_2)}{\epsilon_1(\epsilon_1^2+2\epsilon_2)} \leq a \leq \min\left\{a_*, \frac{N}{\epsilon_1}\right\}$.

If the conditions specified in Theorem 5.1 (with $t = N-1$ and $s = k-1$) are satisfied by the one-way unbalanced model, then clearly we have a scalar multiple of $\mathbf{u}'W^+\mathbf{u}$ better than *every* unbiased IQE of σ_1^2 (in particular, better than the ANOVA estimator and the MINQUE of σ_1^2). On the other hand, if the conditions in Theorem 5.3 hold we have a scalar multiple of SS_{tr} uniformly better than the ANOVA estimator of σ_1^2. In Table 5.1, we have reported the MSE's and squared biases of several competing estimators of the random effect variance component σ_1^2 for the unbalanced one-way model. If $\mathbf{n}$ denotes the vector of treatment replications, the one-way unbalanced models considered in Table 5.1 are given by

$$\begin{aligned} &5(i)\ \mathbf{n} = (3,15,27)';\ 5(ii)\ \mathbf{n} = (3,3,5,5,7,7)';\\ &5(iii)\ \mathbf{n} = (1,1,5,5,9,9)'; 5(iv)\ \mathbf{n} = (1,1,1,1,13,13)';\\ &5(v)\ \mathbf{n} = (3,3,3,5,5,5,7,7,7)' \text{ and}\\ &5(vi)\ \mathbf{n} = (1,1,1,5,5,5,9,9,9)'. \end{aligned} \tag{5.16}$$

In Table 5.1, $\hat{\sigma}_{1A}^2$ and $\hat{\sigma}_{1M}^2$ respectively denote the ANOVA estimator and MINQUE of σ_1^2 and $\hat{\sigma}_{1i}^2$ $(i = 1, 2, 3, 4)$ denote four estimators we have proposed in this section, given by

$$\begin{aligned} \hat{\sigma}_{11}^2 &= \frac{1}{(k+1)}\mathbf{u}'V_1^+\mathbf{u},\\ \hat{\sigma}_{12}^2 &= \begin{cases} 0 & \text{if } k \leq 3\\ \frac{(k-3)\mathbf{u}'V_1^+\mathbf{u}}{(k-1)(k+1)} & \text{if } k \geq 4, \end{cases}\\ \hat{\sigma}_{13}^2 &= \frac{N\epsilon_1}{(\epsilon_1^2+2\epsilon_2)}SS_{tr}, \end{aligned}$$

$$\text{and } \hat{\sigma}_{14}^2 = \begin{cases} 0 & \text{if } \epsilon_1^2 \le 2\epsilon_2 \\ \frac{N(\epsilon_1^2-2\epsilon_2)}{\epsilon_1(\epsilon_1^2+2\epsilon_2)} SS_{tr} & \text{if } \epsilon_1^2 > 2\epsilon_2. \end{cases}$$

In Table 5.1, the squared biases are given (within brackets) only for the four estimators $\hat{\sigma}_{1i}^2$ (i = 1, 2, 3, 4), since $\hat{\sigma}_{1A}^2$ and $\hat{\sigma}_{1M}^2$ are unbiased. We note that the estimators $\hat{\sigma}_{12}^2$ and $\hat{\sigma}_{14}^2$ are respectively of the form $\epsilon_0 \mathbf{u}' V_1^+ \mathbf{u}$ and $a_0 SS_{tr}$, where ϵ_0 is the smallest possible value of ϵ in Theorem 5.1, and a_0 is the smallest possible value of a in Theorem 5.3. The estimators $\hat{\sigma}_{11}^2$ and $\hat{\sigma}_{13}^2$ were included in the numerical comparisons since the coefficient of σ_1^4 in MSE($\epsilon \mathbf{u}' V_1^+ \mathbf{u}$) and MSE($aSS_{tr}$) are respectively minimized by $\epsilon = \frac{1}{(k+1)}$ and $a = \frac{N\epsilon_1}{(\epsilon_1^2+2\epsilon_2)}$. Thus the estimators $\hat{\sigma}_{11}^2$ and $\hat{\sigma}_{13}^2$ are expected to perform well when σ_1^2 is large relative to σ^2. The values of k and the vector $\mathbf{n}$ in Table 5.1 are selected from those considered by Swallow and Monahan (1984). The results in Table 5.1 correspond to $\sigma^2 = 1$. Among the vectors $\mathbf{n}$ we have considered, $\hat{\sigma}_{12}^2$ is obviously zero for the case $k = 3$. It can be verified that the vector $\mathbf{n} = (3, 3, 5, 5, 7, 7)$ satisfies the condition in Theorem 5.1 (*ii*) and the remaining choices do not satisfy this condition. The choices $(3, 15, 27)'$ and $(1, 1, 1, 1, 13, 13)'$ satisfy $\epsilon_1^2 \le 2\epsilon_2$. The remaining choices of $\mathbf{n}$ in Table 5.1 satisfy $\epsilon_1^2 > 2\epsilon_2$ along with the condition in Theorem 5.3 (*ii*). For the values of $\mathbf{n}$ considered in Table 5.1, the significant reduction in MSE achieved by the proposed estimators over $\hat{\sigma}_{1A}^2$ and $\hat{\sigma}_{1M}^2$ is obvious. However, the estimators $\hat{\sigma}_{12}^2$ and $\hat{\sigma}_{14}^2$ have a substantial amount bias. If σ_1^2/σ^2 is not very small, the estimator $\hat{\sigma}_{11}^2$ has a definite edge over the others, in terms of MSE. Furthermore, its bias is not significantly large. It should be noted that the estimation of σ_1^2 is of practical interest only when the ratio σ_1^2/σ^2 is not too small. When this is the case, the estimator we recommend is $\hat{\sigma}_{11}^2$.

Table 5.1. MSE's and squared biases of $\hat{\sigma}_{1A}^2$, $\hat{\sigma}_{1M}^2$, $\hat{\sigma}_{11}^2$, $\hat{\sigma}_{12}^2$, $\hat{\sigma}_{13}^2$ and $\hat{\sigma}_{14}^2$ for the one-way unbalanced models 5(i)–5(vi) in (5.16) with $\sigma^2 = 1$ (the squared biases are given within brackets).

σ_1^2		0.1	0.3	0.5	1.0	4.0	9.0
5(i)	$\hat{\sigma}_{1A}^2$	0.039	0.186	0.447	1.595	23.378	116.413
	$\hat{\sigma}_{1M}^2$	0.043	0.229	0.563	2.044	30.337	151.404
	$\hat{\sigma}_{11}^2$	0.018	0.058	0.138	0.513	8.013	40.513
		(0.000)	(0.006)	(0.031)	(0.183)	(3.714)	(19.600)
	$\hat{\sigma}_{12}^2$	0.010	0.090	0.250	1.000	16.000	81.000
		(0.010)	(0.090)	(0.250)	(1.000)	(16.000)	(81.000)
	$\hat{\sigma}_{13}^2$	0.007	0.052	0.143	0.577	9.338	47.398
		(0.001)	(0.020)	(0.067)	(0.304)	(5.340)	(27.492)
	$\hat{\sigma}_{14}^2$	0.010	0.090	0.250	1.000	16.000	81.000
		(0.010)	(0.090)	(0.250)	(1.000)	(16.000)	(81.000)

(*contd.*)

Table 5.1 (*contd.*)

σ_1^2		0.1	0.3	0.5	1.0	4.0	9.0
5(*ii*)	$\hat{\sigma}_{1A}^2$	0.041	0.109	0.211	0.620	7.656	36.840
	$\hat{\sigma}_{1M}^2$	0.042	0.122	0.246	0.748	9.535	46.174
	$\hat{\sigma}_{11}^2$	0.040	0.063	0.109	0.323	4.609	23.180
		(0.018)	(0.006)	(0.000)	(0.016)	(0.964)	(5.810)
	$\hat{\sigma}_{12}^2$	0.010	0.026	0.075	0.336	6.104	31.718
		(0.002)	(0.006)	(0.036)	(0.226)	(4.792)	(25.465)
	$\hat{\sigma}_{13}^2$	0.031	0.054	0.101	0.325	4.861	24.572
		(0.013)	(0.003)	(0.000)	(0.026)	(1.151)	(6.719)
	$\hat{\sigma}_{14}^2$	0.006	0.027	0.082	0.373	6.702	34.707
		(0.000)	(0.010)	(0.050)	(0.278)	(5.524)	(29.036)
5(*iii*)	$\hat{\sigma}_{1A}^2$	0.046	0.128	0.255	0.775	9.957	48.344
	$\hat{\sigma}_{1M}^2$	0.044	0.151	0.323	1.043	14.005	68.530
	$\hat{\sigma}_{11}^2$	0.165	0.188	0.234	0.448	4.734	23.305
		(0.080)	(0.051)	(0.029)	(0.001)	(0.690)	(5.104)
	$\hat{\sigma}_{12}^2$	0.048	0.050	0.084	0.309	5.859	31.109
		(0.017)	(0.000)	(0.010)	(0.148)	(4.403)	(24.558)
	$\hat{\sigma}_{13}^2$	0.028	0.051	0.103	0.362	5.757	29.387
		(0.010)	(0.008)	(0.002)	(0.052)	(1.756)	(9.953)
	$\hat{\sigma}_{14}^2$	0.003	0.035	0.112	0.509	8.946	46.093
		(0.000)	(0.026)	(0.094)	(0.045)	(8.230)	(42.617)
5(*iv*)	$\hat{\sigma}_{1A}^2$	0.067	0.192	0.401	1.283	17.430	85.677
	$\hat{\sigma}_{1M}^2$	0.059	0.265	0.620	2.156	30.867	152.989
	$\hat{\sigma}_{11}^2$	0.376	0.399	0.445	0.659	4.945	23.516
		(0.217)	(0.167)	(0.124)	(0.044)	(0.420)	(4.134)
	$\hat{\sigma}_{12}^2$	0.115	0.099	0.116	0.297	5.585	30.398
		(0.057)	(0.016)	(0.000)	(0.076)	(3.956)	(23.485)
	$\hat{\sigma}_{13}^2$	0.021	0.046	0.110	0.450	7.827	40.449
		(0.007)	(0.000)	(0.015)	(0.141)	(3.613)	(19.729)
	$\hat{\sigma}_{14}^2$	0.010	0.090	0.250	1.000	16.000	81.000
		(0.010)	(0.090)	(0.250)	(1.000)	(16.000)	(81.000)
5(*v*)	$\hat{\sigma}_{1A}^2$	0.025	0.068	0.132	0.388	4.807	23.142
	$\hat{\sigma}_{1M}^2$	0.026	0.076	0.155	0.472	6.045	29.294
	$\hat{\sigma}_{11}^2$	0.044	0.060	0.092	0.242	3.242	16.242
		(0.026)	(0.015)	(0.007)	(0.000)	(0.384)	(2.623)
	$\hat{\sigma}_{12}^2$	0.019	0.026	0.052	0.206	3.753	19.664
		(0.009)	(0.000)	(0.004)	(0.070)	(2.146)	(12.005)
	$\hat{\sigma}_{13}^2$	0.033	0.049	0.082	0.241	3.451	17.412
		(0.019)	(0.009)	(0.003)	(0.003)	(0.493)	(3.162)
	$\hat{\sigma}_{14}^2$	0.013	0.021	0.052	0.224	4.139	21.635
		(0.005)	(0.000)	(0.010)	(0.099)	(2.581)	(14.130)
5(*vi*)	$\hat{\sigma}_{1A}^2$	0.028	0.078	0.156	0.479	6.191	30.100
	$\hat{\sigma}_{1M}^2$	0.026	0.092	0.198	0.641	8.644	42.335
	$\hat{\sigma}_{11}^2$	0.177	0.193	0.225	0.375	3.375	16.375
		(0.109)	(0.084)	(0.062)	(0.022)	(0.203)	(2.104)
	$\hat{\sigma}_{12}^2$	0.088	0.082	0.095	0.217	3.574	19.168
		(0.049)	(0.020)	(0.004)	(0.019)	(1.790)	(11.141)
	$\hat{\sigma}_{13}^2$	0.030	0.047	0.084	0.270	4.163	21.233
		(0.017)	(0.006)	(0.001)	(0.012)	(0.815)	(4.953)
	$\hat{\sigma}_{14}^2$	0.078	0.020	0.061	0.290	5.440	28.411
		(0.002)	(0.004)	(0.027)	(0.185)	(4.068)	(21.737)

We next deal with the estimation of σ^2, and as in the balanced case, provide two sets of nonnegative estimators which are uniformly better than the biased estimator $\delta_0 = \frac{1}{t-s+2}\mathbf{u}'E_{g+1}\mathbf{u}$, and hence the ANOVA estimator $\hat{\sigma}^2_A$. It is important to observe that the estimators we have proposed here provide uniform improvement only over $\hat{\sigma}^2_A$. These estimators may or may not be better than other unbiased IQE's of σ^2. In fact, the IQE δ_0 does not always dominate $\hat{\sigma}^2_M$. We have not been able to exhibit a nonnegative IQE uniformly better than $\hat{\sigma}^2_M$ or other unbiased estimators of σ^2. This is an important problem since all of them can take negative values. It is worth pointing out that $\hat{\sigma}^2_A$ given in (1.8) is the unique nonnegative unbiased IQE of σ^2 (as a function of the minimal sufficient statistic). Thus any other unbiased IQE of σ^2 is bound to take negative values in the unbalanced case. In particular, $\hat{\sigma}^2_M$ given in (1.9) can be negative.

Theorem 5.4 below, patterned after Stein (1964), exhibits a class of nonquadratic estimators that are uniformly better than δ_0. As mentioned before, such a testimator for σ^2 for the one-way balanced model has been derived by Klotz, Milton and Zacks (1969) and our result in Theorem 5.4 accomplishes the same in the unbalanced case, thus providing a useful generalization. A different class of nonnegative improved estimators, based on the ideas in Strawderman (1974), is presented in Theorem 5.5. Both the theorems are due to Mathew, Sinha and Sutradhar (1992c).

For notational convenience, we shall write $e_i^2 = \mathbf{u}'E_i\mathbf{u}$, $i = 1, 2, \ldots, g+1$. Then the e_i^2's are independent, and $\frac{e_{g+1}^2}{\sigma^2}$ and $\frac{e_i^2}{\sigma^2(1+\theta\lambda_i)}$ $(i = 1, 2, \ldots, g)$ have central chisquare distributions with degrees of freedom $t-s$ and m_i respectively. Here $\theta = \sigma_1^2/\sigma^2$ and λ_i's are defined in (1.6). Let $\lambda^* = \max(\lambda_1, \ldots, \lambda_g)$ and $\lambda_* = \min(\lambda_1, \ldots, \lambda_g)$.

THEOREM 5.4 *The testimator* $\delta_1 = \min\left\{\frac{e_{g+1}^2}{t-s+2}, \frac{\sum_{i=1}^{g+1} e_i^2}{t+2}\right\}$ *is uniformly better than* δ_0.

THEOREM 5.5 *(i) For any* $i = 1, 2, \ldots, g$, *the estimator* $\delta_i^* = \frac{e_{g+1}^2}{t-s+2} \times \left\{1 - \epsilon_i \frac{e_{g+1}^2}{e_i^2 + e_{g+1}^2}\right\}$ *is uniformly better than* δ_0 *whenever* $0 < \epsilon_i \leq \frac{4m_i}{(t-s+4)(t-s+m_i+2)}$.

(ii) The estimator $\delta_2 = \frac{e_{g+1}^2}{t-s+2}\left\{1 - \epsilon\frac{e_{g+1}^2}{\sum_{i=1}^{g+1} e_i^2}\right\}$ *is uniformly better than* δ_0 *whenever* $0 < \epsilon \leq \epsilon_*$, *where* $\epsilon_* = 2\left(\frac{\lambda_*}{\lambda^*}\right)^5 \cdot \frac{(t+4)(t-s+2)}{(t+2)(t-s+4)} \times \left[\frac{(t+2)(t-s+4)}{(t+4)(t-s+2)} - \left(\frac{\lambda^*}{\lambda_*}\right)^5\right]$.

Proof of Theorem 5.4 is based upon some ideas from Stein (1964), Sinha (1976) and Shorrock and Zidek (1976), and depends on estimators of σ^2 of the form

$$\delta = e_{g+1}^2 \phi \left(\sum_{i=1}^{g} \frac{e_i^2}{e_{g+1}^2} \right) \tag{5.17}$$

where ϕ is a nonnegative real valued function. Writing $\tilde{e}_i^2 = e_i^2/\sigma^2$, it follows that the MSE of δ, say $R_\theta(\delta)$, can be written as

$$\begin{aligned} R_\theta(\delta) &= E_\theta \left\{ e_{g+1}^2 \phi \left(\sum_{i=1}^{g} \frac{e_i^2}{e_{g+1}^2} \right) - \sigma^2 \right\}^2 \\ &= \sigma^4 E_\theta \left\{ \tilde{e}_{g+1}^2 \phi \left(\sum_{i=1}^{g} \frac{\tilde{e}_i^2}{\tilde{e}_{g+1}^2} \right) - 1 \right\}^2 \\ &= \sigma^4 E_\theta \left[E_\theta \{ S\phi(U) - 1 \}^2 | U = u \right] \end{aligned} \tag{5.18}$$

where we write

$$S = \tilde{e}_{g+1}^2, \quad U = \sum_{i=1}^{g} \frac{\tilde{e}_i^2}{\tilde{e}_{g+1}^2} \tag{5.19}$$

and the inner expectation in (5.18) is with respect to S conditional on U while the outer expectation is with respect to U. Clearly, the inner conditional expectation, given U, is a minimum when $\phi(\cdot)$ is chosen as

$$\phi(u) = \frac{E_\theta(S|U=u)}{E_\theta(S^2|U=u)} = \phi_\theta(u) \text{ (say)}. \tag{5.20}$$

Mathew, Sinha and Sutradhar (1992c) then prove that for $\theta \geq 0$,

$$\phi_\theta(u) \leq \phi_0(u) \tag{5.21}$$

and

$$\phi_0(u) = \frac{E_0(S|U=u)}{E_0(S^2|U=u)} = \frac{1+u}{t+2}. \tag{5.22}$$

It then follows from the convexity with respect to ϕ of the conditional expectation in (5.18) that given any ϕ and the resultant δ defined in (5.17), $\delta^* = e_{g+1}^2 \phi^* \left(\sum_{i=1}^{g} e_i^2 / e_{g+1}^2 \right)$, with $\phi^*(u) = \min(\phi(u), \phi_0(u))$, has a smaller conditional and hence unconditional MSE than δ. In particular, choosing $\phi = \frac{1}{t-s+2}$ proves the claim.

The proof of Theorem 5.5 is rather involved and is skipped (see Mathew, Sinha and Sutradhar (1992c) for details).

6 Some Open Problems

The results given in the present paper apply only to univariate mixed models. Even in the univariate case, many problems remain to be addressed and furthermore, extension of our results to the multivariate case remains open. Listed below are a few problems that are currently under investigation.

1. Numerical computations for the one-way balanced model indicate that the empirical Bayes estimators of σ_1^2 and σ^2 have substantially smaller MSE's compared to their respective ANOVA estimators, namely, $\hat{\sigma}_{1A}^2$ and $\hat{\sigma}_A^2$. However, a theoretical proof of this is not yet available. Using Theorem 2.2 and Theorem 4.3, it may be possible to theoretically establish the superiority of the empirical Bayes estimators over the ANOVA estimators.

2. Theorem 5.1 provides a sufficient condition for the existence of a nonnegative IQE uniformly better than every unbiased IQE of σ_1^2, and Theorem 5.3 provides a necessary and sufficient condition for the existence of a scalar multiple of the treatment sum of squares uniformly better than the ANOVA estimator of σ_1^2, both in the unbalanced one-way random model. When Theorem 5.1 fails, the problem of deriving conditions for the existence of nonnegative IQE's uniformly better than other specific well known unbiased IQE's of σ_1^2 (for example the MINQUE $\hat{\sigma}_{1M}^2$ of σ_1^2 given in (1.9)) remains open.

3. For the one-way balanced model, Theorem 2.2 and Theorem 4.3 provide classes of estimators $\hat{\sigma}_{1\phi}^2$ and $\hat{\sigma}_\phi^2$ that are uniformly better than $\hat{\sigma}_{1A}^2$ and $\hat{\sigma}_A^2$ respectively. A similar derivation in the unbalanced case remains open.

4. Derivation of satisfactory nonnegative definite estimators of multivariate components of variance in balanced and unbalanced multivariate mixed models is another important open problem, because of its wide applications (see Laird, Lange and Stram (1987)). Some results in the balanced case have recently been obtained by Calvin and Dykstra (1991a, 1991b) and Mathew, Niyogi and Sinha (1992).

Acknowledgment

The authors are grateful to a referee for providing additional relevant references and some helpful comments.

REFERENCES

Box, G. E. P. and Tiao, G. C. (1973) *Bayesian Inference in Statistical Analysis*, Addison-Wesley, Reading, MA.

Broemeling, L. D. (1985) *Bayesian Analysis of Linear Models*, Marcel Dekker, New York.

Calvin, J. A. and Dykstra, R. L. (1991a) Maximum likelihood estimation of a set of covariance matrices under loewner order restrictions with applications to balanced multivariate variance components models, *The Annals of Statistics*, **19**, 850-869.

Calvin, J. A. and Dykstra, R. L. (199b) Least squares estimation of covariance matrices in balanced multivariate variance components models, *Journal of the American Statistical Association*, **86**, 388-395.

Chaubey, Y. P. (1983) A non-negative estimator of variance component closest to the MINQUE, *Sankhya, Series A*, **45**, 201-211.

Chow, S. C. and Shao, J. (1988) A new procedure for the estimation of variance components, *Statistics and Probability Letters*, **6**, 349-355.

Conerly, M. D. and Webster, J. T.(1987) MINQE for the one-way classification, *Technometrics*, **29**, 229-237.

Corbeil, R. R. and Searle, S. R. (1976) A comparison of variance component estimators, *Biometrics*, **32**, 779-791.

Gnot, S., Kleffe, J. and Zmyslony, R. (1985) Nonnegativity of admissible invariant quadratic estimates in mixed linear models with two variance components, *Journal of Statistical Planning and Inference*, **12**, 249-258.

Hartung, J. (1981) Nonnegative minimum biased invariant estimation in variance component models, *The Annals of Statistics*, **9**, 278-292.

Herbach, L. (1959) Properties of type II analysis of variance tests, *The Annals of Mathematical Statistics*, **30**, 939-959.

Hill, B. (1965) Inferences about variance components in the one-way model, *Journal of the American Statistical Association*, **60**, 806-825.

Kelly, R. J. and Mathew, T. (1992) Improved estimators of variance components having smaller probability of negativity, *Journal of the Royal Statistical Society, Series B*, to appear.

Kleffe, J. (1984) Some remarks on improving unbiased estimators by multiplying with a constant, in *Linear Statistical Inference* (eds. T. Calinski and W. Klonecki), New York: Springer-Verlag, 150-161.

Kleffe, J. and Rao, J. N. K. (1986) The existence of asymptotically unbiased nonnegative quadratic estimates of variance components in ANOVA models, *Journal of the American Statistical Association*, **81**, 692-698.

Klotz, J. H., Milton, R. C. and Zacks, S. (1969) Mean square efficiency of estimators of variance components, *Journal of the American Statistical Association*, **64**, 1383-1402.

Kubokawa, T. (1991) A unified approach to improving equivariant estimators, *submitted for publication.*

Laird, N., Lange, N. and Stram, D. (1987) Maximum likelihood computations with repeated measures: application of the EM algorithm, *Journal of the American Statistical Association*, **82**, 97-105.

LaMotte, L. R. (1973) On nonnegative quadratic unbiased estimation of variance components, *Journal of the American Statistical Association*, **68**, 728-730.

LaMotte, L. R. (1979) On admissible invariant quadratic estimators of variance components, in *Variance Components and Animal Breeding, Proceedings of a Conference in Honor of C. R. Henderson*, (eds. L. D. van Vleck and S. R. Searle), Cornell University, 181-188.

LaMotte, L. R. (1985) Admissibility, unbiasedness, and nonnegativity in the balanced random one-way ANOVA model, in *Linear Statistical Inference* (eds. T. Calinski and W. Klonecki), New York: Springer-Verlag, 184-199.

Mathew, T. (1984) On nonnegative quadratic unbiased estimability of variance components, *The Annals of Statistics*, **12**, 1566-1569.

Mathew, T., Niyogi, A. and Sinha, B. K. (1992) Improved nonnegative estimators of variance components in balanced multivariate mixed models, *submitted for publication.*

Mathew. T. and Sinha, B. K. (1988) Optimum tests for fixed effects and variance components in balanced models, *Journal of the American Statistical Association*, **83**, 133-135.

Mathew, T., Sinha, B. K. and Sutradhar, B. (1992a) Nonnegative estimation of random effects variance components in balanced mixed models, in *Nonparametric Statistics and Related Topics* (ed. A. K. Md. Saleh), North-Holland, Amsterdam, pp. 281-295.

Mathew, T., Sinha, B. K. and Sutradhar, B. (1992b) Improved estimation of error variance in general balanced mixed models, *Statistics and Decisions*, **10**, 227-238.

Mathew, T., Sinha, B. K. and Sutradhar, B. (1992c) Nonnegative estimation of variance components in unbalanced mixed models with two variance components, *Journal of Multivariate Analysis*, **42**, 77-101.

Niyogi, A. (1992) *Some Contributions to Nonnegative Estimation of Variance Components*, unpublished Ph.D. dissertation, University of Maryland Baltimore County.

Perlman, M. D. (1972) Reduced mean squared error estimation for several parameters, *Sankhya, Series B*, 89-92.

Portnoy, S. (1971) Formal bayes estimation with application to a random effect model, *The Annals of Mathematical Statistics*, **42**, 1379-1402.

Pukelsheim, F. (1981a) On the existence of unbiased nonnegative estimates of variance covariance components, *The Annals of Statistics*, **9**, 293-299.

Pukelsheim, F. (1981b) Linear models and convex geometry: aspects of nonnegative variance estimation, *Mathematische Operationsforschung und Statistics, Series Statistics*, **12**, 271-286.

Rao, C. R. and Kleffe, J. (1988) *Estimation of Variance Components and Applications*, North-Holland, Amsterdam.

Rao, P. S. R. S. and Chaubey, Y. P. (1978) Three modifications of the principle of MINQUE, *Communications in Statistics, Part A - Theory and Methods*, **7**, 767-778.

Searle, S. R. (1971) *Linear Models*, John Wiley, New York.

Shorrock, R. W. and Zidek, J. V. (1976) An improved estimator of the generalized variance, *The Annals of Statistics*, **4**, 629-638.

Sinha, B. K. (1976) On improved estimators of the generalized variance, *Journal of Multivariate Analysis*, **6**, 617-625.

Stein, C. (1964) Inadmissibility of the usual estimator of the variance of a normal distribution with unknown mean, *Annals of the Institute of Statistical Mathematics*, **16**, 155-160.

Strawderman, W. (1974) Minimax estimation of powers of the variance of a normal population under squared error loss, *The Annals of Statistics*, **2**, 190-198.

Swallow, W. H. and Monahan, J. F. (1984) Monte Carlo comparisons of ANOVA, MIVQUE, REML and ML estimators of variance components, *Technometrics*, **26**, 47-57.

Thompson, W. A. Jr. (1962) The problem of negative estimates of variance components, *The Annals of Mathematical Statistics*, **33**, 273-289.

Tierney, L. and Kadane, J. B. (1986) Accurate approximations for posterior moments and marginal densities, *Journal of the American Statistical Association*, **81**, 82-86.

Department of Mathematics and Statistics
University of Maryland Baltimore County
Baltimore, Maryland 21228
USA

J.K. Ghosh, S.K. Mitra, K.R. Parthasarathy and B.L.S. Prakasa Rao (*eds.*),
Statistics and Probability : A Raghu Raj Bahadur Festschrift
Published by Wiley Eastern Limited, Publishers, (1993), 445 – 461

On the Choice of Ancillary in the Cauchy Location-scale Problem

PETER MCCULLAGH
The University of Chicago

Abstract

In an earlier paper (McCullagh, 1992) it was shown that the configuration ancillary for the Cauchy location-scale problem is not unique and that the choice of ancillary has some effect on conditional probability calculations. The aim in this paper is to clarify the nature and asymptotic order of the dependence on the choice of ancillary. For moderate deviations it is shown that the choice of ancillary has very little effect on probability calculations, the asymptotic effect being of order $O_p(n^{-1})$ at most. In the case of large deviations the picture is more complicated. Our results show that the relative or logarithmic difference between the two conditional densities is $O_p(n^{1/2})$, and unbounded in probability as $n \to \infty$. As a consequence, any large deviation approximation to the conditional density of $\hat{\theta}$, such as Barndorff-Nielsen's (1980, 1983) p^* formula, that has bounded relative error must depend on the choice of ancillary. By contrast, there do exist moderate-deviation approximations with absolute error $O(n^{-1})$ that are independent of the choice of ancillary, given only that $(\hat{\theta}, A)$ is sufficient.

1 Introduction

In non-Bayesian statistical inference, the precision of an estimator $\hat{\theta}$ of θ is sometimes judged by its conditional distribution given the observed value

AMS Subject Classifications : 62 F 12

Key words and Phrases : ancillary statistic, Cauchy configuration statistic, Mobius transformation, large, moderate and intermediate deviations of a suitable ancillary $A = a$. The rationale for conditioning is that two samples of equal size might by chance supply rather different amounts or different kinds of information concerning θ. The ancillary supposedly distinguishes the more informative from the less informative samples, so that any inferences drawn are tailored to the information content of the particular configuration observed. Arguments along these lines can be traced back to Fisher (1934), Pitman (1938) and Cox (1958).

By the term 'ancillary statistic' we mean a statistic whose distribution is the same for all parameter values. Following Cox and Hinkley (1974), we require the ancillary to be a function of the minimal sufficient statistic. Although Fisher did not give a formal definition, the current definition is in agreement with Fisher's usage of the term. One objection to all currently-used definitions is that the ancillary need not be unique. Imposing the restriction that A be a function of the minimal sufficient statistic reduces the degree of non-uniqueness, but does not eliminate it. Further, two ancillaries A_1, A_2 need not be jointly ancillary, so there need not exist a largest ancillary (Basu, 1959).

In this paper, we study a particular example in which $(\hat{\theta}, A_1)$ and $(\hat{\theta}, A_2)$ are alternative representations of the minimal sufficient statistic. The complementary statistics A_1 and A_2 are marginally ancillary but not jointly ancillary. The purpose of the paper is to study the difference between the two conditional densities, in particular to characterize the large-sample behaviour of the difference. The good news is that for moderate deviations in which $\hat{\theta} - \theta$ is $O(n^{-1/2})$ the difference is small and occurs in the $O_p(n^{-1})$ term. By contrast, the difference between the unconditional distribution and any conditional distribution is $O_p(n^{-1/2})$. The bad news is that for large deviations, in which $\hat{\theta} - \theta$ is $O(1)$, the relative difference is $O_p(n^{1/2})$ provided that the two ancillaries are marginally typical, i.e. in the moderate-deviation region. If the ancillaries are marginally atypical, relative differences of order $O(n)$ can be obtained.

We deduce from this that moderate-deviation approximations to the distribution of $\hat{\theta}$, with error $O(n^{-1})$, can be obtained that are independent of the choice of ancillary. This is consistent with McCullagh (1984). It does not appear to be possible to construct a large-deviation approximation for the distribution of $\hat{\theta}$ with bounded relative error that is independent of the choice of ancillary.

2 Conditional inference in location-scale problems

Suppose that $Y_1, \ldots, Y_n$ are independent and identically distributed random variables with density

$$f(y; \mu, \sigma) = \frac{1}{\sigma} f\left(\frac{y - \mu}{\sigma}\right).$$

It is assumed that $f(\cdot)$ is a known function and that the parameter space is the cartesian product of $-\infty < \mu < \infty$ with $\sigma > 0$. Fisher (1934) observed that the configuration statistic A with components $a_j = (Y_j - \hat{\mu})/\hat{\sigma}$ is ancillary in the sense that the distribution of A does not depend on the parameter. This is evident on the grounds of invariance of A under location and positive scale transformations of the data, i.e. $Y_j \mapsto bY_j + d$ with $b > 0$. We take $(\hat{\mu}, \hat{\sigma})$ to be the maximum-likelihood estimator, but in fact any equi-variant estimator is sufficient for the present purposes.

Fisher argued that the precision with which μ and σ are estimated depends on the configuration observed, some configurations being more informative than others. The appropriate distribution to be used for inferential purposes is the conditional distribution of $(\hat{\mu}, \hat{\sigma})$ given the observed value of the configuration ancillary. From this point of view, the choice of estimator is unimportant: two equi-variant estimators differ only by a function of the invariant statistic A, which is now regarded as fixed. In this sense the sample mean and standard deviation are satisfactory choices even for Cauchy samples.

Fisher's arguments are appealing from several points of view. First, among equi-variant functions, all estimators are conditionally equivalent, so questions of efficiency are entirely bypassed. Second, although all inferences are based on sample-space computations, the answers obtained are close in spirit as well as numerically to standard Bayesian calculations. Third, it is rarely feasible to compute the exact unconditional distribution of $(\hat{\mu}, \hat{\sigma})$: even the asymptotic distribution may be excessively complicated in non-standard cases where $f(\cdot)$ has bounded support. By contrast, the exact conditional distribution is surprisingly easy to obtain.

Let $T = (T_1, T_2)$ denote the random variable whose observed value is $(\hat{\mu}, \hat{\sigma})$. On transforming from the vector Y to (T, A) we obtain the conditional density of T given $A = a$ in the form

$$\operatorname{pr}(T \in dt \mid A = a; \mu, \sigma) = c(a)\, t_2^{n-2} \prod_{j=1}^{n} f(t_1 + a_j t_2; \mu, \sigma)\, dt_1\, dt_2 \qquad (1)$$

where $T \in dt$ means $T \in (t_1, t_1 + dt_1) \times (t_2, t_2 + dt_2)$. The normalizing constant $c(a)$ is a function of the configuration: it does not depend on the parameter. Laplace approximation gives

$$1/c(a) \simeq 2\pi J^{-1/2} \prod f(a_j)$$

where, in the Cauchy case, $J = (n^2 - |T^A|^2)/4$, which can be taken as $n^2/4$ without significant loss of accuracy. The quantity T^A occurring in J is a second-order Fourier coefficient defined in section 4.4: it is a function of a of order $O_p(n^{1/2})$.

3 The Cauchy location-scale problem

3.1 Cauchy parameterizations

Following the usual conventions, we say that the random variable Y has the Cauchy distribution with parameters θ_1, θ_2 if the density function has the form

$$f_Y(y;\theta) = \frac{1}{\pi|\theta_2|\left(1+\left(\frac{y-\theta_1}{\theta_2}\right)^2\right)}$$

for $-\infty < y < \infty$. For our present purposes it is helpful to think of θ as a point in the complex plane without the usual restriction that $\theta_2 > 0$. We write $Y \sim C(\theta)$ meaning that Y has the Cauchy density, which can be written in the form

$$f_Y(y;\theta) = \frac{|\theta_2|}{\pi|y-\theta|^2}.$$

The median and probable error are given by the real and imaginary parts of θ as follows:

$$\begin{aligned} \text{median}(Y) &= \Re(\theta) \\ \text{pe}(Y) &= |\Im(\theta)|. \end{aligned}$$

Thus θ and its complex conjugate $\bar{\theta}$ give rise to the same Cauchy density. This duplication of densities in the lower half of the complex plane poses no insuperable difficulties. In practice, we may adopt the convention of taking θ_2 to be non-negative. Alternatively, and perhaps preferably, complex conjugate pairs are identified as equivalent.

The advantage of the complex parameterization is evident from the Möbius transformation property

$$Y^* = \frac{aY+b}{cY+d} \sim C\left(\frac{a\theta+b}{c\theta+d}\right) \tag{2}$$

for all real numbers a, b, c and d with $ad - bc \neq 0$. In other words, not only is the two-parameter Cauchy family closed under affine transformation and reciprocals (the real Möbius group), but the induced transformation on the parameter space has exactly the same form as the transformation on the sample space. Since the transformed parameter can have a negative imaginary part, it is essential in (2) that no restrictions be imposed on θ_2. Without the introduction of complex-valued parameters the essential simplicity of (2) is not so easy to detect. In particular, the median and probable error of Y^* are

$$\begin{aligned} \text{median}(Y^*) &= \Re\left(\frac{a\theta+b}{c\theta+d}\right) = \frac{(a\theta_1+b)(c\theta_1+d)+ac\theta_2^2}{(c\theta_1+d)^2+c^2\theta_2^2} \\ \text{pe}(Y^*) &= \left|\Im\left(\frac{a\theta+b}{c\theta+d}\right)\right| = \left|\frac{(ad-bc)\theta_2}{(c\theta_1+d)^2+c^2\theta_2^2}\right|. \end{aligned}$$

3.2 *The Cauchy configuration statistic*

For a sample of n independent and identically distributed Cauchy observations $y_1, \ldots, y_n$, the likelihood equations are

$$\sum \frac{a_j}{1+a_j^2} = 0, \qquad \sum \frac{1-a_j^2}{1+a_j^2} = 0,$$

where $a_j = (y_j - \hat{\theta}_1)/|\hat{\theta}_2|$ is the jth component of the configuration. Strictly speaking, since the minimal sufficient statistic is the unordered set $\{y_1, \ldots, y_n\}$ rather than the vector $(y_1, \ldots, y_n)$, it is proper to refer to the unordered set $\{a_j\}$ as the configuration statistic. Where no danger of confusion exists, we refer to components of the configuration as if a were a vector. Note that although the sign of $\hat{\theta}_2$ is indeterminate, no similar indeterminacy is permitted in the configuration statistic because otherwise $(\hat{\theta}, A)$ would not be a sufficient statistic.

For a number of reasons, it is helpful to transform the configuration statistic from the real line on to the unit circle in the complex plane. Thus we take $z_j = (1+ia_j)/(1-ia_j)$, so that

$$\Re(z_j) = \frac{1-a_j^2}{1+a_j^2}, \qquad \Im(z_j) = \frac{2a_j}{1+a_j^2}.$$

The likelihood equation is $\sum z_j = 0$: asymptotically the points z_j are independent and uniformly distributed on the unit circle. Given $n-2$ of the points z_j whose sum z_0 satisfies $0 < |z_0| \leq 2$, the remaining two points can be determined uniquely from the likelihood equation. In particular, if $n = 3$, the configuration points are $\{z_1, \omega z_1, \omega^2 z_1\}$, where $\omega = e^{2\pi i/3}$.

3.3 *The reciprocal configuration*

It follows from (2) that $R = 1/Y$ has the Cauchy distribution with parameter $\psi = 1/\theta$. Since the maximum-likelihood estimate is equi-variant under the Möbius group, we have

$$\hat{\psi}_1 = \hat{\theta}_1/|\hat{\theta}|^2 \quad \text{and} \quad \hat{\psi}_2 = \hat{\theta}_2/|\hat{\theta}|^2,$$

the sign of $\hat{\psi}_2$ being regarded as indeterminate. The components of the reciprocal configuration are $b_j = (R_j - \hat{\psi}_1)/|\hat{\psi}_2|$. Evidently $(\hat{\psi}, B)$ is minimal sufficient and B is ancillary. Furthermore, since $\hat{\theta} = 1/\hat{\psi}$, $(\hat{\theta}, B)$ is an equivalent representation of the minimal sufficient statistic. Finally, B is not a function of A, nor is (A, B) jointly ancillary as will be seen below. In the terminology of Basu (1959), A and B are non-equivalent maximal ancillaries, but the pair (A, B) is informative.

The easiest way to understand the relationship between the reciprocal configuration B and the original configuration A is to transform both on to the unit circle. Thus we define $w_j = (1+ib_j)/(1-ib_j)$ for the reciprocal configuration. Each of the transformations $w_j \mapsto b_j \mapsto y_j \mapsto a_j \mapsto z_j$ is a Möbius transformation, and hence the composite transformation is also of the Möbius type. The sequence of transformations together with their associated matrix representations are given below.

$$z_j = \frac{1+ia_j}{1-ia_j} \qquad \begin{pmatrix} i & 1 \\ -i & 1 \end{pmatrix}$$
$$a_j = \frac{y_j - \hat{\theta}_1}{|\hat{\theta}_2|} \qquad \begin{pmatrix} 1 & -\hat{\theta}_1 \\ 0 & |\hat{\theta}_2| \end{pmatrix}$$
$$y_j = \frac{|\hat{\theta}|^2}{\hat{\theta}_1 + |\hat{\theta}_2| b_j} \qquad \begin{pmatrix} 0 & |\hat{\theta}|^2 \\ |\hat{\theta}_2| & \hat{\theta}_1 \end{pmatrix}$$
$$b_j = \frac{-iw_j + i}{w_j + 1} \qquad \begin{pmatrix} -i & i \\ 1 & 1 \end{pmatrix}$$

The matrix representation of the composite transformation is the product of the above matrices, which is

$$\begin{pmatrix} i & 1 \\ -i & 1 \end{pmatrix} \begin{pmatrix} 1 & -\hat{\theta}_1 \\ 0 & |\hat{\theta}_2| \end{pmatrix} \begin{pmatrix} 0 & |\hat{\theta}|^2 \\ |\hat{\theta}_2| & \hat{\theta}_1 \end{pmatrix} \begin{pmatrix} -i & i \\ 1 & 1 \end{pmatrix} = \begin{pmatrix} 0 & 2\hat{\theta} \\ 2\bar{\hat{\theta}} & 0 \end{pmatrix}$$

with the convention that $\hat{\theta}_2 > 0$. Thus $z_j = \hat{\theta}/(\bar{\hat{\theta}} w_j)$. In other words,

$$\arg(z_j) = 2\arg(\hat{\theta}) - \arg(w_j), \tag{3}$$

with the convention that $0 < \arg(\hat{\theta}) < \pi$. Thus, on the unit circle, the reciprocal configuration can be obtained from the original configuration by reflection in the real axis, followed by a rotation through the angle $2\arg(\hat{\theta})$.

From the preceding description it is evident that for sufficiently large n, knowledge of both A and the reciprocal configuration B is sufficient to determine $\arg(\hat{\theta})$. We need only rotate the points $\{\bar{w}_j\}$ until they coincide with the points $\{z_j\}$. For unique determination $n \geq 5$ is typically sufficient. If $n = 3$ the points z_j have angular spacing $2\pi/3$ on the unit circle. In this special case the points $\{\exp(2i\arg(\hat{\theta}))\bar{w}_j\}$ coincide with $\{z_j\}$, but so also do the points $\{\omega \exp(2i\arg(\hat{\theta}))\bar{w}_j\}$ and $\{\omega^2 \exp(2i\arg(\hat{\theta}))\bar{w}_j\}$, where $\omega = e^{2\pi i/3}$. In this case it is not possible to determine $\arg(\hat{\theta})$ uniquely from the pair of sets (a, b).

3.4 *Conditional densities given the configurations*

Let T be the random variable whose observed value is $\hat{\theta}$. Likewise, $P = 1/T$ is the random variable whose observed value is $\hat{\psi} = 1/\hat{\theta}$. In this section, we derive the conditional density of T given $A = a$ and the conditional density of T given $B = b$ demonstrating that the two conditional densities are not the same.

From Fisher's formula (1) we have

$$\mathrm{pr}(T \in dt \mid A = a; \theta) = c(a)\, t_2^{n-2} \prod f(t_1 + a_j t_2; \theta)\, dt_1\, dt_2 \tag{4}$$

where $c(a)$ is the constant of integration, and $f(\cdot;\theta)$ denotes the Cauchy density with parameter θ. Likewise, by the same argument, the conditional density of P given $B = b$ is

$$\mathrm{pr}(P \in dp \mid B = b; \theta) = c(b)\, p_2^{n-2} \prod f(p_1 + b_j p_2; \psi)\, dp_1\, dp_2.$$

Back-transformation from P to $T = 1/P$ yields the conditional density of T given $B = b$ in the form

$$\mathrm{pr}(T \in dt \mid B = b; \theta) = c(b)\, t_2^{n-2} \prod \Big\{ f\Big(\frac{|t|^2}{t_1 + b_j t_2}; \theta\Big) \frac{|t|^2}{(t_1 + b_j t_2)^2} \Big\}\, dt_1\, dt_2. \tag{5}$$

Both (4) and (5) are instances of Barndorff-Nielsen's p^* formula, but with different choices of ancillary. As probability measures on the parameter space, they are exact but not equal.

The ratio of (4) to (5) is the rational function

$$Q(t;\theta) = \frac{c(a)}{c(b)} \prod_j \frac{|t|^4 - 2\theta_1 |t|^2 (t_1 + b_j t_2) + |\theta|^2 (t_1 + b_j t_2)^2}{|t|^2 \{ |\theta|^2 - 2\theta_1 (t_1 + a_j t_2) + (t_1 + a_j t_2)^2 \}}. \tag{6}$$

To the extent that this function is not constant in t, inferences based on the conditional distribution given $A = a$ differ from those based on the conditional distribution given the reciprocal configuration. This lack of uniqueness exposes a flaw in Fisher's edifice. By the symmetry of the derivation, any argument purporting to show that A is preferable as an ancillary to B can be turned around to prove the converse. An essential aspect of this argument is that all functions of θ are to be treated on an equal footing, in particular that no non-trivial function of θ is identified as the component of interest.

In the following section we examine the extent to which (4) and (5) differ in large samples.

4 Asymptotic effect of the ancillary

4.1 General

Ideally, the inferences given by the conditional densities (4) and (5) should be equal and independent of the choice of ancillary. Unfortunately that is not the case, so it is natural to inquire into the magnitude of the difference between the conditional densities (4) and (5) especially when the sample size is large. Ultimately our goal is to describe the order of magnitude in n of the differences

$$\begin{aligned} D_1(S) &= \text{pr}(T \in S \mid A = a) - \text{pr}(T \in S \mid B = b) & (7) \\ D_2(S) &= \log \text{pr}(T \in S \mid A = a) - \log \text{pr}(T \in S \mid B = b) & (8) \end{aligned}$$

where S, a subset of the estimation space, may depend on n. To avoid complications, we do not consider sets that depend on a or b.

4.2 Large deviations

In the case of large deviations, we consider a arbitrary fixed set $S \subset \Theta$ not depending on a, b, or n. To avoid trivialities and unnecessary complications, we take S to be open and not containing the true parameter point. Then the probability that T lies in S is exponentially small for large n and is governed by the points in S that are nearest to the true θ in some appropriate metric. Our aim is to compare the probability content of S under the measure (4) with its content under (5). We do so using the logarithmic difference (8). Since the probability content of S is governed by the points in S that are nearest to θ, we need only examine the function $\log Q(t;\theta)$ for bounded $|t-\theta|$.

Regarding the configuration statistics, A and B, we say that A is marginally typical if its components have the statistical properties of a sample of independent standard Cauchy random variables. Equivalently, A is marginally typical if the Fourier coefficients $\sum z_j^r$ are $O_p(n^{1/2})$ for $r = 2, 3, \ldots$. Clearly, if A is marginally typical so also is B. An extra complication arises here in that the value of (a, b) jointly determines $\hat{\phi} = \arg(\hat{\theta})$. If, in addition to being marginally typical, $\hat{\phi} - \phi = O_p(n^{-1/2})$, we say that A and B are jointly typical. Otherwise, if $\hat{\phi} - \phi = O_p(1)$, we say that A and B are marginally but not jointly typical. In the discussion of the following section we assume that the configurations are marginally typical.

It is important to understand that at no stage of any calculation do we condition on A and B jointly. Further, even though both conditional probabilities in (8) are positive, the two configurations may be such that

$$\text{pr}(T \in S \mid A, B) = 0,$$

for example if the set S contains no points such that $\arg(t) = \hat{\phi}$. There is no contradiction here in considering sets S whose points satisfy $\arg(t) - \arg(\theta) = O(1)$, even though the two configurations are such that $\hat{\phi} - \arg(\theta)$ is $O(n^{-1/2})$.

4.3 *Conditional large deviation rates*

Associated with the density (4) we define the conditional large deviation rate at t by

$$\rho(t;a) = -n^{-1}\log f_{T|A}(t \mid a;\theta).$$

The usual definition of a large deviation rate is the limit as $n \to \infty$ of $\rho(t;a)$, but it is more convenient here not to proceed to the limit. The limit, if it exists, depends on the infinite set a_∞ in addition to the parameter θ. The large deviation rate for density (5) is denoted by $\rho^*(t;b)$. Clearly,

$$\rho(t;a) - \rho^*(t;b) = -n^{-1}\log Q(t;\theta).$$

Thus the two conditional densities have the same limiting large-deviation rate if $\log Q(t;\theta) = o_p(n)$ for bounded $t-\theta$.

In the case of the Cauchy distribution with parameter $\theta = i$ we have

$$\begin{aligned} -\rho(t;a) &= \frac{n-2}{n}\log t_2 - \frac{1}{n}\sum \log\Big(\frac{1+(t_1+a_jt_2)^2}{1+a_j^2}\Big) \\ &\quad + \frac{1}{n}\log\Big(\frac{c(a)}{\prod \pi(1+a_j^2)}\Big). \end{aligned} \tag{9}$$

Since the normalizing constant has the form $c(a) = k_n(1+O(n^{-1}))\prod \pi(1+a_j^2)$, with $k_n = 1/(n\pi)$, the final term is essentially a constant and can be ignored. The summands are bounded, exchangeable, and asymptotically pairwise independent. Hence $\rho(t;a)$ has a probability limit. There is no guarantee of a limit for an arbitrary sequence of constants a_j. Similarly, for the conditional large-deviation rate given B,

$$\begin{aligned} -\rho^*(t;b) &= \frac{n-2}{n}\log t_2 - \frac{1}{n}\sum \log\Big(\frac{|t|^4+(t_1+b_jt_2)^2}{|t|^2(1+b_j^2)}\Big) \\ &\quad + \frac{1}{n}\log\Big(\frac{c(b)}{\prod \pi(1+b_j^2)}\Big). \end{aligned} \tag{10}$$

We now show that the probability limits as $n \to \infty$ of $\rho(t;a)$ and $\rho^*(t;b)$ exist and are equal. First observe that if Y_1 has the standard Cauchy distribution, then

$$Y_2 = \frac{Y_1\cos\phi - \sin\phi}{Y_1\sin\phi + \cos\phi}$$

is also standard Cauchy for all real ϕ. Consequently, the rational functions

$$\frac{c_0+2c_1Y_1+c_2Y_1^2}{1+Y_1^2} \quad\text{and}\quad \frac{d_0+2d_1Y_2+d_2Y_2^2}{1+Y_2^2}$$

are equal in distribution if, for some ϕ, one can be transformed into the other. The conditions for this to occur are

$$d_0+d_2 = c_0+c_2 \quad\text{and}\quad d_0d_2 - d_1^2 = c_0c_2 - c_1^2,$$

i.e. the two matrices must have equal eigenvalues. In particular,

$$\frac{1+(t_1+Y_1t_2)^2}{1+Y_1^2} \sim \frac{|t|^4+(t_1+Y_2t_2)^2}{|t|^2(1+Y_2^2)}, \tag{11}$$

i.e. the ratios have the same distribution for all $t \neq 0$. These ratios occur in (9) and (10) with Y_1 and Y_2 replaced by a_j and b_j respectively. Ignoring terms of order $o(1)$, a_j and b_j are marginally standard Cauchy. To this order of accuracy, the ratios occurring in (9) and (10) have the same distribution. Hence in the limit as $n \to \infty$, ρ and ρ^* are equal with probability one.

In fact, as shown in the Appendix, if the matrix C with components

$$C = \begin{pmatrix} c_0 & c_1 \\ c_1 & c_2 \end{pmatrix}$$

is positive semi-definite, the integral

$$\int_{-\infty}^{\infty} \frac{\log(c_0+2c_1y+c_2y^2)}{(1+y^2)}\,dy = \pi\log(c_0+c_2+2\sqrt{c_0c_2-c_1^2})$$

converges and is equal to $\pi\log(2\bar\lambda+2\dot\lambda)$, where $\bar\lambda$ and $\dot\lambda$ are the arithmetic and geometric means of the eigenvalues of C. Consequently, if Y has the standard Cauchy distribution, we have

$$\begin{aligned} E\log(1+Y^2) &= \log 4 \\ E\log(1+(t_1+t_2Y)^2) &= \log(1+|t|^2+2|t_2|). \end{aligned}$$

Thus, the limiting value of the large deviation rate in (9) and (10) is

$$\rho(t) = \log(1+|t|^2+2t_2) - \log(4t_2)$$

for $t_2 > 0$. The general expression for arbitrary θ is

$$\rho(t) = \log\bigl(1+|t-\theta|^2/(4t_2\theta_2)\bigr) = \log\Bigl(1+\frac{(t-\theta)(\bar t-\bar\theta)}{(t-\bar t)(\theta-\bar\theta)}\Bigr) \tag{12}$$

provided that t and $\bar t$ are chosen such that $\Im(t)\Im(\theta) > 0$. The latter expresion, involving the cross-ratio of the four numbers $(t,\theta,\bar\theta,\bar t)$, makes it clear that the large deviation rate is invariant under Möbius transformation. Thus, the second expression (12) applies also to the circular Cauchy distribution provided that t and θ are either both inside or both outside the unit circle.

Note that $\rho(t) \geq 0$ with equality only at $t = \theta$. However, unlike the large deviation rate for sums of independent random variables, $\rho(t)$ is convex only in the figure-eight region that is the union of $|t-\theta|^2 \leq 2\theta_2^2$ and its complex conjugate.

The second random variable in (11) is undefined for $t = 0$, and both are degenerate for $t = i$. Otherwise they are identically equal if and only if either

$$Y_2 = \frac{t_2 - t_1 Y_1}{t_1 + t_2 Y_1} \quad \text{or} \quad Y_2 = \frac{t_2(1 - |t|^2) + t_1(1 + |t|^2) Y_1}{t_1(1 + |t|^2) - t_2(1 - |t|^2) Y_1}, \tag{13}$$

i.e. the relation must be a Möbius transformation with coefficients depending on t. Otherwise, if Y_2 is a fixed function of Y_1, the difference is $O_p(1)$, bounded with zero mean. In our case, the relation between a_j and b_j is the fixed Möbius transformation (14) depending on $\hat{\phi}$. Hence the average difference, $\rho - \rho^*$, is $O_p(n^{-1/2})$, and consequently $\log Q(t;\theta) = O_p(n^{1/2})$ for bounded $t - \theta$. Simulation results confirm that $n^{-1/2} \log Q(t;\theta)$ has a non-degenerate limiting distribution as $n \to \infty$ for $t \neq \theta$. In particular, although the limiting large-deviation rates are equal, the relative difference between the two conditional densities is unbounded as $n \to \infty$.

The preceding arguments assume that the configurations are either marginally or jointly typical. Although it might not be particularly meaningful from the point of view of statistical inference, we could consider large deviations for the conditioning statistic as well as for the parameter estimate. The above results do not then apply. There is no guarantee that (9) and (10) are equal in the limit for arbitrary sequences $\{a_j\}$ and $\{b_j\}$ satisfying the arithmetical properties, namely (3) and the likelihood equations. In that case it appears that the conditional large deviation rates ρ and ρ^* must have different limits and consequently $\log Q(t;\theta) = O(n)$.

4.4 Moderate deviations

We assume here that the configurations A and B are not just marginally typical, but also jointly typical. Thus $\hat{\phi}$, which is a function of the two configurations, is assumed to satisfy $\hat{\phi} - \arg(\theta) = O(n^{-1/2})$. For moderate-deviation purposes, we consider only points t satisfying $t - \theta = O(n^{-1/2})$, which may be interpreted conditionally on one or other of the configurations, or else unconditionally. A consequence of these assumptions is that, in the region of interest, $\arg(t) - \hat{\phi} = O(n^{-1/2})$.

The relation between a_j and b_j is

$$b_j = \frac{\sin\hat{\phi} - a_j \cos\hat{\phi}}{\cos\hat{\phi} + a_j \sin\hat{\phi}}. \tag{14}$$

Consequently, if $\arg(t) = \hat{\phi}$, the first equation (13) is satisfied with $Y_1 = a_j$ and $Y_2 = b_j$. The random variables at (11) are then identical. More generally, if $\arg(t) - \hat{\phi} = O(n^{-1/2})$, the difference between the variables in (11) is $O_p(n^{-1/2})$ with zero expectation. Thus $\log Q(t;\theta)$, which is the sum of these differences, is $O_p(1)$ for all t in the moderate-deviation region.

Furthermore, since

$$\log\left(\frac{1+(t_1+a_jt_2)^2}{1+a_j^2}\right)-\log\left(\frac{|t|^4+(t_1+b_jt_2)^2}{|t|^2(1+b_j^2)}\right)$$

is $O_p(n^{-1/2})$ with zero expectation for all t, the derivatives are also $O_p(n^{-1/2})$ with zero expectation. Consequently, the derivatives of $\log Q(t;\theta)$ are $O_p(1)$ in the moderate-deviation region, in particular at $t=\theta$. In fact, the first derivative vector is identically zero, a consequence of the likelihood equations. Thus the Taylor expansion of $\log Q(t;\theta)$ is

$$\log Q(t;\theta)=\log Q(\theta;\theta)+O_p(n^{-1})$$

for $t-\theta=O(n^{-1/2})$. Hence $Q(t;\theta)=1+O_p(n^{-1})$, implying that

$$f_{Y|A}(t\,|\,a;\theta)=f_{Y|B}(t\,|\,b;\theta)(1+O_p(n^{-1}))$$

uniformly in t in any $O(n^{-1/2})$ neighbourhood of θ. Hence, for any set S in this neighbourhood, the arithmetic difference

$$\mathrm{pr}(T\in S\,|\,A=a)-\mathrm{pr}(T\in S\,|\,B=b)$$

is $O_p(n^{-1})$. Since sets outside this neighbourhood have exponentially small probability content, the above result can be extended to any set. In other words, for moderate deviations, the choice of ancillary has only a small effect on the conclusions. By contrast, the difference between either conditional distribution and the unconditional distribution is $O_p(n^{-1/2})$.

Direct calculation shows that, for $t-\theta=O(n^{-1/2})$,

$$\log Q(t;\theta)=\tfrac{1}{2}c_{11}(t_1-\theta_1)^2+c_{12}(t_1-\theta_1)(t_2-\theta_2)+\tfrac{1}{2}c_{22}(t_2-\theta_2)^2+O_p(n^{-3/2}),$$

i.e. only the quadratic term contributes to this order. The coefficients, which are $O_p(1)$, depend on the Fourier coefficients $T^A=\sum z_j^2$ or $T^B=\sum w_j^2$, and are given by

$$c_{11}=-c_{22}=\frac{T_0^B(\cos 4\phi-\cos 4\hat{\phi})+T_1^B(\sin 4\phi-\sin 4\hat{\phi})}{2\theta_2^2}.$$

This can also be written in terms of T^A as follows:

$$c_{11}=-c_{22}=\frac{T_0^A\{\cos(4\hat{\phi}-4\phi)-1\}+T_1^A\sin(4\hat{\phi}-4\phi)}{2\theta_2^2}.$$

The coefficient of $(t_1-\theta_1)(t_2-\theta_2)$ is

$$c_{12}=\frac{T_0^B(\sin 4\phi-\sin 4\hat{\phi})+T_1^B(\cos 4\phi-\cos 4\hat{\phi})}{2\theta_2^2}.$$

From the nature of these coefficients, in particular from the fact that the coefficient matrix has zero trace, it is apparent that for certain sets S the difference between the two conditional probabilities is in fact $O_p(n^{-3/2})$ or smaller. For example, if S is approximately elliptical and centered at θ, the difference is $O_p(n^{-3/2})$. In particular, the conditional coverage probabilities of any likelihood-based confidence region differ by $O_p(n^{-3/2})$ or smaller. Also, if S is any half-space such as

$$\{t: a_1t_1 + a_2t_2 \geq a_0\} \quad \text{or} \quad \{t: a_1t_1 + a_2 \log t_2 \geq a_0\},$$

the difference is again $O_p(n^{-3/2})$. These results are consistent with Severini (1991).

4.5 *Intermediate deviations*

There is in fact a smooth transition between the moderate-deviation results and the large-deviation results given in the preceding sections. In this section, we consider the relative difference at intermediate points satisfying $t - \theta = O(n^{-1/2+q/2})$ with $0 \leq q \leq 1$, so that $q = 0$ corresponds to moderate deviations and $q = 1$ to large deviations. Regarding the configurations, it is enough to assume that A and B are marginally typical and that $\hat{\phi} - \arg(\theta)$ is $O(n^{-1/2+q/2})$. Together, these assumptions imply that $\arg(t) - \hat{\phi} = O(n^{-1/2+q/2})$. It follows that the difference between the random variables in (11) is $O_p(n^{-1/2+q/2})$ with zero expectation. Hence, the sum of these differences satisfies $\log Q(t;\theta) = O_p(n^{q/2})$ for t in this intermediate-deviation range. The derivatives are also $O_p(n^{q/2})$ in this region, in particular at $t = \theta$. In the Taylor expansion of $\log Q(t;\theta)$ about $t = \theta$, the first derivative is zero. The quadratic term dominates and is of order $O_p(n^{3q/2-1})$. In other words,

$$\log Q(t;\theta) = O_p(n^{3q/2-1}) \tag{15}$$

for t in this intermediate deviation region. Thus, the relative difference is bounded only for intermediate deviations with index $q \leq 2/3$.

The arithmetic difference (7) remains of order $O(n^{-1})$ for all sets S, whether in the moderate deviation or large deviation region.

5 Unconditional distributions

For large n, the approximate unconditional distribution of T is bivariate Normal with mean θ, variances $2/n$ and zero covariance. However, it appears from the discussion in section 4.3 that a more accurate approximation having the correct large-deviation behaviour is given by

$$f_n(t;\theta) = \frac{n-1}{4\pi t_2^2}\left[\frac{4t_2\theta_2}{|t-\bar{\theta}|^2}\right]^n \tag{16}$$

for t_2 and θ_2 both positive. The constant has been adjusted to make the integral exactly one. Note that this approximation has the form

$$\text{const} \times \text{invariant measure} \times [\text{cross-ratio}]^n,$$

where the invariant cross-ratio is equal to $\exp(-\rho(t))$. The measure $dt_1\, dt_2/t_2^2$ is invariant under the group of Möbius transformations that leaves the real axis invariant, i.e. the Möbius group with real coefficients. On account of its construction, approximation (16) applies equally to the two-parameter circular Cauchy distribution. However, for the Möbius group that leaves the unit circle invariant, the invariant measure is $dt_1\, dt_2/\left|1-|t|^2\right|^2$.

Reverting to the standard case with $\theta = (0, 1)$, we have

$$f_n(t) = \frac{n-1}{4\pi t_2^2}\left[\frac{4t_2}{t_1^2 + (1+t_2)^2}\right]^n.$$

The marginal distribution of T_2, or more generally of T_2/θ_2, is

$$\frac{(n-1)2^{2n} t_2^{n-2}}{4\pi(1+t_2)^{2n-1}} B(\tfrac{1}{2}, n-\tfrac{1}{2}) = \frac{t_2^{n-2}}{(1+t_2)^{2n-1} B(n, n-1)},$$

which is $\frac{n-1}{n} F_{2(n-1),2n}$, with unit expectation, and variance $2/(n-2)$ for $n > 2$. The conditional distribution of T_1 given $T_2 = t_2$ is proportional to Students t:

$$T_1 \mid T_2 = t_2 \sim \frac{1+t_2}{\sqrt{2n-1}}\, t_{2n-1},$$

which has mean 0, conditional variance $(1+t_2)^2/(2n-3)$, and unconditional variance $2/(n-2)$ for $n > 2$. These moments are all consistent with the Normal approximation.

Density (16) has the property that the random variables

$$Z_1 = (T_1 - \theta_1)/T_2 \quad \text{and} \quad Z_2 = T_2/\theta_2$$

are jointly pivotal, i.e. the joint density

$$f_Z(z) = \frac{n-1}{4\pi}\, \frac{4^n z_2^{n-1}}{[z_1^2 z_2^2 + (1+z_2)^2]^n}$$

is independent of θ. The marginal density of Z_1 is given by

$$\frac{(n-1)4^n}{4\pi(1+z_1^2)^{n/2}} \int_0^\infty \frac{u^{n-1}\, du}{[1 + 2u\cos\phi + u^2]^n} \tag{17}$$

where $\cos\phi = (1+z_1^2)^{-1/2}$. Laplace approximation of the integral gives

$$f_Z(z_1) \simeq \frac{n^{1/2}\, 2^n}{2\sqrt{2\pi}(1+z_1^2)^{1/4}(1+\sqrt{(1+z_1^2)})^{n-1/2}}$$

Professor Seshadri has pointed out to me that the integral (17) can be evaluated exactly in terms of the hypergeometric function using formula 3.25.10 of Gradshteyn and Ryzhik (1965).

Simulation results indicate that (16) is quite accurate, particularly if n is replaced by $n-1$. This makes sense because T is not uniquely defined for $n=2$. I had initially thought that one of these forms of (16) might be exact, but regrettably that conjecture proves to be false. For $n=3$ and $n=4$, direct calculation shows that the exact densities of (T_1, T_2) are

$$\frac{3\sqrt{3}}{4t_2^2\pi^2(1+3d+3d^2)} \quad \text{and} \quad \frac{3\log(1+2d)}{t_2^2\pi^3 d(d+1)(2d+1)} \tag{18}$$

respectively, where $d = |t-\theta|^2/(4t_2\theta_2)$ is a form of the cross-ratio. These exact densities have been obtained by transformation and integration, a rather tedious process even for $n=3$. Further, since the invariant measure of the annulus $\{r \leq |t-\theta|^2/(4t_2\theta_2) \leq r+dr\}$ is equal to $4\pi dr$, the marginal density of the random variable $D = |T-\theta|^2/(4T_2\theta_2)$ with respect to Lebesgue measure on R^+ can be obtained from (18) by multiplying by $4\pi t_2^2$.

All attempts to obtain the exact density of T for general n, or even for any $n \geq 5$, have thus far been singularly unsuccessful.

Appendix

Note on the evaluation of a certain contour integral

Consider the complex integral

$$I(r) = \int_{-\infty}^{\infty} \frac{\log(z-r)}{1+z^2}\,dz = \pi\log(\pm i - r),$$

where $r = r_1 + ir_2$ with $r_2 \neq 0$. If $r_2 < 0$, the integral may be obtained by deflecting the contour of integration into the upper half plane, leaving the residue at $z = +i$. This gives $\pi\log(i-r)$. Otherwise, if $r_2 > 0$, we use the residue at $z=-i$, giving $\pi\log(-i-r)$. Hence, $I(r) + I(\bar{r})$ is equal to

$$\begin{aligned} \int_{-\infty}^{\infty} \frac{\log((z-r)(z-\bar{r}))}{1+z^2}\,dz &= \begin{cases} \pi\log((i-r)(-i-\bar{r})) & \text{if } r_2 < 0 \\ \pi\log((i-\bar{r})(-i-r)) & \text{if } r_2 > 0 \end{cases} \\ &= \pi\log(1+2|r_2|+|r|^2). \end{aligned}$$

In particular, we have

$$\int_{-\infty}^{\infty} \frac{\log(x^2 - 2ax\cos\theta + a^2)}{1+x^2}\,dx = \pi\log(1+a^2+2|a\sin\theta|).$$

Gradshteyn and Ryzhik (1965, eqn. 4.296.2), citing Bierens de Haan (1867, eqn. 145.78), give the above integral, but with the absolute value sign missing in the answer. This error is uncorrected in the latest editions. Bierens de Haan (1867), citing Cauchy (1826), gives the same formula, also omitting the absolute value sign. Again, the error is repeated in subsequent editions. In equation (90) of his paper, in the middle of a list of thirty such integrals, Cauchy gives the same formula, omitting the absolute value sign. However, on a preceding page, Cauchy warns that the numerical values of certain constants are not always entirely arbitrary. Presumably Cauchy would have expected the reader to notice that the integral is an even function of a and θ, so that the restrictions $a \geq 0$, $0 \leq \theta \leq \pi$ should be applied. Such restrictions are discussed in the descriptive sections of Cauchy's paper, but are omitted in the listing of particular integrals.

Acknowledgements

I am grateful to O. Barndorff-Nielsen, M. Stein and E.J. Williams for helpful comments on an earlier draft of this paper.

REFERENCES

Barndorff-Nielsen, O.E. (1980) Conditionality resolutions. *Biometrika* **67**, 293–310.

Barndorff-Nielsen, O.E. (1983) On a formula for the distribution of the maximum likelihood estimator. *Biometrika* **70**, 343–65.

Basu, D. (1959) The family of ancillary statistics. *Sankhyā* A **21**, 247–56.

Bierens de Haan, D. (1957) *Nouvelles Tables d'Intégrales Définies.* (Revised version of the 1867 edition). New York, Hafner.

Cauchy, A. (1826) Recherche d'une formule générale qui fournit la valeur de la plupart des intégrales définies connues et celle d'un grand nombre d'autres. *Annales de Mathématiques* **17**, 84–127. Reprinted in *Œuvres Complètes D'Augustin Cauchy*. Series 2, **2**, 353-87.

Cox, D.R. (1958) Some problems connected with statistical inference. *Ann. Math. Statist.* **29**, 357–72.

Cox, D.R. (1971) The choice between alternative ancillary statistics. *J. R. Statist. Soc.* B **33**, 251–5.

Cox, D.R. and Hinkley, D.V. (1974) *Theoretical Statistics.* Chapman and Hall, London.

Fisher, R.A. (1934) Two new properties of mathematical likelihood. *Proc. Roy. Soc.* A **144**, 285–307.

Gradshteyn, I.S. and Ryzhik, I.W. (1965) *Tables of Integrals, Series and Products.* Academic Press, New York.

McCullagh, P. (1984) Local sufficiency. *Biometrika* **71**, 233–44.

McCullagh, P. (1992) Conditional inference and Cauchy models. *Biometrika* **79**, 247–59.

Pitman, E.J.G. (1938) The estimation of the location and scale parameters of a continuous population of any given form. *Biometrika* **30**, 391–421.

Severini, T. (1991) On the relation between Bayesian and non-Bayesian interval estimates. *J. R. Statist. Soc.* B **53**, 611–8.

Department of Statistics
University of Chicago
5734 University Avenue
Chicago, Illinois 60637
USA

J.K. Ghosh, S.K. Mitra, K.R. Parthasarathy and B.L.S. Prakasa Rao (*eds.*),
Statistics and Probability : A Raghu Raj Bahadur Festschrift
Published by Wiley Eastern Limited, Publishers, (1993), 463 – 474

Transformations of the Range Space and the Direct Orthogonal Projector[1]

SUJIT KUMAR MITRA
Indian Statistical Institute, New Delhi

Abstract

Given an orthogonal projector P_1 of order n and a matrix W of commensurate order, the orthogonal projector onto the range of WP_1 is explicitly determined for the following two cases (a) $W = P_2$ another orthogonal projector of order n, (b) $W = \binom{I_n}{b'}$ where b is a given arbitrary n-tuple.

Let $I\!R^n$ denote the vector space of all real n-tuples and $\mathcal{P}$, the collection of all orthogonal projectors defined on $I\!R^n$. Given $P_1, P_2 \in \mathcal{P}$, the object is to find the orthogonal projector onto $\mathcal{R}(P_2P_1)$, the range of P_2P_1. We shall denote this new orthogonal projector by $\Pi(P_2, P_1)$. The following result is well known.

LEMMA 1. P_2P_1 is itself the required orthogonal projector if and only if P_1 and P_2 commute, i.e.,

$$P_1P_2 = P_2P_1. \tag{1}$$

Lemma 2 establishes the same result under a somewhat weaker condition.

[1] Research partly supported by the Council of Scientific and Industrial Research, Government of India under its Emeritus Scientist Scheme

AMS Subject Classifications : 15A09, 62J05

Key words and Phrases : Orthogonal Projector, Löwner Ordering, Shorted Operator, Parallel-sum, Infimum of two projections, BLUE, Regression Diagnostics, Hat Matrix, Adaptive Signal Processing

LEMMA 2. $\Pi(P_2, P_1) = P_2P_1$ if and only if

$$\mathcal{R}(P_2P_1) \subset \mathcal{R}(P_1). \tag{2}$$

PROOF. The necessity of (2) follows from that of (1). For sufficiency, observe that (2) $\Rightarrow$

$$P_1P_2P_1 = P_2P_1. \tag{3}$$

Hence P_2P_1 is symmetric. Also, $P_2P_1P_2P_1 = P_2P_2P_1 = P_2P_1$. ■

The following lemma is a simple consequence of Lemmas 1 and 2. We omit the proof.

LEMMA 3. $\Pi(P_2, P_1) = \Pi(P_1, P_2)$ if and only if (1) holds. The operation $\Pi(,)$ viewed as a mapping from the cartesian product $\mathcal{P} \times \mathcal{P} \to \mathcal{P}$ is thus in general noncommutative. However, the operation is seen to be associative since $\Pi(P_3, \Pi(P_2, P_1)), \Pi(\Pi(P_3, P_2), P_1)$, both represent orthogonal projections to $\mathcal{R}(P_3P_2P_1)$ and hence are equal. The operation is monotone only in the second coordinate under the Löwner ordering (also known as positive semidefinite ordering) that is, $P > P_1 \Rightarrow \Pi(P_2, P) > \Pi(P_2, P_1)$. That $P > P_2$ does not imply $\Pi(P, P_1) > \Pi(P_2, P_1)$ can be seen from the following example. Let P_1 and P_2 be two different orthogonal projectors each of rank 1 such that $P_2P_1 \neq 0$ and let $P = I$. Here $\Pi(P_2, P_1) = P_2$ while $\Pi(P, P_1) = P_1$.

We now recall the Krein - Anderson - Trapp definition of the shorted positive operator [2,6]. Let $\mathcal{C}_n$ denote the cone of real symmetric nonnegative definite (nnd) matrices of order n. Let $V \in \mathcal{C}_n$ and $\mathcal{S}$ be a subspace of $\mathbb{R}^n$. Consider the class $\mathcal{C}$ of matrices defined as follows:

$$\mathcal{C} = \{C : C \in \mathcal{C}_n, V > C, \mathcal{R}(C) \subset \mathcal{S}\}.$$

It was shown that the class $\mathcal{C}$ has a unique maximal element under the Löwner ordering. This maximal element is the shorted operator relative to $\mathcal{S}$ and denoted by $S(V|\mathcal{S})$. The following result was proved in Mitra and Puri ([10], Theorem 2.4).

LEMMA 4 : If $P \in \mathcal{P}$ and $\mathcal{S}$ is a subspace of $\mathbb{R}^n$, then $S(P|\mathcal{S}) \in \mathcal{P}$ and is in fact the orthogonal projector onto $\mathcal{R}(P) \cap \mathcal{S}$.

Observing that $\mathcal{R}(P_2P_1) \subset \mathcal{R}(P_2)$, we have thus

THEOREM 1 : $\Pi(P_2, P_1) = S(P_2|\mathcal{R}(P_2P_1))$.

As a simple corollary of Theorem 1, we have

THEOREM 2 : $\Pi(P_2, P_1) > P_2P_1P_2$.

PROOF : Observe that $P_2P_1P_2 \in \mathcal{C}_n$ and $\mathcal{R}(P_2P_1P_2) \subset \mathcal{R}(P_2P_1)$. Further, $P_2 - P_2P_1P_2 = P_2(I - P_1)P_2 \in \mathcal{C}_n$. Hence $P_2 > P_2P_1P_2$. The definition of the shorted operator and Theorem 1 thus lead to Theorem 2. ■

A pair of matrices A, B of the same order are said to be parallel summable if $A(A+B)^-B$ is invariant under the choice of the g-inverse ([1], [12, p. 188]) in which case $A(A+B)^-B$ is called the parallel sum of A and B and denoted by the symbol $P(A, B)$.

THEOREM 3 : $\Pi(P_2, P_1) > 2P(P_1, P_2)$.

PROOF : Let $P_1 \wedge P_2$ be the infimum of the two projections, which is by definition the orthogonal projection to $\mathcal{R}(P_1) \cap \mathcal{R}(P_2), P_1 > P_1 \wedge P_2$. Therefore, $\Pi(P_2, P_1) > \Pi(P_2, P_1 \wedge P_2) = P_1 \wedge P_2$. Theorem 3 now follows from the expression of $P_1 \wedge P_2$ given by Anderson and Duffin [1]. ■

We need Lemma 5 for proving Theorem 4.

LEMMA 5 : Let $P_1, P_2 \in \mathcal{P}$. Then

$$P_2 \in \{P_1^-\}, P_1 \in \{P_2^-\} \Leftrightarrow P_1 = P_2.$$

PROOF : Since an idempotent matrix is a g-inverse of itself, the '$\Leftarrow$' part is trivial. For the '$\Rightarrow$' part observe that

$$P_1^2 = P_1, P_2^2 = P_2, P_1P_2P_1 = P_1, P_2P_1P_2 = P_2 \Rightarrow$$
$$(P_1 - P_2)^3 = 0,$$

which in turn, on account of symmetry of $P_1 - P_2$ implies

$$P_1 - P_2 = 0 \Rightarrow P_1 = P_2$$ ■

REMARK 1 : Lemma 5 is a simple consequence of Theorem 4 of Rao, Mitra and Bhimasankaram [13] which, for real or complex matrices A and B of the same order, asserts that

$$A^+ \in \{B^-\}, B^+ \in \{A^-\} \Rightarrow A = B.$$

where A^+ and B^+ denote the unique Moore Penrose inverse of A and B respectively. Nevertheless we gave a separate proof of Lemma 5 since this is straightforward and considerably simpler than that of the theorem quoted above.

REMARK 2 : Lemma 5 is not true for oblique projectors. Consider for example the pair

$$P_1 = \begin{pmatrix} 0 & 0 \\ 0 & 1 \end{pmatrix}, P_2 = \begin{pmatrix} 0 & 0 \\ 1 & 1 \end{pmatrix}$$

which satisfies all the assumptions of Lemma 5 other than the crucial symmetry assumption involved in the definition of $\mathcal{P}$.

Let

$$P_{1(m)} = (P_1P_2)^m P_1 = [P_{1(1)}]^m.$$

for a positive integer $m \geq 1$. Similarly $P_{2(m)} = (P_2P_1)^m P_2 = [P_{2(1)}]^m$ for $m \geq 1$. We shall now prove Theorem 4.

THEOREM 4 [VON NEUMANN]: Let $P_1, P_2 \in \mathcal{P}$ and $P_1P_2 \neq P_2P_1$, then for each $i = 1, 2$ (a) $\{P_{i(m)}\}_{m=1,2,\ldots}$ is a strictly decreasing sequence of real symmetric nnd matrices bounded above by P_i and below by $P_1 \wedge P_2$. (b) Further

$$\lim_{m\to\infty} P_{i(m)} = P_1 \wedge P_2.$$

PROOF : Since $P_i = P_iP_i'$, $P_{1(1)} = P_1P_2P_2'P_1'$, $P_{2(1)} = P_2P_1P_1'P_2'$ and in general $P_{1(m)} = [P_{1(1)}]^m$, $P_{2(m)} = [P_{2(1)}]^m$, these are easily seen to be symmetric nnd. That the sequences are nonincreasing and the claim about the lower bound follows from the following expressions for $P_{i(m)} - P_{i(m+1)}$ and $P_{i(m)} - P_1 \wedge P_2$, each trivially seen to be symmetric nnd. We confine our attention to the case $i = 1$. The other case is similarly established.

$$\begin{aligned}
P_{1(m)} - P_{1(m+1)} &= (P_1P_2)^k(I - P_1)(P_2P_1)^k && \text{if } m = 2k-1\\
&= [P_{1(1)}]^k(I - P_2)[P_{1(1)}]^k && \text{if } m = 2k\\
P_{1(m)} - P_1 \wedge P_2 &= (P_1P_2)^k(I - P_1 \wedge P_2)(P_2P_1)^k && \text{if } m = 2k-1\\
&= [P_{1(1)}]^k(I - P_1 \wedge P_2)[P_{1(1)}]^k && \text{if } m = 2k.
\end{aligned}$$

The strict decreasing nature is a consequence of the fact that for $m \geq 1$

$$\begin{aligned}
P_{1(m)} = P_{1(m+1)} &\Rightarrow [P_{1(1)}]^m = [P_{1(1)}]^{m+1},\\
&\Rightarrow P_{1(1)} = [P_{1(1)}]^2\\
&\Rightarrow P_1P_2P_1 = P_1P_2P_1P_2P_1\\
&\Rightarrow P_1P_2(I - P_1)P_2'P_1' = 0 \Rightarrow P_1P_2(I - P_1) = 0\\
&\Rightarrow P_1P_2 = P_1P_2P_1 \Rightarrow P_1P_2 = P_2P_1
\end{aligned}$$

which is excluded by the assumption made in Theorem 4.

Both sequences of symmetric nnd matrices being strictly decreasing and bounded clearly have limits. Let

$$\lim_{m\to\infty} P_{1(m)} = A, \quad \lim_{m\to\infty} P_{2(m)} = B.$$

Since

$$P_{1(m)}P_{1(m)} = P_{1(2m)}, P_{1(m)}P_{2(m)}P_{1(m)} = P_{1(3m+1)}$$

$$P_{2(m)}P_{2(m)} = P_{2(2m)}, P_{2(m)}P_{1(m)}P_{2(m)} = P_{2(3m+1)}$$

taking limits on both sides in all the four expressions, we have

$$A^2 = A, ABA = A; B^2 = B, BAB = B.$$

An appeal to Lemma 5 now shows $A = B$. Since $A = B = \lim_{m\to\infty} P_{i(m)}, i = 1, 2$, it follows from the decreasing nature of both sequences that the range of the common matrix is contained in $\mathcal{R}(P_i), i = 1, 2$. Hence $\mathcal{R}(A) \subset \mathcal{R}(P_1 \wedge P_2)$. The reverse inclusion is trivially seen to be true. Hence $P_1 \wedge P_2$ is the limit for both the sequences. See also Halmos [6, p. 257] which refers to earlier work of von Neumann [14, p 55] in this connection. ■

For a real symmetric nnd matrix A and a positive integer m, let $(A)^{1/m}$ denote the real symmetric nnd mth root of A. We shall prove Theorem 5.

THEOREM 5 : Let $P_1, P_2 \in \mathcal{P}$ and $P_1P_2 \neq P_2P_1$, then (a) $\{(P_1P_2P_1)^{1/m}\}_{m=1,2,\ldots}$ and $\{(P_2P_1P_2)^{1/m}\}_{m=1,2,\ldots}$ are strictly increasing sequences of real symmetric nnd matrices bounded above by P_1 and P_2 respectively. (b) Further

$$\lim_{m\to\infty} (P_1P_2P_1)^{1/m} = \Pi(P_1, P_2),$$

$$\lim_{m\to\infty} (P_2P_1P_2)^{1/m} = \Pi(P_2, P_1).$$

PROOF : While proving Theorem 4 we have seen that

$$P_{1(m)} > P_{1(m+1)}, \text{ i.e. } (P_1P_2P_1)^m > (P_1P_2P_1)^{m+1}.$$

Taking the $m(m+1)$th root on both sides, we conclude that

$$(P_1P_2P_1)^{1/(m+1)} > (P_1P_2P_1)^{1/m}$$

thus establishing the strict increasing nature of both the sequences. One should use this argument judiciously while dealing with real symmetric nnd matrices. The argument is clearly valid for an arbitrary pair of nonnegative real scalars but not so for an arbitrary pair of real symmetric nnd matrices. For example for the pair of matrices

$$A = \begin{pmatrix} 1 & 1 \\ 1 & 1 \end{pmatrix}, B = \begin{pmatrix} 2 & 1 \\ 1 & 1 \end{pmatrix}$$

we have $B > A$. However $B^2 = \begin{pmatrix} 5 & 3 \\ 3 & 2 \end{pmatrix} \not> \begin{pmatrix} 2 & 2 \\ 2 & 2 \end{pmatrix} = A^2$. The reasoning works in the present context since the matrices under comparison are powers of the same real symmetric nnd matrix. Here the spectral theorem for such matrices enables us to transfer the argument to the nonnegative scalars occuring at identical diagonal positions in the pair of matrices under comparison, which are powers of one another.

Let $P_1P_2P_1 = L'DL$ be the spectral representation of $P_1P_2P_1$ where the matrix L is orthogonal and D is a diagonal matrix the diagonal entries being the eigenvalues of $P_1P_2P_1$. The chain of inequalities $P_1P_2P_1 < P_1 < I$ imply that the diagonal entries of D are either 0, 1 or a positive scalar strictly less than 1. Notice now that $(P_1P_2P_1)^{1/m} = L'D^{1/m}L$ and that $\lambda^{1/m} = \lambda$ if $\lambda = 0$ or 1. However if $0 < \lambda < 1$

$$\lambda^{1/m} \uparrow 1$$

as $m \uparrow \infty$. This establishes the (b) part. ■

REMARK 3 : Let us write $P_{1(0)} = (P_1P_2P_1)^{1/2}$ and for a negative integer $-m$, $P_{1(-m)} = (P_1P_2P_1)^{1/(m+2)}$. Similarly define $P_{2(0)} = (P_2P_1P_2)^{1/2}$ and $P_{2(-m)} = (P_2P_1P_2)^{1/(m+2)}$

(a) $\quad ..., P_{1(-2)}, P_{1(-1)}, P_{1(0)}, P_{1(1)}, P_{1(2)}, ...$ and

(b) $\quad ..., P_{2(-2)}, P_{2(-1)}, P_{2(0)}, P_{2(1)}, P_{2(2)}, ...$

are thus two way sequences (matrix valued maps defined on the set $\mathbb{Z}$ of integers) and both are strictly decreasing, when $P_1P_2 \neq P_2P_1$, in the sense that

$$P_{1(u)} > P_{1(u+1)}(P_{1(u)} \neq P_{1(u+1)})$$

$$P_{2(u)} > P_{2(u+1)}(P_{2(u)} \neq P_{2(u+1)}), \text{ for } \forall\, u \in \mathbb{Z}.$$

Further

$$\lim_{u \to -\infty} P_{1(u)} = \Pi(P_1, P_2),$$

$$\lim_{u \to -\infty} P_{2(u)} = \Pi(P_2, P_1)$$

and $\lim_{u\to\infty} P_{1(u)} = \lim_{u\to\infty} P_{2(u)} = P_1 \wedge P_2$. Looking this way one has thus a clear perception of $\Pi(P_1, P_2), \Pi(P_2, P_1), P_1 \wedge P_2$ and their interconnections.

THEOREM 6 : Let $P_{1(u)}$ and $P_{2(v)}$ be arbitrary members of the two sequences. Then $P(P_{1(u)}, P_{2(v)}) = \frac{1}{2}(P_1 \wedge P_2)$.

PROOF : Since we are exclusively dealing with real symmetric nonnegative definite matrices, existence of their parallel sum and shorted versions is always guaranteed. Hence, from Theorem 3.6 of Mitra and Puri [11], it follows that $P(P_1, P_2)$ is the same as $P(S(P_1|\mathcal{R}(P_2)), S(P_2|\mathcal{R}(P_1)))$. However, by Thoerem 2.4 of Mitra and Puri [10], $S(P_1|\mathcal{R}(P_2))$ and $S(P_2|\mathcal{R}(P_1))$ are both equal to $P_1 \wedge P_2$. Hence $P(P_1, P_2) = \frac{1}{2}(P_1 \wedge P_2)$. We now show that $P(P_{1(u)}, P_{2(v)})$ is also equal to $\frac{1}{2}(P_1 \wedge P_2)$, the proof of which will follow from the fact that $S(P_{1(u)}|\mathcal{R}(P_{2(v)}))$ and $S(P_{2(v)}|\mathcal{R}(P_{1(u)}))$ are both equal to $P_1 \wedge P_2$. Since $P_{2(v)} > B$, which we have seen to be equal to $P_1 \wedge P_2$, it follows that $\mathcal{R}(P_1 \wedge P_2) \subset \mathcal{R}(P_{2(v)})$. Also, $P_{1(u)} > A = P_1 \wedge P_2$ implies from the definition of the shorted operator that $S(P_{1(u)}|\mathcal{R}(P_{2(v)})) > P_1 \wedge P_2$. However, $P_1 > P_{1(u)}$ and $\mathcal{R}(P_{2(v)}) \subset \mathcal{R}(P_2)$. Hence, using the monotone property of the shorted operator, $S(P_1|\mathcal{R}(P_2)) > S(P_{1(u)}|\mathcal{R}(P_2)) >$

$S(P_{1(u)}|\mathcal{R}(P_{2(v)}))$ which implies that $P_1 \wedge P_2 > S(P_{1(u)}|\mathcal{R}(P_{2(v)}))$. Hence $S(P_{1(u)}|\mathcal{R}(P_{2(v)})) = P_1 \wedge P_2$. Using a similar argument it follows that

$$S(P_{2(v)}|\mathcal{R}(P_{1(u)}) = P_1 \wedge P_2. \qquad \blacksquare$$

Let us write Q_i for the matrix $I - P_i, i = 1, 2$. Theorem 7 gives an explicit expression for $\Pi(P_2, P_1)$.

THEOREM 7 : $\Pi(P_2, P_1) = P_2P_1P_2$, if P_1 and P_2 commute

$$= P_2P_1P_2 + P_2P_1Q_2(Q_2Q_1Q_2)^{-}Q_2P_1P_2, \text{ otherwise.} \tag{4}$$

PROOF : The first part of the theorem follows from Lemma 1. We shall prove the second part essentially using a statistical argument. Consider a linear model $(\boldsymbol{Y}, \mathcal{S}, V)$ where $\boldsymbol{Y}' = (Y_1, Y_2, ..., Y_n)$ is a vector of random variables such that $E(\boldsymbol{Y}) \in \mathcal{S}$, a given subspace of $\mathbb{R}^n$ and dispersion matrix of $\boldsymbol{Y}, D(\boldsymbol{Y}) = V$. Let D be the dispersion matrix of the BLUE of $E(\boldsymbol{Y})$ under this linear model. It was observed in Mitra and Puntanen [9, page 98] that $D = S(V|\mathcal{S})$. To find $\Pi(P_2, P_1)$ we therefore refer to a linear model $(\boldsymbol{U}, \mathcal{R}(P_2P_1), P_2)$, obtain the BLUE of expectation of $\boldsymbol{U}$ under this model and finally the dispersion matrix of the BLUE.

We initially consider a linear model $(\boldsymbol{Y}, \mathcal{R}(P_1), I)$. It is well known that $P_1\boldsymbol{Y}$ is the BLUE of $E(\boldsymbol{Y})$ in this model and $D(P_1\boldsymbol{Y}) = P_1$. Hence $P_2P_1\boldsymbol{Y}$ is the BLUE based on $\boldsymbol{Y}$ of $E(P_2\boldsymbol{Y})$ and $D(P_2\boldsymbol{Y}) = P_2P_1P_2$. Put $\boldsymbol{U} = P_2\boldsymbol{Y}$. Then, $\boldsymbol{U}$ obeys the linear model $(\boldsymbol{U}, \mathcal{R}(P_2P_1), P_2)$. We have therefore to recompute the BLUE of $E(\boldsymbol{U}) = E(P_2\boldsymbol{Y})$ based on $\boldsymbol{U}$ alone. For a matrix A, let the column vector formed by the jth column of A be denoted by $(A)_j$. The row vector formed by the jth row of A would then be equal to $[(A')_j]'$ which simplifies to $(A)_j'$ when A is symmetric. Since $E[(I - P_1)\boldsymbol{Y}] = 0$, for each choice of the column vector $\boldsymbol{\lambda}, T_{j\boldsymbol{\lambda}}\boldsymbol{Y} = (P_2)_j'P_1\boldsymbol{Y} - \boldsymbol{\lambda}'[Q_2(I - P_1)\boldsymbol{Y}]$ is an unbiased estimator of the jth coordinate $E(P_2\boldsymbol{Y}) = E(\boldsymbol{U})$. Now choose $\boldsymbol{\lambda}$ such that

$$\text{Cov}[T_{j\boldsymbol{\lambda}}\boldsymbol{Y}, Q_2\boldsymbol{Y}] = 0$$

that is,

$$(P_2)_j'P_1Q_2 - \boldsymbol{\lambda}'Q_2(I - P_1)Q_2 = 0. \tag{5}$$

$$\begin{aligned} Q_2(I - P_1)Q_2\boldsymbol{x} = 0 \;&\Rightarrow\; (I - P_1)Q_2\boldsymbol{x} = 0 \Rightarrow P_1Q_2\boldsymbol{x} = Q_2\boldsymbol{x} \\ &\Rightarrow\; (P_2)_j'P_1Q_2\boldsymbol{x} = (P_2)_j'Q_2\boldsymbol{x} = 0. \end{aligned}$$

This shows that the equations (5) are consistent and one choice of $\boldsymbol{\lambda}$ is given by

$$\boldsymbol{\lambda}' = (P_2)_j'P_1Q_2(Q_2Q_1Q_2)^{-} = \boldsymbol{\lambda}_0' \text{ (say)}.$$

The linear estimator $T_j\boldsymbol{Y} = (P_2)_j'P_1\boldsymbol{Y} - \boldsymbol{\lambda}_0'Q_2(I - P_1)\boldsymbol{Y}$ is uncorrelated with $Q_2\boldsymbol{Y}$. It therefore depends on $\boldsymbol{Y}$ only through $\boldsymbol{U}(= P_2\boldsymbol{Y})$. Further,

$(P_2)'_j P_1 \boldsymbol{Y}$ and $Q_2 P_1 \boldsymbol{Y}$ being BLUE of their respective expectations, are uncorrelated with all error functions based on $\boldsymbol{Y}$ including those based exclusively on $\boldsymbol{U}$. The expression $Q_2 \boldsymbol{Y}$ is uncorrelated with $\boldsymbol{U}$ and therefore with every linear functional in $\boldsymbol{U}$, including the error functions. We are thus led to the conclusion that $T_j \boldsymbol{Y}$ is the BLUE of the jth coordinate of $E((P_2\boldsymbol{Y})$ based on $\boldsymbol{U}$. The BLUE of $E(P_2\boldsymbol{Y})$ based on $\boldsymbol{U}$ is

$$T\boldsymbol{Y} = P_2 P_1 \boldsymbol{Y} - P_2 P_1 Q_2 (Q_2 Q_1 Q_2)^- Q_2 Q_1 \boldsymbol{Y}$$

and

$$D(T\boldsymbol{Y}) = P_2 P_1 P_2 + P_2 P_1 Q_2 (Q_2 Q_1 Q_2)^- Q_2 P_1 P_2 \Rightarrow (4).$$

This concludes the proof of Theorem 7. ∎

We may have already exposed ourselves to the criticism that after all these efforts we have only succeeded in deriving a formula for $\Pi(P_2, P_1)$ which possibly involves heavier computations than the expression

$$P_2 P_1 (P_1 P_2 P_1)^- P_1 P_2. \tag{6}$$

Formula (6) can be deduced from the well known expression, already available in literature,

$$X(X'X)^- X' \tag{7}$$

for the orthogonal projector onto the range of the matrix X. We shall answer this criticism later but first let us record the interesting identity

$$P_2 P_1 P_2 = P_2 P_1 (P_1 P_2 P_1)^- P_1 P_2 - P_2 P_1 Q_2 (Q_2 Q_1 Q_2)^- Q_2 P_1 P_2 \tag{8}$$

which can be derived equating the two alternative expressions (4) and (6) for $\Pi(P_2, P_1)$. It should be noted that the equality of expressions (4) and (6) is yet another manifestation of duality aspects of the Gauss-Markov theorem (see [12], Theorem 3.2.4). Formula (6) is a direct consequence of orthogonal projections being viewed as a least squares problem while formula (4) is based upon the dispersion matrix of the BLUE thereby capitalizing on the minimum norm property of these estimators.

As regard the criticism noted above, we wish to record here that none of the two formulae (4) and (6) is really in a finished form for computational purpose.

Let r_i and d_i denote respectively the ranks of P_i and $(I - P_i)$, and let the columns of A_i and B_i constitute their respective orthonormal bases, $i = 1, 2$. Since P_i and $I - P_i$ are orthogonal projectors and are therefore symmetric and idempotent

$$r_i = \text{ tr } P_i, d_i = \text{ tr } (I - P_i).$$

Further A_i and B_i can be obtained applying the Gram-Schmidt algorithm to the columns of P_i and $(I - P_i)$. $P_i = A_i A_i', (I - P_i) = B_i B_i'$ by construction.

Since the range of P_2P_1 is the same as that of P_2A_1, using (7) we have an alternative formula which requires inversion of a real symmetric nnd matrix of order $r_1 \times r_1$

$$P_2A_1(A_1'P_2A_1)^-A_1'P_2. \tag{9}$$

Alternatively, since $A_2'P_1P_2P_1A_2$ has the same rank as $P_1P_2P_1$, it is seen that $A_2(A_2'P_1P_2P_1A_2)^-A_2'$ is a possible choice of a g-inverse of $P_1P_2P_1$. As the expression (6) is invariant under choice of g-inverse of $P_1P_2P_1$, plugging in the expression for the g-inverse just recorded we arrive at another expression for $\Pi(P_2, P_1)$ viz.,

$$P_2P_1A_2[(A_2'P_1A_2)^2]^-A_2'P_1P_2 \tag{10}$$

requiring inversion of a matrix of order $r_2 \times r_2$. Similar arguments when used in respect to formula (4) lead to the following two alternative formulae for $\Pi(P_2, P_1)$.

$$P_2P_1P_2 - P_2P_1B_2(B_2'Q_1B_2)^-B_2'P_1P_2 \tag{11}$$

and

$$P_2P_1P_2 - P_2P_1Q_2B_1[(B_1'Q_2B_1)^2]^-B_1'Q_2P_1P_2 \tag{12}$$

requiring respectively inversion of matrices of order $d_2 \times d_2$ and $d_1 \times d_1$. The matrix to be inverted in each of the four cases (9) - (12), is real symmetric nnd dominated above in the Löwner ordering (positive semidefinite ordering) by the identity matrix of corresponding order. Let $\min(r_1, r_2, d_1, d_2)$ be denoted by δ. If the main consideration is to reduce the order of the matrix to be inverted, then depending on whether $\delta = r_1, r_2, d_1$ or d_2, Table 1 lists the formula appropriate in each case.

Table 1 : Alternative formulae for $\Pi(P_2, P_1)$

δ	Formula for $\Pi(P_2, P_1)$
r_1	$P_2A_1(A_1'P_2A_1)^-A_1'P_2$
r_2	$P_2P_1A_2[(A_2'P_1A_2)^2]^-A_2'P_1P_2$
d_1	$P_2P_1P_2 - P_2P_1Q_2B_1[(B_1'Q_2B_1)^2]^-B_1'Q_2P_1P_2$
d_2	$P_2P_1P_2 - P_2P_1B_2(B_2'QB_2)^-B_2'P_1P_2$

REMARK 4 : Consider the special case where $P_2 = \text{diag}(1, 1, ..., 1, 0)$. Here $B_2 = (0, 0, ..., 0, 1)'$. Substituting these in (11), we are led to the same expression as was derived in [8], corresponding to deletion of the last observation in the linear model $(\boldsymbol{Y}, \mathcal{R}(P_1), I)$, a result used extensively in regression diagnostics. The orthogonal projection matrix is called here the 'hat' matrix [3,5].

The general case requiring deletion of several (say n_2) observations can without any loss of generality be discussed in terms of orthogonal projectors P_1 and P_2, where $P_2 = \text{diag}(\overbrace{1, 1, ..., 1}^{n_1}, \overbrace{0, 0, ..., 0}^{n_2})$, $n_1 + n_2 = n$. The matrix

P_1 was perhaps redefined by permuting the rows and columns of the original P_1 in a symmetric fashion, to fit in the above model for P_2. We have here

$$Q_2 = \begin{pmatrix} 0 & 0 \\ 0 & I_{n_2} \end{pmatrix}, B_2 = \begin{pmatrix} 0 \\ I_{n_2} \end{pmatrix}. \tag{13}$$

Consider the conformable partition for P_1

$$P_1 = \begin{pmatrix} P_{11} & P_{12} \\ P_{21} & P_{22} \end{pmatrix}. \tag{14}$$

From (11), we therefore arrive at the following interesting expression due to Bhimasankaram [4] for the orthogonal projector to the range of the matrix formed by the first n_1 rows of the redefined X matrix

$$P_{11} + P_{12}(I - P_{22})^- P_{21}. \tag{15}$$

We shall consider here another transformation of the range space which corresponds to addition of an independent observation with unit variance to the linear model $(\boldsymbol{Y}, \mathcal{R}(P_1), I)$. This result is relevant in the context of adaptive linear filtering for signal processing. Consider the matrix

$$W = \begin{pmatrix} I_n \\ \boldsymbol{b}' \end{pmatrix}. \tag{16}$$

Given the orthogonal projector P_1 of order n, we are required to find an orthogonal projector to $\mathcal{R}(WP_1)$. We have seen that under the linear model $(\boldsymbol{Y}, \mathcal{R}(P_1), I)$, $P_1\boldsymbol{Y}$ is the BLUE of $E(\boldsymbol{Y})$. We have to consider the additional observation Y_{n+1} with $E(Y_{n+1}) = \boldsymbol{b}'E(\boldsymbol{Y}).\boldsymbol{b}'P_1\boldsymbol{Y}$ is the BLUE of $E(Y_{n+1})$ based on $\boldsymbol{Y}$ and $\text{Var}(\boldsymbol{b}'P_1\boldsymbol{Y}) = \boldsymbol{b}'P_1\boldsymbol{b}$. Combining the estimators Y_{n+1} and $\boldsymbol{b}'P_1\boldsymbol{Y}$ in the best possible manner, we have the BLUE of $E(Y_{n+1})$ as

$$(\boldsymbol{b}'P_1\boldsymbol{b}Y_{n+1} + \boldsymbol{b}'P_1\boldsymbol{Y})/(1 + \boldsymbol{b}'P_1\boldsymbol{b}).$$

To obtain the new BLUE for $E(\boldsymbol{Y})$, we consider the estimator $P_1\boldsymbol{Y} + \boldsymbol{\theta}(Y_{n+1} - \boldsymbol{b}'P_1\boldsymbol{Y})$ and determine $\boldsymbol{\theta}$ so that $\text{Cov}[P_1\boldsymbol{Y} + \boldsymbol{\theta}(Y_{n+1} - \boldsymbol{b}'P_1\boldsymbol{Y}), Y_{n+1} - \boldsymbol{b}'P_1\boldsymbol{Y}] = 0$. This gives $\boldsymbol{\theta} = P_1\boldsymbol{b}/(1 + \boldsymbol{b}'P_1\boldsymbol{b})$. Expressing the BLUE of $\begin{bmatrix} E(\boldsymbol{Y}) \\ E(Y_{n+1}) \end{bmatrix}$ linearly in terms of $\boldsymbol{Y}$ and Y_{n+1} and picking out the coefficients, we have the following expression for the new projection matrix.

$$(1 + \boldsymbol{b}'P_1\boldsymbol{b})^{-1} \begin{bmatrix} P_1(1 + \boldsymbol{b}'P_1\boldsymbol{b}) - P_1\boldsymbol{b}\boldsymbol{b}'P_1 & P_1\boldsymbol{b} \\ \boldsymbol{b}'P_1 & \boldsymbol{b}'P_1\boldsymbol{b} \end{bmatrix}. \tag{17}$$

In the above computation, $E(Y_{n+1})$ is a linear combination of the coordinates of $E(\boldsymbol{Y})$. Suppose $E(Y_i), i = 1, 2, ..., n+1$ are linear functions of the same set of parameters but the coefficients vector for $E(Y_{n+1})$ is linearly independent of those of $E(Y_1), E(Y_2), ..., E(Y_n)$. Then given the orthogonal projector P_1 to the range of the coefficients matrix of the first n observations, the orthogonal projector to the range of the augmented coefficient matrix is given by $\text{diag}(P_1, 1)$.

REMARK 5: P. Bhimasankaram has pointed out that Theorem 7, as well as the expression for the orthogonal projector given in (17) could be derived in a straightforward manner from Lemma 6, using the expression (7) for the orthogonal projector onto $\mathcal{R}(X)$, the range of a matrix X.

LEMMA 6 (Bhimasankaram [4]) Let P_1 be an orthogonal projector and B a matrix such that P_1B is defined. Then

(a) $P_1 + P_1B(I - B'P_1B)^- B'P_1 \in \{(P_1 - P_1BB'P_1)^-\}$

(b) $P_1 - P_1B(I + B'P_1B)^{-1}B'P_1 \in \{(P_1 + P_1BB'P_1)^-\}$.

PROOF: The proof is by direct verification of the condition required to be satisfied by a generalized inverse. ■

Readers unfamiliar with the necessary background material in Statistics will find the purely algebraic approach via Bhimasankaram's lemma more appealing.

For Theorem 7, $X = P_2P_1$. Hence $X'X = P_1P_2P_1 = P_1 - P_1BB'P_1$ where $P_2 = I - BB'$ and $B'B = I$. Here we need Lemma 6(a). For (7) or its extension requiring addition of more than one observation

$$X = \begin{pmatrix} I \\ B' \end{pmatrix} P_1.$$

Here $X'X = P_1 + P_1BB'P_1$ and we use Lemma 6(b).

REFERENCES

[1] Anderson, W.N. and Duffin, R.J. Series and parallel addition of matrices, *J.Math. Anal. Appl.*, **11** (1963), 576-594.

[2] Anderson, W.N. and Trapp, G. Shorted operators II, *SIAM J. Appl. Math.* **28** (1975), 60-71.

[3] Belsley, D.A., Kuh, E. and Welsh, R.E. *Regression Diagnostics : Identifying Influential Observations and Sources of Collinearity*, Wiley, New York (1980).

[4] Bhimasankaram, P., Personal communication (1992).

[5] Chatterjee, S. and Hadi, A.S. *Sensitivity Analysis in Linear Regression*, Wiley, New York(1988).

[6] Halmos, P.R. *A Hilbert Space Problem Book*, Springer-Verlag, Berlin (1974).

[7] Krein, M.G. The theory of self-adjoint extensions of semi-bounded Hermitian operators and its applications, *Mat. Sbornik N.S.*,**20** (62), 431-495 and **21**(63), 365-404 (1947).

[8] Mitra, S.K. Regression post-mortem, *J. Indian Soc. Agricultural Statist.* **43** (1991), 201-210.

[9] Mitra, S.K. and Puntanen, S. The shorted operator statistically interpreted, *Calcutta Statist. Assoc. Bull. H.K. Nandi memorial volume*, **40**(1990-91), 97-102.

[10] Mitra, S.K. and Puri, M.L. Shorted operators and generalized inverse of matrices, *Linear Algebra Áppl.*, **25** (1979), 45-56.

[11] Mitra, S.K. and Puri, M.L. Shorted matrices-an extended concept and some applications *Linear Algebra Appl.*, **42** (1982), 57-79.

[12] Rao, C.R. and Mitra, S.K., *Generalized Inverse of Matrices and its Applications*, Wiley, New York (1971).

[13] Rao, C.R., Mitra, S.K. and Bhimasankaram, P. Determination of a matrix by its subclasses of generalized inverses *Sankhya* **A 34** (1972), 5-8.

[14] Von Neumann, J. *Functional Operators*, Vol. II Princeton University Press, Princeton (1950).

Indian Statistical Institute
Delhi Centre
7, S.J.S. Sansanwal Marg
New Delhi 110 016
INDIA

J.K. Ghosh, S.K. Mitra, K.R. Parthasarathy and B.L.S. Prakasa Rao (*eds.*),
Statistics and Probability : A Raghu Raj Bahadur Festschrift
Published by Wiley Eastern Limited, Publishers, (1993), 475 – 497

A Quantum Probabilistic Analogue of Feller's Condition for the Existence of Unitary Markovian Cocycles in Fock Spaces

A. MOHARI and K.R. PARTHASARATHY
Indian Statistical Institute, New Delhi

Abstract

A rigorous proof of Journé's time reversal principle is presented and applied to find conditions under which a quantum stochastic differential equation of the form $dX = X(\Sigma E_i dA_i^\dagger + \Sigma F_j^i d\Lambda_i^j + \Sigma G_j dA_j + Hdt)$ admits isometric, coisometric and unitary solutions. The problem of uniqueness is also investigated. The conditions on the coefficient operators turn out to be strikingly similar to the resolvent conditions of Feller for the classical Kolmogorov equations in the theory of Markov processes.

1 Introduction

Quantum stochastic differential equations (q.s.d.e's) in the tensor product $h \otimes \Gamma(L^2(\mathbb{R}_+) \otimes k)$ of an initial Hilbert space h and the boson Fock space over $L^2(\mathbb{R}_+) \otimes k$, of the form

$$dX = X(\sum_i E_i dA_i^\dagger + \sum F_j^i d\Lambda_i^j + \sum G_j dA_j + Hdt) \qquad (1.1)$$

AMS Subject Classification : 81 S 25.
Key words and Phrases : quantum stochastic differential equation, boson Fock space, creation, conservation and annihilation processes.

where $E_i, F_j^i, G_j, H (1 \leq i, j \leq \dim k)$ are operators in h and $A_i^\dagger, \Lambda_j^i, A_j (1 \leq i, j \leq \dim k)$ are respectively the basic creation, conservation (gauge or exchange) and annihilation processes in the Fock component, were introduced by Hudson and Parthasarathy [7] and a necessary and sufficient condition for the existence of a unitary solution $U = \{U(t)\}$ with initial condition $U(0) = I$ was obtained when $\dim k < \infty$ and the coefficient operators E_i, F_j^i, G_j, H are bounded. This result was later extended to cover the case $\dim k = \infty$ by Mohari and Sinha [11]. A self-contained and systematic presentation of these results and some of their applications may be found in [14], [10]. When $\dim k < \infty$ and the coefficients are unbounded an investigation of the problem of existence of unitary solutions for (1.1) was initiated by Journé [8] and continued by Fagnola [4], [5], Chebotarev [2] and Vincent-Smith [18]. In order to realise many classical Markov processes through unitary evolutions in a Hilbert space it was already observed in [12, 15, 16, 17] that it is necessary to study (1.1) when $\dim k = \infty$ and the coefficient operators are not necessarily bounded. The notational framework of M.P. Evans [3] developed in the theory of q.s.d.e.'s and used by Mohari and Sinha [11] in the case of infinite degrees of freedom has recently been successfully exploited by Mohari [13] in the study of (1.1) with unbounded coefficients and infinite degrees of freedom. In the present paper we make a relatively "co-ordinate-free" approach to the same problem and find conditions under which (1.1) has an isometric, co-isometric or unitary solution. The problem of uniqueness is also investigated. The conditions on the coefficient operators turn out to be strikingly similar to the resolvent conditions of Feller [6], [9] on the infinitesimal Markov generator to ensure the uniqueness and strict Markov property of the "minimal solution". As an explicit illustration of our main result we provide an example of a q.s.d.e. for which the condition for the uniqueness and existence of a unitary solution leads directly to Feller's condition. This seems to indicate the possibility of a rich boundary theory of Markovian co-cycles associated with a given q.s.d.e. For some interesting applications of these methods in the context of continuous time Markov chains with countable state space we refer to Mohari [13].

2 Preliminaries of quantum stochastic calculus

We shall present a brief summary of quantum stochastic calculus with emphasis on the quantum Ito's formula in a compact notation which seems convenient for handling q.s.d.e.'s with infinite degrees of freedom, especially in the context of Markov processes whose state spaces are not necessarily finite.

All the Hilbert spaces that we deal with are complex and separable with scalar product $< \cdot, \cdot >$ linear in the second variable. If $h_i, i = 1, 2, ..., k$ are Hilbert spaces, $u_i \in h_i$ for each i and there is no confusion we denote the product vector $u_1 \otimes \cdots \otimes u_k$ by $u_1 u_2 \cdots u_k$. If X is an operator in h_{i_0} with domain D for some i_0 we frequently denote by the same letter X any operator with domain including all elements of the form $u_1u_2\cdots u_k$ where $u_{i_0} \in D$ and satisfying $X(u_1u_2\cdots u_k) = u_1u_2\cdots u_{i_0-1}(Xu_{i_0})u_{i_0+1}\cdots u_k$.

Choose and fix two Hilbert spaces h, k. Fix an orthonormal basis $\{e_i, i \in S\}$ in k. Denote by $\mathcal{M}_S \subset L^2(\mathbb{R}_+) \otimes k$ the dense linear manifold of all elements u satisfying $< e_i, u(\cdot) >_k = 0$ for all but a finite number of $i \in S$ when u is looked upon as a k-valued function on $\mathbb{R}_+$ which is integrable in norm square. Imbed k in the larger Hilbert space

$$\hat{k} = \mathbb{C}e_{-\infty} \oplus k \oplus \mathbb{C}e_{\infty} \tag{2.1}$$

for which $\{e_i, i \in \overline{S}\}$ with $\overline{S} = S \cup \{-\infty, \infty\}$ is an orthonormal basis. Introduce the basic Hilbert spaces

$$\mathcal{H} = h \otimes \Gamma(L^2(\mathbb{R}_+) \otimes k) \tag{2.2}$$

$$\hat{\mathcal{H}} = h \otimes \hat{k} \otimes \Gamma(L^2(\mathbb{R}_+) \otimes k) \tag{2.3}$$

where for any Hilbert space $\mathcal{V}$, $\Gamma(\mathcal{V})$ denotes the boson Fock space over it. Denote by F the unique unitary operator (called the *flip* exchanging $e_{-\infty}$ and e_∞) in $\hat{\mathcal{H}}$ which satisfies

$$Ffe_j\psi = \begin{cases} fe_j\psi & \text{if } j \in S, \\ fe_\infty\psi & \text{if } j = -\infty, \\ fe_{-\infty}\psi & \text{if } j = +\infty \end{cases} \tag{2.4}$$

for all $f \in h, \psi \in \Gamma(L^2(\mathbb{R}_+) \otimes k)$. Note that, according to our conventions, F also denotes the flip in $\hat{k}$ as well as $h \otimes \hat{k}$.

We now modify the Evans' notation [3] and introduce the basic integrator processes $\{\Lambda^i_j, i, j \in \overline{S}\}$ of quantum stochastic calculus by putting

$$\begin{aligned}
\Lambda^i_{-\infty}(t) &= a(I_{[0,t]} \otimes e_i), \quad i \in S \\
\Lambda^\infty_j(t) &= a^\dagger(I_{[0,t]} \otimes e_j), \quad j \in S \\
\Lambda^i_j(t) &= \lambda(I_{[0,t]} \otimes |e_j >< e_i|), \quad i, j \in S \\
\Lambda^\infty_{-\infty}(t) &= tI, \\
\Lambda^{-\infty}_i(t) &= \Lambda^j_\infty(t) = 0, i, j \in \overline{S}
\end{aligned}$$

where, for any Hilbert space $\mathcal{V}, a(v)$ and $a^\dagger(v)$ denote respectively the annihilation and creation operators in $\Gamma(\mathcal{V})$ associated with $v \in \mathcal{V}, \lambda(T)$ is the conservation (or gauge) operator in $\Gamma(\mathcal{V})$ associated with the operator T in $\mathcal{V}$ and $I_{[0,t]}$ is either the indicator function or multiplication by the indicator function of the interval $[0, t]$. (Meyer [10] in his notes calls $\Lambda^i_j, i, j \in S$ as exchange operator processes. We may say that exchange from i to $-\infty$ is 'annihilation' and that from ∞ to i is 'creation' whereas an exchange from $-\infty$ to i or from i to ∞ is impossible.)

For any Hilbert space $\mathcal{V}$ denote by $\mathcal{B}(\mathcal{V})$ the algebra of all bounded operators on $\mathcal{V}$ and for any $v \in \mathcal{V}$ let $e(v)$ be the exponential or coherent vector associated with v in the boson Fock space $\Gamma(\mathcal{V})$ over $\mathcal{V}$. Let

$$e(v) = \sum_{n=0}^{\infty} v^{(n)}, v^{(0)} = 1 \tag{2.5}$$

be the finite particle vector expansion of $e(v)$ where $v^{(n)} = (n!)^{-\frac{1}{2}} v^{\otimes^n}$. The fact that $e(zv) = \sum_{n=0}^{\infty} z^n v^{(n)}$ is a convergent power series in the complex variable z will play an important role in this paper. (See the proofs of Proposition 4.3 and Theorem 4.4).

Let $D \subset h$ be a dense linear manifold in the initial Hilbert space. Associated with D are the dense linear manifolds $D_S, \mathcal{D}_S$ and $\hat{\mathcal{D}}_S$ in $h \otimes \hat{k}, \mathcal{H}$ and $\hat{\mathcal{H}}$ generated by $\{fe_j | f \in D, j \in \overline{S}\}, \{fe(u) | f \in D, u \in \mathcal{M}_S\}$ and $\{fe_j e(u) | f \in D, j \in \overline{S}, u \in \mathcal{M}_S\}$ respectively. (See (2.1) - (2.3)). By a $(D_S, \mathcal{M}_S)$ - *adapted process* L in $\hat{\mathcal{H}}$ we mean a family $\{L(t), t \geq 0\}$ of operators in $\hat{\mathcal{H}}$ satisfying the following:

(i) $\hat{\mathcal{D}}_S$ is contained in the domain of $L(t)$ for each t;

(ii) The map $t \to L(t)\xi e(u)$ is measurable for each $\xi \in D_S, u \in \mathcal{M}_S$;

(iii) The element $L(t)\xi e(u)$ admits the factorization

$$L(t)\xi e(u) = \{L(t)\xi e(I_{[0,t]}u)\} \otimes e(I_{(t,\infty)}u)$$

for every $\xi \in D_S, u \in \mathcal{M}_S, t \geq 0$ where the first factor on the right hand side belongs to $h \otimes \hat{k} \otimes \Gamma(L^2[0,t] \otimes k)$ and the second is an element of $\Gamma(L^2(t,\infty) \otimes k)$.

The adapted process L as defined above is said to be *regular* if, in addition, the following holds:

(iv) The map $t \to L(t)\xi e(u)$ is continuous for every fixed $\xi \in D_S, u \in \mathcal{M}_S$.

Similarly, $(D, \mathcal{M}_S)$-adapted and regular processes in $\mathcal{H}$ are defined when $h \otimes \hat{k}$ and D_S are replaced by h and D respectively.

Denote by $\mathcal{I}(D_S, \mathcal{M}_S)$ the family of all $(D_S, \mathcal{M}_S)$-adapted regular processes L in $\hat{\mathcal{H}}$ satisfying the following:

$$L(t) f e_{-\infty} e(u) = 0; \tag{2.6}$$

$$< fe_\infty e(u), L(t) g e_j e(v) >= 0 \tag{2.7}$$

for all $f, g \in D, u, v \in \mathcal{M}_S, j \in \overline{S}$.

The regularity condition on any $L \in \mathcal{I}(D_S, \mathcal{M}_S)$ implies, in particular, that

$$\int_0^t \|L(s)\xi e(u)\|^2 ds < \infty \text{ for all } \xi \in D_S, u \in \mathcal{M}_S, t \geq 0. \tag{2.8}$$

Every $L \in \mathcal{I}(D_S, \mathcal{M}_S)$ defines the family $\{L^i_j, i, j \in \overline{S}\}$ of regular $(D, \mathcal{M}_S)$-adapted processes in $\mathcal{H}$ through the relations:
$< fe(u), L^i_j(t)ge(v) >$
$= < fe_ie(u), L(t)ge_je(v) >, \; f, g \in D, u, v \in \mathcal{M}_S.$

Then (2.8) can be expressed as

$$\int_0^t \sum_{i \in \overline{S}} \|L^i_j(s)ge(v)\|^2 ds < \infty, g \in D, v \in \mathcal{M}_S, t \geq 0 . \tag{2.9}$$

Indeed,

$$\|L(t)ge_je(v)\|^2 = \sum_{i \in \overline{S}} \|L^i_j(t)ge(v)\|^2 \tag{2.10}$$

for every $j \in \overline{S}, g \in D, v \in \mathcal{M}_S$. Further, $L^\infty_j = L^i_{-\infty} = 0$ for all $i, j \in \overline{S}$.

PROPOSITION 2.1: Let X_0 be any operator in h with domain D and let $L \in \mathcal{I}(D_S, \mathcal{M}_S)$. Then there exists a unique $(D, \mathcal{M}_S)$-adapted regular process X in $\mathcal{H}$ satisfying

$$X(t) = X_0 + \int_0^t \sum_{i,j \in \overline{S}} L^i_j(s) d\Lambda^j_i(s) \tag{2.11}$$

on the domain $\mathcal{D}_S$.

PROOF : In view of the fact that $L^\infty_j = L^i_{-\infty} = 0, \Lambda^j_\infty = \Lambda^{-\infty}_i = 0$ for all $i, j \in \overline{S}$ and (2.9) this is just a restatement of Proposition 27.1 in [14]. See also [11]. ∎

For any $L \in \mathcal{I}(D_S, \mathcal{M}_S)$ we write

$$\Lambda_L(t) = \int_0^t \sum_{i,j \in \overline{S}} L^i_j(s) d\Lambda^j_i(s)$$

and express (2.11) in the language of quantum stochastic differentials :

$$dX = \sum_{i,j \in \overline{S}} L^i_j d\Lambda^j_i, X(0) = X_0$$

on the domain $\mathcal{D}_S$. In particular, Λ_L is a $(D, \mathcal{M}_S)$-adapted process in $\mathcal{H}$ satisfying

$$d\Lambda_L = \sum_{i,j \in \overline{S}} L^i_j d\Lambda^j_i, \Lambda_L(0) = 0$$

on $\mathcal{D}_S$.

PROPOSITION 2.2 (Quantum Ito's formula): Let $L, M \in \mathcal{I}(D_S, \mathcal{M}_S)$ and let

$$dX = d\Lambda_L, X(0) = X_0,$$

$$dY = d\Lambda_M, Y(0) = Y_0$$

on $\mathcal{D}_S$. Then, for all $f, g \in D, u, v \in \mathcal{M}_S$ the following holds:

(i) $< fe(u), X(t)ge(v) > \;=\; < fe(u), X_0 ge(v) >$

$$+ \int_0^t < f(e_{-\infty} + u(s))e(u), L(s)g(v(s) + e_\infty)e(v) > ds;$$

(ii) $< X(t)fe(u), Y(t)ge(v) > \;=\; < X(0)fe(u), Y(0)ge(v) >$

$$+ \int_0^t \{< (X(s) + FL(s)F)f(e_{-\infty} + u(s))e(u), (Y(s) + M(s))g(v(s) + e_\infty)e(v) > - < X(s)f(e_{-\infty} + u(s))e(u), Y(s)g(v(s) + e_\infty)e(v) >\}ds.$$

PROOF : This is just a restatement of the first and second fundametal formulae of quantum stochastic calculus expressed in the new notations introduced here. See, in particular, Corollary 27.2 of [14]. See also [7], [11], [12]. ■

It may be noted that in (i) of Proposition 2.2 f, u can be arbitrary elements of $h, L^2(\mathbb{R}_+) \otimes k$ respectively.

COROLLARY 2.3 : Let $L \in \mathcal{I}(D_S, \mathcal{M}_S)$,

$$dX = d\Lambda_L, X(0) = X_0$$

on $\mathcal{D}_S$. For any $u \in L^2(\mathbb{R}_+) \otimes k$ introduce

$$\nu_u(t) = \int_0^t (1 + \|u(s)\|^2)ds. \tag{2.12}$$

Then

$$\|[X(t_2) - X(t_1)]fe(u)\|^2$$

$$\leq 2e^{\nu_u(t_2)} \int_{t_1}^{t_2} \|L(s)f(u(s) + e_\infty)e(u)\|^2 ds \tag{2.13}$$

for all $f \in D, u \in \mathcal{M}_S, 0 \leq t_1 \leq t_2 < \infty$.

PROOF : In (ii), Proposition 2.2 choose $L = M, X = Y, f = g, u = v$ and proceed along the same lines as in the proof of Proposition 27.1 by using Proposition 25.5 in [14]. ■

When $D = h$, a $(D, \mathcal{M}_S)$-adapted process $X = \{X(t)\}$ in $\mathcal{H}$ is called *contractive* , *isometric, co-isometric* or *unitary* according as all the operators $X(t), t \geq 0$ are contractive, isometric, co-isometric or unitary.

Suppose $L \in \mathcal{I}(D_S, \mathcal{M}_S)$ but $L(t)$ is independent of t. Then L can be identified with the operator $L = L(0)$ in $h \otimes \hat{k}$ with domain D_S and satisfying the conditions: $Lfe_{-\infty} = 0, < fe_{\infty}, Lge_j >= 0$ for all $f, g \in D, j \in \overline{S}$. Our aim is to examine q.s.d.e.'s of the form

$$dX = Xd\Lambda_L, X(0) = X_0 \tag{2.14}$$

and find conditions on L so that a contractive, isometric, co-isometric or unitary $(h, \mathcal{M}_S)$-adapted regular solution $X = \{X(t)\}$ satisfying (2.14) on $\mathcal{D}_S$ may exist and be unique. In this context we introduce the basic algebra $\mathbb{B} = \mathbb{B}(h, k) \subset \mathcal{B}(h \otimes \hat{k})$ associated with the pair (h, k) of Hilbert spaces:

$$\mathbb{B} = \{L | L \in \mathcal{B}(h \otimes \hat{k}), Lfe_{-\infty} = L^* fe_{\infty} = 0 \text{ for all } f \in h\}. \tag{2.15}$$

The algebra $\mathbb{B}$ admits the involution : $L \to L^b$ where $L^b = FL^*F, F$ being the unitary flip operator exchanging $e_{-\infty}$ and e_{∞}. We call the algebra $\mathbb{B}$ with the involution b the *Belavkin algebra* associated with the pair (h, k). When L varies in $\mathbb{B}$ the quantum Ito's formula of Proposition 2.2 takes the form :

(i) Λ_L and Λ_{L^b} are adjoint to each other on the domain $\mathcal{D}_S$ with $D = h$;

(ii) $d\Lambda_L d\Lambda_M = d\Lambda_{LM}$.

It is clear that the map $L \to \Lambda_L$ is linear in L. (See [1], [16]).

3 A class of q.s.d.e.'s with bounded coefficients

Let h, k be fixed Hilbert spaces and let $\mathbb{B}$ be the Belavkin algebra with involution b associated with the pair (h, k). For any $L \in \mathbb{B}, u, v \in k$ define the bounded operators $L_{u,v}$ in h by the relations

$$< f, L_{u,v} g > \quad = \quad < f(e_{-\infty} + u), Lg(v + e_{\infty}) > \tag{3.1}$$

for all f, g in h. Then

$$\|L_{u,v}\| \leq \|L\| \{(1 + \|u\|^2)(1 + \|v\|^2)\}^{1/2}. \tag{3.2}$$

For any $s \geq 0$ we introduce the operators θ_s, ρ_s in $L^2(\mathbb{R}_+) \otimes k$ and their respective second quantizations S_s, R_s in $\Gamma(L^2(\mathbb{R}_+) \otimes k)$ by putting

$$(\theta_s u)(t) = \begin{cases} 0 & \text{if } t < s, \\ u(t-s) & \text{if } t \geq s, \end{cases} \tag{3.3}$$

$$(\rho_s u)(t) = \begin{cases} u(s-t) & \text{if } t < s, \\ u(t) & \text{if } t \geq s, \end{cases} \tag{3.4}$$

$$S_s e(u) = e(\theta_s u), \tag{3.5}$$

$$R_s e(u) = e(\rho_s u), \tag{3.6}$$

for all $u \in L^2(\mathbb{R}_+) \otimes k$. The operators θ_s, S_s are isometric whereas ρ_s, R_s are unitary. We have $\theta_s \theta_t = \theta_{s+t}, S_s S_t = S_{s+t}$ and $R_s^2 = I$. θ_s and S_s are 'time shifts' whereas ρ_s and R_s are 'time reversals in the interval $[0, s]$'

According to the conventions we mentioned at the beginning of Section 2, we shall view S_s and R_s also as operators in both the Hilbert spaces $\mathcal{H}$ and $\hat{\mathcal{H}}$ defined by (2.2) and (2.3).

PROPOSITION 3.1 : For any $L \in \mathbb{B}$ and $s \geq 0$ there exists a unique regular $(h, \mathcal{M}_S)$-adapted process $X(s,.)$ in the interval $[s, \infty)$ satisfying the q.s.d.e.

$$dX(s,t) = d\Lambda_L(t)X(s,t), X(s,s) = I \tag{3.7}$$

for all $t \geq s$ in $\mathcal{D}_S$ with $D = h$. Furthermore

$$S_h^* X(s+h, t+h)S_h = X(s,t) \tag{3.8}$$

in $\mathcal{D}_S$ for all $0 \leq s \leq t < \infty, h \geq 0$.

PROOF : We view L as a 'constant' adapted process in $\hat{\mathcal{H}}$, define L_j^i as in Proposition 2.1 and observe that (2.10) holds. Thus

$$\sum_{i \in \overline{S}} \|L_j^i f\|^2 = \|Lfe_j\|^2 \leq \|L\|^2 \|f\|^2$$

for all $f \in h$. It now follows from Proposition 27.5, [14] (See also [11]) that the process $X(s,.)$ satisfying (3.7) is uniquely determined. Equation (3.8) is immediate from the fact that L_j^i are independent of t, the uniqueness of the solution of (3.7) and the property

$$S_h^*[\Lambda_j^i(t+h) - \Lambda_j^i(s+h)]S_h = \Lambda_j^i(t) - \Lambda_j^i(s).$$

■

PROPOSITION 3.2: Let $\{X(s,t), 0 \leq s \leq t < \infty\}$ be defined as in Proposition 3.1. Suppose that $u, v \in \mathcal{M}_S$ satisfy the boundedness condition

$$\sup_{0 \leq t \leq a} (\|u(t)\| + \|v(t)\|) < \infty \text{ for every } a > 0. \tag{3.9}$$

Then there exists a family $\{N(s,t), 0 \leq s \leq t < \infty\}$ of bounded operators on h satisfying the following:

(i) $< fe(u), X(s,t)ge(v) > \; = \; < f, N(s,t)g >$ for all $f, g \in h$;

(ii) $\frac{\partial}{\partial t} N(s,t) = L_{u(t),v(t)} N(s,t)$ a.e. t,

where $L_{u(t),v(t)}$ is defined by (3.1) and differentiation is in the operator norm sense;

(iii) $N(s,t) =< e(u), e(v) > \{I+$

$$\sum_{n=1}^{\infty} \int_{s<t_1<\cdots<t_n<t} L_{u(t_n),v(t_n)} L_{u(t_{n-1}),v(t_{n-1})} \cdots L_{u(t_1),v(t_1)} dt_1 dt_2 \ldots dt_n\}$$

where the right hand side converges in operator norm ;

(iv) $\frac{\partial N(s,t)}{\partial s} = -N(s,t)L_{u(s),v(s)}$ a.e. s.

PROOF : Combining Proposition 3.1 with the first part of Proposition 2.2 we have
$< fe(u), X(s,t)ge(v) > \ = \ < fe(u), ge(v) >$
$+ \int_s^t < f(e_{-\infty} + u(\tau))e(u), LX(s,\tau)g(v(\tau) + e_\infty)e(v) > d\tau$
$= \ < fe(u), ge(v) >$

$$+ \int_s^t < fe(u), L_{u(\tau),v(\tau)}X(s,\tau)ge(v) > d\tau. \tag{3.10}$$

Now consider the ordinary differential equation

$$\frac{\partial N(s,t)}{\partial t} = L_{u(t),v(t)}N(s,t), t \geq s \tag{3.11}$$

with the initial condition $N(s,s) =< e(u), e(v) > I$ in $\mathcal{B}(h)$. By condition (3.9) and the inequality (3.2) it follows that (3.11) admits a unique solution given by the infinite series in (iii) of the proposition. Now a comparison with (3.10) shows that property (i) of the proposition holds. (iv) follows from (iii) by straightforward differentiation with respect to s. ■

THEOREM 3.3: Let $L \in \mathbb{B}$. In order that there may exist a family $U = \{U(s,t), 0 \leq s \leq t < \infty\}$ of isometric operators in $\mathcal{H}$ such that $U(s,.)$ is a regular $(h, \mathcal{M}_S)$-adapted process in $[s, \infty)$ for every s, satisfying

$$dU(s,t) = d\Lambda_L(t)U(s,t), U(s,s) = I \tag{3.12}$$

in $\mathcal{D}_S$ with $D = h$ it is necessary and sufficient that

$$L + L^b + L^b L = 0. \tag{3.13}$$

In such a case the family U is unique. Furthermore

$$U(r,t) = U(s,t)U(r,s),$$

$$S_h^* U(s+h, t+h)S_h = U(s,t)$$

for all $0 \leq r \leq s \leq t < \infty, h \geq 0$.

PROOF : This is already essentially proved in [11], [14] but, nevertheless, we shall illustrate how it works in the new notations introduced in Section 2. By Proposition 3.1 there exists a unique regular $(h, \mathcal{M}_S)$-adapted solution $U = \{U(s,t)\}$ satisfying (3.12) for a given $L \in \mathbb{B}$. By Proposition 2.2, (ii)
$< U(s,t)fe(u), U(s,t)ge(v) > - < fe(u), ge(v) >$

$= \int_s^t < U(s,\tau)f(e_{-\infty} + u(\tau))e(u),$

$$\{(I + L^b)(I + L) - I\}U(s,\tau)g(v(\tau) + e_\infty)e(v) > d\tau \tag{3.14}$$

for all $f, g \in h, u, v \in \mathcal{M}_S$. Suppose that $U(s,t)$ extends to an isometry on $\mathcal{H}$ for all $0 \leq s \leq t < \infty$. Then (3.14) implies that the integrand in the right hand side vanishes for almost all τ in $[s, \infty)$. By the regularity of the process $\{U(s, \cdot)\}$, choosing u and v to be simple and continuous at s and letting $\tau \to s$ we conclude that

$$< f(e_{-\infty} + u), (L + L^b + L^b L)g(v + e_\infty) >= 0$$

for all $f, g \in h$ and $u, v \in k$. Since $L \in I\!\!B$ we have $Le_{-\infty} = L^b e_{-\infty} = 0$. Thus (3.13) holds and necessity is proved.

To prove sufficiency note that (3.13) and (3.14) imply that $U(s,t)$ is isometric on the total set $\{fe(u), f \in h, u \in \mathcal{M}_S\}$ and hence extends to an isometry on $\mathcal{H}$. This proves the first part.

To prove the second part note that $U(r, \cdot)$ and $U(s, \cdot)U(r,s)$ are both solutions of

$$dX(r,t) = d\Lambda_L(t)X(r,t)$$

in the interval $[s, \infty)$ with initial condition $X(r,s) = U(r,s)$. It now follows from the uniqueness property that $U(r,t) = U(s,t)U(r,s)$ for all $t \geq s \geq r$.

The last part is immediate from property (3.8) in Proposition 3.1. ∎

In order to arrive at the co-isometric counter-part of Theorem 3.3 we formulate a definition. A family $X = \{X(s,t), 0 \leq s \leq t < \infty\}$ of contraction operators in $\mathcal{H}$ is called a *contractive right cocycle* if the following are fulfilled.

(a) $X(s,s) = I$ and $X(s, \cdot)$ is a regular $(h, \mathcal{M}_S)$-adapted process in $[s, \infty)$ for every s;

(b)$X(r,t) = X(s,t)X(r,s)$ for all $0 \leq r \leq s \leq t < \infty$;

(c) $S_h^* X(s+h, t+h)S_h = X(s,t)$ for all $h \geq 0, s \leq t$.

A family $Y = \{Y(s,t), 0 \leq s \leq t < \infty\}$ of contraction operators in $\mathcal{H}$ is called a contractive *left cocycle* if their adjoints $\{Y(s,t)^*, 0 \leq s \leq t < \infty\}$ constitute a right cocycle.

THEOREM 3.4 : (Journé's time reversal principle [8])

Let $L \in I\!\!B$ and let $X = \{X(s,t)\}$ be a contractive right cocycle satisfying

$$dX(s,t) = d\Lambda_L(t)X(s,t), X(s,s) = I$$

for $t \geq s$ in $\mathcal{D}_S$ with $D = \mathcal{H}$. Define

$$\tilde{X}(s,t) = R_t X(0, t-s)^* R_t$$

where R_t is given by (3.6). Then $\tilde{X}$ is a contractive right cocycle satisfying

$$d\tilde{X}(s,t) = d\Lambda_{L^b}(t)\tilde{X}(s,t), \tilde{X}(s,s) = I$$

for $t \geq s$ in $\mathcal{D}_S$ with $D = h$.

PROOF : Choose and fix $T > 0$. Define

$$Y(s,t) = R_T X(T-t, T-s)^* R_T, \ 0 \leq s \leq t \leq T. \tag{3.15}$$

Then $Y(s,t)$ is a contraction, $Y(s,s) = I$ and for $0 \leq r \leq s \leq t \leq T$

$$Y(r,t) = Y(s,t)Y(r,s).$$

Let $Z = \{Z(s,t), \ 0 \leq s \leq t < \infty\}$ be the unique solution of

$$dZ(s,t) = d\Lambda_{L^b}(t)Z(s,t), \ Z(s,s) = I$$

in $t \geq s$ given by Proposition 3.1. Fix u, v satisfying (3.9) and define the operators $M(s,t), N(s,t)$ in h by the relations

$$< f, M(s,t)g > \ = \ < fe(u), Y(s,t)ge(v) >, \tag{3.16}$$

$$< f, N(s,t)g > \ = \ < fe(u), Z(s,t)ge(v) > \tag{3.17}$$

for all $f, g \in h$. By Proposition 3.2 (iv), $N(s,t)$ is a bounded operator and

$$\frac{dN(s,t)}{ds} = -N(s,t)L^b_{u(s),v(s)} \text{ a.e. } s. \tag{3.18}$$

The definitions of X and Y, (3.16), and Proposition 2.2 (i) imply, through a change of variables.

$$\begin{aligned} < M(s,t)g, f > \ &= \ \overline{< fe(u), Y(s,t)ge(v) >} \\ &= \ < ge(v), fe(u) > \\ &\quad + \int_{T-t}^{T-s} < g(e_{-\infty} + (\theta_T v)(\tau))e(\theta_T v), \\ &\quad LX(T-t,\tau)f((\theta_T u)(\tau) + e_\infty)e(\theta_T u) > d\tau \\ &= \ < ge(v), fe(u) > \\ &\quad + \int_s^t < g(e_{-\infty} + v(\tau))e(\theta_T v), \\ &\quad LX(T-t, T-\tau)f(u(\tau) + e_\infty)e(\theta_T u) > d\tau \end{aligned}$$

for all $0 \leq s \leq t \leq T$. Thus

$$\begin{aligned} < f, M(s,t)g > \ &= \ < fe(u), ge(v) > \\ &\quad + \textstyle\int_s^t < f(e_{-\infty} + u(\tau))e(u), Y(\tau,t)L^b g(v(\tau) + e_\infty)e(v) > d\tau \\ &= < fe(u), ge(v) > + \textstyle\int_s^t < f, M(\tau,t)L^b_{u(\tau),v(\tau)}g > d\tau. \end{aligned}$$

In other words

$$\frac{dM(s,t)}{ds} = -M(s,t)L^b_{u(s),v(s)} \text{ a.e. } s, 0 \le s \le t \le T.$$

Comparing with (3.18) we conclude that $M(s,t) = N(s,t)$ for all $0 \le s \le t \le T$. Since u, v are arbitrary but subject to (3.9) only, it follows that $Y(s,t) = Z(s,t)$ on $\mathcal{D}_S$ for all $0 \le s \le t \le T$. Letting T descend to t we see that $\tilde{X}$ satisfies the required conditions. ■

THEOREM 3.5 : Let $L \in \mathbb{B}$. In order that there may exist a family $U = \{U(s,t), 0 \le s \le t < \infty\}$ of co-isometric operators in $\mathcal{H}$ such that $U(s,\cdot)$ is a regular $(h, \mathcal{M}_S)$-adapted process in $[s,\infty)$ for every s satisfying

$$dU(s,t) = d\Lambda_L(t)U(s,t), U(s,s) = I$$

in $\mathcal{D}_S$ with $D = h$ it is necessary and sufficient that

$$L + L^b + LL^b = 0 \tag{3.19}$$

In such a case the family U is unique. Furthermore U is a co-isometric right cocycle.

PROOF : To prove necessity define

$$\tilde{U}(s,t) = R_t U(0, t-s)^* R_t.$$

Then by Theorem 3.4 $\tilde{U}$ is an isometric cocycle satisfying

$$d\tilde{U}(s,t) = d\Lambda_{L^b}(t)\tilde{U}(s,t), \quad t \ge s.$$

By Theorem 3.3 it follows that (3.19) holds.

To prove sufficiency, once again by Theorem 3.3 there exists an isometric right cocycle V such that

$$dV(s,t) = d\Lambda_{L^b}(t)V(s,t), \quad t \ge s$$

since L^b satisfies $L + L^b + (L^b)^b L^b = 0$. Now define

$$U(s,t) = R_t V(0, t-s)^* R_t.$$

Once again by Theorem 3.4, U satisfies the requirements of the theorem. ■

THEOREM 3.6 : Let $L \in \mathbb{B}$. In order that there may exist a unitary right cocycle $U = \{U(s,t), 0 \le s \le t < \infty\}$ satisfying

$$dU(s,t) = d\Lambda_L(t)U(s,t), t \ge s$$

in $\mathcal{D}_S$ with $D = h$ it is necessary and sufficient that

$$L + L^b + L^b L = L + L^b + LL^b = 0.$$

PROOF : This is immediate from Theorem 3.3 and 3.5. ■

THEOREM 3.7 : Let $L \in \mathbb{B}$. Consider the q.s.d.e.

$$dU(s,t) = U(s,t)d\Lambda_L(t),\ U(s,s) = I, t \geq s \tag{3.20}$$

in $\mathcal{D}_S$ with $D = h$, where $U(s,\cdot)$ is a regular $(h, \mathcal{M}_S)$-adapted process in $[s,\infty)$ for every s. Then the following holds:

(a) An isometric solution of (3.20) exists if and only if $L + L^b + L^b L = 0$.

(b) A co-isometric solution of (3.20) exists if and only if $L + L^b + LL^b = 0$.

(c) A unitary solution of (3.20) exists if and only if $L + L^b + LL^b = L + L^b + L^b L = 0$.

Furthermore, in each of these cases, the solution is unique and U is a left cocycle.

PROOF : In (a) the necessity part is straightforward from Proposition 2.2. To prove sufficiency consider the unique co-isometric cocycle V satisfying

$$dV(s,t) = d\Lambda_{L^b}(t)V(s,t), \quad t \geq s.$$

Indeed, by Theorem 3.5, such a V exists. By Journé's time reversal principle both the maps $t \to V(s,t)$ and $t \to V(s,t)^*$ are strongly continuous. In particular $V(s,\cdot)^* L$ is stochastically integrable. By (i) of Proposition 2.2, $U(s,t) := V(s,t)^*$ satisfies (3.20).

We omit the proofs of (b) and (c) since they follow by similar arguments. The last part is immediate. ■

4 A class of q.s.d.e.'s with unbounded coefficients

Let (h,k) be a fixed pair of Hilbert spaces where k and $\hat{k}$ defined by (2.1) are equipped with the orthonormal bases $\{e_i, i \in S\}$ and $\{e_i, i \in \overline{S}\}$ respectively, $\overline{S}$ being $S \cup \{-\infty, \infty\}$. Our aim is to study the q.s.d.e.

$$dV = V d\Lambda_L, V(0) = I$$

where L is an operator on $h \otimes \hat{k}$ with domain D_S being the dense linear manifold generated by $\{fe_j, f \in D, j \in \overline{S}\}$ and D is a fixed dense linear manifold in h. We say that L satisfies condition (A) if the following holds:

(A) There exists a sequence $\{L_n\}$ in the Belavkin algebra $\mathbb{B}$ such that

(i) $\lim_{n\to\infty} L_n fe_j = Lfe_j$ for all $f \in D, j \in \overline{S}$; (4.1)

(ii) There exists a regular $(h, \mathcal{M}_S)$-adapted contractive process $\{V_n\}$ in $\mathcal{H}$ satisfying

$$dV_n = V_n d\Lambda_{L_n}, V_n(0) = I \tag{4.2}$$

for every n.

PROPOSITION 4.2 : Let L satisfy condition (A) so that (i) and (ii) of (A) hold for a sequence $\{L_n\}$ in $\mathbb{B}$ and contractive processes $\{V_n\}$. Then there exists a regular $(h, \mathcal{M}_S)$-adapted contractive process V in $\mathcal{H}$ satisfying the following:

(i) There exists a subsequence $\{V_{n_k}\}$ such that

$$\text{w.lim}_{k\to\infty} V_{n_k}(t) = V(t) \text{ for all } t \geq 0;$$

(ii) For any $\psi \in \mathcal{H}, f \in D, u \in \mathcal{M}_S$ and $T > 0$

$$\lim_{k\to\infty} \sup_{t\leq T} |< \psi, [V_{n_k}(t) - V(t)]fe(u) > | = 0;$$

(iii) For each $f \in D, u \in \mathcal{M}_S$ and $T > 0$ there exists a constant $c = c(f, u, T)$ such that

$$\|[V(t) - V(s)]fe(u)\|^2 \leq c\,|\nu_u(t) - \nu_u(s)|$$

where ν_u is given by (2.12);

(iv) $dV = V d\Lambda_L, V(0) = I$ in $\mathcal{D}_S$, the linear manifold generated by $\{fe(u), f \in D, u \in \mathcal{M}_S\}$.

PROOF : Equation (4.2) can be expressed as

$$dV_n = d\Lambda_{V_n L_n}$$

according to the conventions of Section 2. By Corollary 2.3 and the hypothesis that $V_n(t)$ is a contraction we have

$$\begin{aligned} &\|[V_n(t_2) - V_n(t_1)]fe(u)\|^2 \\ &\leq\; 2e^{\nu_u(t_2)} \int_{t_1}^{t_2} \|V_n(s)L_n f(u(s) + e_\infty)e(u)\|^2 ds \\ &\leq\; 2e^{\nu_u(t_2)+\|u\|^2} \int_{t_1}^{t_2} \|L_n f(u(s) + e_\infty)\|^2 ds \end{aligned} \tag{4.3}$$

for all $0 \leq t_1 \leq t_2 < \infty, f \in h, u \in \mathcal{M}_S$. By (4.1), (4.3) and the definition of $\mathcal{M}_S$ we conclude that, for any $f \in D, u \in \mathcal{M}_S$ and $T > 0$, there exists a constant $c = c(f, u, T)$ such that

$$\begin{aligned} &\sup_n \|[V_n(t_2) - V_n(t_1)]fe(u)\|^2 \\ &\leq\; c[\nu_u(t_2) - \nu_u(t_1)] \quad \text{for all } 0 \leq t_1 \leq t_2 \leq T\,. \end{aligned} \tag{4.4}$$

Let now $\psi \in \mathcal{H}$ be fixed. Then for any $f \in D, u \in \mathcal{M}_S$,

$$\sup_n | < \psi, [V_n(t_2) - V_n(t_1)]fe(u) > |$$

$$\leq \|\psi\| c^{1/2} [\nu_u(t_2) - \nu_u(t_1)]^{1/2} \text{ for } 0 \leq t_1 \leq t_2 \leq T \tag{4.5}$$

and

$$\sup_n \sup_t | < \psi, V_n(t) fe(u) > | \leq \|\psi\| \, \|f\| \, \|e(u)\| \,. \tag{4.6}$$

The regularity of $V_n(t)$, inequalities (4.5), (4.6) and the Arzela-Ascoli theorem imply that the sequence $\{< \psi, V_n(\cdot)fe(u) >\}$ of continuous functions on $\mathbb{R}_+$ admits a subsequence which converges uniformly over every bounded interval in $\mathbb{R}_+$. Using the separability of the Hilbert spaces involved and the standard diagonalisation procedure we conclude the existence of a subsequence $\{V_{n_k}\}$ satisfying properties (i) and (ii) of the proposition for some family $\{V(t)\}$ of contraction operators in $\mathcal{H}$. Property (i) implies that $V = \{V(t)\}$ is $(h, \mathcal{M}_S)$-adapted. (ii) and (4.5) imply (iii). In particular, V is a strongly continuous contraction-valued adapted process. Thus $V(t)L$ can be viewed as an element of $\mathcal{I}(\mathcal{D}_S, \mathcal{M}_S)$. By Proposition 2.2

$$\begin{aligned} < fe(u), V_n(t)ge(v) > &= < fe(u), ge(v) > \\ &+ \int_0^t < f(e_{-\infty} + u(s))e(u), V_n(s)L_n g(v(s) + e_\infty)e(v) > ds. \end{aligned}$$

Using property (i) of the proposition and (4.1) we now substitute $n = n_k$, let $k \to \infty$ and conclude that, for all $g \in D, u, v \in \mathcal{M}_S$ and $f \in h$

$$\begin{aligned} &< fe(u), V(t)ge(v) > = < fe(u), ge(v) > \\ &+ \int_0^t < f(e_\infty + u(s))e(u), V(s)Lg(v(s) + e_\infty)e(v) > ds. \end{aligned}$$

This proves (iv). ■

Let L be as in Proposition 4.2. For any u, v in the linear manifold generated by $\{e_i, i \in \overline{S}\}$ define the operator L_v^u in h with domain D by

$$< f, L_v^u g > \;=\; < fu, Lgv > \text{ for all } f \in h, g \in D. \tag{4.7}$$

Define

$$L_\infty^{-\infty} = L_{e_\infty}^{e_{-\infty}}, L_v^{-\infty} = L_v^{e_{-\infty}}, L_\infty^u = L_{e_\infty}^u. \tag{4.8}$$

We say that L satisfies condition (B) if the following holds:

(B) The operator $L_\infty^{-\infty}$ with domain D extends to the infinitesimal generator of a contraction semigroup of operators in h for which D is a core.

PROPOSITION 4.3 : In Proposition 4.2 let L satisfy conditions (A) and (B). Then the regular $(h, \mathcal{M}_S)$- adapted contractive solution V of

$$dV = V d\Lambda_L, \; V(0) = I$$

in $\mathcal{D}_S$ is unique. The sequence $\{V_n\}$ in condition (A) satisfies the property

$$\text{w.lim}_{n\to\infty} V_n(t) = V(t) \text{ for all } t \geq 0.$$

PROOF : Suppose that $V' = \{V'(t)\}$ is another solution of

$$dV' = V' d\Lambda_L, V'(0) = I.$$

with the same properties as V. Define $X(t) = V(t) - V'(t)$. Then X satisfies

$$dX = X d\Lambda_L, \quad X(0) = 0, \quad ||X(t)|| \leq 2$$

with respect to the same domain $\mathcal{D}_S$. Fix $u, v \in \mathcal{M}_S$ and define the bounded operators $M(t)$ in h by the relations:

$$< f, M(t)g > = < fe(u), X(t)ge(v) > \text{ for all } f, g \in h. \tag{4.9}$$

By Proposition 2.2 (i) we have

$$< f, M(t)g > = \int_0^t < f(e_{-\infty} + u(s))e(u), X(s)Lg(v(s) + e_\infty)e(v) > ds \tag{4.10}$$

for all $f \in h,\ g \in D$. Using (4.7) - (4.9) we express this as

$$\frac{dM(t)}{dt} = M(t)\{L_\infty^{-\infty} + L_{v(t)}^{-\infty} + L_\infty^{u(t)} + L_{v(t)}^{u(t)}\} \tag{4.11}$$

in the strong sense on the domain D. Changing u to xu, v to yv in (4.10), using (2.5) and identifying coefficients of $x^m y^n$ on both sides of (4.11) we obtain for the operators $M(t, m, n)$ defined in h by the relations

$$< f, M(t, m, n)g > \quad = \quad < fu^{(m)}, X(t)gv^{(n)} >$$

the following:

$$\begin{aligned} \frac{dM(t,m,n)}{dt} \quad &= \quad M(t,m,n)L_\infty^{-\infty} \\ &\quad + M(t,m,n-1)L_{v(t)}^{-\infty} + M(t,m-1,n)L_\infty^{u(t)} \\ &\quad + M(t,m-1,n-1)L_{v(t)}^{u(t)} \end{aligned} \tag{4.12}$$

in the strong sense on the domain D. First we claim that $M(t, 0, 0) = 0$. Indeed, (4.12) implies

$$\frac{dM(t,0,0)}{dt} = M(t,0,0)L_\infty^{-\infty} .$$

Thus, for any $\lambda > 0$, an integration by parts shows that

$$\lambda \int_0^\infty e^{-\lambda t} M(t,0,0)dt = \int_0^\infty e^{-\lambda t} M(t,0,0)L_\infty^{-\infty} dt$$

on D. Putting

$$R(\lambda) = \int_0^\infty e^{-\lambda t} M(t,0,0)dt$$

we conclude that

$$R(\lambda)(L_\infty^{-\infty} - \lambda I)f = 0, \quad f \in D.$$

By condition (B), $(L_\infty^{-\infty} - \lambda I)D$ is dense in h. Since $R(\lambda)$ is bounded we conclude that $R(\lambda) = 0$. By uniqueness of Laplace transforms $M(t,0,0) = 0$ and the claim is proved. Now we employ induction. Suppose that

$$M(t,k,\ell) = 0 \text{ whenever } k + \ell \leq n.$$

If $k + \ell = n + 1$ then (4.12) implies

$$\frac{dM(t,k,\ell)}{dt} = M(t,k,\ell)L_\infty^{-\infty}$$

on D. By repeating the earlier argument we conclude that $M(t,k,\ell) = 0$. In other words $M(t) = 0$. Since u, v are arbitrary in $\mathcal{M}_S$,(4.9) implies $X(t) = 0$. ∎

THEOREM 4.4: Let L satisfy the conditions (A) and (B). In order that the unique contraction porocess V satisfying the q.s.d.e.

$$dV = Vd\Lambda_L, V(0) = I$$

may be isometric it is necessary that

$$< \xi, L\eta > + < FLF\xi, \eta > + < FLF\xi, L\eta >= 0 \text{ for all } \xi, \eta \in D_S \quad (4.13)$$

and together with (4.13) it is sufficient that for some positive λ the space

$$\begin{aligned} \mathcal{B}_\lambda &= \{B \in \mathcal{B}(h) | B \geq 0, \\ &\quad < fe_{-\infty}, BLge_\infty > + < Lfe_\infty, Bge_{-\infty} > \\ &\quad + < FLFfe_{-\infty}, BLge_\infty >= \lambda < f, Bg > \\ &\quad \text{for all } f, g \in D\} \\ &= \{0\}. \end{aligned}$$

PROOF : By Proposition 2.2 we have for all $f, g \in D, u, v \in \mathcal{M}_S$

$$\begin{aligned} &< V(t)fe(u), V(t)ge(v) > - < fe(u), ge(v) > \\ &= \int_0^t \{< V(s)(I + FLF)f(e_{-\infty} + u(s))e(u), \\ &\quad V(s)(I + L)g(v(s) + e_\infty)e(v) > \\ &- < V(s)f(e_{-\infty} + u(s))e(u), V(s)g(v(s) + e_\infty)e(v) >\}ds \quad (4.14) \end{aligned}$$

Suppose $V(t)$ is isometric for every t. Then (4.14) yields

$$< (I + FLF)f(e_{-\infty} + u), (I + L)g(v + e_\infty) > \\ - < f(e_{-\infty} + u), g(v + e_\infty) >= 0 \quad (4.15)$$

for all $f, g \in D$ and u, v in the linear manifold generated by $\{e_i, i \in S\}$. By condition (A) (i) we have

$$Lfe_{-\infty} = 0, \quad < fe_\infty, L\xi >= 0, \quad \xi \in D_S$$

Thus (4.15) becomes (4.13), proving the necessity part.

In order to prove the second part of the theorem let us suppose that $V(t)$ is not isometric for some t. Put

$$X(t) = I - V(t)^* V(t).$$

Then $X(t)$ is a positive operator and for every $\lambda > 0$ the operator

$$Y_\lambda = \int_0^\infty e^{-\lambda t} X(t) dt$$

satisfies $Y_\lambda \neq 0$ and $Y_\lambda \geq 0$. Owing to (4.13) equation (4.14) becomes

$$\begin{aligned} &< fe(u), X(t)ge(v) > \\ &= \int_0^t \{< FLFf(e_{-\infty} + u(s))e(u), X(s)g(v(s) + e_\infty)e(v) > \\ &\quad + < f(e_{-\infty} + u(s))e(u), X(s)Lg(v(s) + e_\infty)e(v) > \\ &\quad + < FLFf(e_{-\infty} + u(s))e(u), X(s)Lg(v(s) + e_\infty)e(v) >\} ds \end{aligned} \quad (4.16)$$

As in the proof of Proposition 4.3 we change u to xu, v to yv and identify the coefficients of $x^m y^n$ on both sides of (4.17). Let n_0 be the smallest integer for which

$$X(t)fu^{(n_0)} \neq 0 \text{ for some } f \in D, u \in \mathcal{M}_S.$$

Then we obtain

$$\begin{aligned} &< fu^{(n_0)}, X(t)gu^{(n_0)} > \\ &= \int_0^t \{< FLFfe_{-\infty}u^{(n_0)}, X(s)ge_\infty u^{(n_0)} > \\ &\quad + < fe_{-\infty}u^{(n_0)}, X(s)Lge_\infty u^{(n_0)} > \\ &\quad + < FLFfe_{-\infty}u^{(n_0)}, X(s)Lge_\infty u^{(n_0)} >\} ds. \end{aligned}$$

Multiply both sides of this equation by $e^{-\lambda t}$, integrate with resprect to t in $[0, \infty)$ and on the right hand side do an integration by parts. Using the fact

$X(0) = 0$ and the identity (4.16) define the bounded positive operator B_λ in h by

$$< f, B_\lambda g > = < fu^{(n_0)}, Y_\lambda g u^{(n_0)} >$$

and conclude that, for $f, g \in D$,

$$\begin{aligned} < f, B_\lambda g > &= \lambda^{-1}\{[< FLFfe_{-\infty}, B_\lambda g e_\infty > \\ &+ < fe_{-\infty}, B_\lambda L g e_\infty > + < FLFfe_{-\infty}, B_\lambda L g e_\infty >\} \end{aligned}$$

Since $B_\lambda \neq 0$ we obtain a contradiction. ■

THEOREM 4.5 : Let $D \subset h$ be a dense linear manifold and let $L, \tilde{L}$ be operators in $h \otimes \hat{k}$ with domain D_S satisfying the following conditions:

(i) There exists a sequence $\{L_n\}$ in $I\!B$ such that

$$\begin{aligned} \lim_{n\to\infty} L_n f e_j &= L f e_j, \\ \lim_{n\to\infty} L_n^b f e_j &= \tilde{L} f e_j \text{ for all } f \in D, j \in \overline{S} . \end{aligned}$$

(ii) For each n there exists a contractive left cocycle V_n satisfying

$$dV_n(s,t) = V_n(s,t) d\Lambda_{L_n}(t), \quad t \geq s$$

(iii) L and $\tilde{L}$ satisfy condition (B).

Then there exists a unique pair of contractive left cocycles $V, \tilde{V}$ and $\mathcal{H}$ satisfying

$$\begin{aligned} dV(s,t) &= V(s,t) d\Lambda_L(t), \\ d\tilde{V}(s,t) &= \tilde{V}(s,t) d\Lambda_{\tilde{L}}(t), \quad t \geq s \end{aligned}$$

in the domain $\mathcal{D}_S$. Furthermore

$$\tilde{V}(s,t) = R_t V(0, t-s)^* R_t \text{ for all } 0 \leq s \leq t < \infty \tag{4.17}$$

where R_t is the time reversal operator in (3.6). In particular V is a co-isometric cocycle if and only if $\tilde{V}$ is an isometric cocycle.

PROOF : The existence and uniqueness of V follow from Proposition 4.2, 4.3. Define

$$\tilde{V}_n(s,t) = R_t V_n(0, t-s)^* R_t \, . \tag{4.18}$$

By "the left analogue" of Theorem 3.4 we have

$$d\tilde{V}_n(s,t) = \tilde{V}_n(s,t) d\Lambda_{L_n^b}(t) \, .$$

Now another application of Proposition 4.2 and 4.3 yields the existence and uniqueness of $\{\tilde{V}(s,t)\}$. (4.18) follows from (4.19) in the limit. The last part is immediate. ■

5 An example

Let h, k be Hilbert spaces with orthonormal bases $\{f_i, i \in \mathbb{Z}\}, \{e_i, i \in \mathbb{Z}\}$ so that $S = \mathbb{Z}$. Let $S_i, i \in \mathbb{Z}$ denote the unitary shift operators in h satisfying

$$S_i f_k = f_{k+i},\ i,\ k \in \mathbb{Z}. \tag{5.1}$$

Let

$$A = ((\alpha_{ij}))_{i,j\in\mathbb{Z}} \tag{5.2}$$

where α_{ij} are complex scalars satisfying

$$\sum_j |\alpha_{ij}|^2 = \pi_i < \infty \tag{5.3}$$

for every $i \in \mathbb{Z}$. Let $Z_i, i \in \mathbb{Z}$ be the normal operator in h satisfying

$$Z_i f_k = \alpha_{k,k+i} f_k,\ i, k \in \mathbb{Z}\,. \tag{5.4}$$

Denote

$$P_n = \sum_{|k|\le n} |f_k >< f_k| \tag{5.5}$$

which is the projection on the linear span of $\{f_k, |k| \le n\}$. Let D be the dense linear manifold generated by $\{f_i, i \in \mathbb{Z}\}$ so that $D_{\mathbb{Z}}$ is the dense linear manifold generated by $\{f_i e_j, i \in \mathbb{Z}, j \in \bar{\mathbb{Z}}\}$ in $h \otimes \hat{k}$ where $\bar{\mathbb{Z}} = \mathbb{Z} \cup \{-\infty, \infty\}$. Define the operators $L, \tilde{L}$ with domain $D_{\mathbb{Z}}$ by

$$\begin{aligned} L &= -\sum_{i\in\mathbb{Z}} S_i Z_i \otimes |e_i >< e_\infty| \\ &\quad + \sum_{j\in\mathbb{Z}} Z_j^* S_j^* \otimes |e_{-\infty} >< e_j| \\ &\quad -\frac{1}{2}(\sum_{k\in\mathbb{Z}} Z_k Z_k^*) \otimes |e_{-\infty} >< e_\infty|, \end{aligned} \tag{5.6}$$

$$\begin{aligned} \tilde{L} &= -\sum_{i\in\mathbb{Z}} Z_i^* S_i^* \otimes |e_{-\infty} >< e_i| \\ &\quad + \sum_{j\in\mathbb{Z}} S_j Z_j \otimes |e_j >< e_\infty| \\ &\quad -\frac{1}{2}(\sum_{k\in\mathbb{Z}} Z_k Z_k^*) \otimes |e_{-\infty} >< e_\infty|. \end{aligned} \tag{5.7}$$

Define the bounded operators $\{L_n, n = 1, 2, \ldots\}$

$$\begin{aligned} L_n &= -\sum_{i\in\mathbb{Z}} S_i Z_i P_n \otimes |e_i >< e_\infty| \\ &\quad + \sum_{j\in\mathbb{Z}} P_n Z_j^* S_j^* \otimes |e_{-\infty} >< e_j| \\ &\quad -\frac{1}{2}(\sum_{k\in\mathbb{Z}} Z_k Z_k^*) P_n \otimes |e_{-\infty} >< e_\infty| \end{aligned} \tag{5.8}$$

so that

$$\begin{aligned} L_n^b &= -\sum_{i\in\mathbb{Z}} P_n Z_i^* S_i^* \otimes |e_{-\infty}><e_i| \\ &+\sum_{j\in\mathbb{Z}} S_j Z_j P_n \otimes |e_j><e_\infty| \\ &-\frac{1}{2}(\sum_{k\in\mathbb{Z}} Z_k Z_k^*)P_n \otimes |e_{-\infty}><e_\infty|. \end{aligned} \tag{5.9}$$

Indeed, the boundedness of L_n for each n is a consequence of Schwarz's inequality and (5.3). A straightforward computation shows that

$$L_n + L_n^b + L_n^b L_n = L_n + L_n^b + L_n L_n^b = 0. \tag{5.10}$$

Furthermore

$$\begin{aligned} \lim_{n\to\infty} L_n f_i e_j &= L f_i e_j, \\ \lim_{n\to\infty} L_n^b f_i e_j &= \tilde{L} f_i e_j, \quad i\in\mathbb{Z}, j\in\bar{\mathbb{Z}} \end{aligned}$$

and the operators $L_\infty^{-\infty}$ and $\tilde{L}_\infty^{-\infty}$ are the restrictions to D of the selfadjoint operator C satisfying

$$C f_k = -\frac{1}{2}\pi_k f_k, \quad k\in\mathbb{Z} \tag{5.11}$$

Owing to Theorem 3.7, (5.10) and (5.11) all the conditions (i) - (iii) of Theorem 4.5 are fulfilled. Thus there exist contractive left cocycles $V = \{V(s,t)\}$, $\tilde{V} = \{V(s,t)\}$ satisfying

$$\begin{aligned} dV(s,t) &= V(s,t)d\Lambda_L(t), \\ d\tilde{V}(s,t) &= \tilde{V}(s,t)d\Lambda_{\tilde{L}}(t), \quad 0\le s\le t<\infty \end{aligned} \tag{5.12}$$

in the domain $\mathcal{D}_{\mathbb{Z}}$ which is the linear manifold generated by $\{f_k e(u), k\in\mathbb{Z}, u\in\mathcal{M}_{\mathbb{Z}}\}$. Furthermore

$$\tilde{V}(s,t) = R_t V(0,t-s)^* R_t, \quad 0\le s\le t<\infty$$

where R_t is the time reversal operator defined by (3.6). We shall now apply Theorem 4.4 in order to find conditions under which the cocycle V satisfying (5.12) may be isometric or co-isometric. Since L_n and L_n^b converge to L and $\tilde{L}$ respectively on $D_{\mathbb{Z}}$ as $n\to\infty$ it follows that condition (4.13) is fulfilled by L and $\tilde{L}$.

Now consider the set $\mathcal{B}_\lambda$ defined in Theorem 4.4 for some $\lambda>0$. Suppose there exists $B\in\mathcal{B}_\lambda, B\ne 0$. Define $p_k =<f_k, Bf_k>, k\in\mathbb{Z}$. Then $0\le p_k\le\|B\|$ and $\underline{p}=\{p_k\}$ is a non zero element of $\ell^\infty(\mathbb{Z})$. The definition of $\mathcal{B}_\lambda$ implies

$$\begin{aligned} \lambda p_k &= <f_k e_{-\infty}, BLf_k e_\infty> + <FLFf_k e_{-\infty}, Bf_k e_\infty> \\ &+ <FLFf_k e_{-\infty}, BLf_k e_\infty>, \quad k\in\mathbb{Z}. \end{aligned} \tag{5.13}$$

By (5.1) - (5.4) and (5.6)

$$Lf_k e_\infty = -\sum_{i \in \mathbb{Z}} \alpha_{k\ k+i} f_{k+i} e_i - \frac{1}{2}\pi_k f_k e_{-\infty}.$$

Substituting this in (5.13) we obtain

$$\lambda p_k = \sum_j |\alpha_{kj}|^2 (p_j - p_k). \tag{5.14}$$

Similarly, suppose that there exists a $\tilde{B} \in \tilde{\mathcal{B}}_\lambda$ where $\tilde{\mathcal{B}}_\lambda$ is $\mathcal{B}_\lambda$ with L replaced by $\tilde{L}$ and $\tilde{B} \neq 0$. Defining $\tilde{p}_k = < f_k, \tilde{B} f_k >$ and employing the same arguments using (5.7) we obtain (5.14) with p_k replaced by $\tilde{p}_k$ for every $k \in \mathbb{Z}$. Thanks to Theorem 4.4, 4.5 we now obtain the following theorem.

THEOREM 5.1: Let $A = ((\alpha_{ij})), i, j \in \mathbb{Z}$ be any matrix with complex entries satisfying

$$\sum_j |\alpha_{ij}|^2 < \infty \quad \text{for every} \quad i \in \mathbb{Z}.$$

Let $\Omega = ((q_{ij}))$ be the infinitesimal Markov matrix defined by

$$q_{ij} = \begin{cases} |\alpha_{ij}|^2 & \text{if} \quad i \neq j, \\ -\sum_{k \neq i} |\alpha_{ik}|^2 & \text{if} \quad i = j \end{cases}$$

Suppose that for some $\lambda > 0$ there does not exist a non zero sequence $\underline{p} = \{p_k\}, k \in \mathbb{Z}$ such that $p_k \geq 0, \underline{p} \in \ell^\infty(\mathbb{Z})$ and

$$(\Omega - \lambda)\underline{p} = 0$$

where $\underline{p}$ is viewed as a column vector. Then there exists a unique unitary cocycle $\{V(s,t), 0 \leq s \leq t < \infty\}$ in $\mathcal{H}$ satisfying

$$dV(s,t) = V(s,t) d\Lambda_L(t), \quad 0 \leq s \leq t < \infty$$

in the domain $D_{\mathbb{Z}}$ where L is defined by (5.1) - (5.4) and (5.6).

REFERENCES

[1] Belavkin, V.P.: A new form and $*-$ algebraic structure of quantum stochastic integrals in Fock space, Rendiconti del Seminario Matematico e Fisico di Milano , Vol. LVIII, 177-193 (1988).

[2] Chebotarev, A.N. : The theory of conservative dynamical semigroups and its applications, Preprint N 1 , M.I.E.M. Moscow (1990).

[3] Evans, M.P.: Existence of quantum diffusions, Probab. Th. Rel. Fields 81, 473-483 (1989)

[4] Fagnola, F.: On quantum stochastic differential equations with unbounded coefficients, Preprint, University of Trento (1989)

[5] Fagnola, F.: Pure birth and pure death processes as quantum flows in Fock space, Preprint, University of Trento (1990), To appear in Sankhya.

[6] Feller, W.: On the integro-differential equations of purely discontinuous Markov processes, Trans. Amer. Math. Soc. 48, 488-515 (1940).

[7] Hudson R.L. and Parthasarathy, K.R.: Quantum Ito's formula and stochastic evolutions, Commun. Math. Phys. 93, 301-323 (1984)

[8] Journé, J.L.: Structure de cocycles Markoviens sur l'espace de Fock, Probab. Th. Rel. Fields. 75, 291 - 316 (1987)

[9] Ledermann, W. and Reuter, G.E.H. : On the differential equations for the transition probabilities of Markov processes with enumerably many states, Proc. Camb. Philos. Soc. 49(1963) 247-262.

[10] Meyer, P.A. : Elements de probabilités quantiques, Seminaire de Probabilités, Springer LNM 1204, 186-312 (1986), 1247, 34-80 (1987), 1321, 101 - 128 (1988), 1372, 175-185 (1989)

[11] Mohari, A. and Sinha, K.B.: Quantum stochastic flows with infinite degrees of freedom and countable state Markov processes, Sankhya, Ser. A. Part 1, 43-57 (1990)

[12] Mohari, A. and Parthasarathy, K.R.: On a class of generalised Evans-Hudson flows related to classical Markov processes, Preprint, Indian Statistical Institute, Delhi Centre 1991 (Submitted to Proc. Quantum Probability & Applications, Delhi).

[13] Mohari, A.: Quantum stochastic differential equations with unbounded coefficients and dilations of Feller's minimal solution, Indian Statistical Institute, Delhi Centre Preprint 1991.

[14] Parthasarathy, K.R.: An Introduction to Quantum Stochastic Calculus, Birkhauser Verlag, Basel (1992).

[15] Parthasarathy, K.R.: Realisation of a class of Markov processes through unitary evolutions in Fock space, Seminaire de Probabilités XXV, Springer LNM 1485, 31-36 (1991) (Ed. Azèma, Meyer and Yor).

[16] Parthasarathy, K.R.: An additional remark on unitary evolutions in Fock space, Seminaire de Probabilités XXV, Springer LNM 1485, 36-80 (1991) (Ed. Azéma, Meyer and Yor).

[17] Parthasarathy, K.R. and Sinha, K.B. : Markov Chains and Evans-Hudson diffusions in Fock space, Seminaire de Probabilités XXIV 1988/89, Springer LNM 1426, 362-369 (1990) (Ed. Azema, Meyer and Yor).

[18] Vincent-Smith. Private communication.

Indian Statistical Institute
7, S.J.S. Sansanwal Marg
New Delhi - 110 016

J.K. Ghosh, S.K. Mitra, K.R. Parthasarathy and B.L.S. Prakasa Rao (*eds.*),
Statistics and Probability: A Raghu Raj Bahadur Festschrift
Published by Wiley Eastern Limited, Publishers, (1993), 499 – 509

Bernstein-type Inequality for Associated Sequences

B.L.S. PRAKASA RAO
Indian Statistical Institute

Abstract

A version of Bernstein-type inequality applicable for finite sequences of associated random variables is derived. Inter alia, some inequalities on covariance of functions of a pair of associated random variables are obtained. Some special cases are discussed.

1 Introduction

Let X be a random variable such that $E[e^{kX}] < \infty$ for some $k > 0$. Bernstein inequality states that

$$P(X \geq t) \leq e^{-kt} E[e^{kX}], t \in R\,.$$

It is easy to see that this inequality is a simple consequence of Chebyshev's inequality. Bennett (1962) and Hoeffding (1963) extended this inequality for sums of independent random variables (cf. Prakasa Rao (1983), p. 183). Steiger (1970) generalized the inequality to martingales. Collomb (1984) extended this result to ϕ-mixing sequences (see Lemma 1, p. 449) in his study of the almost sure convergence of a nonparametric predictor in the time series analysis. The inequality has also been generalized by others

AMS Subject Classifications : Primary 60E 15, 60F10; secondary 60G99
Key words and Phrases : Associated Sequences, Bernstein-type inequality.

to dependent as well as independent sequences of random elements taking values in a Banach space. It is found to be extremely useful in obtaining large deviation type results and in the study of almost sure convergence properties for sums of random variables.

Our aim is to obtain a generalization of the inequality for associated sequences. We follow the technique of Collomb (1984). The need for such an investigation arose in the study of nonparametric estimation of density for stationary associated sequences. Section 2 contains some inequalities for covariance of functions of a pair of associated random variables. A version of Bernstein-type inequality for associated sequences is given in Section 3. Some special cases are discussed in Section 4.

2 Inequalities for Covariances of Functions

Suppose (X, Y) is a bivariate random vector with $E(X^2) < \infty$ and $E(Y^2) < \infty$. It is known that (cf. Lehmann (1966))

$$\mathrm{Cov}(X, Y) = \int_{-\infty}^{\infty}\int_{-\infty}^{\infty} H(x, y)\, dx dy \tag{2.1}$$

where

$$\begin{aligned} H(x, y) &= P(X > x, Y > y) - P(X > x)P(Y > y) \\ &= P(X \leq x, Y \leq y) - P(X \leq x)P(Y \leq y). \end{aligned} \tag{2.2}$$

In general, for any two functions $f(\cdot)$ and $g(\cdot)$ with $E[f(X)]^2 < \infty$ and $E[g(Y)]^2 < \infty$ and finite derivaties $f'(\cdot)$ and $g'(\cdot)$.

$$\mathrm{Cov}(f(X), g(Y)) = \int_{-\infty}^{\infty}\int_{-\infty}^{\infty} f'(x)g'(y)\, H(x, y)\, dx dy \tag{2.3}$$

(cf. Newman (1980)).

A pair of random variabes $\{X, Y\}$ is said to be *associated* if

$$\mathrm{Cov}[f(X, Y), g(X, Y)] \geq 0 \tag{2.4}$$

for any real nondecreasing (coordinate wise) functions f and g on R^2. In particular, if (X, Y) is associated, then, for any pair of nondecreasing functions f and g on R,

$$\mathrm{Cov}[f(X), g(Y)] \geq 0\,. \tag{2.5}$$

It can be shown that the condition (2.5) is equivalent to the condition

$$H(x, y) \geq 0, \quad x, y \in R\,. \tag{2.6}$$

In the following discussion, we assume that the random vector (X,Y) satisfies (2.6). Further let f and g be nondecreasing differentiable functions on R. Since f' and g' exist and f and g are nondecreasing, it follows that f' and g' are both non-negative on R. Relations (2.3), (2.4) and (2.6) imply that

$$\begin{aligned} 0 &\le \operatorname{Cov}[f(X), g(Y)] \\ &\le \text{ess-sup}_x[f'(x)]\ \text{ess-sup}_y[g'(y)] \int_{-\infty}^{\infty}\int_{-\infty}^{\infty} H(x,y)dxdy \\ &= \text{ess-sup}_x[f'(x)]\ \text{ess-sup}_y[g'(y)]\ \operatorname{Cov}(X,Y) \end{aligned} \tag{2.7}$$

whenever ess-sup$_y[g'(y)]$ and ess-sup$_x$ $[f'(x)]$ are finite. This inequality is a special case of Lemma 3 in Newman (1980). On the other hand, if $\sup_y\ g'(y)$ is finite , then

$$\begin{aligned} 0 &\le \operatorname{Cov}[f(X), g(Y)] \\ &\le \text{ess-sup}_y[g'(y)] \int_{-\infty}^{\infty}\int_{-\infty}^{\infty} f'(x)H(x,y)dxdy \qquad (2.8) \\ &= \text{ess-sup}_y[g'(y)]\ \operatorname{Cov}[f(X), Y]\ . \qquad (2.9) \end{aligned}$$

The last equality follows from (2.3). Applying Cauchy-Schwartz inequality, we have

$$0 \le \operatorname{Cov}[f(X), g(Y)] \le \text{ess-sup}_y[g'(y)]\ [\operatorname{Var}[f(X)]\ \operatorname{Var}(Y)]^{1/2}\ . \tag{2.10}$$

An alternate inequality for $\operatorname{Cov}[f(X), g(Y)]$ can be obtained using (2.8). Note that

$$\begin{aligned} 0 &\le \operatorname{Cov}[f(X), g(Y)] \\ &\le \text{ess-sup}_y[g'(y)] \int_{-\infty}^{\infty}\int_{-\infty}^{\infty} f'(x)H(x,y)dxdy \\ &= \text{ess-sup}_y[g'(y)] \int_{-\infty}^{\infty}\int_{-\infty}^{\infty} f'(x)H^{1/2}(x,y)H^{1/2}(x,y)\ dxdy \\ &\le \text{ess-sup}_y[g'(y)]\{\int_{-\infty}^{\infty}\int_{-\infty}^{\infty} [f'(x)]^2 H(x,y)\ dxdy\}^{1/2} \\ &\qquad \times\ \{\int_{-\infty}^{\infty}\int_{-\infty}^{\infty} H(x,y)dxdy\}^{1/2}\ . \end{aligned} \tag{2.11}$$

Let $h(x)$ be a function such that $h'(x) = [f'(x)]^2$. Then

$$0 \le \operatorname{Cov}[f(X), g(Y)] \le \text{ess-sup}_y[g'(y)]\ \{\operatorname{Cov}[h(X), Y]\ \operatorname{Cov}[X, Y]\}^{1/2}. \tag{2.12}$$

It is easy to see that all the inequalities derived above are sharp in the sense that, for appropriate choices of f and g, equality holds. Let us now consider some special functions f and g

$$f(x) = e^{rx} \quad \text{and} \quad g(y) = e^{sy}$$

where $s \geq 0$ and $r \geq 0$. Assume that $E[e^{2rX}] < \infty$ and $E[e^{2sY}] < \infty$. Further assume that $|Y| \leq d$ a.s. It is easy to see that

$$\begin{aligned} 0 \leq \text{Cov}[e^{rX}, e^{sY}] &\leq \text{ess-sup}[se^{sY}]\{\text{Cov}(\frac{r}{2}e^{2rX}, Y)\text{Cov}(X,Y)\}^{1/2} \\ &= se^{sd}(\frac{r}{2})^{1/2}\{\text{Cov}(e^{2rX}, Y)\text{Cov}(X,Y)\}^{1/2} \qquad (2.13) \end{aligned}$$

from (2.12). On the other hand, it can be checked that

$$\begin{aligned} 0 \leq \text{Cov}(e^{rX}, e^{sY}) &= \int_{-\infty}^{\infty} re^{rx}se^{sy}H(x,y)dxdy \\ &\leq rse^{sd}\int_{-\infty}^{\infty}\int_{-\infty}^{\infty} e^{rx}H(x,y)dxdy \\ &= rse^{sd}\,\text{Cov}[e^{rX}, Y] \qquad (\text{By } (2.3)) \\ &= rse^{sd}E[\{e^{rX} - E(e^{rX})\}\{Y - E(Y)\}] \\ &\leq 2rse^{sd}dE|e^{rX} - E(e^{rX})| \\ &\leq 4rse^{sd}dE[e^{rX}] \\ &= 4\|sY\|_\infty \|e^{sY}\|_\infty E[e^{rX}]. \qquad (2.14) \end{aligned}$$

The inequality (2.13) involves $E[e^{2rX}]$ whereas (2.14) involves $E[e^{rX}]$.

3 Bernstein - type inequality

For every $n \geq 1$, let $Z_i^{(n)}, 1 \leq i \leq l_n$ be associated random variables such that $E(Z_i^{(n)}) = 0, |Z_i^{(n)}| \leq d_n \leq \infty$. Define

$$\psi^{(n)}(m) = \sup_{1 \leq i_1, i_2 \leq l_n} \sum_{\substack{i_1 \leq j,k \leq i_2 \\ 0 < |j-k| \leq m}} \text{Cov}(Z_j^{(n)}, Z_k^{(n)}). \qquad (3.1)$$

Let

$$S_n = \sum_{i=1}^{l_n} Z_i^{(n)}. \qquad (3.2)$$

Define $1 \leq m_n \leq l_n$ and α by the relation

$$\alpha m_n d_n \leq \frac{1}{4}. \qquad (3.3)$$

THEOREM 3.1 : For any $\varepsilon < 0$.

$$P(|S_n| \geq \epsilon) \leq 2\,\exp\{6\alpha^2 l_n C_{1n} + \frac{3}{4}\sqrt{e}\,\alpha^{-2} l_n m_n^{-2} C_{1n}^{-1} - \alpha\epsilon\} \qquad (3.4)$$

where

$$C_{1n} = (d_n^2 + \psi^{(n)}(m_n)). \qquad (3.5)$$

PROOF: Let N_n be the integer defined by the inequalities

$$2m_n(N_n - 1) \leq l_n \leq 2m_n N_n. \tag{3.6}$$

Define

$$A_{j,k}^{(n)} = \sum_{i=i_1^{(n)}}^{i=i_2^{(n)}} Z_i^{(n)}, 1 \leq k \leq N_n, j = 1, 2 \tag{3.7}$$

where

$$i_1^{(n)} = \inf\,[(2k+j-3)m_n + 1, l_n] \quad \text{and} \quad i_2^{(n)} = \inf\,[i_1^{(n)} + m_n - 1, l_n]. \tag{3.8}$$

Let

$$B_{j,k}^{(n)} = \sum_{t=1}^{k} A_{j,t}^{(n)} \quad \text{with} \quad B_{j,0}^{(n)} = 0, 1 \leq k \leq N_n, j = 1, 2. \tag{3.9}$$

Choose $\alpha > 0$ satisfying

$$\alpha m_n d_n \leq 1/4 \tag{3.10}$$

as specified earlier by (3.3). Then

$$\begin{aligned} E[e^{\alpha S_n}] &= E[\exp\{\alpha B_{1N_n}^{(n)} + \alpha B_{2N_n}^{(n)}\}] \\ &= E[\exp(\alpha B_{1N_n}^{(n)})\exp(\alpha B_{2N_n}^{(n)})] \\ &\leq \frac{1}{2}(E[\exp(\beta B_{1N_n}^{(N)}] + E[\exp(\beta B_{2N_n}^{(n)})]) \end{aligned} \tag{3.11}$$

where $\beta = 2\alpha$ by the elementary inequality

$$(ab)^\alpha \leq \frac{a^{2\alpha} + b^{2\alpha}}{2} \quad \text{whenever} \quad a \geq 0 \quad \text{and} \quad b \geq 0.$$

It is easy to check that $B_{j,k-1}^{(n)}$ and $A_{j,k}^{(n)}$ are associated for every $1 \leq k \leq N_n$ and $j = 1, 2$ since $\{Z_i^{(n)}, 1 \leq i \leq l_n\}$ are associated. Furthermore

$$|A_{j,k}^{(n)}| \leq m_n d_n \tag{3.12}$$

and hence

$$\|\beta A_{j,k}^{(n)}\|_\infty \leq \beta m_n d_n \leq 1/2 \tag{3.13}$$

by (3.10). Applying the inequality (2.14), we have

$$\begin{aligned} 0 &\leq \operatorname{Cov}(e^{\beta B_{j,k-1}^{(n)}}, e^{\beta A_{j,k}^{(n)}}) \\ &\leq 4\|\beta A_{j,k}^{(n)}\|_\infty \|e^{\beta A_{j,k}^{(n)}}\|_\infty\, E[e^{\beta B_{j,k-1}^{(n)}}] \qquad (3.14) \\ &\leq 4(\frac{1}{2})e^{1/2} E[e^{\beta B_{j,k-1}^{(n)}}] \qquad (3.15) \end{aligned}$$

by the inequality (3.13). Hence

$$\begin{aligned}E[\exp\{\beta B_{j,k}^{(n)}\}] &= E[\exp\{\beta B_{j,k-1}^{(n)}\}\exp\{\beta A_{j,k}^{(n)}\}] \\ &\leq E[\exp\{\beta B_{j,k-1}^{(n)}\}]E[\exp\{\beta A_{j,k}^{(n)}\}] \\ &\quad +2e^{1/2}E[\exp\{\beta B_{j,k-1}^{(n)}\}] \\ &= E[\exp\{\beta B_{j,k-1}^{(n)}\}](E[\exp\{\beta A_{j,k}^{(n)}\}] + 2\sqrt{e}). \quad (3.16)\end{aligned}$$

Following the inequality

$$e^x \leq 1 + x + x^2 \quad \text{for} \quad |x| \leq 1/2\,,$$

it follows that

$$\exp\{\beta A_{j,k}^{(n)}\} \leq 1 + \beta A_{j,k}^{(n)} + \beta^2 [A_{j,k}^{(n)}]^2 \tag{3.17}$$

from (3.13). Hence

$$E[\exp\{\beta A_{j,k}^{(n)}\}] \leq 1 + \beta^2 E[A_{j,k}^{(n)}]^2 \tag{3.18}$$

since $E[A_{j,k}^{(n)}] = 0$. Furthermore

$$1 + \beta^2 E[A_{j,k}^{(n)}]^2 \leq e^{\beta^2 [A_{j,k}^{(n)}]^2} \tag{3.19}$$

since $1 + x \leq e^x$ for $x \geq 0$. Combining the above inequalities, we obtain that

$$E[\exp\{\beta A_{j,k}^{(n)}\}] \leq e^{\beta^2 E[A_{j,k}^{(n)}]^2}. \tag{3.20}$$

Let us now estimate $E[A_{j,k}^{(n)}]^2$. Note that

$$\begin{aligned}E[A_{j,k}^{(n)}]^2 &= \sum_{i=i_1^{(n)}}^{i=i_2^{(n)}} E[Z_i^{(n)}]^2 + \sum_{i=i_1^{(n)}}^{i_2^{(n)}} \sum_{i'=i_1^{(n)}}^{i_2^{(n)}} E[Z_i^{(n)} Z_{i'}^{(n)}] \\ &\leq m_n d_n^2 + \sum_{i=i_1^{(n)}}^{i_2^{(n)}} \sum_{i'=i_1^{(n)}}^{i_2^{(n)}} \operatorname{Cov}(Z_i^{(n)}, Z_{i'}^{(n)}) \\ &= m_n d_n^2 + \sum_{i=i_1^{(n)}}^{i_2^{(n)}} \{ \sum_{\substack{i'=i_1^{(n)} \\ 0<|i'-i|\leq m_n}}^{i_2^{(n)}} \operatorname{Cov}(Z_i^{(n)}, Z_{i'}^{(n)})\} \\ &\leq m_n d_n^2 + \psi^{(n)}(m_n) \\ &= m_n[d_n^2 + m_n^{-1}\psi^{(n)}(m_n)] \\ &= m_n C_{1n} \quad \text{(say)} \qquad (3.21)\end{aligned}$$

where

$$C_{1n} = d_n^2 + m_n^{-1}\psi^{(n)}(m_n).$$

Combining relations (3.18) to (3.21), we have

$$E[e^{\beta A^{(n)}_{J,k}}] \leq e^{\beta^2 m_n C_{1n}} \tag{3.22}$$

Relations (3.16) and (3.22) prove that

$$\begin{aligned} E[e^{\beta B^{(n)}_{j,k}}] &\leq E[e^{\beta B^{(n)}_{j,k-1}}](e^{\beta^2 m_n C_{1n}} + 2\sqrt{\epsilon}) \\ &= E[e^{\beta B^{(n)}_{j,k-1}}]e^{\beta^2 m_n C_{1n}}(1 + 2\sqrt{\epsilon}\, e^{-\beta^2 m_n C_{1n}}) . \end{aligned} \tag{3.23}$$

Repeated application of this inequality for $1 \leq k \leq N_n$ leads to the inequality (note that $B^{(n)}_{j,0} = 0$),

$$\begin{aligned} E[e^{\beta B^{(n)}_{j,N_n}}] &\leq \exp\{N_n\beta^2 m_n C_{1n}\}(1 + 2\sqrt{\epsilon}\, e^{-\beta^2 m_n C_{1n}})^{N_n} \\ &= \exp\{N_n\beta^2 m_n C_{1n} + N_n \log(1 + 2\sqrt{\epsilon}\, e^{-\beta^2 m_n C_{1n}})\} \\ &\leq \exp\{N_n\beta^2 m_n C_{1n} + N_n 2\sqrt{\epsilon}\, e^{-\beta^2 m_n C_{1n}}\} \\ &\quad (\text{since} \quad \log(1+x) \leq x \quad \text{for} \quad x \geq 0) \\ &\leq \exp\{\frac{3}{2} l_n \beta^2 C_{1n} + \frac{3}{2}\frac{l_n}{m_n} 2\sqrt{\epsilon}\, e^{-\beta^2 m_n C_{1n}}\} \\ &\quad (\text{since} \quad N_n \leq \frac{3l_n}{2m_n}) \\ &\leq \exp\{\frac{3}{2} l_n \beta^2 C_{1n} + \frac{3}{2}\frac{l_n}{m_n} 2\sqrt{\epsilon}(\beta^2 m_n C_{1n})^{-1}\} \\ &\quad (\text{since} \quad e^{-x} \leq \frac{1}{x} \quad \text{for} \quad x > 0) \\ &= \exp\{\frac{3}{2} l_n \beta^2 C_{1n} + 3\sqrt{\epsilon}\frac{l_n}{m_n}(\beta^2 m_n C_{1n})^{-1}\} \end{aligned} \tag{3.24}$$

for $j = 1, 2$. Inequalities (3.11) and (3.24) together prove that

$$\begin{aligned} E[e^{\alpha S_n}] &\leq \exp\{\frac{3}{2} l_n (2\alpha)^2 C_{1n} + 3\sqrt{\epsilon}\frac{l_n}{m_n}(4\alpha^2 m_n C_{1n})^{-1}\} \\ &= \exp\{6\alpha^2 l_n C_{1n} + \frac{3}{4}\sqrt{\epsilon}\alpha^{-2} l_n m_n^{-2} C_{1n}^{-1}\} \end{aligned} \tag{3.25}$$

whenever $\alpha > 0$ and $\alpha m_n d_n \leq \frac{1}{2}$. Note that, for any $\epsilon > 0$,

$$\begin{aligned} P(|S_n| \geq \epsilon) &\leq P(S_n \geq \epsilon) + P(-S_n \geq \epsilon) \\ &\leq e^{-\alpha\epsilon} E[e^{\alpha S_n}] + e^{-\alpha\epsilon} E[e^{-\alpha S_n}]. \end{aligned} \tag{3.26}$$

Observe that $\{-Z_i^{(n)}, 1 \leq i \leq l_n\}$ satisfies the same conditions as $\{Z_i^{(n)}, 1 \leq i \leq l_n\}$ does. Applying the inequality (3.25) for both the sequences $\{Z_i^{(n)}, 1 \leq i \leq l_n\}$ and $\{-Z_i^{(n)}, 1 \leq i \leq l_n\}$, we have

$$P(|S_n| \geq \epsilon) \leq 2e^{-\alpha\epsilon} \exp\{6\alpha^2 l_n C_{1n} + \frac{3}{4}\sqrt{\epsilon}\, \alpha^{-2} l_n m_n^{-2} C_{1n}^{-1}\} \tag{3.27}$$

$$= 2\exp\{6\alpha^2 l_n C_{1n} + \frac{3}{4}\sqrt{\epsilon}\, \alpha^{-2} l_n m_n^{-2} C_{1n}^{-1} - \alpha\epsilon\} \tag{3.28}$$

provided $\alpha > 0$ and $0 \le \alpha m_n d_n \le \frac{1}{2}$ where $C_{1n} = d_n^2 + m_n^{-1}\psi^{(n)}(m_n)$. This completes the proof of Theorem 3.1.

4 Special Cases

Let us now consider some special cases of Theorem 3.1. Choose $\alpha = \frac{1}{2}m_n^{-1}d_n^{-1}$ and replace ϵ by $\gamma_n\epsilon$ where $\gamma_n \to \infty$ as $n \to \infty$. The inequality (3.30) reduces to

$$\begin{aligned}
\log P(|S_n| \ge \gamma_n\epsilon) - \log 2 & \\
\le \; & 6(\frac{1}{2}m_n^{-1}d_n^{-1})^2\ell_n C_{1n} + \frac{3}{4}\sqrt{e}\,(\frac{1}{2}m_n^{-1}d_n^{-1})^{-2}\ell_n m_n^{-2}C_{1n}^{-1} \\
& \quad -\frac{1}{2}m_n^{-1}d_n^{-1}\gamma_n\varepsilon \\
= \; & 3\sqrt{e}\,\ell_n d_n^2 C_{1n}^{-1} + \frac{3}{2}\ell_n m_n^{-2}d_n^{-2}C_{1n} - \frac{1}{2}m_n^{-1}d_n^{-1}\gamma_n\varepsilon \\
\le \; & 3\sqrt{e}\,\ell_n d_n^2 C_{1n}^{-1} + \frac{3}{2}\ell_n m_n^{-2}d_n^{-2}(d_n^2 + m_n^{-1}\psi^{(n)}(m_n)) \\
& \quad -\frac{1}{2}m_n^{-1}d_n^{-1}\gamma_n\varepsilon \qquad (4.1)
\end{aligned}$$

where

$$C_{1n} = d_n^2 + m_n^{-1}\psi^{(n)}(m_n)\,. \qquad (4.2)$$

SPECIAL CASE I

Supppose $\ell_n = n, m_n = n^p$ where $0 \le p \le 1, \gamma_n = n^q$ and $d_n = n^{-\tau}, \tau \ge 0$. Since $|S_n| \le \ell_n d_n = n^{1-\tau}$, we assume that $q \le 1-\tau$ as otherwise $P(|S_n| \ge \gamma_n\varepsilon) = 0$ for large n. Further assume that

$$0 < \overline{\lim_{n}}\ \frac{\psi^{(n)}(m_n)}{hm_n^s} < \infty \qquad (4.3)$$

for some $h > 0$ and $s \ge 0$. This condition essentially implies that there exist $k_1 > 0$ and $k_2 > 0$ such that

$$k_1 m_n^s \le \psi^{(n)}(m_n) \le k_2 m_n^s \qquad (4.4)$$

for large n. In particular, it follows that

$$k_1 n^{ps} \le \psi^{(n)}(m_n) \le k_2 n^{ps} \qquad (4.5)$$

for large n. Then, for $0 \le q \le 1-\tau$,

$$\begin{aligned}
\log\ P(|S_n| \ge n^q\varepsilon) \; & - \; \log 2 \\
& \le \; 3\sqrt{e}\,n^{1-2\tau}(n^{-2\tau} + n^{-p}(\psi^{(n)}(m_n))^{-1} + \frac{3}{2}n^{1-2p}
\end{aligned}$$

$$
\begin{aligned}
&+\frac{3}{2}n^{1-3p+2\tau}\psi^{(n)}(m_n)-\frac{1}{2}n^{q-p+\tau}\varepsilon\\
\le\;& 3\sqrt{e}\,n^{1-2\tau+p}[\psi^{(n)}(m_n)]^{-1}+\frac{3}{2}n^{1-2p}\\
&+\frac{3}{2}n^{1-3p+2\tau}\psi^{(n)}(m_n)-\frac{1}{2}n^{q-p+\tau}\varepsilon\\
\le\;& 3\sqrt{e}\,k_1n^{1-2\tau+p-ps}+\frac{3}{2}n^{1-2p}\\
&+\frac{3}{2}n^{1-3p+2\tau+ps}-\frac{1}{2}n^{q-p+\tau}\varepsilon
\end{aligned}
\tag{4.6}
$$

for large n. The last term in the above expression dominates the first three terms provided

$$
\begin{gathered}
1-2\tau+p-ps\le q-p+\tau \quad\text{i.e.,}\quad 1-3\tau+2p-ps\le q,\\
1-2p\le q-p+\tau \quad\text{i.e.,}\quad 1-p-\tau\le q,\\
1-3p+2\tau+ps\le q-p+\tau \quad\text{i.e.,}\quad 1-2p+\tau+ps\le q.
\end{gathered}
\tag{4.7}
$$

Suppose the condition (4.3) holds with $s<2$. In other words we assume that

$$
0<\overline{\lim_n}\,\frac{\psi^{(n)}(m_n)}{hm_n^s}<\infty \tag{4.8}
$$

for some $0\le s<2$. Choose $p=\frac{2\tau}{2-s}$. Note that $0\le q\le 1-\tau$ by assumption. Then inequalities (4.7) can be written in the form

$$
-ps+1-3\tau+2p\le q\le 1-\tau,\quad 1-p-\tau\le q\le 1-\tau, \tag{4.9}
$$

and

$$
ps+1-2p+\tau\le q\le 1-\tau,\quad p=\frac{2\tau}{2-s}.
$$

It is easy to see that $q=1-\tau$ for the above choice of p and

$$
\log P(|S_n|\ge n^{1-\tau}\varepsilon)-\log 2\le 3\sqrt{e}\,k_1\,n^{1-p}+\frac{3}{2}n^{1-2p}+\frac{3}{2}n^{1-p}-\frac{1}{2}n^{1-p}\varepsilon\,.
$$

Therefore, for any $\varepsilon>0$, there exists a constant $C_\varepsilon>0$ such that

$$
\log P(|S_n|\ge n^{1-\tau}\varepsilon)\le -C_\varepsilon n^{1-p}\varepsilon
$$

or equivalently

$$
P(|S_n|\ge n^{1-\tau}\varepsilon)\le \exp\{-C_\varepsilon n^{1-p}\varepsilon\} \tag{4.10}
$$

for large n.

SPECIAL CASE II

If $\text{Cov}(Z_i^{(n)}, Z_{i'}^{(n)}) = 0$ for all $i \neq i'$, then $\psi^{(n)}(m_n) = 0$ for all n and hence the inequality (4.1) reduces to

$$\begin{aligned}\log\ P(|S_n| \geq \gamma_n\varepsilon) - \log\ 2 \ &\leq\ 3\sqrt{e}\,\ell_n + \frac{3}{2}\ell_n m_n^{-2} - \frac{1}{2}m_n^{-1}d_n^{-1}\gamma_n\varepsilon \\ &=\ 3\sqrt{e}\,n + \frac{3}{2}n^{1-2p} - \frac{1}{2}n^{q-p+\tau}\varepsilon\ \log\ n \end{aligned} \tag{4.11}$$

when

$$\ell_n = n, \gamma_n = n^q(\log\ n)^{-1}, d_n = n^{-\tau}(\log\ n)^{-2} \text{ and } m_n = n^p, 0 \leq p \leq 1, \tau \geq 0\,.$$

In this case, the last term dominates the first two terms only when

$$\frac{1}{2}\,n^{q-p+\tau}\ \varepsilon\ \log\ n > 3\sqrt{e}\,n + \frac{3}{2}n^{1-2p} \tag{4.12}$$

which in turn holds for large n provided $q-p+\tau \geq 1$ and $q-p+\tau \geq 1-2p$.

Choose $p = 0$ and $q = 1-\tau$. Then (4.12) reduces to

$$\frac{1}{2}\,n\varepsilon \log\ n > 3\sqrt{e}\,n + \frac{3}{2}\,n. \tag{4.13}$$

It is clear that (4.13) holds for large n. Hence,

$$\log\ P\{|S_n| \geq n^{1-\tau}(\log\ n)^{-1}\varepsilon\} \leq \{-(n\ \log\ n)C_\varepsilon\} \tag{4.14}$$

where $C_\varepsilon > 0$ depending on ε. Therefore

$$P(|S_n| \geq n^{1-\tau}(\log\ n)^{-1}\varepsilon) \leq e^{-(n\ \log\ n)C_\varepsilon} \tag{4.15}$$

for large n. However, in this case, since $\{Z_i^{(n)}, 1 \leq i \leq \ell_n\}$ is associated sequence satisfying the condition $\text{Cov}(Z_i^{(n)}, Z_{i'}^{(n)}) = 0$ for $i \neq i'$, it follows that $\{Z_i^{(n)}, 1 \leq i \leq \ell_n\}$ are independent.

REFERENCES

Bennett, G. (1962) Probability inequalities for the sum of independent random variables, *J. Amer. Statist. Assoc.* **57**, 33-45.

Collomb, G. (1984) Propertes de convergence presque complete du predicteur a noyau, *Z. Wahrscheinlichkeitstheorie verw. Gebiete* **66**, 441-460 .

Hoeffding, W. (1963) Probability inequalities for sums of bounded random variables, *J. Amer. Statist. Assoc.* **58**, 13-30.

Lehmann, E.L. (1966) Some concepts of dependence, *Ann. Math. Statist.* **37**, 1137-1153.

Newman, C.M. (1980) Normal fluctuations and the FKG inequalities, *Commun. Math. Phys.* **74** , 119-128.

Prakasa Rao, B.L.S. (1983) *Nonparametric Functional Estimation*, Academic Press, Orlando.

Steiger, W.L. (1970) Bernstein's inequality for martingales, *Z. Wahrscheinlichkeitstheorie verw. Gebiete* **16**, 104-106.

Indian Statistical Institute
203, B.T Road
Calcutta 700 035
INDIA

J.K. Ghosh, S.K. Mitra, K.R. Parthasarathy and B.L.S. Prakasa Rao (*eds.*),
Statistics and Probability : A Raghu Raj Bahadur Festschrift
Published by Wiley Eastern Limited, Publishers, (1993), 511 – 530

The Metaplectic Representation in the Fock Model

R. RANGA RAO
University of Illinois, Urbana

This paper is dedicated to Professor R.R. Bahadur.

Abstract

The operators of the metaplectic representation in the Fock model have been determined by Bargmann and others upto an ambiguous phase factor. In this paper it is determined completely based on a similar determination in the case of Schrödinger model carried out earlier in [RR].

1 Introduction

Over the past two decades the central role played by the Heisenberg group (and the associated projective representation of the symplectic group) in many problems in analysis, geometry, and number theory (not to mention their role in mathematical physics) has been generally accepted and indeed has attracted a great deal of attention (see the papers of R. Howe for beautiful expositions, or the book [Fo] for references). The Heisenberg group has a unique unitary irreducible representation associated to a nontrivial central character according to a celebrated result of Stone and von Neumann. There are three important models which exhibit this representation and bring out various geometrical aspects. These are the Schrödinger, Fock and lattice models. The Schrödinger model plays an important role in a

AMS Subject Classifications : 22E45 22E41
Key words and Phrases : Heisenberg group, metaplectic representation, symplectic group, Maslov index

group theoretic understanding of the symbol calculus, Hermitian geometry is central in the Fock model, and the lattice model is ideally suited to understanding the ideas where theta functions enter prominently. In all this, the symplectic group and the associated projective representation play an important supportive role. A lot is known about this projective representation. That it actually lifts to an ordinary representation of the twofold cover of the symplectic group was observed in the first papers; hence the name – the metaplectic representation. Another name popularized by Howe is that of oscillator representation – based on its origins in mathematical physics. The fundamental paper of Weil discusses the above representation in the context of Heisenberg groups associated to general locally compact abelian groups. He also gives the operators of the metaplectic representation, upto a phase factor, for certain members of the symplectic group. The same thing was done in the case of Heisenberg group associated to an infinite dimensional Hilbert space by Segal and Shale. The Fock model originated in mathematical physics and their mathematical importance was brought out in the works of Segal and Bargmann. For the lattice model, see Cartier, Igusa, the paper of Weil, and Mumford [M]. In the Schrödinger model, the cocycle associated to the projective representation is closely related to a geometric invariant, known as the Maslov index. In an earlier paper [RR] we expressed this cocycle explicitly as a coboundary on the simply connected covering group and this leads to an explicit description of the operators of the metaplectic representation without any ambiguous phase factors. In this paper, using the above results in the Schrödinger model we do the same for the Fock model. We have included a brief exposition of some of the ideas that arise here naturally, such as vacuum vectors, etc. The literature on the subject is vast and growing and we have mentioned only some here. Some other relevant ones, although not discussed or used here, are included in the list of references (see [B-W], [I], [K-V], [Ma], [L-V], [St], [Car], [Sa]).

This paper is dedicated to one of my teachers, Professor R.R. Bahadur. In the early days of our growing up mathematically in Calcutta, he was one of the people who influenced me greatly, and I am thankful to him for all the kindness he has shown to me.

2 Notation and preliminaries

Let X be a real vector space of dimension $2n$ and Ω a nondegenerate skew-symmetric form on X. Let $e_1, ..., e_n, e_1^*, e_2^*, ..., e_n^*$ be a fixed symplectic basis, i.e., $\Omega(e_i, e_j^*) = \delta_{ij}, \Omega(e_i, e_j) = \Omega(e_i^*, e_j^*) = 0$. Let $J_0 \in \text{End}\, X$, such that $J_0 e_j = -e_j^*, J_0 e_j^* = e_j$ or J_0 has the matrix $= \begin{pmatrix} 0 & I \\ -I & 0 \end{pmatrix}$ on the symplectic basis. Let $V = \sum \mathbb{R} e_j, V^* = \sum \mathbb{R} e_j^*$.

Let $X^{\mathbb{C}} = X \otimes \mathbb{C}$ denote the complexification of X and let $\xi \to \bar{\xi}$ be the conjugation of $X^{\mathbb{C}}$ defined by X. Extend Ω by $\mathbb{C}$ bilinearity to $X^{\mathbb{C}}$. Define $h(\xi, \eta) = \frac{1}{2i}\Omega(\xi, \bar{\eta}), \xi, \eta \in X^{\mathbb{C}}$. Then $h(\cdot, \cdot)$ is a Hermitian form on

$X^{\mathbb{C}}$, nondegenerate and of maximal Witt index. In fact, the symplectic basis $e_1, ..., e_n, e_1^*, ..., e_n^*$ is a (multiple) of a hyperbolic basis for h, i.e., $h(e_r, e_s) = 0 = h(e_r^*, e_s^*)$, and $h(e_r, e_s^*) = 2i\delta_{rs}$. Let $G^{\mathbb{C}} = SpX^{\mathbb{C}}$ be the symplectic group and let $G = SpX = \{\sigma \in G^{\mathbb{C}} : \sigma X = X \text{ or } \sigma \text{ is real, i.e., } \bar{\sigma} = \sigma\}$. Then G also leaves the Hermitian form $h(\cdot,\cdot)$ invariant. Moreover, the action of G in $X^{\mathbb{C}}$ being irreducible the form $h(\cdot,\cdot)$ is the unique (upto a scalar multiple) G-invariant Hermitian form on $X^{\mathbb{C}}$. It is easy to check that

$$G = \{\sigma \in G^{\mathbb{C}} | \sigma \text{ leaves } h \text{ invariant}\}.$$

2.1 Let Lag $X^{\mathbb{C}}$ denote the set of all Lagrangian subspaces of $X^{\mathbb{C}}$, i.e., $\mathbb{C}$-linear totally isotropic subspaces of $X^{\mathbb{C}}$ of dimension n. Clearly $\lambda \in$ Lag $X^{\mathbb{C}}$ implies that $\bar{\lambda} \in$ Lag $X^{\mathbb{C}}$, where $\bar{\lambda} = \{\bar{\xi} : \xi \in \lambda\}$. Call λ real if $\lambda = \bar{\lambda}$, totally complex if $\lambda \cap \bar{\lambda} = (0)$. Note $h(\bar{\xi}, \bar{\eta}) = -h(\eta, \xi)$ and this implies that $\lambda \cap \bar{\lambda} = (\lambda \cap X)^{\mathbb{C}} = \{\xi \in \lambda : h(\xi, \lambda) = 0\}$ the radical of the quadratic space (λ, h).

LEMMA 2.1. *Let $\lambda \in$ Lag $X^{\mathbb{C}}$. Then there exist orthogonal symplectic subspaces X_1, X_2 of X, with $X = X_1 + X_2$, and $\lambda = \lambda_1 + \lambda_2$ with $\lambda_j \in$ Lag $X_j^{\mathbb{C}}$. λ_1 is real and λ_2 is totally complex.*

PROOF: Now $\lambda \cap \bar{\lambda} = (\lambda \cap X)^{\mathbb{C}}$. Let X_1 be any symplectic subspace of X containing $\lambda \cap X$ and of dimension 2 dim $\lambda \cap X$. Such subspaces exist. Let $X_2 = X_1^{\perp}$. Then X_2 is also a symplectic subspace and $X = X_1 + X_2$. Also $\lambda \cap X_1^{\mathbb{C}}$ contains $(\lambda \cap X)^{\mathbb{C}}$ and from dimension considerations it follows that $\lambda \cap X_1^{\mathbb{C}}$ is actually $= (\lambda \cap X)^{\mathbb{C}}$. Since $X^{\mathbb{C}} = X_1^{\mathbb{C}} + X_2^{\mathbb{C}}$, direct sum, if $\xi \in \lambda$ and $\xi = x_1 + x_2, x_j \in X_j^{\mathbb{C}}$, then both $x_1, x_2 \in \lambda$. From this it follows that $\lambda = \lambda_1 + \lambda_2$, with $\lambda_j = \lambda \cap X_j^{\mathbb{C}}$. Clearly $\lambda_1 = \lambda \cap \bar{\lambda}$ is real, and $\lambda_2 \cap \bar{\lambda}_2 = (0)$ or λ_2 is totally complex.

LEMMA 2.2. *Let $\lambda \in$ Lag $X^{\mathbb{C}}$ be totally complex. Let $u_1, ..., u_n$ be a frame for (λ, h), i.e., $u_1, ..., u_n$ is a basis for λ such that $h(u_r, u_s) = \epsilon_r \delta_{rs}$ where $\epsilon_r = \pm 1$. Let $u_r = x_r - iy_r, x_r, y_r \in X$. Then $x_1, ..., x_n, \epsilon_1 y_1, ..., \epsilon_n y_n$ is a symplectic basis of X. In particular, if λ, λ' are both totally complex, and $\alpha : \lambda \to \lambda'$ is a $\mathbb{C}$-linear isomorphism which is also an isometry of the quadratic spaces (λ, h) and (λ', h'), then there exists $\sigma \in G = Sp\ X$, such that $\sigma|\lambda = \alpha$.*

PROOF. Note that $\Omega(x_r, y_s) - i\Omega(x_r, x_s) = h(u_r, u_s) = h(-iu_r, -iu_s) = \Omega(y_r, x_s) - i\Omega(y_r, y_s)$. Thus $\Omega(x_r, x_s) = 0 = \Omega(y_r, y_s)$ and $\Omega(x_r, y_s) = \epsilon_r \delta_{rs}$. This implies the first part. If $u'_r = \alpha(u_r)$, then u'_r is a frame for (λ', h) and $h(u'_r, u'_s) = \epsilon_r \delta_{rs}$ and if $u'_r = x'_r - iy'_s$, then $x'_1, ..., x'_n, \epsilon_1 y'_1, ..., \epsilon_n y'_n$ is also a symplectic basis for X. If you define $\sigma x_r = x'_r, \sigma y_r = y'_r$, then $\sigma \in G$ and $\sigma u_r = u'_r$ or $\sigma|\lambda = \alpha$.

Next let H^+ denote the subspace spanned by $e_j^* + ie_j, 1 \leq j \leq n$ and $H^- = (H^+)^-$. Then $H^+ \in \text{Lag } X^{\mathbb{C}}$ and $h > 0$ on H^+ and $h < 0$ on H^-. For any subset $S \subset \{1, 2, ..., n\}$ let $V_S = \sum_{j \in S} \mathbb{R} e_j$. Similarly define $V_S^*, X_S, H_S^{\pm}$. Then combining the above two lemmas we have the following (the proofs are omitted).

LEMMA 2.3. *Let S_1, S_2, S_3 be a partition of $\{1, 2, ..., n\}$. Then $X = X_{S_1} + X_{S_2} + X_{S_3}$ is an orthogonal direct sum of symplectic subspaces. Let $L = H_{S_1}^+ + H_{S_2}^- + V_{S_3}$. Then $L \in$ Lag $X^{\mathbb{C}}$ and the Hermitian quadratic space (L, h) has the generalized signature (n_1, n_2, n_3) with $n_j = |S_j|$. Any $\lambda \in$ Lag $X^{\mathbb{C}}$ is in the G-orbit of one of these L's. In particular, Lag $X^{\mathbb{C}}$ splits into a finite number of G-orbits, the isometry class of (λ, h) being a complete invariant.*

In connection with these results see Howe's article [H] or the memoir [R-Ro]. The space Lag $X^{\mathbb{C}}$ is a compact complex manifold which is a homogeneous space of $G^{\mathbb{C}}$ and is of complex dimension $\frac{1}{2}n(n+1)$. For a totally complex λ the stabilizer $G_\lambda = \{\sigma \in G | \sigma \cdot \lambda = \lambda\}$ is easily seen to be the unitary group of the Hermitian quadratic space (λ, h) and thus $\dim_{\mathbb{R}} G_\lambda = n^2$. Thus the orbits $G \cdot \lambda$, for λ totally complex have real dimension equal to $2n^2 + n - n^2 = n(n+1)$ and so are open in Lag $X^{\mathbb{C}}$. In general, one computes without difficulty $\dim_{\mathbb{R}} G_\lambda = n^2 + \frac{1}{2}r(r+1)$, where $r = \dim_{\mathbb{C}} \lambda \cap \bar{\lambda}$, giving that the open orbits are those of totally complex λ.

Next an element $\lambda \in$ Lag $X^{\mathbb{C}}$ is said to be positive if h is positive definite on λ and all the positive elements form an open subset denoted by $\text{Lag}^+ X^C$. There is a standard way of parametrizing this open set by means of the generalized upper half-plane. In fact, let λ be a positive Lagrangian subspace. Then λ is totally complex and moreover $\lambda \cap V_{\mathbb{C}} = \lambda \cap V_{\mathbb{C}}^* = (0)$. Thus the projection map of λ into $V_{\mathbb{C}}^*$ (using the decomposition $X^{\mathbb{C}} = V_{\mathbb{C}} + V_{\mathbb{C}}^*$) is a linear isomorphism. Thus there is a well defined and unique ordered basis $w_1(\lambda), ..., w_n(\lambda)$ of λ such that the projection of $w_j(\lambda)$ into $V_{\mathbb{C}}^*$ is e_j^*, i.e.,

$$w_j(\lambda) = \sum z_{rj} e_r + e_j^*, \qquad 1 \leq j \leq n, \ z_{rj} \in \mathbb{C}. \tag{2.1}$$

The matrix $(z_{rs}) = Z_\lambda$ is the matrix associated to λ. Note that $\Omega(w_r, w_s) = z_{rs} - z_{sr}$ and

$$h(w_r, w_s) = \text{Im}\, z_{rs} = \text{Imaginary part of } z_{rs}. \tag{2.2}$$

Thus $\lambda \in \text{Lag}^+ X^{\mathbb{C}}$, if and only if, the corresponding matrix Z_λ belongs to the Siegel upper half-plane

$$\mathcal{H}_n = \{Z : Z \text{ is } n \times n \text{ complex symmetric matrix with positive definite imaginary part } \}.$$

Clearly the map $\lambda \to Z_\lambda$ is a bijection. Note if we write $u_Z = \begin{pmatrix} I & Z \\ 0 & I \end{pmatrix}$ in the sympletic basis, then $u_Z \in \mathrm{Sp}\, X^{\mathbb{C}}$, and $\lambda = u_Z \cdot V^*_{\mathbb{C}}$, if $Z = Z_\lambda$. Also if $\sigma \in \mathrm{Sp}\, X^{\mathbb{C}}$ is such that $\sigma = \begin{pmatrix} \alpha & \beta \\ \gamma & \delta \end{pmatrix}$ in the symplectic basis, then

$$\sigma \cdot V^*_{\mathbb{C}} = \lambda \Rightarrow \det \delta \neq 0 \text{ and } \beta = Z_\lambda \delta. \tag{2.3}$$

This comes out of the fact that the stabilizer of $V^*_{\mathbb{C}}$ is the subgroup of matrices of the form $\begin{pmatrix} * & 0 \\ * & * \end{pmatrix}$ in $\mathrm{Sp}\, X^{\mathbb{C}}$. From this it follows easily that $Z_{\sigma \cdot \lambda} = \sigma \cdot Z_\lambda$, where

$$\sigma \cdot Z = (\alpha Z + \beta)(\gamma Z + \delta)^{-1}$$

for $Z \in \mathcal{H}_n$ and $\sigma = \begin{pmatrix} \alpha & \beta \\ \gamma & \delta \end{pmatrix}$ in G=Sp X. We will also use the properties of the function trace log or simply written as tr-log on $\mathcal{H}_n$ introduced and studied in [RR]. The function tr-log is the unique holomorphic function on $\mathcal{H}_n$ such that (i) exp tr-log Z = det Z and tr-log $(iI_n) = in\pi/2$. It is actually the trace of the principal logarithm of Z (see [RR]). Later we also introduce and use another parametrization of $\mathrm{Lag}^+ X^{\mathbb{C}}$ through the generalized unit disc (see Section 5).

Next consider complex structures J on X, i.e., real linear endomorphisms J of X such that $J^2 = -id$. Let $X_J^+ = \{\xi \in X^{\mathbb{C}} : J\xi = i\xi\}$ and $X_J^- = \{\xi \in X^{\mathbb{C}} : J\xi = -i\xi\}$. Clearly $X_J^\pm = \{x \mp iJx : x \in X\}$ and X_J^- is the complex conjugate $(X_J^+)^-$ of X_J^+. Also $\dim X_J^\pm = n$. One checks easily that $J \in G = \mathrm{Sp}\, X$ if and only if X_J^+ and X_J^- belong to Lag $X^{\mathbb{C}}$. Notice in this case $X_J^\pm$ are both totally complex. Conversely given any totally complex $\lambda \in \mathrm{Lag}\, X^{\mathbb{C}}$, the maps $\xi \to \mathrm{Re}\, \xi, \xi \to \mathrm{Im}\, \xi$ of λ into $X (\xi \in \lambda \subset X^{\mathbb{C}})$ are $\mathbb{R}$-linear isomorphisms. Thus $J_\lambda : \mathrm{Re}\, \xi \to -\mathrm{Im}\, \xi, \xi \in \lambda$, is a well defined endomorphism of X with $J_\lambda^2 = -id$. Moreover, $J_\lambda \in G$ and λ is recovered as $\lambda = X_J^+$. Note for any complex structure J in G, one has

$$h(x - iJx, y - iJy) = \Omega(x, Jy) - i\Omega(x, y) \tag{2.4}$$

for all $x, y \in X$. We will call J positive if $\Omega(x, Jx) \geq 0$ for all $x \in X$ (note that the symmetric bilinear form $\Omega(x, Jy)$ is nondegenerate owing to the nondegeneracy of Ω). Equivalently from (2.2), it follows that J is positive if and only if X_J^+ is a positive Lagrangian subspace.

3 Vacuum distributions, vacuum vectors, and the system of coherent vectors for the Schrödinger representation

In this section we review some basic notions associated with the Schrödinger representation. Consider the Heisenberg group associated to (X, Ω), i.e., let $E = X \times \mathbb{R}$ with group law $(x, t) \cdot (y, t') = (x + y, t + t' + \frac{1}{2}\Omega(x, y))$. For this group there is a unique (upto unitary equivalence) irreducible unitary representation associated to central character $(0,t) \to e^{2\pi it}$. We use the following construction and description. The subgroup V.Cent E is a maximal abelian subgroup of E and let ν denote the representation obtained by inducing the character $(p,t) \to e^{2\pi it}$ of V.Cent E. Since $V^* \cdot V$. Cent $E = E$, this representation is realized in $L^2(V^*)$ and is given as

$$\nu(p+q,t)f \cdot (x) = f(x-q)\exp\left\{2\pi it + 2\pi i\left(p, x - \frac{1}{2}q\right)\right\} \tag{3.1}$$

with $p \in V, q \in V^*$. We also write $\nu(p+q) = \nu(p+q,0)$ (for all this see [RR] or [Fo]). For the Heisenberg group E, the same space serves as the Lie algebra, with the identity map serving as the exponential map. As a Lie algebra, the Lie bracket in E is $[(x,t),(y,t')] = (0, \Omega(x,y))$. Next we consider the associated Lie algebra representation $\dot{\nu}$. It is known that the space of C^∞ vectors of the representation, denoted by $C^\infty(\nu)$ is the Schwartz space of V^* and the Lie algebra representation on $C^\infty(\nu)$ may be described as ($\dot{\nu}\ (x) = \dot{\nu}\ (x,0)$)

$$\dot{\nu}\ (p+q) = -\partial(q) + 2\pi i\Omega(p,\cdot). \tag{3.2}$$

Here $\partial(q)f(x) = (d/dt)_0(f(x+tq))$ is the directional derivative in the direction $q \in V^*$. Note that group operators $\nu(x,t)$ leave the Schwartz space invariant and so act on tempered distributions by duality. To clarify what we mean by this, for any $\alpha \in \text{End}\ \mathcal{S}$, we will say α has a formal adjoint $\alpha^* \in \text{End}\ \mathcal{S}$, if

$$(\alpha f, g) = (f, \alpha^* g)$$

for all $f, g \in \mathcal{S}$ = the Schwartz space of V^*. Here, (f,g) denotes the inner product in $L^2(V^*)$. For any tempered distribution $u : \mathcal{S} \to \mathbb{C}$, we write $(u,f) = u(\bar{f})$. Then the dual action of endomorphisms possessing a formal adjoint on tempered distributions is defined as $(\alpha \cdot u, f) = (u, \alpha^* f)$. A similar definition works for distributions on V^* as well. Note that the Lie algerba operators $\dot{\nu}\ (x)$ being differential operators have formal adjoints even in the context of $C_c^\infty(V^*)$ and so act on distributions by duality as above (see [Hor]). With these preliminary remarks we now define vacuum distributions etc., motivated by the terminology used in mathematical physics.

DEFINITION 3.1. *Let $\lambda \in$Lag $X^{\mathbb{C}}$. A distribution u on V^* is said to be a vacuum distribution for λ or λ-vacuum distribution if $\dot{\nu}\ (\xi)u = 0$ for all $\xi \in \lambda$. (Note that $\dot{\nu}\ (\xi)$ is defined by $\mathbb{C}$-linear extension of $\dot{\nu}\ (x), x \in X$, i.e., $\dot{\nu}\ (x_1 + ix_2) = \dot{\nu}\ (x_1) + i\,\dot{\nu}\ (x_2)$, with $x_1, x_2 \in X$).*

PROPOSITION 3.2. *Let $I(\lambda)$ = set of all λ-vacuum distributions. Then $\dim_{\mathbb{C}} I(\lambda) = 1$. Moreover, these vacuum distributions are tempered if and only if (λ, h) is positive semidefinite, i.e., $h \geq 0$ on λ. Also these vacuum distributions u are vacuum vectors, i.e., $(u,\varphi) = (f,\varphi)$ for some*

$f \in L^2(V^*)$ and all $\varphi \in C_c^\infty(V^*)$ if and only if λ is positive or h is positive definite on λ. In this case the vacuum vectors are all in $C^\infty(\nu)$ – the Schwartz space of V^*.

PROOF : Let $M = \{\sigma \in G : \sigma V \subset V, \sigma V^* \subset V^*\}$. Then the operator $r(\sigma)f\cdot(x) = |\det(\sigma|V^*)|^{-1/2} f(\sigma^{-1}x)$ has the property that $r(\sigma)\nu(x)r(\sigma)^{-1} = \nu(\sigma\cdot x)$ for all $x \in X$. This implies that $r(\sigma)\ \dot\nu\ (\xi)r(\sigma)^{-1} = \dot\nu\ (\sigma\cdot\xi)$ for all $\xi \in X^{\mathbb{C}}$. Note that the operator $r(\sigma)$ for $\sigma \in M$, leaves both $C_c^\infty(V^*)$ and the Scharwtz space invariant. Thus $u \in I(\lambda)$ implies that $r(\sigma)u \in I(\sigma\cdot\lambda)$. Thus to prove the result we may pick one λ from each orbit of M in Lag $X^{\mathbb{C}}$. Let $U_k = \{\lambda \in \text{Lag } X^{\mathbb{C}} : \dim \lambda \cap V_{\mathbb{C}} = k\}$. Clearly Lag $X^{\mathbb{C}} = \bigcup_{k=0}^n U_k$ and each U_k is M invariant. If $\sigma \in M$, $\sigma\cdot\lambda\cap V_{\mathbb{C}} = \sigma(\lambda\cap V_{\mathbb{C}})$ and since M acts transitively on subspaces of dimension k, it follows that $U_k = M\cdot U_{k,S}$, where S is the subset $= \{1, 2, ..., k\}$ of $\{1, 2, ..., n\}$ and $U_{k,S} = \{\lambda \in U_k : \lambda\cap V_{\mathbb{C}} = (V_S)_{\mathbb{C}}\}$. Consider $\lambda \in U_{k,S}$. Then $V_S = \sum_{j=1}^{k} \mathbb{R}e_j \subset \lambda$. If π is the projection onto $V_{\mathbb{C}}^*$ defined by the decomposition $X^{\mathbb{C}} = V_{\mathbb{C}} + V_{\mathbb{C}}^*$ it follows that $\dim \pi(\lambda) = n - k$ and $\Omega(V_S, \pi(\lambda)) = 0$. Thus $\pi(\lambda) = \sum_{j>k} \mathbb{C}e_j^*$. Thus there exists a basis for λ say $w_j, 1 \le j \le n$ of the form $w_j = e_j, 1 \le j \le k$ and $w_j = \sum_{r>k} z_{rj}e_r + e_j^*, k+1 \le j \le n$. If $u \in I(\lambda)$, then the equations $\dot\nu\ (\xi)u = 0$, for all $\xi \in \lambda$, are just $x_j u = 0, 1 \le j \le k$ and

$$\left\{-\frac{\partial}{\partial x_j} + 2\pi i \sum_{r>k} z_{rj}x_r\right\} u = 0, \qquad k+1 \le j \le n.$$

From these equations it is clear that $I(\lambda) = \mathbb{C}u_0$, where u_0 is the distribution

$$(u_0, f_1 \otimes f_2) = (\delta_0, f_1)(\varphi_0, f_2)$$

where $f_1 \in C_c^\infty(V_S^*)$, $f_2 \in C_c^\infty(V_{S'}^*)$, δ_0 – is the Dirac distribution on V_S^*, and φ_0 is the function

$$\varphi_0(x_{S'}) = \exp i\pi \left(\sum z_{rs}x_r x_s\right)$$

and $x_{S'}$ is the $V_{S'}^*$ component of x. It is clear that u_0 is tempered if and only if Im $Z > 0$. Now a direct calculation shows that $h(w_j, w_{j'}) = (\text{Im } Z)_{j'j}$ for all $j, j' > k$ and 0 otherwise. Thus u_0 is tempered if and only if $h \ge 0$ on λ. It is also clear that u_0 is an element of $L^2(V^*)$ if and only if $k = 0$ and h is positive definite on λ. This completes the proof.

Next we define coherent systems formally.

DEFINITION 3.3. *Let $\lambda \in Lag^+ X^{\mathbb{C}}$. A λ-coherent system is a set $= \{\varphi_{\lambda,\alpha} \in L^2(V^*) : \alpha \in \lambda\}$ such that the following conditions hold:*

(i) $\varphi_{\lambda,\alpha} \in C^\infty(\nu) = \mathcal{S}(V^*)$ *– – the Schwartz space of* V^*;

(ii) $\dot{\nu}(\xi)\varphi_{\lambda,\alpha} = -2\pi h(\xi,\alpha)\varphi_{\lambda,\alpha}$ *for all* ξ *and* $\alpha \in \lambda$;

(iii) $(\varphi_{\lambda,\alpha}, \varphi_{\lambda,\beta}) = exp\ \pi h(\beta,\alpha)$ *for all* $\alpha, \beta \in \lambda$.

REMARK : Note that the above conditions imply that $\varphi_{\lambda,0}$ is a vacuum vector of unit length for λ. Thus, for the above definition to make sense λ has to be positive.

LEMMA 3.4. *The λ-coherent system is unique upto a common constant phase factor. A λ-coherent system satisfying $\varphi_{\lambda,0}(0) > 0$ is unique. If $Z(= Z_\lambda)$ is the matrix in the upper half-plane corresponding to λ or $w_j(\lambda) = \sum z_{rj}e_r + e_j^*$ is a basis of λ, then the λ-coherent system satisfying $\varphi_{\lambda,0}(0) > 0$ is given by*

$$\varphi_{\lambda,\alpha}(x) = \varphi_\lambda(x,y) = 2^{n/4}(det\, B)^{1/4}\ exp\ Q$$

where $Q = i\pi Z[x] + 2\pi\, {}^t x B\bar{y} - \frac{\pi}{2}B[\bar{y}]$. *Here* $Z = A + iB, \alpha = Zy + y = \sum y_j w_j(\lambda) \in \lambda$, *and* $x \in V^*, y \in V^*_{\mathbb{C}}$, *and* $Z[x] = {}^t x \cdot B \cdot x$.

PROOF : Note that $\dot{\nu}(w_j) = -\frac{\partial}{\partial x_j} + 2\pi i \sum z_{rj}x_r$. From this it follows that the solution space of the equations (ii) of Definition 3.3 is one dimensional. So if $\varphi'_{\lambda,\alpha}$ is another λ-coherent system, then $\varphi'_{\lambda,\alpha} = c(\alpha)\varphi_{\lambda,\alpha}$ for some constant $c(\alpha)$. Then the condition (iii) of Definition 3.3 forces $c(\alpha)\overline{c(\beta)} = 1$ for all $\alpha, \beta \in \lambda$. Thus $c(\alpha)$ is independent of α and $|c(\alpha)| = 1$. The verification, that the actual formula for $\varphi_{\lambda,\alpha}$ given above, satisfies the conditions (i), (ii), (iii) of Definition 3.3, is left to the reader.

We next describe the action of the metaplectic or oscillator representation on the λ-coherent system. First we recall a few facts about the metaplectic representation. For notation, further details, and references, see [RR]. Let $\tilde{G}$ denote the simply connected covering group of the symplectic group $G = \mathrm{Sp}\, X$, and let $\tilde{\sigma} \to \sigma$ denote the covering homomorphism. Then there exists a unique set of unitary operators $r(\tilde{\sigma})$ in $L^2(V^*)$ such that

(i) $\tilde{\sigma} \to r(\tilde{\sigma})$ is a representation of G, and

(ii) $r(\tilde{\sigma})\nu(x)r(\tilde{\sigma})^{-1} = \nu(\sigma \cdot x)$ for all $x \in X$.

To proceed further recall the construction of $\tilde{G}$. Let $j(\sigma, Z) = \det(\gamma Z + \delta)$, where $Z \in \mathcal{H}_n$ – the Siegel upper half-plane and σ has the matrix representation $\begin{pmatrix} \alpha & \beta \\ \gamma & \delta \end{pmatrix}$ in the decomposition $X = V + V^*$. Let $\tilde{G} = \{(\sigma,\psi) : \sigma \in G, \psi$ holomorphic function on $\mathcal{H}_n$ such that $\exp \psi(Z) = j(\sigma, Z)$ for all $Z\}$. With group law defined as $(\sigma_1,\psi_1)\cdot(\sigma_2,\psi_2) = (\sigma_1\sigma_2,\psi)$ where $\psi(Z) = \psi_1(\sigma_2 \cdot Z) + \psi_2(Z)$, $\tilde{G}$ becomes the simply connected covering group, with $(\sigma,\psi) \to \sigma$ as the covering homomorphism. Note for example that inverse of (σ,ψ) is $(\sigma^{-1}, -\psi o \sigma^{-1})$.

THEOREM 3.5. *For $\lambda \in Lag^+X^{\mathbb{C}}$ and $\alpha \in \lambda$, let $\varphi_{\lambda,\alpha}$ denote the unique λ-coherent system with $\varphi_{\lambda,0}(0) > 0$ (see Lemma 3.4). Then*

$$r(\tilde{\sigma})\varphi_{\lambda,\alpha} = \left\{\exp - \frac{i}{2} Im\psi\,(Z_\lambda)\right\} \varphi_{\sigma\cdot\lambda,\sigma\cdot\alpha}.$$

Here Z_λ is the unique element of the Siegel upper half-plane corresponding to λ, and $\tilde{\sigma} = (\sigma, \psi) \in \tilde{G}$.

PROOF : To start with, the intertwining property of $r(\tilde{\sigma})$ implies that it leaves the space of C^∞ vectors $C^\infty(\nu)$ invariant and we also have the infinitesimal intertwining property — namely $r(\tilde{\sigma})\,\dot{\nu}\,(\xi)r(\tilde{\sigma}^{-1}) = \dot{\nu}\,(\sigma\cdot\xi)$ for all $\xi \in X^{\mathbb{C}}$. This shows that $r(\tilde{\sigma})\varphi_{\lambda,\alpha}$ is a coherent system for $\sigma\cdot\lambda$. In fact

$$\begin{aligned} \dot{\nu}\,(\sigma\cdot\xi)r(\tilde{\sigma})\varphi_{\lambda,\alpha} &= r(\tilde{\sigma})\,\dot{\nu}\,(\xi)\varphi_{\lambda,\alpha} \\ &= -2\pi h(\xi,\alpha)r(\tilde{\sigma})\varphi_{\lambda,\alpha} \\ &= -2\pi h(\sigma\cdot\xi,\sigma\cdot\alpha)r(\tilde{\sigma})\varphi_{\lambda,\alpha} \end{aligned}$$

since $h(\cdot,\cdot)$ is G-invariant. The other property of coherent system follows from unitarity of $r(\tilde{\sigma})$ and the invariance of $h(\cdot,\cdot)$ under G. Thus from the uniqueness property (see Lemma 3.4), there exists a constant phase factor $c, |c| = 1$, such that

$$r(\tilde{\sigma})\varphi_{\lambda,\alpha} = c\varphi_{\sigma\cdot\lambda,\sigma\cdot\alpha}.$$

Since $\varphi_{\sigma\cdot\lambda,0}(0) > 0$, the constant c is just the phase factor in $(r(\tilde{\sigma})\varphi_{\lambda,0})(0)$. But this has been evaluated in [*RR*], Proposition 4.2. In fact if $F_Z(x) = \exp i\pi Z[x]$, then

$$r(\tilde{\sigma})F_Z = \{\exp - \frac{1}{2}\psi(Z)\}F_{\sigma\cdot Z}.$$

Note that $\varphi_{\lambda,0}$ is a positive constant multiple of F_Z. It follows that $c =$ phase factor in $(r(\tilde{\sigma})F_Z)(0) =$ phase factor in $\exp -\frac{1}{2}\psi(Z)$ and thus $c = \exp\{-\frac{i}{2}\mathrm{Im}\psi(Z)\}$. Here $Z = Z_\lambda$ is the point in the Siegel upper half-plane corresponding to λ.

REMARK. The set of λ-coherent vectors $\{\varphi_{\lambda,\alpha} : \alpha \in \lambda\}$ is actually a total subset of $L^2(V^*)$, i.e., their linear span is dense in $L^2(V^*)$. This is easily checked, for example for the case $\lambda = \lambda_0 = H^+$ and is valid for all λ. The implication is that the operators $r(\tilde{\sigma})$ are completely determined by their action on a λ-coherent system. Thus the above simple formula also determines the operators $r(\tilde{\sigma})$ completely.

4 Holomorphic induction procedure and the Fock representation

In this section we recall the method of holomorphic induction that leads to the construction of the Fock model. Consider the Heisenberg group $E = X \times R$ introduced in Section 3. Consider the induced module obtained from the character $(0,t) \to \exp 2\pi i t$ of the subgroup, Cent E. This module is realized in the space of sections of a homogeneous line bundle L_χ for E, over $X = E/$ Cent E. This line bundle L_χ is easily described as $L_\chi = X \times \mathbb{C}$, with E-action on it described as $(x,t)\cdot(y,z) = (y+x, z \exp(2\pi i t + i\pi\Omega(x,y)))$, where $(x,t) \in E, y \in X, z \in \mathbb{C}$. If $s_f : x \to (x, f(x))$ is a section of the bundle, then the E-module action corresponds to the following action on the function f

$$(x,t)\cdot f : y \to f(y-x)\exp(2\pi i t + i\pi\Omega(x,y)).$$

Now let $J \in \text{End}\, X, J^2 = -id$ be a complex structure on X. Since E-action on $X = E/$ Cent E is just translation, they are holomorphic. Consider those J for which there also exists a compatible complex (manifold) structure on L_χ, which is E-invariant and in which it becomes a holomorphic line bundle over (X, J). One can show that this is possible if and only if the subspaces $X_J^\pm = \{\xi \in X^{\mathbb{C}} : J\xi = \pm i\xi\}$ are Lagrangian or $\in$ Lag $X^{\mathbb{C}}$. This is equivalent to the condition that $J \in SpX(= G)$. In this case such a structure is easily constructed on L_χ by considering the complexified Heisenberg group $E_{\mathbb{C}} = X^{\mathbb{C}} \times \mathbb{C}$. In fact $\mathfrak{b} = X_J^- +$ Cent $E_{\mathbb{C}}$ is a closed complex Lie subgroup of $E_{\mathbb{C}}$ and one checks that $E_{\mathbb{C}}/\mathfrak{b} \simeq (X, J)$ as complex manifolds. One may thus identify L_χ as a $E_{\mathbb{C}}$ homogeneous holomorphic line bundle over (X, J) corresponding to the quasi character $(\xi, t) \to \exp 2\pi i t$ of $\mathfrak{b}$. With this structure one shows that the section s_f is holomorphic if and only if f is C^∞ on X and satisfies the equation

$$\{\partial(\xi) - i\pi\Omega(y,\xi)\} f\cdot(y) = 0 \tag{4.1}$$

for all $y \in X$ and all $\xi \in X_J^-$. Here $\partial(\xi) = \partial(x_1) + i\partial(x_2)$ if $\xi = x_1 + ix_2, x_1, x_2 \in X$, and $\partial(x)$ is the usual directional derivative differential operator. One checks easily that this condition (4.1) is equivalent to the condition that the function

$$F(y) = f(y)\exp\left\{\frac{\pi}{2}\Omega(y, Jy)\right\} \tag{4.2}$$

be holomorphic on (X, J). Thus the space of L^2-holomorphic sections (denoted by $L^2(X, \Omega, J)$ consist of all holomorphic functions F on (X, J) such that $\int |F(y)|^2 \exp(-\pi\Omega(y, Jy))d_X y < \infty$.

LEMMA 4.1. *The space $L^2(X, \Omega, J) \neq (0)$ if and only if $\Omega(y, Jy) \geq 0$ for all $y \in X$ or equivalently X_J^+ is a positive Lagrangian subspace.*

Note the bilinear form $\Omega(x, Jy)$ is symmetric and nondegenerate and so the condition $\Omega(y, Jy) \geq 0$ is equivalent to the condition that the form $\Omega(x, Jy)$ is positive definite on X. Also $X_J^+ = \{x - iJx : x \in X\}$ and $\Omega(x, Jy) =$ real part of $h(x - iJx, y - iJy)$. Thus the above condition is equivalent to saying that $h > 0$ on X_J^+. The lemma itself now follows from the following elementary result (the proof is omitted).

LEMMA 4.2. *Let M be a finite dimensional complex vector space and $h(\cdot,\cdot)$ a Hermitian form on M. Then there exists a non-zero holomorphic function F on M such that*

$$\int_M |F(z)|^2 e^{-\pi h(z,z)} d_M z < \infty,$$

if and only if h is positive definite on M.

Note that E-module action on functions F given by (4.2) is

$$(x,t)\cdot F : y \to F(y-x)\exp Q \tag{4.3}$$

where $Q = 2\pi i t + \pi\{\Omega(y,Jx) - i\Omega(y,x)\} - \frac{\pi}{2}\Omega(x,Jx)$. This can be written in standard form as follows. Let $\lambda = X_J^+$ and $\mathbf{H}(\lambda) = \{\varphi : \varphi$ holomorphic on λ and $\int_\lambda |\varphi(u)|^2\{\exp -\pi h(u,u)\} d_\lambda u < \infty\}$. Let E_λ denote the Heisenberg group constructed as follows: $E_\lambda = \lambda \times R$ with group law defined as

$$(\xi,t)\cdot(\xi',t') = \left(\xi+\xi', t+t' - \frac{1}{2}\,\mathrm{Im}\ h(\xi,\xi')\right).$$

Then $\mathbf{H}(\lambda)$ becomes an E_λ-module for the action defined as follows:

$$(z,t)\cdot\varphi : u \to \varphi(u-z)\exp\left\{2\pi i t + \pi h\left(u - \frac{1}{2}z, z\right)\right\}. \tag{4.4}$$

Note that the map $(x,t) \to (x - iJx, t)$ is a group isomorphism of E with E_λ and the isomorphism $F \to \varphi$, where $\varphi(y - iJy) = F(y), y \in X$ is an isomorphism of the E-module $L^2(X,\Omega,J)$ with the E_λ-module $\mathbf{H}(\lambda)$. The representation given here in $\mathbf{H}(\lambda)$ is known as the Fock representation, and in the above paragraph, we have tried to indicate that it fits into a general constructive procedure known as the method of holomorphic induction.

We next construct the Bargmann transform. Consider the matrix entry $\Phi(f_1,f_2) : y \to (f_1, \nu(y)f_2)$ of the Schrödinger representation. It is known that $\Phi(f_1,f_2) \in L^2(X)$ and we have

$$\|\Phi(f_1,f_2)\|^2_{L^2(X)} = \|f_1\|^2_{L^2(V^*)}\|f_2\|^2_{L^2(V^*)} \tag{4.5}$$

(see [Fo], p. 31). Here the Haar measures are the standard Lebesgue measures in X and V^* via identification with R^{2n} and R^n using the symplectic basis. When f_1 or f_2 is a C^∞-vector for ν, one checks by differentiation that $\Phi(f_1,f_2)$ is C^∞ on X and moreover

$$\begin{aligned}\partial(\xi)\Phi(f_1,f_2) &= i\pi\Omega(y,\xi)\Phi(f_1,f_2) + \Phi(f_1,\dot{\nu}(\bar{\xi})f_2)\\ &= i\pi\Omega(y,\xi)\Phi(f_1,f_2) - \Phi(\dot{\nu}(\xi)f_1,f_2).\end{aligned} \tag{4.6}$$

Thus when $f_2 = \varphi_\lambda (= \varphi_{\lambda,0}$ see Lemma 3.4) the λ-vacuum vector of unit length with $\varphi_\lambda(0) > 0$, the function $\Phi(f,\varphi_\lambda)$ is in $C^\infty(X)$ and satisfies

$$\{\partial(\xi) - i\pi\Omega(y,\xi)\}\Phi(f,\varphi_\lambda) = 0$$

for all $y \in X, \xi \in X_J^- = \bar{\lambda}$. Thus if we define for $u = y - iJy \in \lambda$,

$$B_\lambda f : u \to \{\exp \frac{\pi}{2}\Omega(y, Jy)\}(f, \nu(y)\varphi_\lambda)_{L^2(V^*)} = \left\{\exp \frac{\pi}{2} h(u,u)\right\}(f, \nu(y)\varphi_\lambda), \tag{4.7}$$

then $B_\lambda f \in \boldsymbol{H}(\lambda)$. One checks that (4.5) implies that B_λ is a unitary isomorphism and intertwines the E-module actions. This is called the Bargmann transform in [Fo], see also [Ba]. From (4.6), it follows easily that $\partial(\xi)B_\lambda\varphi_\lambda = 0$, for all $\xi \in \lambda$, and so $B_\lambda\varphi_\lambda$ is a constant and in fact $B_\lambda\varphi_\lambda = 1$, by evaluating at $y = 0$. As a consequence of this we have

LEMMA 4.3. *Let φ_λ be a λ-vacuum vector of unit length in $L^2(V^*)$ and satisfying $\varphi_\lambda(0) > 0$. Then the λ-coherent system can also be described as*

$$\varphi_{\lambda,z} = \left\{\exp \frac{\pi}{2} h(z,z)\right\} \nu(y)\varphi_\lambda$$

where $y \in X, z = y - iJy$ and $\lambda = X_J^+$. In particular

$$B_\lambda f \cdot (z) = (f, \varphi_{\lambda,z})$$

for all $f \in L^2(V^*)$.

PROOF : A simple calculation shows that

$$\nu(y)^{-1} \dot{\nu}(\xi)\nu(y) = \dot{\nu}(\xi) + 2\pi i\Omega(\xi, y)$$

for all $y \in X$ and $\xi \in X^{\mathbb{C}}$. From this it follows that for all $\xi \in \lambda$, $\dot{\nu}(\xi)\nu(y)\varphi_\lambda = 2\pi i\Omega(\xi,y)\nu(y)\varphi_\lambda$. Now $2\pi i\Omega(\xi, y) = -4\pi h(\xi, y) = -2\pi h(\xi, y - iJy)$ for $\xi \in \lambda$. Here note that, $h(\xi, y + iJy) = 0$, for all $\xi \in \lambda = X_J^+$. Thus, $\dot{\nu}(\xi)\nu(y)\varphi_\lambda = -2\pi h(\xi, z)\nu(y)\varphi_\lambda$. Moreover,

$$\begin{aligned}(\nu(y')\varphi_\lambda, \nu(y)\varphi_\lambda) &= \{\exp i\pi\Omega(y',y)\}(\varphi_\lambda, \nu(y'-y)\varphi_\lambda) \\ &= \{\exp i\pi\Omega(y',y)\} \exp -\frac{\pi}{2}h(z'-z, z'-z) \\ &= \exp\{-\frac{\pi}{2}h(z',z') - \frac{\pi}{2}h(z,z) + \pi h(z,z')\}.\end{aligned}$$

This gives the result that $(\varphi_{\lambda,z'}, \varphi_{\lambda,z}) = \exp \pi h(z, z')$. The result on B_λ now follows from the definition (4.7).

REMARK : It follows from the above lemma, and may also be checked directly, that $B_\lambda\varphi_{\lambda,z} = \exp \pi h(\cdot, z)$ and thus the positive λ-coherent system in $\boldsymbol{H}(\lambda)$ is $\{\exp \pi h(\cdot, z) : z \in \lambda\}$.

Next we consider coherent systems in $\boldsymbol{H}(\lambda)$. Let $\mu \in \mathrm{Lag}^+ X^{\mathbb{C}}$ and let $\{f_{\lambda,\mu,\alpha} : \alpha \in \mu\}$ be the positive μ - coherent system in $\boldsymbol{H}(\lambda)$, i.e., they satisfy

(i) $f_{\lambda,\mu,\alpha}$ is a C^∞-vector of the representation $\boldsymbol{H}(\lambda)$ and $f_{\lambda,\mu,0}(0) > 0$;

(ii) If ν_λ denotes the representation of E in $\boldsymbol{H}(\lambda)$, $\{\dot{\nu}_\lambda(\xi) + 2\pi h(\xi,\alpha)\} f_{\lambda,\mu,\alpha} = 0$, for all ξ and $\alpha \in \mu$;

(iii) $(f_{\lambda,\mu,\alpha}, f_{\lambda,\mu,\beta})_{\boldsymbol{H}(\lambda)} = \exp \pi h(\beta,\alpha)$ for all $\alpha, \beta \in \mu$.

Here the inner product in $\boldsymbol{H}(\lambda)$ is defined by

$$(F_1, F_2)_{\boldsymbol{H}(\lambda)} = \int_\lambda F_1(u)\overline{F_2(u)}(\exp -\pi h(u,u))d_\lambda u$$

and the Haar measure is normalized by the condition $\int_\lambda (\exp -\pi h(u,u))d_\lambda u = 1$. Note that the μ-coherent system defined above is uniquely defined. We will give them explicitly later but first we note the following.

LEMMA 4.4. *The system $\{B_\lambda \varphi_{\mu,\alpha} : \alpha \in \mu\}$ where $\varphi_{\mu,\alpha}$ is given in Lemma 3.4, is a μ-coherent system and we have*

$$B_\lambda \varphi_{\mu,\alpha} = a(\lambda,\mu) f_{\lambda,\mu,\alpha}$$

where the phase factor $a(\lambda,\mu)$ is given by

$$a(\lambda,\mu) = \exp\left\{\frac{in\pi}{4} - \frac{i}{2} Im\ tr\text{-}log(Z_\mu - \bar{Z}_\lambda)\right\}$$

where Z_μ, Z_λ are the points in the Siegel upper half-plane corresponding to μ and λ.

PROOF. Since B_λ is an intertwining operator, it follows that $B_\lambda(\varphi_{\mu,\alpha})$ is a μ-coherent system, and from uniqueness (see Lemma 3.4), $a(\lambda,\mu)$ is just the phase factor in $(B_\lambda, \varphi_{\mu,0})(0)$. From the formula for $\varphi_{\lambda,\alpha}$ in Lemma 3.4 and the definition of B_λ, it follows that $a(\lambda,\mu)$ is the phase factor in the integral

$$\int (\exp i\pi(Z_\mu - \bar{Z}_\lambda)[x])dx.$$

The value of the above integral is easily evaluated (in this connection see Proposition 4.2 in [RR]) and is equal to $\exp\{\frac{in\pi}{4} - \frac{1}{2}\text{tr-log}(Z_\mu - \bar{Z}_\lambda)\}$. The lemma follows from this.

Let $R_\lambda(\tilde{\sigma}) = B_\lambda r(\tilde{\sigma}) B_\lambda^{-1}$ denote the metaplectic representation in $\boldsymbol{H}(\lambda)$. Then these operators can be completely described without any ambiguity in phase factors. In fact we have

THEOREM 4.4. *If $f_{\lambda,\mu,\alpha}$ is the positive μ-coherent system in $\boldsymbol{H}(\lambda)$ introduced above, then*

$$R_\lambda(\tilde{\sigma}) f_{\lambda,\mu,\alpha} = \left(\exp -\frac{i}{2} Im\ c(\lambda,\mu,\tilde{\sigma})\right) f_{\lambda,\sigma\cdot\mu,\sigma\cdot\alpha}$$

where

$$c(\lambda,\mu,\tilde{\sigma}) = \psi(Z_\mu) + tr\text{-}log(\sigma\cdot Z_\mu - \bar{Z}_\lambda) - tr\text{-}log(Z_\mu - \bar{Z}_\lambda).$$

Here Z_μ and Z_λ are the points of the Siegel upper half-plane associated to μ and λ respectively.

PROOF. Using Lemma 4.3

$$\begin{aligned} R_\lambda(\tilde{\sigma})f_{\lambda,\mu,\alpha} &= a(\lambda,\mu)^{-1}R_\lambda(\tilde{\sigma})B_\lambda\varphi_{\mu,\alpha} \\ &= a(\lambda,\mu)^{-1}B_\lambda r(\tilde{\sigma})\varphi_{\mu,\alpha} \\ &= a(\lambda,\mu)^{-1}\left(\exp -\frac{i}{2}\mathrm{Im}\psi(Z_\mu)\right)B_\lambda\varphi_{\sigma\cdot\lambda,\sigma\cdot\alpha}, \quad \text{(by Theorem 3.5)} \\ &= a(\lambda,\mu)^{-1}\left(\exp -\frac{i}{2}\mathrm{Im}\psi(Z_\mu)\right)a(\lambda,\sigma\cdot\mu)f_{\lambda,\sigma\cdot\mu,\sigma\cdot\alpha} \end{aligned}$$

and the theorem follows by substituting for $a(\lambda,\mu)$ from Lemma 4.3.

5 The generalized disc parametrization and the metaplectic representation

Fix $\lambda \in \mathrm{Lag}^+X^{\mathbb{C}}$ and let Z_λ be the point of the Siegel upper half-plane corresponding to it. Let $w_j = w_j(\lambda)$ be the canonical basis of λ introduced in (2.1). Then $X^{\mathbb{C}} = \bar{\lambda} + \lambda$ and

$$\bar{w}_1, \bar{w}_2, \ldots, \bar{w}_n, w_1, \ldots, w_n \tag{5.1}$$

is a basis of $X^{\mathbb{C}}$. As observed earlier the group $G = SpX$ is just the subgroup of $GL(X^{\mathbb{C}})$ which leave the forms Ω and $h(\cdot,\cdot)$ invariant. Note that the matrices of the forms $\Omega(\cdot,\cdot)$ and $h(\cdot,\cdot)$ in the basis (5.1) are $2i\begin{pmatrix} 0 & -B \\ B & 0 \end{pmatrix}$ and $\begin{pmatrix} -B & 0 \\ 0 & B \end{pmatrix}$ respectively, (see (2.2)). From this one gets the following easily.

LEMMA 5.1. *Let $\sigma \in GL(X^{\mathbb{C}})$ have the matrix $\begin{pmatrix} \alpha_0 & \beta_0 \\ \gamma_0 & \delta_0 \end{pmatrix}$ in the basis (5.1), arranged in blocks accoding to the decomposition $X^{\mathbb{C}} = \bar{\lambda} + \lambda$. Then $\sigma \in G$ if and only if $\alpha_0 = \bar{\delta}_0, \gamma_0 = \bar{\beta}_0$, ${}^t\delta_0 B\beta_0$ is symmetric and $\delta_0^* B\delta_0 - \beta_0^* B\beta_0 = B$, where $B = ImZ_\lambda$.*

We omit the details of verification. A basic reference is [S]. Next we note the relationship between the two matrix presentations of $\sigma \in G$. Now the transition matrix from the symplectic basis to the basis (5.1) is $L = \begin{pmatrix} \bar{Z}_\lambda & Z_\lambda \\ I & I \end{pmatrix}$ where Z_λ is the point of $\mathcal{H}_n$ corresponding to λ (i.e., $Le_r = \bar{w}_r, Le_r^* = w_r$, etc.). The inverse of L is easily checked to be

$L^{-1} = (2i)^{-1}\begin{pmatrix} -B^{-1} & B^{-1}Z_\lambda \\ B^{-1} & -B^{-1}\bar{Z}_\lambda \end{pmatrix}$. Thus the two matrix presentations of σ are related by

$$\begin{pmatrix} \alpha_0 & \beta_0 \\ \gamma_0 & \delta_0 \end{pmatrix} = L^{-1}\begin{pmatrix} \alpha & \beta \\ \gamma & \delta \end{pmatrix}L.$$

We summarize the details of this calculation.

LEMMA 5.2. *With notation as above* ($B = ImZ_\lambda$)

$$\begin{aligned} 2i\delta_0 &= B^{-1}\{(\alpha Z_\lambda + \beta) - \bar{Z}_\lambda(\gamma Z_\lambda + \delta)\} \\ &= B^{-1}(\sigma \cdot Z_\lambda - \bar{Z}_\lambda)(\gamma Z_\lambda + \delta) \\ 2i\beta_0 &= -B^{-1}\{(\alpha Z_\lambda + \beta) - Z_\lambda(\gamma Z_\lambda + \delta)\} \\ &= -B^{-1}(\sigma \cdot Z_\lambda - Z_\lambda)(\gamma Z_\lambda + \delta). \end{aligned}$$

The verification is straightforward and so omitted. Next we introduce the generalized disc parametrization of $Lag^+X^{\mathbb{C}}$ relative to λ. Note that $h > 0$ on λ and so $h < 0$ on $\bar{\lambda}$, as $h(\xi_1, \xi_2) = -h(\xi_2, \xi_1)$. Thus if $\mu \in Lag^+X^{\mathbb{C}}$, then $\mu \cap \bar{\lambda} = (0)$. This follows from the observation that $h > 0$ on μ and $h < 0$ on $\bar{\lambda}$. Thus the projection map of μ into λ defined by the decomposition $X^{\mathbb{C}} = \bar{\lambda} + \lambda$ is a linear isomorphism. Let $v_j = v_j(\mu)$ be the unique element of μ such that its projection into λ is w_j, i.e.,

$$v_j(\mu) = \sum c_{rj}\bar{w}_r + w_j, \quad j = 1, 2, \ldots, n. \tag{5.2}$$

Then $C_\mu = (c_{rs})$ is called the matrix associated to μ, in the generalized unit disc. From (2.2), one calculates that $\Omega(v_r, v_s) = (2i)\{(BC)_{rs} - (BC)_{sr}\}$ and

$$h(v_r, v_s) = (B - {}^tCB\bar{C})_{rs} \tag{5.3}$$

(note $B = ImZ_\lambda$). Thus $\mu \in Lag^+X^{\mathbb{C}}$ implies that $C_\mu \in \mathcal{D}(\lambda) = \{C : C$ is an $n \times n$ complex matrix such that (i) BC is symmetric and (ii) $B - {}^tCB\bar{C}$ is positive definite $\}$. Conversely if $C \in \mathcal{D}(\lambda)$ and μ is the subspace of $X^{\mathbb{C}}$ defined by the basis v_j given in (5.2), then $\mu \in Lag^+X^{\mathbb{C}}$. Thus $\mu \to C_\mu$ is a parametrization. Note that $\mathcal{D}(\lambda)$ depends only on B. Also if $C_1 = B^{1/2}CB^{-1/2}$ then $C \in \mathcal{D}(\lambda)$ is equivalent to $C_1 \in \mathcal{D}(\lambda_0)$, where $Z_{\lambda_0} = iI_n$, i.e., $\mathcal{D}(\lambda_0) = \{C_1 : C_1$ symmetric and $I_n - C_1\bar{C}_1 > 0\}$. This latter one is usually called the generalized unit disc (see [S]). For any $n \times n$ complex matrix C, let k_C denote the element of $GL(X^{\mathbb{C}})$, whose matrix in the basis $\bar{w}_1, \ldots, w_n$ is $\begin{pmatrix} I & C \\ 0 & I \end{pmatrix}$. Then, one checks directly tht $k_C \in SpX^{\mathbb{C}}$, if and only if BC is symmetric. Arguing exactly as in (2.3), if $\sigma \in SpX^{\mathbb{C}}$ and $\sigma = \begin{pmatrix} \alpha_0 & \beta_0 \\ \gamma_0 & \delta_0 \end{pmatrix}$ in the above basis, then $\sigma \cdot \lambda = \mu$ if and only if $\det \delta_0 \neq 0$ and $\beta_0 = C_\mu \cdot \delta_0$. From this one deduces again, that $C_{\sigma \cdot \mu} = \sigma \cdot C_\mu$, where

$$\sigma \cdot C = (\alpha_0 C + \beta_0)(\gamma_0 C + \delta_0)^{-1}$$

for all $\sigma \in G$ and $C \in \mathcal{D}(\lambda)$. In fact if we write simply $C = C_\mu$ and $C' = C_{\sigma\cdot\mu}$, then $k_{C'} \cdot \lambda = \sigma \cdot k_C \cdot \lambda$ or $k_{C'}^{-1}\sigma k_C \cdot \lambda = \lambda$ or the matrix of $k_{C'}^{-1}\sigma \cdot k_C$ is of the form $\begin{pmatrix} * & 0 \\ * & * \end{pmatrix}$ and $\in SpX^{\mathbb{C}}$. This actually gives that $(\alpha_0 C + \beta_0) - C'(\gamma_0 C + \delta_0) = 0$ and $\gamma_0 C + \delta_0$ is invertible, and hence $C' = (\alpha_0 C + \beta_0)(\gamma_0 C + \delta_0)^{-1}$.

Next consider the relationship between the parametrizations $\mu \to Z_\mu$ and $\mu \to C_\mu$. In fact suppose $\mu = u_Z \cdot V^*_{\mathbb{C}} = k_C \cdot \lambda = k_C \cdot u_{Z_\lambda} \cdot V^*_{\mathbb{C}}$. Now the matrix of k_C in the symplectic basis is $Lk_C L^{-1}$, where $L = \begin{pmatrix} \bar{Z}_\lambda & Z_\lambda \\ I & I \end{pmatrix}$. Thus the matrix of $k_C u_{Z_\lambda}$ in the symplectic basis is

$$\frac{1}{2i}\begin{pmatrix} \bar{Z}_\lambda & Z_\lambda \\ I & I \end{pmatrix}\begin{pmatrix} I & C \\ 0 & I \end{pmatrix}\begin{pmatrix} -B^{-1} & B^{-1}Z_\lambda \\ B^{-1} & -B^{-1}\bar{Z}_\lambda \end{pmatrix}\begin{pmatrix} I & Z_\lambda \\ 0 & I \end{pmatrix}$$

and this simplifies to $\begin{pmatrix} * & \bar{Z}_\lambda C + Z_\lambda \\ * & I + C \end{pmatrix}$. From (2.3), it now follows that $I + C$ is invertible and $Z = (\bar{Z}_\lambda C + Z_\lambda)(I + C)^{-1}$. Substituting, one computes $C = (Z_\lambda - Z)(Z - \bar{Z}_\lambda)^{-1}$. We summarize these in the following.

LEMMA 5.3. *If $\mu = u_Z . V^*_{\mathbb{C}} = k_C \cdot \lambda$, with Z and C belonging to the generalized upper half-plane and disc respectively, then $Z = (\bar{Z}_\lambda C + Z_\lambda)(I + C)^{-1}$ and $C = (Z_\lambda - Z)(Z - \bar{Z}_\lambda)^{-1}$.*

Next we consider the function $j_0(\sigma, C) = det(\gamma_0 C + \delta_0)$ for $\sigma \in G$ and $C \in \mathcal{D}(\lambda)$. Here $\sigma \doteq \begin{pmatrix} \alpha_0 & \beta_0 \\ \gamma_0 & \delta_0 \end{pmatrix}$ in the basis (5.1). We have seen earlier that $j_0(\sigma, C) \neq 0$ for all σ and C. Also one checks directly that it satisfies the cocycle identity $j_0(\sigma_1\sigma_2, C) = j_0(\sigma_1, \sigma_2 \cdot C) j_0(\sigma_2, C)$. Recall that $j(\sigma, Z) = det(\gamma Z + \delta)$ is a similar cocycle, based on the generalized upper half-plane. The relation between these two is given in the following

LEMMA 5.4. *With notation as above*

$$\gamma_0 C + \delta_0 = B^{-1}(\sigma \cdot Z - \bar{Z}_\lambda)(\gamma Z + \delta) B (Z - \bar{Z}_\lambda)^{-1}$$

and

$$j_0(\sigma, C) = j(\sigma, Z) det(\sigma \cdot Z - \bar{Z}_\lambda)\{det(Z - \bar{Z}_\lambda)\}^{-1}$$

and

$$j(\sigma, Z) = j_0(\sigma, C)\{det(I + \sigma \cdot C)\}^{-1} det(I + C)$$

where C and Z correspond to each other as in Lemma 5.3.

The verification is straightforward using Lemma 5.2 and 5.3 and so will be left to the reader.

We now construct a realization of the simply connected covering group based on the generalized unit disc. This is analogous to the construction of $\tilde{G}$ discussed in [RR]. For each $\sigma \in G$, let

$$\Gamma_{\lambda,\sigma} = \{\eta : \eta \text{ a holomorphic function on } \mathcal{D}(\lambda) \text{ such that}$$
$$\exp \eta(C) = j_0(\sigma, C) \text{ for all } C \in \mathcal{D}(\lambda)\}.$$

Note that if $\eta_1, \eta_2 \in \Gamma_{\lambda,\sigma}$, then the difference $\eta_1 - \eta_2$ is a constant and equal to an integral multiple of $2\pi i$. Also the cocyle property implies that if $\eta_1 \in \Gamma_{\lambda,\sigma_1}$ and $\eta_2 \in \Gamma_{\lambda,\sigma_2}$, then $\eta_1 o \sigma_2 + \eta_2 \in \Gamma_{\lambda,\sigma_1\sigma_2}$. From this it follows that

$$\tilde{G}^\lambda = \{(\sigma, \eta) : \sigma \in G, \eta \in \Gamma_{\lambda,\sigma}\}$$

is a group, with group law defined as $(\sigma_1, \eta_1) \cdot (\sigma_2, \eta_2) = (\sigma_1\sigma_2, \eta_1 o \sigma_2 + \eta_2)$. This group has $(Id, 0)$ as identity and $(\sigma^{-1}, -\eta o \sigma^{-1})$ serves as the inverse of (σ, η). The map $(\sigma, \eta) \to \sigma$ is a homomorphism with kernel $\{(0, 2\pi i m) : m \in \mathbb{Z}\} = \Gamma_{\lambda,1}$. Topologized using the product topology, this becomes a topological group, which is a covering group of G by a discrete subgroup isomorphic to $\mathbb{Z}$.

To establish an explicit isomorphism between $\tilde{G}^\lambda$ and the group $\tilde{G}$ (based on the Siegel upper half-plane), recall that Γ_σ was defined in [RR, Section 3] as $\Gamma_\sigma = \{\psi : \psi \text{ holomorphic on } \mathcal{H}_n, \text{ and } \exp \psi(Z) = j(\sigma, Z) \text{ for all } Z \in \mathcal{H}_n\}$.

LEMMA 5.5. *If C and Z are related as in Lemma 5.3, the map $\psi \to \eta$ of Γ_σ into $\Gamma_{\lambda,\sigma}$, defined as $\eta(C) = \psi(Z) +$ tr-log$(\sigma \cdot Z - \bar{Z}_\lambda) -$ tr-log$(Z - \bar{Z}_\lambda)$ is a bijection and moreover the inverse map is*

$$\psi(Z) = \eta(C) - \textit{tr-log}(I + \sigma \cdot C) + \textit{tr-log}(I + C).$$

If η_j correspond to $\psi_j \in \Gamma_{\sigma_j}$ as above, then $\psi_1 o \sigma_2$ corresponds to $\eta_1 o \sigma_2 + \eta_2$. Thus the map $(\sigma, \psi) \in \tilde{G} \to (\sigma, \eta)$, where η and ψ are related as above, is an isomorphism of the topological groups $\tilde{G}$ and $\tilde{G}^\lambda$.

This is clear from Lemma 5.4, and the fact that the map $C \leftrightarrow Z$ preserves G action, i.e., $\sigma \cdot C$ corresponds to $\sigma \cdot Z$. Again we omit the details.

REMARK : Note that the stabilizer G_λ of λ in G is the unitary group of the hermitian quadratic space (λ, h) and is thus the maximal compact subgroup of G and moreover if $\sigma \in G_\lambda$, then $\sigma \cdot \bar{\lambda} = \bar{\lambda}$ also and so the matrix of σ is $\begin{pmatrix} \alpha_0 & 0 \\ 0 & \delta_0 \end{pmatrix}$ and thus $j_0(\sigma, C) = \det \delta_0$. Also $\delta_0^* B \delta_0 = B$ implies that $|\det \delta_0| = 1$. Thus $\eta \in \Gamma_{\lambda,\sigma}$ is a constant and moreover this constant is pure imaginary or $\eta \in i\mathbb{R}$. Note that if $J_\lambda \in G$ is the complex structure corresponding to λ, i.e., $\lambda = X^+_{J_\lambda}$, then $J_\lambda = i.id$ on λ, and elements sitting above J_λ are $(J_\lambda, 2\pi i l)$. We define $\tilde{J}_\lambda = (J_\lambda, 0) \in \tilde{G}^\lambda$.

We will now express the metaplectic representation naturally in terms of the group $\tilde{G}^\lambda$. This seems a natural thing to do when looking at the realization in the Fock space $\boldsymbol{H}(\lambda)$. In fact if we write $W_\lambda(z)$ for the representation of E_λ in $\boldsymbol{H}(\lambda)$, i.e.,

$$W_\lambda(z)F(u) = F(u-z)\exp \pi h\left(u - \frac{1}{2}z, z\right), \quad F \in \boldsymbol{H}(\lambda)$$

and let $W_\lambda((\sigma,\eta))$ denote the unique representation of $\tilde{G}^\lambda$ in $\boldsymbol{H}(\lambda)$ such that $W_\lambda((\sigma,\eta)) = R_\lambda((\sigma,\psi))$ where $(\sigma,\eta) \to (\sigma,\psi)$ is the isomorphism, then

$$W_\lambda((\sigma,\eta))W_\lambda(z)W_\lambda((\sigma,\eta))^{-1} = W_\lambda(\gamma_0\bar{z} + \delta_0 z)$$

and moreover

$$W_\lambda((\sigma,\eta))f_{\lambda,\mu,\alpha} = \left(\exp -\frac{i}{2}\operatorname{Im}\eta(C_\mu)\right) f_{\lambda,\sigma\cdot\mu,\sigma\cdot\alpha}$$

where C_μ is the point in the generalized unit disc $\mathcal{D}(\lambda)$ corresponding to μ and $\alpha \in \mu$. Here we have used Lemma 5.5 and Theorem 4.4. When $\sigma \in G_\lambda$, it was observed by Segal and Bargmann (and checked rather easily) that if $U_\sigma f : u \to f(\sigma^{-1}\cdot u), u \in \lambda, f \in \boldsymbol{H}(\lambda)$, then U_σ is a unitary operator in $\boldsymbol{H}(\lambda)$ and

$$U_\sigma W_\lambda(z)U_\sigma^{-1} = W_\lambda(\sigma\cdot z).$$

From this one checks easily that $f_{\lambda,\mu,\alpha}o\sigma^{-1}$ is a positive $\sigma\cdot\mu$-coherent system. Thus from uniqueness $f_{\lambda,\sigma\cdot\mu,\sigma\cdot\alpha} = f_{\lambda,\mu,\alpha}o\sigma^{-1}$ for all $\sigma \in G_\lambda$. Since a μ-coherent system is a total subset of $\boldsymbol{H}(\lambda)$, it follows that

$$W_\lambda((\sigma,\eta))f = \left(\exp -\frac{1}{2}\eta\right) fo\sigma^{-1}$$

for all $\sigma \in G_\lambda, (\sigma,\eta) \in \tilde{G}^\lambda$, and $f \in \boldsymbol{H}(\lambda)$.

Finally we give an explicit formula for the positive μ-coherent system in $\boldsymbol{H}(\lambda)$. Note that the differential equations defining the μ-coherent system are

$$\{\dot{\nu}_\lambda(\xi) + 2\pi h(\xi,\alpha)\}F = 0, \qquad \text{for all } \xi \in \mu.$$

Now $v_j(\mu)$ defined in (5.2) is a basis of μ. Here we coordinatize λ by using the basis $w_1,\ldots,w_n$ (see 2.1). Now $\dot{\nu}_\lambda(\xi) = -\partial(\xi - iJ_\lambda\xi) + \pi h(\cdot, \bar{\xi} - iJ_\lambda\bar{\xi})$. Also when $\xi = v_j, \xi - iJ_\lambda\xi = 2w_j, \bar{\xi} - iJ_\lambda\bar{\xi} = \sum \bar{c}_{rj}\cdot w_r$ and using $h(w_r,w_s) = b_{rs}, h(v_r,v_s) = (B - {}^tCB\bar{C})_{rs}$ (see (5.3)), the differential equations become

$$\left\{-\frac{\partial}{\partial z_j} + \pi\sum(BC)_{sj}z_s + \pi\sum(B - {}^tCBC)_{sj}\bar{y}_s\right\}F = 0$$

where $\alpha = \sum y_jv_j \in \mu$. Clearly a solution of this is a multiple of

$$K_y(z) = \exp\{\frac{\pi}{2}BC[z] + \pi{}^tz(B - {}^tCB\bar{C})\bar{y}\}.$$

Next

$$(K_y, K_{y'})_{\boldsymbol{H}(\lambda)} = \int_\lambda (\exp\, Q)d_\lambda z$$

where

$$Q = \frac{\pi}{2}BC[z] + \pi^t z(B -^t CB\bar{C})\bar{y} + \pi^t \bar{u}(B -^t \bar{C}BC)y' + \frac{\pi}{2}B\bar{C}[\bar{z}] - \pi^t zB\bar{z}$$

and $d_\lambda z$ means (det B) times the standard Lebesgue measure defined by the basis $w_1, ..., w_n$ of λ. This last condition arises because of the normalization $\int_\lambda (\exp -\pi h(z,z)) d_\lambda z = 1$. If we now use the change of variable $z = B^{-1/2}u$ and the Gaussian integral formula in ([Fo], page 258) then the integral is equal to

$$(\det B)(\det(B -^t CB\bar{C}))^{-1/2} \exp Q_1$$

where

$$Q_1 = \frac{\pi}{2}(B -^t CB\bar{C})\bar{C}[\bar{y}] + \frac{\pi}{2}(B -^t \bar{C}BC)[y'] + \pi^t y'(B -^t CB\bar{C})\bar{y}.$$

From this we easily get the formula

$$f_{\lambda,\mu,\alpha}(z) = (detB)^{-1/2}(\det(B -^t CB\bar{C}))^{1/4} \exp Q$$

where

$$Q = -\frac{\pi}{2}(B -^t CB\bar{C})\bar{C}[\bar{y}] + \frac{\pi}{2}BC[z] + \pi^t z(B -^t CBC)y.$$

Here $C = C_\mu$ and $B = \operatorname{Im} Z_\lambda$ and $\alpha = \sum y_j v_j(\mu)$.

REFERENCES

[Ba] Bargmann, V., Group representations on Hilbert spaces of analytic functions, pp. 27-63, in *Analytical Methods in Mathematical Physics* (R.P. Gilbert and R.G. Newton, eds.), Gordon and Breach, New York, 1970.

[B-W] Borel, A. and Wallach, N., Continuous cohomology of discrete subgroups and representations of reductive groups, *Ann. Math. Studies*, Princeton University Press (1980).

[Car] Carmona, J., Representations du groupe de Heisenberg dans les espaces de (o,q)-formes, *Math. Annalen*, **205** (1973) 89-112.

[C] Cartier, P., Quantum mechanical commutation relations and theta functions, *AMS Proc. Symp. Pure Math.* **9** (1966), pp. 361-383.

[Fo] Folland, G.B., Harmonic analysis in phase space, *Ann. Math. Studies*, Princeton University Press, 1989.

[H] Howe, R., The oscillator semigroup, *AMS Proc. Symp. Pure. Math. Studies* **48** (1988).

[Hor] Hormander, L., *Analysis of Linear Partial Differential Operators*, Volume I, Springer-Verlag, New York.

[I] Itzykson, C., Remarks on boson commutation rules, *Comm. Math. Phys.* **4** (1967), pp. 92-122.

[Ig] Igusa, J., *Theta functions*, Springer-Verlag, New York, 1973.

[K-V] Kashiwara, M. and Vergne, M., On the Segal Shale-Weil representation and harmonic polynomials, *Invent. Math.* **44** (1978), 1-47.

[L-V] Lion, G. and Vergne, M., The Weil representation, Maslov index and theta series, *Progress in Mathematics* **6** (1982), Birkhauser, Boston.

[M] Mumford, D. (with Madhav Nori and Peter Norman) Tata Lectures on Theta III, Birkhauser, Boston, 1992.

[Ma] Magneron, B., Une extension de la notion d'indices de Maslov, *C.R. Acad. Sci. Paris* Ser A-B **289** (1979) no. 14, pp. A683-A686.

[R-Ro] Rawnsley and Robertson, The metaplectic representation, M_p^c structures and geometric quantization, *Memoirs Amer. Math. Soc.* **410** (1989).

[RR] Rao, R.R., The Maslov index on the simply connected covering group and the metaplectic representation, *J. Functional Analysis*, **107** (1992), 211.

[RR-2] Rao, R.R., On some explicit formulas in the theory of Weil representation (Preprint 1978) (to appear in *Pacific Journal of Mathematics*).

[Sa] Satake, I., Unitary representations of a semi-direct product of Lie groups on $\bar{\partial}$-cohomology spaces, *Math. Annalen* **190** (1970) 177-202.

[S] Siegel, C.L., *Symplectic Geometry*, Academic Press, New York, 1964.

[Sh] Shale, D., Linear symmetries of free boson fields, *Trans. Amer. Math. Soc.* **103** (1962), pp. 149-162.

[St] Sternberg, S., The pairing method and bosonic anomalies, pp. 261-293, in *Differential Geometric Methods in Theoretical Physics* (K. Bleuler and M. Werner, eds.), Kluwer Academic Publishers, 1988.

[V] Vergne, M., Seconde quantification et groupe symplectique, *C.R. Acad. Sci. Paris* **285** (1977), pp. A191-A194.

[W] Weil,A., Sur certain groupes des operateurs unitaires, *Acta Math.* **111** (1964), pp. 143-211.

Department of Mathematics
University of Illinois
Urbana, IL 61801
U.S.A.

J.K. Ghosh, S.K. Mitra, K.R. Parthasarathy and B.L.S. Prakasa Rao (*eds.*),
Statistics and Probability : A Raghu Raj Bahadur Festschrift
Published by Wiley Eastern Limited, Publishers, (1993), 531 – 537

Paradoxes in Robustness [1]

KESAR SINGH
Rutgers University, New Brunswick

Abstract

This note describes some paradoxes related to the concept of breakdown in robustness.

In real life, all real-valued random variables (when $\pm\infty$ are not part of the data) are bounded, including the ones generated by the computers. The bound may be an astronomical number (say, 10^{1000}), nonetheless they are bounded. Often the bounds arise from the precision limit of the measuring instrument. For example, consider $\log T$ or $1/T$, where T is some measured time. Suppose the smallest time that can be measured by the device used is 10^{-20} sec; then both these variables are bounded in absolute value by 10^{20}, even though in theory, they are both unbounded. It is established in this note that in the bounded situations the notion of breakdown in robustness can produce some results of paradoxical nature.

The concept of breakdown is attributable to Hampel (1971, 1974). This concept is perhaps the most popular notion of robustness in the modern

[1] Research supported by NSF Grant DMS 88-02558.
AMS Subject Classification: 62G20.
Key words and Phrases: Breakdown, Trimmed means.

statistical literature. We have included a few selected references. Suppose we have a random sample $X_1, X_2, \ldots, X_n$ from F supported on the interval $[L, M]$. Let T_n be a statistic whose robustness is being studied using the concept of breakdown point. A common definition of the sample version of the breakdown point d is as follows:

Let m be the minimum number of X data that should be replaced by the worst possible outliers in order to cause a breakdown of T_n. The breakdown point b is defined as $b = \frac{m}{n}$. The larger the value of b, the more robust T_n is. When the possible outliers are unbounded and T_n is a location statistic, a breakdown of T_n usually means $T_n \to \pm\infty$. In the bounded case, a reasonable definition should be: a breakdown occurs when T_n is pulled out of the convex hull of the original data, and yet it is within the bounds L and M. Let r be a fraction such that $X_{(n)} \leq rM \leq M$ where $X_{(i)}$'s are the order statistics of the X-data. Assume $M > X_{(n)}$, an event which has probability one or near one, typically. For the class of trimmed mean, a popular class of location statistics, we study here the value of b and point out some paradoxes that could occur when the definition of breakdown is as follows: A breakdown occurs when $T_n \geq rM$. A similar theory can be presented for the "left side" breakdown. We now define the class of trimmed means.

For a fraction α such that $\frac{\alpha n}{2}$ is a nonnegative integer and $\alpha n \leq n-1$, let

$$T_\alpha = \frac{1}{n(1-\alpha)} \sum_{n\frac{\alpha}{2}+1}^{n-n\frac{\alpha}{2}} X_{(i)}.$$

One can define T_α for all $0 \leq \alpha \leq 1$, but we do not do so for the sake of simplicity. For the smallest value of α (i.e., $\alpha = 0$), T_α is the arithmetic mean and for the largest value of α allowed, T_α is the usual sample median. Let b_α denote the breakdown point of T_α.

We look at b_α in the unbounded case first. If the outliers are not bounded, clearly $m = \frac{\alpha}{2}n$ and $b_\alpha = \frac{\alpha}{2}$. This is a very well known result on breakdown and indeed it agrees very well with our common understanding of the concept of robustness. It confirms that the more we trim, the more robust result we get. The sample mean is the least robust and the sample median is the most robust member and other trimmed means are in between.

Next, we state and prove a general formula for b_α in the bounded setup. First, a couple of notations follow: Let us define Y_t as the mean of top tn values of the X-data. It is assumed that tn is an integer between 0 and n, and $Y_0 = X_{(n)}$. Also, let

$$g(t) = (1-t)(M - Y_{1-t}).$$

Notice that Y_t is a nonincreasing function of t and hence, so is $g(t)$.

THEOREM. *Under the above setup,* $b_\alpha =$ min d *such that*

$$g(\frac{\alpha}{2} + d) \leq M(1-r)(1-\alpha).$$

PROOF: In the framework that we are working with, the worst outliers are $= M$, and they replace the lowest values of the original data. Out of dn outliers which are $= M$, $\frac{\alpha}{2}n$ are used up for upper trimming. The remaining $(d-\frac{\alpha}{2})n$ are used in the trimmed mean. Thus, after replacement, T_α can be regarded as a convex combination of M and $Y_{1-\frac{\alpha}{2}-d}$, since the top $n(1-\frac{\alpha}{2}-d)$ values of X appear in T_α. In the convex combination, the coefficient of $Y_{1-\frac{\alpha}{2}-d}$ is $p = (1-\frac{\alpha}{2}-d)/(1-\alpha)$ and that for M is $1-p$. Thus

$$T_\alpha = pY_{1-\frac{\alpha}{2}-d} + (1-p)M.$$

Consequently, it follows that $T_\alpha \geq Mr$ if and only if

$$p \leq \frac{M - Mr}{M - Y_{1-\frac{\alpha}{2}-d}}$$

which is equivalent to the statement

$$g(\frac{\alpha}{2} + d) \leq M(1-r)(1-\alpha).$$

Thus, the formula follows.

From this general formula, we shall deduce the familiar result for the case $M = \infty$ and other quite unfamiliar results which defy our common understanding of robustness.

COROLLARY 1. *For a fixed* α, *for which* T_n *is defined,* $b_\alpha \to \alpha/2$ *as* $M \to \infty$, *and* $r \to 0$.

Thus, Corollary 1 includes the case when the breakdown means "pulling T_α out of the data cloud" i.e. $rM = X_{(n)}$.

COROLLARY 2. *When* $r = 1$ *(and* M *is fixed),*

$$b_\alpha = 1 - \frac{\alpha}{2}.$$

Notice the paradoxical nature of this b_α. It is a decreasing function of α. It makes the mean most robust and the median least robust in the class of T_α.

COROLLARY 3. *In general,*

$$\frac{\alpha}{2} \leq b_\alpha \leq 1 - \frac{\alpha}{2}.$$

These bounds can be established by substituting $d = \alpha/2$ and $1-\alpha/2$ for d in the general formula. These bounds reveal some interesting facts about b_α. The standard result (i.e., $b_\alpha = \alpha/2$) turns out to be just the lower bound, but as $\alpha \to 1$, both the bounds merge and approach 1/2. Thus $b_\alpha \approx 1/2$ for $\alpha \approx 1$, i.e. when T_α is nearly median. In fact, $|b_{med} - 1/2| \leq 1/2n$. Notice that this fact does not depend upon the choice of r. When $\alpha \approx 0$, the two bounds are most far apart. As mentioned in the next corollary, b_α can very well be $> 1/2$ for any r in the range allowed.

COROLLARY 4. *If*

$$2M(1-r) < M - Y_{(1-\alpha)/2},$$

then $b_\alpha > 1/2$.

This is deduced by letting $d = 1/2$ in the general formula.

In the next corollary, we provide a condition under which $b_{\alpha+\epsilon} < b_\alpha$ i.e., additional trimming results in reduction in the breakdown point.

COROLLARY 5. *Let α and ϵ be such that T_α and $T_{\alpha+\epsilon}$ are defined. Assume $\epsilon/2 \geq 1/n$. Then $b_{\alpha+\epsilon} < b_\alpha$ if the following condition holds:*

$$\epsilon M(1-r) \leq (M - X_{(\ell)})(\frac{\epsilon}{2} - \frac{1}{n})$$

with $\ell = n(b_\alpha + \frac{\alpha}{2} + \frac{\epsilon}{2} - \frac{1}{n})$, provided the latter is $< n$. If not, the inequality $b_{\alpha+\epsilon} < b_\alpha$ holds automatically.

PROOF: By Corollary 3,

$$b_{\alpha+\epsilon} \leq 1 - \frac{\alpha+\epsilon}{2}.$$

Hence, if $1 - \frac{\alpha}{2} - \frac{\epsilon}{2} \leq b_\alpha - \frac{1}{n}$, then one clearly has $b_{\alpha+\epsilon} < b_\alpha$. Thus, one can assume, W.L.O.G., that

$$n(b_\alpha + \frac{\alpha}{2} + \frac{\epsilon}{2} - \frac{1}{n}) < n.$$

Now, recall the function $g(t) = (1-t)(M - Y_{1-t})$. Let δ be such that $(t+\delta)n$ be an integer $\leq n$ We claim the following bound:

$$g(t) - g(t+\delta) \geq \delta(M - X_{(n(t+\delta))}).$$

In order to prove this claim, note that

$$g(t) - g(t+\delta) = (1-t)(Y_{1-t-\delta} - Y_{1-t}) + \delta(M - Y_{1-t-\delta}).$$

Furthermore,

$$Y_{1-t} \leq \{(1-t-\delta)Y_{1-t-\delta} + \delta X_{((t+\delta)n)}\}/(1-t).$$

Combining the above two claims, the bound follows after some cancellations.

Finally, to deduce the corollary, one takes $t = b_\alpha + \frac{\alpha}{2}$ and $\delta = \frac{\epsilon}{2} - \frac{1}{n}$. Then, under the assumed condition, it follows that

$$g(b_\alpha - \frac{1}{n} + \frac{\alpha + \epsilon}{2}) \leq M(1-r)(1-\alpha-\epsilon).$$

This means, according to the formula in the theorem, that

$$b_{\alpha+\epsilon} \leq b_\alpha - \frac{1}{n}. \quad \square$$

REMARK 1. It is worth mentioning here that if $b_0 \leq b$, then a sufficient condition required for $b_\epsilon < b_0$ is

$$\epsilon M(1-r) \leq (\frac{\epsilon}{2} - \frac{1}{n})(M - X_{((b+\frac{\epsilon}{2}-\frac{1}{n})n)})$$

provided $b + \frac{\epsilon}{2} \leq 1$.

COROLLARY 6. *If $M - Mr < Mr - X_{(n)}$, i.e. Mr is more than the average of $X_{(n)}$ and M, then $b_\alpha > \frac{1}{2}$ for all α and b_α is a nonincreasing function of α, i.e. additional trimming can only decrease the breakdown point.*

PROOF: The claim $b_\alpha > \frac{1}{2}$ is a simple consequence of Corollary 3. To establish the second claim, we modify the proof of the previous corollary to deduce the following

$$b_{\alpha+\epsilon} \leq b_\alpha$$

provided either $\ell = n(b_\alpha + \frac{\alpha}{2} + \frac{\epsilon}{2})$ is $\geq n$ or $2M(1-r) \leq M - X_{(\ell)}$. This will always hold under the assumed condition.

REMARK 2. In the extreme case, when some of the data values are equal to M, the value of d should be modified to d^* defined as follows

$$d^* = \max(d - p, 0)$$

where p is the number of X_i's equal to M. In such a case, $r = 1$ in our framework. If one goes to the farthest extreme, i.e. when all the data values are $= M$, then as a referee has correctly pointed out, one has $d^* \equiv 0$; in which case, some of the statements made above are not valid anymore.

EXAMPLES: We close this note with some examples where the condition required for $b_0 > 1/2$ hold, with the most lenient definition of breakdown in the class considered here. So, let $rM = X_{(n)}$, i.e. breakdown occurs if T_α is pulled out of the data cloud in the upper side.

Let X-data represent the height measurement of adult men included in a sample drawn from a particular region. This example seems to be a typical case of all biological measurements. Using the knowledge of men's height in the region, suppose that one takes $M = 96''$. If $X_{(n)} = 84''$ and $Y_{1/2} = 70''$, then Corollary 4 entails $b_0 > 1/2$, i.e. breakdown for near mean T_α is higher than that of near median T_α. The figures taken seem to be quite plausible for many different regions of the world.

Another general example that we could consider is Huber's contamination model with a very reasonable truncation. Let $H = .9F + .1G$ where F is $N(0, \sigma)$ and G is $N(\mu, \sigma)$, both truncated at $\pm 6\sigma$ from their respective means. This truncation hardly affects the model. Let $\mu \geq 3\sigma$. Here $M = \mu + 6\sigma$. Suppose that $X_{(n)} = \mu + 2.5\sigma$, a reasonable extreme value on this model. With $rM = X_{(n)}$, we have $M(1-r) = 3.5\sigma$. The expected value of $Y_{1/2}$ is close to $.64\sigma + .2\mu$. Therefore, $M - E(Y_{1/2}) \approx .8\mu + 5.36\sigma > 7.76\sigma$. Thus the condition $(M - Y_{1/2}) > 2M(1 - r)$ is expected to hold. Even if $b_0 > 1/2$ does not occur, it is quite possible that the condition in Corollary 5 will hold, resulting in $b_\epsilon < b_0$.

CONCLUDING REMARKS: It is by no means concluded here that the classical notion of breakdown is a meaningless concept. In fact, in a great many situations, the outliers can be large enough that the classical breakdown represents the robustness quite accurately; especially if the human error in data entry is allowed to be a source of outliers. However, if such errors are ruled out using some computer based checking device and moderate bounds exist for the underlying data, then this notion of breakdown seems to miss the essence of robustness and produce quite meaningless results. This is the main point of this note. One natural remedy should be to have a definition to the effect that a breakdown occurs if an estimator T_n is pulled out of the "central body" of the data in the location case. A concept of data depth like Tukey's depth can be used to define centrality in the multivariate cases.

ACKNOWLEDGEMENT:
I wish to thank a referee for a very elaborate and interesting report.

REFERENCES

Donoho, D. L. (1982). Breakdown properties of multivariate location estimators. Ph.D. Qualifying Paper, Department of Statistics, Harvard University.

Donoho, D. L. and Huber, P. J. (1983). The notion of breakdown points. In *A Festschrift for Erich L. Lehmann*. 157-184. Wadsworth, Belmont, CA.

Hampel, F. R. (1971). A general qualitative definition of robustness. *Ann. Math. Statist.* **42**, 1887-1896.

Hampel, F. R. (1974). The influence curve and its role in robust estimation. *JASA* **69**, 383-393.

Stahel, W. A. (1981). *Breakdown of covariance estimators.* Research Report No. 31, E.T.H., Zurich.

Rutgers University
New Brunswick
New Jersey 08903
USA

J.K. Ghosh, S.K. Mitra, K.R. Parthasarathy and B.L.S. Prakasa Rao (*eds.*),
Statistics and Probability : A Raghu Raj Bahadur Festschrift
Published by Wiley Eastern Limited, Publishers, (1993), 539 – 550

On the Speed of Convergence to a Class of Max-Stable Laws

M. SREEHARI
M.S. University of Baroda, Baroda

Abstract

Let $\{X_n\}$ be a sequence of independent identically distributed random variables. Let $M_n = \max(X_1, \cdots, X_n)$. The class of limit distributions (extreme value distributions) of M_n, linearly normalised, and the speed of convergence in the corresponding limit theorems are well discussed in the literature. Pancheva (1984) investigated the class of possible limit distributions of M_n using power normalization. Ravi and Mohan (1989) investigated the inter-relationship between the two classes of limit distributions. In this paper we study the speed of convergence in the power normalization set up and compare it with the speed to the extreme value distributions.

1 Introduction

Let $\{X_n\}$ be a sequence of independent random variables with common distribution function (d.f.) F. Denote $\max(X_1, \ldots, X_n)$ by M_n. Then we say that F belongs to the max-domain of attraction of l-max stable law G if there exist norming constants $a_n > 0$ and b_n real such that as $n \to \infty$

$$P(M_n \leq a_n x + b_n) \to G(x), \qquad x \in \mathcal{L}(G)$$

AMS Subject Classification : 60F99
Key words and Phrases: l-max stable laws, p-max stable laws, domain of attraction, linear normalisation, power normalisation, speed of convergence, normal and exponential distributions.

or equivalently

$$F^n(a_n x + b_n) \to G(x), \qquad x \in \mathcal{L}(G) \tag{1}$$

where $\mathcal{L}(G)$ denotes the set of continuity points of G. We shall then write $F \in \mathcal{D}_l(G)$. Further if there exist positive norming constants α_n and β_n such that

$$P\left[|M_n/\alpha_n|^{1/\beta_n} \text{ sgn } (M_n) \le x\right] \to H(x), \qquad x \in \mathcal{L}(H)$$

or equivalently

$$F^n(\alpha_n|x|^{\beta_n} \text{ sgn } x) \to H(x), \qquad x \in \mathcal{L}(H) \tag{2}$$

as $n \to \infty$, we say that F belongs to the max-domain of attraction of p-max stable law H signifying the power normalisation that is used in (2). Here sgn $x = -1$, 0 or 1 according as $x < 0$, $= 0$ or > 0. Exact forms of G (Galambos (1978) or Resnick (1987)) and H (Pancheva (1984)) are known and are given in the following theorems.

THEOREM A. The d.f. G in (1) is one of the following three types

$$\Phi_\alpha(x) = \begin{cases} 0 & \text{if } x < 0 \\ \exp\{-x^{-\alpha}\} & \text{if } x \ge 0,\ \alpha > 0 \end{cases}$$

$$\Psi_\alpha(x) = \begin{cases} \exp\{-(-x)^{\alpha}\} & \text{if } x \le 0,\ \alpha > 0 \\ 1 & \text{if } x > 0 \end{cases}$$

and

$$\Lambda(x) = \exp\{-e^{-x}\}, \qquad -\infty < x < \infty.$$

THEOREM B. The d.f. H in (2) is one of the following six p-types

$$H_{1,\alpha}(x) = \begin{cases} 0 & \text{if } x < 1 \\ \exp\{-(\log x)^{-\alpha}\} & \text{if } x \ge 1,\ \alpha > 0 \end{cases}$$

$$H_{2,\alpha}(x) = \begin{cases} 0 & \text{if } x \le 0 \\ \exp\{-|\log x|^{\alpha}\} & \text{if } 0 < x < 1 \\ 1 & \text{if } x \ge 1 \end{cases}$$

$$H_{3,\alpha}(x) = \begin{cases} 0 & \text{if } x \le -1 \\ \exp\{-|\log|x||^{-\alpha}\} & \text{if } -1 < x < 0,\ \alpha > 0 \\ 1 & \text{if } x \ge 0 \end{cases}$$

$$H_{4,\alpha}(x) = \begin{cases} \exp\{-(\log|x|)^{\alpha}\} & \text{if } x < -1,\ \alpha > 0 \\ 1 & \text{if } x \ge -1 \end{cases}$$

$$H_5(x) = \Phi_1(x)$$

$$H_6(x) = \Psi_1(x)$$

where Φ_1 and Ψ_1 are as defined in Theorem A and we say the d.f.s. F_1 and F_2 are of the same p-type if $F_1(x) = F_2(a|x|^b \text{ sgn } x)$, $-\infty < x < \infty$, for some positive numbers a and b.

Mohan and Ravi (1989) have discussed the inter-relationship between the max domains of attraction of l-max and p-max stable laws. Since Φ_1 and Ψ_1 arise in both the limiting results at (1) and (2) it is necessary to distinguish between l-max domain and p-max domain of attraction of a $l(p)$ max stable law. Mohan and Ravi have proved the following

THEOREM C. Let F be a d.f. and $r(F) = \sup\{x|F(x) < 1\}$. Then

$$
\begin{array}{lll}
(a) & \left.\begin{array}{l}(i) F \in \mathcal{D}_l(\Phi_\alpha) \\ (ii) F \in \mathcal{D}_l(\Lambda),\ r(F) = \infty\end{array}\right\} & \Longrightarrow F \in \mathcal{D}_p(\Phi_1) \\
(b) & F \in \mathcal{D}_l(\Lambda),\ 0 < r(F) < \infty & \Longleftrightarrow F \in \mathcal{D}_p(\Phi_1),\ 0 < r(F) < \infty \\
(c) & F \in \mathcal{D}_l(\Lambda),\ r(F) < 0 & \Longleftrightarrow F \in \mathcal{D}_p(\Psi_1),\ r(F) < 0 \\
(d) & \left.\begin{array}{l}(i) F \in \mathcal{D}_l(\Lambda);\ r(F) = 0 \\ (ii) F \in \mathcal{D}_l(\Psi_\alpha),\ r(F) = 0\end{array}\right\} & \Longrightarrow F \in \mathcal{D}_p(\Psi_1) \\
(e) & F \in \mathcal{D}_l(\Psi_\alpha),\ r(F) > 0 & \Longleftrightarrow F \in \mathcal{D}_p(H_{2,\alpha}), \\
(f) & F \in \mathcal{D}_l(\Psi_\alpha),\ r(F) < 0 & \Longleftrightarrow F \in \mathcal{D}_p(H_{4,\alpha}).
\end{array}
$$

The two classes of distributions, viz. types of G and types of H, are well connected and X follows a G type distribution if and only if either $Y = \exp(X)$ or $Z = -\exp(-X)$ follows a H type distribution.

Full value of a limit theorem can be realised only if the speed of convergence in the limit theorem is available. In this context we consider

$$
\begin{aligned}
\Delta_n(x) &= |F^n(a_n x + b_n) - G(x)| && (3)\\
\text{and } \Delta_n^*(x) &= \left|F^n(\alpha_n |x|^{\beta_n} \text{ sgn } x) - H(x)\right| && (4)
\end{aligned}
$$

where G and H are as given in Theorems A and B.

Preliminary results concerning uniform and non-uniform bounds on $\Delta_n(x)$ are available in Galambos (1978). For standard normal F, Hall (1979) proved that $\sup_x \Delta_n(x) \sim (\log n)^{-1}$, whereas in the case of exponential, $\sup \Delta_n(x) \sim n^{-1}$ (Hall and Wellner (1979)). Cohen (1982) considered two classes of distributions, N (containing the normal) and E (containing the exponential), which are subclasses of $\mathcal{D}_l(\Lambda)$ and obtained the speed of convergence of $\sup_x \Delta_n(x)$ to zero. Cohen also discussed situations where a Φ_α type distribution or a Ψ_α type distribution is a better approximation for the expression $F^n(a_n x + b_n)$ even when $F \in \mathcal{D}_l(\Lambda)$.

Smith (1982) considered the speed problem when $F \in \mathcal{D}_l(\Phi_\alpha)$ (as also Ψ_α).

Balkema and De Haan (1990) investigated the problem of finding upper and lower bounds for $\sup_x \Delta_n(x)$. All the above results depend heavily on the properties of regularly varying functions.

Falk (1986) obtained similar estimates for $\Delta_n(x)$ using the fact that F can be approximated in an appropriate way by a generalised Pareto distribution.

It has been pointed by Mohan and Ravi that there are distributions $F \in \mathcal{D}_p(H)$ for some H but $F \notin \mathcal{D}_l(G)$ for any G. That is, there are situations where only power normalization approximation is possible for the d.f. of M_n. The remark following Theorem C suggests that for such distributions , a transformed distribution will bring us into the domain of attraction of l-max stable law. In view of Theorem C there are d.f.s F in the domain of attraction of a G as well as a H. The question is : does a H type distribution provide a better approximation than a G to the distribution of M_n? In this paper we prove that it is not so.

Statements and proofs of the main results are given in Section 3. A number of lemmas required for this investigation are given in Section 2; some of them are of independent interest also and may have been known. In the following C denotes a positive constant which may change from step to step.

2 Lemmas

Let $u(x,\theta) = (x^\theta - 1)/\theta$, $x > 0$, $\theta > 0$. Proofs of main results depend on bounds on and asymptotic behaviour of the following four expressions as $\theta \to 0$.

$$\begin{array}{lll} (i) & \exp\{-e^{-u(x,\theta)}\} - e^{-1/x} & \text{for } x > 0 \\ (ii) & \exp\{-e^{u(-x,\theta)}\} - e^{x} & \text{for } x < 0 \\ (iii) & \exp\{-(-u(x,\theta))^\alpha\} - e^{-(-\log x)^\alpha} & \text{for } 0 < x < 1 \\ (iv) & \exp\{-(u(-x,\theta))^\alpha\} - e^{-(\log(-x))^\alpha} & \text{for } x < -1. \end{array}$$

Write $v(x,\theta) = u(x,\theta) - \log x$ and $w(x,\theta) = v(x,\theta)/\theta$. Note that $u(x,\theta) < 0$ for $0 < x < 1$, $u(1,\theta) = 0$ and $u(x,\theta) > 0$ for $x > 1$. Also $v(x,\theta)$ is a decreasing function of x in $(0,1)$ and increases with x in $(1,\infty)$. Thus $v(x,\theta) \geq 0$. Let $\delta > 0$ be fixed.

LEMMA 1.

$$0 < v(x,\theta) < \{1 + x^\delta\} \, |\log x| \text{ for } x > 0 \text{ and } 0 < \theta \leq \delta.$$

PROOF. Write, for $x > 1$, $0 < \theta \leq \delta$

$$v(x,\theta) = \int_1^x \frac{y^\theta - 1}{y} \, dy < \int_1^x y^{\theta-1} \, dy < x^\delta \log x.$$

For $0 < x < 1$, $v(x,\theta) < -\log x$. Hence the result.

LEMMA 2. For $x > 0$ and $0 < \theta \leq \delta$

$$0 \leq w(x,\theta) \leq \frac{(\log x)^2}{2} + w^*(x)$$

where

$$w^*(x) = \begin{cases} 0 & \text{if } 0 < x \le 1 \\ \delta^{-2}\, x^\delta & \text{if } x > 1. \end{cases}$$

PROOF. Write

$$\gamma(x,\theta) = w(x,\theta) - \frac{(\log x)^2}{2}.$$

Then

$$\frac{\partial}{\partial x}\gamma(x,\,\theta) = \frac{v(x,\theta)}{x} \ge 0,$$

so that $\gamma(x,\theta) < 0$ for $0 < x < 1$ and $\gamma(x,\theta) > 0$ for $x > 1$.
Thus for $0 < x < 1$

$$0 < w(x,\theta) < (\log x)^2/2.$$

Consider next $x > 1$ and $0 < \theta \le \delta$. Then $w(x,\theta)$ can be proved to be an increasing function of θ for fixed x.

To see this write

$$\begin{aligned} \frac{\partial}{\partial\theta}\, w(x,\theta) &= \frac{1}{\theta^3}\left[\theta\, x^\theta \log x + \theta \log x - 2(x^\theta - 1)\right] \\ &= I_x(\theta)/\theta^3, \text{ say.} \end{aligned}$$

Now $I'_x(\theta) = \log x\; J_x(\theta)$ where $J'_x(\theta) = \theta\, x^\theta (\log x)^2 > 0$.

Since $J_x(\theta)$ increases with θ and $J_x(0) = 0$ we get $I_x\,(\theta)$ is also increasing with θ. Since $I_x(0) = 0$, $w(x,\theta)$ increases with θ.

Then for $0 < \theta \le \delta$ and $x > 1$

$$\begin{aligned} \gamma(x,\theta) &\le \frac{x^\delta - 1 - \delta\log x}{\delta^2} - \frac{(\log x)^2}{2} \\ &< x^\delta\,\delta^{-2}, \end{aligned}$$

completing the proof of the Lemma.

LEMMA 3. For $x > 0$ and $0 < \theta \le \delta$

$$0 \le e^{-\log x} - e^{-u(x,\theta)} \le \frac{1}{x}\left[1 - e^{-(1+x^\delta)\,|\log x|}\right]$$

and

$$0 \le e^{u(x,\theta)} - e^{\log x} \le x\left[e^{(1+x^\delta)|\log x|} - 1\right].$$

PROOF. Since $u(x,\theta) \ge \log x$ the first part of each inequality is true. Next consider

$$\begin{aligned} e^{u(x,\theta)} - e^{\log x} &= x\left[e^{v(x,\theta)} - 1\right] \\ &\le x\left[e^{(1+x^\delta)|\log x|} - 1\right] \end{aligned}$$

by Lemma 1. Similarly we get the other inequality.

LEMMA 4. For fixed y, $0 < y < 1$ and $\alpha > 0$

$$f(y,\alpha) = \left(\frac{-\log y}{1-y}\right)^{\alpha-1} \left[-\frac{\alpha+1}{2} \log y - 1\right]$$

is an increasng function of α.

PROOF. Writing $a(y) = (-\log y)/2$, $b(y) = -1 + a(y)$ and $C(y) = (-\log y)/(1-y)$ we note that

$$\frac{\partial f}{\partial \alpha} = [C(y)]^{\alpha-1} [\{a(y)\alpha + b(y)\} \log C(y) + a(y)] > 0$$

if and only if $g_y(\alpha) = [a(y)\alpha + b(y)] \log C(y) + a(y) > 0$. Since $g'_y(\alpha) = a(y) \log C(y) > 0$ it follows that $g_y(\alpha) > 0$ (and hence f is an increasing function of α) if $g_y(0) > 0$. Now, to prove that $g_y(0) > 0$ for all y in $(0,1)$ we consider the three cases

(i) $0 < y \le e^{-2}$

(ii) $e^{-2} < y < e^{-1}$

(iii) $e^{-1} \le y < 1$ separately.

Note first that $C(y)$ decreases to 1 as y increases to 1. In case (i), $0 \le -\log y - 2 < \infty$ so that

$$g_y(0) = \frac{1}{2} [-\log y + \{-\log y - 2\} \log C(y)] > 0.$$

In case (ii), $1 < -\log y < 2$ and

$$2g_y(0) = z + (z-2) \log \frac{z}{1-e^{-z}} = h(z),$$

say, where $z = -\log y$.

Now for $1 < z < 2$,

$$h(z) = z^2 - z + (z-2) \log \frac{z}{e^z - 1} > z(z-1) > 0$$

because $e^z > 1 + z$.

Finally in case (iii), $0 < z < 1$ and

$$h(z) = (2-z) \left[\frac{z(z-1)}{2-z} + \log\{(e^z-1)/z\}\right] > 0$$

if and only if $\left[\frac{z(z-1)}{2-z} + \log\{(e^z-1)/z\}\right] > 0$ or equivalently $K(z) = \frac{e^z - 1}{z} \exp\left\{\frac{z(z-1)}{2-z}\right\} > 1.$

But

$$
\begin{aligned}
K(z) &= \frac{1}{z}\left[e^{z/(2-z)} - e^{z(z-1)/(2-z)}\right] \\
&= \frac{1}{z}\left[z + \sum_{2}^{\infty} \frac{z^k}{(2-z)^k k!}\{1-(z-1)^k\}\right] \\
&> 1, \text{ because } 0 < z < 1.
\end{aligned}
$$

This completes the proof of the Lemma.

Lemma 5. The function $f(y,\alpha)$ defined in Lemma 4 satisfies

$$f(y,\alpha) + y > 0 \text{ for } 0 < y < 1 \text{ and } \alpha > 0.$$

Proof. In view of Lemma 4, it is sufficient to show that, for $0 < y < 1$, $Q(y) = f(y,0) + y > 0$. Since $Q(1) = 0$, we shall show that $Q(y)$ is a decreasing function of y, which proves the result. Now

$$
\begin{aligned}
Q(y) &= \frac{1-y}{\log y}\left[1 + \frac{1}{2}\log y\right] + y \\
&= \frac{1+y}{2} + \frac{1-y}{\log y},
\end{aligned}
$$

and,

$$Q'(y) = \frac{y(\log y)^2 - 2y\log y - 2(1-y)}{2y(\log y)^2}.$$

Hence, it is sufficient to show that for $0 < y < 1$

$$R(y) = y(\log y)^2 - 2y\log y - 2(1-y) < 0.$$

Since $R(1) = 0$ and $R'(y) = (\log y)^2 > 0$, the proof of the Lemma is complete.

Lemmas 3, 4 and 5 will be used in proving our final lemma.

Lemma 6. For $0 < x < 1$, $\alpha > 0$, $\theta > 0$,

$$0 \le (-\log x)^\alpha - \left(\frac{1-x^\theta}{\theta}\right)^\alpha \le (-\log x)^\alpha$$

and

$$0 \le \frac{(-\log x)^\alpha - \left(\dfrac{1-x^\theta}{\theta}\right)^\alpha}{\theta} \le \frac{\alpha}{2}(-\log x)^{\alpha+1} \qquad (5)$$

Proof. The first result is obvious because

$$0 \le \frac{1-x^\theta}{\theta} \le (-\log x).$$

In order to prove (5), write

$$\begin{aligned} T(x,\theta,\alpha) &= \frac{(-\log x)^\alpha - \left(\frac{1-x^\theta}{\theta}\right)^\alpha}{\theta} - \frac{\alpha}{2}(-\log x)^{\alpha+1} \\ &= \frac{1}{\theta^{\alpha+1}}\left[(-\theta \log x)^\alpha - (1-x^\theta)^\alpha - \frac{\alpha}{2}(-\theta \log x)^{\alpha+1}\right] \\ &= \frac{1}{\theta^{\alpha+1}}\left[(-\log y)^\alpha - (1-y)^\alpha - \frac{\alpha}{2}(-\log y)^{\alpha+1}\right] \end{aligned}$$

where $y = x^\theta$. We shall show that for $0 < y < 1$, $\alpha > 0$,

$$M(y,\alpha) = (-\log y)^\alpha - (1-y)^\alpha - \frac{\alpha}{2}(-\log y)^{\alpha+1} < 0.$$

Since $M(1,\alpha) = 0$ this follows if $M(y,\alpha)$ is an increasing function of y. But

$$\frac{\partial M}{\partial y} = \frac{\alpha}{y}(1-y)^{\alpha-1}\left[f(y,\alpha) + y\right]$$

where $f(y,\alpha)$ is as defined in Lemma 4. The result (5) now follows from Lemma 5.

3 Main Results

Throughout the following discussion we assume that $F \in \mathcal{D}_l(G)$. By considering various possible d.f.s G given in Theorem A, we determine the d.f. H so that $F \in \mathcal{D}_p(H)$ using Theorem C. We assume that the uniform rate of convergence to zero of the expression at (3) is known to be, say, θ_n. It is well-known that the sharpness of θ_n depends on the appropriate choice of the normalising constants a_n and b_n. For finding the rate of convergence in the p-max stable limit theorem at (2), we consider the norming constants α_n and β_n in terms of a_n and b_n so that the rate θ_n can be used. We thus assume

$$\sup_x |F^n(a_n x + b_n) - G(x)| = 0(\theta_n) \tag{6}$$

where G is as given in Theorem A.

THEOREM 1. Suppose (6) holds with $G = \Phi_\alpha$. For $\alpha_n = a_n$ and $\beta_n = 1/\alpha$

$$\sup_x \left|F^n(\alpha_n |x|^{\beta_n} \text{ sgn } x) - \Phi_1(x)\right| = 0(\theta_n).$$

PROOF. This follows from (6) because for $x \geq 0$,

$$\Phi_\alpha(\alpha^{1/\alpha}) = \Phi_1(x)$$

and

$$\Phi_\alpha(-(-x)^{1/\alpha}) = 0 = \Phi_1(x) \text{ for } x < 0.$$

It may be noted that $\beta_n \to 0$ as $n \to \infty$ in the following theorems except in Theorem 4.

THEOREM 2. Suppose (6) holds with $G = \Lambda$ and $0 < r(F) \le \infty$.

Then for $\alpha_n = b_n$ and $\beta_n = a_n/b_n$

$$\sup_x \left|F^n(\alpha_n |x|^{\beta_n} \text{ sgn } x) - \Phi_1(x)\right| = 0\,(\max(\theta_n, \beta_n, F^n(0))).$$

PROOF. For $x > 0$

$$\begin{aligned}&\left|F^n(\alpha_n x^{\beta_n}) - \Phi_1(x)\right| = |F^n(a_n\, u(x,\beta_n) + b_n) - \Phi_1(x)|\\ &= |F^n(a_n u(x,\beta_n) + b_n) - \Lambda(u(x,\beta_n)) + \Lambda(u(x,\beta_n)) - \Phi_1(x)|\\ &\le 0(\theta_n) + |\Lambda(u(x,\beta_n)) - \Phi_1(x)|\,. \end{aligned} \tag{7}$$

Now

$$\begin{aligned}\Lambda(u(x,\beta_n)) - \Phi_1(x) &= \exp\left\{-e^{-u(x,\beta_n)}\right\} - e^{-1/x}\\ &= e^{-1/x}\, u(e, p(x,\beta_n))p(x,\beta_n)\end{aligned}$$

where

$$p(x,\beta_n) = e^{-\log x} - e^{-u(x,\beta_n)} \to 0 \text{ as } n \to \infty$$

for fixed $x > 0$. Since

$$p(x,\beta_n) = \frac{1}{x}\left[\frac{1 - e^{-v(x,\beta_n)}}{v(x,\beta_n)}\right]\frac{v(x,\beta_n)}{\beta_n}\beta_n$$

where

$$v(x,\beta_n) = u(x,\beta_n) - \log x \to 0 \text{ as } n \to \infty,$$

using Lemmas 1-3 and the fact that $u(e,\theta)$ is bounded for θ small, we get for n large and $0 < a < x < b < \infty$

$$\begin{aligned}&|\Lambda(u(x,\beta_n)) - \Phi_1(x)|\\ &\le C\beta_n e^{-1/x}\frac{1}{x}\left[1 + (1 + x^{\delta})|\log x|\right]\left[(\log x)^2 + \delta^{-2}x^{\delta}\right].\end{aligned}$$

The fact that $e^{-1/x} \to 0$ exponentially fast as $x \to 0$ and δ may be taken less than 1 gives

$$\sup_{x>0} |\Lambda(u(x,\beta_n)) - \Phi_1(x)| < C\beta_n. \tag{8}$$

For $x \le 0$,

$$F^n(\alpha_n |x|^{\beta_n} \text{ sgn } x) \le F^n(0) \to 0$$

as $n \to \infty$ so that the result now follows from (7) and (8).

THEOREM 3. Suppose (6) holds with $G = \Lambda$ and $r(F) \le 0$.
Then with $\alpha_n = -b_n$ and $\beta_n = -a_n/b_n$

$$\sup_x \left|F^n(\alpha_n|x|^{\beta_n} \text{ sgn } x - \Psi_1(x)\right| = 0(\max(\theta_n, \beta_n)).$$

PROOF. For $x < 0$,

$$\begin{aligned}\left|F^n(\alpha_n|x|^{\beta_n} \text{ sgn } x) - \Psi_1(x)\right| &= |F^n(-a_n u(-x, \beta_n) + b_n) \\ &\quad -\Lambda(-u(-x, \beta_n)) + \\ &\quad \Lambda(-u(-x, \beta_n)) - \Psi_1(x)| \\ &\le 0(\theta_n) + |\Lambda(-u(-x, \beta_n)) \\ &\quad -\Psi_1(x)|. \end{aligned} \tag{9}$$

Proceeding as in Theorem 2 we get the result from (9) and Lemmas 1-3. (Note that for $x > 0$, left side of (9) is exactly zero).

THEOREM 4. Suppose (6) holds with $G = \Psi_\alpha$ and $r(F) = 0$.
Then with $\alpha_n = a_n$ and $\beta_n = 1/\alpha$

$$\sup_x \left|F^n(\alpha_n|x|^{\beta_n} \text{ sgn } x) - \Psi_1(x)\right| = 0(\theta_n).$$

PROOF. For $x < 0$

$$\left|F^n(\alpha_n|x|^{\beta_n} \text{ sgn } x) - \Psi_1(x)\right| =$$

$$\left|F^n(-a_n(-x)^{1/\alpha}) - \Psi_\alpha(-(-x)^{1/\alpha})\right|.$$

Hence we get the result from (6).

THEOREM 5. Suppose (6) holds with $G = \Psi_\alpha$ and $r(F) > 0$.
Then with $\alpha_n = b_n$ and $\beta_n = a_n/b_n$

$$\sup_x \left|F^n(\alpha_n|x|^{\beta_n} \text{ sgn } x) - H_{2,\alpha}(x)\right| = 0(\max(\theta_n, \beta_n, F^n(0))). \tag{10}$$

PROOF. Left side of (10)

$$\le 0(\theta_n) + \sup_{0<x<1} |\Psi_\alpha(u(x, \beta_n)) - \Psi_\alpha(\log x)| + F^n(0).$$

Recall that $\alpha_n = b_n = r(F)$ so that $\sup_{x\ge 1} \left|F^n(\alpha_n x^{\beta_n}) - 1\right| = 0.$

Proceeding as in Theorem 2 and using Lemma 6 we get for $0 < a < x < b < 1$ and n large

$$\begin{aligned}&|\Psi_\alpha(u(x, \beta_n)) - \Psi_\alpha(\log x)| \le \\ &C\, e^{-(-\log x)^\alpha} \beta_n(-\log x)^{\alpha+1} [1 + (-\log x)^\alpha]\end{aligned}$$

so that (10) holds.

Finally, we have

THEOREM 6. Suppose (6) holds with $G = \Psi_\alpha$ and $r(F) < 0$.

Then with $\alpha_n = -b_n$ and $\beta_n = -a_n/b_n$

$$\sup_x \left|F^n(\alpha_n|x|^{\beta_n} \text{ sgn } x) - H_{4,\alpha}(x)\right| = 0(\max(\theta_n, \beta_n)). \tag{11}$$

PROOF. For $x < -1$,

$$\begin{aligned} &\left|F^n(\alpha_n|x|^{\beta_n} \text{ sgn } x) - H_{4,\alpha}(x)\right| \leq \\ &0(\theta_n) + |\Psi_\alpha(-u(-x, \beta_n)) - \Psi_\alpha(-\log(-x))|. \end{aligned} \tag{12}$$

Proceeding as in Theorems 2 and 5 and using Lemma 6, the second term on the right side of (12) is seen to be less than

$$C\,\beta_n\, e^{-[\log(-x)]^\alpha}\, [\log(-x)]^{\alpha+1}\, [1 + (\log(-x))^\alpha] \tag{13}$$

and hence (11) holds as the left side of (12) is zero for $x \geq -1$.

4 Examples

(a) Hall (1979) proved that when F is the standard normal d.f. then for appropriate choice of a_n and b_n, the expression at (3) is of the order $1/\log n$. It may be recalled that $G = \Lambda$ in this case. In this case $\beta_n = 1/\log n$ and hence the speed of convergence in (2) is of the order $1/\log n$. Thus both the approximations are of the same order. (Sreehari (1990)).

(b) If F is the standard exponential d.f. for appropriate choice of a_n and b_n, $\theta_n = 1/n$. However β_n being of the order $1/\log n$, the speed of convergence in (2) is of the order of $1/\log n$. Thus result (1) gives a better approximation for the d.f. of M_n (Sreehari (1990)).

REMARK. As noted earlier the speed of convergence in (1) depends on the choice of the normalizing constants a_n and b_n. The same is true in the case of (2). It therefore follows that for a different choice of α_n and β_n the speed of convergence in (2) may be faster.

Acknowledgements

The author is thankful to Professor N.R. Mohan for his comments on an earlier version and Dr. J.R. Patadia for a discussion.

REFERENCES

Balkema, A.A. and De Haan, L. (1990) A convergence rate in extreme value theory, *J. Appl. Prob.*, **27**, 577-585.

Cohen, J.P. (1982) Convergence rates for the ultimate and Penultimate approximations in extreme value theory, *Adv. Appl. Prob.*, **14**, 833-854.

Falk, M. (1936) Rates of uniform convergence of extreme order statistics, *Ann. Inst. Statist. Math.* **38** Part A, 245-262.

Galambos, J. (1978) The Asymptotic Theory of Extreme Order Statistics, John Wiley and Sons, N.Y.

Hall, P. (1979) On the rate of convergence of normal extremes, *J. Appl. Prob.* **16**, 433-439.

Hall, W.J. and Wellner, J.A. (1979) The rate of convergence in law of maximum of an exponential sample, *Statistica Neerlandica*, **33**, 151-154.

Mohan, N.R. and Ravi, S. (1991) On max domains of attraction of univariate and multivariate p-max stable laws, To appear in *Theoriya Veroyatnostei i ee Premeneniya.*

Pancheva, E. (1984) *Limit Theorems for Extreme Order Statistics under Nonlinear Normalization*, Lecture notes in Math. 1155, Springer Verlag, 284-309.

Resnick, S.I. (1987) *Extreme values, Regular Variation and Point Processes*, Springer Verlag.

Smith, R.L. (1982) Uniform rates of convergence in extreme value theory *Adv. Appl. Prob.*, **14**, 600-622.

Sreehari, M. (1990) On the speed of convergence in a max stable limit theorem, presented at the International Seminar on Recent Advances in Probability and Statistics, New Delhi.

Department of Statistics
Faculty of Science
The Maharaja Sayajirao University of Baroda
Lokmanya Tilak Road
Baroda 390 002
INDIA

J.K. Ghosh, S.K. Mitra, K.R. Parthasarathy and B.L.S. Prakasa Rao (*eds.*),
Statistics and Probability : A Raghu Raj Bahadur Festschrift
Published by Wiley Eastern Limited, Publishers, (1993), 551 - 564

Simultaneous Selection and Estimation of a Simple Normal Model

J.H. VENTER and S.J. STEEL
Potchefstroom University, Potchefstroom

Abstract

We consider model selection and estimation of the relevant parameters for the simple case where the data consists of a single observed X and the competing models for the distribution of X are $N(0;1)$ and $N(\mu;1)$ with $\mu \neq 0$ respectively. Various approaches which try to achieve a suitable balance between the selection and estimation aspects of the problem are compared. This culminates in a simple estimator which may be described as approximately minimizing the probability of not selecting the simpler model when it is true within the class of minimax estimators with respect to the squared error with complexity cost loss.

1 Introduction

This paper discusses various approaches to the following simple canonical simultaneous model selection and estimation problem: we observe X which is assumed to be $N(\mu;1)$-distributed, and we must choose between two models regarding the possible values of μ, namely $M_0 : \mu = 0$ and $M_1 : \mu$ has some non-zero value. Our inference objective is to select one of these

AMS Subject Classifications: 62 F 03; 62 F 30
Key words and Phrases: Model selection; simultaneous selection and estimation; pre-test; minimax; complexity cost

models and to estimate μ accordingly, i.e. we must supply an estimator $\delta(X)$ of μ with the interpretation that $\delta(X) = 0$ implies that model M_0 is selected (and μ is estimated to be zero), and $\delta(X) \neq 0$ implies that model M_1 is selected (and μ is estimated to be $\delta(X)$). This objective is often a realistic alternative to testing the hypothesis that $\mu = 0$ against $\mu \neq 0$. Cohen (1965), Nagata (1983), Bickel (1984) and Kempthorne (1987a,b) discuss contexts of which this problem is a special case.

Model selection criteria such as AIC, C_p, etc. can be applied to this problem. The argument leading to the C_p estimator is as follows: assume squared error (SE) loss; if we select model M_0 and estimate μ by $\delta(X) = 0$ **for all** X, the risk is μ^2 (which is unknown but is estimated unbiasedly by $X^2 - 1$); if we select model M_1 and estimate μ by $\delta(X) = X$ **for all** X, the risk has the known value 1; hence we select M_0 if $X^2 - 1 \leq 1$ and M_1 otherwise, leading to the final estimator $\delta(X) = X\mathbf{I}(|X| > \sqrt{2})$, where $\mathbf{I}(A)$ denotes the indicator function of the event A. Apart from the uncritical use of the unbiased estimator $X^2 - 1$ for μ^2 in this argument, we see that the final estimator arrived at is neither 0 **for all** X nor X **for all** X, whereas the derivation implied that the choice was between these two estimators. Thus it seems that the C_p estimator rests on a somewhat shaky foundation. Similar comments apply to estimators based on many other selection criteria (cf. Linhart and Zucchini (1986)).

The **pre-testing approach**, using a test of size α for the hypothesis $\mu = 0$, selects M_0 if $|X| \leq a$ where

$$a = a(\alpha) = \Phi^{-1}(1 - \tfrac{1}{2}\alpha), \tag{1.1}$$

where Φ denotes the $N(0;1)$-distribution function. Otherwise M_1 is selected and μ is estimated to be X. Thus the pre-test estimator is

$$\delta_a(X) = X\mathbf{I}(|X| > a). \tag{1.2}$$

The special case $a = \sqrt{2}$ of (1.2) gives the C_p estimator and this corresponds to $\alpha = 0.1573$. An important consideration in model selection is expressed by the **principle of parsimony** (cf. Kotz and Johnson (1985), p.578), according to which the simplest adequate model should be selected. In the pre-testing approach we accommodate this principle to some extent by making α small (implying a large) so that the probability of not selecting the simpler model M_0 when it is correct becomes small. However, this may affect the accuracy of estimation of μ adversely, and it seems clear that an approach is needed which balances the selection and estimation aspects of the problem.

Bickel (1984) formulated the **Problem P approach**: among all estimators δ whose SE risk $E[\delta(X)-\mu]^2$ is at most $1+\epsilon$, find that one with minimal SE risk at $\mu = 0$. Here $\epsilon \geq 0$ is some prescribed number. This approach provides for the model selection aspect of the problem in the way in which the SE risk of the estimator is appraised. Parsimony is accommodated by being as accurate as possible in the estimation of μ when the simpler model is correct and overall accuracy is maintained by restricting attention to estimators whose risks are suitably limited everywhere. While the exactly optimal solution to Problem P is complicated, the limited translation (LT) estimator

$$\delta(X) = (X - \sqrt{\epsilon}\,\mathrm{sgn}(X))\mathbf{I}(|X| > \sqrt{\epsilon}) \tag{1.3}$$

is nearly optimal. An intuitively unappealing feature of this estimator is its behaviour when $|X|$ is large. In that case it seems that μ should be estimated by X (or something close to X), whereas (1.3) uses $X - \sqrt{\epsilon}\mathrm{sgn}(X)$. This suggests that ϵ must be taken small (which is also desirable to limit the maximal risk), but then the probability of selecting M_0 when it is correct also becomes small. Although one may argue that for any given ϵ, (1.3) is in a sense doing (approximately) as well as is possible, the actual choice of ϵ in practice seems difficult.

Kempthorne (1983, 1987a) discusses what we will refer to as the **Problem K approach**: let D_α denote the class of estimators δ which do not select M_0 when it is correct with probability at most α; then find that $\delta \in D_\alpha$ which minimizes

$$\sup_{\mu\in\Re} E[\delta(X) - \mu]^2 . \tag{1.4}$$

This formulation seems more appealing than the Problem P approach. For given α the optimal solution to Problem K is doing as well as possible in terms of maximal SE risk, while the more readily interpretable quantity α may be adjusted to accommodate the principle of parsimony. Kempthorne (1983, 1987a) shows that the optimal solution has the form

$$\delta(X) = b(X)\mathbf{I}(|X| > a) \tag{1.5}$$

with a as in (1.1) and $b(X)$ the posterior expectation of μ corresponding to a suitably chosen discrete symmetric prior for μ. In general this estimator is difficult to work with. In §2 we show that a simple close to optimal approximate solution is given by

$$\delta_{a,c,d}(X) = c\,\mathrm{sgn}(X)\mathbf{I}(a < |X| \le c+d) + (X - d\,\mathrm{sgn}(X))\mathbf{I}(|X| > c+d) \tag{1.6}$$

where a is as in (1.1) and where $c > a$ and $d \geq 0$ are chosen to minimize the maximal SE risk of (1.6).

Meeden and Arnold (1979) accommodate the principle of parsimony by using the loss function

$$L(\delta, \mu) = (\delta - \mu)^2 + s^2\mathbf{I}(\delta \neq 0) \tag{1.7}$$

where the quantity s^2 denotes the cost entailed when the more complex model M_1 rather than the simpler model M_0 is selected. We refer to (1.7) as the **squared error with complexity cost** (SEWCC) loss function.

Meeden and Arnold (1979) show that the pre-test estimator (1.2) with $a = s$ is admissible under this loss function. Kempthorne (1987b) shows that the estimator X itself is inadmissible but minimax with constant risk $1 + s^2$. In §3 we show that no estimator of the form (1.2) can be minimax. We also present numerical evidence indicating that the simple estimator

$$\delta_{a,c}(X) = c\,\mathrm{sgn}(X)\mathbf{I}(a < |X| \leq c) + X\mathbf{I}(|X| > c) \qquad (1.8)$$

dominates X if the constants a and c are chosen appropriately. We then argue that this estimator may be regarded as an approximately optimal solution to what we will refer to as **Problem M**: among all SEWCC minimax estimators, find that one which minimizes the probability of not selecting M_0 if it is correct. This estimator seems preferable to the others considered in this paper.

For the sake of completeness, two other approaches may be mentioned. Cohen (1965) considers a loss function that is similar to (1.7) but penalizes for complexity only when complexity is inappropriate (i.e. when $\mu = 0$). Inagaki (1977) introduces a "loss function" that attempts to combine model selection and estimation errors and Nagata (1983) and Nagata and Inaba (1985) discuss related results for this "loss function". However, this "loss function" has unusual features in that it depends on the observed X also in addition to μ and δ and it may be negative. We do not discuss these approaches further in this paper.

In §4 we conclude with a brief indication of extensions of the results.

2 An approximately optimal solution to Problem K

Throughout this paragraph we work with the SE loss function unless otherwise stated. Kempthorne (1987a) approximates the solution to problem K as follows: for $0 \leq m < \infty$, restrict μ to $[-m; m]$ and find the estimator which minimizes

$$\sup_{\mu \in [-m;m]} E[\delta(X) - \mu]^2 \qquad (1.9)$$

among all estimators $\delta \in D_\alpha$. The solution to this problem depends on m and α, and it is given by $\overline{\delta}_m(X) = b(X)\mathbf{I}(|X| > a)$, with a as in (1.1) and with

$$b(X) = \frac{\sum p_i t_i[\phi(X - t_i) - \phi(X + t_i)]}{\sum p_i[\phi(X - t_i) + \phi(X + t_i)]} \qquad (1.10)$$

In (1.10) ϕ is the standard normal density function and $0 < t_1 < t_2 < \cdots$ are the support points of a symmetric discrete prior for μ that places probability p_i at $\pm t_i, i = 1, 2, \ldots$. For given α and m, the t_i's and p_i's have to be determined numerically and Kempthorne (1987a) describes an algorithm for this purpose. Kempthorne (1983) and (1987a) give their values for a range of α-values for $m = 6$ and $m = 7$. An inspection of the graph of $\overline{\delta}_7(X)$ suggests that the optimal solution for $m = 7$ may be approximated rather accurately by the simple estimator given in (1.6). In addition to a of (1.1), this estimator depends on c and d and these may be determined to minimize the maximal risk of (1.6) as follows. The risk of $\delta_{a,c,d}$ is given by

$$\begin{aligned}
& E[\delta_{a,c,d}(X)-\mu]^2 \\
= \; & \mu^2[\Phi(a-\mu)+\Phi(a+\mu)-1]+(c-\mu)^2[\Phi(c+d-\mu)-\Phi(a-\mu)]+ \\
& (c+\mu)^2[\Phi(c+d+\mu)-\Phi(a+\mu)]+ \qquad (1.11) \\
& (1+d)^2[2-\Phi(c+d-\mu)-\Phi(c+d+\mu)]+ \\
& (c-d-\mu)\phi(c+d-\mu)+(c-d+\mu)\phi(c+d+\mu)
\end{aligned}$$

and this is a symmetric function of μ so that we can limit attention to the case $\mu \geq 0$. Clearly (1.11) approaches $1+d^2$ as $\mu \to \infty$. Numerical work showed that for given $c > a$ and $d > 0$, (1.11) has a minimum at $\mu = 0$, reaches a local maximum, say $r(a,c,d)$, at some finite value of $\mu > 0$, and eventually increases monotonically to $1+d^2$ as $\mu \to \infty$. We used IMSL routine DZBREN to determine the value of $d = d(c)$ for which $r(a,c,d) = 1+d^2$. In this process we used IMSL routine DUVMIF to determine $r(a,c,d)$ numerically. These values of c and $d = d(c)$ balance the local maximum $r(a,c,d)$ and the maximum $1+d^2$ at $\mu \to \infty$. Finally we varied $c > a$ until c and d were found for which the common value of $r(a,c,d)$ and $1+d^2$ is a minimum. Columns 3 and 4 of Table 1.1 give c and d for a range of α-values and the corresponding maximal risks are given in column 5. Column 2 gives a obtained from (1.1). To see how close to optimal $\delta_{a,c,d}$ is, define

$$R(\delta) = \sup_{\mu\in\Re} E[\delta(X)-\mu]^2 \text{ and } \overline{R}(\alpha) = \inf_{\delta\in D_\alpha} R(\delta) \qquad (1.12)$$

and define R_m and $\overline{R}_m$ similarly with $\mu \in [-m;m]$. Then since $\delta_{a,c,d} \in D_\alpha$ for $a = a(\alpha)$,

$$R(\delta_{a,c,d}) \geq \overline{R}(\alpha) \geq \overline{R}_m(\alpha) \qquad (1.13)$$

We computed the optimal $\overline{\delta}_8$ and $\overline{R}_8(\alpha)$ using IMSL routine DNCONF to do the multivariate optimization required and show the results in column 6 of Table 1.1. Since the entries of columns 5 and 6 are close to each other, it follows from (1.13) that $\delta_{a,c,d}$ comes very close to being an optimal solution of Problem K. Further evidence to this effect is provided by Figure 1 which shows $R(\delta_{a,c,d})$ together with $\overline{R}_m(\alpha)$ for $m = 4$ and 8 as functions of $\alpha \in [0.01; 0.25]$.

In Figure 2 we show the risks of $\delta_{a,c,d}$ and $\overline{\delta}_8$ as functions of μ for $\alpha = 0.10$. The closeness of these curves again confirms the approximate optimality of $\delta_{a,c,d}$ as a solution of Problem K. Analytical support for this finding analogous to the way in which Bickel (1984) shows (1.3) to be an approximately optimal solution to Problem P would be nice but have not been found.

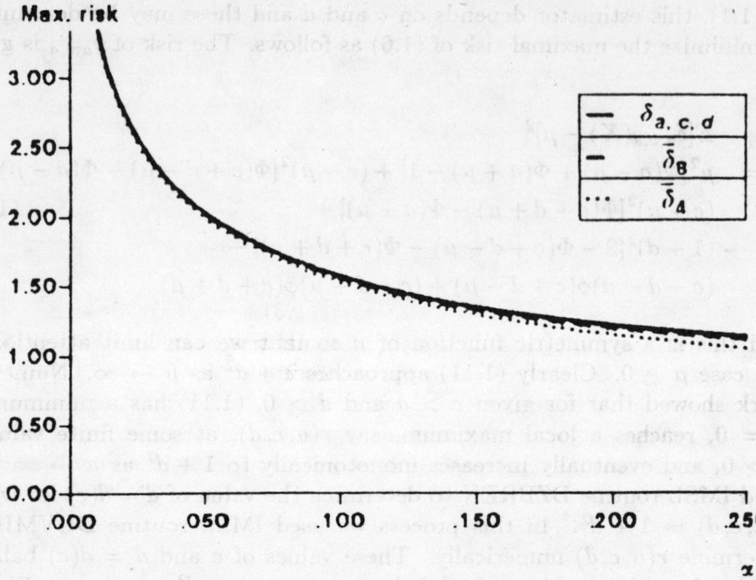

Figure 1 Maximal risks of $\delta_{a,c,d}$, $\overline{\delta}_4$ and $\overline{\delta}_8$

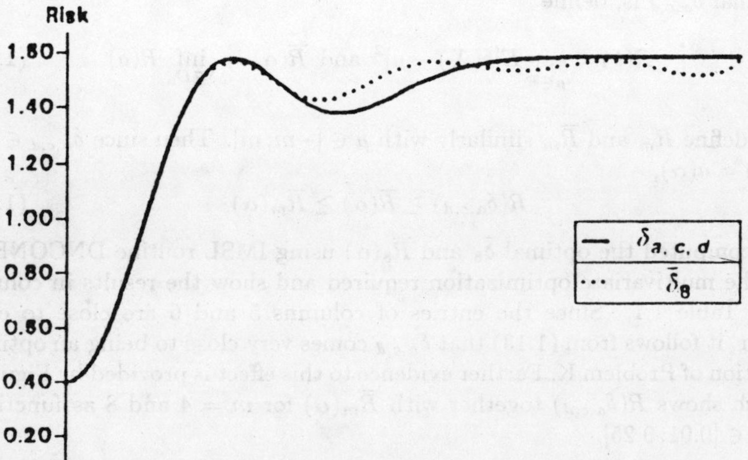

Figure 2 Risks of $\delta_{a,c,d}$ and $\overline{\delta}_8(\alpha = 0.10)$

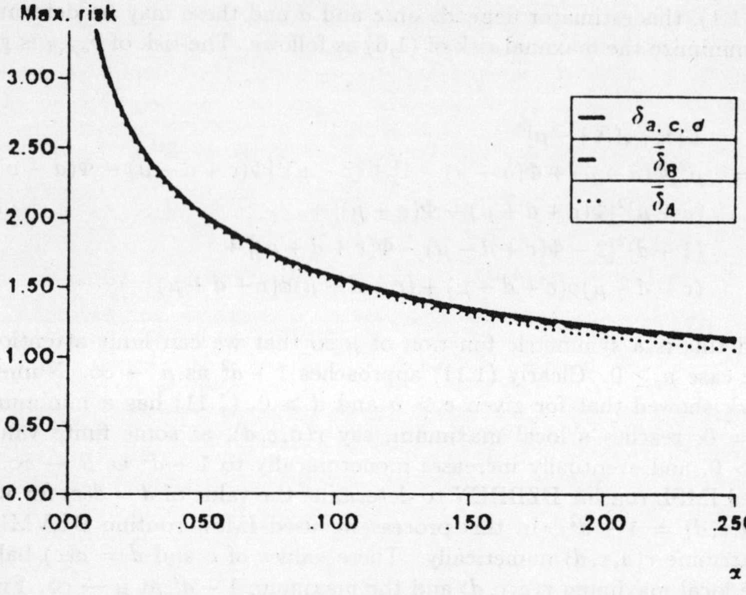

Figure 1 Maximal risks of $\delta_{a,c,d}$, $\overline{\delta}_4$ and $\overline{\delta}_8$

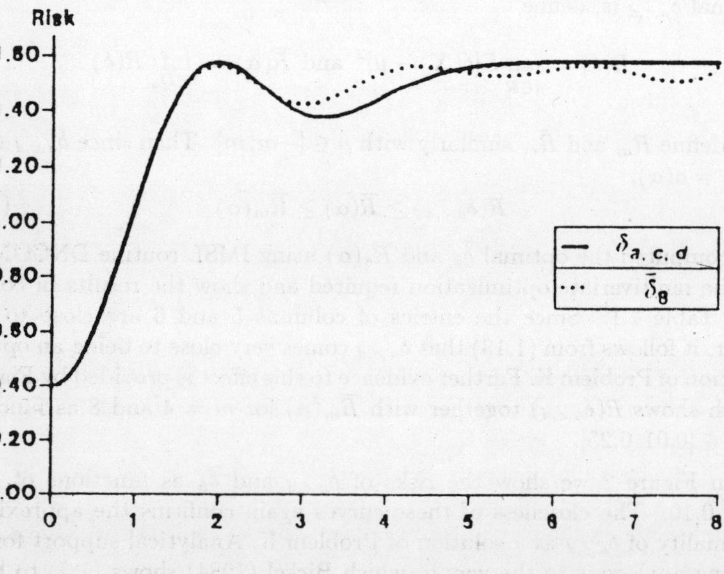

Figure 2 Risks of $\delta_{a,c,d}$ and $\overline{\delta}_8(\alpha = 0.10)$

Table 1.1 Approximately optimal solution to Problem K

α	a	c	d	$R(\delta_{a,c,d};\alpha)$	$\bar{R}_8(\alpha)$
0.01	2.576	2.576	1.530	3.340	3.332
0.05	1.960	2.156	1.018	2.036	2.021
0.10	1.645	1.965	0.758	1.574	1.559
0.15	1.440	1.841	0.590	1.348	1.331
0.20	1.282	1.739	0.464	1.215	1.196
0.25	1.150	1.644	0.363	1.132	1.108

Finally note that the LT estimator (1.3) is in the class D_α if we take $\epsilon = a^2$ with a as in (1.1). Then its maximal risk is $1 + a^2$. This is typically much larger than $1 + d^2$ which is the maximal risk of $\delta_{a,c,d}$. Thus from the point of view of Problem K, $\delta_{a,c,d}$ is much preferable over the LT estimator. It still has the unappealing feature that for large $|X|$ it does not reduce to X but to $X - d\text{sgn}(X)$ but since $d < \sqrt{\epsilon}$ it behaves better than the LT estimator in this repect.

3 Squared error with complexity cost loss

Throughout this paragraph we work with the SEWCC loss given by (1.7); unless otherwise stated terms such as risk, minimax, etc. now refer to this loss. In this loss s is supposed to be a given positive number. We comment briefly on its choice at the end of §3. To begin with, consider the pretest estimators δ_a of (1.2). With the choice $a = s$, δ_s becomes admissible (Meeden and Arnold (1979)); however, as we now show, neither it nor any other δ_a with $a > 0$, is minimax. It is known that the minimax value for estimating μ is $1 + s^2$ (Kempthorne (1987b)). The risk of δ_a is given by

$$\begin{aligned} g(\mu) &= E[X\mathbf{I}(|X| > a) - \mu]^2 + s^2 E\mathbf{I}(|X| > a) \\ &= 1 + s^2 + (\mu^2 - s^2 - 1)\int_{-a}^{a} \phi(x-\mu)dx + \qquad (1.14) \\ & \quad (a-\mu)\phi(a-\mu) + (a+\mu)\phi(a+\mu) \end{aligned}$$

Evidently $g(\mu) \to 1 + s^2$ as $|\mu| \to \infty$. Also, differentiating (1.14) with respect to μ we find that

$$\begin{aligned} & g\prime(\mu) \\ = \ & 2\mu \int_{-a}^{a} \phi(x-\mu)dx - a(2\mu - a)\phi(a-\mu) - \\ & a(2\mu + a)\phi(a+\mu) + s^2[\phi(a-\mu) - \phi(a+\mu)] \qquad (1.15) \\ = \ & 4a\mu e^{-\frac{1}{2}\mu^2}\cosh(a\mu)\phi(a)\left[\frac{\int_0^a \cosh(\mu x)\phi(x)dx}{a\cosh(a\mu)\phi(a)} + \frac{a^2+s^2}{2a\mu}\tanh(a\mu) - 1\right] \end{aligned}$$

By dominated convergence the first term in brackets tends to 0 as $|\mu| \to \infty$, the second clearly also goes to 0 and the factor between brackets is negative for all $|\mu|$ large enough. Therefore (1.14) eventually decreases monotonically to $1+s^2$ and we must have $\sup_\mu g(\mu) > 1+s^2$. Hence no member of (1.2) can be minimax, and in particular, although δ_s is admissible, it is not minimax. Numerical work showed that the maximal risk of δ_s exceeds $1+s^2$ by at most 8% over the range $0 < s \leq 5$. Therefore we can say that δ_s is not far from minimaxity. It also turns out that the maximal risk of δ_a decreases as we decrease a (keeping s fixed) and for $a = 0$ we get $\delta_0(X) = X$ which is minimax. However, X is not a satisfactory estimator since it has probability 1 of never selecting the simpler model M_0. Also it is known that X is inadmissible (Kempthorne (1987b)), but no estimator dominating X seems to have been proposed in the literature.

Next consider **Problem M**: among the class of minimax estimators, find that one which minimizes the probability of not selecting M_0 when it is correct (i.e. $\mu = 0$). To study the solution of this problem, consider the equivalent of Problem K for the SEWCC risk: for given α and s, find the estimator $\delta \in D_\alpha$ which minimizes the maximal risk. Let $\tilde{\delta}_\alpha$ be the optimal solution of this problem and $\tilde{R}(\alpha)$ its maximal risk. Evidently $\tilde{R}(\alpha) \geq 1+s^2$ for all α, $\tilde{R}(1) = 1+s^2$ and $\tilde{R}(\alpha)$ decreases as α increases. Let $\alpha = \alpha(s)$ be the smallest α such that $\tilde{R}(\alpha) = 1+s^2$. Then $\tilde{\delta}_{\alpha(s)}$ is the optimal solution to Problem M. Unfortunately, exact computation of $\tilde{\delta}_{\alpha(s)}$ is again difficult, but we can proceed by approximation as before. Let $\tilde{\delta}_{m,\alpha}$ and $\tilde{R}_m(\alpha)$ be as above but with μ restricted to $[-m, m]$. These quantities can be computed numerically by extending the methods of Kempthorne (1987a) to the SEWCC case. In particular, $\tilde{\delta}_{m,\alpha}(X)$ again has the form (1.5) with $a = a(\alpha)$ of (1.1). Define $\alpha_m(s)$ as the smallest α such that $\tilde{R}_m(\alpha) = 1+s^2$. Then $\alpha_m(s)$ increases as m increases and $\Delta_{m,s} = \tilde{\delta}_{m,\alpha_m(s)}$ gives an approximation to the optimal solution $\tilde{\delta}_{\alpha(s)}$ of Problem M.

Figure 3 illustrates numerical work implementing this process for the case $s = 1.5$. The three lower curves in Figure 3 show $\tilde{R}_m(\alpha)$ as functions of α for $m = 4, 8$ and 12 and they cross the level $1+s^2 = 3.25$ at $\alpha_4(1.5) = 0.084$, $\alpha_8(1.5) = 0.120$ and $\alpha_{12}(1.5) = 0.135$ respectively. For $s = 1.5$ it is quite difficult to compute $\alpha_m(1.5)$ accurately beyond $m = 12$; for $s < 1.5$ even smaller m's are difficult while for $s > 1.5$ we can go beyond $m = 12$. Columns 2 and 3 of Table 1.2 list the largest m and $\alpha_m(s)$-values which we were able to compute for some values of s. We believe that these values of $\alpha_m(s)$ are reasonably close lower bounds for $\alpha(s)$. Column 4 of Table 1.2 gives $a = a(\alpha_m(s))$ from (1.1). For $|X| \leq a(\alpha_m(s))$, $\Delta_{m,s}(X) = 0$ while otherwise $\Delta_{m,s}(X)$ is a Bayes estimator with respect to a suitably chosen discrete prior. The broken curves in Figure 4 show $\Delta_{m,s}(X)$ as functions of $|X|$ for $m = 8$ and $m = 12$ and $s = 1.5$ (of course, the vertical sections are not part of these functions but are shown to make the functions more visible). As before these estimators are tedious to deal with in practice and so we look for simpler approximations. The solid curve in Figure 4 shows the function $\delta_{a,c,0}(X) = \delta_{a,c}(X)$ of (1.6) with $a = 1.3202$ and $c = 2.295$ (the reason for this choice will become clear below) and it clearly captures the main features of $\Delta_{m,s}$. This suggests that we may look for simple minimax estimators within the class (1.6) which approximately solves Problem M.

Table 1.2 Approximating the optimal solution to Problem M

s	m	$\alpha_m(s)$	$a(\alpha_m(s))$	$\overline{\alpha}(s)$	$a(\overline{\alpha}(s))$	$c(\overline{\alpha}(s))$
0.5	9	0.3213	0.9918	0.5344	0.6213	1.136
1.0	10	0.2236	1.2170	0.3202	0.9942	1.824
1.5	12	0.1349	1.4951	0.1868	1.3202	2.295
2.0	12	0.0655	1.8416	0.0957	1.6659	2.800
2.5	14	0.0277	2.2018	0.0407	2.0468	3.250
3.0	14	0.0093	2.6008	0.0142	2.4530	3.660

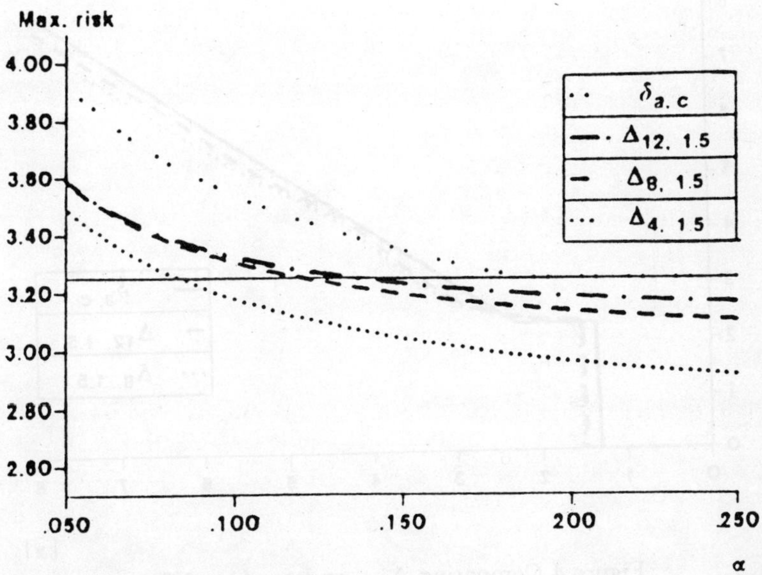

Figure 3 Approximating the optimal solution to Problem M (s=1.5)

The risk of $\delta_{a,c,d}$ is easily seen to approach $1+s^2+d^2$ as $|\mu| \to \infty$. Hence we must take $d = 0$ to get a minimax member from this family and we may restrict attention to $\delta_{a,c}$ of (1.8). Numerical work showed that it is indeed possible to choose many combinations of a and c to make $\delta_{a,c}$ minimax. In the spirit of the approximate solution of Problem M we may proceed as follows. For given α, choose $a = a(\alpha)$ by (1.1) so that $\delta_{a(\alpha),c} \in D_\alpha$; then compute the maximal risk of $\delta_{a(\alpha),c}$ and choose $c = c(\alpha)$ to minimize this maximal risk. Now find the smallest $\alpha = \overline{\alpha}(s)$ such that the maximal risk of $\delta_{a(\alpha),c(\alpha)}$ is still $1+s^2$. The top curve in Figure 3 shows the maximal risk of $\delta_{a(\alpha),c(\alpha)}$ as a function of α and the smallest α for which this is still $1+s^2 = 3.25$ is $\overline{\alpha}(1.5) = 0.187$. Columns 5, 6 and 7 of Table 1.2 list values of $\overline{\alpha}(s)$, $a(\overline{\alpha}(s))$ and $c(\overline{\alpha}(s))$ for other values of s. Of course we must have $\alpha_m(s) \leq \alpha(s) \leq \overline{\alpha}(s)$. For small values of s the difference between $\overline{\alpha}(s)$ and $\alpha_m(s)$ is rather large so that we cannot be certain that $\overline{\alpha}(s)$ is close to $\alpha(s)$. As s becomes larger $\overline{\alpha}(s)$ and $\alpha_m(s)$ become close and then $\overline{\alpha}(s)$ must be close to $\alpha(s)$ so that $\delta_{a(\overline{\alpha}(s)),c(\overline{\alpha}(s))}$ provides a solution to Problem M which is both simple and close to optimal. By using a family of functions more flexible than $\delta_{a,c}$ it is no doubt possible to find simple solutions to Problem M which is also close to optimal for small s but we will not pursue this here.

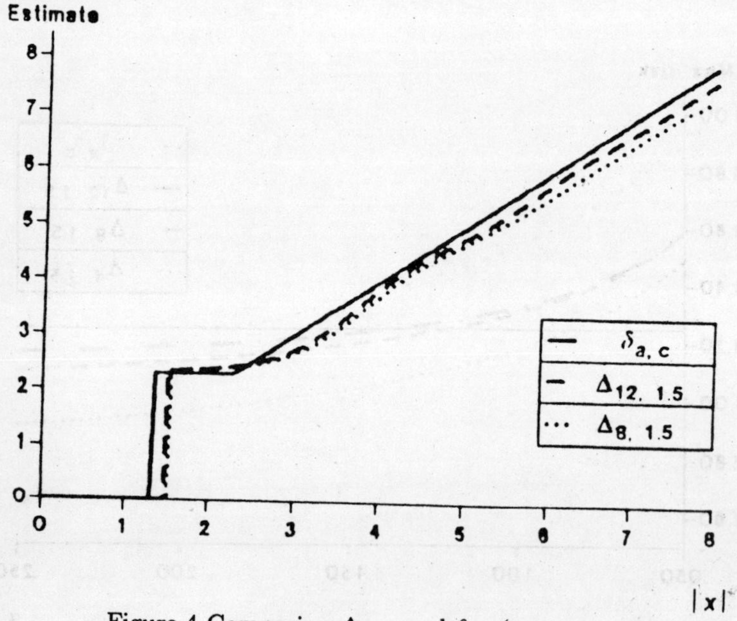

Figure 4 Comparing $\Delta_{m,s}$ and $\delta_{a,c}$ $(s = 1.5)$

The $s = 1.5$ row in Table 1.2 gives the values of a and c used for $\delta_{a,c}$ in Figure 4. Continuing with this case, Figure 5 shows the risk of $\delta_{a,c}$ and it clearly demonstrates the minimaxity of $\delta_{a,c}$ in this case. We also show separately the two components making up this risk, namely the SE risk $E[\delta_{a,c}(X)-\mu]^2$ and the complexity cost (CC) risk $s^2P(|X|>a)$. It is clear that $\delta_{a,c}$ allows its SE risk to exceed 1 (the SE risk of X) for small μ in exchange for low CC risk there. We also note that the SE risk of $\delta_{a,c}$ quickly approaches 1 as $|\mu|$ increases. This contrasts sharply with the behaviour of the LT estimator and $\delta_{a,c,d}$ of §2 whose risks approach $1+\epsilon$ and $1+d^2$ respectively. It reflects the fact that $\delta_{a,c}(X)$ does not continue to shrink X as $|X| \to \infty$. On balance we feel that the SEWCC loss approach leads to estimators having more reasonable behaviour than the estimators obtained from the Problem P and K approaches.

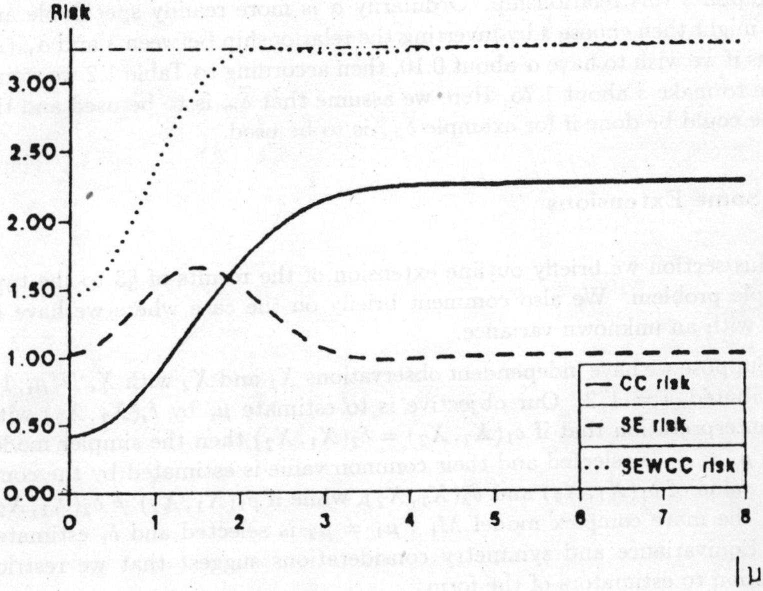

Figure 5 Risks of $\delta_{a,c}$ (s=1.5)

An interesting property of these estimators is illustrated in Figure 4. For values of $|X|$ smaller than some critical value, the simple model M_0 is selected and μ is estimated to be 0. As $|X|$ increases, this decision eventually changes and the more complex model M_1 is selected instead. For a range of $|X|$- values following the point at which this change occurs, μ is estimated by a value further away from 0 than X. Thus when these estimators find model M_0 no longer tenable, they make a rather radical choice for model M_1. Their implied discontinuities are even larger than for the ordinary pre-test estimator (1.2). From a purely estimation point of view this behaviour may seem somewhat strange. However, selecting one model in preference to another is a radical decision so that such discontinuous behaviour is perhaps to be expected in the context of simultaneous model selection and estimation.

Finally, some comments on the choice of s for the SEWCC loss function: clearly the role of s is to strike a balance between the mean squared error $E(\delta(X)-\mu)^2$ and the probability of not selecting the simpler model $P(\delta(X) \neq 0)$. Since these two quantities reflect quite different aspects of the estimator δ it would usually be hard to decide on a suitable value for s. A possible guideline is provided by the monotone relationship between s and $\alpha(s)$, which is the minimal probability of not selecting M_0 when M_0 is correct that can be achieved by a minimax estimator. Unfortunately $\alpha(s)$ is hard to compute exactly, but as discussed above $\alpha_m(s)$ provides a reasonably close approximation and the corresponding entries in Table 1.2 exemplifies this relationship. Ordinarily α is more readily specifiable and one might then choose s by inverting the relationship between s and $\alpha_m(s)$. Thus if we wish to have α about 0.10, then according to Table 1.2 we would have to make s about 1.75. Here we assume that $\overline{\delta}_m$ is to be used and the same could be done if for example $\delta_{a,c}$ is to be used.

4 Some Extensions

In this section we briefly outline extension of the results of §3 to the two-sample problem. We also comment briefly on the case where we have to deal with an unknown variance.

Suppose we have independent observations X_1 and X_2 with X_i $N(\mu_i; 1)$-distributed, $i = 1, 2$. Our objective is to estimate μ_i by $\delta_i(X_1, X_2)$ with the interpretation that if $\delta_1(X_1, X_2) = \delta_2(X_1, X_2)$ then the simpler model $M_0 : \mu_1 = \mu_2$ is selected and their common value is estimated by the common value of $\delta_1(X_1, X_2)$ and $\delta_2(X_1, X_2)$, while if $\delta_1(X_1, X_2) \neq \delta_2(X_1, X_2)$ then the more complex model $M_1 : \mu_1 \neq \mu_2$ is selected and δ_i estimates μ_i. Equivariance and symmetry considerations suggest that we restrict attention to estimators of the form

$$\delta_1(X_1, X_2) = X_1 + h(X_2 - X_1)\ , \quad \delta_2(X_1, X_2) = X_2 + h(X_1 - X_2) \quad (1.16)$$

for some function h. Put $Y = (X_2 - X_1)/\sqrt{2}$ and $\nu = (\mu_2 - \mu_1)/\sqrt{2}$, so that Y is $N(\nu; 1)$-distributed. Then by using the arguments in Venter and Steel (1990) it is easily shown that the total SE risk of (1.16) is given by

$$\begin{aligned} r(h, \nu) &= E[\delta_1(X_1, X_2) - \mu_1]^2 + E[\delta_2(X_1, X_2) - \mu_2]^2 \\ &= 1 + \tfrac{1}{2}E[Y - \nu - \sqrt{2}h(\sqrt{2}Y)]^2 + \tfrac{1}{2}E[Y - \nu + \sqrt{2}h(-\sqrt{2}Y)]^2. \end{aligned} \quad (1.17)$$

We assume that $h(-Y) = -h(Y)$, since if this were not the case we can put $\tilde{h}(Y) = \frac{1}{2}[h(Y) - h(-Y)]$ and then (1.16) with $\tilde{h}$ replacing h will dominate (1.16) in the sense that $r(h, \nu) - r(\tilde{h}, \nu) \geq 0$ for all ν with equality holding if and only if $h(-Y) = -h(Y)$. With this assumption (1.17) becomes

$$r(h, \nu) = 1 + E[\delta(Y) - \nu]^2 \text{ where } \delta(Y) = Y - \sqrt{2}h(\sqrt{2}Y) \quad (1.18)$$

is an estimator of ν. The CC loss is $s^2\mathbf{I}(\delta_1 \neq \delta_2)$ and the CC risk of (1.16) is seen to reduce to $s^2P(\delta(Y) \neq 0)$. Hence the total SEWCC risk of (1.16) becomes

$$1 + E[\delta(Y) - \nu]^2 + s^2P(\delta(Y) \neq 0) \quad (1.19)$$